Introduction to VLSI D

Chip designing is a complex task that requires an in-depth understanding of VLSI design flow, skills to employ sophisticated design tools, and keeping pace with the bleeding-edge semiconductor technologies. This lucid textbook is focused on fulfilling these requirements for students, as well as a refresher for professionals in the industry.

The book consists of four parts. The first part describes foundational concepts related to VLSI design flow and integrated circuits. It also gives an overview of the design, verification, and test methods employed in a typical VLSI design flow. The second part of the book describes the logic implementation and verification steps such as simulation, static timing analysis, and formal methods. It also explains the modelling of hardware using Verilog and logic synthesis; technology libraries; and timing constraints along with logic, power, and timing optimization techniques. The third part of the book describes the design for test (DFT) methods for digital circuits. The fourth and final part describes physical design methods and physical verification. All the physical design implementation steps such as floorplanning, placement, clock-tree synthesis, and routing are described in this part. Moreover, physical verification steps, such as parasitic extraction, design rule checks (DRCs), electrical rule checks (ERCs), layout versus schematic (LVS) checks, and post-silicon validation are explained.

Illustrations and pictorial representations are used liberally to simplify the explanation. Additionally, activities are suggested at the end of relevant chapters to help readers gain a practical understanding of VLSI design flow. Review questions and problems are given at the end of each chapter to revise the concepts. Recent trends and references are listed at the end of each chapter for further reading.

Sneh Saurabh is Associate Professor in the Department of Electronics and Communication Engineering, Indraprastha Institute of Information Technology, New Delhi, India. He has rich experience in the semiconductor industry, having spent 16 years working for industry leaders such as Cadence Design Systems, Synopsys India, and Magma Design Automation. His research interests include VLSI design and automation, nanoelectronics, and energy-efficient systems.

Introduction to VLSI Design Flow

Sneh Saurabh

CAMBRIDGE
UNIVERSITY PRESS

University Printing House, Cambridge CB2 8BS, United Kingdom

One Liberty Plaza, 20th Floor, New York, NY 10006, USA

477 Williamstown Road, Port Melbourne, vic 3207, Australia

314 to 321, 3rd Floor, Plot No.3, Splendor Forum, Jasola District Centre, New Delhi 110025, India

103 Penang Road, #05–06/07, Visioncrest Commercial, Singapore 238467

Cambridge University Press is part of the University of Cambridge.

It furthers the University's mission by disseminating knowledge in the pursuit of education, learning and research at the highest international levels of excellence.

www.cambridge.org
Information on this title: www.cambridge.org/9781009200813

First published 2023

Printed in India by Nutech Print Services, New Delhi 110020

A catalogue record for this publication is available from the British Library

Library of Congress Cataloging-in-Publication Data
Names: Saurabh, Sneh, author.
Title: Introduction to VISI design flow / Sneh Saurabh.
Description: Cambridge, United Kingdom ; New York, NY, USA : Cambridge University Press, 2023. | Includes bibliographical references and index.
Identifiers: LCCN 2023004179 (print) | LCCN 2023004180 (ebook) | ISBN 9781009200813 (paperback) | ISBN 9781009200790 (ebook)
Subjects: LCSH: Integrated circuits–Very large scale integration.
Classification: LCC TK7874.75 .S38 2023 (print) | LCC TK7874.75 (ebook) | DDC 621.3815–dc23/eng/20230208
LC record available at https://lccn.loc.gov/2023004179
LC ebook record available at https://lccn.loc.gov/2023004180

ISBN 978-1-009-20081-3 Paperback

To Saraswati, the Goddess of Learning

Contents

Preface

Modern integrated circuits are very complex and can have billions of transistors crammed onto a few square centimeters of silicon. To ensure that the manufactured chips produce the desired results, a designer needs to perform tasks related to design implementation, verification, and testing. These tasks consist of several individual design steps that are accomplished with the help of sophisticated electronic design automation (EDA) tools. Besides knowing the details of these design steps, a designer should be familiar with the capabilities, assumptions, limitations, and framework of various EDA tools employed in a design flow. Moreover, these design steps are interrelated. Therefore, a designer must develop a holistic view of a design flow to utilize EDA tools efficiently and obtain a high-quality design.

Most books on very large scale integration (VLSI) focus on specific aspects of design flow, such as register transfer level (RTL) modeling, logic synthesis, formal verification, physical design, or testing. These books provide excellent study material and references for individual design tasks. However, they become overwhelming for undergraduate and postgraduate students.

An undergraduate or postgraduate student needs to develop concepts on VLSI design flow in a limited time frame, typically one or two semesters. The existing books on these topics allow students to cover only a few design tasks in great detail while leaving out many others. We believe a student should devote limited academic time to acquiring the broadest base of knowledge rather than prematurely specializing in one domain. This textbook is written to fulfill this objective. It explains the fundamental aspects of VLSI design flow for the senior undergraduate and postgraduate students. It covers the design flow extensively, taking a holistic view of the VLSI design and technology. Further, it emphasizes the basic principles of design flow that are expected to be helpful in the long-term for a VLSI design engineer.

OVERVIEW OF THE BOOK

We have divided this book into four parts:

1. **Overview of VLSI design flow:** It introduces integrated circuit concepts and gives an overview of the design, verification, and test methods employed in a typical design flow.
2. **Logic design:** It explains RTL modeling using Verilog. Further, it describes the logic synthesis, technology libraries, and timing constraints along with logic, power, and timing optimization techniques. It also explains verification steps such as simulation, static timing analysis, and formal methods.
3. **Design for testability (DFT):** It describes basic concepts of DFT, scan design methodology, test pattern generation, and built-in self-test techniques.
4. **Physical design:** It describes physical design tasks such as chip planning, placement, clock tree synthesis, and routing. It also describes physical verification steps such as parasitic extraction,

design rule checks (DRCs), electrical rule checks (ERCs), layout versus schematic (LVS) checks, signoff, and post-silicon validation.

TO TEACHERS

We have written this textbook for a foundation course in VLSI design flow. We have intentionally covered more breadth and depth than what can be covered in a one-semester course. Therefore, we encourage you to select topics based on your interests, expertise, and students' requirements. The topics we have previously covered in a one-semester VLSI design flow course are listed on the book's website. A two-semester course on VLSI design flow can also be designed using this book. An indicative list of topics for a two-semester course on VLSI design flow is also available on the book's website. We would be happy to receive your questions, suggestions, comments, and feedback on the course design based on this book.

We have taken an approach of explaining concepts using examples. We have found that this approach is highly effective for introductory-level courses. The examples presented in this book illustrate the motivation, optimization principles, assumptions, limitations, and side effects of a design task. We encourage you to discuss these examples in the classroom to augment the learning process.

VLSI design flow is incomplete without hands-on experience with the EDA tools. We have provided activities at the end of each major chapter. We encourage you to ask students to carry out these activities independently. We have intentionally avoided suggesting any commercial tools to perform these activities. In the past, we have used these activities as weekly and fortnightly assignments in the introductory course on VLSI design flow. Students have used open-source tools to carry out these activities and have found them to be highly beneficial in augmenting their understanding of the design flow. Since these activities are typically performed on small designs, the students can conduct rigorous experiments and carry out in-depth analysis with the help of EDA tools. The solutions for some activities using open-source tools will be made freely available. We will also be happy to provide on request the solutions to the numerical problems appearing at the end of the chapters.

Finally, we would love to receive your suggestions, questions, and feedback on this book. Please feel free to reach out to us.

TO STUDENTS

This book is intended to introduce VLSI design flow concepts to undergraduate and postgraduate students. We have attempted to present topics that allow students to comprehend the subject independently. The preliminary concepts required for understanding this book are described in Chapter 1 ("Foundation"). You may like to review this chapter before proceeding further.

In the first part of the book, we have presented an overview of the design flow. The topics described in this part of the book will help you develop a holistic view of the design flow. We encourage you to go through this part of the book first. The rest of the chapters are laid out according to the sequence of tasks performed in a typical design flow. You may follow the same sequence because it allows you to read the chapters and perform design activities side by side. In the past, students have found this sequence to be highly effective in building both theoretical and practical design flow concepts. To make the topics easy to understand, we have tried to explain the concepts using examples, illustrations, and figures as much as possible. Moreover, the content of this book has gone through extensive revision based on the students' feedback. We hope the book has become more readable and students can independently master the concepts. However, the proof of the pudding is in the eating. Therefore, we will be happy to receive your feedback, questions, and suggestions on the readability of the book. We will try to incorporate your feedback in future editions as much as possible.

The content of this book has been reviewed extensively by industry experts and academicians. We have tried our best to cover topics relevant to the current industry requirements. We have intentionally avoided discussing tool-specific features and commands in this book. We believe that specific EDA tools can be mastered easily after understanding the principles of VLSI design flow and the associated techniques. We strongly recommend you to run EDA tools to gain a hands-on experience of the design flow after covering each chapter. The activities given at the end of each major chapter can be their starting points. If you have access to the commercial EDA tools, you can use them to perform these activities. You can also use freely available tools to perform most of these activities (we have added some references of the open-source tools at appropriate places). The results obtained for some of these activities using open-source tools will be made freely available. Nevertheless, we encourage you to carry out these activities on your own. A practical experience with EDA tools will be invaluable in your future VLSI design projects and career.

TO PRACTICING ENGINEERS

We have covered VLSI design concepts comprehensively in this book. We have also added extensive references for delving deeper into these topics. Therefore, this book can serve you as a ready reference in real design projects. Moreover, we have tried to highlight good engineering practices, typical assumptions and limitations of EDA tools, and common pitfalls and fallacies. We believe that these ideas can be extremely helpful in avoiding and fixing challenging design problems. The book also highlights various trade-offs needed in a typical design flow. These concepts can guide you in judiciously making tough engineering decisions. We have tried our best to make this book relevant for practicing engineers with the help of inputs from the industry experts. We will be happy to receive your suggestions and feedback to make this book more useful in real design projects.

TYPOGRAPHIC CONVENTIONS

No.	Typeface	Purpose	Examples
1	Normal	Normal text	This is normal
2	`Monospaced font`	Keywords of languages such as Verilog, Library format, Constraint format	`always`, `class`, `cell`, `pin`, `create_clock`
3	*Italic*	Design object names,	*Clk*, *MyInst*, *net_1*
		Introduce technical terms,	... is known as *placement*
		Emphasize	... the *next* clock edge

ERRORS

We have made all efforts to ensure the correctness of the material presented in this book. Despite all our efforts, some errors might have escaped our scrutiny. In case there are any errors, we apologize for them in advance. Further, we will be grateful to the readers if they inform us about any errors. We will make such findings available to all the readers before we get an opportunity to make corrections in the future edition of this book.

AUTHOR CONTACT AND BOOK'S WEBSITE

Send an email to vlsi.design.flow@gmail.com.
The website for the book is mentioned on the back cover of the book.

Acknowledgments

Writing this book has been a long-term project spanning several years, and many people have helped me in this endeavor. I want to acknowledge their contributions and express my deepest gratitude.

First, I would like to thank my teachers and mentors for helping me gain sufficient knowledge and skills to write this book. I am incredibly grateful to Prof. P.R. Panda (IIT Delhi) for teaching the rudiments of electronic design automation and providing valuable suggestions throughout this project. I would like to extend my deepest gratitude to Prof. G.S. Visweswaran (IIT Delhi), Prof. P. Jalote (IIIT Delhi), and Prof. M. Jagadesh Kumar (IIT Delhi) for constant encouragement, guidance, and support in writing this book. I would like to thank Prof. S.K. Lahiri (IIT Kharagpur), Prof. B. Bhaumik (IIT Delhi), and Prof. Jayadeva (IIT Delhi) for introducing me to the VLSI design, tools, and technology. I am deeply indebted to Fr. Varghese Panangatt (SJS Bhagalpur) for helping me in many ways, including honing my writing skills, without which I would not have ventured into book writing.

I would like to express my deepest gratitude to all my mentors and colleagues in the industry who have helped me appreciate the practical aspects of VLSI design flow. In particular, I would like to thank Sanjay Churiwala, Manish Goel, Shailendra Pokhriyal, and P.K. Prasoon for their help and guidance during my initial days in the semiconductor industry.

While writing this book, I have taken help from several of my colleagues, friends, and mentors from academics and industry. They have helped me review the book chapters, find errors, and provide insightful suggestions. I am highly grateful to them for taking time from their busy schedule and helping me on this project at a personal level. In this regard, I would specially thank Alwin Gupta, Anil Mishra, Anshumani, Anuj Grover, Chandan Kumar, Gaurav Gupta, Hitesh Shrimali, Indrajit Dutta, Mahesh Chandra, Manish Goel, Md. Shaukat Ullah, Naresh Kumar, N. Kannan, N. Nallam, Prashant Sethia, P.R. Panda, P.K. Prasoon, Praveen Chhabra, Rajat Bishnoi, R.S. Saxena, Sanjay Churiwala, Sujay Deb, and Sumit Darak.

The students at IIIT Delhi who have taken my course *VLSI Design Flow* have contributed immensely to this book. I would like to thank them for providing constructive feedback, asking questions, doing short- and long-term projects, and participating in classroom activities. The book's content has refined significantly over the years due to the active involvement of the students, including TAs. I would like to especially thank the following students for directly helping me in this project by reviewing chapters, improving the readability, carrying out activities with the electronic design automation (EDA) tools, and verifying the solutions: Abhinav Gupta, Abhishek TC, Amina Haroon, Arushi Vasudevan, Gagandeep, Gyan Deep, Jasmine Kaur, Pooja Beniwal, Pranav Jain, Shakti Shrey, Sonali Jain, and Swapnil Bansal. I would also like to thank IIIT Delhi for providing an excellent academic environment where intellectual activity, such as book writing, can thrive and be easily pursued.

I would also like to gratefully acknowledge the help of the publishing team at Cambridge University Press, especially Mr. Agnibesh Das, for reaching out to me for this project and constantly supporting me in this endeavor.

I would like to express my deepest gratitude to my parents, Dr. S.P. Sah and Mrs. Sunita Sah, for their love, care and teaching values of continuously pursuing and sharing knowledge. I would also thank my sisters and extended family members for providing rock-solid support, especially during challenging times.

This book could have never taken the current shape without the love and unwavering support of my wife Onam and my kids, Vibhu and Ishaan. They have backed me with immaculate patience for years, and their strength and encouragement have taken me to the finish line of this book-writing marathon.

Finally, I would like to thank my friends, colleagues, collaborators, well-wishers, and everyone who has helped me directly or indirectly in writing this book.

PART ONE

Overview of VLSI Design Flow

In this part of the book, we will introduce integrated circuits (ICs) concepts and give a top-level view of the entire VLSI design flow. In the remaining portion of this book, we will discuss each task of the VLSI design flow in detail. We have taken this approach to present the topics because an early introduction to the entire design flow eases building context for individual tasks. It allows us to relate one task with another in the remaining portion of this book. We can appreciate the interaction between them and understand how decisions of one task impact other tasks. As a result, we can easily comprehend challenges and opportunities in optimizations and develop a solid understanding of the entire flow.

In the first chapter ("Foundation"), we will briefly describe the concepts of digital circuits, devices, CMOS inverters, data structures, and algorithms. Since readers would have been exposed to these topics previously, this chapter will serve as a refresher. Additionally, this chapter will be a handy reference if a reader needs to recall some of these concepts. In the second chapter ("Introduction to Integrated Circuits"), we will introduce the concepts related to ICs, their history, fabrication, economics, and figures of merit. In the third ("Pre-RTL Methodologies") and fourth ("RTL to GDS Implementation Flow") chapters, we will describe the implementation of a design from concept to the layout. In the fifth ("Verification Techniques") and sixth ("Testing Techniques") chapters, we will discuss basic concepts related to the verification and testing of ICs. In the seventh chapter ("Post-GDS processes"), we will explain processes involved in fabricating chips from a given layout.

1 Foundation

To look at a thing is very different from seeing a thing. One does not see anything until one sees its beauty...

—Oscar Wilde, "The Decay of Lying," *Intentions*, 1891

In this chapter, we review some of the basic concepts of electronics and computer science. These concepts are required in understanding subsequent chapters of this book. We expect that readers are already familiar with these concepts. Therefore, the purpose of this chapter is to review them for the sake of completeness. If required, readers can refer to the dedicated textbooks on these topics. We have provided a few references at appropriate places inside the chapter.

1.1 DIGITAL SYSTEMS

A digital system is designed to accomplish some tasks by manipulating digital signals. These tasks can be related to digital signal processing, computing, data transfer, data storage, and data retrieval. Examples of digital systems are computers, mobiles, hard disks, video game consoles, electrocardiograph (ECG) machines, the internet, and robots. Most electronic appliances now depend on digital signal manipulations due to digital systems' inherent robustness and flexibility.

1.1.1 Analog Signal to Digital Signal

Analog signals can take any value within some range. Naturally occurring signals are analog. For example, the speech signal that gets generated when we talk. It is composed of air pressure that varies with time and can take any value within some range. Analog signals can appear as variations in a physical quantity such as pressure, light intensity, and temperature.

We convert a naturally occurring analog signal to an *electrical signal* using a *transducer* for digital signal processing. An electrical signal is a variation in an electrical quantity such as voltage or current. For example, we convert a sound signal to an electrical signal using a microphone. Subsequently, we convert the analog electrical signal to a *digital signal* by *sampling* at discrete intervals and *quantizing* the sampled signal within a discrete set of values. Figure 1.1 illustrates the process of extracting a digital signal from an analog signal.

We typically represent a digital signal as a sequence of binary digits (bits or 0/1). For example, we can represent a digital signal with a magnitude of 24 in 8-bits as $(0001\,1000)_2$. We convert

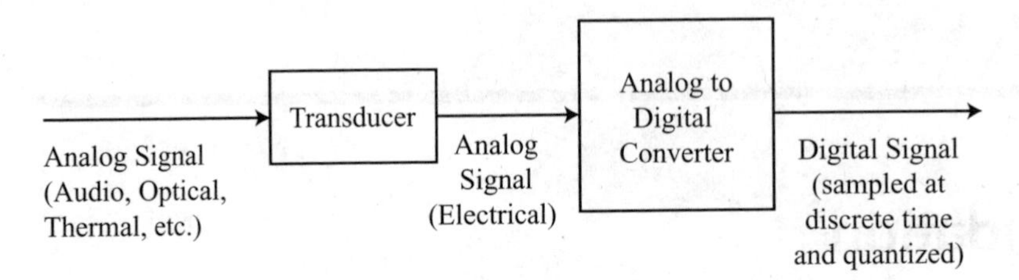

Figure 1.1 Conversion of an analog signal to a digital signal

an analog signal to the corresponding digital signal using an *analog-to-digital converter* (ADC). We divide the full analog signal range of an ADC into a finite number of discrete values. Then, we map an analog signal that has a value within that range to an appropriate discrete value. We explain it in the following example.

Example 1.1 Assume that an ADC produces a digital signal with 8-bits. Therefore, it can produce a digital signal from $(0000\ 0000)_2$ to $(1111\ 1111)_2$, i.e., 256 distinct values. Consequently, we map the full analog signal range of the ADC to 256 distinct values. Further, assume that the full range of an ADC is 0–1 V, as shown in Figure 1.2(a).

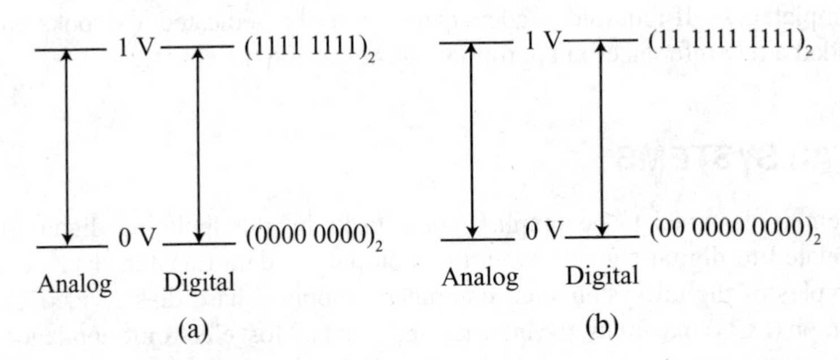

Figure 1.2 Analog to digital conversion (a) with 8-bits and (b) with 10-bits

Therefore, we can represent 0 V as $(0000\ 0000)_2$, 1 V as $(1111\ 1111)_2$, and any intermediate voltage level proportionally between $(0000\ 0000)_2$ and $(1111\ 1111)_2$. In this case, each bit represents 1 V/256 $\approx$ 3.9 mV. To make a 1-bit change, we need to change the analog signal by 3.9 mV.

We can increase the accuracy of an ADC by increasing the number of bits in the digital representation. For example, if we increase the number of bits from 8-bit to 10-bit, as shown in Figure 1.2(b), we obtain a 1-bit change even for a 1 V/1024 $\approx$ 0.98 mV change in the analog signal.

1.1.2 Robustness of Digital Systems

A digital system operates on *bits*. A bit can take only one of the two discrete values: zero (0) or one (1). Physically, we represent these two values as two levels of voltages (0 and supply voltage V_{DD}). While interpreting voltage levels as bits, we add appropriate margins to the voltage levels.

Example 1.2 Consider a system with the mapping of voltage levels to the bit value as shown in Figure 1.3. We have inserted δ_1 and δ_2 voltage margins to the interpretation of bit 0 and 1, respectively. Hence, the voltage in the interval 0 to $(0 + \delta_1)$ is taken as 0. Similarly, the voltage in the interval $(V_{DD} - \delta_2)$ to V_{DD} is taken as 1.

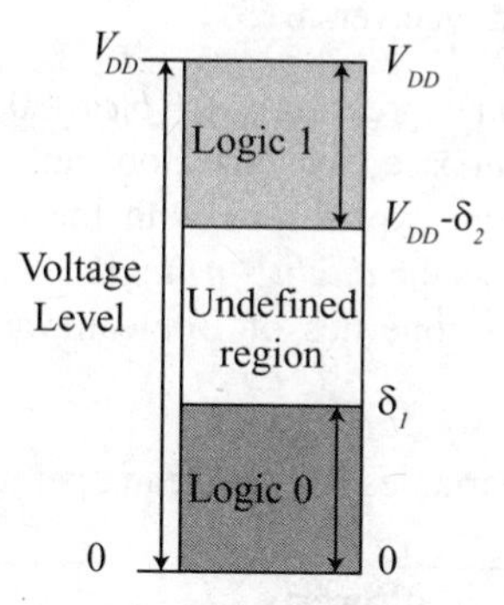

Figure 1.3 Interpretation of voltage levels as logic values

Thus, due to noise or other reasons, even when the voltage of 0 bit degrades by δ_1V, the interpretation of that bit is still correct. Similarly, we can tolerate the degradation of δ_2V in interpreting 1 bit. These margins make digital systems more robust.[1]

1.1.3 Boolean Algebra

Since a digital circuit operates on bits, we can design and analyze digital circuits using *binary Boolean algebra.* Binary Boolean algebra operates on variables that can take values 0 or 1. These variables are known as *Boolean variables.* There are two basic operations defined on Boolean variables:

1. **AND:** It is denoted by "." (dot). It is also called *product* or *conjunction.* The AND of two Boolean variables a and b is 1 if $a = b = 1$, and is 0 otherwise. The identity element for the AND operation is 1, i.e., for any Boolean variable a, $a.1 = a$.
2. **OR:** It is denoted by "+" (plus). It is also called *sum* or *disjunction.* The OR of two Boolean variables a and b is 0 if $a = b = 0$, and 1 otherwise. The identity element for the OR operation is 0, i.e., for any Boolean variable a, $a + 0 = a$.

Complement: For any Boolean variable a, a *complement* denoted as a' exists, such that if $a = 0$ then $a' = 1$, and if $a = 1$ then $a' = 0$. A Boolean variable and its complement satisfy: $a + a' = 1$ and $a.a' = 0$. Taking complement of a Boolean variable is also referred to as NOT operation.

Boolean expressions: We can form an infinite number of Boolean or logic *expressions* using AND, OR, and NOT operators. The precedence of these operators (starting with the highest precedence) is: NOT, AND, and OR. We can use parentheses to specify the order of operations in

[1] The robustness of a logic gate to *noise* is measured by a quantity called *noise margin.* Interested readers can refer to [1] for the mathematical definition of noise margins in digital circuits.

a Boolean expression. Some examples of Boolean expressions are: $((a + b).c + d'f).a'$, $a.b + b.c + a.c$, and $(a + b + c).(e + f)$, where a, b, c, d, e, and f are Boolean variables.

Dual of an expression: We obtain a dual of a given Boolean expression as follows:

1. Replace all the AND operators with the OR operators and vice versa.
2. Replace all the 0's with 1's and vice versa.

For example, the dual of the Boolean expression $a + ((b.c) + 0)$ is $a.((b + c).1)$.

Duality principle: If a statement based on some Boolean expressions is true, then the statement obtained by substituting the Boolean expressions with the corresponding duals of the Boolean expressions is also true. It is known as the *duality principle*.

Properties: We list out some properties of Boolean variables and their dual properties in Table 1.1.

Table 1.1 Properties of Boolean variables and Boolean operators

No.	Name	Property	Dual
1	Commutative	$a + b = b + a$	$a.b = b.a$
2	Associative	$a + (b + c) = (a + b) + c$	$a.(b.c) = (a.b).c$
3	Distributive	$a.(b + c) = a.b + a.c$	$a + (b.c) = (a + b).(a + c)$
4	Absorption	$a + a.b = a$	$a.(a + b) = a$
5	DeMorgan's Theorem	$(a + b)' = a'.b'$	$(a.b)' = a' + b'$

1.1.4 Realization of a Digital Circuit

We can classify digital circuits and circuit elements as combinational and sequential.

1.1.5 Combinational Circuits

A circuit in which the output value depends *only on the present value* of the signals at the inputs is known as a *combinational circuit*. It implies that its output does not depend on the past sequence of inputs. We describe some combinational circuit elements below.

Logic Gates

A logic gate has one or more inputs and one output. A logic gate can implement basic Boolean functions such as NOT, AND, and OR. Physically, we implement a logic gate by connecting components such as transistors, diodes, and resistors. The schematic representation of basic logic gates is shown in Figure 1.4(a).

We can implement a complicated Boolean function by connecting logic gates and obtaining a *logic circuit*. For example, Figure 1.4(b) shows the logic circuit for the Boolean function $y = a + b'c$ implemented using AND, OR, and NOT logic gates.

Although basic logic gates are sufficient to implement any logic function, there are some other Boolean functions that we can efficiently implement using transistors. For example, we can efficiently implement gates such as NAND, NOR, XOR, XNOR, AND, AND-OR-Invert (AOI), and OR-AND-Invert (OAI) using transistors. We often use these gates in digital circuit implementations. Figure 1.5 shows the schematic symbol and the functionality of some commonly used logic gates.

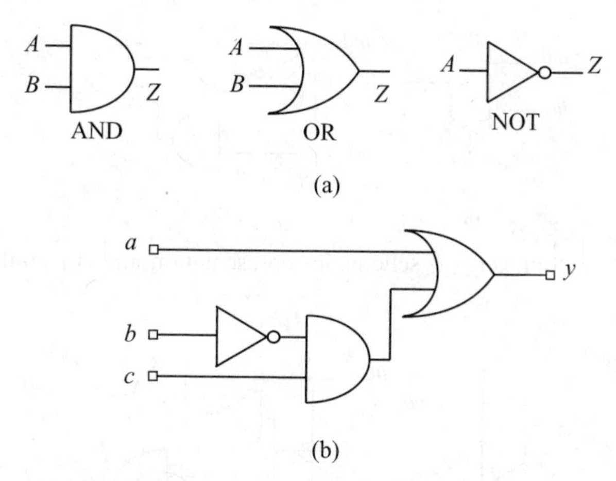

Figure 1.4 Schematic representation of (a) AND, OR, and NOT gates and (b) logic circuit for $y = a + b'c$

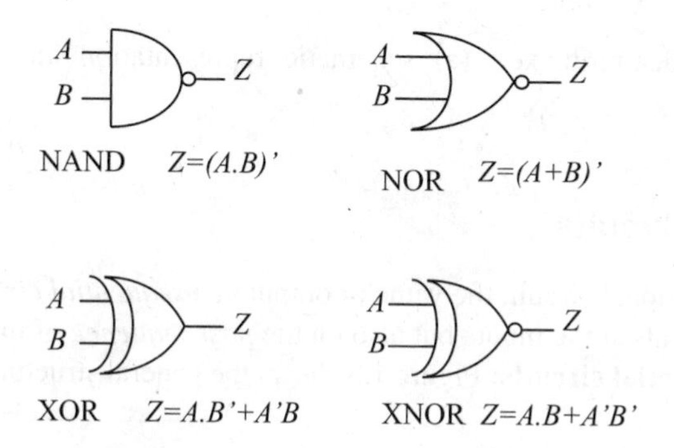

Figure 1.5 Simple logic gates

Multiplexers and Demultiplexers

A multiplexer is a circuit element having multiple *data* and *select* pins, and only one output pin. Depending on the value of signals at the select pins, one of the data pin signals is passed to the output pin. Figure 1.6(a) shows a schematic representation of a 2-to-1 multiplexer. The pins *in0* and *in1* are the data pins, the pin *s* is the select pin and the pin *y* is the output pin. The value at *y* is *in0* when $s = 0$ and is *in1* when $s = 1$. The Boolean function representing the output can be written as $y = s'.in0+s.in1$. Therefore, a 2-to-1 multiplexer can be implemented as shown in Figure 1.6(b).

A demultiplexer performs the functionality that is the opposite of a multiplexer. It has one input data pin and many select and output pins. Depending on the value of signals at the select pins, the data pin signal is passed to one of the output pins, while the value at other output pins remains 0. Figure 1.7(a) shows a schematic representation of a 1-to-2 demultiplexer. The pin *in* is the data pin, the pin *s* is the select pin, and the pins *y0* and *y1* are the output pins. When *s*=0, the value of *y0*=*in* and *y1*=0. When *s*=1, the value of *y0*=0 and *y1*=*in*. The Boolean function representing the outputs can be written as *y0*=*s'*.*in* and *y1*=*s*.*in*. Therefore, a 1-to-2 demultiplexer can be implemented as shown in Figure 1.7(b).

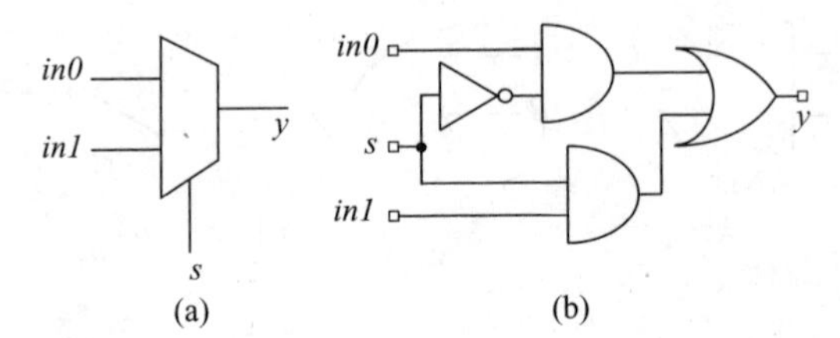

Figure 1.6 A 2-to-1 multiplexer: (a) schematic representation and (b) implementation in terms of basic logic gates

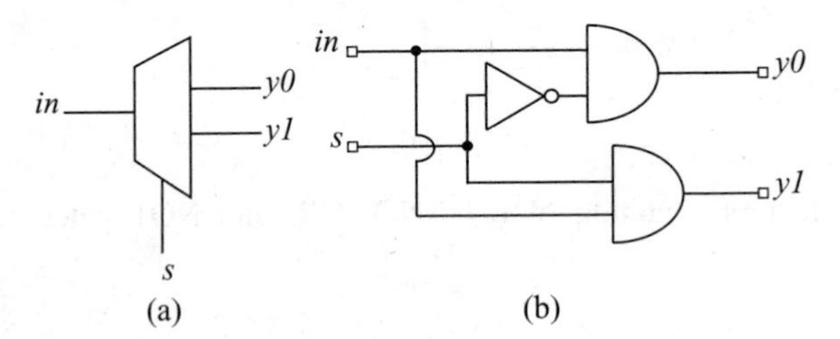

Figure 1.7 A 1-to-2 demultiplexer: (a) schematic representation and (b) implementation in terms of basic logic gates

1.1.6 Sequential Circuits

In contrast to a combinational circuit, the value of output in a *sequential circuit* depends not only on the present value of signals at the inputs but also on the *past sequence* of inputs.

Structure of sequential circuits: Figure 1.8 shows the general structure of a sequential circuit.

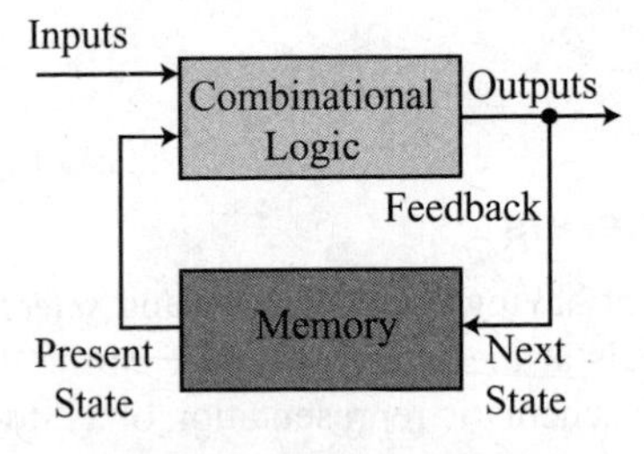

Figure 1.8 General structure of a sequential circuit

A sequential circuit includes some *memory* elements to remember the past behavior, in addition to some combinational circuits. The memory elements can store binary information (0 or 1). The value stored in the memory elements at a given time represents the *present state* of the sequential circuit. As the input of the sequential circuit changes, the state of the memory element can change and attain the *next state*. The *next state* and the *output* depend on the present state and the input values, as shown in Figure 1.8.

In general, there are two types of sequential circuits: *synchronous sequential circuits* and *asynchronous sequential circuits*.

Synchronous sequential circuits: In a synchronous sequential circuit, the state of the circuit changes only at discrete time intervals. The discrete-time interval is defined using a signal called *clock*. A clock signal repeats regularly after a specific time interval known as the *clock period*, as shown in Figure 1.9. Within a clock period, a clock signal toggles from 0 → 1 and 1 → 0. In a synchronous circuit, the clock signal controls the read and write operations in memory elements. As data updates in memory elements, other combinational circuit elements also update their output values. Thus, in a synchronous circuit, the activities are controlled and synchronized by the clock signal.

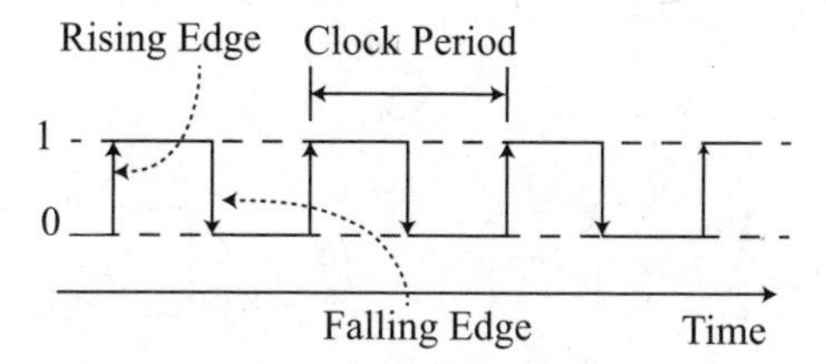

Figure 1.9 A clock signal

Asynchronous sequential circuits: In an asynchronous sequential circuit, the *state* or the stored data can change at any time due to the change in the input values. We can store data with the help of combinational circuit elements having an appropriate *feedback* mechanism. However, problems such as instability and race conditions can occur due to the feedback action. Therefore, designing and ensuring the correctness of an asynchronous circuit is challenging. In contrast, since the state of a synchronous circuit changes only at discrete time intervals, its timing analysis needs to be done only at discrete time steps. It simplifies the design and verification of a synchronous circuit. Therefore, synchronous circuits are more widely employed.

In this book, we will consider primarily synchronous circuits.

Cross-coupled Inverter Pair

We can realize a rudimentary memory element using a cross-coupled inverter pair, as shown in Figure 1.10.

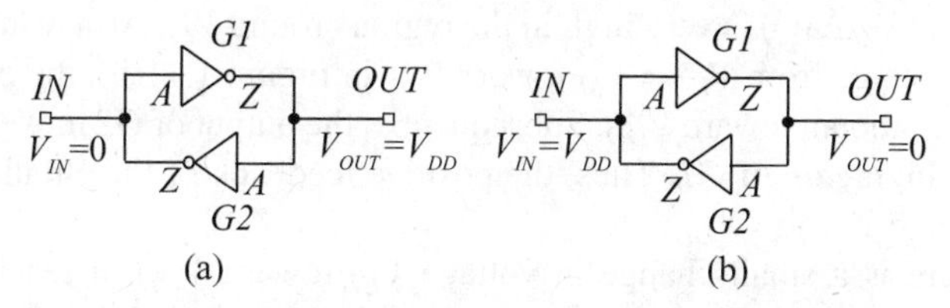

Figure 1.10 A cross-coupled inverter pair: (a) state 1 and (b) state 0

When the input *IN* is at logic 0 (V_{IN}=0 V), the output of the inverter *G1* is at logic 1 ($V_{OUT} = V_{DD}$), as shown in Figure 1.10(a). Since the inverter *G2* is cross-coupled with *G1*, the input of *G2* receives logic 1 and produces logic 0. The output of *G2* reinforces a 0 at input *IN*. Thus, the circuit operates in a positive feedback mode and attains a stable state with *IN*=0 and *OUT*=1 (equivalently V_{IN}=0 *V* and $V_{OUT} = V_{DD}$, ignoring voltage margins in the logic abstraction).

Similarly, the circuit can attain another stable state with *IN* = 1 and *OUT* = 0, as shown in Figure 1.10(b) (equivalently $V_{IN} = V_{DD}$ and $V_{OUT} = 0$ *V*).

Stable states and memory action: The above two states are stable. When a circuit reaches any of the above two states, it can remain in that state indefinitely. Due to a small noise in the circuit, the voltage level V_{IN} and V_{OUT} can change slightly. However, the positive feedback of the circuit takes back the voltage levels to stable values. Thus, we can say that the cross-coupled inverter remembers its old state and implements a rudimentary memory action.

Metastable states: For practical inverters, there is a continuous input–output voltage relationship similar to Figure 1.11(a). Let us understand the behavior of a cross-coupled inverter pair that has inverters with characteristics shown in Figure 1.11(a).

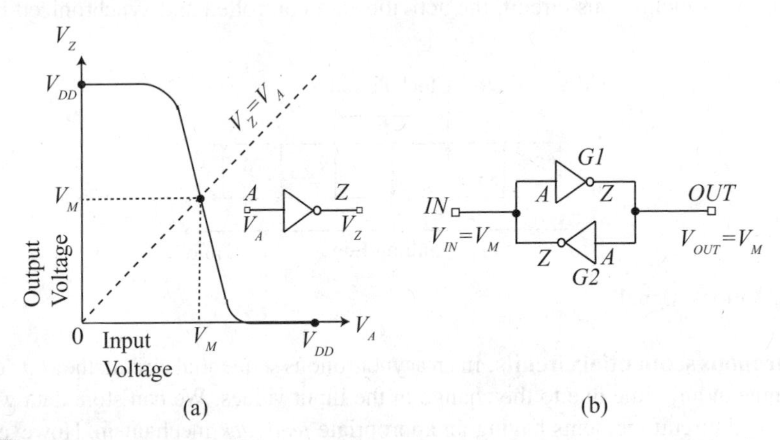

Figure 1.11 (a) Input–output voltage relationship of an inverter implemented in Complementary MOSFET (CMOS) technology and (b) metastable state

For a low input voltage ($V_A \approx 0$ V), the output voltage is high ($V_Z \approx V_{DD}$) and for a high input voltage ($V_A \approx V_{DD}$), the output voltage is low ($V_Z \approx 0$ V). These are stable states as described above.

Next, let us define a voltage V_M such that the input voltage and output voltage are the same for the inverter, i.e., $V_A = V_Z = V_M$ as shown in Figure 1.11(a). At this voltage, the cross-coupled inverter pair can reach another state with voltage $V_{IN} = V_{OUT} = V_M$, as shown in Figure 1.11(b).

It is worthy to note from Figure 1.11(a) that around the voltage V_M, the output voltage changes (ΔV_Z) by a large amount due to a small change in the input voltage (ΔV_A). In other words, the *gain* of the inverter (*gain*=$\Delta V_Z/\Delta V_A$) is very high in the region around V_M. As a result, if there is a small change in the input voltage from $V_{IN} = V_M$ toward 0 V in Figure 1.11(b), the output voltage of *G1* will change by a large amount toward V_{DD}. Subsequently, the output of *G2* moves by a larger amount toward 0 V due to a high *gain* of *G1*. Thus, the positive feedback of the circuit takes it toward the stable state $V_{IN} = 0$ and $V_{OUT} = V_{DD}$.

Similarly, if there is a small change in voltage V_{IN} toward V_{DD}, it reaches the other stable state $V_{IN} = V_{DD}$ and $V_{OUT} = 0$. Thus, from the state shown in Figure 1.11(b), it reaches one of the stable states due to a small noise perturbation. We call the state that is shown in Figure 1.11(b) as *metastable state*.

Non-deterministic behavior: Given a metastable state, we cannot predict which of the two stable states will finally be reached due to noise perturbations. Moreover, if a circuit goes into a metastable state, we cannot predict the duration for which it will remain in that state. Thus, metastable states can create *non-deterministic* behavior in a circuit. Therefore, we design digital circuits such that the probability of reaching a metastable state is practically zero.

A cross-coupled inverter pair can illustrate many concepts of memory elements. However, it is not practically useful because there is no mechanism to change its state. This difficulty is overcome by *latches* and *flip-flops*. These circuit elements have extra pins that allow us to change states and store them across *clock cycles*.

Latches and Flip-flops

Latch: The schematic symbol of a simple latch is shown in Figure 1.12(a). It consists of two input pins *D* and *EN* and one output pin *Q*. We connect the EN-pin normally to a periodic signal called *clock*. When *EN*=1, the value at the D-pin of a latch gets propagated to the Q-pin. We say that when *EN*=1, the latch has become *transparent* and it works similar to a buffer or a combinational circuit element. Thus, a latch is a *level-triggered* device. However, when *EN*=0, the value at the Q-pin remains unchanged or latched and it remembers the D-pin value when *EN* was 1. It reflects the memory action of a latch. We refer to the value at the Q-pin of a latch as its *state*.

Figure 1.12 Schematic symbol of (a) latch and (b) flip-flop

Flip-flop: The schematic symbol of a simple flip-flop is shown in Figure 1.12(b). It consists of two input pins *D* and *CP* and one output pin *Q*. The CP-pin is normally connected to a clock signal and is called the *clock pin*. A flip-flop is an *edge-triggered* device. When a *clock edge* arrives at the clock pin, the D-pin value propagates to the Q-pin. A flip-flop can be *positive* edge-triggered, meaning that the D-pin value propagates when the clock signal transitions from 0 → 1. Similarly, it can be *negative* edge-triggered, meaning that the D-pin value propagates when the clock signal transitions from 1 → 0.

Differences between a latch and a flip-flop: A latch is a level-triggered device, and a flip-flop is an edge-triggered device. We illustrate it in the following example.

Example 1.3 Assume that we apply the clock signal shown in Figure 1.13 to the EN-pin of a latch and CP-pin of a positive edge-triggered flip-flop. Further, assume that we apply the data signal to their D-pins, as shown in the figure.

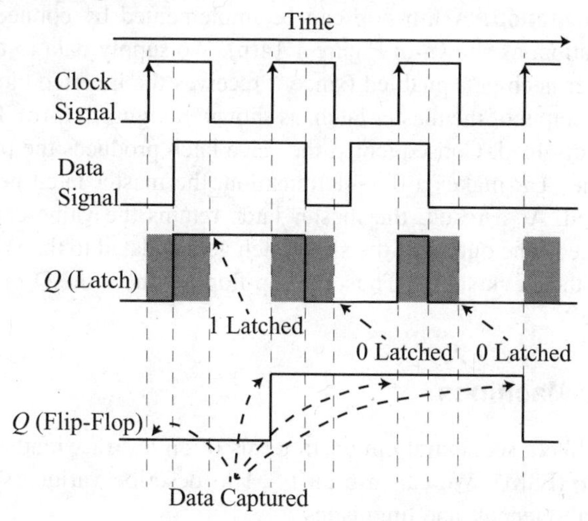

Figure 1.13 Comparison of behavior of a latch and a flip-flop

When *EN*=1 for a latch, it becomes transparent (the value at the input propagates to the Q-pin), as shown by the shaded region. When *EN*=0, the value at the Q-pin remains latched to the last input value. In contrast, a flip-flop captures data only at the positive clock edges.

Latch implementation: We can implement a latch using a multiplexer, as shown in Figure 1.14(a).

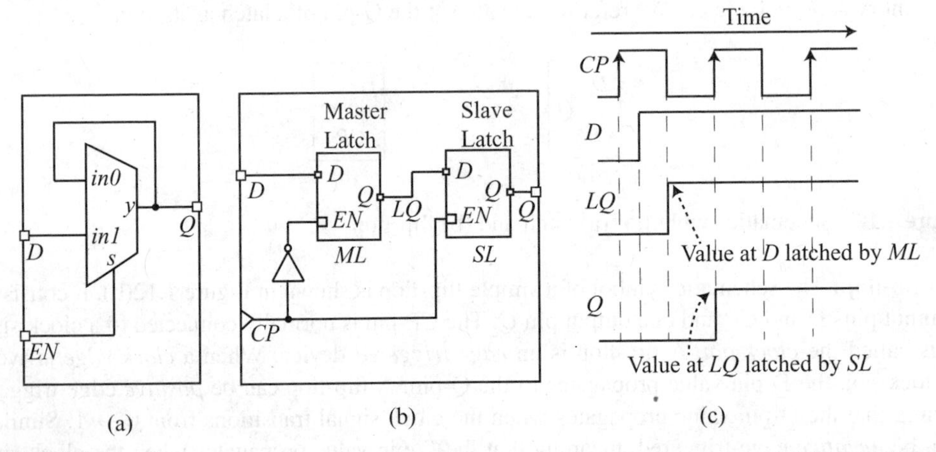

Figure 1.14 Latch and flip-flop: (a) latch implementation using multiplexer, (b) positive edge-triggered flip-flop implementation using master–slave latches, and (c) waveform in a flip-flop

When *EN*=1, the pin *in1* of the multiplexer is selected. As a result, the value at the D-pin of the latch propagates to the Q-pin. When *EN*=0, the pin *in0* of the multiplexer is selected. Since the pin *in0* is connected to the output pin *y*, the output simply recirculates through the pin *in0*. As a result, the Q-pin retains its old value.

Flip-flop implementation: A flip-flop can be implemented by connecting two latches in a master–slave configuration, as shown in Figure 1.14(b). We supply data to the master latch (*ML*). When *CP*=0, the master latch gets enabled (since it receives the inverted clock signal). Therefore, the data appears at the output of the master latch, as shown in Figure 1.14(c). However, when *CP*=0, the slave latch (*SL*) is disabled. Consequently, the slave latch produces the previously stored value till *CP* remains 0. When *CP* makes a 0 → 1 transition, the master latch gets disabled, while the slave latch gets enabled. As a result, the master latch retains the value captured just before the 0 → 1 transition occurred. The output of the slave latch gets updated to the value of the master latch on receiving 0 → 1 of the clock signal. Thus, the flip-flop produces the D-pin value at the Q-pin at the positive clock edge.

1.1.7 Finite State Machines

We can formally describe a sequential circuit in terms of an abstract mathematical model known as *finite state machine* (FSM). We can use an FSM to describe various systems in electronics, computation, biological sciences, and linguistics.

An FSM consists of a finite *set of states, inputs, outputs*, conditions for *transitioning* from one state to another, and a given *initial state*. In a sequential circuit, we use flip-flops to model and store different states. The function describing how an FSM changes from one state to another is known as the *transition function*. The transition function of an FSM is modeled using a combinational circuit and is a function of the present state and inputs.

We can model the output of an FSM in two ways:

1. We consider output as a function of *only* states. We call such FSMs as *Moore machine*.
2. We consider output as a function of both states and inputs. We call such FSMs as *Mealy machine*.

State diagram: We can represent FSMs pictorially using a *directed graph* known as *state diagram*. The states of an FSM form the vertices of the state diagram. We represent the transition function using edges in the state diagram. An edge starts from a present state and ends in the next state. We mark the input values that lead to a particular transition on the edges. We mark the output inside the state for the Moore machine and at the edges for the Mealy machine.

Figure 1.15 shows two states *S1* and *S2* for a Moore machine and a Mealy machine. The machine transitions from *S1* to *S2* on receiving an input *I1*. The Moore machine has the output *O1* in the state *S1* and *O2* in the state *S2*. The Mealy machine produces an output *O1* when it transitions from *S1* to *S2* on receiving the input *I1*.

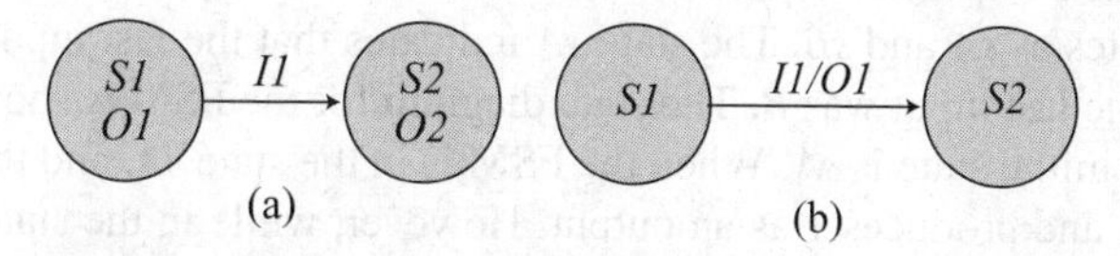

Figure 1.15 State diagram for (a) Moore machine and (b) Mealy machine

Example 1.4 We illustrate two examples of FSM.

1. **Moore machine:** Consider a fan regulator. Let us represent the four different settings of the fan regulator as the four states: *s0*, *s1*, *s2*, and *s3*. Let there be two inputs to the fan regulator: rotate clockwise represented as *C* and rotate anticlockwise represented as *A*.

   Assume that the initial state is *s0*. From the state *s0*, when the input *C* is applied successively, the states

   $$s1 \rightarrow s2 \rightarrow s3 \rightarrow s0 \rightarrow s1 \rightarrow s2 \rightarrow s3 \rightarrow s0 \rightarrow \ldots$$

   are reached in a circular fashion. Similarly, from the state *s0*, when the input *A* is applied successively, the states

   $$s3 \rightarrow s2 \rightarrow s1 \rightarrow s0 \rightarrow s3 \rightarrow s2 \rightarrow s1 \rightarrow s0 \rightarrow \ldots$$

   are reached in a circular fashion. The speed of the fan would be different under these four settings. Let the output speed be represented as *speed0*, *speed1*, *speed2*, and *speed3*, corresponding to the states *s0*, *s1*, *s2*, and *s3*. We show the state diagram for the fan regulator in Figure 1.16(a). Since the output speed depends just on the state of the fan regulator, this is an example of a Moore machine.

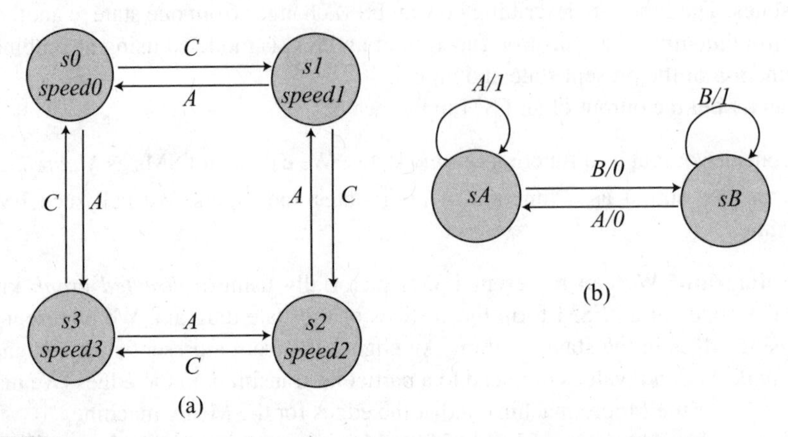

Figure 1.16 State diagram for (a) fan regulator and (b) input repeat detector

2. **Mealy machine:** Consider an FSM that detects whether the input is repeated. Suppose the input can be either *A* or *B*. Let the output be 1 when the input is repeated, and 0 otherwise. Let us represent the states as *sA* and *sB*. The state *sA* indicates that the last input was *A*. The state *sB* represents that the last input was *B*. The state diagram for the FSM is shown in Figure 1.16(b). Assume that the initial state is *sA*. When the FSM is in the state *sA*, and the input *A* arrives, the state remains *sA* and produces 1 as an output. However, while in the state *sA*, if *B* arrives, the state changes to *sB* and produces a 0 as an output. Similarly, we can determine the transitions from *sB* and the output values. In this case, since the output depends on the state and the input values, the FSM is a Mealy machine.

In this section, we described the basic concepts related to a digital system. For more discussions on these topics, readers can refer to textbooks on digital circuit design such as [2, 3].

1.2 DEVICES AND CIRCUITS

We use transistors to implement digital circuits. In this section, we review concepts related to transistors.

1.2.1 P–N Junction

The theory of p–n junction is essential in understanding semiconductor devices. A p–n junction gets formed when regions of p-type (rich in holes) and n-type (rich in electrons) are adjacent. Holes diffuse from the p-type region to the n-type region due to a density gradient. It leaves negatively charged acceptor atoms in the p-type region. Similarly, electrons diffuse from the n-type region to the p-type region due to a density gradient. It leaves positively charged donor atoms in the n-type region. Thus at the junction of the p-type region and n-type region, the material becomes devoid of mobile carriers, and this region is known as the *depletion region*.

The separation of charges at the junction leads to an electric field that opposes further diffusion of electrons from the n-type region and holes from the p-type region. As a result, a potential difference exists across the depletion region. This potential difference is known as the *built-in potential barrier*.

Reverse-biased mode: When we apply a bias such that the p-type region is at a lower potential than the n-type region, we say that the p–n junction is *reverse biased*. The potential barrier at the junction further increases when a p–n junction is reverse-biased. Therefore, negligible current flows through the p–n junction in the reverse-biased mode. With the change in the reverse bias voltage, the charge inside the depletion region changes. Hence, we can define *junction capacitance* (C_j) as the incremental charge change (ΔQ) in the depletion region with the change in the reverse bias voltage (ΔV_R) (i.e., $C_j = \Delta Q/\Delta V_R$).

Forward-biased mode: When we apply a bias such that the p-type region is at a higher potential than the n-type region, we say that the p–n junction is *forward-biased*. In the forward-biased mode, the potential barrier reduces. Consequently, electrons from the n-type region and holes from the p-type region can flow freely across the junction. It establishes a forward-biased current which is significantly higher than the reverse-biased current. Hence, the behavior of a p–n junction is different in the forward-biased and reverse-biased modes.

1.2.2 MOSFETs

There are two types of *metal oxide semiconductor field-effect transistors* (MOSFETs): *n-channel MOSFET* (NMOS) and *p-channel MOSFET* (PMOS). In an NMOS, the current conduction is primarily due to *electrons*, while in a PMOS, the current conduction is primarily due to *holes*.

Structure: Figure 1.17(a) shows the structure of a typical NMOS. It consists of n^+-doped source and drain. The substrate is p-type silicon. A p–n junction gets formed at the source–channel and drain–channel junctions because the dopings are of the opposite type. There is a gate oxide (insulator) sandwiched between the gate electrode and the silicon body. The gate oxide material can be silicon dioxide or a high-k dielectric material such as HfO_2. The gate electrode can be a metal or polycrystalline silicon. The metallic gate electrode (M), insulating oxide (O), and the silicon semiconductor (S) form the *MOS* structure in a MOSFET. The source and the body terminals are often tied together in a MOSFET.

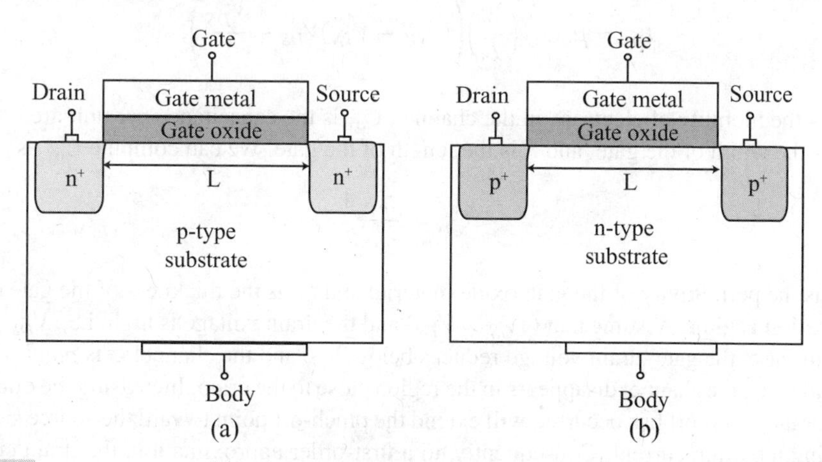

Figure 1.17 Cross-sectional view of a typical (a) n-channel MOSFET and (b) p-channel MOSFET

A PMOS also consists of a MOS structure. However, it has p^+-doped source and drain. The substrate is n-type silicon, as shown in Figure 1.17(b). The gate electrode in a PMOS is metal or polycrystalline silicon, similar to an NMOS. However, typically, it has a different work function than NMOS.

Working principle: The heart of a MOSFET is the MOS structure that works similarly to a parallel-plate capacitor. The gate electrode and the semiconductor act as two electrodes of a parallel-plate capacitor. The gate oxide works as the dielectric between the parallel plates. When voltage is applied to the gate electrode, the electron and hole concentrations change inside the semiconductor due to the *field effect*.

In an NMOS, if the gate–source voltage (V_{GS}) is sufficiently high (higher than the threshold voltage (V_{TN})), we obtain a high concentration of electrons in a thin layer inside the semiconductor just below the gate. This highly conducting layer of electrons is referred to as the *channel*. When the channel gets formed, if we apply a drain–source voltage (V_{DS}), current (I_{DS}) will flow through them. We name the terminal that supplies electrons to the channel as the *source*, and the one which takes them away as the *drain*. The current I_{DS} depends on the voltage V_{DS}. The gate–source voltage V_{GS} also controls I_{DS} by modulating the channel's charge and conductivity.

A PMOS works analogously; however, the current flow is due to holes instead of electrons.

Operation: First, let us consider the operation of an NMOS. For simplicity, let us assume that we have connected the source and the body terminals of the NMOS to the ground or 0 V.

Subthreshold region: When $V_{GS} < V_{TN}$, the potential barriers between the source/drain and the channel block the current. The MOSFET is said to operate in the subthreshold region or the OFF-state. Ideally, I_{DS} should be zero in the OFF-state. However, due to the statistical thermal energy distribution (Fermi-Dirac energy distribution), some electrons possess enough energy to surmount the potential barrier and move from the source to the drain. It constitutes the *subthreshold current* of the device.

Linear region: When $V_{GS} > V_{TN}$, a conducting channel gets formed just below the gate. Let us assume that we have applied a low drain voltage, i.e., $0 < V_{DS} < (V_{GS} - V_{TN})$. In this case, an NMOS behaves as a simple resistor. Hence, it is said to operate in the *linear region*. We can compute the drain current I_{DS} as follows:

$$I_{DS} = \mu_n C_{ox} \left(\frac{W}{L}\right)\left((V_{GS} - V_{TN})V_{DS} - \frac{V_{DS}^2}{2}\right), \tag{1.1}$$

where μ_n is the mobility of electrons in the channel, C_{ox} is the capacitance per unit area of the gate oxide, W is the width of the gate, and L is the length of the gate. We can compute C_{ox} as

$$C_{ox} = \frac{\epsilon_{ox}}{t_{ox}}, \tag{1.2}$$

where ϵ_{ox} is the permittivity of the gate oxide material and t_{ox} is the thickness of the gate oxide.

Saturation region: Assume that ($V_{GS} > V_{TN}$) and the drain voltage is high, i.e., $V_{DS} \geq (V_{GS} - V_{TN})$. In this case, the gate–drain voltage reduces below V_{TN} and the channel gets *pinched off* from the drain side, i.e., the channel disappears in the region close to the drain. Increasing the drain voltage further after the pinch-off has occurred will extend the pinch-off point toward the source side instead of increasing the drain current. Consequently, to a first-order approximation, the drain current I_{DS} becomes independent of the applied drain voltage V_{DS}, and the I_{DS} reaches a constant value known as the *saturation current*. This region of operation is known as the *saturation region* and the current I_{DS} can be given as

$$I_{DS} = \frac{\mu_n C_{ox}}{2}\left(\frac{W}{L}\right)(V_{GS} - V_{TN})^2. \tag{1.3}$$

Channel length modulation: If we consider second-order effects in the saturation region, then the drain voltage also impacts the drain current. As we increase V_{DS}, the pinch-off point extends toward the source. As a result, the effective length of the channel decreases. This phenomenon is known as *channel length modulation*. Consequently, the drain current increases slightly with the drain voltage in the saturation region also. We can model this effect as follows:

$$I_{DS} = \frac{\mu_n C_{ox}}{2}\left(\frac{W}{L}\right)(V_{GS} - V_{TN})^2(1 + \lambda V_{DS}), \tag{1.4}$$

where λ is the channel length modulation parameter and is inversely proportional to the gate length.

Short-channel effects: With the advancement in technologies, the transistor sizes and channel lengths get reduced. Consequently, the gate electrode can control a smaller fraction of the total channel charge. Moreover, the distance between the source and drain reduces. As a result, the drain can have increased control over the drain current compared to the gate. Therefore, V_{TN} of a MOSFET decreases on increasing V_{DS}. This phenomenon is known as *Drain-Induced Barrier Lowering* (DIBL). Moreover, due to the less control of the gate, the V_{TN} decreases with the decrease in the gate length. This effect is known as *threshold voltage roll-off*. Since these effects become prominent at small channel lengths, these effects are referred to as *short-channel effects*.

PMOS behavior: A PMOS behaves analogously to an NMOS. Nevertheless, some prominent differences are as follows. We connect the source and the body terminals of a PMOS to the supply voltage instead to the ground. The voltage polarity and the current direction are opposite in a PMOS. Additionally, we should consider the mobility of holes instead of electrons in the current equations.

Switch model: In this book, we are primarily considering digital systems. In digital systems, for simplicity, we can model MOSFETs as a switch. The drain and the source terminals are the contacts. The gate is the controlling terminal. The current flows between the contacts when the controlling terminal closes the switch.

For an NMOS, when $V_G \approx 0\ V$ the switch is open, and when $V_G \approx V_{DD}$ it is closed (assuming $V_S = 0\ V$). For a PMOS, when $V_G \approx 0\ V$ the switch is closed, and when $V_G \approx V_{DD}$ it is open (assuming $V_S = V_{DD}$).

In this section, we described the basic concepts of MOSFETs that are important in understanding subsequent chapters. However, for an in-depth understanding of MOSFETs, the readers can refer to textbooks on semiconductor devices such as [4, 5].

1.2.3 CMOS Inverter

A *Complementary MOSFET* (CMOS) inverter is the simplest digital circuit that we can implement in the CMOS technology. A good understanding of a CMOS inverter builds the foundation for digital VLSI design. Therefore, we explain some of the basic concepts related to a CMOS inverter in the following paragraphs.

Static Characteristics

A CMOS inverter consists of a PMOS and an NMOS, as shown in Figure 1.18(a). The static characteristic of a CMOS inverter is shown in Figure 1.11(a). We show it again in Figure 1.18(b) for convenience. Let us understand how a CMOS inverter produces this characteristic.

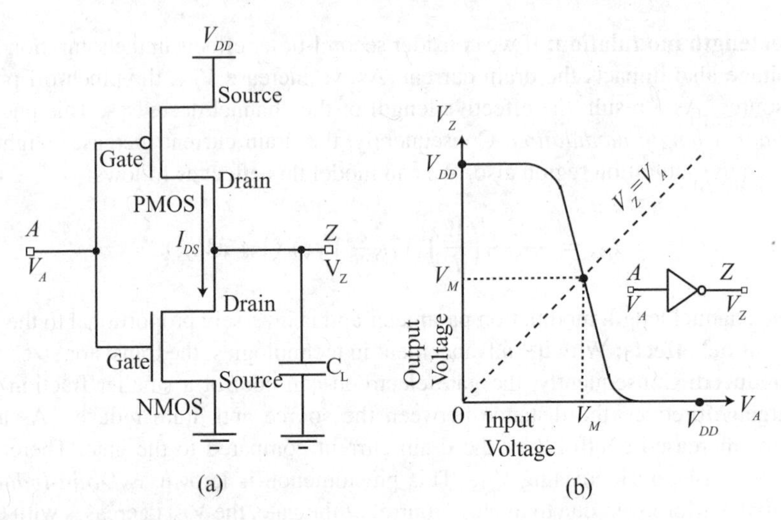

Figure 1.18 CMOS inverter: (a) circuit diagram and (b) static characteristic

First, let us consider MOSFETs as switches. When $V_A \approx 0$, the NMOS is switched-off, while the PMOS is switched-on. Therefore, the output Z gets connected to the power supply V_{DD} through the PMOS. As a result, the capacitor C_L gets charged to V_{DD}. Next, let us consider $V_A \approx V_{DD}$. The PMOS is switched-off, while the NMOS is switched-on. The output Z gets connected to the ground (0 V) through the NMOS. As a result, the capacitor C_L gets discharged to the ground potential. This explains the CMOS inverter characteristic for $V_A \approx 0$ and $V_A \approx V_{DD}$.

In the intermediate region, as the input voltage increase from 0 V to V_{DD}, the NMOS transistor moves from the cutoff to the saturation and finally to the linear region. Similarly, the PMOS moves from the linear to the saturation and finally to the cutoff region. The NMOS and the PMOS behave differently in these regions. Nevertheless, in the steady state, the drain current flowing through both these transistors should be the same. Note that no current flows through the capacitor C_L in the steady state. Therefore, both transistors are forced to keep the drain voltage such that the same I_{DS} flows through PMOS and NMOS. It makes the output voltage reach a steady-state voltage somewhere between V_{DD} and 0 V.

High gain at V_M: There is a sharp fall in the output voltage V_Z when input is $V_A = V_M$, as shown in Figure 1.18(b). Under this condition, *both* the transistors are operating in their respective saturation regions. In the saturation region, the drain current is ideally independent of the drain voltage. It only depends on the gate voltage. As a result, the drain voltage (V_Z) of both transistors can vary independently of their gate voltages (V_A). Thus, the output voltage V_Z can change arbitrarily with respect to the input voltage. Therefore, ideally, the gain of the inverter ($= \Delta V_Z / \Delta V_A$) will be infinite at $V_A = V_M$. However, the drain current increases gradually in the saturation region due to channel length modulation, and we observe a finite gain at $V_A = V_M$.

Adjusting V_M for noise margin: To obtain a high noise margin, we should keep the point V_M close to $V_{DD}/2$. However, the mobility of the electrons is higher compared to the mobility of holes in silicon. Therefore, we keep the PMOS width more than the NMOS width. It compensates for the mobility differences, and we can obtain V_M close to $V_{DD}/2$.

Complementary configuration: In general, a CMOS circuit combines a pull-up and pull-down network in a complementary symmetry configuration. The pull-up network pulls the output node to V_{DD}, while the pull-down network takes it to the ground. The complementary configuration ensures that under *static conditions*, the output node is connected either to V_{DD} or ground through a low resistance path for any combination of bits at the inputs. It ensures that the output voltage of the circuit is more resilient to noise. Furthermore, since there is always at least one MOSFET in the switched-off state in the path from V_{DD} to ground, a CMOS circuit draws negligible power from the battery under the static condition.

Delay

Let us understand the delay characteristics of a CMOS inverter using the switch model of the MOSFETs.

High-to-low transition: Let the input make a high-to-low transition. We can assume that the MOSFETs change their ON/OFF states instantaneously. Consequently, the PMOS gets switched-on, while the NMOS gets switched-off instantaneously. In the switched-on state, let us assume that the PMOS offers an average resistance R_P, as shown in Figure 1.19(a). The capacitor load C_L gets charged by the power supply through the PMOS, and the output makes a low-to-high transition with some delay. The delay depends on the time taken to charge the capacitor C_L through the resistance R_P. It is proportional to R_PC_L. To reduce the delay of the inverter, we need to reduce C_L. Furthermore, we can reduce the delay by reducing R_P. We can accomplish this by increasing the W/L ratio of the PMOS.

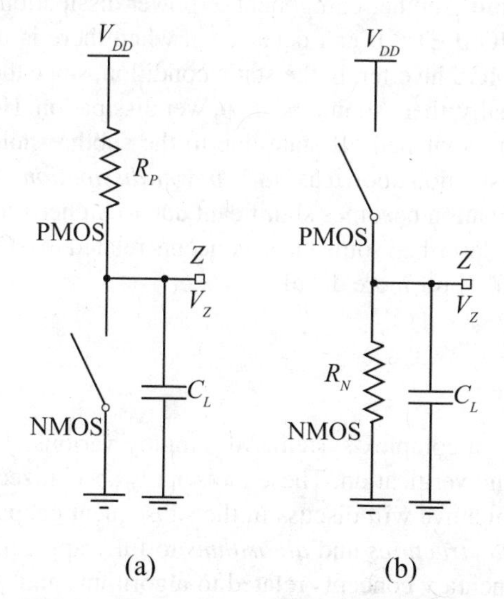

Figure 1.19 Simplistic model of a CMOS inverter making a (a) high-to-low transition at the input and (b) low-to-high transition at the input

Low-to-high transition: The equivalent circuit for the case when the input makes a low-to-high transition is shown in Figure 1.19(b). The load capacitor C_L discharges from V_{DD} to ground potential through the NMOS. Let us assume that the NMOS offers an average resistance of R_N. The delay depends on the time taken to discharge the capacitor C_L through R_N. It is proportional to $R_N C_L$. We can decrease it by increasing the W/L ratio of the NMOS and reducing C_L.

Power Dissipation

There are mainly three kinds of power dissipation in a CMOS inverter.

Switching power dissipation: During a low-to-high transition at the output, the capacitor C_L gets charged from 0 V to V_{DD} through the PMOS. While charging, some energy gets dissipated, while some energy gets stored in the capacitor C_L. When the output makes a high-to-low transition, the stored energy in the capacitor C_L gets dissipated. The power dissipated by the CMOS inverter during charging or discharging of the load capacitor is known as the *switching power dissipation*. If an inverter switches with a frequency of f, then the switching power dissipation P_D can be computed as follows:

$$P_D = C_L V_{DD}^2 f \tag{1.5}$$

The above equation suggests that we can reduce P_D by decreasing the load capacitance C_L, the supply voltage V_{DD}, and the switching frequency f.

Short circuit power dissipation: When the input of an inverter changes slowly, both its transistors are switched on for some time. During this time, there is a direct path from V_{DD} to the ground. As a result, a high transient current flows through these transistors. This current spike is responsible for power dissipation, known as *short circuit power dissipation*. We can reduce it by decreasing the input transition time.

Static power dissipation: Another component of power dissipation in a CMOS inverter is due to the leakage current of MOSFETs. It can occur even when there is no switching activity at the input of the inverter. In a CMOS inverter, in the static condition, since there is at least one transistor in the switched-off state, ideally, there should be no power dissipation. However, some current flows through the transistors in the switched-off state due to the subthreshold conduction. This current gives rise to some power dissipation known as *static power dissipation*. At the advanced technology nodes, the static power dissipation becomes significant due to higher subthreshold conduction.

In this section, we have described some basic concepts related to a CMOS inverter. Readers can refer to textbooks such as [6, 7] for more details.

1.3 ALGORITHMS

An integrated circuit (IC) is a complex system. We employ various electronic design automation (EDA) tools for its design and verification. These tools apply specialized algorithms and techniques to solve EDA problems, which we will discuss in the subsequent chapters. We need to understand some basic concepts of *data structures* and *algorithms* to fully appreciate them. In the rest of this chapter, we review the elementary concepts related to algorithms and data structures. Readers can refer to specialized textbooks, such as [8], for a detailed discussion on these topics.

1.3.1 Computational Problems

Algorithms are tools for solving *computational problems*. We can define a computational problem as a relationship between a given set of inputs and the desired outputs. For example, *sorting* is a widely studied computational problem. In a sorting problem, a sequence of N numbers is the given input. The output is another sequence in which we arrange the given numbers in a non-decreasing manner.

Problem instance: A given set of inputs for a computational problem is known as the *instance* of a problem. For example, for the sorting problem, an instance of the sorting problem is ⟨20, 2, 65, 32, 12⟩ and the output is ⟨2, 12, 20, 32, 65⟩.

Algorithm: An algorithm is a sequence of computational steps that solves a given computational problem. It transforms the inputs of a problem instance into the desired output. Some popular algorithms for solving the sorting problem are bubble sort, insertion sort, quick sort, merge sort, and heap sort.

1.3.2 Computational Effort

The computational effort of an algorithm depends on the size of a given problem instance. The number of items in a given input set typically quantifies the size of the problem instance. For example, if N is large for the sorting problem, the computational effort will be more.

Computational resource requirements: For a given computational problem, there can be several algorithms that can produce the desired output. However, for the same problem, different algorithms can consume different amounts of computational resources. The computation resources can be *memory* and running time. We call the running time of an algorithm as *runtime* in short. Typically, the runtime is the most critical efficiency parameter for an algorithm. We quantify runtime by the number of *elementary operations* executed by the algorithm for a given problem instance. The elementary operations are the resource-consuming steps performed by an algorithm and are easily identifiable for a given algorithm. For example, in the sorting algorithm, elementary operations are comparing two numbers, indexing them into an array and assigning a value to a variable.

Average case and worst-case behavior: While designing an algorithm, it is often necessary to analyze the *computational effort* for various problem instances. The computational effort can be runtime or some quantity relevant to a given problem. It is worth mentioning that, even for the same problem instance size, the computational effort required by an algorithm can be different for different inputs. For example, some sorting algorithms can take less runtime for a sorted input compared to reverse-sorted input. Therefore, it is desirable to estimate the *average* computational effort of an algorithm by considering all possible inputs for a given problem instance size. However, it is often difficult to analyze the average case behavior of an algorithm. It involves considering the probability distribution of the inputs, which is typically unknown and challenging to model. In practice, finding the problem instance that results in the *worst*-case behavior of an algorithm is comparatively easy. Therefore, typically, we do the worst-case analysis of an algorithm and try to improve it. We hope that improving the worst-case behavior will also improve the average case behavior of the algorithm.

1.3.3 Asymptotic Analysis of an Algorithm

In the analysis of an algorithm, we often determine the computational effort's *growth rate* as a function of the input size rather than determining the worst-case computational effort. Moreover, we carry out *asymptotic analysis* of an algorithm and focus on the algorithm's behavior for large problem instance size. It allows us to represent the computational effort only in terms of the *dominating* factors and simplifies our analysis of the algorithms. Typically, we represent the asymptotic behavior of an algorithm using *big-O notation*.

Big-O notation: Suppose there is a problem instance of size n. Let us measure the computational effort by the number of elementary operations carried out by the algorithm. Let us denote the number of elementary operations carried out by the algorithm in solving this problem instance by $f(n)$. If there exist some constants c and n_0 such that for all $n \geq n_0$,

$f(n) \leq cg(n)$,

then we say that the *order* of the function $f(n)$ is $g(n)$, and we write

$f(n) = O(g(n))$.

Example 1.5 Let $f(n) = 2n^2 + 5n + 12$.

We can write $f(n) \leq 2n^2 + 5n^2 + 12n^2$, for $n \geq 1$. Therefore, $f(n) \leq 19n^2$, and we can say $f(n) = O(n^2)$, for $c = 19$.

The order of the function $f(n)$ says that the growth rate of $f(n)$ is no more than $g(n) = n^2$. The effect of lower-order terms (such as n) and constants (such as 2, 5, and 12) get abstracted out in the asymptotic analysis. However, large constant factors can also impact the computational effort if the problem instance size is small.

Note that, we can also write, $f(n) \leq 2n^3 + 5n^3 + 12n^3$ for $n \geq 1$ and $f(n) \leq 19n^3$. Therefore, we can write $f(n) = O(n^3)$, as well. However, in general, we use big-O notation to characterize an upper bound on a function as tightly as possible. Therefore, we prefer $O(n^2)$ over $O(n^3)$.

1.3.4 "Hard" Problems

A few examples of $g(n)$ that we commonly use in the big-O notation are: $O(1)$, $O(n)$, $O(n^2)$, $O(log(n))$, $O(nlog(n))$, $O(2^n)$, and $O(n^n)$. When an algorithm takes constant effort for all problem instance sizes, we denote its complexity as $O(1)$. In general, algorithms with an order of *polynomial* ($g(n) = n^k$, where k is a constant) or the ones that grow slower than a polynomial are desirable.

NP-hard problems: There is a class of problems whose worst-case behavior is exponential (such as $O(2^n)$ and $O(n^n)$), and we do not know of any polynomial complexity algorithm that can solve them. Moreover, if any problem in this class can be solved using a polynomial complexity algorithm, we can solve all problems in this class with a polynomial complexity algorithm. Consequently, researchers have put considerable effort into finding a polynomial complexity algorithm that can solve any problem in this class of problems but have been unsuccessful. Therefore, it is widely believed that no such algorithm exists, though it has not yet been proven that such an algorithm cannot exist. This class of problems is known as *NP-hard* problems.

NP-complete problems: It has been shown that some of the NP-hard problems can be solved using algorithms of polynomial complexity running on *non-deterministic* machines (hypothetical machines that can make guesses and that do not exist). These problems are known as *NP-complete* problems. Unfortunately, there are many problems in VLSI design and verification that are NP-hard or NP-complete. We need to tackle them using some heuristics or approximation techniques.

1.4 DATA STRUCTURES

A *data structure* is a method of organizing data in a computer. It allows efficient computational resource usage. We employ different kinds of data structures based on data types, the given computational problem, and the computational resource constraints.

In VLSI design, voluminous data is generated and consumed routinely. Therefore, we need to often choose data structures that allow efficient storage, easy manipulation, and are suitable for implementing EDA algorithms.

1.4.1 Arrays

An array is a data structure in which data items of the same type are stored contiguously in the memory. Consequently, any data item can be referred to in the memory by its *index*. Given an index, we can locate a data item in the memory by simply adding an offset to the base location, as shown in Figure 1.20. Thus, given an index, we can access a data item in an array in constant time or with a complexity of $O(1)$.

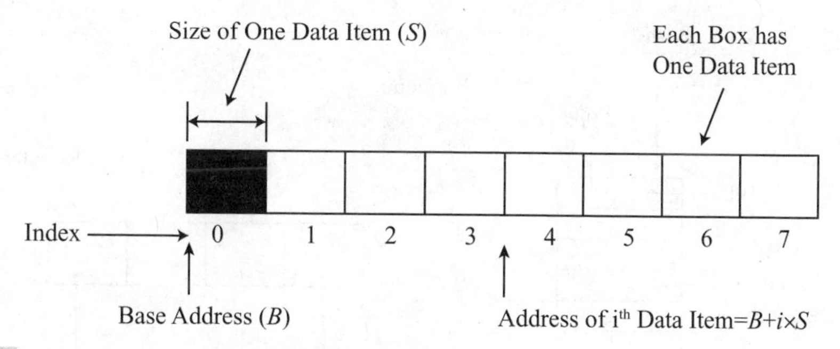

Figure 1.20 Schematic representation of an array

1.4.2 Linked Lists

A linked list consists of entities known as *nodes*. A node contains a data item and a *link* (or pointer or reference) to the *next* node, as shown in Figure 1.21. Using the link to the next node, we can traverse a linked list from the *head* to the last element. We indicate the end of a linked list by pointing the next link to 0 or some sentinel element.

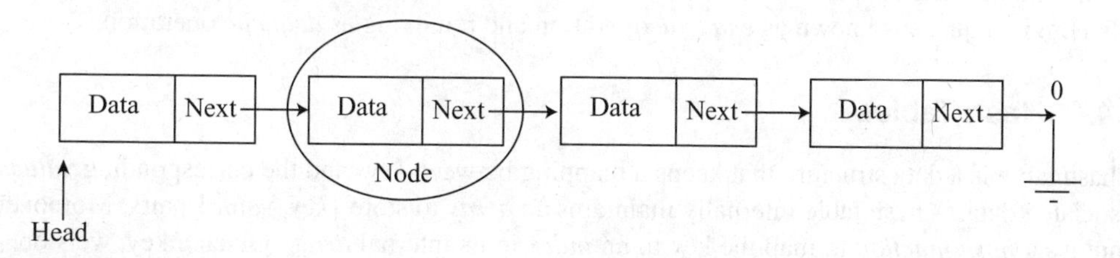

Figure 1.21 Schematic representation of a linked list

In contrast to an array, the data items are not stored contiguously in the memory in a linked list. Therefore, inserting or removing a data item from somewhere in the middle of a linked list can be done in constant time by making updates to the existing link and creating new ones. However, these operations in an array will be more complex because they can require complete restructuring.

The disadvantage of a linked list is that we cannot randomly access an *i th* element, in contrast to an array. To access an *i th* element of a linked list, we must traverse all previous $(i - 1)$ elements sequentially.

1.4.3 Stacks and Queues

Stacks and queues are sets in which we can insert and remove elements dynamically. In a stack, the element that we inserted most recently is removed first. Therefore, a stack follows the *last-in-first-out* (LIFO) policy. A stack can be visualized as shown in Figure 1.22(a). The insertion and removal of a data item are carried out at the top of the stack. The process of inserting a data item in a stack is known as *push* operation and removing as *pop* operation.

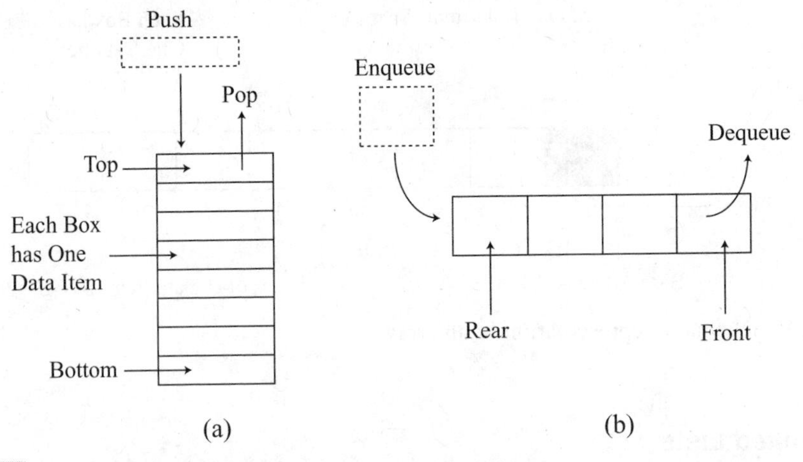

Figure 1.22 Schematic representation of a (a) stack and (b) queue

On the other hand, in a queue, the element that we inserted first is removed first. Therefore, the queue follows a *first-in first-out* (FIFO) policy. A queue can be visualized as shown in Figure 1.22(b). We insert a data item at the rear of a queue and remove it from the front. The process of inserting a data item in a queue is known as *enqueue* operation and removing as *dequeue* operation.

1.4.4 Hash Tables

A hash table is a data structure that keeps a mapping between *keys* and the corresponding *values* or associated data. A hash table internally maintains an *array* to store {*key*, *value*} pairs. Moreover, it employs a *hash function* to map the key to an *index* in its internal *array*. Given a key, we store the value and retrieve it from the internal array at the index obtained by the hash function.

Example 1.6 Consider a hash table that internally maintains an array of 11 elements. Let the hash function for the hash table be *index=key*%11, where % denotes modulus operation.

Let us insert the following {*key*, *value*} pairs in the hash table: {27,91}, {125,85}, {22,68}, and {65,79}. For the keys 27, 125, 22, and 65, the indices can be computed using the hash function as 5, 4, 0, and 10, respectively. Hence, we can visualize the hash table as shown in Figure 1.23. The values 91, 85, 68, and 79 are stored at indices 5, 4, 0, and 10, respectively.

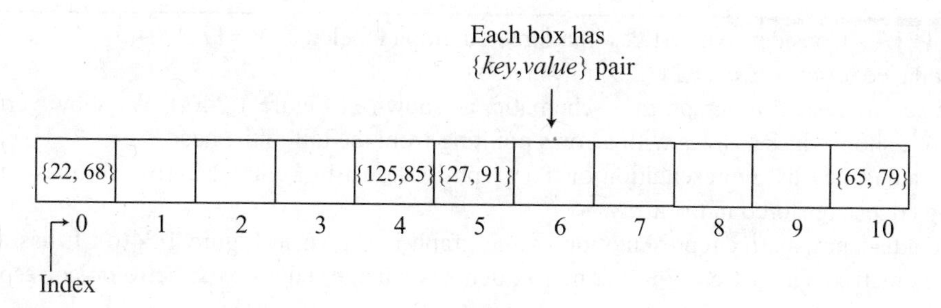

Figure 1.23 Schematic representation of a hash table

Suppose we need to find a value for the *key*=65. We compute the index using the hash function (65%11=10). Then, we retrieve the value 79 from the 10th index in the array. Thus, we can insert and retrieve data in a hash table in constant time or with a complexity of $O(1)$.

Ideally, the hash function should map each key to a unique index in the internal array. However, in practice, a hash function can generate the same index for multiple keys. For example, index 0 will be generated by the hash function *key*%11 for all keys 0, 11, 22, 33, and other multiples of 11. This phenomenon is known as *collision*.

We can handle collision by maintaining a list of {*key*, *value*} pairs at each slot in the array instead of a single {*key*, *value*} pair. Note that the worst-case complexity of insertion and retrieval from a hash table will be worse than $O(1)$ in this case.

1.4.5 Graphs

Graphs are widely used in EDA algorithms because we can model many EDA problems as well-known graph problems.

A graph consists of a finite set of *vertices* (V) and a finite set of *edges* (E). A vertex is also called a *node*. An edge is a connecting link between a pair of vertices.

Directed and undirected graphs: There are two types of graphs: *directed* graph and *undirected* graph. In a directed graph, an edge has a direction associated with it, and we represent it by an ordered pair of vertices. Let us denote two vertices in a graph as u and v. Then, an edge from the vertex u to v is represented as (u, v). The starting vertex u is called the *tail* of the edge and the ending vertex v is called the *head* of the edge.

In an undirected graph, edges do not have any direction associated with them. Therefore, an edge from u to v is the same as an edge from v to u. We represent an undirected edge by an unordered pair $\{u, v\}$.

Representation: In a computer, we typically represent a graph using one of the two methods: *adjacency list* or *adjacency matrix*.

In adjacency list representation, each vertex is stored in an array V. For each vertex $u \in V$, a list is maintained that contains vertices v such that $(u, v) \in E$.

In adjacency matrix representation, we maintain a matrix A of size $V \times V$. Furthermore, each vertex $u \in V$ is associated with a number in $[1, |V|]$. Therefore, each vertex can be written as v_i where $1 \leq i \leq |V|$. An element $a_{ij} \in A$ is 1 if $(v_i, v_j) \in E$, else it is 0. For a sparsely connected graph, the adjacency list representation is compact and preferred.

Example 1.7 Consider five vertices in a directed graph labeled as $V = \{1,2,3,4,5\}$.

Let the edge set be $E=\{(1,2),(2,3),(3,4),(1,3),(1,5),(5,4)\}$.

We can represent the graph in a schematic, as shown in Figure 1.24(a). We show vertices as circles. We show edges as lines with arrows pointing from the tail to the head.

The adjacency list representation of the graph is shown in Figure 1.24(b). Vertices and links denoting edges are stored in the array.

The adjacency matrix representation of the graph is shown in Figure 1.24(c). It has a 5×5 adjacency matrix. The 0/1 entry in the matrix denotes whether edges exist between corresponding vertices.

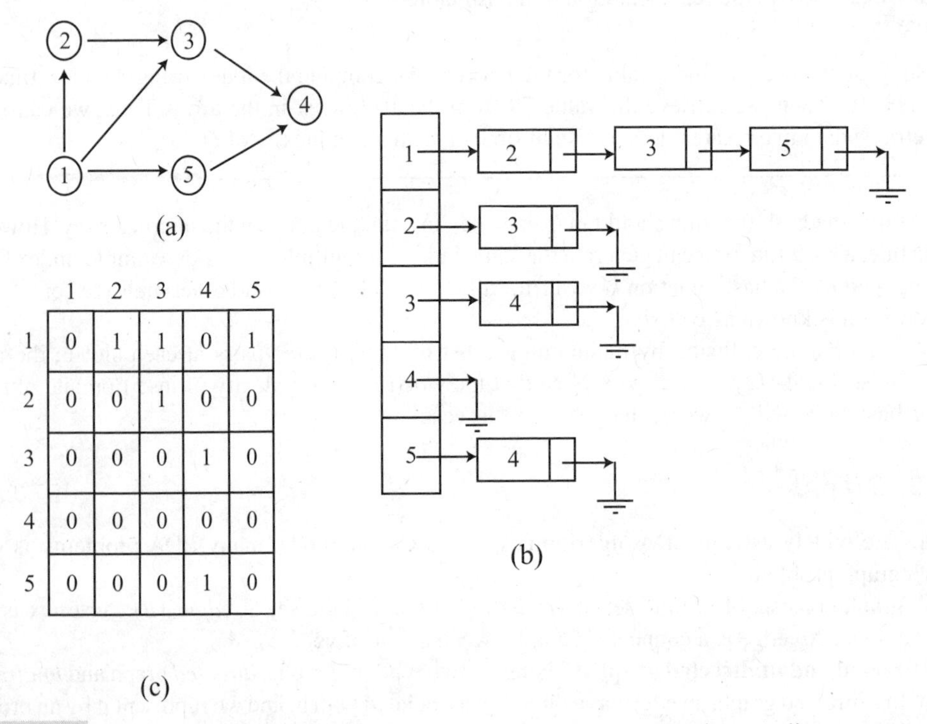

	1	2	3	4	5
1	0	1	1	0	1
2	0	0	1	0	0
3	0	0	0	1	0
4	0	0	0	0	0
5	0	0	0	1	0

Figure 1.24 An example of a graph: (a) schematic representation, (b) adjacency list representation, and (c) adjacency matrix representation

Next, we look at a few important definitions related to graphs. These definitions will be used in the subsequent chapters. We explain them with the help of an undirected graph, shown in Figure 1.25. We can easily extend these definitions to directed graphs.

Walk: A *walk* in a graph is an alternating sequence of vertices and edges in that graph. A walk can have repeated vertices and edges. A few examples of walks are: $\{v_1, e_{12}, v_2, e_{24}, v_4, e_{45}, v_5\}$, $\{v_1, e_{12}, v_2, e_{24}, v_4, e_{34}, v_3, e_{13}, v_1\}$, and $\{v_1, e_{12}, v_2, e_{12}, v_1, e_{13}, v_3\}$. A walk in which the start and end vertices are the same is considered *closed*. For example, $\{v_1, e_{12}, v_2, e_{24}, v_4, e_{34}, v_3, e_{13}, v_1\}$ is a closed walk. A walk in which the start and end vertices are distinct is considered as *open*.

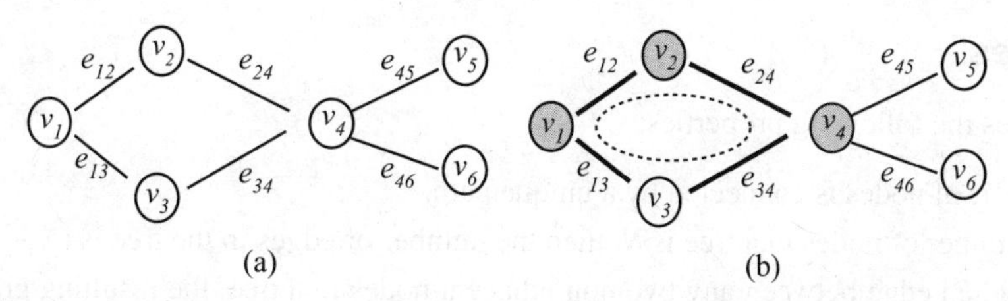

Figure 1.25 An undirected graph: (a) schematic representation and (b) a cycle shown with shaded vertices and thicker edges

Trail: A walk with all distinct edges is called a *trail*. The walks $\{v_1, e_{12}, v_2, e_{24}, v_4, e_{45}, v_5\}$ and $\{v_1, e_{12}, v_2, e_{24}, v_4, e_{34}, v_3, e_{13}, v_1\}$ are trails, while the walk $\{v_1, e_{12}, v_2, e_{12}, v_1, e_{13}, v_3\}$ is not a trail since the edge e_{12} is repeated.

Path: A trail with all distinct vertices is called a *path*. The trail $\{v_1, e_{12}, v_2, e_{24}, v_4, e_{45}, v_5\}$ is a path, since all the vertices are distinct. A graph in which a path exists for each pair of vertices is known as a *connected graph*. The graph as shown in Figure 1.25(a) is a connected graph.

Cycle: A closed trail in which only the start and the end vertices are repeated is called a *cycle*. For example, the trail $\{v_1, e_{12}, v_2, e_{24}, v_4, e_{34}, v_3, e_{13}, v_1\}$ is a cycle but not a path. Cycles appear as *loops* in the schematic diagram of a graph, as illustrated in Figure 1.25(b).

Acyclic graph: A graph in which no cycle exists is known as an *acyclic graph*. Since the graph as shown in Figure 1.25(b) contains cycles, it is not an acyclic graph.

Acyclic graphs are important from a graph algorithm's perspective since several efficient algorithms can work on acyclic graphs but not on graphs with cycles. Moreover, we can make a graph acyclic by removing certain edges from the cycles in that graph. For example, we can make the graph shown in Figure 1.25 acyclic by removing the edge e_{24}, as shown in Figure 1.26(a).

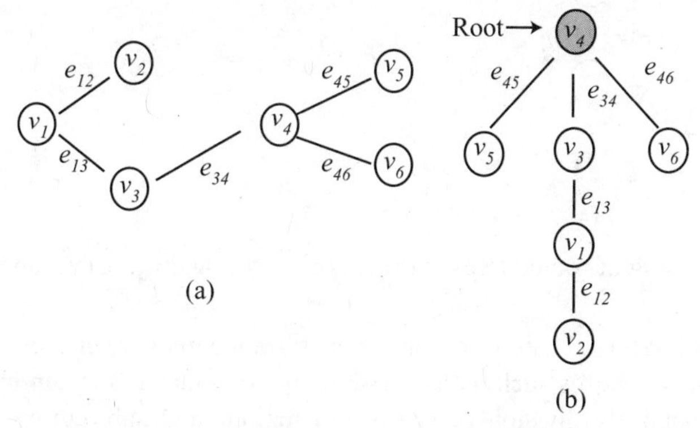

Figure 1.26 Modified graph of Figure 1.25: (a) cycle removed and (b) rooted tree

Tree: An acyclic graph that is connected is known as a *tree*. The graph as shown in Figure 1.26(a) is connected and acyclic and is an example of a tree. The vertices of a tree are also called *nodes*.

Trees are useful data structures, and we often employ them in VLSI design and verification for various purposes.

1.4.6 Trees

A tree satisfies the following properties:

1. Every pair of nodes is connected by a unique path.
2. If the number of nodes in a tree is N, then the number of edges in the tree is $(N-1)$.
3. If we add an edge between any two non-adjacent nodes in a tree, the resulting graph contains exactly one cycle.

Rooted tree: If we designate a vertex of a tree as the *root* of the tree, we obtain a *rooted tree.* For the tree shown in Figure 1.26(a), if we designate v_4 as the root, we obtain a rooted tree shown in Figure 1.26(b). Note that, in Figure 1.26(b), we have just redrawn the tree shown in Figure 1.26(a) with the root at the top.

Subtrees: We can partition the nodes other than the root node of a tree into n *disjoint* set of rooted trees $T_1, T_2, \ldots, T_n$. The rooted trees $T_1, T_2, \ldots, T_n$ are called *subtrees* of the root. The subtrees T_1, T_2, and T_3 of the rooted tree at v_4 are shown in Figure 1.27(a). We can further partition the rooted subtrees recursively until $n = 0$. Figure 1.27(b–d) illustrates the recursive partitioning of T_1, T_2, and T_3 into subtrees until $n = 0$.

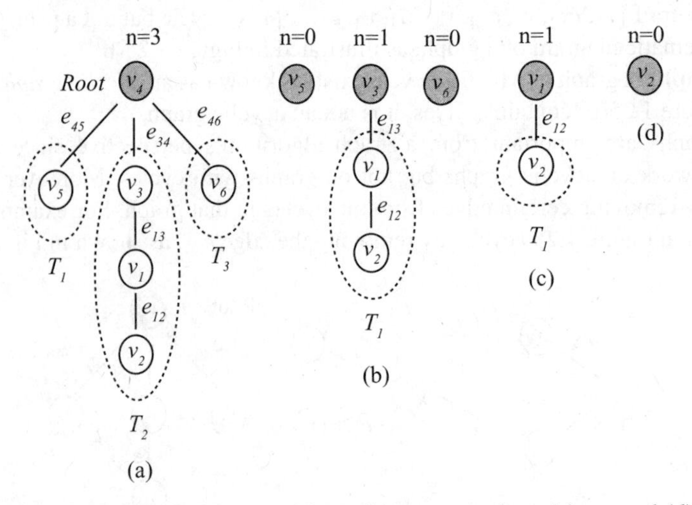

Figure 1.27 Partitioning of rooted trees at (a) v_4; (b) v_5, v_3, and v_6; (c) v_1; and (d) v_2

In contrast to arrays, linked lists, stack, and queues, rooted trees are *hierarchical data structures.* Therefore, hierarchical patterns such as file systems in computers, Boolean expressions and sub-expressions, and recursively division of space into regions and sub-regions can be represented conveniently using rooted trees. The hierarchical nature of rooted trees allows efficient processing such as accessing a node and comparison of nodes.

Parent and children: We consider the nodes connected to the root as the *children* of the root node, and the root is considered the *parent* of these nodes. The root node does not have any *parent.* All other tree nodes have exactly one parent (the first node on the path from that node to the root). All nodes in a rooted tree can have zero or more children. A node that does not have any child is called a *leaf node.*

Binary tree: A type of rooted tree in which each node can have a maximum of two children is known as a *binary tree*. The two children are referred to as the *left child* and the *right child*. If a binary tree is constructed with the following constraints, we obtain a *binary search tree*:

1. The parent's key is greater than or equal to the key of the left child.
2. The parent's key is less than or equal to the key of the right child.

By adding the above constraints while building a tree, we can efficiently search for a key in a binary search tree.

REFERENCES

[1] J. R. Hauser. "Noise margin criteria for digital logic circuits." *IEEE Transactions on Education* 36, no. 4 (1993), pp. 363–368.

[2] M. M. Mano and M. D. Ciletti. *Digital Design: With an Introduction to the Verilog HDL, VHDL, and System Verilog*. Pearson, 6th ed., 2018.

[3] S. Brown and Z. Vranesic. *Fundamentals of Digital Logic With Verilog Design*. McGraw Hill Education, 2nd ed., 2017.

[4] S. M. Sze and K. K. Ng. *Physics of Semiconductor Devices*. John Wiley & Sons, 2006.

[5] D. A. Neamen. *Semiconductor Physics and Devices: Basic Principles*. McGraw-Hill, 2012.

[6] J. M. Rabaey, A. P. Chandrakasan, and B. Nikolic. *Digital Integrated Circuits*, vol. 2. Prentice Hall Englewood Cliffs, 2002.

[7] N. H. Weste and D. Harris. *CMOS VLSI Design: A Circuits and Systems Perspective*. Pearson Education India, 2015.

[8] T. H. Cormen, C. E. Leiserson, R. L. Rivest, and C. Stein. *Introduction to Algorithms*. MIT Press, 2009.

2 Introduction to Integrated Circuits

...sources of excellence in the work produced by machinery depend on a principle which pervades a very large portion of all manufactures, and is one upon which the cheapness of the articles produced seems greatly to depend. The principle alluded to is that of COPYING, taken in its most extensive sense.

—Charles Babbage, *On the Economy of Machinery and Manufactures*, Chapter 11, 1832

To understand the VLSI design flow, we should be familiar with some basic concepts related to integrated circuits (IC). In this chapter, we explain them and introduce some terminologies that we use throughout this book.

2.1 VERY LARGE SCALE INTEGRATION (VLSI): AN HISTORICAL PERSPECTIVE

We connect various active and passive components such as transistors, diodes, resistors, and capacitors in an electronic circuit. However, with increasing complexity, interconnecting or assembling discrete components become expensive, time-consuming, and unreliable [1–3]. Therefore, the technology to build *monolithic silicon chips* containing several electronic components was invented in the early 1960s [1, 3]. These monolithic silicon chips are known as *integrated circuits* (IC).

IC technology: The invention of *IC technology* resulted in a tremendous decrease in the cost of electronic products and fueled the current *information technology revolution* [2, 4]. During the initial development of IC technology, the following problems were challenging to solve [1, 3]:

1. Isolating separate components electrically on a monolithic semiconductor was a challenge. We could accomplish this by employing *reverse-biased p–n junctions* that can block the unwanted flow of current through the semiconductor [5]. The technology to diffuse dopants directly into a semiconductor at precise locations using silicon dioxide (SiO_2) mask was critical in solving this problem [4, 6].
2. Obtaining sufficiently high *yield* to be profitable was a challenge. Yield is the percentage of good ICs among all the fabricated ICs. A technique known as *photolithography* was critical in solving this problem. We will describe photolithography in detail in the following sections.

Many of the IC technology advancements relied on the discovery that the native oxide of silicon has an excellent interface property. These advancements culminated in developing the *planar process* of transistor fabrication [7]. A planar process of fabrication is simpler and scalable to smaller dimensions [6]. Moreover, it allows interconnections using a thin film of metal over a semiconductor [4]. Thus, we could make internal interconnections, and we could achieve more complex integration in a chip. Moreover, due to improved reliability, higher yield, and lower cost, we could employ the IC technology profitably for mass production.

Increasing integration: With continued advancement in IC technologies, the integration evolved from *small scale integration* (SSI) to *very large scale integration* (VLSI) with more than a million transistors on a single IC [1, 8]. Gordon E. Moore observed the trend of increasing integration in 1965 and made the now famous *Moore's prediction* [9].

2.2 MOORE'S PREDICTION

In the original paper, Moore predicted that the number of components in an IC realized at *minimum cost* will *double every year* [9]. In 1975, he revised this prediction to *double every two years* [10–13]. These predictions are popularly known as *Moore's law* [10].

The exponential growth in the number of components per IC, as predicted by Moore, has continued unabated for more than five decades. The unprecedented success of exponential Moore's prediction for more than five decades has been driven by both technological innovations and pulls from economic incentives. Furthermore, Moore's prediction has often been treated as production goals. Therefore, profit-making incentives to achieve that goal have ensured that technology advancement follows the predicted trajectory [12].

Shrinking transistors and improving characteristics: Moore's prediction has been primarily enabled by the shrinking sizes of transistors (MOSFETs). The successive technology nodes such as 90 nm, 65 nm, 45 nm, 32 nm, 22 nm, and 16 nm represent reducing transistor sizes. The reduction in transistor sizes results in improving the characteristics such as speed and energy efficiency and reducing the cost per transistor [12]. However, with scaling, in general, designing an IC becomes more complicated and challenging. We will discuss these challenges and the techniques to tackle them throughout this book.

2.3 PHOTOLITHOGRAPHY

The fundamental technology that has enabled low-cost mass production of ICs and allowed a successive reduction in the feature size is *photolithography* [10]. It is the process by which we transfer shapes or features to thin films on a silicon wafer. We mark features on a glass plate with opaque chrome thin films, as shown in Figure 2.1. These glass plates are known as *masks* or *photomasks* or *reticles*. A thin transparent covering known as a *pellicle* is added to the mask to prevent contamination of the chrome and the glass plate.

We fabricate an IC by a sequence of steps consisting of photolithography, ion implantation, oxidation, and deposition. We prepare different masks to define features on different layers. Subsequently, we carry out photolithography separately for different layers.

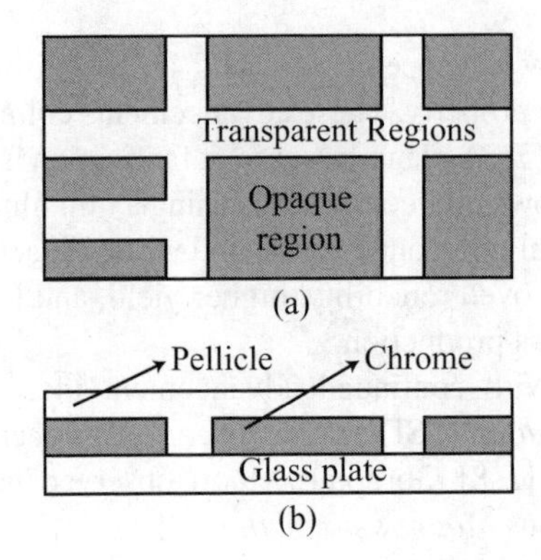

Figure 2.1 Schematic representation of a mask: (a) top view and (b) side view

The major steps involved in photolithography are as follows [14, 15]:

1. **Preparation:** Before starting photolithography, we deposit a layer of thin film that needs to be patterned, as shown in Figure 2.2(a).

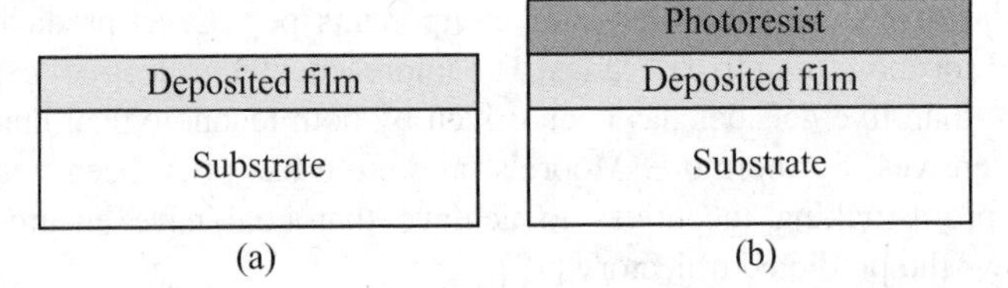

Figure 2.2 Photolithography: (a) deposition and (b) photoresist application

   The deposited film can be silicon dioxide, silicon nitride, polysilicon, or other materials. After deposition, we clean the film and dehydrate it. These steps ensure that the chemicals used in subsequent processing adhere well to the surface of the film.
2. **Photoresist application:** After the thin film is ready, we apply a *photoresist* layer to the film, as shown in Figure 2.2(b). A *photoresist* is a chemical that is sensitive to light. The solubility of the photoresist in a special kind of solution increases when exposed to light.[1] Such solutions are known as *developer solutions*. The photoresist and a suitable solvent are poured onto the wafer and spin-coated. The solvent allows the photoresist to form thin layers over the substrate by spinning. We control the photoresist thickness on the wafer by adjusting its volume, composition, and spin speed.

   After applying the photoresist, we remove the excess solvent using a *post-apply bake process* with the help of a convection oven or a hot plate. The baking process increases the photoresist's stability at room temperature and is critical for the subsequent steps.
3. **Exposure and development:** After baking, we expose the wafer to the light through the mask, as shown in Figure 2.3(a). By shining light through the patterned mask, we obtain the required spatial variation in the photoresist solubility. Consequently, the mask features get transferred to the photoresist layer during exposure.

[1] This kind of photoresist is a *positive photoresist*. There is another type of photoresist whose solubility decreases on light exposure. Such photoresists are known as *negative photoresists*.

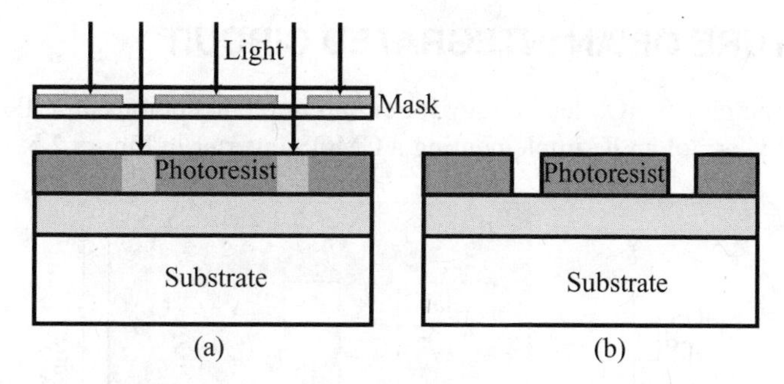

Figure 2.3 Photolithography: (a) exposure and (b) development (continued from Figure 2.2)

We expose the wafer in a *step and repeat* process. In this process, we shine a light on a small portion of the wafer and repeat it to cover the entire wafer. It helps in attaining a high *resolution*, i.e., the ability to distinguish closely spaced features.

The wavelength of light and the optical instruments determine the resolution of the image obtained on the wafer. Currently, we commonly use the wavelength of 193 nm in photolithography.

Post-exposure baking: After exposure, we carry out *post-exposure baking* (PEB). The PEB step reduces the unwanted ridges obtained in the sidewall of the photoresist. These ridges appear due to constructive and destructive interference of the incident light.

Development: After PEB, we develop the photoresist by immersing the wafer in an aqueous developer solution such as tetramethylammonium hydroxide (TMAH). Consequently, the soluble part of the photoresist layer dissolves and gets removed, as shown in Figure 2.3(b). After development, we bake the wafer again so that the photoresist can withstand the harsh environment of subsequent steps.

4. **Etching and photoresist removal:** After development, the mask features appear on the photoresist layer. As a result, when we etch the deposited film covered with photoresist, the etching is non-uniform. The regions covered by the photoresist withstand etching, while the unprotected regions etch easily, as shown in Figure 2.4(a).

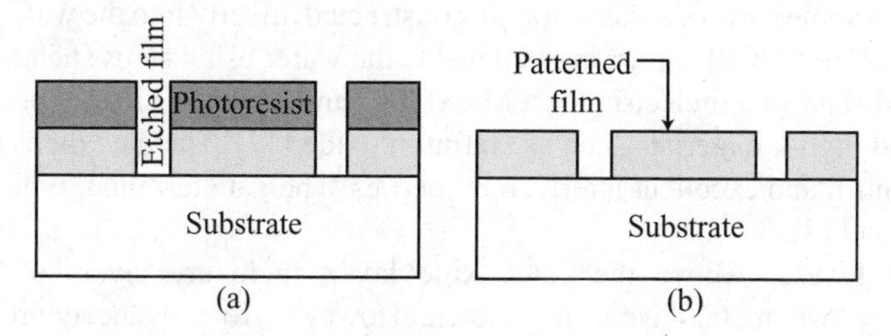

Figure 2.4 Photolithography: (a) etching and (b) photoresist removal (continued from Figure 2.3)

We can carry out etching using wet chemicals such as sulfuric acid or in a dry environment using plasma etching. After etching, we remove the leftover photoresist layer. Thus, we obtain the required features on the deposited layer, as shown in Figure 2.4(b).

2.4 STRUCTURE OF AN INTEGRATED CIRCUIT

We predominantly employ CMOS technology for circuit implementations. For illustration, we show the cross-sectional view[2] of an IC implementing a CMOS inverter in Figure 2.5.

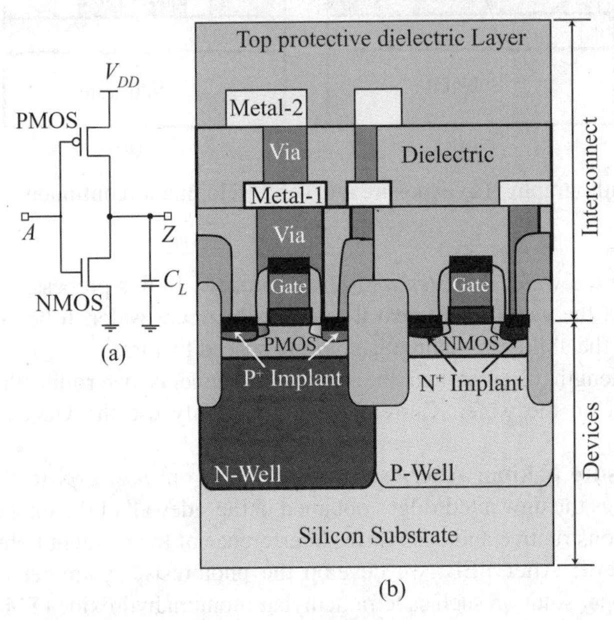

Figure 2.5 A CMOS inverter: (a) schematic and (b) IC implementation [14]

Note that the structure of an IC can vary across fabrication technologies. Nevertheless, there are some features that are common to all. We discuss them in the following paragraphs.

Layered structure: A salient feature of an IC is that it consists of different layers. The bottom-most layer consists of the silicon wafer that provides support to all other structures lying above it. In the lower layers of an IC, devices or transistors are built. In the upper layers, interconnects or wires are built, as shown in Figure 2.5(b).[3]

Device layers: Some parts of a transistor are constructed directly into the wafer. For example, the drain and the source of a MOSFET are created inside the wafer using steps such as ion implantation and annealing [14]. The gate dielectric for a MOSFET can be silicon dioxide or a stack containing silicon dioxide and high-k materials such as Hafnium oxide [17]. The gate dielectric layer requires tight thickness control and excellent interface properties. The gate terminal, typically, consists of a metal or polysilicon [18].

Interconnect layers: Above the gate oxide layer, there are layers of interconnects. In Figure 2.5(b), only two metal layers are shown. However, for advanced processes and high-performance ICs, there can be more than ten metal layers [19, 20]. An interconnect can be a *local interconnect* or a *global interconnect*. A local interconnect is a connection between components that are in close vicinity. Local interconnects are normally realized using polysilicon. Above the

[2] Another fast and easy way to draw and capture various layer information of an IC is using *stick diagram*. Readers can refer to [16] for this.

[3] In this book, we have used the words *interconnect* and *wire* synonymously to describe a physical entity that makes connections within an IC.

local interconnect, there are multiple layers of global interconnects. They connect different pins or provide a network of global signals such as clocks and resets. The highest interconnect layers have low resistance because they are thicker and use low resistivity material [20, 21]. We typically allocate them for power and ground lines.

As the number of pins in a circuit increases and the connectivity becomes more complex, we require more interconnect layers. In a given circuit, the possibility of making a valid physical connection depends on the number of available interconnect layers. We illustrate it in the following example.

Example 2.1 Suppose there are four points A1, A2, B1, and B2 in a plane, as shown in Figure 2.6(a). Assume that we need to make two connections, A1–A2 and B1–B2, without short-circuiting these connections.

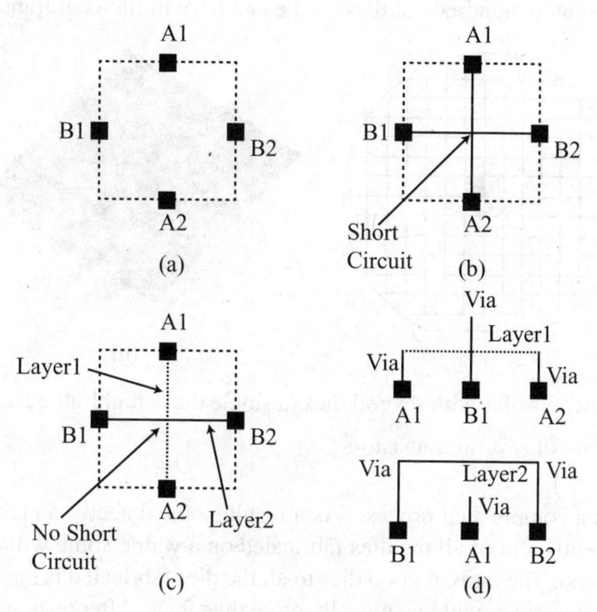

Figure 2.6 Making connections using layers (a) given point A1, A2, B1, and B2; (b) wires in the same plane and within the square; (c, d) wires allowed to go to different planes: ((c) top view and (d) side views)

If the wires are constrained to be within the square in Figure 2.6(a) and in the same plane, no connection is possible without short circuiting, as shown in Figure 2.6(b).

However, if two wires are allowed to run in two different planes, we can connect, as shown in Figure 2.6(c) and (d). The wire from A1 takes off out of the plane, runs vertically in Layer1, and then goes into the plane to connect A2. The wire from B1 takes off out of the plane, runs horizontally in Layer2, and then goes into the plane to connect B2.

Thus, if the wires can run in different planes (layers), we can make complicated connections. Therefore, employing multiple layers of interconnects eases making connections.

2.5 REALIZING AN IC ON SILICON

An integrated circuit, such as shown in Figure 2.5, is realized using a sequence of processes. These processes build structures layer by layer on a *silicon wafer*.

Silicon wafer: A silicon wafer is a thin piece of silicon that we obtain by slicing *silicon ingots*. A silicon ingot is large monocrystalline silicon prepared widely using the *Czochralski (CZ) process*. In this process, we pull out a pure seed crystal of appropriate orientation from highly pure silicon melted at around 1425°C [22]. Subsequently, monocrystalline and nearly defect-free silicon wafers are sliced out from the silicon ingots using diamond-edged saws. The silicon wafers with 300 mm diameters are most widely used for IC fabrication.

Dies and chips: We fabricate several rectangular ICs simultaneously on a silicon wafer, as shown in Figure 2.7. Each rectangular silicon chunk on the wafer is a self-contained IC, such as a processor, ethernet chip, or microcontroller. These rectangular ICs fabricated on the wafer are known as *dies*. A wafer can contain hundreds of dies. A die can have millions of transistors.

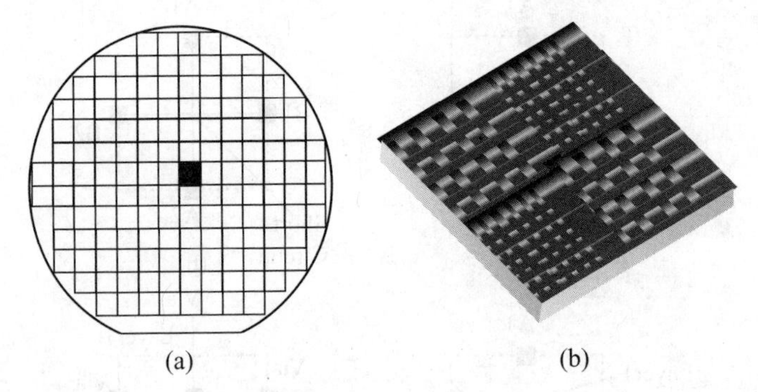

Figure 2.7 (a) A silicon wafer with several dies (a single die is highlighted as black rectangle) and (b) single die consisting of several transistors

IC fabrication is a complicated process. As a result, some defects can creep into an IC during fabrication. Consequently, out of all the dies fabricated on a wafer, some will have defects and will not be functionally good. The ratio of good dies to all the dies fabricated is known as the *yield* of the process. For a matured process, yield is typically more than 90%. After testing, the good dies are cut out from the wafer and appropriately packaged to obtain a *chip*.

2.6 DESIGNING VERSUS FABRICATION

An IC is hardware or physical entity that realizes a given functionality such as a microprocessor, graphics controller, voice recorder, MP3 decoder, and bitcoin miner. A packaged IC hardware is popularly known as a *chip*. We can divide the process of implementing the given functionality and obtaining a chip into two distinct phases: (a) designing and (b) fabrication, as shown in Figure 2.8.

Designing: It involves determining the parameters and composition of a circuit that can achieve the desired functionality. At the end of the design process, we obtain the layout of a circuit. The circuit layout defines the geometrical patterns on various mask layers that can be used in photolithography.

Fabrication: Post-designing, we fabricate an IC. The fabrication involves creating an IC physically using various processes such as photolithography, oxidation, etching, and ion implantation. Note that the mask layers, which are the output of the design process, serve as input for fabrication. We fabricate ICs in *semiconductor foundries*, typically in large volumes.

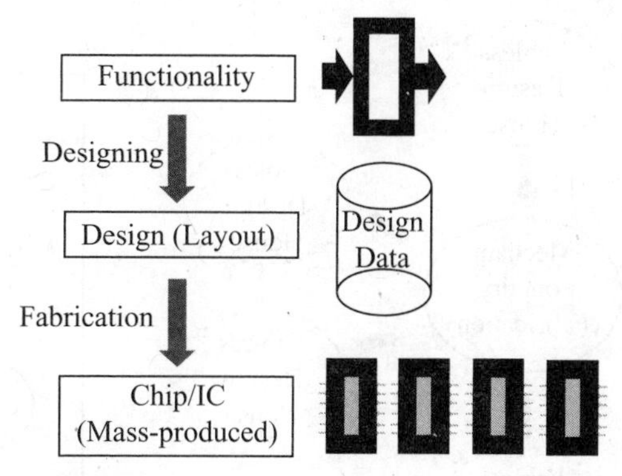

Figure 2.8 Designing and fabrication

2.7 SEMICONDUCTOR FOUNDRY

A semiconductor foundry is a manufacturing plant for an IC. It is also called a semiconductor *fabrication plant* or *fab*. A foundry contains sophisticated instruments for various fabrication tasks such as cleaning, ion implantation, etching, deposition, and photolithography.

A *cleanroom* is a critical part of a semiconductor foundry. We control the environment in a cleanroom tightly and avoid contaminants such as dust particles. It is essential to avoid them because dust particles and features on ICs are of similar sizes. Therefore, a dust particle can make a fabricated circuit faulty.

The requirement of a highly controlled environment in a foundry and costly instruments make the cost of setting up and maintaining a foundry very high. Therefore, a semiconductor foundry is sustainable only when we utilize it to its *full potential*. Otherwise, the cost of setting up a foundry cannot be recovered before it being obsolete.

To ensure the profitability of foundries, a business model consisting of semiconductor foundries and *fabless design companies* has evolved such that both the foundries and the fabless design companies can run profitably. We illustrate it in Figure 2.9.

Fabless design companies: A fabless design company carries out only designing and gets the chips fabricated using *merchant foundries* [23]. The fabless design companies can be profitable since they do not need to invest heavily in the fabs. A few examples of fabless design companies are Qualcomm, Broadcom, Nvidia, and Apple.

Merchant foundries: On the other hand, merchant foundries carry out fabrication for other companies. They draw business from a large number of design companies and can keep working to their *full potential*. Therefore, merchant foundries can also be profitable in this business model. A few examples of merchant foundries are TSMC, UMC, and Globalfoundries.

Integrated device manufacturers: Some companies such as Intel and Samsung carry out both designing and fabrication. These companies are known as *integrated device manufacturers* (IDMs) and can be more efficient by integrating and controlling the entire design and fabrication tasks.

Consistency between design and fabrication: It is worth pointing out that, though we keep designing and fabrication processes separate, they are intricately related. For example, the properties of a transistor and interconnects strongly depend on the fabrication process. Therefore, a designer needs to know these properties while designing an IC. Moreover, a designer should make a design that can be fabricated in a given technology available at the foundry. Therefore, despite separating the designing and fabrication processes, we need to maintain consistency between these two processes.

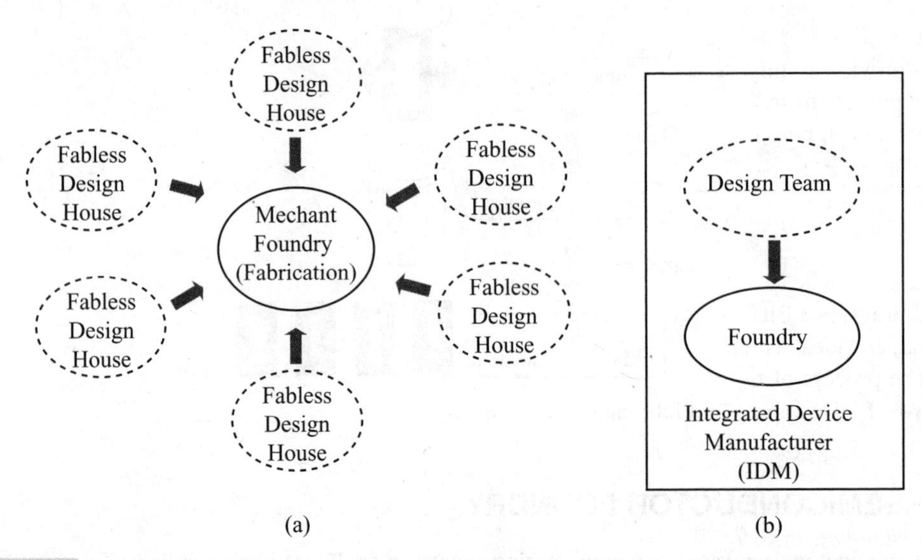

Figure 2.9 Business models of designing and fabrication: (a) fabless design house and merchant foundry and (b) integrated device manufacturer (IDM)

Process design kit: We can establish consistency between designing and fabrication processes by sharing information. A foundry shares a bundle of data known as *process design kit* (PDK) with the design team, as illustrated in Figure 2.10.

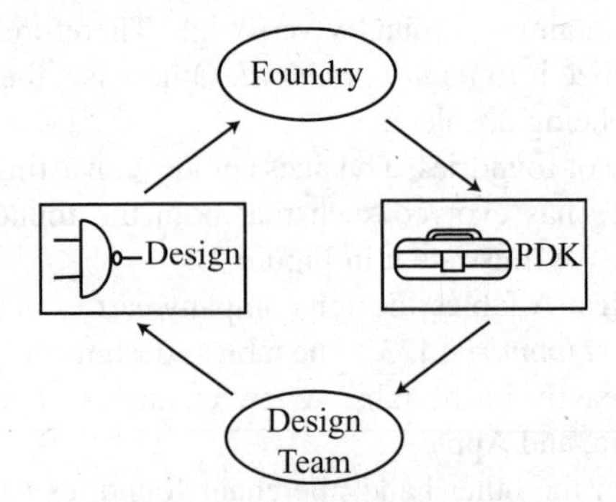

Figure 2.10 Sharing of information between a foundry and a design team

A PDK contains a basic circuit element called *parameterized cell* (PCell)[4] and the electrical models of resistors, capacitors, inductors, and transistors. These models can be used by circuit simulators such as *simulation program with integrated circuit emphasis* (SPICE). A PDK also contains technology and process information, such as the number of interconnect layers allowed in a process and their attributes. Additionally, it contains information and rules for *physical extraction*, *layout versus schematic* (LVS) checks, *design rule check* (DRC), and *electrical rule check* (ERC).[5]

[4] PCell is a circuit element whose structure depends on one or more parameters. For example, a resistor with length and width as parameters.

[5] We will discuss these tasks in detail in the subsequent chapters.

By obeying these rules in a design, we can ensure that we can fabricate that design profitably. Thus, PDKs provide an important link between a foundry and a designer. It also shields the complexities of one process from the other. For academic and research purposes, we can employ freely available PDKs for designing [24].

2.8 REALIZING AN "IDEA" USING INTEGRATED CIRCUIT

An IC is a vehicle to realize an "idea." An "idea" can be to build a system that computes, compresses video, provides a Bluetooth interface, carries out some signal processing task, or controls a robot. In general, an "idea" is realized using a combination of ICs, software, and other components.

The process of transforming an "idea" into a final system is complicated. It requires making complex tradeoffs between multiple *objectives* (such as speed and power dissipation) and fulfilling several *constraints* (such as cost and manufacturability) [25]. We simplify this process by breaking it down into several smaller tasks. The output of one task becomes the input for the next task. This sequence of tasks forms a part of the complete process known as *VLSI design flow*. We also refer to it as *design flow* in short.

During a design flow, an "idea" goes through a series of transformations and representations. The flow starts with an abstract representation of an "idea," possibly as input/output functional description. As the flow progresses, we add details to the representation and the "idea" takes a concrete form. Among various representations, the following representations are of particular significance:

1. **Register transfer level (RTL):** An RTL describes how signals flow from *register to register* in an IC. Generally, an RTL is written in *Hardware Description Languages (HDLs)* such as Verilog or VHDL [26–28].
2. **Graphical database system (GDS):** A GDS represents the layout of an IC. It contains details of the set of masks required for photolithography during IC fabrication. Typically, we write GDS in Graphic Design System II (GDSII) or Open Artwork System Interchange Standard (OASIS) formats [29]. These are binary formats, and tools extract different mask layers by reading a GDS file.

Based on the above representations, the transformation of an "idea" to the final chip can be divided into three major parts, as shown in Figure 2.11:

1. **"Idea" to RTL flow or pre-RTL flow:** In this part of designing, the hardware portion of an "idea" gets transformed into an RTL representation. We group all the design processes before obtaining an RTL as *pre-RTL methodologies*. We also refer to them as *system-level design* and describe them in Chapter 3 ("Pre-RTL methodologies").
2. **RTL to GDS flow:** In this part of the design flow, we take an RTL through a sequence of logical and physical design processes. Finally, we obtain a layout from the RTL, which we write in the GDS format. The RTL to GDS flow is the theme of this book. We discuss it briefly in Chapter 4 ("RTL to GDS implementation flow") and in detail in Parts II–IV of this book.
3. **GDS to chip flow or post-GDS flow:** After obtaining a GDS, we fabricate a given design and produce a finished chip, typically in a high volume. We describe the post-GDS processes briefly in Chapter 7 ("Post-GDS processes").

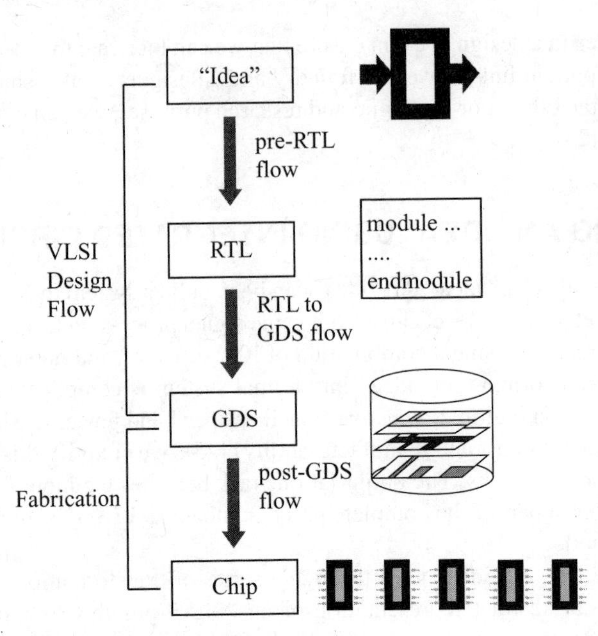

Figure 2.11 Transforming an "Idea" to final chip

Note that the design flow or the design methodology, especially after obtaining an RTL, is mostly the same for ICs with different functionality. Therefore, once we have developed a design flow for a technology node, we can reuse that design flow for many ICs with diverse functionality. Thus, creating a robust design flow reduces subsequent design effort and pays in the long run. However, design flows are different for different types of ICs. Therefore, it is crucial to be familiar with the different types of ICs, which we describe in the next section.

2.9 TYPES OF INTEGRATED CIRCUITS

We can classify ICs based on the scope of applications and the design styles [16, 30].

2.9.1 Based on Scope of Applications

Based on the scope of application, ICs are classified as *application specific integrated circuit* (ASIC) and *general-purpose ICs*. We design ASICs for a specific end application [30]. For example, ICs for video compression, digital camera, Bluetooth connectivity, and bitcoin mining. However, we design general-purpose ICs to perform some generic functions. For example, microprocessor, memory, and FPGA. We customize a general-purpose IC to fulfill application-specific requirements. We can customize by running relevant software, providing appropriate external connections, and biasing some control ports with pertinent voltages. Since we can employ general-purpose ICs for many applications, we typically fabricate these ICs in a large volume.

2.9.2 Based on Design Styles

Based on the design styles, we classify ICs as full-custom design, standard cell-based design, gate array design, and field programmable gate array (FPGA)-based design.

Full-custom Design

In a full-custom design, a designer defines the layout of each transistor and interconnects in a circuit. Thus, a designer works at the transistor level in full-custom design and has more freedom in making design decisions. As a result, higher performance and lower power dissipation can be achieved in a full-custom design, though at the cost of high design effort. Therefore, we undertake full-custom design only for circuits requiring very high performance [31]. For example, we can undertake a full-custom design for a few critical paths in a high-performance microprocessor. Moreover, we undertake a full-custom design in mixed-signal or analog designs, where designing at the transistor level gives better electrical characteristics than automated design styles [32].

Standard Cell-based Design

Standard cells are combinational circuit elements such as logic gates, multiplexers, adders, and multipliers, and sequential circuit elements such as flip-flops and latches.

Standard cell library: A standard cell is optimally designed at the transistor level to obtain a high speed, low power dissipation, and less area overhead. We measure the electrical characteristics of a standard cell using circuit simulation. Subsequently, we store these characteristics of each standard cell in some files. These files are known as *technology libraries* or *standard cell libraries*. After creating a technology library, the internal structure of each standard cell gets fixed.

Design using standard cells: A designer works at the cell level and uses cells existing in a given set of technology libraries in a standard cell-based design. A designer cannot change the internal details of the standard cells. However, the desired circuit function can be obtained by choosing different cells from the given standard cell libraries and varying their interconnections. Thus, we can create many designs with varying functions using the same set of technology libraries.

Rows of standard cells: In a standard cell-based design, we can vary the location of standard cells on the layout. Typically, we arrange standard cells in multiple rows of fixed height, as shown in Figure 2.12(a). This task gets facilitated by keeping the *same height* of all the standard cells in a technology library [33].

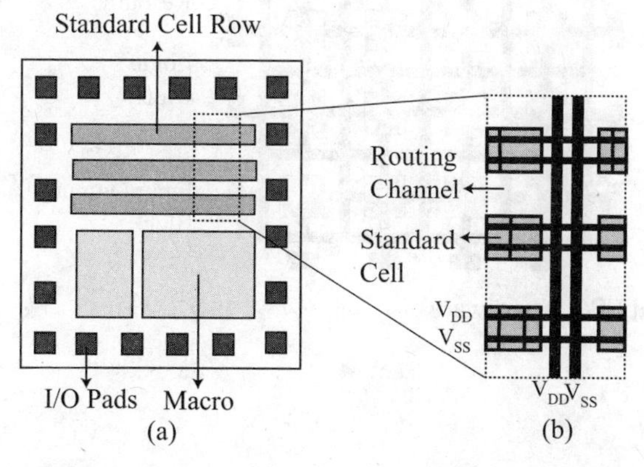

Figure 2.12 Standard cell-based design: (a) layout and (b) an expanded view of standard cell rows

We perform routing[6] using interconnect layers *over-the-cell* (OTC). We can also allocate extra space for routing between the standard cell rows, as shown in Figure 2.12(b). We make the power (V_{DD}) and ground (V_{SS}) connections by running these lines over the standard cells. A standard cell design can also contain *macros*. A macro is a larger design entity such as a multiplier and memory block.

Automation: The well-organized structure of a standard cell-based design allows a high degree of automation in creating layouts. Based on the design requirements, *computer-aided design* (CAD) tools can choose appropriate standard cells and arrange them in standard cell rows. Therefore, all masks, including the bottom device layers and top interconnect layers, are design specific in a standard cell-based design.

In this book, our focus is on standard cell-based design flow.

Gate Array Design

In a gate array design, transistors are predefined on the IC in the form of a gate array. The smallest repeated element that forms the gate array is called the *base cell* or the *primitive cell*. A designer customizes a gate array only by defining the interconnection between base cells. Therefore, only the top-most layers of the mask that define the interconnections between base cells are design specific.

Note that the functionality of a base cell is fixed. Therefore, implementing some functions, such as memory, can be inefficient or challenging. This problem is solved in some gate array ICs by embedding custom blocks that implement memory and microcontroller functionality.

FPGA-based Design

In an FPGA-based design, the hardware is fixed. The hardware consists of an array of *configurable logic blocks*, interconnections, and input/output cells, as shown in Figure 2.13 [34]. The configurable logic blocks can realize a wide variety of combinational and sequential functions. The signals are routed in an FPGA using configurable interconnect switches and routing wires. A commercial FPGA board or system can also have embedded microprocessors, analog components, and signal processing units. They enable us to accomplish a wide variety of tasks efficiently [35].

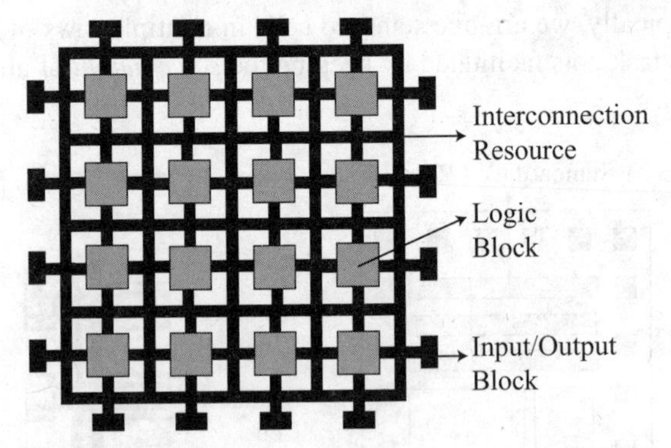

Figure 2.13 A typical FPGA hardware

[6] Routing involves making connections between cells or pins on the layout.

Configuring an FPGA: We configure the functionality of an FPGA hardware with the help of tools provided by the FPGA vendor. First, we define the required functionality in an appropriate format and give it as an input to the FPGA tool. Subsequently, the FPGA tool configures or sets up the FPGA hardware to deliver the required functionality. A vendor can make the hardware configurable using several technologies, such as SRAM-controlled pass transistors, flash-controlled signal routing, and anti-fuses that make the direct connection when programmed.

Comparison with other design styles: An FPGA has fixed hardware. After implementing a given function, some blocks (hardware) of FPGA can be left unused. Furthermore, we incur some overhead in providing the field programmability to an FPGA chip. Therefore, in general, there is a significant area penalty in an FPGA-based design compared to a full-custom or standard cell-based design. Additionally, performance and power dissipation for an FPGA-based design are worse than a full-custom design and standard cell-based design [34, 35]. Nevertheless, design effort is lower for FPGA-based design. Furthermore, the risk of failure is less in an FPGA-based design since the problem can be fixed by reconfiguring.

We can implement a given function in different types of ICs. However, economics plays a crucial role in choosing an IC type, as described in the following section.

2.10 ECONOMICS OF INTEGRATED CIRCUIT

We can break the cost of an IC into two components:

1. **Fixed cost:** The *fixed cost* is independent of the number of chips we manufacture. For instance, we incur the cost only once in designing a chip and preparing its masks. We can employ the same masks for manufacturing any volume of chips. Therefore, we can consider the cost of designing and mask preparation as fixed costs [22]. Similarly, software (such as CAD tools), hardware (such as computing resources), and other capital costs (such as buildings) are fixed costs for a chip.
2. **Recurring cost:** The *recurring cost* or *variable cost* depends on the volume of the chips we manufacture. For example, the fabrication cost depends on the number of chips that we manufacture. The fabrication cost, such as the cost of chemicals used in processing, increases if we process extra wafers. Similarly, the cost of individual dies (that depends on the size of the die and the yield) is a recurring cost since we incur them for individual chips.

The fixed cost and the variable cost are different for different IC types. Figure 2.14 shows a trend of change in the total cost of a chip with the volume for standard-cell-based and FPGA technologies. The fixed costs for standard-cell-based design are higher due to the higher design effort (design cost) and design-specific masks. For FPGA, since the hardware is the same, the cost of hardware design and mask preparation is distributed over several designs [35]. Moreover, the cost involved in customizing an FPGA is significantly lower than making a standard-cell-based design. Therefore, the fixed costs of an FPGA-based chip are lower. However, the per unit cost of FPGA is higher than standard-cell-based design. As a result, there is a break-even volume below which FPGA technology is more profitable than standard-cell-based technology. Therefore, for low-volume products, we prefer FPGA.

In addition to the cost, various figures of merit (FOMs) play a crucial role in determining the preferred IC type for a given application. We discuss this aspect of IC in the next section.

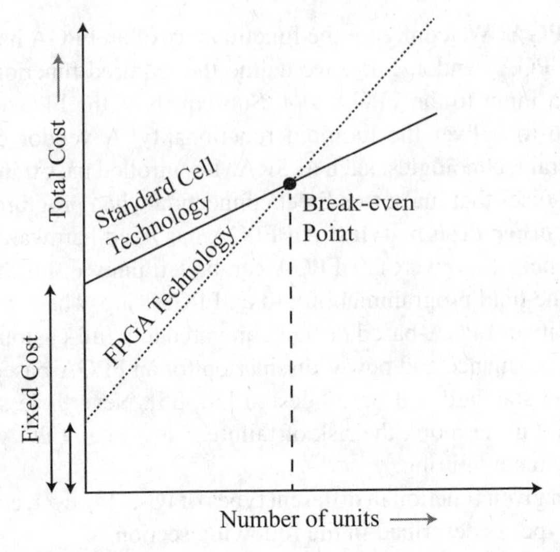

Figure 2.14 A simplistic model for the total cost vs volume for a standard cell technology and FPGA technology [35]

2.11 FIGURES OF MERIT (FOMS)

We refer to the quantities that we can use to assess the "goodness" of an IC as *figures of merit* (FOMs). Some of the important FOMs are shown in Figure 2.15 and described below.

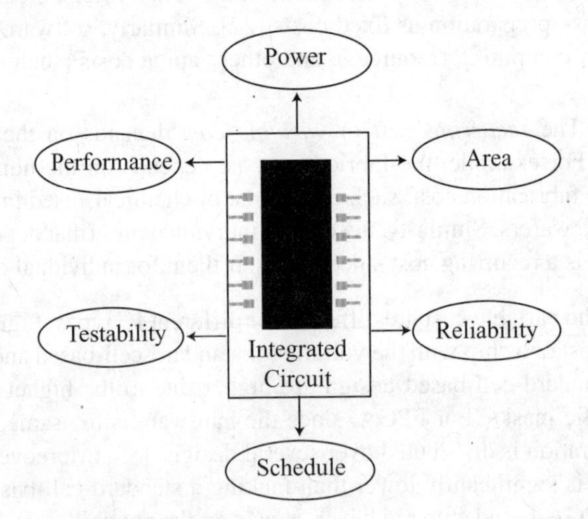

Figure 2.15 Figures of merit (FOMs) for an IC

1. **Performance:** We typically quantify the *speed* or *performance* of a synchronous circuit by the maximum clock frequency at which it can operate. To ensure that a circuit can operate at a given clock frequency, we often need to reduce the propagation delay of the internal circuit elements.

2. **Power:** We compute or estimate the *power* dissipated in a circuit and ensure that it should be within some limit. We typically consider two components of power dissipation: *dynamic power dissipation* (power dissipated when the circuit is computing actively) and *static power dissipation* (power dissipated when it is in an idle state). Increased power dissipation in an IC has several harmful effects, including reduced battery life for portable devices.
3. **Area:** The die *area* determines the variable cost of a chip. We should implement a given functionality in the smallest possible die area to reduce its variable cost. However, during the implementation of a cell-based design, we often treat the die area as a given constraint. Therefore, we attempt to meet the die area constraint by reducing the cell size and packing the interconnects compactly. The package size also puts an additional constraint on the die area.
4. **Reliability:** We want the functionality of an IC not to degrade significantly over time. We quantify this ability of an IC using *reliability* measure. We ensure that the reliability of an IC is acceptable under the stated condition for the product's lifetime.
5. **Testability:** It refers to the property of a circuit by which the *faults* associated with manufacturing defects can be easily detected, diagnosed, and localized. Testability helps us in preventing a defective IC from reaching the end user and maintaining the product's perceived quality.
6. **Schedule:** It refers to the time spent in designing an IC. The time spent in designing increases the fixed cost of a chip. Additionally, it can determine the success or failure of a product in a highly competitive semiconductor industry. Therefore, the schedule is a critical FOM for an IC.

Tradeoffs: The challenges in VLSI design flow often originate due to the requirement of *simultaneously* maximizing or minimizing *multiple* FOMs. However, maximizing one FOM can have a deleterious effect on others. For example, to improve the performance, we can increase the sizes of the transistors in a circuit. However, increasing transistor sizes increases the area and power dissipation in a circuit. Therefore, making intelligent *tradeoffs* between different FOMs is an integral and the most challenging part of a design flow.

PPA and QoR: We often refer to the power (P), performance (P), and area (A) of an IC together as *PPA*. We also refer to the metrics such as power, performance, area, reliability, and testability obtained during designing collectively as the *quality of results* (QoR) for a design. The dependence of these metrics on circuit parameters and their interdependence is complicated. Therefore, the *mathematical optimum QoR* attainable for a given design is rarely known or achieved. In practice, the goal of a design flow is to find one of the *feasible solutions* with *acceptable QoR*.

2.12 RECENT TRENDS

The miniaturization of transistors has continued for more than fifty years and has been the principal driver for the semiconductor industry. In general, it facilitates realizing electronic products with a higher speed (clock frequency), lower power dissipation, and incurring a lower cost per elementary function. With the increased availability of transistors in an IC, realizing a complete system consisting of microprocessors, memory and peripherals became possible. It has enabled implementation of complex system-on-chip (SoC) and interconnections with billions of components [36–39]. This miniaturization trend is sustaining and being driven by newer applications such as AI, big data, IoT, and cloud computing. However, power dissipation and non-scalability of the supply voltage are the biggest challenges for miniaturization in the future.

In recent times, we have started exploiting the third dimension of an IC [40]. Instead of pursuing a more challenging 2D reduction in transistor sizes, layers of devices are built one on top of another using *monolithic 3D integration*. It allows packing more components in an IC, reducing wire lengths, and alleviating interconnect bottlenecks [40]. It has been shown that a monolithic 3D IC can deliver the same performance as a 7 nm technology node, even when implemented using an old technology node such as 90 nm [41].

REVIEW QUESTIONS

2.1 What are the advantages of an integrated circuit over assembling discrete components?

2.2 What is the significance of silicon dioxide in realizing integrated circuits?

2.3 How does the wavelength of the light source impact photolithography?

2.4 Why are multiple layers of interconnects required in an IC?

2.5 What is the difference between a silicon wafer and a die?

2.6 For a wafer of diameter 300 mm, how many dies of size 20 mm × 20 mm can be obtained (Assuming there is no wafer material waste)? For a wafer of diameter 450 mm, how many dies of size 20 mm × 20 mm can be obtained (Assuming there is no waste of wafer material)? Assume that the cost of fabricating a wafer of diameter 300 mm is $1000 and a wafer of diameter 450 mm is $1200. What is the cost per die for the 300 mm technology and the 450 mm technology?

2.7 What is the difference between designing an IC and fabricating an IC?

2.8 The cost of setting up a foundry is very high. For profitability, we should utilize a foundry to its full potential. How does the business model of fabless design companies and merchant foundries help to achieve it?

2.9 What information does a GDS file contain?

2.10 What is the merit of using standard cells of fixed height in a standard cell-based design?

2.11 What are the advantages and disadvantages of FPGA technology over standard cell-based technology?

2.12 How can monolithic 3D integration improve the QoR of a design?

2.13 For a certain standard cell-based technology, the total cost of producing N chips is given as:

$$cost = 10000 + 0.02 \times N$$

For producing an IC with the same functionality in FPGA technology, the cost of producing N chips is given as:

$$cost = 1000 + 0.08 \times N$$

(a) Find N at which the total cost of chip produced in standard cell-based technology and FPGA technology is the same.

(b) For $N = 10000$, which technology is cheap?

(c) For $N = 1000000$, which technology is cheap?

REFERENCES

[1] G. O'Regan. "The invention of the integrated circuit and the birth of silicon valley." In *Introduction to the History of Computing*. Springer, Cham, 2016.

[2] S. Chih-Tang. "Evolution of the MOS transistor-from conception to VLSI." *Proceedings of the IEEE* 76 (Oct. 1988), pp. 1280–1326.

[3] J. S. Kilby. "The integrated circuit's early history." *Proceedings of the IEEE* 88 (Jan. 2000), pp. 109–111.

[4] W. F. Brinkman, D. E. Haggan, and W. W. Troutman. "A history of the invention of the transistor and where it will lead us." *IEEE Journal of Solid-State Circuits* 32 (Dec. 1997), pp. 1858–1865.

[5] R. N. Noyce. "Microelectronics." *Scientific American* 237, no. 3 (1977), pp. 62–69.

[6] R. R. Schaller. "Moore's law: Past, present and future." *IEEE Spectrum* 34 (Jun. 1997), pp. 52–59.

[7] J. A. Hoerni. "Planar silicon diodes and transistors." *1960 International Electron Devices Meeting* (Oct. 1960), pp. 50–50.

[8] G. Moore. "Solid state: VLSI: Some fundamental challenges: Defining and designing the products made possible by very-large-scale integration are first on the list of priority tasks." *IEEE Spectrum* 16 (Apr. 1979), pp. 30–30.

[9] G. E. Moore. "Cramming more components onto integrated circuits." *Electronics* 87 (Apr. 1965), pp. 114–117.

[10] G. E. Moore. "Lithography and the future of Moore's law." In *Integrated Circuit Metrology, Inspection, and Process Control IX*, vol. 2439, pp. 2–18. International Society for Optics and Photonics, 1995.

[11] G. E. Moore et al. "Progress in digital integrated electronics." *Electron Devices Meeting* 21 (1975), pp. 11–13.

[12] C. Mack. "The multiple lives of Moore's law." *IEEE Spectrum* 52 (Apr. 2015), pp. 31–31.

[13] A. Huang. "Moore's law is dying (and that could be good)." *IEEE Spectrum* 52, no. 4 (2015), pp. 43–47.

[14] J. D. Plummer. *Silicon VLSI Technology: Fundamentals, Practice and Modeling*. Pearson Education India, 2009.

[15] R. F. Pease and S. Y. Chou. "Lithography and other patterning techniques for future electronics." *Proceedings of the IEEE* 96 (Feb. 2008), pp. 248–270.

[16] N. H. Weste and D. Harris. *CMOS VLSI Design: A Circuits and Systems Perspective*. Pearson Education India, 2015.

[17] E. Gusev, D. Buchanan, E. Cartier, A. Kumar, D. DiMaria, S. Guha, A. Callegari, S. Zafar, P. Jamison, D. Neumayer, et al. "Ultrathin high-K gate stacks for advanced CMOS devices." *International Electron Devices Meeting. Technical Digest (Cat. No. 01CH37224)* (2001), pp. 20–1, IEEE.

[18] K. Mistry, C. Allen, C. Auth, B. Beattie, D. Bergstrom, M. Bost, M. Brazier, M. Buehler, A. Cappellani, R. Chau, et al. "A 45nm logic technology with high-k+ metal gate transistors, strained silicon, 9 Cu interconnect layers, 193nm dry patterning, and 100% Pb-free packaging." *2007 IEEE International Electron Devices Meeting* (2007), pp. 247–250, IEEE.

[19] C.-H. Jan, U. Bhattacharya, R. Brain, S.-J. Choi, G. Curello, G. Gupta, W. Hafez, M. Jang, M. Kang, K. Komeyli, et al. "A 22nm SoC platform technology featuring 3-D tri-gate and highk/metal gate, optimized for ultra low power, high performance and high density SoC applications." *2012 International Electron Devices Meeting* (2012), pp. 3–1, IEEE.

[20] S. Natarajan, M. Agostinelli, S. Akbar, M. Bost, A. Bowonder, V. Chikarmane, S. Chouksey, A. Dasgupta, K. Fischer, Q. Fu, et al. "A 14nm logic technology featuring 2nd-generation FinFET, air-gapped interconnects, self-aligned double patterning and a 0.0588 μm 2 SRAM cell size." *2014 IEEE International Electron Devices Meeting* (2014), pp. 3–7, IEEE.

[21] D. Edelstein, J. Heidenreich, R. Goldblatt, W. Cote, C. Uzoh, N. Lustig, P. Roper, T. McDevitt, W. Motsiff, A. Simon, et al. "Full copper wiring in a sub 0.25 μm CMOS ULSI technology." *International Electron Devices Meeting. IEDM Technical Digest* (1997), pp. 773–776, IEEE.

[22] W. Zulehner. "Historical overview of silicon crystal pulling development." *Materials Science and Engineering: B* 73, no. 1 (2000), pp. 7–15.

[23] D. B. Fuller. "Chip design in China and India: Multinationals, industry structure and development outcomes in the integrated circuit industry." *Technological Forecasting and Social Change* 81 (2014), pp. 1–10.

[24] M. Martins, J. M. Matos, R. P. Ribas, A. Reis, G. Schlinker, L. Rech, and J. Michelsen. "Open cell library in 15nm FreePDK technology." *Proceedings of the 2015 Symposium on International Symposium on Physical Design* (2015), pp. 171–178.

[25] R. K. Brayton, G. D. Hachtel, and A. L. Sangiovanni-Vincentelli. "A survey of optimization techniques for integrated-circuit design." *Proceedings of the IEEE* 69, no. 10 (1981), pp. 1334–1362.

[26] G. D. Micheli. *Synthesis and Optimization of Digital Circuits*. McGraw-Hill Higher Education, 1994.

[27] S. Palnitkar. *Verilog HDL: A Guide to Digital Design and Synthesis*. Pearson Education India, 2003.

[28] Z. Navabi. *VHDL: Analysis and Modeling of Digital Systems*. McGraw-Hill, Inc., 1997.

[29] Y. Chen, A. B. Kahng, G. Robins, A. Zelikovsky, and Y. Zheng. "Evaluation of the new OASIS format for layout fill compression." *Proceedings of the 2004 11th IEEE International Conference on Electronics, Circuits and Systems, 2004. ICECS 2004* (2004), pp. 377–382, IEEE.

[30] M. J. S. Smith. *Application-Specific Integrated Circuits*, vol. 7. Addison-Wesley Reading, MA, 1997.

[31] P. E. Gronowski, W. J. Bowhill, R. P. Preston, M. K. Gowan, and R. L. Allmon. "High-performance microprocessor design." *IEEE Journal of Solid-State Circuits* 33, no. 5 (1998), pp. 676–686.

[32] R. A. Rutenbar. "Analog design automation: Where are we? Where are we going?" *Proceedings of IEEE Custom Integrated Circuits Conference-CICC'93* (1993), pp. 13–1, IEEE.

[33] A. E. Dunlop and B. W. Kernighan. "A procedure for placement of standard cell VLSI circuits." *IEEE Transactions on Computer-Aided Design* 4, no. 1 (1985), pp. 92–98.

[34] I. Kuon, R. Tessier, et al. "FPGA architecture: Survey and challenges." *Foundations and Trends® in Electronic Design Automation* 2, no. 2 (2008), pp. 135–253.

[35] S. M. Trimberger. "Three ages of FPGAs: A retrospective on the first thirty years of FPGA technology." *Proceedings of the IEEE* 103, no. 3 (2015), pp. 318–331.

[36] G. Martin and H. Chang. "System-on-chip design." *ASICON 2001. 2001 4th International Conference on ASIC Proceedings (Cat. No. 01TH8549)* (2001), pp. 12–17, IEEE.

[37] T. A. Claasen. "An industry perspective on current and future state of the art in system-on-chip (SoC) technology." *Proceedings of the IEEE* 94, no. 6 (2006), pp. 1121–1137.

[38] L. Benini and G. De Micheli. "Networks on chips: A new SoC paradigm." *Computer* 35, no. 1 (2002), pp. 70–78.

[39] S. Markidis, S. W. Der Chien, E. Laure, I. B. Peng, and J. S. Vetter. "Nvidia tensor core programmability, performance & precision." *2018 IEEE International Parallel and Distributed Processing Symposium Workshops (IPDPSW)* (2018), pp. 522–531, IEEE.

[40] E. P. DeBenedictis, M. Badaroglu, A. Chen, T. M. Conte, and P. Gargini. "Sustaining Moore's law with 3D chips." *Computer* 50, no. 8 (2017), pp. 69–73.

[41] S. K. Moore. "3 Directions for Moore's law: The last few months have sent mixed signals about where chips are headed." *IEEE Spectrum* 55 (Nov. 2018), pp. 14–15.

3 Pre-RTL Methodologies

True brevity of expression consists in everywhere saying only what is worth saying, and in avoiding tedious detail about things which everyone can supply for himself. This involves correct discrimination between what is necessary and what is superfluous.

—Arthur Schopenhauer (from "The art of literature" in *The Essays of Arthur Schopenhauer* [translated by T. Bailey Saunders], On Style, 1891)

Designing an electronic system is a complicated process. We simplify the design process by decomposing it into multiple tasks. As we carry out these tasks, an abstract "idea" of a system gradually transforms into a design and finally into the desired product. During this transformation, creating a *register transfer level* (RTL) model is a milestone. However, before RTL model creation, several tasks need to be carried out. In this book, we refer to these tasks collectively as *pre-RTL methodologies*.

Pre-RTL methodologies involve determining various components and their interactions that achieve the desired functionality at a high level of abstraction. These components can be hardware or software. The hardware components can be processors, memories, peripherals, sensors, data converters, and signal processing blocks. The software can be firmware, device drivers, operating systems, and application programs. Since pre-RTL methodologies involve taking decisions at the system level, we also refer to the pre-RTL design activities collectively as *system-level design*.

At the system level, we have only a few necessary constraints of the system. Therefore, we have a greater freedom in choosing components, and many feasible solutions are possible. However, evaluating the impact of different choices is challenging because we have limited *implementation details*. Nevertheless, choices made during system-level design strongly impact the quality of the final product. Therefore, system-level design is critical, as well as challenging.

We have been doing IC-related system-level design for the last several decades. However, due to diverse application-specific issues, system-level design methodologies could not be standardized [1]. Till date, system-level design consists of widely varying tools and technologies [2–8]. For example, system-level design methodologies can differ between a signal processing chip (such as a video encoder and decoder) and a multicore processor. Therefore, in this chapter, we describe a few methodologies that are applicable in diverse scenarios. Subsequently, readers can apply the concepts explained in this chapter to different applications and platforms. To know more about system-level design, we suggest that readers refer to dedicated books on system-level design such as [1, 9–12].

3.1 A SYSTEM PERSPECTIVE

Figure 3.1 illustrates the process of transformation of an "idea" into a final product. We discuss it in more detail in the following paragraphs.

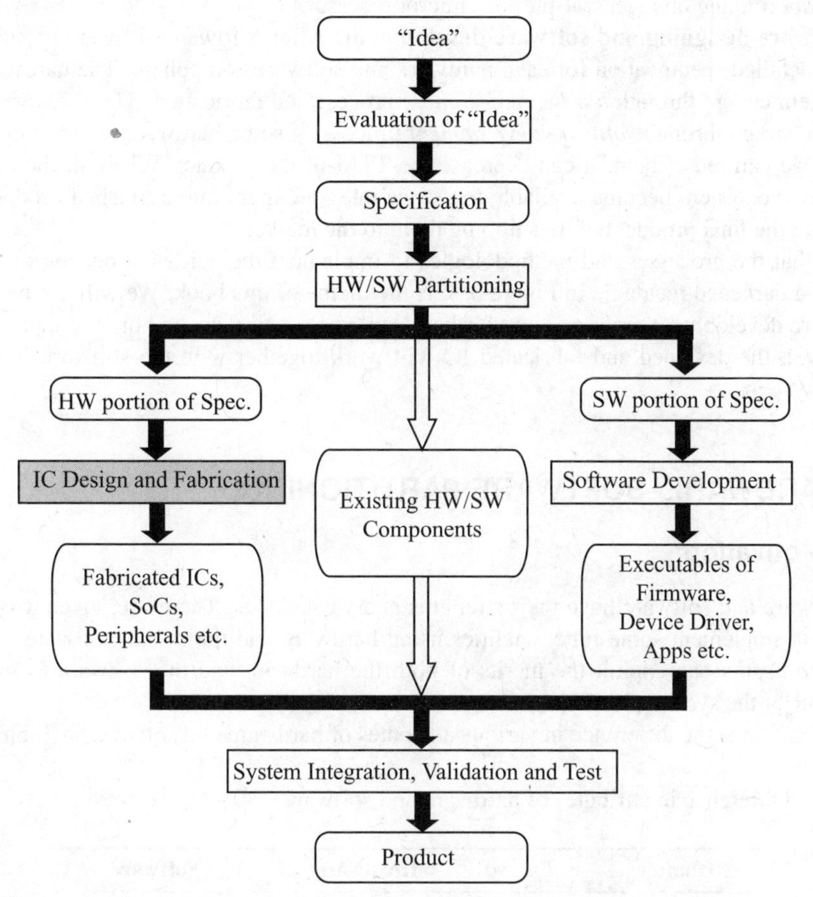

Figure 3.1 A system perspective of transforming an "idea" into the final product

Evaluation of an idea: Once we conceive an "idea" of a product, we first evaluate it for market needs, economic viability, and technical feasibility. Typically, we gather information about market requirements and competitive products. Subsequently, we assess the viability of the product and the risks involved in undertaking the venture.

Preparing specifications: After ascertaining that the product is viable, we prepare the top-level specifications of the product. The specifications can describe the desired functionality or features of the product, expected performance in terms of clock frequency, maximum power consumption, and product cost. It is important to note that, with cut-throat competition existing in the semiconductor market, the *time-to-market* (TTM) is a crucial parameter in product design. The time elapsed between conceiving a product and the product being available in the market is considered the TTM. Quite often, the commercial success of a product depends on the TTM. Therefore, decisions at the system level like choosing the implementation platform and system components can be guided by the available TTM for the product.

Hardware–software partitioning: Once we have made the top-level specification, we *partition* the system and identify its various components. Specifically, we decide whether to implement each component as hardware or software. This task is known as *hardware–software partitioning*. We can implement the hardware as a dedicated *chip* or a *block* inside a chip. We can implement the software as a *program* running on a general-purpose microprocessor.

Hardware designing and software development: After hardware–software partitioning, we prepare a detailed specification for each hardware and software component. The hardware portion of the system can go through *hardware designing* process and fabrication. The software portion of the system can go through *software development* process. If some hardware or software is already available, we can reuse them. It can decrease the TTM of the product. When all the components required by the system become available to us, we integrate them into a finished product. We test and validate the final product before shipping them to the market.

Note that the processes and methodologies to implement the hardware portion of the system (shown as a darkened rectangle in Figure 3.1) is the theme of this book. We will not be discussing the software development processes any further. Nevertheless, it must be borne in mind that, at the system level, the designed and fabricated IC will work together with the software to realize the complete system.

3.2 HARDWARE–SOFTWARE PARTITIONING

3.2.1 Motivation

Both hardware and software have their strengths and weaknesses. Therefore, given a system, it is beneficial to implement some functionalities using hardware and the rest in software. *Hardware–software co-design* can exploit the merits of both the hardware and the software to improve the overall QoR of the system [13–15].

We summarize the difference in various attributes of hardware and software in Table 3.1.

Table 3.1 Difference in attributes of hardware and software

Attribute	Hardware	Software
Speed	High	Low
Development time	High	Low
Cost	High	Low
Risk due to bug	High	Low
Customization	Low	High

The speed of the hardware is typically higher than the software. We can implement hardware as *parallel circuits* that can carry out the computation *concurrently*. On the other hand, instructions in a software program execute *sequentially* on a microprocessor or similar hardware. Therefore, it can be beneficial to implement the performance-critical part of a system in hardware. However, developing hardware is more time-taking, is costly, and involves more risks due to bugs. The software is easier to customize and more flexible in functionality. Therefore, it can be advantageous to implement the parts that need to be flexible and are not performance-critical in software.

Example 3.1 Assume that a system needs to compress video. We can divide the algorithm of video compression into two parts:

1. Computing Discrete Cosine Transform (DCT)
2. Handling different frames and performing other computations

The algorithm that computes DCT is executed many times in video compression. Assume that DCT is the bottleneck in the overall algorithm.

In this case, it is prudent to implement the DCT algorithm in hardware. We can design the hardware to have parallel circuits and enable it to do multiple computations concurrently. Therefore, it can compute several orders of magnitude faster than the software running on a general-purpose microprocessor. We can also optimize the dedicated hardware that performs DCT for other QoR metrics such as power dissipation.

We can implement the algorithm that handles frames and performs other computations as software. It can offer flexibility, enable easy bug fixing, and reduce the overall cost of the system.

3.2.2 Techniques

For hardware–software partitioning, we specify the functionality of the system at an algorithmic level. Subsequently, we map the functionality either to the hardware or the software.

Architecture: For hardware–software partitioning, we typically assume the architecture shown in Figure 3.2 [1, 16, 17].

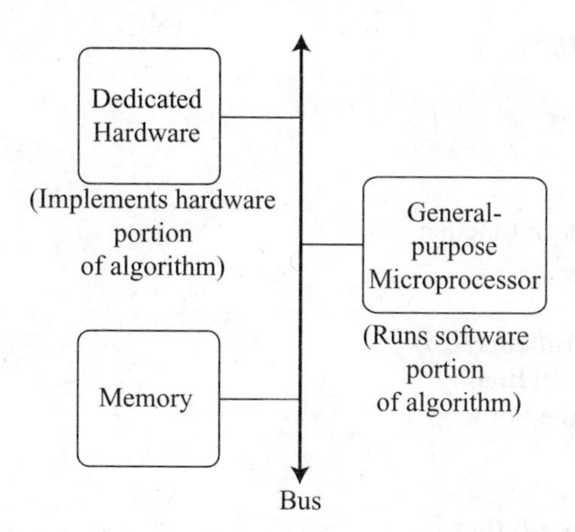

Figure 3.2 A typical architecture assumed in hardware–software partitioning [1, 16, 17]

A common *bus* connects the *dedicated hardware* with the *general-purpose microprocessor*. The dedicated hardware is also referred to as *hardware accelerator*. It increases the speed of the computation. The general-purpose microprocessor runs the software portion of the system. We decide the type of general-purpose processor before hardware–software partitioning.[1] It enables us

[1] One of the critical aspects of system-level design is deciding the family of processors required for the system. The choice of processor impacts optimal hardware–software partitioning. Hence, the task of choosing processor and hardware–software partitioning can be iterative.

to estimate the performance of the software during the partitioning. The bus shares information between the hardware accelerator and the software.

The architecture shown in Figure 3.2 is ideal for FPGA platforms [1]. The hardware accelerator can be implemented by programming FPGA fabric, while the processor available on the FPGA platforms can run the software. Note that, Figure 3.2 is a simplistic representation of a system. It can comprise multiple hardware accelerators and general-purpose processors. Moreover, the communication links between them can be of different topologies [16].

Approach: We present one of the approaches to carry out hardware–software partitioning in Algorithm 3.1.

Algorithm 3.1 PARTITION_HW_SW

Input:

- Given algorithm that is implemented entirely in software S
- Acceptable performance P
- Number of bottleneck functions tried for hardware acceleration in each iteration N

Output:

- Returns set of functions H to be implemented in hardware
- Returns *status* whether solution is found

Steps:

1: $H \leftarrow \{\}$
2: $speed \leftarrow Evaluate(H,S)$
3: $status = true$
4: **while** $(speed < P)$ **do**
5: $\quad Profile(H,S)$
6: $\quad$ **for** $i \leftarrow 1$ to N **do**
7: $\quad\quad f_i \leftarrow i^{th}$ bottleneck function
8: $\quad\quad S \leftarrow (S - f_i)$
9: $\quad\quad H \leftarrow (H \cup f_i)$
10: $\quad\quad newspeed \leftarrow Evaluate(H,S)$
11: $\quad\quad$ **if** $(newspeed \geq P)$ **then**
12: $\quad\quad\quad$ **break** from the For loop
13: $\quad\quad$ **end if**
14: $\quad$ **end for**
15: $\quad$ **if** $(newspeed < speed)$ **then**
16: $\quad\quad status \leftarrow false$
17: $\quad\quad$ **break** from the While loop
18: $\quad$ **end if**
19: $\quad speed \leftarrow newspeed$
20: **end while**
21: **return** $H, status$

It works on the assumption that the objective of partitioning is to design a system with acceptable performance P. Furthermore, since the hardware is typically costlier than software, we implement only a minimal set of functions using dedicated hardware.[2] Additionally, we allow the algorithm to try a maximum of N functions to move to the hardware in each iteration.

First, the given system is implemented entirely in software S running on a pre-decided general-purpose microprocessor. We estimate its *speed* using a function *Evaluate*(H, S). If the *speed* is acceptable, then we do not require any dedicated hardware. Otherwise, we need to identify and move some function(s) to dedicated hardware. We execute the software on a set of representative inputs and carry out dynamic program analysis using *software profilers* [18]. A software profiler measures the frequency or duration of each function call in a program. This task is referred to as *profiling*. We represent profiling as *Profile*(H, S) in Algorithm 3.1.

Using the results of the profiler, we identify i^{th} most severe *bottleneck function* f_i in the program. A bottleneck function consumes a large fraction of the total runtime of the program. After identifying the bottleneck function f_i, the function f_i is considered to be implemented in hardware.

We expect that the implementation of f_i on dedicated hardware will be faster than the corresponding software implementation. Therefore, the overall system speed is expected to increase. However, we should consider the communication overhead between the dedicated hardware and the software also. If the communication overhead is significant, the overall system speed can decrease. Therefore, we re-evaluate the overall system speed *newspeed* after moving a function from the software to hardware.[3]

After moving the function f_i to hardware, if the overall system speed does not become acceptable, we keep trying the N most severe bottleneck function. If the overall system speed becomes acceptable, we produce the computed partition. Otherwise, we repeat identifying the next bottleneck function(s) in the software that can be moved to dedicated hardware. At any stage, even after moving N bottleneck functions, if we cannot increase the speed, we cannot produce a solution. The algorithm terminates with a *status* = *false*.

Challenges: The above technique to obtain hardware–software partitioning is simple and makes a *greedy search* on the solution space. In practice, we face the following challenges:

1. **Performance estimation:** During hardware–software partitioning, the implementation of the dedicated hardware does not exist. Therefore, we evaluate system performance by assuming some timing model of the dedicated hardware. The past implementations of similar dedicated hardware can help us in building the model and estimating the performance. We can also perform *behavioral synthesis* (described later in this chapter) to estimate the performance of the dedicated hardware. Nevertheless, it is challenging to have a decent performance estimate of the dedicated hardware during hardware–software partitioning.
2. **Verification:** After hardware–software partitioning, we must verify that the partitioned system realizes the given functionality. To achieve this, we give inputs to the system consisting of both hardware and software and observe the output response. This task is referred to as *hardware–software co-simulation* [20]. The output response observed during hardware–software co-simulation must match the expected response. However, carrying out hardware–software co-simulation is challenging because of the non-existence of dedicated hardware. In practice, some abstract model of the dedicated hardware, such as *instructor set model*, is combined

[2] In practice, we need to honor the hardware-resource constraint.

[3] The data dependency can limit the overall system speedup. For example, if the hardware needs to wait for the input data, we will not achieve the expected parallelization-induced speedup [19].

with the software to carry out hardware–software co-simulation. Another strategy is to use an *emulation-based methodology*. In this methodology, we quickly implement a *prototype* of the dedicated hardware in FPGA. Then, we combine the prototype with the software at the system level and carry out the hardware–software co-simulation.

Most formulations of hardware–software partitioning are NP-hard[4] [21]. Therefore, researchers have mainly focused on developing efficient heuristics for hardware–software partitioning [13–15, 17, 21–25]. The problem of defining *computational models* for combined hardware–software systems is crucial and being researched. It becomes particularly challenging in systems with *heterogeneous integration* of hardware, processors, memories, and communication links.

3.3 FUNCTIONAL SPECIFICATION TO RTL

Once we have done the hardware–software partitioning, the functionality of the hardware part of the system becomes well-defined. For some applications, the hardware–software partitioning can be evident and not necessary. In either case, at the system level, we can specify the functionality of the hardware as algorithms using *high-level languages* such as C, C++, SystemC, or MATLAB.

Specification at a higher level of abstraction: Making specifications of the hardware at a *higher level of abstraction* has several advantages. It facilitates easy design-space exploration by quickly changing the algorithm and evaluating the effects on the QoR. Since the system contains less detail at a higher level of abstraction, it allows easier analysis, verification, and debugging. It also provides a golden reference for the subsequent tasks of the design flow. However, making specification at a higher level of abstraction opens up an *implementation gap* [3]. We need to subsequently transform the hardware description written in *high-level languages* into an RTL model and fill this implementation gap. It is a non-trivial task.

Adding details: A high-level algorithmic description does not carry the timing information of the circuit. However, an RTL model describes the data flow from register to register at different time instances or clock cycles. Thus, an RTL description of a design carries timing information of the circuit. Therefore, we need to add the timing information to the algorithmic model to obtain an RTL model. It involves making crucial decisions for a circuit such as *scheduling* operations in different clock cycles, determining the depth of the *pipeline*, and computational *resource allocation*. It is challenging to perform these tasks *optimally*.

RTL model: For brevity, we often refer to an *RTL model* simply as *RTL*. The schematic of a typical RTL is shown in Figure 3.3.

It consists of registers driving the output ports of the circuit. The signals propagate to the registers through computational elements or arithmetic logical units (ALU). The results of the computation are steered by the multiplexers to the registers using control signals. A finite state machine (FSM) generates the control signals. In general, we can divide an RTL into two parts:

1. **Data path:** The data path consists of computational elements such as adders, multipliers, and other arithmetic logic units (ALUs). The shared computational elements have their inputs multiplexed. The *select* signals for the multiplexers are generated in the control path.
2. **Control path:** The control signals steer the data appropriately in the data path. The control signals are generated by an FSM and control input ports.

[4] For NP-hard, see Section 1.3.4, "Hard problems".

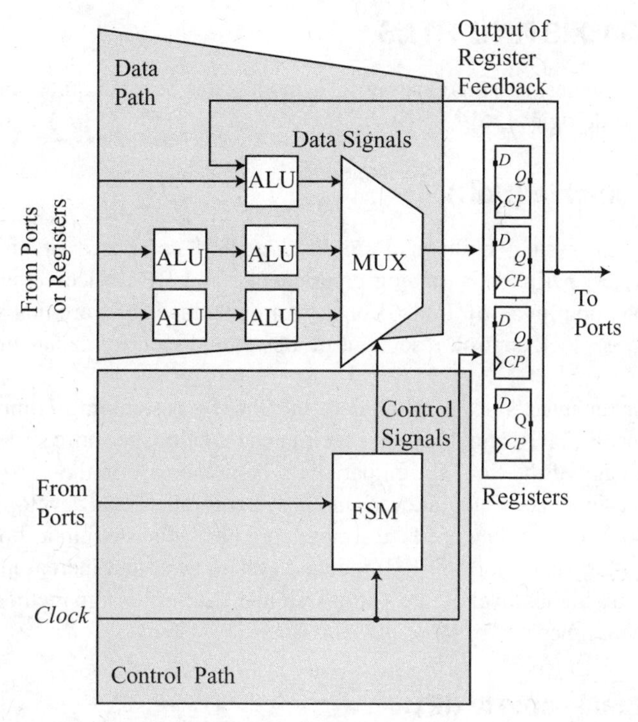

Figure 3.3 Structure of an RTL model

It is evident from Figure 3.3 that, in an RTL, computations are done both *concurrently* and *sequentially*. For example, the computation in parallel data paths is done concurrently, while in the same data path, one computation can be sequentially followed by another. Therefore, *hardware description languages* (HDLs) having constructs to describe both concurrent and sequential operations are a natural choice for RTL modeling.

From a given functional or algorithmic description, we can obtain RTL using the following approaches:

1. **Manually coding RTL:** Given a high-level description of a design, we can manually code the corresponding RTL.[5] However, this process is time-consuming, tedious, and prone to human errors. After manual RTL coding, we should verify that the coded RTL description is functionally correct.
2. **Reusing existing RTL:** To reduce the TTM, we can reuse existing RTL models. This approach is especially taken in *system-on-chip* (SoC) designs. We can also purchase RTL models in a packaged form from *intellectual property* (IP) vendors.
3. **Behavioral synthesis:** In recent times, techniques to automatically transform a given algorithm to an RTL model are becoming popular. These techniques are called *behavioral synthesis* or *high-level synthesis*. These techniques can alleviate some of the problems faced in the manual coding of RTL.

We will discuss the above methods in the following sections.

[5] We will discuss how to create an RTL model using Verilog in Chapter 8 ("Modeling hardware using Verilog").

3.4 REUSING EXISTING RTLS

During the last two decades, reusing existing RTL models has become quite popular, especially in the SoC design methodologies [26].

3.4.1 System-on-chip (SoC)

An *SoC* is a complete system built on a *single chip*. It is composed of processors, hardware accelerators, memories, peripherals, analog components, and RF devices connected using some structured communication links [6]. It also consists of embedded software that supports functions, such as boot sequence, clock control, reset control, power and security management, and low-level drivers.

Earlier, we implemented systems similar to an SoC by assembling components on boards. However, advancements in IC technology have facilitated realizing the entire system on a single chip.

A crucial advantage of the SoC design paradigm is an increase in the *designer's productivity*. The role of an SoC designer is to integrate various subsystems rather than develop them from scratch. Consequently, an SoC designer can produce a more complicated system in a shorter time [27]. The SoC design methodology has shorter TTM, reduced system cost, and increased ability to perform complex tasks (due to a higher level of integration). Hence, the SoC design methodology has become quite popular in recent times.

3.4.2 Intellectual Property (IP)

An SoC design methodology uses predesigned and pre-verified subsystems or blocks. These subsystems or blocks are called *intellectual properties* (IPs) or *IP cores*. An IP can be developed internally by a design company. It can also be purchased from third-party IP vendors.

Content: An IP can be a hardware block such as a processor, peripherals, memory, analog components, or interface. An IP can also contain software components such as real-time operating systems (RTOS) and device drivers. Some IPs, especially related to communication protocols, such as USB and Ethernet, can have an associated *verification IP* (VIP). A VIP eases verification effort of an IP user by encapsulating information related to communication protocol verification and other complex functions within an IP.

Sharing information: The information provided in an IP relates to the structure, configurability, and interfaces of the subsystem. A critical challenge in developing an IP block is to correctly *package* this information and allow a seamless integration at the SoC level. However, quite often, the IP packaging team and the SoC integration team are in different organizations. It creates a communication gap between the IP creator and the SoC designer. The problem aggravates due to non-uniform packaging approaches followed by various vendors and groups. Therefore, we need to be careful that IPs carry adequate knowledge from the IP creator to the SoC designer. It helps in avoiding the hassles of fixing integration issues at the SoC level.

3.4.3 Integration of IPs

The primary task in SoC design is to assemble and connect different IPs. It involves *instantiating* various IP blocks and making connections at the SoC level.

Generator tools: Earlier, we assembled IPs manually. However, with the increased complexity of SoCs, manual assembly is error-prone. Therefore, now we use *generator tools* to assemble IPs. We define the top-level IP models, bus interfaces, ports, registers, and the required configuration.

This information is referred to as the *metadata*. We can specify metadata in various formats such as IP-XACT, SystemRDL, XML, or spreadsheet [1]. A generator tool employs metadata to produce an SoC-level RTL with instantiated IPs. A generator tool can also produce a verification environment and low-level software drivers.

Configuring IPs: Some IPs are highly *configurable* and can have several configuration parameters such as bus width, power modes, and communication protocols. The task of IP assembly involves choosing the set of configuration parameters that achieves the required functionality. Some of the parameters of one IP can conflict with the parameters of another IP. Therefore, while assembling, we should check for consistency of configuration parameters among IPs.

Communication links: We face another challenge in connecting IPs to the *communication links* inside an SoC. Earlier, we employed *ad hoc bus-based* approaches to connect IPs. However, structured links borrowed from the network domain are also employed in SoCs to tackle the increased complexity of connections. These structured connecting links between IPs are known as *networks-on-chip* (NoC) [6, 27].

Verification: A crucial task in SoC design is to develop *verification methodologies* for the overall system. Multiple IPs and their communication links present a huge functional space for verification. Therefore, SoC verification is challenging. Moreover, functional verification of the software portion of an SoC exacerbates this problem.

3.5 BEHAVIORAL SYNTHESIS

The framework for the behavioral synthesis is shown in Figure 3.4. We provide a design or an algorithm specified in a high-level language such as C, C++, SystemC, and MATLAB as input. Additionally, we provide *constraints* also as input. The constraints can impose restrictions on resource usage, delay, latency, and power dissipation. It guides the behavioral synthesis tool in choosing one of the several possible RTL implementations. Moreover, we should also provide a *library* of circuit elements as an input to the behavioral synthesis tool. A library provides information about the area, delay, power, and other attributes for various computational elements such as combinational gates, flip-flops, adders, and multipliers. The output of the behavioral synthesis tool is an RTL model, typically written in Verilog or VHDL. The generated RTL is optimized to reduce *cost metrics* such as area, delay, latency, and power dissipation.

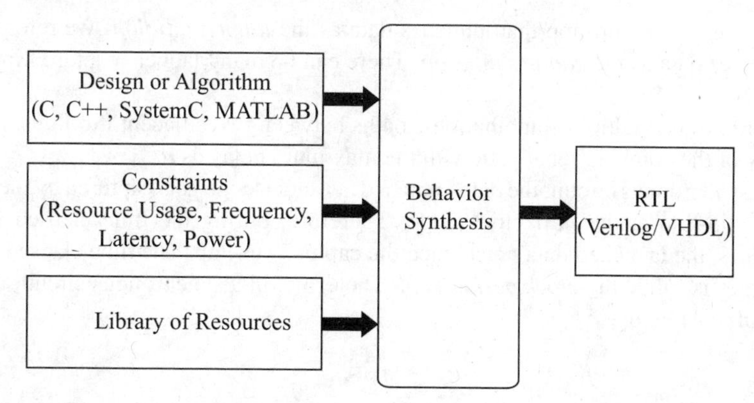

Figure 3.4 Framework of behavioral synthesis

3.5.1 Cost Metrics

Some of the cost metrics that a behavioral synthesis tool can consider in its optimization are as follows:

1. **Area:** During behavioral synthesis, we compute the total area as the sum of areas of individual circuit elements. To simplify the problem, we can ignore the area due to wiring and unused space on the die. To reduce the area cost, we should reduce the number of circuit elements, especially of large computing entities such as adders and multipliers. We can achieve this by *sharing* or reutilizing a circuit element for multiple computations.
2. **Delay of the critical path:** To appreciate the timing behavior of a circuit, we need to be familiar with the following concepts of timing analysis.

   Let us consider a synchronous circuit.[6] For simplicity, let us assume that it consists of combinational circuit elements and flip-flops. These circuit elements are connected by *wires* or interconnects. We define a *path* as a *sequence* of pins through which a signal can propagate. We consider a path as a *combinational path* if it does not contain any sequential circuit element such as a flip-flop. Thus, a combinational path can have combinational circuit elements connected with wires.

   In a circuit, if the output of one flip-flop is fed as an input to the other flip-flop through a combinational path, we say that these two flip-flops are *sequentially adjacent*. A pair of sequentially adjacent flip-flops is shown in Figure 3.5.

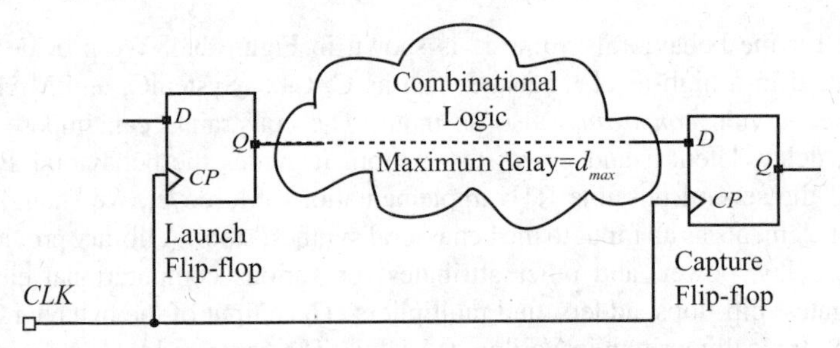

Figure 3.5 A pair of sequentially adjacent flip-flops in a synchronous circuit

   We refer to the flip-flop that launches data as the *launch flip-flop*. We refer to a flip-flop that captures data as the *capture flip-flop*. There can be many launch–capture flip-flop pairs in a circuit.

   There can be multiple combinational paths between two adjacent flip-flops. Let us denote the delay of the combinational path with the maximum delay as d_{max}.

   In a synchronous circuit, the data launched at a clock edge gets captured by the sequentially adjacent flip-flop at the *next* clock edge. Therefore, during the time elapsed between two clock edges, the launched data must reach the capture flip-flop. The time elapsed between two clock edges is called the *clock period* (T_p). Therefore, in a synchronous circuit, the following constraint must hold:

$$T_p > d_{max} \tag{3.1}$$

[6] Industrial designs are mostly synchronous or their major portion works synchronously. In this book, we will primarily consider synchronous circuits.

The clock frequency (f) is the reciprocal of the clock period, i.e., $f = 1/T_p$. Therefore, the maximum clock frequency f_{max} corresponds to the minimum clock period. Using Eq. 3.1, we can write f_{max} as follows:

$$f_{max} = 1/d_{max} \tag{3.2}$$

The combinational path that determines the *maximum clock frequency* (f_{max}) at which a circuit can operate is called the *critical path* of that circuit. For estimating f_{max}, we can consider the combinational path that offers the maximum delay d_{max} as the critical path.

Note that the above formalism of f_{max} is simplistic.[7] Nevertheless, at a higher level of abstraction, such as during behavioral synthesis, the above formalism is often adequate.

3. **Latency:** We can define the latency of a circuit as the maximum time elapsed after the application of input for the result to appear at the corresponding output. For a synchronous circuit, we can measure the latency in terms of the number of clock cycles. To reduce the latency, we should reduce the number of flip-flops in the *pipeline* (from the input to the output).

Example 3.2 Assume that we provide the following equation to a behavioral synthesis tool for implemention in RTL:

$$Y = a + b + c$$

Further assume that the library contains the circuit elements shown in Table 3.2.

Table 3.2 Resources to be used in behavioral synthesis

Resource Name	Area (μ^2)	Delay (ns)
Inverter	1	1
Multiplexer	6	12
Adder	110	140
Flip-flop	14	0

The given equation does not contain any timing information about the circuit. The behavioral synthesis tool can implement it in several ways.

Three possible implementations are shown in Figure 3.6. Let us examine each of them.

a) One possible implementation of $Y = a + b + c$ is shown in Figure 3.6(a). In this implementation, we compute $(a + b + c)$ using two adders $A1$ and $A2$. The result computed by the adders is assigned to the output port Y at the positive clock edge. Thus, the latency of the circuit is one clock cycle. The combinational path with the maximum delay consists of two adders.

b) Another possible implementation of $Y = a + b + c$ is shown in Figure 3.6(b). We introduce a flip-flop $F1$ between the two adders $A1$ and $A2$. Consequently, the area increases. The intermediate result $(a + b)$ is computed by $A1$ and is captured by the flip-flop $F1$ at the positive clock edge. Subsequently, the result $(a + b + c)$ is computed by $A2$. The final result becomes

[7] We will derive f_{max} more accurately in Chapter 14 ("Static timing analysis").

available at the output port Y at the next positive clock edge. Thus, the latency of the circuit is two clock cycles. However, the combinational path with the maximum delay consists of just one adder. Therefore, the maximum operating frequency increases by trading off area and latency.

c) Another implementation of $Y = a + b + c$ is shown in Figure 3.6(c). In this implementation, the adder *A1* performs different computations in different clock cycles. Since an adder consumes a large area, employing only one adder reduces the area cost.

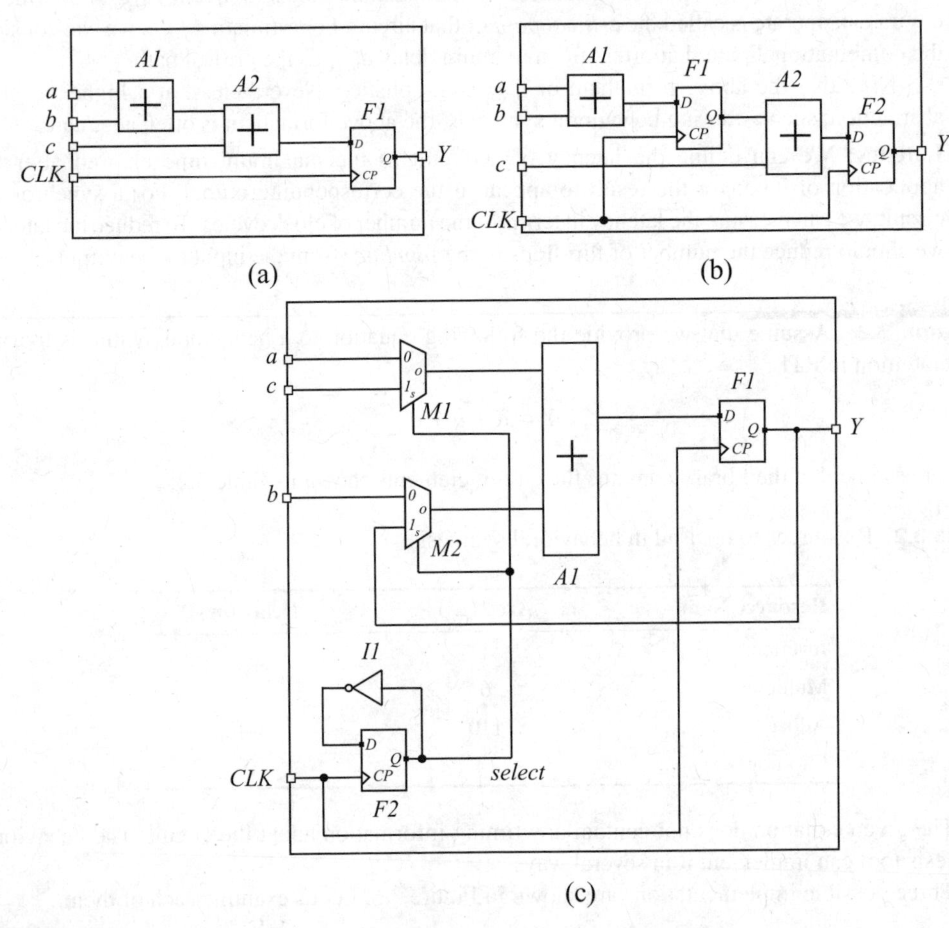

Figure 3.6 Three implementations of given function $Y = a + b + c$

The flip-flop *F2* generates the *select* signal. The input pin *D* of *F2* is connected to the output pin *Q* through an inverter *I1*. Consequently, the *select* signal toggles between 0 and 1 in alternate clock cycles.

When *select* = 0, the adder *A1* computes $(a + b)$. The result is captured at port *Y* at the positive clock edge. When *select* becomes 1 in the next clock cycle, then the adder *A1* computes $(Y + c)$. It is the same as $(a + b + c)$ since the value of *Y* was computed as $(a + b)$ in the previous cycle. Thus, the value at the output port *Y* is $(a + b)$ and $(a + b + c)$ in alternate clock cycles.

Since the result $(a + b + c)$ appears after two clock cycles, the latency is two. The critical path in this circuit comprises a multiplexer and an adder.

The cost metrics for the three implementations are shown in Table 3.3.

Table 3.3 Cost metrics for various circuits

Circuit	Latency (cycle)	Area (μ^2)	Critical Path Delay (ns)
Figure 3.6(a)	1	234	280
Figure 3.6(b)	2	248	140
Figure 3.6(c)	2	151	152

A behavioral synthesis tool can represent the circuits shown in Figure 3.6 as RTL models. There can be several other RTL models for this functionality. A behavioral synthesis tool can generate one of these implementations based on the *constraints* or *objectives*. For example:

1. If the objective is to attain *minimum area*, the tool can choose the circuit shown in Figure 3.6(c).[8]
2. If the objective is to attain *maximum operable frequency*, the tool can choose the circuit shown in Figure 3.6(b).
3. If the objective is to attain *minimum latency*, the tool can choose the circuit shown in Figure 3.6(a).

The above example illustrates that a behavioral synthesis enables automatic design-space exploration and can reduce the design effort significantly [28–31].

3.5.2 Challenges

Though behavioral synthesis is a powerful technique, there are many challenges associated with it:

1. **Physical design:** A behavioral synthesis tool does not have information about the physical design, such as the location of logic gates, wire length, and congestion. Therefore, we can sometimes face problems during physical design in using behavioral synthesis-generated RTL [29].

   For example, if registers or other resources are highly *shared* in the generated RTL, it can produce complicated connections around that shared resource. Consequently, when we make connections in the layout, we can face *congestion* around that highly shared resource. Note that the congestion problem occurs when many wires are laid out within a small region.[9]
2. **Incremental changes:** Sometimes, we need to make small changes in a behavioral model. It can be due to *engineering change order* (ECO) or some bug found in the behavioral model late in the design flow. However, making a slight change in a behavioral model can lead to significant changes in the generated RTL [32]. Such large changes can have a cascading effect on the design flow. Consequently, the overall design effort can increase. It can impact the design schedule adversely. The problem gets aggravated because the machine-generated RTL lacks

[8] Another cost metric at the system level is *throughput*. It can be measured as the number of computations per clock cycle. This implementation has the disadvantage of lower throughput than the others.

[9] We will discuss congestion in detail in Chapter 28 ("Routing").

human readability. In contrast, it is often easy to make localized incremental changes in a hand-coded RTL model.

3. **Verification:** We must verify the equivalence of the generated RTL model with the original behavioral model. It is a challenging task and an active area of research. Additionally, debugging an automatically-generated RTL is difficult. For easy debugging, a behavioral synthesis tool can provide some *back-referencing mechanism* between the behavioral model and the generated RTL [33].

3.6 RECENT TRENDS

With the increasing demand for computing, delivering complicated functionality and meeting stringent TTM becomes critical. Consequently, hardware–software co-design, IP integration, and high-level synthesis are now widely employed for complex chips and SoCs. Other techniques and models are also adopted that integrate architectural exploration, design implementation, software development, and verification tasks. For example, abstract models based on *transaction-level modeling* (TLM) and *unified modeling language* (UML) are now widely used for SoCs. These models deliver significant speedup in verification and greatly enhance productivity by providing a common reference [34, 35].

The advancements in the tools and models for system-level design have enabled us to progress from *integrated circuits* to *integrated systems* [28, 36–40]. In the future, we expect that the system complexities will increase further. Hence, there are merits in adopting a more structured approach to pre-RTL methodologies, standardization, and greater EDA tool support for system-level design.

REVIEW QUESTIONS

3.1 What are the advantages and disadvantages of designing at a higher level of abstraction?

3.2 What are the challenges in standardizing pre-RTL methodologies?

3.3 What is time-to-market? How do pre-RTL methodologies help in decreasing the time to market for a product?

3.4 How can hardware–software partitioning help in improving the quality of a system?

3.5 A software was profiled, and the % of time spent (on an average) in different functions are shown below.

Table 3.4 Profiling information

Function Name	% Time
F1	5
F2	3
F3	15
F4	5
F5	10
F6	30
F7	7
F8	25

The cost of a product entirely implemented with the above software running on a general-purpose microprocessor is \$100. We can implement each function of the software using separate dedicated hardware. The runtime of any function implemented with dedicated hardware is 1/10th of that implemented in software. However, when we implement any function with a dedicated hardware, the product cost increases by \$40 per function, while the software cost remains unchanged.

a) Assume that we need to keep the product cost below \$200. How can the software–hardware partitioning be performed such that we obtain the minimum runtime meeting the cost constraint? What is the estimated cost of the system? Assume that a given application consumed 1000s when entirely implemented in software. What is the estimated runtime for the designed system?

b) Assume that a given application takes a runtime of 1000s when entirely implemented in software.. However, we need to reduce the runtime of this application to below 350s by running some functions on a dedicated hardware. How can hardware–software partitioning be performed to obtain the minimum cost and meet the runtime constraint? What is the estimated runtime for the application and the cost of this system?

3.6 Why is an algorithm written in C language considered untimed while the corresponding design in Verilog obtained after high-level synthesis is considered timed?

3.7 How does SoC-based design improve the productivity of a designer?

3.8 What is the significance of verification IPs in an SoC-based design?

3.9 What are the challenges faced in integrating IPs in an SoC-based design? What are the possible solutions for these challenges?

3.10 How does behavioral synthesis enable automated design-space exploration?

3.11 Suggest a few techniques to increase the maximum operable frequency of a design.

3.12 Suggest a few techniques to account for the physical aspect of a design during behavioral synthesis.

TOOL-BASED ACTIVITY

Take any high-level synthesis tool (HLS) available to you and carry out the following activities. You can use open-source tools also [41].

```
#include <stdio.h>
#include <stdbool.h>
long myMacro(int,int,bool);
main()
{
    int j=1; int k=2;
    bool c= false;
    int res = myMacro(j,k,c);
    return 0;
}
```

```
long myMacro(int j, int k, bool c)
{
    int i=0;
    if(c){
        i = j + k;
    }else {
        i = j - k;
    }
    return i;
}
```

1. Obtain the RTL for the above C code using an HLS tool. You need to implement the function *myMacro* in RTL.
2. Explain the control path in the generated RTL model. Draw the state diagram if the control path contains an FSM.
3. Explain the function of different modules in the generated RTL. Explain how + and – operations are carried out on the data path.

REFERENCES

[1] L. Scheffer, L. Lavagno, and G. Martin. *EDA for IC System Design, Verification, and Testing*. CRC Press, 2016.

[2] M. Itoh, Y. Takeuchi, M. Imai, and A. Shiomi. "Synthesizable HDL generation for pipelined processors from a micro-operation description." *IEICE Transactions on Fundamentals of Electronics, Communications and Computer Sciences* 83, no. 3 (2000), pp. 394–400.

[3] H. Nikolov, T. Stefanov, and E. Deprettere. "Systematic and automated multiprocessor system design, programming, and implementation." *IEEE Transactions on Computer-Aided Design of Integrated Circuits and Systems* 27, no. 3 (2008), pp. 542–555.

[4] B. Khailany, E. Krimer, R. Venkatesan, J. Clemons, J. S. Emer, M. Fojtik, A. Klinefelter, M. Pellauer, N. Pinckney, et al. "A modular digital VLSI flow for high-productivity SoC design." *2018 55th ACM/ESDA/IEEE Design Automation Conference (DAC)* (2018), pp. 1–6, IEEE.

[5] P. Coussy, A. Baganne, and E. Martin. "A design methodology for integrating IP into SoC systems." *Proceedings of the IEEE 2002 Custom Integrated Circuits Conference (Cat. No. 02CH37285)* (2002), pp. 307–310, IEEE.

[6] L. Benini and G. De Micheli. "Networks on chips: A new SoC paradigm." *Computer*, 35, no. 1 (2002), pp. 70–78.

[7] N. Ohba and K. Takano. "An SoC design methodology using FPGAs and embedded microprocessors." *Proceedings of the 41st Annual Design Automation Conference* (2004), pp. 747–752, ACM.

[8] W. Kruijtzer, P. Van Der Wolf, E. De Kock, J. Stuyt, W. Ecker, A. Mayer, S. Hustin, C. Amerijckx, S. De Paoli, and E. Vaumorin. "Industrial IP integration flows based on IP-XACT standards." *2008 Design, Automation and Test in Europe* (2008), pp. 32–37, IEEE.

[9] A. Gerstlauer, R. Dömer, J. Peng, and D. D. Gajski. *System Design: A Practical Guide with SpecC*. Springer Science & Business Media, 2001.

[10] D. D. Gajski, N.D. Dutt, A. C. Wu, and S. Y. Lin. *High-Level Synthesis: Introduction to Chip and System Design*. Springer Science & Business Media, 2012.

[11] R. Camposano and W. Wolf. *High-level VLSI Synthesis*, vol. 136. Springer Science & Business Media, 2012.

[12] P. Michel, U. Lauther, and P. Duzy. *The Synthesis Approach to Digital System Design*, vol. 170. Springer Science & Business Media, 2012.

[13] R. K. Gupta and G. De Micheli. "Hardware-software cosynthesis for digital systems." *IEEE Design & Test of Computers* 10, no. 3 (1993), pp. 29–41.

[14] R. Ernst, J. Henkel, and T. Benner. "Hardware-software cosynthesis for microcontrollers." *IEEE Design & Test of Computers* 10, no. 4 (1993), pp. 64–75.

[15] G. De Michell and R. K. Gupta. "Hardware/software co-design." *Proceedings of the IEEE* 85, no. 3 (1997), pp. 349–365.

[16] W. H. Wolf. "An architectural co-synthesis algorithm for distributed, embedded computing systems." *IEEE Transactions on Very Large Scale Integration (VLSI) Systems* 5, no. 2 (1997), pp. 218–229.

[17] W. Wolf. "A decade of hardware/software codesign." *Computer* no. 4 (2003), pp. 38–43.

[18] D. C. Suresh, W. A. Najjar, F. Vahid, J. R. Villarreal, and G. Stitt. "Profiling tools for hardware/software partitioning of embedded applications." In *ACM SIGPLAN Notices* 38 (2003), pp. 189–198, ACM.

[19] R. Kastner, J. Matai, and S. Neuendorffer. "Parallel programming for FPGAs." *arXiv preprint arXiv:1805.03648*, 2018.

[20] D. Becker, R. K. Singh, and S. G. Tell. "An engineering environment for hardware/software co-simulation." *Proceedings 29th ACM/IEEE Design Automation Conference* (1992), pp. 129–134, IEEE.

[21] P. Arató, Z. Á. Mann, and A. Orbán. "Algorithmic aspects of hardware/software partitioning." *ACM Transactions on Design Automation of Electronic Systems (TODAES)* 10, no. 1 (2005), pp. 136–156.

[22] J. Henkel and R. Ernst. "An approach to automated hardware/software partitioning using a flexible granularity that is driven by high-level estimation techniques." *IEEE Transactions on Very Large Scale Integration (VLSI) Systems* 9, no. 2 (2001), pp. 273–289.

[23] M. López-Vallejo and J. C. López. "On the hardware-software partitioning problem: System modeling and partitioning techniques." *ACM Transactions on Design Automation of Electronic Systems (TODAES)* 8, no. 3 (2003), pp. 269–297.

[24] W. Zuo, L.-N. Pouchet, A. Ayupov, T. Kim, C.-W. Lin, S. Shiraishi, and D. Chen. "Accurate high-level modeling and automated hardware/software co-design for effective soc design space exploration." *Proceedings of the 54th Annual Design Automation Conference 2017* (2017), p. 78, ACM.

[25] X.-H. Yan, F.-Z. He, and Y.-L. Chen. "A novel hardware/software partitioning method based on position disturbed particle swarm optimization with invasive weed optimization." *Journal of Computer Science and Technology* 32, no. 2 (2017), pp. 340–355.

[26] M. Keating and P. Bricaud. *Reuse Methodology Manual for System-on-a-Chip Designs: For System-on-a-Chip Designs*. Springer Science & Business Media, 2002.

[27] R. Saleh, S. Wilton, S. Mirabbasi, A. Hu, M. Greenstreet, G. Lemieux, P. P. Pande, C. Grecu, and A. Ivanov. "System-on-chip: Reuse and integration." *Proceedings of the IEEE* 94 (June 2006), pp. 1050–1069.

[28] A. Takach. "High-level synthesis: Status, trends, and future directions." *IEEE Design & Test* 33, no. 3 (2016), pp. 116–124.

[29] J. Zhao, T. Liang, S. Sinha, and W. Zhang. "Machine Learning Based Routing Congestion Prediction in FPGA High-Level Synthesis." *2019 Design, Automation & Test in Europe Conference & Exhibition (DATE)* (2019), pp. 1130–1135, IEEE.

[30] S. Lahti, P. Sjövall, J. Vanne, and T. D. Hämäläinen. "Are we there yet? A study on the state of high-level synthesis." *IEEE Transactions on Computer-Aided Design of Integrated Circuits and Systems* 38, no. 5 (2018), pp. 898–911.

[31] L. Piccolboni, G. Di Guglielmo, and L. P. Carloni. "KAIROS: Incremental Verification in High-Level Synthesis through Latency-Insensitive Design." *2019 Formal Methods in Computer Aided Design (FMCAD)* (2019), pp. 105–109, IEEE.

[32] Q. Wang, Y. Kimura, and M. Fujita. "Methods of equivalence checking and ECO support under C-based design through reproduction of C descriptions from implementation designs." *2017 18th International Symposium on Quality Electronic Design (ISQED)* (2017), pp. 432–437, IEEE.

[33] J. S. Monson and B. L. Hutchings. "Using source-level transformations to improve high-level synthesis debug and validation on FPGAs." *Proceedings of the 2015 ACM/SIGDA International Symposium on Field-Programmable Gate Arrays* (2015), pp. 5–8, ACM.

[34] V. Jain, A. Kumar, and P. Panda. "Exploiting UML based validation for compliance checking of TLM 2 based models." *Des Autom Embed Syst* 16 (2012), pp. 93–113.

[35] L. Cai and D. Gajski. "Transaction level modeling: an overview." *First IEEE/ACM/IFIP International Conference on Hardware/Software Codesign and Systems Synthesis (IEEE Cat. No. 03TH8721)* (2003), pp. 19–24, IEEE.

[36] W. Chen, S. Ray, J. Bhadra, M. Abadir, and L.-C. Wang. "Challenges and trends in modern SoC design verification." *IEEE Design & Test* 34, no. 5 (2017), pp. 7–22.

[37] R. Nane, V.-M. Sima, C. Pilato, J. Choi, B. Fort, A. Canis, Y. T. Chen, H. Hsiao, S. Brown, et al. "A survey and evaluation of FPGA high-level synthesis tools." *IEEE Transactions on Computer-Aided Design of Integrated Circuits and Systems* 35, no. 10 (2015), pp. 1591–1604.

[38] Q. Zhu and M. Tatsuoka. "High quality IP design using high-level synthesis design flow." *2016 21st Asia and South Pacific Design Automation Conference (ASP-DAC)* (2016), pp. 212–217, IEEE.

[39] S. Rigo, R. Azevedo, and L. Santos. *Electronic System Level Design: An Open-Source Approach*. Springer Science & Business Media, 2011.

[40] M. Goli and R. Drechsler. *Automated Analysis of Virtual Prototypes at the Electronic System Level: Design Understanding and Applications*. Springer Nature, 2020.

[41] C. Pilato and F. Ferrandi. "Bambu: A modular framework for the high level synthesis of memory-intensive applications." *2013 23rd International Conference on Field programmable Logic and Applications* (2013), pp. 1–4, IEEE.

4 RTL to GDS Implementation Flow

In our own reflection abstraction is a throwing off of useless baggage for the sake of more easily handling the knowledge which is to be compared, and has therefore to be turned about in all directions...

—Arthur Schopenhauer, *The World as Will and Idea* (translated from the German by R. B. Haldane and J. Kemp), 1909

VLSI design flow often starts with an RTL model and goes through a sequence of logical and physical design processes. Finally, we obtain a layout typically represented in the GDS format.

In this chapter, we briefly explain the tasks involved in transforming an RTL model into GDS. We will describe these tasks in detail in the subsequent chapters.

4.1 ABSTRACTION IN VLSI DESIGN FLOW

We can represent a design in several ways. For example, we can describe it as an algorithm, RTL model, connection of logic gates, and layout. These descriptions differ in (a) the level of *abstraction* and (b) the design *view* or the design domain.

Abstraction: Abstraction means hiding lower-level details in a description. For example, an algorithm does not contain the timing information of a design, while an RTL model typically describes the flow of data at different clock cycles. Therefore, an algorithm is a more abstract description than an RTL model.

View: A view refers to the aspect of a design that we capture in the description. It can be a design's behavior, structure, or physical form. For example, we can describe a design by an algorithm (behavioral view), the connection of logic gates (structural view), or the layout (physical view).

4.1.1 Gajski-Kuhn's Y-chart

We can depict different abstract representations of a design and visualize their transformations using Gajski-Kuhn's *Y-chart*, shown in Figure 4.1 [1–3].

The concentric circles show different levels of abstraction. As we move from an outer circle to an inner one, the abstraction level decreases, and more details get added to a design. For example, the abstraction decreases as a design evolves as follows: system specification → algorithm → RTL specification → Boolean equation → differential equation.

The three different views of a design are shown on three axes forming a "Y": *behavioral*, *structural*, and *physical*. The behavior describes the functionality, the structure describes the interconnections, and the physical design view describes the components' shape, size, and location.

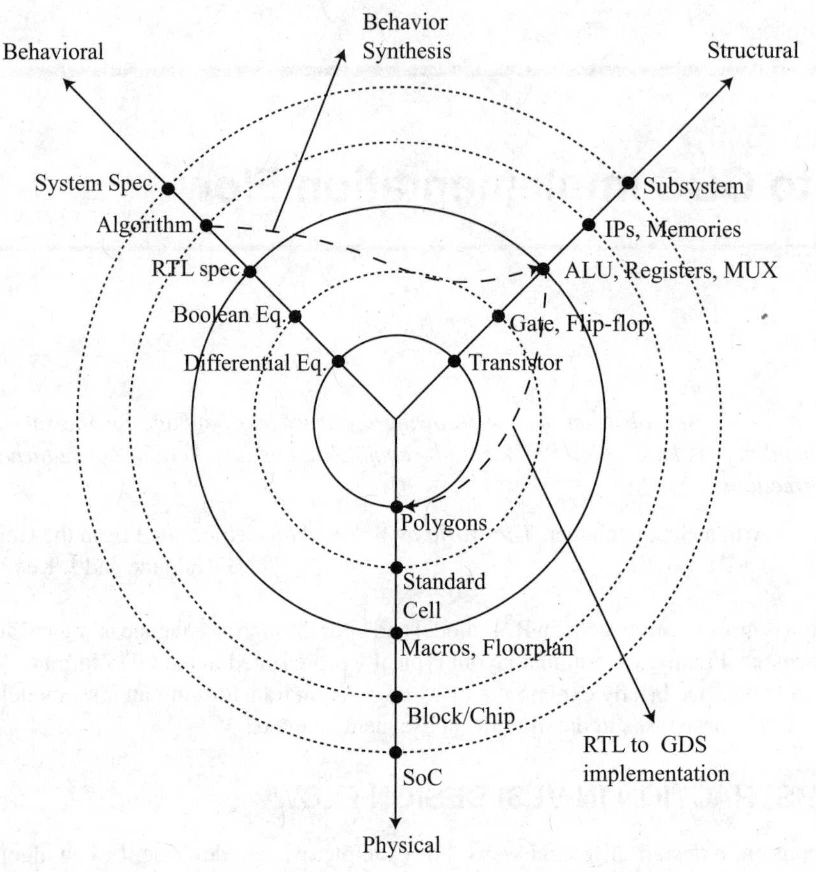

Figure 4.1 Gajski-Kuhn's Y-chart [1–3]

The intersection of abstraction circles and the domain axes are shown as blackened circles and marked by the corresponding design elements. For example, consider a high level of abstraction (second circle from outside). We describe behavior by a set of algorithms, structure by connections of IPs and memories, and the physical form using a layout of blocks in a design. Next, let us consider a low level of abstraction (innermost circle). We describe behavior by a set of differential equations, structure by connections of transistors, and the physical form by polygons, typically in the GDS format.

4.1.2 Design Transformations

We can visualize a design implementation task as a transformation of the following types:

1. **Refinement:** It transforms a design from a high level of abstraction to a low level of abstraction by adding details. In the Y-chart, *refinement* involves moving from an outer circle toward an inner circle.
2. **Domain transformation:** It transforms a design from one view to another. For example, from behavioral view to structural view or from structural view to physical view.

3. **Optimization:** It does not change the abstraction level or the view of a design. It improves some QoR measures, typically by trading off other QoR measures. For example, *optimization* can improve timing by incurring some area costs.

In the Y-chart, we can depict design transformations by an arrow. As an illustration, we have denoted *behavioral synthesis* by an arrow in Figure 4.1. Behavioral synthesis transforms an algorithm into an RTL model. As discussed in the previous chapter, we can describe an RTL model *structurally* as an interconnection of ALUs, flip-flops, and multiplexers. Thus, behavioral synthesis involves refinement, domain transformation, and optimization. Since optimizations do not change an abstraction level or a view, we do not show them by an arrow on the Y-chart.

In Figure 4.1, we also represent the RTL to GDS implementation flow by an arrow. It transforms a design from RTL to polygon representation in the GDS format. It involves refinement, domain transformation, and optimizations.

4.1.3 Logic Synthesis and Physical Design

We carry out multiple tasks sequentially in the RTL to GDS flow. We divide these tasks into two stages: *logic synthesis* and *physical design*, as shown in Figure 4.2.

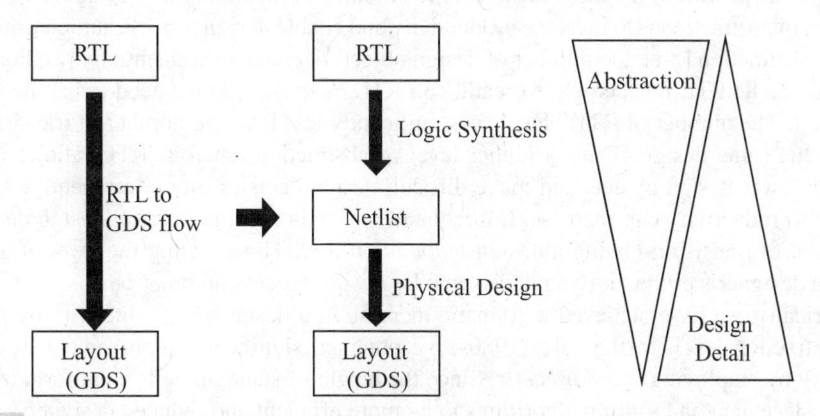

Figure 4.2 Dividing RTL to GDS flow into logic synthesis and physical design

1. **Logic synthesis:** It involves transforming an RTL to its corresponding *netlist*. A netlist is a network or interconnection of logic gates or standard cells. A netlist implements the required *logical functionality* of a design. We will describe the components of a netlist in detail in the following sections.
2. **Physical design:** It involves transforming a netlist into a layout that can be represented in the GDS format. Thus, it takes a design from the structural domain to the physical domain.

As a design flow progresses, the abstraction decreases, while the details in a design increase. Let us understand the motivation for employing such a design flow.

4.1.4 Motivation for Abstraction

During the early phases of a design flow, we represent designs at a higher level of abstraction. At a higher level of abstraction, there is less detail. When there is less detail, it is easier to modify a design.

Therefore, we can analyze and try out many solutions in less time. Consequently, optimization and *design-space exploration* become easier at a higher level of abstraction. Later in the design flow, details get added to a design by refinement. It makes modifying a design difficult.

Example 4.1 Consider a function $F = (A + B)'$ that we implement using logic gates in a netlist. It is easy to transform the implementation from NOR gate to a combination of AND and NOT gates ($F = A'.B'$) in the netlist.

Next, consider that we have implemented $F = (A + B)'$ using a standard cell delivering NOR function, and we have already created a layout. Now, changing the NOR cell with a combination of AND and NOT cells is more complicated. We need to add new cells to the layout. Hence, we need to find suitable locations for these cells on the layout and make changes to the layout of the wires also. These changes will require more design and computational effort.

Thus, modifying a design is more convenient at a higher level of abstraction. Therefore, we make large transformations early in the design flow when the abstraction is high. Subsequently, as abstraction decreases, we make smaller design modifications.

Designer's productivity and automation: Another motivation for a designer to work at a higher level of abstraction is to increase productivity and enable design process automation [2, 3]. By raising the abstraction level, the number of design objects decreases exponentially [2]. For example, we need to handle RTL objects while creating an RTL model, while we need to handle transistors in the layout. The number of RTL objects is significantly less than the number of transistors in the layout for the same design. Thus, a higher level of abstraction (such as RTL) allows a designer to focus on fewer design objects and make difficult design decisions more efficiently. As a result, a designer's productivity can increase. Subsequently, we can add the missing design details (such as the layout of transistors) using a design *automation tool*. Thus, raising the level of abstraction enhances a designer's productivity and also enables design process automation.

Historically, we have achieved a dramatic increase in a designer's productivity by raising the design abstraction level. In the early 1980s, we obtained significant improvement in designers' productivity by employing *standard cells*. Since the height of standard cells is uniform in a library, it allowed placement and routing algorithms to be more efficient and reduced designers' effort. We achieved a similar kind of productivity gains in the early 1990s by employing *logic synthesis tools*. It raised the abstraction level from logic gates to RTL objects [3]. In the early 2000s, SoC-based design and IP assembly methods further delivered productivity gains by raising the abstraction levels.[1]

4.2 LOGIC SYNTHESIS

Logic synthesis converts an RTL model to a functionally equivalent netlist.

4.2.1 The Framework

The inputs and output for logic synthesis are shown in Figure 4.3.

[1] For details on SoC design methodology, see Section 3.4, "Reusing existing RTL." SoC design methodology allowed designers to work at the IP level.

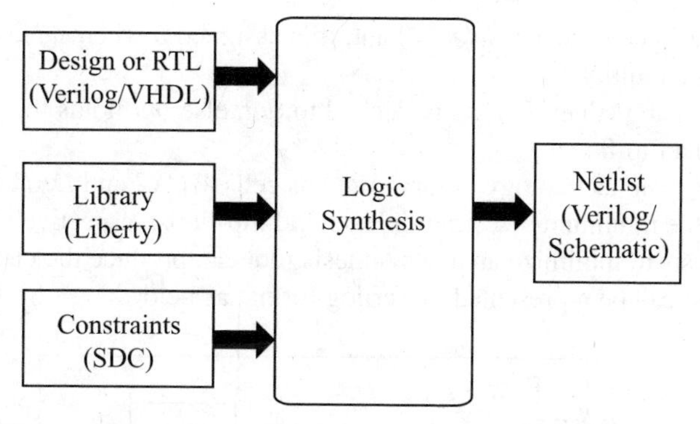

Figure 4.3 Inputs and output of logic synthesis

It requires the following inputs:

1. **Design:** We provide an RTL model to a logic synthesis tool. The RTL model is typically represented in Verilog, VHDL, or a mix of Verilog and VHDL.
2. **Technology library:** A logic synthesis tool creates netlist by choosing *standard cells* from the given *technology library*. We also refer to the technology library as *library* in short. A library is modeled in Liberty format [4, 5]. It describes the functionality and attributes such as delay, power, and area of each standard cell.[2]
3. **Constraints:** We define our design goals using *constraints*. The design goals are related to the expected timing behavior, such as maximum operable frequency. We also use constraints to convey information about the design environment. For example, we can specify the attributes of the incoming signal and the generated output signal. Typically, we specify constraints in Synopsys Design Constraint (SDC) format [4, 6].[3]

The process of translating an RTL to an equivalent circuit composed of logic gates is straightforward. We explain it in the following example.

Example 4.2 Consider the Verilog code shown below.[4] Let us examine how a logic synthesis tool can translate it to netlist.

```
module top(a, b, clk, select, out);
    input a, b, clk, select;
    output out;
    reg out;
    wire y;
    assign y = (select) ? b : a;
    always @(posedge clk)
    begin
        out <= y;
    end
endmodule
```

[2] We will explain technology libraries in Chapter 13 ("Technology library").

[3] We will explain constraints in Chapter 15 ("Constraints").

The signal *y* is assigned *a* when *select*=0 and it is assigned *b* when *select*=1. This behavior is similar to a two-to-one multiplexer.

When there is a rising edge of *clk*, *y* is latched to *out*, else *out* holds the previous value. This behavior is similar to a flip-flop.

Assume that the given technology library contains cells *MUX2* and *DFF*. Further, assume that *MUX2* and *DFF* are the minimum-sized multiplexer and flip-flops, respectively, in the given library. If we perform synthesis to minimize area, a synthesis tool can produce the circuit shown in Figure 4.4. The output netlist can be represented in Verilog format as below.

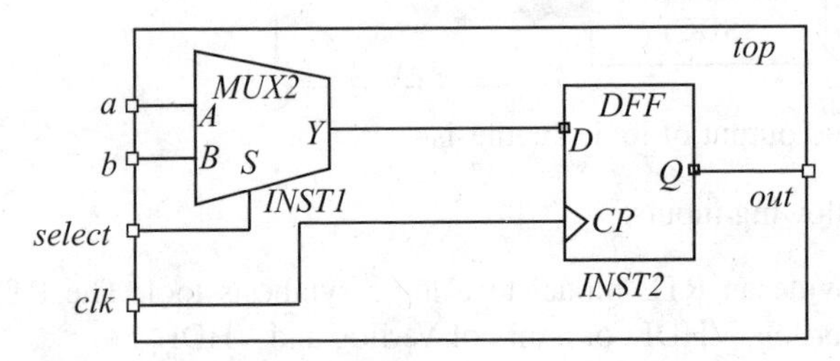

Figure 4.4 Schematic of the synthesized netlist

```
module top(a, b, clk, select, out);
    input a, b, clk, select;
    output out;
    wire y;
    MUX2 INST1(.A(a), .B(b), .S(select), .Y(y));
    DFF INST2(.D(y), .CP(clk), .Q(out));
endmodule
```

The instance names (*INST1* and *INST2*) are arbitrary names given to the *instances* by the synthesis tool. The names *MUX2* and *DFF* are the cell names in the given technology library.

The above example illustrates that translating an RTL to an equivalent netlist is straightforward. However, a naïve translation of an RTL to logic gates can lead to unacceptable QoR measures such as timing, area, and power. Therefore, a logic synthesis tool needs to generate a netlist, meeting the given constraints and *optimizing* some QoR measures such as area, timing, or power. This task of logic *optimization* is challenging for a logic synthesis tool.

4.2.2 Technique

Logic synthesis is a mature technology. Designers have been using it since the early 1990s. In a typical logic synthesis system, we decompose the synthesis task into two distinct phases [7–9]:

---

[4] This example is just to familiarize readers with a Verilog model. We will discuss modeling RTL using Verilog in detail in Chapter 8 ("Modeling hardware using Verilog").

1. **Technology-independent phase:** In this phase, a synthesis tool does not consider the cells of the given technology library. It generates a netlist composed of *generic gates*. A generic gate has a well-defined Boolean function. It can deliver the function of a combinational gate such as AND, NAND, XOR, multiplexer, and demultiplexer. It can also deliver sequential logic functions such as of latches and flip-flops. Therefore, a *generic gate netlist* can define the *design functionality* unambiguously. However, a generic gate does not have a fixed transistor-level implementation. Hence, it does not have a well-defined area, delay, and power attributes. Therefore, it is difficult to estimate the QoR of a generic gate netlist. As a result, logic optimizations performed during this phase are based on crude metrics such as the number of logic gates and logic depth.
2. **Technology-dependent phase:** In this phase, a synthesis tool considers the information contained in the given technology library. Therefore, the attributes such as timing and area of the *cells* guide the second phase of logic synthesis. The second phase of logic synthesis starts with the *mapping* of generic gates in a netlist to the *library cells* in the given library. This step is known as *technology mapping*.[5] After performing technology mapping, a synthesis tool carries out further logic optimization to improve the QoR of a design.

4.2.3 Netlist and Associated Terminologies

A netlist describes *connections* between various *components* in a design. We can represent a netlist using a schematic or a Verilog file. There are many kinds of entities in a netlist. In this section, we describe terminologies associated with these entities. These terminologies are used commonly in EDA tools, referred to in literature, and will frequently appear in subsequent chapters. Therefore, we suggest that readers get familiarized with these terminologies.

Consider the netlist and its corresponding library shown in Figure 4.5(a) and 4.5(b), respectively. We can identify the following entities in the netlist.

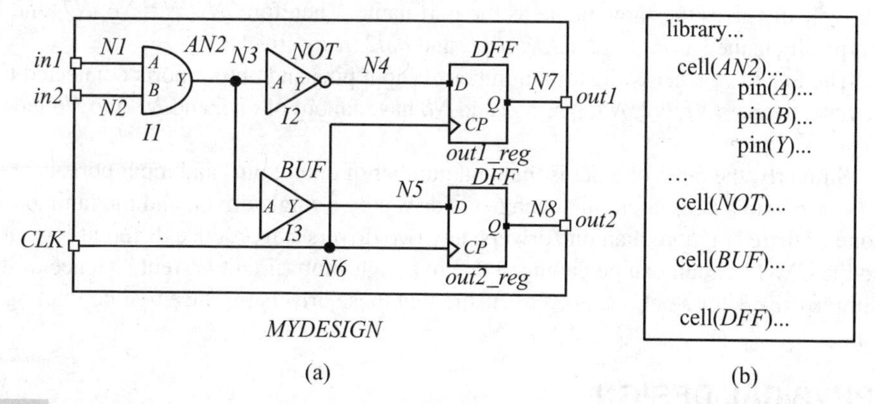

Figure 4.5 Illustrating entities in a netlist: (a) schematic and (b) corresponding library

1. **Design:** We refer to the top-level entity as a *design*. The name of the design in Figure 4.5 is *MYDESIGN*.

[5] We will discuss technology mapping in Chapter 16 ("Technology mapping").

2. **Port:** The interface of a design through which it communicates with the external world is known as a *port*. The ports of *MYDESIGN* are *in1*, *in2*, *CLK*, *out1*, and *out2*. The port through which a signal enters a design is known as an *input port* and the port through which a signal leaves a design is known as an *output port*. The ports *in1*, *in2*, and *CLK* are the input ports, and the ports *out1* and *out2* are the output ports. An input port is also referred to as *primary input* (PI) and an output port as *primary output* (PO).
3. **Cell:** The basic entity delivering a combinational or sequential logic function in a technology library is known as a *library cell* or *cell*. In Figure 4.5(b), the cells are *AN2*, *NOT*, *BUF*, and *DFF*.
4. **Instance:** When a library cell is used inside a design, it is called an *instance*. The process of using a cell in a design is known as *instantiation*. The instances in *MYDESIGN* are *I1*, *I2*, *I3*, *out1_reg*, and *out2_reg*. Note that the same cell can be instantiated multiple times, with different instance names (*out1_reg* and *out2_reg* are instances of the same cell *DFF*).
5. **Pin:** An interface of a *library cell* or *instance* through which it communicates with the other components is called a *pin*. For example, *A*, *B*, and *Y* are pins of the cell *AN2* and instance *I1*.

   A pin that brings a signal into a cell or instance is known as an *input pin*. A pin that sends a signal out from a cell or instance is known as an *output pin*. Pins *A* and *B* are input pins, while the pin *Y* is an output pin.

   It is often clear from the context whether we are describing a library cell's pin or an instance's pin. Nevertheless, when we want to be explicit, we refer to the pin of a library cell as *library pin* and the pin of an instance as *instance pin*.

   We can refer to the instance pin as a combination of instance and pin names. For example, we can refer to the pin *A* of the instance *I1* as *I1/A*, where / is a separator between names. We have followed this naming convention for instance pins throughout this book.
6. **Net:** A wire that connects different pins and ports in a design is called a *net*. *N1*, *N2*, *N3*, *N4*, *N5*, *N6*, *N7*, and *N8* are nets in *MYDESIGN*. For nets that are connected to a port, a synthesis tool typically gives the same name as the port name. Therefore, *N1*, *N2*, *N6*, *N7*, and *N8* will be typically named as *in1*, *in2*, *CLK*, *out1*, and *out2*, respectively.

   The fanout of a net is the total number of input pins and output ports connected to it. For example, the nets *N1*, *N2*, *N4*, *N5*, *N7*, and *N8* have fanout of one, and *N3* and *N6* have fanout of two.

   Similarly, the fanin of a net is the total number of output pins and input ports connected to it. Typically, in a CMOS circuit, each net is driven by a single driver, and the fanin of each net is one. If there are more than one drivers, any two drivers can produce different logical values, and the CMOS circuit can be damaged due to a high short-circuit current.[6] Hence, if there are multiple drivers for a net, we need to ensure that these drivers produce non-conflicting values.

4.3 PHYSICAL DESIGN

Physical design is the process of deriving a layout from a given netlist. We represent the layout in GDS or geometrical patterns that can be written on masks for IC fabrication.

[6] A CMOS logic gate has a low resistance path from the output to the ground while driving 0 and to V_{DD} while driving 1. Hence, V_{DD} gets connected to the ground through a low resistance path for conflicting logic values at a net.

4.3.1 The Framework

The inputs and output for physical design are shown in Figure 4.6. It takes the following inputs:

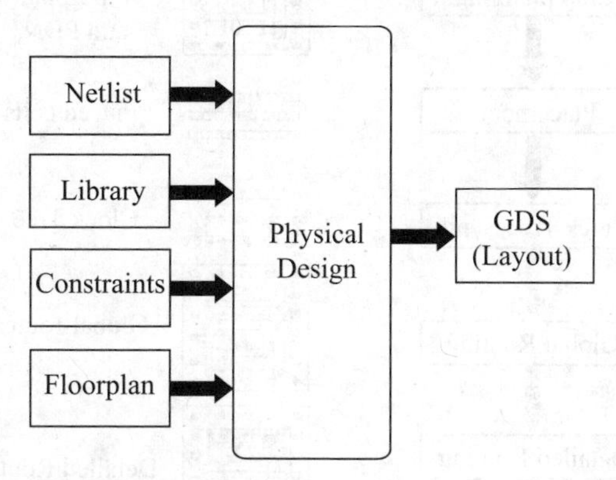

Figure 4.6 Framework of physical design

1. **Netlist:** We provide the netlist generated by the logic synthesis tool as an input to the physical design tool.
2. **Library:** We provide the library that contains the cells instantiated in the above netlist. A physical design tool can compute delay and other attributes for cell instances with the help of this library.

   A physical design tool also needs to know some information about the library cell layout, such as the bounding box of cells and the location of pins. It also needs to know process-dependent parameters such as the number of metal layers supported by the technology and the resistivity of the metal layers. This information is contained in *physical libraries*, and we need to provide them to the physical design tool.
3. **Constraints:** We provide constraints to a physical design tool defining the design goals and the external environment. We can provide the same constraints that we used in logic synthesis after making some modifications related to wires and physical design.
4. **Floorplan:** A floorplan captures the designer's intent about the physical design. It defines the size and aspect ratio of a die. It contains information such as the location of hard macros, the shape and location of various blocks, spaces for the rows of standard cells, and the location of input/output (I/O) pads. A physical design tool uses a floorplan as a constraint in placing various components and creating a wire layout.

Using the above inputs, a physical design tool determines the *location* of all instances in a design. It decides the *clocking structure* and layout of wires through which the clocked elements such as flip-flops receive the clock signal. It builds the topology of the metal and via layers for all nets in a design. Finally, a physical design tool produces a layout represented in GDS format which contains all the information required for fabricating a chip.

4.3.2 Major Tasks

The major tasks in a physical design flow are shown in Figure 4.7. We explain them briefly in the following paragraphs.

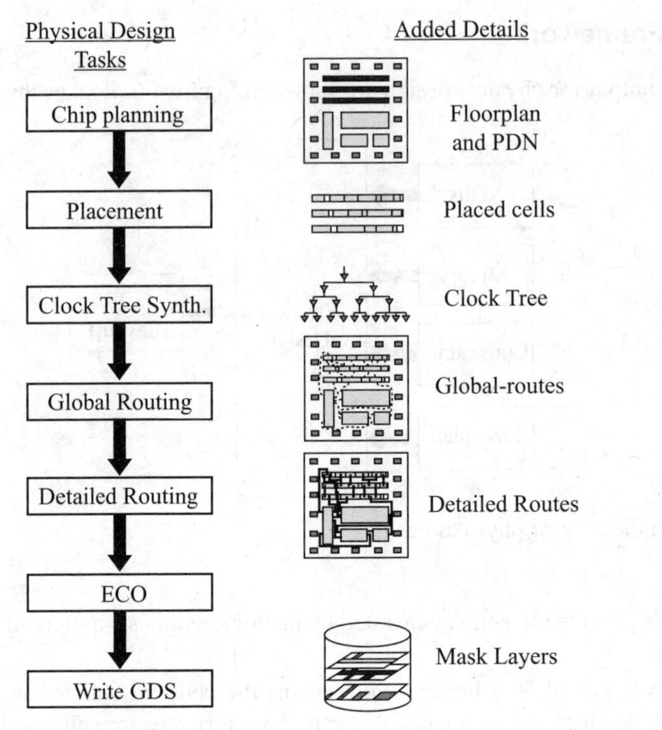

Figure 4.7 Major tasks in physical design

1. **Chip planning:** We start the physical design task with chip planning.[7] We make some crucial decisions about the design layout at this stage. For creating a layout for industrial and large designs, we first partition it into smaller *blocks* or *subsystems* that can be managed easily. Note that blocks are large entities and can contain more than a million standard cells.

   After partitioning, we decide the location of each block on the layout. While creating the layout, we need to consider the physical constraints such as die size, aspect ratio, and the location of fixed entities such as input/output ports. Some blocks are flexible, and we can change their shapes. We also allocate some area for the standard cells and specify the location of input/output pads. This task is known as *floorplanning*.

   A key consideration in floorplanning is to avoid *congestion* (too many wires going through a small region). An inappropriate location or shape of blocks can lead to congestion, and it will be discovered later in the design flow. Congestion in a design leads to timing and routing issues. Therefore, floorplanning tools employ various techniques to predict and avoid this situation.

   During chip planning, we also carry out *power planning*. We decide the topology of the *power delivery network* (PDN) for a design. We build structures, typically using top metal layers, that can deliver sufficient power to all the cell instances and active elements in a design. A key consideration in power planning is to avoid supply voltage from dropping below a threshold at each cell instance in a design. Additionally, the current in the network should remain sufficiently low to avoid reliability issues.

[7] We will explain chip planning in detail in Chapter 25 ("Chip planning").

2. **Placement:** We decide the location of bigger entities such as blocks and macros during chip planning. However, a design also has numerous standard cells, often more than a million. We need to decide their location on the layout also. This task is accomplished by *placement*.[8] The overwhelming number of standard cells and their connections make the placement problem challenging. We heavily rely on an automatic placement tool for this task. In contrast, we often make manual interventions in floorplanning because it involves fewer entities.

   The primary objective used by placement tools is to minimize the *total wire length* of a design. Since the wires are still not laid out during placement, some *estimate* of wire length is used during placement. Additionally, a placement tool tries to ensure that the design works at the given clock frequency. Therefore, placement tools are often *timing-driven*. They attempt to improve the performance by placing the cells on the critical path nearby. Placement tools also consider congestion problems and move out cells from the regions where estimated congestion is high.

3. **Clock tree synthesis:** After placement, we determine the topology of the clock network and decide how a clock signal reaches each clocked element in a design. This task is known as *clock tree synthesis* (CTS).[9] The primary objective of CTS is to minimize the *clock skew*. The clock skew is the *difference* between the arrival time of the clock signal at two clocked elements. We typically choose a symmetric architecture for the clock network to minimize the clock skew. To reduce the clock skew, we also need to insert clock buffers of appropriate sizes and at optimal locations.

   CTS is also responsible for building the complete wire layout for the clock network. Note that clock signals are the most critical signal of a synchronous design. We should avoid *detours* or long routes for the clock signal. Therefore, in physical design, we give higher priority to the clock signals while allocating routing resources. Hence, we carry out CTS before routing data signals. The other objective of CTS is to decrease the power dissipation in a clock network. It is crucial because a clock network consumes a significant fraction of the total dynamic power. Therefore, CTS inserts special circuit elements, such as *clock gaters*, to reduce power dissipation in a clock network.

4. **Global routing and detailed routing:** After CTS, we create a wire layout for all nets in the given netlist. This task is known as *routing*.[10] Note that the routing of the power delivery network and the clock signal is already done in previous design steps. Hence, we route only the signals that are not yet routed. Since creating a wire layout for numerous connections is complicated and time-consuming, we break the overall routing into two processes: *global routing* and *detailed routing*.

   (a) **Global routing:** It is the *planning* phase of routing. It does not create the actual layout for nets. It simply allocates the region on the die for each net through which we can make its connection subsequently. For global routing, the entire routing region is divided into rectangular sections known as *global bins*. For a given net, global routing assigns a set of global bins that will be used for making connections.

   (b) **Detailed routing:** It decides the layout of each net in the pre-assigned global bins determined by the global router. It allocates routing tracks on the metal layers and employs vias wherever needed to change layers for a net.

[8] We will describe placement in detail in Chapter 26 ("Placement").

[9] We will describe CTS in detail in Chapter 27 ("Clock tree synthesis").

[10] We will discuss routing in detail in Chapter 28 ("Routing").

5. **Engineering change order:** Engineering change order (ECO) is needed to make minor fixes in a design. These fixes may be related to timing violations or minor changes in the functionality. Since ECO fixes are done at the end of a design flow, we must verify them carefully. Any inadvertent bug introduced by ECO can be detrimental to a chip.
6. **GDS writing and tape-out:** After thorough verification, we write a design in the GDS format. We send the GDS file to the foundry for fabrication. This task is referred to as *tape-out*.

It is worth pointing out that after each major physical design task, we also make some *incremental changes* in a design to fix timing violations or improve some figures of merit. We should apply these changes incrementally because a large change can inadvertently disrupt the entire design and create issues in achieving design closure.

4.3.3 Challenges

In this chapter, we have discussed only the implementation aspect of a design. However, we need to carry out *verification* tasks throughout the design flow. Among the verification tasks, ensuring that a circuit meets the given timing constraints is critical. If the final design meets them, we say that *timing closure* has been attained for the design. However, achieving timing closure is challenging for real designs because of the following reasons:

1. **Poor estimates:** At any given stage of a design flow, it is difficult to predict the problems that might occur down the flow. The origin of these problems lies in the abstraction or the lack of details in a design. Tools need to make an *estimate* for the non-existent design information. However, these estimates can be of poor quality, and the decisions taken based on them can be incorrect or inferior. Such decisions cause problems in achieving timing closure down the flow.

   For example, it is challenging to predict routing problems during placement. The problem occurs because, during placement, a design does not contain wire layouts. Therefore, a placement tool estimates the wire length and the wire delay based on some heuristics. However, the estimated wire length can widely differ from the actual implementation. Therefore, wrong decisions taken by a placement tool can cause timing problems during routing.
2. **Interdependency of timing paths:** In a design, multiple paths can have *slack* close to the critical path slack. The slack is the extra margin that we have in the delay before timing violations show up. Sometimes, speeding up the critical path can impact the timing of other paths. For example, moving the cells on the critical path can also displace cells on other paths and impact their timing. Hence, improving the timing of the critical path can create timing violations in some other paths that have slacks close to the critical path slack. Thus, the improvement of the critical path can be nullified, and achieving timing closure becomes difficult.

Therefore, we often need to repeat some design implementation tasks to obtain timing closure. It becomes unavoidable when fixing a problem requires retracting some decisions of the preceding tasks. For example, assume that during global routing, we encounter heavy congestion. Moreover, assume that after investigating, we found that the root problem is the inappropriate placement of some of the cells. Therefore, we might need to re-do the placement for those cells to fix the problem.

The repetition of design implementation tasks creates loops in a design flow and increases design effort and time. Therefore, the goal of a design flow is to avoid repetitions or iterations in achieving design closure.

4.4 RECENT TRENDS

A key challenge in design implementation flow is to reduce the design effort. Two techniques are commonly employed to decrease the design effort: (a) divide and conquer and (b) raising the abstraction level of design elements [3, 10]. In recent times, machine learning is also being explored to reduce design effort and achieve faster design closure using better prediction of the downstream flow problems [11].

Another challenge faced in designing is obtaining the full QoR benefits of scaling and technology advancements. By reducing design effort, we can free up some resources that can be utilized in careful design exploration and achieve better QoR at advanced technologies [12]. However, the rising cost of designing at advanced process nodes, shortage of expertise, and risk barrier block designers from using advanced technologies [13]. Fully automated and open-source toolchains for RTL to GDS flow are being developed and employed to tackle these problems [13–17]. In the future, open-source collaborative effort and machine learning can dramatically change the EDA landscape.

REVIEW QUESTIONS

4.1 What are the advantages of designing at a higher level of abstraction?

4.2 Why is it more difficult to make changes in a design at a lower level of abstraction?

4.3 Can we represent a netlist as a graph? If yes, represent the circuit shown in Figure 4.5(a) as a graph.

4.4 Why do we carry out CTS before global routing?

4.5 What is the difference between global routing and detailed routing?

4.6 Why is it difficult to achieve timing closure at advanced technology nodes?

4.7 What are the advantages and disadvantages of open-source EDA tools?

REFERENCES

[1] D. D. Gajski and R. H. Kuhn. "New VLSI tools." *Computer* no. 12 (1983), pp. 11–14.

[2] A. Gerstlauer and D. D. Gajski. "System-level abstraction semantics." *Proceedings of the 15th International Symposium on System Synthesis* (2002), pp. 231–236, ACM.

[3] A. Hemani. "Charting the EDA roadmap." *IEEE Circuits and Devices Magazine* 20, no. 6 (2004), pp. 5–10.

[4] H. Bhatnagar. *Advanced ASIC Chip Synthesis*. Springer, 2002.

[5] Synopsys Inc. "Liberty." https://www.synopsys.com/community/interoperability-programs/tap-in.html. Last accessed on January 6, 2023.

[6] Synopsys Inc. "Synopsys design constraints (SDC)." https://www.synopsys.com/community/interoperability-programs/tap-in.html. Last accessed on January 6, 2023.

[7] R. K. Brayton, G. D. Hachtel, and A. L. Sangiovanni-Vincentelli. "Multilevel logic synthesis." *Proceedings of the IEEE* 78, no. 2 (1990), pp. 264–300.

[8] R. Rudell and A. Sangiovanni-Vincentelli. "Logic synthesis for VLSI design." PhD thesis, University of California, Berkeley, 1989.

[9] G. D. Micheli. *Synthesis and Optimization of Digital Circuits*. McGraw-Hill Higher Education, 1994.

[10] B. Khailany, E. Krimer, R. Venkatesan, J. Clemons, J. S. Emer, M. Fojtik, A. Klinefelter, M. Pellauer, N. Pinckney, et al. "A modular digital VLSI flow for high-productivity SoC design." *2018 55th ACM/ESDA/IEEE Design Automation Conference (DAC)* (2018), pp. 1–6, IEEE.

[11] A. B. Kahng. "Machine learning applications in physical design: Recent results and directions." *Proceedings of the 2018 International Symposium on Physical Design* (2018), pp. 68–73, ACM.

[12] A. B. Kahng. "Reducing time and effort in IC implementation: A roadmap of challenges and solutions." *2018 55th ACM/ESDA/IEEE Design Automation Conference (DAC)* (2018), pp. 1–6, IEEE.

[13] T. Ajayi, V. A. Chhabria, M. Fogaça, S. Hashemi, A. Hosny, A. B. Kahng, M. Kim, J. Lee, U. Mallappa, et al. "Toward an open-source digital flow: First learnings from the OpenROAD project." *Proceedings of the 56th Annual Design Automation Conference 2019* (2019), p. 76, ACM.

[14] E. Alon, K. Asanović, J. Bachrach, and B. Nikolić. "Open-source EDA tools and IP: A view from the trenches." *2019 56th ACM/IEEE Design Automation Conference (DAC)* (2019), pp. 1–3, IEEE.

[15] T.-W. Huang, C.-X. Lin, G. Guo, and M. D. Wong. "Essential building blocks for creating an open-source EDA project." *DAC* (2019), pp. 78–1.

[16] WOSET. "Repo for articles in workshop on open-source EDA technology (WOSET)." https://woset-workshop.github.io/. Last accessed on January 6, 2023.

[17] R. Friesenhahn and J. York. "ARL: UT's experiences in the free open-source VLSI EDA landscape." 2018. https://woset-workshop.github.io/PDFs/2018/a8.pdf. Last accessed on January 6, 2023.

5 Verification Techniques

Take nothing on its looks; take everything on evidence. There's no better rule.

—Charles Dickens, *Great Expectations*, Chapter 40, 1861

In the previous chapter, we discussed *design implementation*: from RTL to the final layout. We carry out design implementation with the help of sophisticated EDA tools. These tools need manual inputs to gather design information, set tool options, and obtain hints to guide their algorithms. As a design gets transformed and moves through various EDA tools and design teams, *bugs* can creep into it. Some frequent bug sources are human error, miscommunication, inappropriate usage of EDA tools, and unexpected EDA tool behavior. These bugs can make a design erroneous. Therefore, it is critical to detect bugs or design problems using some *verification techniques*.

If design verification fails, we need to take some remedial actions. Therefore, it is prudent to verify a design as soon as we make some non-trivial design changes so that *debugging* and *fixing* effort is minimized. Hence, we need to verify a design multiple times during a design flow. Consequently, during designing, we spend significant effort on verification alone [1–3]. In the future, with the increase in design complexity, the verification effort is expected to increase further.

During various stages of a design flow, we employ different verification techniques. These techniques differ in exhaustiveness, computational resource requirement, and designer effort. In this chapter, we will briefly discuss some of the commonly used verification techniques. We will discuss these techniques in detail in Part II and Part IV of this book.

5.1 SIMULATION

During the early stages of a design flow, we represent the functionality of a design using an RTL model. Given a functionality, we can manually develop RTL models, reuse existing RTL, or generate them using a behavioral synthesis tool. Irrespective of the source, we need to verify that the RTL functionality matches the specification. Typically, we employ *simulation* to verify the functional correctness of an RTL model [4].

Using simulation, we mathematically compute the *response* of a given design for a given input or *stimulus*. We specify stimulus by a sequence of bits (0 or 1) at the input ports. We also specify the *timing* information such as *when* does $0 \rightarrow 1$ or $1 \rightarrow 0$ transition occur. A set of input stimuli applied to a circuit is commonly referred to as *test vectors* or *test patterns*. The response of a digital circuit is also a sequence of bits (0 or 1) at the output ports and its associated timing attributes.

Framework: We illustrate one of the simulation-based verification frameworks in Figure 5.1.

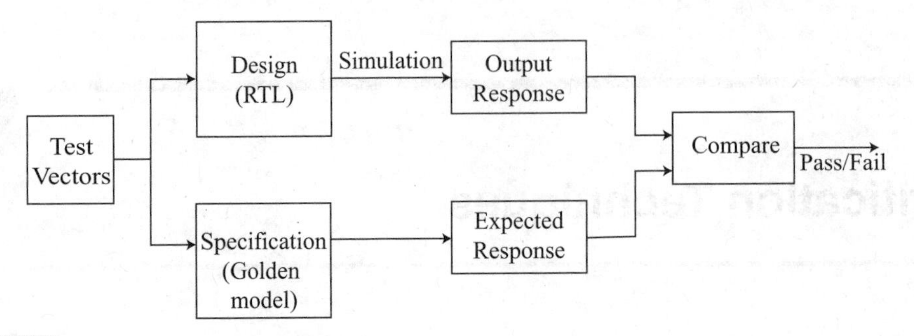

Figure 5.1 A simulation framework

For a given design and a set of test vectors, a *simulator* computes the resultant output response. For the same test vectors, we compute the expected response using a *golden model*. The golden model describes the required functionality using high-level languages such as C, PERL, Python, or MATLAB. The output of the golden model is generated by running corresponding executables or scripts. Next, we compare the simulator-computed outputs of the RTL model with the expected response. If they match, we say the verification has passed; else, it has failed. When it fails, we require to debug and fix design problems.[1]

Speed and versatility: Simulation is an easy and the most popular technique for functional verification. It can quickly discover bugs, especially in the early stages of designing. Additionally, it is a versatile technique. We can employ it for various purposes at different abstraction levels [5]. Besides verifying an RTL model, we can use the simulation for hardware–software partitioning, developing software when the targeted hardware is unavailable, and predicting and fine-tuning the performance of a hardware model. In the later stages of a design flow, we can use simulation to check whether a design meets the expected power and performance targets.

Incompleteness: We need to provide a set of test vectors to a simulator for functional verification. For realistic designs, too many test vectors are possible, and simulating all of them is not feasible. Therefore, we choose only a subset of possible test vectors for simulation. Typically, we choose test vectors to *cover* key design features or portions of code that are vulnerable to bugs. Thus, simulation-based verification is not *complete* and can miss detecting some functional issues. Nevertheless, because of its speed and versatility, it is an essential component of a design flow.

5.2 PROPERTY CHECKING

Another technique that verifies whether an RTL meets the specification is *property checking*. Based on the specification, we can identify properties for a design. For example, consider a traffic light controller. We can identify the following property that must be valid for it:

Not more than one light should be green.

Similarly, consider a resource-sharing environment. We can identify the following property that must hold:

Request for a resource should be granted eventually.

[1] We will discuss simulation-based verification in detail in Chapter 9 ("Simulation-based verification").

After identifying a set of properties that we expect to be valid for a given design, we can verify whether it holds using a *property checking tool*.

Mechanism of property checking: For a given design, the number of possible test stimuli increases exponentially with the number of inputs and the state elements (flip-flops). Therefore, verifying a given property rigorously using simulation is not feasible. Hence, a property checking tool employs *formal verification techniques* to verify if the given property holds [6]. Formal verification techniques establish the *proof* of a given property using formal mathematical tools such as deductions. Once we have proven a property mathematically, it is guaranteed to hold for all test stimuli. Thus, formal verification implicitly covers all test stimuli and is a *complete* verification technique.[2] However, formal verification is computationally challenging and cannot replace simulation everywhere in the design flow. In practice, we employ hybrid verification techniques that combine the benefits of both simulation and formal verification [7].

5.3 COMBINATIONAL EQUIVALENCE CHECKING

A given design gets transformed multiple times during a design flow. Some of these changes are big. For example, an RTL model gets transformed to its corresponding netlist during logic synthesis. Nevertheless, the functionality of the RTL model and the netlist must be *equivalent*.

Functional equivalence: Note that the functionality of an RTL and its corresponding netlist must be *equivalent*, they need not be *equal*. For example, in the RTL description, some signals can be specified as don't care (x). However, a logic synthesis tool can assign either 0 or 1 to them depending on which value reduces the cost of the solution. Such *refinements* in functionality are acceptable. Thus, we need to check for functional equivalence rather than functional equality.

During physical design also, a tool can make inadvertent functional changes. A physical design tool can restructure logic, resize cells, insert buffers and inverters, and change cell connections. Ideally, these changes should not modify the functionality of a design. Thus, we need to establish the functional equivalence of the transformed design at various stages of a design flow. We can accomplish this by a technique known as *combinational equivalence checking* (CEC). In general, after each non-trivial design transformation, we should verify the *functional equivalence* of the models, as shown in Figure 5.2.

Mechanism of CEC: To perform CEC, we provide two models for checking to the CEC tool. These models can be RTL or netlist. First, a one-to-one correspondence between the ports of the two models is determined. The one-to-one correspondence between the sequential circuit elements of the two models is also determined. The sequential circuit elements are typically flip-flops. Then, the tool determines a correspondence between *combinational logic cones* in two models by treating ports and flip-flops as invariant points. Subsequently, the tool checks the equivalence of the corresponding combinational logic cones using *formal techniques* [8]. If all the corresponding logic cones in the two models are equivalent, we say the given models are equivalent. Note that a CEC tool does not check *all* possible input cases by *simulation*. However, it covers them implicitly by a formal technique.[3]

[2] We will describe the formal verification methodology in detail in Chapter 11 ("Formal verification").

[3] We will discuss CEC in detail in Chapter 11 ("Formal verification").

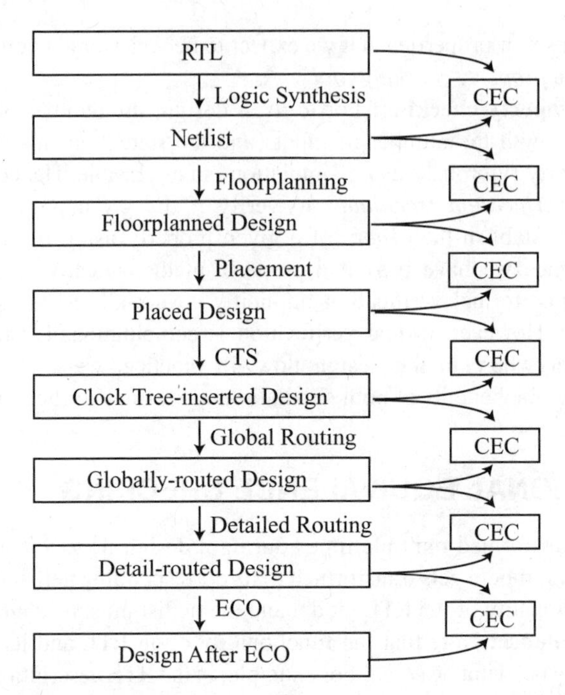

Figure 5.2 Applying CEC in VLSI design flow

5.4 STATIC TIMING ANALYSIS

For a synchronous design, we need to ensure the *synchronicity* of data transfer between flip-flops. We ensure it by performing *static timing analysis* (STA). An STA tool considers pairs of *launch* and *capture* flip-flops for verifying the timing of a synchronous circuit.[4] The flip-flop that produces data at the Q-pin is known as the *launch flip-flop*. This data propagates through a combinational circuit and reaches the D-pin of another flip-flop. This flip-flop is known as the *capture flip-flop*. An STA tool makes the following checks on each launch-capture flip-flop pair.

1. **Late checks:** The data produced by the launch flip-flop at a particular clock edge is read by the capture flip-flop on the *next* clock edge. To ensure this, the delay of the combinational gates between the launch flip-flop and the capture flip-flop should be *less than* some *maximum value*. This maximum value is known as the *setup constraint*. If the setup constraint gets violated, the signal starting from the launch flip-flop will reach the capture flip-flop too late. It will disrupt the synchronous behavior of the circuit.[5] These types of checks are known as *late checks* or *setup checks* or *max checks*.

   The setup constraint depends on the clock period. Since the time elapsed between two active clock edges is equal to the specified clock period, the setup constraint becomes stricter when we reduce the clock period. The setup constraint also depends on the *clock skew*. The clock skew is the difference between the clock arrival time at the launch flip-flop and the capture flip-flop. If the clock signal reaches late at the launch flip-flop, the setup constraint becomes

[4] Readers can refer to Figure 3.5 to get a pictorial view of the launch and capture flip-flops.

[5] We will explain how it will disrupt the synchronous behavior in Chapter 14 ("Static timing analysis").

stricter. Additionally, the setup constraint should have a sufficient margin for the setup time of the capture flip-flop.

2. **Early checks:** The data produced by the launch flip-flop should not be read by the capture flip-flop in the *same* clock cycle. To ensure this, the delay of the combinational gates between the launch flip-flop and the capture flip-flop should be *more* than some *minimum value*. This minimum value is known as the *hold constraint*. If the hold constraint gets violated, the signal from the launch flip-flop will reach the capture flip-flop too early. It will disrupt the synchronous behavior of the circuit.[5] These types of checks are known as *early checks* or *hold checks* or *min checks*. The hold constraint depends on the clock skew and the hold time of the capture flip-flop.

Constraints: For STA tools to carry out the verification for a design, we need to provide them with the corresponding *constraints*. A constraint defines the design intent, such as *maximum operable clock frequency*, and the *environment* under which a circuit is expected to work. The environment relates to the delay and the transition time of the incoming signals, the expected load to be driven by the output ports, and the constant logic values that may be present at some input ports during circuit operation.[6]

Temporal safety: An STA tool does not consider all *test vectors* during timing analysis. It also does not perform any dynamic simulation to compute the delays of circuit elements. However, it calculates delays by assuming *the worst-case* timing behavior of a circuit. Furthermore, it checks all possible paths in a circuit without determining whether a signal can propagate through them. Thus, the goal of the STA is to ensure the *temporal safety* of a circuit under a given set of constraints, rather than determining the best possible timing performance of the circuit. Ensuring temporal safety for a given set of constraints is an easier problem, allowing STA to employ efficient algorithms and produce results quickly.

STA in design flow: We perform STA after each major task in the design flow, as shown in Figure 5.3.

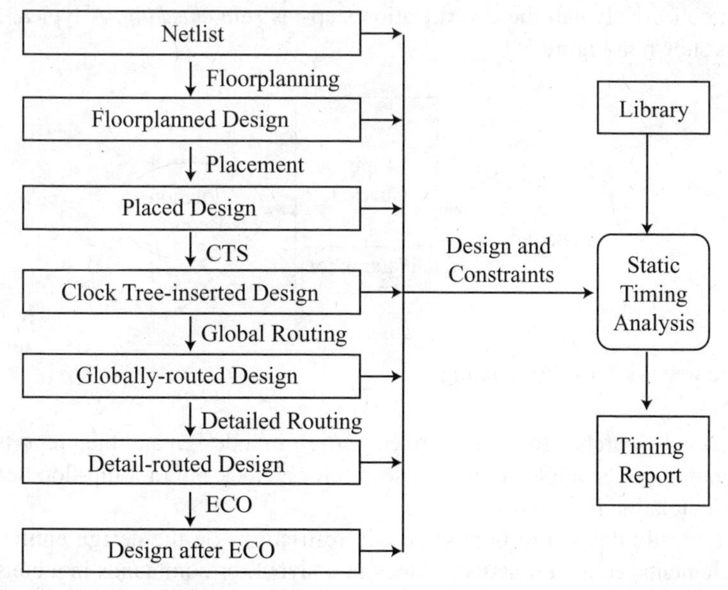

Figure 5.3 Using STA in the design flow

[5] We will explain how it will disrupt the synchronous behavior in Chapter 14 ("Static timing analysis").

[6] We will discuss constraints in detail in Chapter 15 ("Constraints").

If we find timing constraint violations, we might need to make changes to the circuit. We can employ various techniques to fix timing violations depending on the design stage in which the violations get detected. Note that STA is a verification technique. It carries out timing analysis and reports timing violations. It does not modify a design. We need to fix timing violations using appropriate design implementation tools.

Abstract delay and interconnect models: An STA tool can typically handle different levels of abstraction in a design flow. With a change in abstraction level, an STA tool can employ different *delay calculators*. For example, just after the logic synthesis, it can calculate delay by assuming *zero wire load*. After placement, it can calculate delay based on *semi-perimeter wire length estimation*. After detailed routing, it can use the *extracted parasitics* of the wires to compute delay.[7]

As we add details to a design, the estimates of the timing attribute become more realistic. For example, after building a *power distribution network*, we can carry out a *rail analysis*. By rail analysis, we get the information on the voltage drop in the supply/ground rails (wires) of a circuit. Using this information, we can compute realistic delay and ensure that the voltage drops in the supply rails are within acceptable limits [9, 10]. Similarly, after creating a layout for the interconnects, we can extract the associated interconnect parasitics (resistance and capacitance) and compute delays more accurately. Additionally, we can perform *signal integrity* (SI) analysis using the extracted interconnect parasitics. The SI analysis ensures that the coupling of signals between different wires does not cause extra timing violations or disrupt the functionality of a design [11].

Power analysis: Similar to STA, we carry out the *power analysis* of a design at different stages of a design flow to ensure that the dissipated power is within an acceptable limit.[8]

5.5 RULE CHECKING

In VLSI design flow, there are several occasions when we check or verify some characteristics of a design. We can collectively call these verification steps as *rule checking*. A typical framework for rule checking is shown in Figure 5.4.

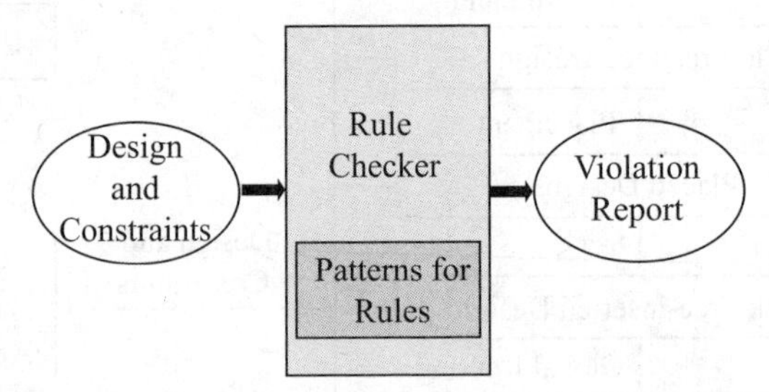

Figure 5.4 Framework for rule checking

In general, a *rule* searches for a pre-defined *pattern* in a design and then reports a violation if that pattern is found. For example, a rule can search for a clock-pin of a flip-flop tied to constant 0 or 1 and report a violation if it finds it.

Rules are typically defined to impose certain restrictions on the design entities. The entities can be RTL statements, cells of a netlist, shapes in a layout, or commands in a constraint file. We

[7] We will discuss these interconnect models in Part IV of this book.

[8] We will describe power analysis in Chapter 18 ("Power analysis").

impose such restrictions typically to avoid potential issues later in the design flow. For example, at the RTL level, we can define rules to detect simulation-synthesis mismatch, *race conditions* in simulation, and inference of latch in a flip-flop-based design.[9] We can identify testability issues, incorrect clock signal usage, and connectivity problems at the netlist level. Once a rule checker reports a rule violation, we need to analyze it. If the violation is tolerable, we waive it; otherwise, we make design changes to fix the problem.

5.6 PHYSICAL VERIFICATION

During physical design, we need to carry out some verification tasks. These tasks ensure that the layout does not have manufacturing and connectivity issues and the yield for a design remains high.

Some of the physical verification tasks are the following: *design rule checks* (DRC), *electrical rule checks* (ERC), and *layout versus schematic* (LVS) checks. The DRCs are defined by the foundries and depend on the manufacturing technology. A design must be made free of all the DRC violations before sending it to the foundry for fabrication. The ERCs ensure that the connectivity of a design is proper. For example, there should not be a short circuit between separate signal lines. The LVS checks ensure that the layout is functionally equivalent to the original netlist.[10]

5.7 RECENT TRENDS

Verification is gradually becoming a bottleneck in terms of effort and cost in a design flow. We can reduce this effort by targeting verification to critical design features and specific needs [3]. We can decrease verification execution runtime by parallelizing the corresponding algorithms [12]. Moreover, data analytics and machine learning can improve verification efficiency [13]. Additionally, we can reduce verification effort by adopting verification-driven design methodologies [14–16].

REVIEW QUESTIONS

5.1 Why is verification required in a design flow?

5.2 What are the limitations of simulation-based verification?

5.3 How does property checking help in solving the limitations of simulation-based verification?

5.4 Why is combinational equivalence checking required in a design flow?

5.5 What is the significance of the word "static" in static timing analysis (STA)?

5.6 A synchronous design violates the late check constraint. What can be its consequences?

5.7 A synchronous design violates the early check constraint. What can be its consequences?

5.8 How can rule checking help improve the productivity of a designer?

5.9 What can be the possible consequences of manufacturing a design that violates some DRCs?

5.10 What can be the possible consequences if a design violates some LVS checks?

---

[9] We will discuss the significance of these rules in Chapter 9 ("Simulation-based verification") and Chapter 10 ("RTL synthesis").

[10] We will describe physical verification in Chapter 29 ("Physical verification and signoff").

REFERENCES

[1] D. MacMillen, R. Camposano, D. Hill, and T. W. Williams. "An industrial view of electronic design automation." *IEEE Transactions on Computer-Aided Design of Integrated Circuits and Systems* 19, no. 12 (2000), pp. 1428–1448.

[2] R. E. Bryant, K.-T. Cheng, A. B. Kahng, K. Keutzer, W. Maly, R. Newton, L. Pileggi, J. M. Rabaey, and A. Sangiovanni-Vincentelli. "Limitations and challenges of computer-aided design technology for CMOS VLSI." *Proceedings of the IEEE* 89, no. 3 (2001), pp. 341–365.

[3] W. Chen, S. Ray, J. Bhadra, M. Abadir, and L.-C. Wang. "Challenges and trends in modern SoC design verification." *IEEE Design & Test* 34, no. 5 (2017), pp. 7–22.

[4] B. Wile, J. Goss, and W. Roesner. *Comprehensive Functional Verification: The Complete Industry Cycle*. Morgan Kaufmann, 2005.

[5] K. Olukotun, M. Heinrich, and D. Ofelt. "Digital system simulation: Methodologies and examples." *Proceedings 1998 Design and Automation Conference. 35th DAC.(Cat. No. 98CH36175)* (1998), pp. 658–663, IEEE.

[6] J. R. Burch, E. M. Clarke, K. L. McMillan, D. L. Dill, and L.-J. Hwang. "Symbolic model checking: 1020 states and beyond." *Information and Computation* 98, no. 2 (1992), pp. 142–170.

[7] J. Bhadra, M. S. Abadir, L.-C. Wang, and S. Ray. "A survey of hybrid techniques for functional verification." *IEEE Design & Test of Computers* 24, no. 2 (2007), pp. 0112–122.

[8] E. I. Goldberg, M. R. Prasad, and R. K. Brayton. "Using SAT for combinational equivalence checking." *Proceedings Design, Automation and Test in Europe. Conference and Exhibition 2001* (2001), pp. 114–121, IEEE.

[9] H. H. Chen and D. D. Ling. "Power supply noise analysis methodology for deep-submicron VLSI chip design." *Proceedings of the 34th Annual Design Automation Conference* (1997), pp. 638–643.

[10] Y.-M. Jiang and K.-T. Cheng. "Analysis of performance impact caused by power supply noise in deep submicron devices." *Proceedings of the 36th Annual ACM/IEEE Design Automation Conference* (1999), pp. 760–765.

[11] R. Arunachalam, K. Rajagopal, and L. T. Pileggi. "TACO: timing analysis with coupling." *Proceedings of the 37th Annual Design Automation Conference* (2000), pp. 266–269.

[12] Y. Zhu, B. Wang, and Y. Deng. "Massively parallel logic simulation with GPUs." *ACM Transactions on Design Automation of Electronic Systems (TODAES)* 16, no. 3 (2011), pp. 1–20.

[13] M. Arar, M. Behm, O. Boni, R. Gal, A. Goldin, M. Ilyaev, E. Kermany, J. Reysa, B. Saleh, K.-D. Schubert, et al. "The verification Cockpit——Creating the dream playground for data analytics over the verification process." *Haifa Verification Conference* (2015), pp. 51–66, Springer.

[14] R. Drechsler, M. Soeken, and R. Wille. "Formal specification level: Towards verification-driven design based on natural language processing." *Proceeding of the 2012 Forum on Specification and Design Languages* (2012), pp. 53–58, IEEE.

[15] J. Urdahl, S. Udupi, T. Ludwig, D. Stoffel, and W. Kunz. "Properties first? A new design methodology for hardware, and its perspectives in safety analysis." *2016 IEEE/ACM International Conference on Computer-Aided Design (ICCAD)* (2016), pp. 1–8, IEEE.

[16] T. Ludwig, M. Schwarz, J. Urdahl, L. Deutschmann, S. Hetalani, D. Stoffel, and W. Kunz. "Property-driven development of a RISC-V CPU." *Proceedings of the Design and Verification Conference United States (DVCON-US)* (2019), pp. 865–898.

6 Testing Techniques

...I won't descend to that. I'll be bad; but anyway not a liar, a cheat.

—Leo Tolstoy, *Anna Karenina* (translated by Constance Garnett), Chapter 35, 1878

We fabricate ICs using sophisticated technology and under tight process control. Nevertheless, during fabrication, defects can sneak in, and the functionality of the *fabricated IC* can differ from its *design* [1]. Therefore, post-fabrication, we should detect defective ICs and prevent them from reaching the end-user. This is the primary purpose of *testing* an IC.

At the outset, it is important to understand the difference between a *verification process*[1] and a *testing process*. A verification process ensures that the functionality of a *design* meets its corresponding *specification*. We carry out verification processes multiple times in a design flow to ensure the *design correctness*. Therefore, verification processes work on design representations such as RTL, netlist, layout, or GDS.

In contrast, we carry out testing on a *fabricated* IC. The objective of testing is to ensure that the IC has been fabricated as per the given design. Therefore, testing ensures that the functionality of the fabricated IC meets the given design. To achieve this, we need to detect *defects* introduced during manufacturing.

There are several steps in a design flow that enable or reduce the complexities of test processes. Therefore, even though we conduct testing after fabrication, we can enhance its efficiency and effectiveness by considering testing during the design phase. Consequently, test-related design steps are now an integral part of the VLSI design flow [2].

In this chapter, we introduce the testing techniques. We will explain the details of the testing techniques in Part III of this book.

6.1 MANUFACTURING DEFECTS

6.1.1 Origin of Defects

The fabrication of an IC is a complex process. It requires an extremely clean environment and tight process control. In addition to particulate and gaseous contaminants, we need to tightly control factors such as vibration, static electricity, and operational practices. We refer to the tightly controlled environment of IC manufacturing as *cleanrooms* [3].

[1] Verification is explained in Chapter 5 ("Verification techniques").

Defects: Despite tight process control, some physical imperfections inevitably get introduced into a wafer due to the statistical deviations in the properties of materials used in fabrication, finite tolerances of process control such as the furnace temperature, and random fluctuations in the process conditions such as the turbulent gas flows. Imperfections can also appear due to deviations in the mask features and deposition of undesired chemicals and airborne particles. These physical imperfections that are permanent are known as *defects*.

Spot defects: Some large-area defects can also be generated by wafer mishandling, mask misalignment, over-etching, and under-etching. These large-area defects or global defects are simpler to minimize and typically eliminated in matured fabrication processes [4]. However, small-area defects or *spot defects* are random in nature and inevitable in IC fabrication [5]. Furthermore, the expected number of spot defects increases with the area of the die. Spot defects are primarily responsible for the functional failures of ICs. Therefore, spot defects are of greater significance from the perspective of testing [1].

Location of defects: A manufacturing defect can occur in different structural parts of an IC. Since gate oxides are the smallest features realized on an IC, gate oxides are more vulnerable to manufacturing defects. The thickness of gate oxide is only a few atomic layers. Therefore, controlling the thickness of gate oxide during fabrication is challenging. During manufacturing, defects can be generated in the gate oxide due to insufficient oxidation, crystal defects, and chemical contamination. Defects can also be generated in the metal, diffusion, or polysilicon layers due to *missing* or *extra* material created by particulate deposition during photolithography. Similarly, defects can be generated in vias or between the metal layers due to local contaminants such as dust particles [4].

6.1.2 Manifestation of Defects

There can be several electrical manifestations of defects in an IC.

Short circuit and open circuit: Some defects can lead to a *short* circuit due to extra material, while others can lead to an *open* circuit due to missing material on the die.

Example 6.1 Consider three signal lines *A*, *B*, *C*, and the ground line *GND* running parallelly on a good die, as shown in Figure 6.1(a).

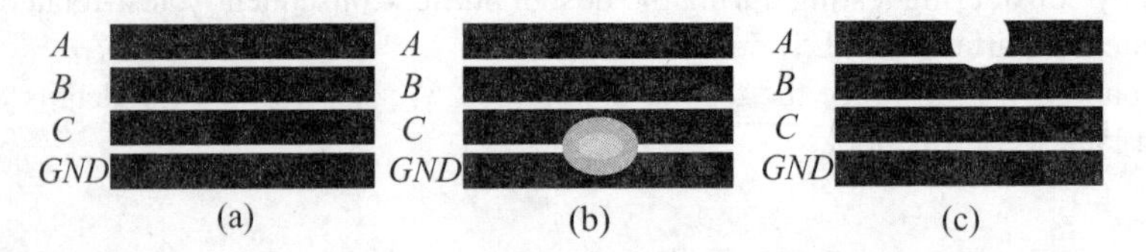

Figure 6.1 Layout for a: (a) good die, (b) short circuit defect, and (c) open circuit defect

Assume that, during fabrication, a conducting particle gets permanently lodged between the line *C* and the ground, as shown in Figure 6.1(b). It will create a short circuit between the line *C* and the ground. The signal *C* will be *stuck* to 0 under all circuit conditions. Thus, we have a *fatal defect* in the signal line due to a short circuit. It can lead to improper functionality of the fabricated IC.

Assume that there is an opening in a wire due to missing conducting material, as shown in Figure 6.1(c). It can occur due to the deposition of non-conducting particles. As a result, we observe an *open circuit defect*. Its impact on the electrical characteristics depends on its size, location, and the other structures in its vicinity [6].

A conducting particle can also get lodged between two signal lines. These types of defects are known as *bridges*. For instance, a bridge can form between signal lines A and B. If the logic value of two signals A and B differ, an IC can exhibit improper functionality or can even be damaged due to high current flow depending on the strength of the two signals and the circuit topology [7].

Fault model: In the analysis and simulation of circuits, we account for the impact of different types of defects using some *fault models*. A fault model represents a *physical* defect by a simple *logical* or *electrical* model. The resultant abstraction obtained using a fault model is known as *fault*. For example, we can represent the defect in Figure 6.1(b) using *stuck-at fault model*. In this model, we assign a logic 0 to the signal C. The resultant abstraction is known as *stuck-at-0 fault* (SA0).

Parametric fault: Note that some defects can cause a change in a continuous parameter of a circuit. These defects can be modeled as *parametric faults* [7]. For example, some defects can increase the propagation time of signals in a faulty circuit. These types of parametric faults are known as *delay faults*. Similarly, some defects can increase the quiescent power-supply current (I_{DDQ}) for some test patterns. These parametric faults are called I_{DDQ} *faults*.

6.1.3 Distortions and Inconsequential Flaws in Manufacturing

With the shrinking device sizes, it is difficult to create features on a die that are replicas of what was designed. Due to optical effects such as diffraction, photolithography can produce *distorted* features on a die. Some resolution enhancement techniques, such as *optical proximity corrections* (OPC) and *multi-patterning*, reduce such distortions.[2] Despite these techniques, features created on a die can differ from the designed features on the masks. Nevertheless, the logic function of a *die with distortions* can still match the original design. Therefore, such distortions are harmless and the goal of testing is not to detect such distortions.

Besides distortions due to photolithography, there can be other deviations of a fabricated circuit from the ideal. These deviations are called *flaws* [7]. If a flaw does not cause any *measurable* change in the circuit, that flaw is an *inconsequential flaw*. In Figure 6.1(b) and (c), if the sizes of the defects are too small with respect to the dimensions of the wires, their existence will not be electrically noticeable in a defective circuit. Therefore, we can ignore inconsequential flaws during testing [8]. In general, the probability that a defect will lead to a fault depends on its size relative to the patterns in the circuit layout [8].

6.2 YIELD

Yield is defined as the percentage of *good* die on a wafer. A good die is one with no fatal manufacturing defect.

Example 6.2 Assume that there are 400 dies manufactured on a single wafer. Further, assume that out of these 400 dies, 28 dies are defective.

[2] We will explain resolution enhancement techniques in Chapter 7 ("Post-GDS processes").

We can compute the yield as:

$$Yield = \frac{(400 - 28)}{400} \times 100 = 93\%.$$

6.2.1 Yield Learning

Yield quantifies the quality of a manufacturing process. It is a function of the complexity and the maturity of the process. In the initial stages of advanced process nodes, yield can be as low as 20% [9]. However, by identifying critical issues, eliminating the sources of defects, and rectifying the process, we can increase the yield substantially [9, 10]. This yield-enhancing process is known as *yield learning* [11]. Since higher yield leads to higher profitability, yield should be close to 100%. For a matured process, it is typically more than 90%.

6.2.2 Factors Affecting Yield

Yield strongly depends on the attributes of a manufacturing process, especially on the *defect density* and the *defect distribution* [1, 12].

Defect density: *Defect density* is defined as the average number of defects per unit area. It depends on the manufacturing technology, process complexity, and the minimum design feature size [13]. An increase in defect density reduces the yield.

Defect distribution: Another important factor that determines the yield of a process is the spatial distribution of defects. Fortunately, defects tend to *cluster* on a wafer during IC fabrication [13]. The clustering of defects means that defects are generated close to each other on a wafer. Note that one fatal defect on a die is sufficient to make it non-functional. More than one fatal defect existing on a die do not contribute to extra yield loss. As a result, for the same number of defects on a wafer, less number of dies become non-functional if the defects are clustered [12, 14].

Example 6.3 Assume that there are 32 dies manufactured on a single wafer. Assume that there are 10 defects on the wafer.

If the distribution of defects is as shown in Figure 6.2(a), 10 dies will be defective.

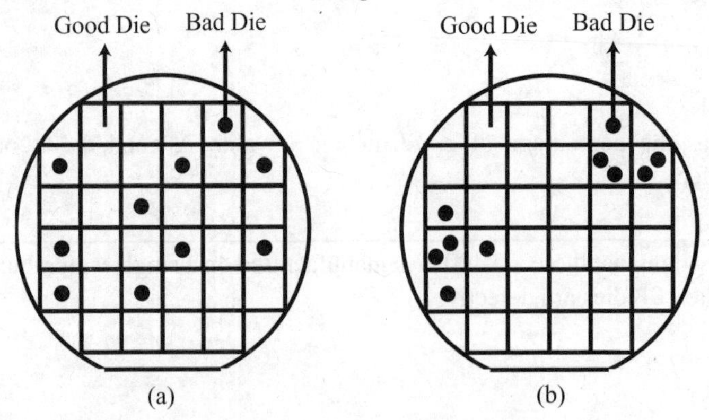

Figure 6.2 Defects distributed on wafer (each dark circle represents one defect): (a) defects not clustered and (b) defects clustered

Hence, the yield is

$$Yield = \frac{(32 - 10)}{32} \times 100 = 68.75\%.$$

Next, assume that the 10 defects are distributed as shown in Figure 6.2(b). Due to clustering, only 7 dies become defective. Hence, the yield is

$$Yield = \frac{(32 - 7)}{32} \times 100 = 78.13\%.$$

Thus, we obtain more yield when defects are clustered.

Design attributes: Yield also depends on the attributes of a design such as a die area [1]. Since spot defects are the dominant source of yield loss, increasing the die area increases the probability of their occurrence. Therefore, in general, yield decreases with an increase in the die area. If the die area is made too large, then it can reduce to such an extent that its manufacturing can become non-profitable even in a matured process [15]. Therefore, we should minimize the die area or design area.

It is worth noting that, even for dies with the same area, yield can show variations because of the variation in the density of circuit elements and layout [12]. Therefore, to improve yield, we need to also consider the arrangement of circuit elements within a given die area [8].

6.2.3 Yield Estimation and Modeling

Yield has a substantial impact on the profitability of producing an IC. Therefore, it is essential to *estimate the expected yield* for an IC before starting the venture [13, 15]. We use *yield model* to make this estimate.

Yield models: Several yield models have been developed over the years with varying accuracy, complexity, and assumptions on the spatial distribution of defects [1, 12–14, 16]. Due to clustering, we can assume that the probability of having a defect in a given area increases linearly with the number of defects already present in that area. This assumption leads to the popular *negative binomial distribution*-based yield model, which has shown an excellent match to the experimental yield data. We can describe it mathematically as follows:

$$Yield = (1 + Ad/\alpha)^{-\alpha} \times 100\% \tag{6.1}$$

where A is the area of the die, d is the defect density, and α is the clustering parameter.

The clustering parameter α represents the spatial distribution of defects on a wafer and takes a positive value ($0 < \alpha < \infty$). A high value of α indicates a weak clustering of defects. The practical range of α is between 0.3 and 5.0 [12].

Example 6.4 Assume that $Ad = 1$. From Eq. 6.1, we can write

$$Yield = (1 + 1/\alpha)^{-\alpha} \times 100\%$$

We plot the above equation in Figure 6.3.

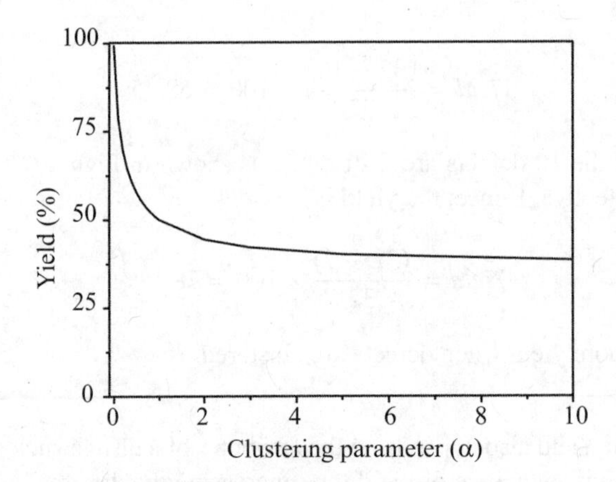

Figure 6.3 Variation in yield with clustering parameter (α) for $Ad = 1$

When α tends to zero, defects get localized to a small area on the wafer, and only a small number of dies are affected by defects. As a result, the yield becomes close to 100%.

When α tends to infinity, clustering disappears, and defects are distributed uniformly over the wafer. Consequently, yield decreases to $1/e \times 100 = 36.8\%$.

Note that a lower limit on yield obtained for $\alpha \to \infty$ does not imply that a lower limit on the yield exists for the negative binomial distribution-based yield model. We obtained the above lower limit because we have assumed $Ad = 1$. If Ad increases, the yield would decrease further.

Yield estimation: Using yield data of previously manufactured chips and Eq. 6.1, we can derive the defect density d and the clustering parameter α (e.g., by curve fitting). Then, given the area A of the die for the proposed chip, we can estimate its yield using Eq. 6.1 and the derived values of d and α.

6.3 TESTING METHODOLOGY

At the beginning of this chapter, we explained that the primary goal of testing is to detect defects in a fabricated IC and prevent it from reaching the end-user. This section describes a widely employed IC testing framework illustrated in Figure 6.4(a). It requires applying *test patterns* and measuring *response* using an *automatic test equipment (ATE)*.

6.3.1 Test Patterns and Response

We test a fabricated die using a set of *test patterns*. A test pattern is a bit pattern or a bit sequence that we can apply at the inputs of a circuit to *differentiate* between a faulty circuit and a good one. By applying a test pattern at the inputs of a *device under test* (DUT) and observing the output response, we can detect a fault. When we are performing wafer-level testing, a DUT refers to the die on the wafer. When we are doing package-level testing, a DUT refers to a packaged die or a chip.

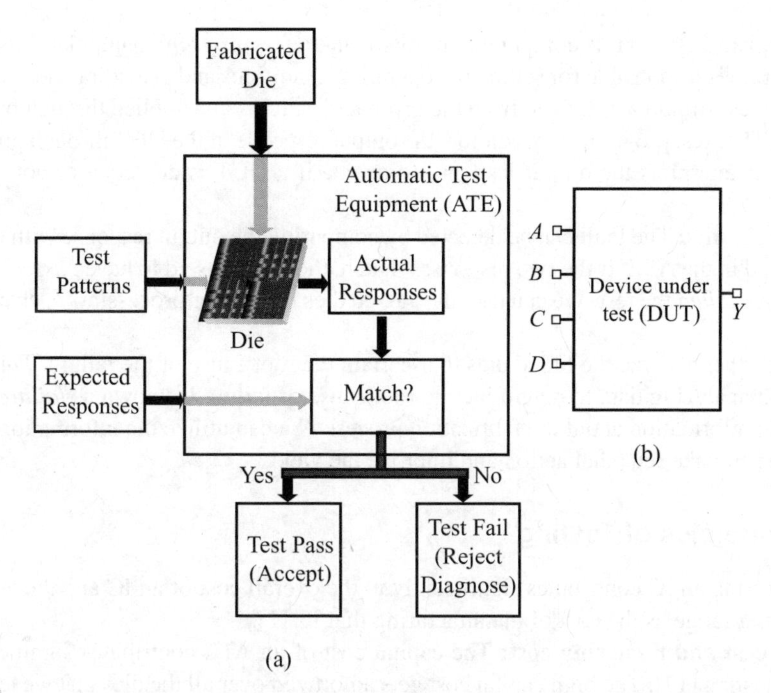

Figure 6.4 (a) A popular testing framework and (b) a device under test (DUT)

Example 6.5 Consider a simple circuit with four input ports *A*, *B*, *C*, and *D*, as shown in Figure 6.4(b). Assume that a bit sequence {0111} is a test pattern to detect some faults in this circuit. It implies that when we apply the corresponding stimulus at the input ports (A=0 and B=C=D=1), the value at the output *Y* will be different for a good circuit and the circuit having that fault. Therefore, we can detect a fault by applying a test pattern to the inputs of the circuit.

For a given IC, we generate a set of test patterns and their expected responses during the *design phase* using *automatic test pattern generation* (ATPG) tools.[3] These test patterns and the expected response help us to detect faulty dies or chips. We also refer to a test pattern as a *test vector*.

6.3.2 Automatic Test Equipment (ATE)

An *automatic test equipment* (ATE) is the most critical component of the testing framework illustrated in Figure 6.4(a). It is responsible for performing actual electrical tests on a DUT to determine the existence of a fault.

Test head: An ATE consists of one or more *test heads* that hold channels that interface with a DUT using a *probe card*. A probe card consists of several microscopic *probe needles*. The needles make contact with the test pads on a DUT and provide electrical paths between an ATE and a DUT. Before carrying out electrical tests, a set of test patterns and the expected responses for a DUT are loaded and stored inside an ATE, as illustrated in Figure 6.4(a).

[3] We will describe ATPG in detail in Chapter 22 ("Automatic test pattern generation (ATPG)").

Test program: An ATE is computer controlled, and a *test program* manipulates its operations. The test program is responsible for setting the operating conditions and providing the test stimulus to the DUT. The test stimuli are derived from the given test patterns and applied through probe needles onto the DUT. The test program also monitors the output response of the DUT through probe needles. Furthermore, it interprets the output response to decide if a DUT is defective or not for each test pattern.

Detecting faults: The fault can be detected by comparing the output response with the expected response stored in the ATE. If the two responses match, the DUT is said to have *passed* the test, else it is said to have *failed* the test. We can take the passed dies for further processing, such as packaging and binning.[4]

We can further analyze the failed dies to ascertain the root cause of the failure. This process of collecting and analyzing data to determine the root cause of failure is known as *failure analysis*. It gives valuable information about the fabrication process. We can utilize the information gathered in failure analysis to take remedial action and improve the yield.

6.3.3 Economics of Testing

The cost of testing an IC contributes significantly to the overall cost of an IC and the testing cost is now in a similar range as the cost of manufacturing that IC [2].

Capital cost and recurring cost: The capital cost of an ATE contributes significantly to the fixed cost of testing [17]. The high capital cost gets amortized over all the dies that we test. The time spent by an IC on ATE or the *test time* contributes significantly to the recurring cost of testing [18]. If the test time per die increases, we can test fewer dies in a given time, and the test cost increases.

Test time: The test time depends on the volume of test data transferred from ATE to DUT, which further depends on the number of test patterns. Therefore, various techniques such as *test compression* are adopted to reduce the test data volume and the test time [19, 20]. The number of ports on a DUT, the nature of DUT, and the test methodology also impact the test time.

6.4 FAULT COVERAGE AND DEFECT LEVEL

The quality of testing strongly depends on the *exhaustiveness* of the test patterns in detecting faults in an IC. Ideally, the applied set of test patterns should detect all possible faults in an IC. However, this is not practically possible because many different types of faults can occur in an IC, and finding all of them will require an enormous number of test patterns. It can make the test time and the cost of testing prohibitively high. Even if we consider only a few common fault types, it is computationally difficult to find test patterns to detect all such faults. Despite ATPG tools becoming quite efficient over the years, they can take unacceptably high runtime to generate an exhaustive set of test patterns. Therefore, we need to use a non-exhaustive set of test patterns for testing. As a result, some faults can remain untested or uncovered during testing.

Fault coverage: The quality of testing is quantified using a metric known as *fault coverage*. For a given fault model, we define fault coverage as follows:

$$\textit{Fault coverage} = \frac{\textit{Number of faults detected}}{\textit{Number of faults possible}} \times 100\% \qquad (6.2)$$

[4] Packaging and binning will be described in Chapter 7 ("Post-GDS processes").

Fault coverage indicates the capability of a given set of test patterns in detecting faults for a given fault model. A fault coverage of less than 100% suggests that some faults can go undetected by employing the given set of test patterns. However, it is worth pointing out that there are some faults, known as *redundant faults*, for which no test pattern exists that can distinguish a faulty and a fault-free circuit. Redundant faults can decrease the fault coverage, though their non-detection has no bearing on the test quality because their existence on a fabricated die does not impact functionality.

If fault coverage is low, then some faults can go undetected and faulty products can reach end-users. It will negatively impact the perception of a product among end-users. Thus, the perceived quality of a product depends heavily on the fault coverage. Therefore, we try to keep the fault coverage as high as possible.

Defect level: The quality of a product or a chip is quantified using a metric known as *defect level*. We define the defect level (DL) of a product as the fraction of faulty products among the products that have passed the tests:

$$DL = \frac{\textit{Number of faulty products that passed the test}}{\textit{Total number of products that passed the test}} \tag{6.3}$$

It indicates a fraction of chips that escaped testing. We typically measure defect level in *parts per million (ppm)*. If we make testing more rigorous by increasing the fault coverage, fewer faulty products can escape testing. Thus, the defect level decreases with the increasing fault coverage.

Models of defect level: Since defect level is an important metric for assessing testing quality, we often need to model and estimate it. One of the simplistic models for the defect level (DL in ppm) is as follows [21, 22]:

$$DL = \left(1 - \frac{(\alpha + TAd)^{\alpha}}{(\alpha + Ad)^{\alpha}}\right) \times 10^6 \tag{6.4}$$

where T is the fault coverage expressed as a fraction, α is the clustering parameter, A is the area of the die, and d is the defect density.

Example 6.6 Assume that $Ad = 1$. From Eq. 6.4, we can write

$$DL = \left(1 - \frac{(\alpha + T)^{\alpha}}{(\alpha + 1)^{\alpha}}\right) \times 10^6$$

We plot the above equation for varying fault coverage (T) and clustering parameter (α) in Figure 6.5.

When fault coverage is 100% or $T = 1$, no faulty product can escape testing. Thus, the defect level reduces to zero. When fault coverage decreases, more faulty products go undetected, and the defect level increases.

When defects get clustered in a small region (α is low), fewer faulty products are made during manufacturing. Therefore, the defect level decreases. The defect level also decreases if we reduce the die area or the defect density.

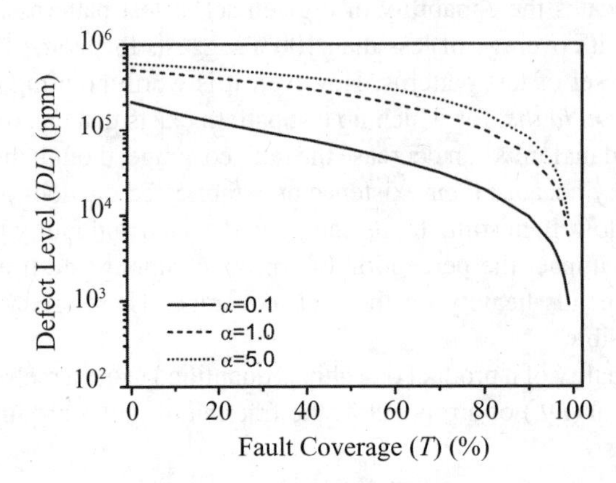

Figure 6.5 Variation in defect level (*DL*) with fault coverage (*T*) (It is assumed that $Ad = 1$.)

6.5 DESIGN FOR TESTABILITY

To carry out testing efficiently, we need to perform several tasks *during the design phase*. These tasks are collectively called *design for testability* (DFT). In this section, we introduce some concepts related to DFT. We will describe them in detail in Part III of this book.

6.5.1 Testability of a Circuit

A fault can occur at any *internal* pin or net of a circuit. However, we need to detect this fault by applying test patterns *only* at the *input ports* or the external interfaces of the circuit. Moreover, we can observe the output response for a given test pattern *only* at the *output ports* or the external interfaces of the circuit. For example, in the circuit shown in Figure 6.4(b), while the fault can occur at any internal pin, we can apply the test patterns only at the ports *A*, *B*, *C*, and *D*, and observe the output response only at the port *Y*. However, ensuring that a circuit is testable by accessing *only* its external interfaces is challenging. We refer to the ease of detecting and diagnosing a fault in a circuit as *testability*.

Controllability and observability: For testing a fault at an internal pin, we need to write logic values (0/1) at that pin. Therefore, a circuit should be designed such that it is possible to obtain an arbitrary logic value (0/1) at an internal pin just by applying some test patterns at the input ports. This property of a circuit is known as the *controllability* of a pin. Similarly, to observe a fault at an internal pin, we need to read logic values (0/1) from that pin to some output port. This property of a circuit is known as the *observability* of a pin.

We employ several techniques during designing to make a circuit easily testable. These techniques enhance the controllability and observability of the internal pins. Among them, the most popular is implementing *scan chain*-based testing. Though there are *overheads* associated with scan chain-based testing, it makes testing highly efficient [2].

6.5.2 Test Pattern Generation

As discussed earlier, during testing, ATE applies a set of test patterns to a DUT. We obtain the test pattern set using ATPG tools during the design phase. Ideally, we want that the test pattern set should deliver 100% fault coverage. Moreover, to reduce the test time, we want this set to contain the minimum number of test patterns for a given fault coverage. ATPG tools try to achieve these goals.

6.5.3 Self-testing

ATE-based testing is costly. Therefore, alternative test strategies, such as self-testing, that eliminate the dependency on ATE and reduce the test cost have become popular. Self-testing allows testing to be done outside the production environment and also during the use of a product. However, additional hardware and software need to be added to a design to enable the self-testing of circuits. The self-testing technique is known as *built-in self-test* (BIST).

6.6 RECENT TRENDS

With the wide-scale adoption of DFT and progress in test technologies, testing has now evolved from being a *necessary evil* to an integral part of VLSI design flow [2]. With the advent of new technologies, such as 3D ICs, IoT, and neuromorphic computing, we face new test challenges, and we need to develop their solutions [23–28].

In recent times, test technologies are leveraging machine learning and big data analytics in various areas such as yield improvement and test cost reduction [29]. In the future, we expect this trend to continue.

In the future, there can be a paradigm shift in testing. For example, some imperfections could be desirable for some systems, such as memristor-based neuromorphic computing [2]. Additionally, instead of testing a large number of individual IoT devices, we could allow them to test each other and self-organize into a reliable system [2].

REVIEW QUESTIONS

6.1 What are the differences between verification and testing processes?

6.2 How do defects get introduced into a manufactured chip?

6.3 What are the advantages of a high yield learning rate?

6.4 For a given process, the yield is modeled by the following equation:

$$Yield = (1 + Ad/\alpha)^{-\alpha} \times 100\%$$

where A is the die area, d is the defect density, and α is the clustering parameter. The wafer size is 300 mm, and the die size is 25 mm^2. Assume that the cost of fabricating a wafer is \$100, and there is no wastage of material in creating dies out of the wafer. Assume defect density is 0.5 defect/cm^2 and the clustering parameter is 0.5.

(a) Estimate the yield and the cost per die.

(b) Due to yield learning, defect density decreases to 0.1 defect/cm^2. Estimate the new yield and the cost per die.

(c) Assume that defect density remains as 0.1 defect/cm^2, but the die area is increased to 100 mm^2. Compute the new yield and the cost per die?

6.5 What should we do with a die that has failed the ATE-based test?

6.6 Why should the set of test patterns used for ATE-based testing be small in size, and why can it not be made arbitrarily small?

6.7 A die had passed the ATE-based testing, though it had a manufacturing fault. Comment on how is it possible.

6.8 Why should an internal pin of a circuit have high controllability and high observability?

REFERENCES

[1] B. T. Murphy. "Cost-size optima of monolithic integrated circuits." *Proceedings of the IEEE* 52, no. 12 (1964), pp. 1537–1545.

[2] K. Chakrabarty. "Quo vadis test? The past, the present, and the future: No longer a necessary evil." *IEEE Design & Test* 34, no. 3 (2017), pp. 93–95.

[3] M. Kozicki. *Cleanrooms: Facilities and Practices*. Springer Science & Business Media, 2012.

[4] I. Koren and Z. Koren. "Defect tolerance in VLSI circuits: Techniques and yield analysis." *Proceedings of the IEEE* 86, no. 9 (1998), pp. 1819–1838.

[5] I. Koren and A. D. Singh. "Fault tolerance in VLSI circuits." *Computer* 23, no. 7 (1990), pp. 73–83.

[6] C. F. Hawkins, J. M. Soden, A. W. Righter, and F. J. Ferguson. "Defect classes-an overdue paradigm for CMOS IC testing." *Proceedings International Test Conference* (1994), pp. 413–425, IEEE.

[7] F. J. Ferguson, M. Taylor, and T. Larrabee. "Testing for parametric faults in static CMOS circuits." *Proceedings International Test Conference 1990* (1990), pp. 436–443, IEEE.

[8] V. K. R. Chiluvuri and I. Koren. "Layout-synthesis techniques for yield enhancement." *IEEE Transactions on Semiconductor Manufacturing* 8, no. 2 (1995), pp. 178–187.

[9] A. Oberai and J.-S. Yuan. "Efficient fault localization and failure analysis techniques for improving IC yield." *Electronics* 7, no. 3 (2018), p. 28.

[10] C. Terwiesch and R. E. Bohn. "Learning and process improvement during production rampup." *International Journal of Production Economics* 70, no. 1 (2001), pp. 1–19.

[11] C. Weber. "Yield learning and the sources of profitability in semiconductor manufacturing and process development." *IEEE Transactions on Semiconductor Manufacturing* 17, no. 4 (2004), pp. 590–596.

[12] W. Kuo and T. Kim. "An overview of manufacturing yield and reliability modeling for semiconductor products." *Proceedings of the IEEE* 87, no. 8 (1999), pp. 1329–1344.

[13] A. V. Ferris-Prabhu. "A cluster-modified Poisson model for estimating defect density and yield." *IEEE Transactions on Semiconductor Manufacturing* 3, no. 2 (1990), pp. 54–59.

[14] B. El-Kareh, A. Ghatalia, and A. Satya. "Yield management in microelectronic manufacturing." *1995 Proceedings 45th Electronic Components and Technology Conference* (1995), pp. 58–63, IEEE.

[15] J. A. Cunningham. "The use and evaluation of yield models in integrated circuit manufacturing." *IEEE Transactions on Semiconductor Manufacturing* 3, no. 2 (1990), pp. 60–71.

[16] A. Ferris-Prabhu. "Modeling the critical area in yield forecasts." *IEEE Journal of Solid-State Circuits* 20, no. 4 (1985), pp. 874–878.

[17] J. Gatej, L. Song, C. Pyron, and R. Raina. "Evaluating ATE features in terms of test escape rates and other cost of test culprits." *Proceedings International Test Conference* (2002), pp. 1040–1049, IEEE.

[18] I. D. Dear, C. Dislis, A. P. Ambler, and J. Dick. "Economic effects in design and test." *IEEE Design Test of Computers* 8 (Dec 1991), pp. 64–77.

[19] P. Girard. "Survey of low-power testing of VLSI circuits." *IEEE Design & Test of Computers* 19, no. 3 (2002), pp. 82–92.

[20] A. Chandra and K. Chakrabarty. "System-on-a-chip test-data compression and decompression architectures based on Golomb codes." *IEEE Transactions on Computer-Aided Design of Integrated Circuits and Systems* 20, no. 3 (2001), pp. 355–368.

[21] J. T. de Sousa and V. D. Agrawal. "Reducing the complexity of defect level modeling using the clustering effect." *Proceedings Design, Automation and Test in Europe Conference and Exhibition 2000 (Cat. No. PR00537)* (2000), pp. 640–644, IEEE.

[22] M. Bushnell and V. Agrawal, *Essentials of Electronic Testing for Digital, Memory and Mixed-Signal VLSI Circuits*, vol. 17. Springer Science & Business Media, 2004.

[23] H.-H. S. Lee and K. Chakrabarty. "Test challenges for 3D integrated circuits." *IEEE Design & Test of Computers* 26, no. 5 (2009), pp. 26–35.

[24] E. J. Marinissen, Y. Zorian, M. Konijnenburg, C.-T. Huang, P.-H. Hsieh, P. Cockburn, J. Delvaux, V. Rožić, B. Yang, et al. "IoT: Source of test challenges." *2016 21th IEEE European Test Symposium (ETS)* (2016), pp. 1–10, IEEE.

[25] T.-C. Huang and J. Schroff. "Precompensation, BIST and analogue Berger codes for self-healing of neuromorphic RRAM." *2018 IEEE 27th Asian Test Symposium (ATS)* (2018), pp. 173–178, IEEE.

[26] A. Koneru, S. Kannan, and K. Chakrabarty. "A design-for-test solution based on dedicated test layers and test scheduling for monolithic 3D integrated circuits." *IEEE Transactions on Computer-Aided Design of Integrated Circuits and Systems* (2018).

[27] A. Koneru and K. Chakrabarty. "Test and design-for-testability solutions for monolithic 3D integrated circuits." *Proceedings of the 2019 on Great Lakes Symposium on VLSI* (2019), pp. 457–462, ACM.

[28] E. J. Marinissen, T. McLaurin, and H. Jiao. "IEEE Std P1838: DfT standard-under-development for 2.5 D-, 3D-, and 5.5 D-SICs." *2016 21th IEEE European Test Symposium (ETS)* (2016), pp. 1–10, IEEE.

[29] L.-C. Wang. "Experience of data analytics in EDA and test–principles, promises, and challenges." *IEEE Transactions on Computer-Aided Design of Integrated Circuits and Systems* 36, no. 6 (2016), pp. 885–898.

7 Post-GDS Processes

Good sentences, and well pronounced.
They would be better if well followed.

—William Shakespeare, *The Merchant of Venice*, Act 1, Scene 2, 1596

In the previous chapters, we discussed the tasks required for *designing* an IC. Once we have obtained the final layout, the design process is complete. Subsequently, the GDS file corresponding to the final layout is employed for making a chip using tasks such as mask preparation, wafer fabrication, testing, and packaging. In this book, we have grouped all chip-making tasks carried out after obtaining the GDS file as *post-GDS processes*.

Though post-GDS processes are not directly related to the *design* flow, we need to understand them to appreciate the challenges of fabrication, and possibly address some of these challenges during the design phase. Therefore, we briefly explain post-GDS processes in this chapter. For a detailed understanding of post-GDS processes, readers can refer to dedicated books on these topics such as [1–5].

7.1 MASK FABRICATION

We have discussed in Chapter 2 ("Introduction to integrated circuits") that an essential step in IC fabrication technology is *photolithography*. It involves transferring the patterns in a layout for a given layer to the silicon wafer. We carry out photolithography separately for each layer.

To start with, we create a replica of the pattern of a given layer on a substrate such as glass. This replica of the pattern is known as *mask* or *reticle*. After creating a mask, we use it many times for carrying out photolithography during high-volume manufacturing [4].

We can fabricate a mask using several techniques. Nevertheless, a typical mask fabrication flow consists of the following steps:

1. Data preparation
2. Mask writing and chemical processing
3. Quality checks and adding protections

We explain these steps briefly in the following paragraphs.

7.1.1 Data Preparation

First, we prepare the given layout data for mask writing. We translate the GDS-specified mask information to a format comprehended by a *mask writing tool*. It involves converting complicated polygons to simpler rectangles and trapeziums. This process is popularly known as *fracturing*. It simplifies the task for the mask writing hardware. Additionally, data preparation involves augmenting the mask data to enhance the resolution. We will describe some of these techniques later in this chapter.

7.1.2 Mask Writing and Chemical Processing

The steps carried out in creating a mask are shown in Figure 7.1. The starting material for mask fabrication is known as *photomask blank*. It consists of a highly transparent substrate such as glass or fused silica over which we coat a thin opaque layer of chromium and a photoresist material, as shown in Figure 7.1(a). Subsequently, we write the required pattern on the photoresist, as shown in Figure 7.1(b). Typically, we write patterns on a mask by focusing an electron or LASER beam precisely at the required locations on the substrate.[1] The photoresist material is sensitive to the electron or the LASER beam. As a result, its property changes wherever it is exposed to the beams. Therefore, the light-exposed portion of the photoresist gets easily eliminated when we develop it, as shown in Figure 7.1(c).

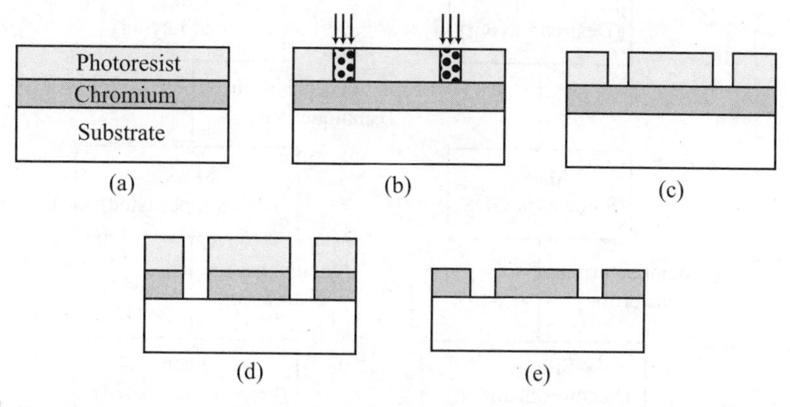

Figure 7.1 Steps to create a mask: (a) chromium and photoresist coated on a glass substrate, (b) pattern written on photoresist, (c) photoresist developed, (d) chromium etched, and (e) photoresist stripped

After development, chromium on the mask is etched. The photoresist material is resistant to the etchant. Therefore, chromium gets etched at locations where the photoresist was removed during development, as shown in Figure 7.1(d). Finally, we strip off the remaining photoresist material. Thus, we obtain a mask shown in Figure 7.1(e).

[1] Electron beam lithography has a high spatial resolution but is extremely slow compared to optical lithography.

7.1.3 Quality Checks and Protection

After creating a mask, we inspect for defects and make quality assurance checks [6]. We can find a defect in a mask by scanning its surface and comparing it with the reference image. If a defect is beyond tolerance, we repair it with the help of LASER, focused ion beams, or atomic force microscope-based nano-machining techniques [4].

After quality checks and cleaning, we apply a protective covering to the mask. The protective cover is a thin transparent membrane stretched over a metal frame. This cover is known as *pellicle*. It keeps unwanted particles away from the mask surface and avoids printing them on the wafer during photolithography.

7.2 RESOLUTION ENHANCEMENT TECHNIQUES

Currently, we carry out photolithography using a light source, typically with a wavelength of 193 nm. However, we require to print features on wafers much smaller than this wavelength. If we fabricate a mask with the same shapes we desire on the silicon, the final layout obtained on silicon can be pretty different from what we want, as illustrated in the flow diagram of Figure 7.2(a). These distortions are due to physical effects such as diffractions.

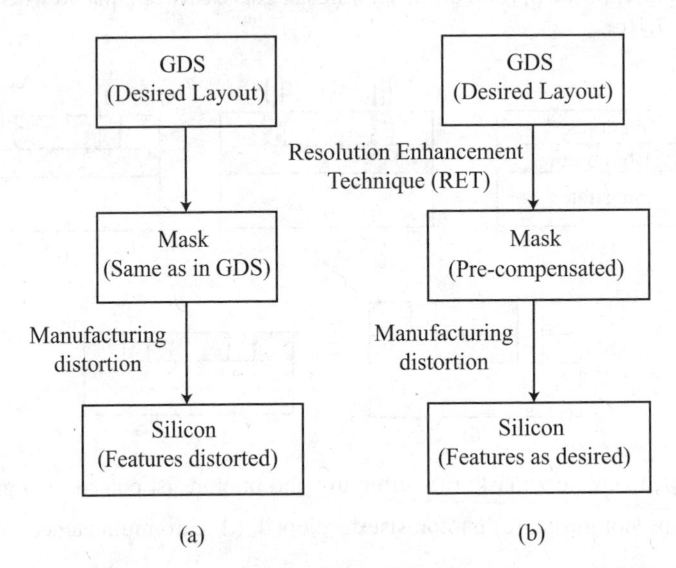

Figure 7.2 Fabrication of features on silicon using different masks: (a) mask has the same layout as in GDS and (b) layout in the mask is pre-compensated using RET

We can reduce these distortions but can never eliminate them, especially when the feature size is very small. Moreover, with decreasing feature size and the advancement in technology nodes, these distortions become more prominent, and we need to tackle them using some techniques.

One of the techniques to alleviate the problem of manufacturing distortions is by pre-compensating the layout, and the corresponding mask, using *resolution enhancement techniques* (RET), as shown in the flow diagram of Figure 7.2(b). If we use appropriately pre-compensated masks in photolithography, we can obtain the desired features on the silicon. We explain some of these techniques in the following paragraphs.

7.2.1 Optical Proximity Correction

If we print a mask pattern smaller than the light wavelength, we obtain significant distortions [7]. There can be rounding of corners and *line-end pullback*. Consequently, the electrical behavior of the fabricated circuit can deviate appreciably from the designed one. We can reduce these distortions by adding appropriate serifs, hammerheads, and mouse bites to the mask. It improves the resolution of photolithography. The technique of adding such features in the layout is known as *optical proximity correction* (OPC).

Example 7.1 Assume that we want to print the feature on the wafer as shown in Figure 7.3(a). Let us first use the same pattern on the mask, as shown in Figure 7.3(b).

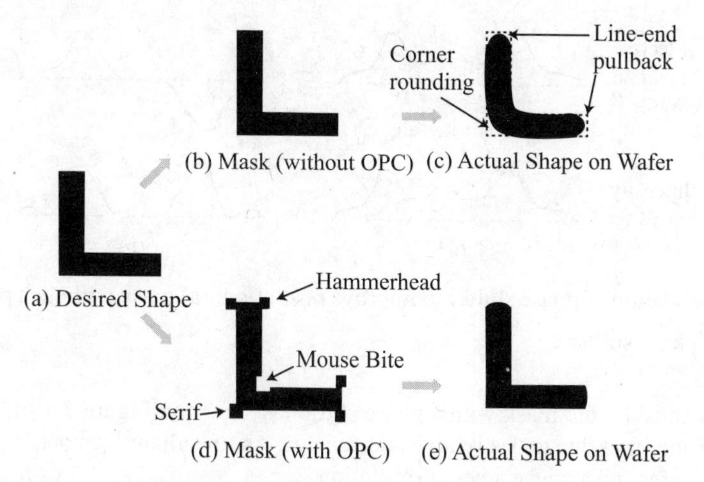

Figure 7.3 Illustration of optical proximity correction: (a) desired pattern, (b) mask same as the desired pattern, (c) distorted pattern obtained on the wafer, (d) mask augmented with the OPC features, and (e) pattern obtained on the wafer with the OPC-augmented mask [7]

If the pattern size is smaller than the wavelength of light, we will observe rounding of corners and line-end pullback, as shown in Figure 7.3(c).

We can add serifs, hammerheads, and mouse bites to the mask, as shown in Figure 7.3(d) [7]. As a result, we will obtain features similar to the layout. Thus, the added features improve the resolution of photolithography.

OPC features are minute and make the mask more complicated. Thus, the mask production tools need to print finer features to introduce OPC.

7.2.2 Introduction of Phase Shift

When there are closely spaced small apertures on the mask, we can enhance the resolution by introducing phase shifts [8]. To accomplish this, we adjust the optical phase difference of the waves transmitted through the adjacent apertures to 180°. Consequently, some diffraction effects get canceled due to destructive interference between waves. Thus, we can resolve closely spaced features.

Example 7.2 Let us consider a mask without a phase shifter as shown in Figure 7.4(a). The light intensity gets blurred between the apertures due to the interference and diffraction effects.

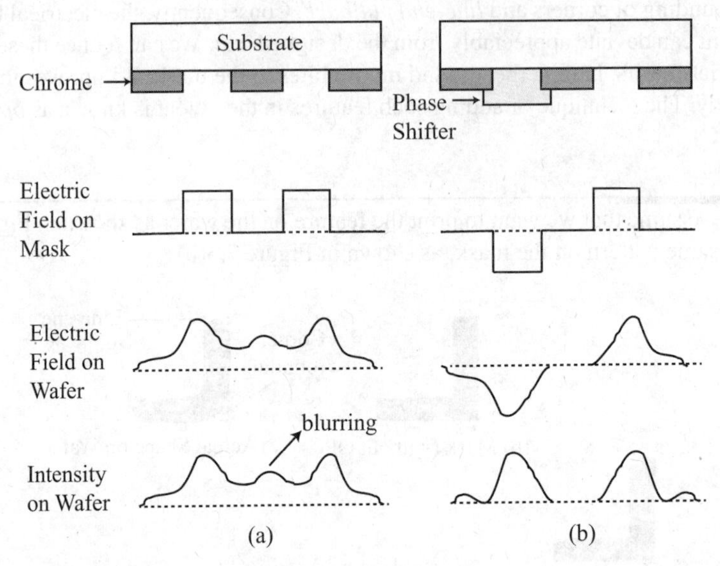

Figure 7.4 Application of phase shifter to improve resolution: (a) mask without a phase shifter and (b) mask with a phase shifter

Next, let us consider the mask with a phase shifter as shown in Figure 7.4(b). The intensity is low between the apertures due to destructive interference. As a result, we can obtain the desired light intensity on the wafer and achieve a better resolution.

We can introduce a phase shift on masks in several ways. For example, we can etch a rectangular groove on the mask substrate. Alternatively, we can add a transparent layer of appropriate refractive index and thickness.

7.2.3 Double Patterning or Multi-patterning

Due to the limited resolution of photolithography, we face problems in printing closely spaced features on a die. If there are inadequate gaps between features, they can overlap while fabricating, as explained above. We can solve the problem by increasing the spacing between features. We can decompose a closely spaced layout into two or more layouts. As a result, in each layout, the spacing between features increases. We use *different masks* and *different exposures* for each layout. Thus, each exposure needs a lower resolution due to decreased feature or pattern density. However, the combined effect of all the masks delivers the required high resolution.

Example 7.3 Assume that we need to print the features as shown in Figure 7.5(a).

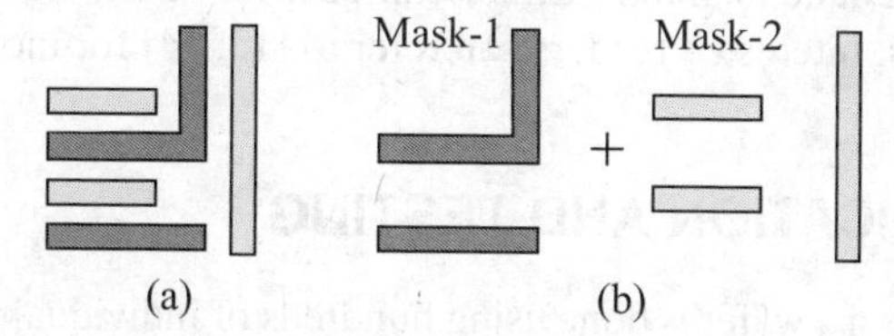

Figure 7.5 Double patterning: (a) given layout and (b) features decomposed into two masks

We decompose it into two masks as shown in Figure 7.5(b). As a result, the spacing between features on each mask increases. We use different exposures for each mask. Thus, each exposure needs a lower resolution, and we achieve an overall high resolution.

We can decompose a layout by identifying features that are closer than some predefined spacing. Subsequently, we assign different masks to them. We often refer to partitioning a layout's features to different masks as assigning different *colors* to features. Sometimes, we might need to divide a single feature (such as a rectangle) into different colors to meet the feature spacing constraint [9]. Alternatively, we can undertake triple patterning and quadruple patterning to improve the resolution [10, 11]. It is noteworthy that multiple patterning makes the process flow complicated and requires tighter overlay control between different exposures.

We can implement double-pattern lithography in many ways. Two popular techniques are *litho-etch-litho-etch* technique (LELE) and *self-aligned double patterning* (SADP) technique [12–14].

We illustrate the LELE technique in Figure 7.6. We employ a *hard mask* over the target layer as an intermediate. In the first exposure, we print the features of the first mask to the hard mask with the help of a photoresist, as illustrated in Figure 7.6(a–c). Then, we re-apply the photoresist layer over the wafer. We print the second mask features on the photoresist in the following exposure, as shown in Figure 7.6(d). Then, we etch the target layer through the openings left by the hard mask and the photoresist (the etchant is chosen such that it etches the material of the target layer selectively). Thus, we obtain the features corresponding to the two masks on the target layer, as shown in Figure 7.6(e). However, we need to accurately control the alignment of the two masks to avoid overlay errors.

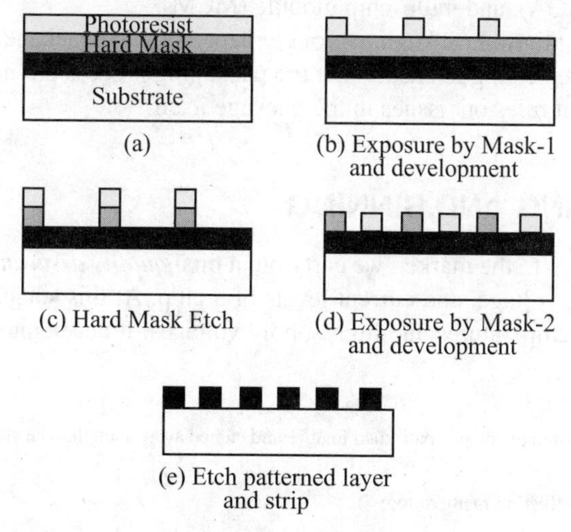

Figure 7.6 Illustration of processing steps for double patterning

The SADP technique generates finer features with the help of *spacers* that are self-aligned.[2] It is a more complicated technique than the LELE technique. However, the self-alignment allows it to alleviate the overlay errors. Interested readers can refer to [13, 14] for more details on SADP.

7.3 WAFER FABRICATION AND TESTING

The fabrication of circuits on a wafer is done using hundreds of individual process steps. These steps are carried out in a sequence and build features on a wafer layer-wise, starting from the bottom layer.

First, we fabricate circuit elements such as resistors, capacitors, diodes, and transistors on the lower layers of the wafer using a set of processes known as the *front end of the line* (FEOL) processes. Subsequently, we make interconnections using metallic layers at the top of the wafer using a set of fabrication steps known as the *back end of the line* (BEOL) processes. All these processes involve some form of photolithographic patterning. We will discuss these steps in more detail in Chapter 24 ("Basic concepts for physical design").

After we have fabricated an IC over a silicon wafer, we carry out *die-level testing*.[3] If we find that a die is faulty, we put an identification mark on it. Subsequently, we discard the faulty ones and diagnose the root cause of their failure, if needed. We proceed with the good dies for packaging.

7.4 PACKAGING

Once a die passes testing, we cut it from the wafer and encapsulate it in a supporting case known as a *package* to form a chip. A package provides pins for external connection to the die. The package pins impact the attributes of signals entering or leaving a chip, such as a delay and transition time. Therefore, we need to design the package for a chip carefully. The package is also responsible for drawing heat from a die and dissipating it into the environment. Therefore, thermal design and the surface area of a package are critical considerations in designing or choosing a package for a chip. A package also protects the die from mechanical damage and corrosion.

We can use different types of materials, such as ceramic and plastic, for packages. We consider the cost and the material's electrical, thermal, and mechanical properties while choosing a material for packaging. Packages can be of different types, such as dual in-line package (DIP), pin grid array (PGA), ball grid array(BGA), and multi-chip module (MCM).

The steps carried out during packaging strongly depend on the package type. After packaging, we perform package-level testing to ensure that the packaging process did not introduce any defect in the die. Additionally, it rules out issues in the package itself.

7.5 FINAL TESTING AND BINNING

Before shipping out a chip to the market, we carry out a final *quality assurance* testing. We measure the performance, power, voltage, and current levels of a chip. At this stage, we can also carry out *burn-in testing* at high temperatures and high supply voltages. It accelerates latent defect-induced

[2] Spacer is a layer that is deposited over a pre-patterned feature and etched away such that the sidewall spacer remains around the pre-patterned feature.

[3] For details, see Section 6.3 ("Testing methodology").

infant mortalities, and thus avoids shipping a product that fails in the early stages of operation [15, 16]. However, burn-in testing can take additional test time and increase the cost of a product.

During final testing, we can also perform *binning* [17, 18]. The binning of chips becomes necessary because statistical process variations result in a large spread in the operating frequency. We subject a fabricated chip to some functional or structural tests at *different* frequencies during binning. Using these test results, we determine the highest performance that a chip can deliver [17]. A special on-chip delay measurement circuitry inserted in the chip enables these measurements [18]. After binning, we can assign performance-based price points to different bins.

7.6 RECENT TRENDS

With the advancement in technology, manufacturing processes have undergone several changes. The smallest feature size obtained using photolithography is related to the wavelength of employed light. We can attain a smaller feature size by employing light of a reduced wavelength. Hence, *extreme ultraviolet* (EUV) lithography that uses a 13.5 nm wavelength light source is being used in advanced process nodes [19–22]. However, we need to make massive changes in photoresist materials and light sources for transitioning from 193 nm lithography to EUV lithography [23].

With the continued miniaturization of the transistors, controlling their electrostatics becomes challenging. Therefore, the MOSFET architecture has been evolving over the years: from the 2D planar structure to FinFET and *gate-all-around* (GAA) architecture implemented using nanosheets or nanowires [24]. Moreover, newer materials, such as those based on III–V semiconductors, that can deliver better performance can be integrated into the mainstream CMOS technology [25]. Additionally, transistors based on newer mechanisms such as tunneling and spintronics are being widely researched for future commercial applications [26].

REVIEW QUESTIONS

7.1 While making masks, we typically write the pattern on the photoresist using an electron beam or LASER. Why is the same technique not used for mass photolithography on wafers?

7.2 What is the purpose of a pellicle on a mask?

7.3 The adoption of OPC techniques at advanced process nodes has made the requirement of mask writing more stringent. Comment.

7.4 Explain the technique of double patterning.

7.5 Explain how we can decompose a layout for double patterning?

7.6 What are the challenges of EUV lithography?

REFERENCES

[1] J. D. Plummer. *Silicon VLSI Technology: Fundamentals, Practice and Modeling*. Pearson Education India, 2009.

[2] P. Van Zant. *Microchip Fabrication*. McGraw-Hill, Inc., 2004.

[3] W. M. Moreau. *Semiconductor Lithography: Principles, Practices, and Materials*. Springer Science & Business Media, 2012.

[4] S. Rizvi. *Handbook of Photomask Manufacturing Technology*. CRC Press, 2005.

[5] R. Tummala, E. J. Rymaszewski, and A. G. Klopfenstein. *Microelectronics Packaging Handbook: Semiconductor Packaging, Part II, Second Edition*. Springer Science & Business Media, 1997.

[6] J. H. Bruning, M. Feldman, T. Kinsel, E. K. Sittig, and R. L. Townsend. "An automated mask inspection system (AMIS)." *IEEE Transactions on Electron Devices* 22, no. 7 (1975), pp. 487–495.

[7] R. Seisyan. "Nanolithography in microelectronics: A review." *Technical Physics* 56, no. 8 (2011), p. 1061.

[8] M. D. Levenson, N. Viswanathan, and R. A. Simpson. "Improving resolution in photolithography with a phase-shifting mask." *IEEE Transactions on Electron Devices* 29, no. 12 (1982), pp. 1828–1836.

[9] A. B. Kahng, C.-H. Park, X. Xu, and H. Yao. "Layout decomposition for double patterning lithography." *2008 IEEE/ACM International Conference on Computer-Aided Design* (2008), pp. 465–472, IEEE.

[10] B. Yu, K. Yuan, D. Ding, and D. Z. Pan. "Layout decomposition for triple patterning lithography." *IEEE Transactions on Computer-Aided Design of Integrated Circuits and Systems* 34, no. 3 (2015), pp. 433–446.

[11] M. Neisser and S. Wurm. "ITRS lithography roadmap: 2015 challenges." *Advanced Optical Technologies* 4, no. 4 (2015), pp. 235–240.

[12] P. Zimmerman. "Double patterning lithography: Double the trouble or double the fun?" *SPIE Newsroom* 20 (2009).

[13] J. Hwang, J. Seo, Y. Lee, S. Park, J. Leem, J. Kim, T. Hong, S. Jeong, K. Lee, et al. "A middle-1X nm NAND flash memory cell (M1X-NAND) with highly manufacturable integration technologies." *2011 International Electron Devices Meeting* (2011), pp. 9–1, IEEE.

[14] Y. Ma, J. Sweis, H. Yoshida, Y. Wang, J. Kye, and H. J. Levinson. "Self-aligned double patterning (SADP) compliant design flow." In *Design for Manufacturability Through Design-Process Integration VI*, vol. 8327, pp. 49–61, SPIE, 2012.

[15] D. Appello, C. Bugeja, G. Pollaccia, P. Bernardi, R. Cantoro, M. Restifo, E. Sanchez, and F. Venini. "An optimized test during burn-in for automotive SoC." *IEEE Design & Test* 35, no. 3 (2018), pp. 46–53.

[16] M. F. Zakaria, Z. A. Kassim, M.-L. Ooi, and S. Demidenko. "Reducing burn-in time through high-voltage stress test and Weibull statistical analysis." *IEEE Design & Test of Computers* 23, no. 2 (2006), pp. 88–98.

[17] B. D. Cory, R. Kapur, and B. Underwood. "Speed binning with path delay test in 150-nm technology." *IEEE Design & Test of Computers* 20, no. 5 (2003), pp. 41–45.

[18] A. Raychowdhury, S. Ghosh, and K. Roy. "A novel on-chip delay measurement hardware for efficient speed-binning." *11th IEEE International On-Line Testing Symposium* (2005), pp. 287–292, IEEE.

[19] C. A. Mack. "Reducing roughness in extreme ultraviolet lithography." *Journal of Micro/Nanolithography, MEMS, and MOEMS* 17, no. 4 (2018), p. 041006.

[20] S. K. Moore. "3 Directions for Moore's Law: The last few months have sent mixed signals about where chips are headed." *IEEE Spectrum* 55 (Nov. 2018), pp. 14–15.

[21] T. Song, J. Jung, W. Rim, H. Kim, Y. Kim, C. Park, J. Do, S. Park, S. Cho, et al. "A 7nm FinFET SRAM using EUV lithography with dual write-driver-assist circuitry for low-voltage applications." *2018 IEEE International Solid-State Circuits Conference-(ISSCC)* (2018), pp. 198–200, IEEE.

[22] G. Yeap, S. Lin, Y. Chen, H. Shang, P. Wang, H. Lin, Y. Peng, J. Sheu, M. Wang, et al. "5nm CMOS Production Technology Platform featuring full-fledged EUV, and High Mobility Channel FinFETs with densest 0.021 μm 2 SRAM cells for Mobile SoC and High Performance Computing Applications." *2019 IEEE International Electron Devices Meeting (IEDM)* (2019), pp. 36–7, IEEE.

[23] O. Kostko, B. Xu, M. Ahmed, D. S. Slaughter, D. F. Ogletree, K. D. Closser, D. G. Prendergast, P. Naulleau, D. L. Olynick, et al. "Fundamental understanding of chemical processes in extreme ultraviolet resist materials." *The Journal of Chemical Physics* 149, no. 15 (2018), p. 154305.

[24] D. Nagy, G. Espineira, G. Indalecio, A. J. Garcia-Loureiro, K. Kalna, and N. Seoane. "Benchmarking of FinFET, nanosheet, and nanowire FET architectures for future technology nodes." *IEEE Access* 8 (2020), pp. 53196–53202.

[25] N. Collaert. *High Mobility Materials for CMOS Applications*. Woodhead Publishing, 2018.

[26] S. Manipatruni, D. E. Nikonov, C.-C. Lin, T. A. Gosavi, H. Liu, B. Prasad, Y.-L. Huang, E. Bonturim, R. Ramesh, and I. A. Young. "Scalable energy-efficient magnetoelectric spin–orbit logic." *Nature* 565, no. 7737 (2019), pp. 35–42.

PART TWO

Logic Design

Earlier, we discussed dividing the RTL to GDS implementation flow into two parts: logic synthesis and physical design. In this part of the book, we will discuss logic synthesis. In Part IV, we will discuss physical design.

Logic design involves transforming a high-level functional description to a netlist of standard cells and macros. It takes a design from the functional domain to the structural domain. The primary task of logic design is to decide the logic elements that will deliver the required functionality. Additionally, we need to ensure that the design metrics such as area, performance, power, and testability meet the given requirements.

To ensure that the logic design meets the above requirements, we interleave implementation and verification tasks in a design flow. We have arranged implementation and verification-related chapters similarly. However, note that we carry out some of the verification tasks, such as combinational equivalence checking, timing analysis, and power analysis, multiple times in a design flow.

We will explain implementation of the logical design in Chapter 8 ("Modeling Hardware using Verilog"), Chapter 10 ("RTL Synthesis"), Chapter 12 ("Logic Optimization"), Chapter 16 ("Technology Mapping"), Chapter 17 ("Timing-driven Optimization"), and Chapter 19 ("Power-driven Optimization"). We will discuss verification aspects for a design in Chapter 9 ("Simulation-based Verification"), Chapter 11 ("Formal Verification"), Chapter 14 ("Static Timing Analysis"), and Chapter 18 ("Power Anlaysis"). We will present the information that is used both in implementation and verification in Chapter 13 ("Library") and Chapter 15 ("Constraints").

It is worthy to point out that the primary objective of these chapters is to build a foundation for logic design. Therefore, we explain essential concepts and principles governing it. We have attempted to provide explanations not based on any specific logic synthesis tool or proprietary data format. Therefore, a reader can apply these concepts to any tool s/he chooses for logic design. Additionally, note that these chapters build a foundation for logic design. To gain more depth on these topics, we encourage readers to refer to standard textbooks on them. We have provided references to those textbooks at appropriate places in each chapter.

8 Modeling Hardware Using Verilog

'When I use a word,' Humpty Dumpty said in rather a scornful tone,
'it means just what I choose it to mean—neither more nor less.'
'The question is,' said Alice, 'whether you can make words mean different things.'
'The question is,' said Humpty Dumpty, 'which is to be master—that's all.'

—Lewis Carroll, *Through the Looking-Glass*, Chapter 6, 1871

Obtaining a register transfer level (RTL) model for a design is often the starting point of a design flow. We model RTL using hardware description languages (HDLs), such as Verilog and VHDL. We synthesize an RTL model to obtain a netlist. Subsequently, we use this netlist in the design flow.

In this chapter, we will describe modeling of hardware using Verilog. We will primarily focus on the language constructs and explain their application in creating hardware models. We will describe the impact of Verilog constructs on simulation and synthesis in Chapter 9 ("Simulation-based verification") and Chapter10 ("RTL synthesis"), respectively.

This chapter is an introduction to Verilog. It will enable readers to understand various constructs in a given Verilog code, develop Verilog models, and subsequently use the Verilog models in a design flow. For a deeper understanding, we advise readers to refer to dedicated resources on Verilog and the *IEEE Std 1364-2001* [1–3].

8.1 HARDWARE DESCRIPTION LANGUAGES

HDLs are created to describe hardware easily and realistically. They enable designers to express the design intent and functionality in a precise manner. Moreover, they allow electronic design automation (EDA) tools to extract design information and process them efficiently. Therefore, HDLs have several features distinct from conventional programming languages such as FORTRAN, C, and C++. Some of the distinctive features of HDLs are:

1. **Concurrency:** Hardware can compute *concurrently*. Different components in a circuit such as flip-flops, adders, and multipliers can compute in *parallel*.

Example 8.1 Consider two adder circuit elements *I1* and *I2*. We can connect them as shown in Figure 8.1(a). In this case, the adders can operate concurrently. We can also connect these two adders as shown in Figure 8.1(b). The adder *I2* uses the result of *I1*. Therefore, the operation of *I2* can be completed only after *I1* has completed its operation. Hence, these adders operate sequentially.

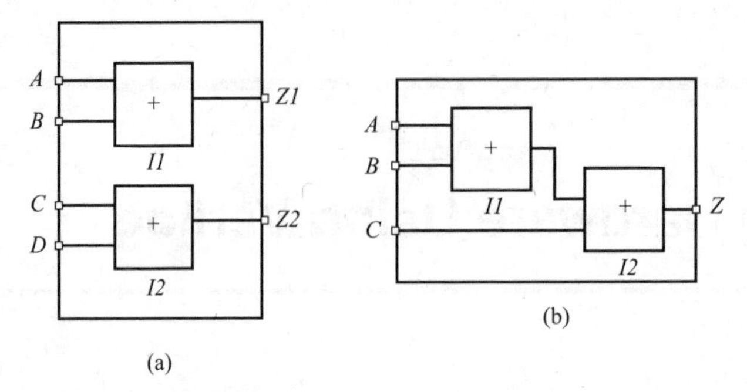

Figure 8.1 Two adders: (a) concurrent operation and (b) sequential operation

The above example illustrates that we need to describe the sequence of operations in the hardware model. HDLs need to provide features to support both parallel and sequential operations. Additionally, it should enable EDA tools to distinguish between these two types of operations unambiguously.

2. **Notion of time:** Delay is a fundamental concept for hardware. Additionally, we need to capture the attributes of the periodic signals, such as the clock, and also model the timing sequence of operations. Therefore, HDLs have in-built notion of *time*.
3. **Electrical characteristics:** HDLs need to support features to describe the electrical characteristics of circuit elements. For example, a driver of a signal can be weak or strong. When there is contention (more drivers are driving a signal), the signal can resolve based on the *drive strength* of the drivers.
4. **Bit-true data types:** HDLs should provide a mechanism to model the behavior of buses such as 32-bit data buses or 64-bit address buses. A bus describes a group of signals that are treated as parallel lines and are operated together. Moreover, in the hardware models, we sometimes need to describe the behavior of individual bits or a part of a bus. Therefore, HDLs need to support multiple levels of signal abstraction.

Thus, HDLs need to support many features which are absent in the traditional programming languages. Therefore, concerted efforts have been made since the 1970s to develop languages with HDL features [4]. Eventually, two languages that became popular for modeling hardware were Verilog and VHDL.

8.1.1 Verilog and SystemVerilog

The Verilog language was created at Gateway Design Automation in 1983–84. The word Verilog was derived by combining the words *verification* and *logic* [5]. Initially, Verilog gained popularity due to the high-speed simulator Verilog-XL (1987) and the support for logic synthesis by Design Compiler (1987) [4]. The simplicity of Verilog was a major attraction for hardware designers.

Verilog was standardized by IEEE in 1995 (*IEEE Standard 1364-1995*) and 2001 (*IEEE Standard 1364-2001*). Subsequently, features got added to support robust verification methodologies such as coverage-directed constrained random verification and assertion-based verification. As a result, a superset of Verilog known as SystemVerilog was developed. The merged Verilog and

SystemVerilog was standardized in 2009 (*IEEE Standard 1800-2009*) [6]. SystemVerilog 2009 also ensures interoperability with other HDLs such as SystemC and VHDL. The unified Verilog or SystemVerilog is now widely used and supported by EDA tools to design, verify, synthesize, and test hardware. In 2018, about 80% of integrated circuit (IC) design teams worldwide used Verilog or SystemVerilog [4].

8.1.2 VHDL

VHDL stands for *VHSIC HDL*, where VHSIC is an acronym for *very high speed integrated circuit*. VHDL started as a documenting language for ICs and was first standardized by IEEE in 1987 (*IEEE 1076-1987*). Later, major revisions were made in 1993 (*IEEE 1076-1993*) and 2019 (*IEEE 1076-2019*) [7]. VHDL is a verbose language and performs strict type checking. It was popular in the 1990s and is still in use. It is based on the Ada programming language. In contrast, Verilog has syntax similar to the C programming language. Therefore, designers who are already familiar with C find Verilog easier to learn and use.

8.2 BASIC CONSTRUCTS OF VERILOG

In this section, we briefly describe basic concepts and constructs of Verilog based on *IEEE Standard 1364-2001* [1].

8.2.1 Lexical Conventions

The lexical convention in Verilog is similar to the C programming language. A Verilog file consists of stream of lexical *tokens*. A token can be: *white space*, *identifier*, *keywords*, *operators*, *numbers*, *string*, and *comment*. Let us understand each of them.

1. **White space:** A white space consists of the characters for spaces, tabs, newlines, and formfeeds. These characters are used as separators for other tokens.
2. **Identifier:** An identifier is a unique name given to an object so that we can refer to it in the Verilog code. The objects can be modules, ports, nets, registers, functions, etc. An identifier must begin with an alphabetic character or an underscore (a–z A–Z _). However, subsequent characters can be alphanumeric, underscore, or dollar (a–z A–Z 0–9 _, $). The maximum character length allowed for an identifier is 1024. However, a tool can support more than 1024 characters also. The identifiers are *case-sensitive*. Thus, *net_a* and *Net_A* are treated as two distinct identifiers.

   We can create an *escaped identifier* by preceding with a backslash "\" and ending with a white space. An identifier can contain any printable ASCII character between the "\" and the white space. We do not consider the leading backslash and the terminating white space part of the identifier name. Thus, an escaped identifier $\backslash net\_(a+b)*c$ is treated as $net\_(a+b)*c$.
3. **Keywords:** They are reserved words of Verilog that define various constructs of the language. For example, `module`, `input`, `begin`, `end`, `initial`, and `always`. All the keywords of Verilog are only in lower case.
4. **Operators:** They are predefined sequences of one, two, or three characters used in an expression. For example, `!`, `+`, `-`, `&&`, `==`, `!==`, etc.

5. **Numbers:** We can specify *integers* or *real* numbers in Verilog code. Though we can specify a number in multiple ways, we typically choose a convenient and readable representation. Nevertheless, note that tools internally represent these numbers simply as a sequence of bits.

   We can define integers as traditional decimal numbers. For example, 169, –123, etc. Alternatively, we can define an integer in the following format:

   –⟨*size*⟩'⟨*base*⟩⟨*value*⟩

   where –(minus) is an optional negative sign, an optional ⟨*size*⟩ specifies the number of bits in the number, ⟨*base*⟩ is a single letter that specifies the base of the number, and ⟨*value*⟩ specifies the value of the integer. The above integer definition obeys the following rules:

   (a) The valid letters for ⟨*base*⟩ are `b/B` for binary, `o/O` for octal, `d/D` for decimal, and `h/H` for hexadecimal.

   (b) For hexadecimal, octal, and binary constants, `x/X` represent the unknown value and `z/Z/?` represent the high-impedance value. When the high-impedance value is a don't-care, we prefer using `?` instead of `z/Z` to improve readability.

   (c) We can add underscore (_) in ⟨*value*⟩ to improve readability. These underscores are ignored in the internal representation of the number.

   (d) If ⟨*size*⟩ is not big enough to accommodate ⟨*value*⟩, the leftmost bits of ⟨*value*⟩ are truncated.

   (e) If ⟨*size*⟩ is more than ⟨*value*⟩ in an unsigned number, the leftmost bits are filled with `0`, `x`, or `z`, depending on whether the leftmost bits in ⟨*value*⟩ is `0/1`, `x`, or `z`, respectively.

   (f) Negative numbers are internally represented in two's complement form.

   A few examples of integer constants are given in Table 8.1.

Table 8.1 Example of integer constants

Verilog Representation	Comments	Internal Bit Representation
`'h48F`	Unsized hexadecimal	`0000 0000 0000 0000 0000 0100 1000 1111`[a]
`4'b110x`	Least significant bit unknown	`110x`
`8'b1001_1010`	Underscore ignored	`1001 1010`
`6'h8F`	Truncation of 2 leftmost bits	`00 1111`
`8'b11`	Extension to 8 bits	`0000 0011`
`8'bz1`	Extension to 8 bits	`zzzz zzz1`
`-8'd6`	Two's complement of 6 in 8 bits	`1111 1010`

[a]The number of bits in an unsized number is at least 32.

Note that for integers specified as simple decimal number or when ⟨*size*⟩ is not specified, the number of bits in the internal representation is at least 32.

We can define real numbers in decimal notation (e.g., 3.14159) or scientific notation (e.g., 2.99E8). Internally, real numbers are represented in IEEE standard for double-precision floating-point numbers [8].

6. **String:** A string is a sequence of characters. It is enclosed by double quotes and is contained on a single line. Each character is represented by its corresponding 8-bit ASCII value.[1] For example, "Hello."
7. **Comment:** We can add comments in a Verilog code, as shown below.

```
// This is a single-line comment.
/* This is a
multi-line
comment. */
```

A one-line comment starts with a double back-slash // and ends with a newline. A block comment is enclosed between /* and */. A *nested* block comment is not allowed.

8.2.2 Data Types, Vectors, and Arrays

To model digital hardware, Verilog supports following four *data values* for signals:

1. `0`: represents logic *false* or Boolean *zero*.
2. `1`: represents logic *true* or Boolean *one*.
3. `x`: represents unknown value.
4. `z`: represents high-impedance state.

Additionally, Verilog supports two main groups of data types: *nets* and *variables*.

1. **Nets:** A net represents physical connections between structural entities, such as gates and instances. It cannot store value, rather it takes value from its driver. We commonly declare nets using the statement `wire`, as shown below.

```
wire w1, w2;
wire w3=1'b1;
supply0 gnd;
supply1 vdd;
```

If there is no driver connected to a `wire`, it takes a value *z*. Additionally, tools assume an undeclared signal in the Verilog code as a 1-bit `wire`. Some other types of nets supported in Verilog are `supply0`, `supply1`, `wand`, and `wor`. We use `supply0` net type to model ground connections or logic `0`. We use `supply1` net type to model connections to power supplies or logic `1`. The `wand` create wired AND type net configurations, such that if any driver is 0, the value of that net becomes 0. Similarly, the `wor` create wired OR type net configurations, such that if any driver is 1, the value of that net becomes 1.

[1] ASCII stands for American Standard Code for Information Interchange. It is a popular character encoding standard.

2. **Variables:** A variable is an abstraction for an element that stores data. A variable retains its already assigned value until it gets changed by another *assignment*. We commonly declare variables using the statement `reg`, as shown below.

```
reg r1, r2;
```

The initial value for the `reg` type data is `x` (unknown value). We assign values to a `reg` type data using *procedural assignments* (will be discussed later in detail). Since a `reg` type variable holds its value between assignments, we can use it to model hardware registers (flip-flops and latches). However, a `reg` type data need not always represent a hardware storage element. It can also represent combinational logic elements, as will be explained in the later sections. To ease hardware modeling, Verilog supports other types of variables also, such as `integer`, `real`, and `time`.

The nets and variables are 1-bit wide or *scalar* by default. However, we can declare them as *vectors* by preceding the declaration with vector definition in the following format: [⟨*left_range*⟩:⟨*right_range*⟩]. The ⟨*left_range*⟩ is the *most significant bit* (MSB) and ⟨*right_range*⟩ is the *least significant bit* (LSB). We can select a bit of a vector net or vector variable by specifying the address within the square bracket (`[]`). It is referred to as *bit-select*. Similarly, if we want to select a portion of a vector, we need to specify the range of MSB and LSB separated by a colon (`:`). It is referred to as *part-select*. If the specified address is illegal or wrong, we obtain `x` for that selection.

Example 8.2 Consider the following piece of Verilog code.

```
wire [31:0]databus;
reg [7:0]addressbus;
```

It declares a 32-bit wide net *databus* with MSB=31 and LSB=0. In the next statement, it declares a 8-bit wide variable *addressbus* with MSB=7 and LSB=0. If we want to select the bit at index 4 in *databus*, we write:

```
databus [4] = 1'b0; // arbitrary RHS just for illustration
```

If we want to select the first four lower bits of *addressbus*, we write:

```
addressbus [3:0] = 4'b1001; // arbitrary RHS just for illustration
```

Sometimes we need to group elements into multidimensional objects. We can use *arrays* for grouping together objects. We define the address range [⟨*left_range*⟩:⟨*right_range*⟩] for the array *after* the declared identifier. For each dimension, we need to define the address range.

Example 8.3 A few examples of arrays are shown below.

```
reg r[15:0];                // an array of 16 scalar reg
wire matrix[9:0][9:0];      // a 2D 10x10 array of wires
reg [7:0]mbyte[0:127];      // an array of 128 eight-bit register
reg [31:0]mword[0:1023];    // an array of 1024 32-bit register
```

Note the difference in the declaration of arrays and vectors. Arrays are useful in modeling memories.

8.2.3 Modules, Ports, Instantiation, and Parameters

Modules are the basic building blocks of a Verilog design. They fully describe their interfaces and their functions in structural, behavioral, or mixed form. A module definition is enclosed between the keywords `module` and `endmodule`. The identifier immediately following the keyword `module` is the module name.

The interfaces of a module are known as *ports*. A port can be of type `input`, `output`, or `inout`. Ports are defined in the module declaration. We can specify a port as a bus or a group of bits by preceding the port name with the vector specification, similar to nets and variables. We can also define the direction, width, and type of ports in the module declaration. It eliminates some duplication in defining the ports of the module.

Example 8.4 Consider the following piece of Verilog code.

```
module flipflop(d, clk, rst, q);
        input d, clk, rst;
        output q;
        ...
endmodule
module top(data, cntrl, clk, rst, result);
        input [7:0]data, cntrl, clk, rst;
        output [3:0]result;
endmodule
```

It declares two modules named *flipflop* and *top*. The module *flipflop* has input ports *d*, *clk*, and *rst* and an output port *q*. The module *top* has an input port *data* of 8 bits and *cntrl*, *clk*, and *rst* of 1 bit each. It has also an output port *result* of 4 bits. We can also declare module and ports together, as shown in the following piece of code.

```
module top(input  [7:0]  data,  // direction and width defined
       input cntrl, clk, rst,  // shared definition
       output reg  [3:0]  result);  // reg type also declared
```

We cannot *nest* module definitions, i.e., we cannot define another module within a pair of `module–endmodule`. However, we can allow one module to use a copy of another module by *instantiation*. We create one or more named *instances* of a module using an instantiation statement. We can specify the connections between instances in two ways:

1. **Order-based connections:** We simply specify the connections in the instance in the order in which ports are declared in the master module. The tools automatically make connections based on the order of port declaration of the master module. Since order-based connections are implicit, debugging can be difficult when there are many ports in a module.
2. **Name-based connections:** We can specify connections by explicitly naming the ports of the master module. In this case, we can set the connections in arbitrary order.

We illustrate instantiation in the following example.

Example 8.5 Consider the following piece of Verilog code.

```
module flipflop(d, clk, q);
    input d, clk;
    output q;
    ...
endmodule
module middle(D, CLK, Q);
    input D, CLK;
    output Q;
    flipflop F1(D, CLK, Q);
endmodule
module Top(data, clk, result);
    input data, clk;
    output result;
    wire w1, w2;
    flipflop I1(data, clk, w1);
    flipflop I2(.d(w1), .clk(clk), .q(w2));
    middle I3(.D(w2), .Q(result), .CLK(clk));
endmodule
```

We have instantiated module *flipflop* with the instance name *F1* in the module *middle*. We refer to the module which is instantiated as the *master module*. Thus, the master module of the instance *F1* is *flipflop*. Further, the module *Top* contains instances *I1* and *I2* of the master module *flipflop* and *I3* of the master module *middle*. Note that we can create as many instances of a module in any other module. These instantiations create a design hierarchy, which we can represent as in Figure 8.2(a). Often EDA tools show these design hierachies as a schematic or a hierarchy tree.

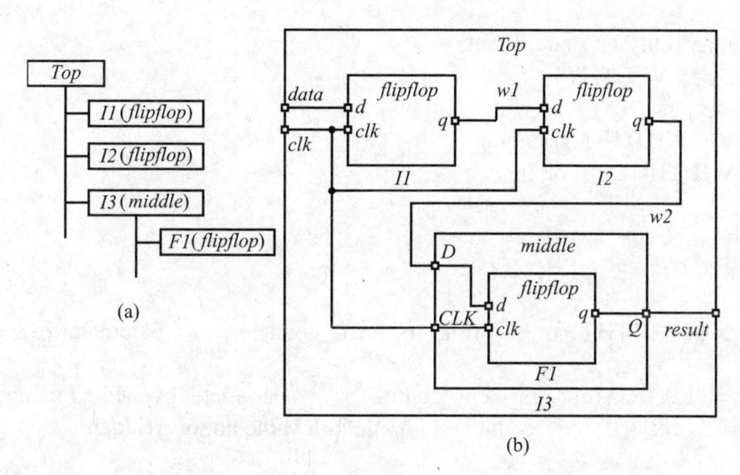

Figure 8.2 An illustration of (a) design hierarchy; (b) instances and connections

We have specified order-based connections for the instance *F1* consistent with the master module *flipflop*. Similarly, we have specified the order-based connection of the instance *I1* in the module *Top*. Tools automatically infer connections based on the order of port declaration of the master module, as shown in Figure 8.2(b). We have created name-based connections for the instances *I2* and *I3*. Note that, in the name-based connection, we can specify the connection in an arbitrary order.
Remark: While writing a Verilog code, we should prefer *name-based connections* because it avoids inadvertent errors in defining connections due to order mismatch. It also makes the code readable since the connections are explicit and helps in debugging.

Verilog allows definition of constants for a *module* using the keyword `parameter`. We declare and provide a *default value* for the parameter in the module. The module in which a parameter is defined is referred to as a *parameterized module*. The scope of the parameter is the module in which it is defined. After providing a default value to a parameter, we cannot change it further inside the parameterized module. However, we can override the default value of the parameter during instantiation by providing non-default values within `#()`. Thus, parameters provide a mechanism to obtain a different implementation of a module based on instantiations.[2]

[2] We can also modify the value of a parameter using `defparam` statement. Verilog also supports `localparam`. They work the same as parameters, but we cannot directly modify their values in module instantiation or by `defparam` statements.

Example 8.6 Suppose we need counters of various widths and the same functionality in a design. We can accomplish this by designing a counter in which *WIDTH* is defined as a parameter with some default value. We describe the functionality of the counter using the parameter *WIDTH*. To obtain counters of different widths, we instantiate the parameterized module with the appropriate parameter value. One possible Verilog implementation is shown below.

```
module counter(clk, rst, en, count);
    parameter WIDTH=4;
    input clk, rst, en;
    output [WIDTH-1:0] count;
    reg [WIDTH-1:0] count;
    ...
endmodule
module top();
    ...
    counter #(.WIDTH(16)) C1(clk, rst, en, count1); // parameter overriding using
name
    counter #(128) C2(clk, rst, en, count2); // parameter overriding using position
    counter C3(clk, rst, en, count3); // default value not overridden
    ...
endmodule
```

The instances *C1*, *C2*, and *C3* have WIDTH defined as 16, 128, and 4, respectively.

8.2.4 Built-in Gates

To allow easy hardware modeling, Verilog inherently supports some logic gates and switches. It supports multi-input gates AND, NAND, NOR, OR, XNOR, and XOR using keywords `and`, `nand`, `nor`, `or`, `xnor`, and `xor`, respectively. The first pin is the output pin, and the rest are input pins. Additionally, Verilog supports multi-output buffers and inverters using keywords `buf` and `not`, respectively. The last pin is the input pin for buffers and inverters, and the rest are output pins. We can instantiate and use these built-in gates to model various Boolean functions. Additionally, Verilog supports various types of tristate gates, pullup/pulldown gates, MOS switches, and bidirectional switches [1].

Example 8.7 The following piece of Verilog code illustrates the usage of built-in gates.

```
module myfunction(a, b, s, out1, out2);
    input a, b, s;
    output out1, out2;
    wire sbar, w1, w2, o;
    not N1(sbar, s); // sbar is output, s is input
    and A1(w1, a, sbar); // w1 is output, a and sbar are inputs
```

```
    and A2(w2, b, s);
    or O1(o, w1, w2); // MUX o=as'+bs
    buf B1(out1, out2, o); // out1, out2 are outputs, o is input
endmodule
```

N1, *A1*, *A2*, *O1*, and *B1* are instances of build-in gates.

8.2.5 Operators, Operands, and Expressions

Verilog supports *operators* with symbols similar to the C programming language. For example, arithmetic operators (`+`, `-`, `*`, `/`, `**`, `%`), logical operators (`!`, `&&`, `||`), bit-wise operators (`~`, `&`, `|`, `^`, `~^`), shift operators (`>>`, `<<`, `>>>`, `<<<`), conditional write operator (`? :`), and relational operators (`>`, `>=`, `<`, `<=`) are supported in Verilog.

Operators act on *operands* to produce results. For each operator, the number of operands it takes is defined by the language. For example, logical negation (e.g., `!`*a*) takes one operand, less than operator (e.g., *a*<*b*) takes two operands, and conditional write operator (e.g., *c* `?` *a* `:` *b*) takes three operands. An operand can be of various data types such as net, variable, bit-select, part-select, or array element.

There are some special operators in Verilog that are especially useful in hardware modeling:

1. **Equality operators:** There are two equality operators `==` and `===`. The result of *a*==*b* is `x` if either operand is `x` or `z`. If both *a* and *b* are Boolean (`0` or `1`), the result is `1` if *a* and *b* are equal, and `0` otherwise. The result of *a*===*b* is always Boolean (`0` or `1`) and is obtained by comparing four possible values of *a* and *b*.
2. **Reduction operators:** The reduction operators perform a bit-wise operation on a *single operand* to produce a single bit result. Verilog supports reduction operators `&` (AND), `~&` (NAND), `|` (OR), `~|` (NOR), `^` (XOR), and `~^` (XNOR). The results of the reduction operators NAND, NOR, and XNOR are obtained by first computing reduction AND, OR, and XOR, respectively, on the given operand and finally taking the complement.

Example 8.8 Assume that a 4-bit register *a* holds `4'b1011`. Let us compute `&`*a* (reduction AND on *a*).

The computation starts from the LSB and moves toward the MSB, as shown below:

$$
\begin{aligned}
\&a &= \&\,(1011)\\
&= \&\,(10(1 \text{ AND } 1))\\
&= \&\,(101)\\
&= \&\,(1(0 \text{ AND } 1))\\
&= \&\,(10)\\
&= 1 \text{ AND } 0\\
&= 0
\end{aligned}
$$

The final result is `0`.

3. **Concatenation and replication operators:** The concatenation operator (`{...,...}`) allows joining together the bits of two or more data objects. For example, assume that a =`2'b00`, b =`4'b1111`, and c =`1'b0`. Then, `{a, b, c}` is `7'b0011110`. If we want to perform repetitive concatenations of an object y a constant (N) number of times, we use the replication operator $N\{y\}$. For example, `{4{a}, b, 2{c}}` is `14'b00000000111100`.

We form *expressions* by combining operators with operands that yield some *result*. The result is determined by the operands and the semantic meaning of the operator. Any legal operand (such as net or variable) without any operator is also considered an expression. Moreover, Verilog defines operator precedence rules similar to the C programming language. We can also use parentheses for sub-expressions to specify the desired precedence.

8.2.6 Control Statements

We can group a bunch of Verilog statements between the keywords `begin` and `end`. This group of statements forms a *block*. Optionally, we can give a name to a block, as shown below.

```
begin : myblock
        statement-1
        statement-2
        statement-3
        ...
end
```

The statements within a `begin–end` block are executed in the order in which they are specified in the code. Thus, in the above code *statement-1* will be executed first, then *statement-2*, etc. Therefore, the block of code within `begin–end` is referred to as a *sequential block*.

Verilog supports another grouping of statements using keywords `fork` and `join`. The statements within a `fork–join` block are executed concurrently or in parallel.

```
fork : myblock
        statement-1
        statement-2
        statement-3
        ...
join
```

In the above code *statement-1*, *statement-2*, etc. will execute parallelly. Therefore, the block of code within `fork–join` is referred to as a *parallel block*. We can embed sequential and parallel blocks within each other to model complex control structure of the hardware.

We can control the sequence of execution of statements using traditional programming constructs such as `if-else`, `for`, `case`, `while`, and `repeat` in the blocks.

Example 8.9 The following piece of Verilog code illustrates branching using the `if-else` statements.

```
// if-else example, used for two-way branching
if ( select == 1'b0) begin
     x=a;
end else begin
     x=b;
end
```

The following piece of code illustrates branching using the `case` statements.

```
// case example, used for multi-way branching
case ( sel ) // sel is of two bits
     2'b00: begin z=a; w=b; end
     2'b11: begin z=b; w=a; end
     default: begin z=1'b0; w =1'b1; end // when none of the above cases match
endcase
```

Verilog supports two variants of the `case` statement to capture dont't-care conditions in the case comparison: `casez` and `casex`. The `casez` statement treats bit positions with `z` as don't-cares. Note that we can denote `z` as `?` also in the code to make don't-care bits more readable. The `casex` statement treats bit positions with both `z` or `x` as don't-cares.

Example 8.10 The following piece of Verilog code explains the usage of the `casez` statements.

```
// casez example
casez ( instruct ) // instruct is of 8 bits
     2'b0???????: begin cntrl=2'b00; end // matches when MSB is 0, other bits
are don't-cares
     2'b11??????: begin cntrl=2'b11; end
     2'b101????1: begin cntrl=2'b01; end
     default: begin cntrl=2'b10;
end endcase
```

We can use `for`, `while`, and `repeat` to create loops in a Verilog code.

Example 8.11 The following pieces of Verilog code illustrate how we can interchangeably use `for`, `while`, and `repeat` constructs to create loops.

```
// for example
for(i = 1; i < 8; i = i + 1 ) begin // three parts: initialize, repeat until the given condition
is false, update value of loop control
        state[i] = 1'b1;
        y[i] = c[i];
end
```

```
// while example, i is initially 1
while (i < 8) begin // executes until the given condition becomes false
        y[i] = c[i];
        i = i + 1;
end
```

```
// repeat example, i is initially 1
repeat (8) begin // executes fixed number of times
        y[i] = c[i];
        i = i + 1;
end
```

8.2.7 Structured Procedures

Verilog supports *structured procedures* using four constructs: *initial block*, *always block*, *function*, and *task*.

Initial and Always Blocks

The blocks of codes associated with the keyword `initial` and `always` are known as initial block and always block, respectively. These blocks are enabled when the simulation begins (time is 0). An initial block executes only once and stops when it reaches the end of the block. However, the always block executes repeatedly throughout the simulation.

Example 8.12 The following piece of Verilog code illustrates initial block and always block.

```
module myTop(datain, dataout);
        input datain;
        output dataout;
```

```
        reg clock, counter, dataout;
        // initial block
        initial begin
             clock = 1'b0;
             counter = 1'b0;
        end
        // always block with delay control
        always begin
             #10 clock = ~clock;
        end
endmodule
```

The registers *clock* and *counter* get initialized in the initial block. Thus, *clock* and *counter* have values `1'b0` after the initial block is executed. After the initial block is executed once, it never executes. In the always block, the register *clock* is assigned the complement of its value after 10 time units. Note that `#10` denotes a delay of 10 time units in assigning the value to the register. Since the always block executes throughout the simulation, the above code implies that *clock* toggles *every* 10 time units.

Note that always block is useful only with *timing control*. An always block with no timing control will, creates an infinte loop (deadlock) in simulation. For example, the following code will create infinite loop in simulation and is of no use:

```
// always block with infinite loop in simulation
always clock = ~clock;
```

We can specify timing control in Verilog in three ways: *delay control*, *event control*, and *wait statement*.

1. **Delay control:** The delay control is provided by the constructs (`#<d> rega = regb;`). It delays the execution of the assignment by the amount of simulation time specified by `<d>`, as demonstrated in the previous example.
2. **Event control:** We can synchronize a procedural statement by providing event control. The controlling event can be a value change on a net or a variable.

Example 8.13 Consider the following piece of Verilog code.

```
// always block with event control
always @(en) begin
     rega = regb; // triggered when en changes
end
```

The assignment gets triggered by a change in the value of the net *en*. We can use such timing controls to model the combinational circuit or latches.

We can also define event based on the *direction* of change of a net or variable. The transition toward 1 is known as *positive edge* and specified with the keyword `posedge`. For four valued logic in Verilog, the following transitions are considered as a positive edge:

(a) from `0/x/z` to `1`

(b) from `0` to `x/z`.

Similarly, the following transitions are considered as a *negative edge*:

(a) from `1/x/z` to `0`

(b) from `1` to `x/z`.

We specify negative edge with the keyword `negedge`.

Example 8.14 Consider the following piece of Verilog code.

```
// always block with edge event control
always @(posedge clock)
begin // triggered at positive edge of clock
        dout = din;
end
```

The register *dout* is assigned at the positive edge of the *clock*. We use edge-triggered event to model flip-flops.

We can specify the logical OR of multiple events such that any one of the events can trigger the associated block or statement. The keyword `or`/comma (`,`) is used to specify logical OR of event. The `or`-separated list of triggering events is referred to as the *sensitivity list*.

Example 8.15 Consider the following piece of Verilog code.

```
// always block with multiple event control
always @(posedge clock or negedge reset)
begin
        dout = din;
end
```

The assignment gets triggered when either the positive edge of *clock* or the negative edge of *reset* occurs.

We often need to put all the signals being read in an always block in the sensitivity list. For example, in modeling a combinational circuit using an always block, we should put all the input signals in the sensitivity list. However, for complicated combinational blocks, the sensitivity list can contain many signals. Therefore, manually adding all the signals in the sensitivity list can be error-prone. Verilog provides a shorthand notation `@*` to include all nets and variables being read in the sensitivity list. It is known as *implicit event expression*.

Example 8.16 Consider the following piece of Verilog code.

```
// always block with implicit event expression
// avoids cumbersome always@(cin1 or cin2 or din1 or din2 or f1 or f2 or g1 or g2)
always @* begin
      dout1 = (cin1 + din1 - f1*g1 );
      dout2 = (cin2 - din2 - f2*g2 );
end
```

The always block gets triggered when a change occurs to any of the signal *cin1*, *cin2*, *din1*, *din2*, *f1*, *f2*, *g1*, or *g2*.

3. **Wait statement:** We use the Verilog keyword `wait` with some conditional expression *expr* to delay a procedural statement until *expr* becomes true. Note that the wait statement is *level-sensitive*. It implies that the corresponding procedural statement keeps being executed while *expr* remains true. It is in contrast to the event-driven control provided by @. We illustrate this in the following example.

Example 8.17 Consider the following piece of Verilog code.

```
// always block with wait control
always begin
      wait (enable) #10 clk = ~clk;
end
```

The always block is continuously executed. When *enable* is true, the value of *clk* is toggled after 10 time units, else the execution remains at the wait statement. Note that the wait statement is level-sensitive. It implies that if *enable* remains true, the value of *clk* keeps toggling every 10 time units.

Next, consider the following piece of Verilog code.

```
// always block with wait control
always @(enable)
begin
      #10 clk = ~clk;
end
```

The execution enters the `begin–end` block only when there is a change in value of *enable*. Whenever *enable* changes, *clk* toggles once. After that, *clk* will remain constant until *enable* changes again.

Usage of initial and always blocks: We typically use an initial block in test benches to initialize our design for simulation. On the other hand, we use an always block as an essential element for RTL and behavioral modeling. We can define any number of initial and always blocks in a module. All these blocks get executed at the beginning of the simulation time. Note that there is no predefined order in starting the execution of these initial and always blocks. Thus, an initial block need not get executed before an always block.

The initial and always blocks provide a mechanism to model the concurrency of the hardware. We can think of these blocks as different hardware working in parallel and whose order of starting the execution is not within our control (a simulation tool is free to choose any arbitrary order).

Functions and Tasks

We can use functions and tasks to reuse repeated codes and improve the readability of a hardware description. Verilog supports functions and tasks using keywords `function`, `endfunction` and `task`, `endtask`, respectively. We use functions when the response to the given inputs is a single value. Therefore, we can use a function as an operand in the expression. The value returned by the function is taken as the value of the operand. We use a task to calculate multiple outputs and support multiple goals. The major differences between tasks and functions are summarized in Table 8.2.

Table 8.2 Differences in functions and tasks

Attribute	Functions	Tasks
Arguments	Can have one or more `input` type and no `output` or `inout` type	Can have zero or multiple `input`, `output`, and `inout` types
Return value	One return value	No return value
Timing control	Not allowed	Allowed
Simulation	Zero delay in simulation	Delay as defined in the code
Calls	Can call other functions but not tasks	Can call other functions and tasks

We illustrate the usage of a function and a task in the following example.

Example 8.18 Consider the following piece of Verilog code.

```
module top(a, b, c, d, out1, out2) ; // illustrates function
    input a, b, c, d;
    output out1, out2;
    // function definition
    function myfunc;
```

```
        input x, y, z;
        begin
            myfunc = x-y+z;
        end
    endfunction;
    // function calls
    assign out1 = myfunc (a, b, c) ; // returned value assigned to out1
    assign out2 = myfunc (b, c, d) ; // returned value assigned to out2
endmodule
```

It illustrates the definition and calling of functions. Note that a function definition implicitly declares a variable with the same name as the function name (*myfunc* in the above case). This variable is assigned the return value of the function.

The following piece of Verilog code illustrates the usage of a task named *mytask*.

```
module top(a, b, c, s1, c1, s2, c2) ; // illustrates function
        input a, b, c;
        output s1, c1, s2, c2;
        // task definition
        task mytask;
            input x, y;
            output sum, carry;
            begin
                sum = x ^y;
                carry = x & y;
            end
        endtask;
        // task invokations
        mytask (a, b, s1, c1) ;
        mytask (b, c, s2, c2) ;
    endmodule
```

8.2.8 Assignments

We provide values to the nets and variables using *assignments*. There are two basic forms of assignments: *continuous assignment* and *procedural assignment*. The continuous assignments provide values to the nets, while procedural assignments provide values to the variables.

Continuous Assignment

Verilog supports continuous assignment using the keyword `assign`. It drives values to nets continuously without any sensitivity list. Whenever the value on the assignment's right-hand side (RHS) changes, the left-hand side (LHS) is updated. Thus, we can use continuous assignments to model combinational logic functions as expressions rather than the interconnection of logic gates.

Example 8.19 Consider the following piece of Verilog code.

```
module mymux(a, b, s, out);
    input a, b, s;
    output out;
    assign out = (s) ? a : b; // continuous assignment
endmodule
```

It models a 2-to-1 multiplexer using the continuous assignment.

Procedural Assignments

We use procedural assignments in the structured procedures (initial block, always block, functions, and tasks). They assign values to variables (such as `reg` or `integer` types) and they do not change their values until the next procedural assignment is made. Note that nets (`wire` type signals) cannot store value. Therefore, we cannot use simple nets in the procedural assignments.

Verilog supports two types of procedural assignments: *blocking assignment* and *nonblocking assignment*.

1. **Blocking assignment:** A blocking assignment blocks the execution of the next statement until it is executed when it appears in a sequential block (`begin–end` block). In other words, a blocking assignment is executed before executing statements following it in a sequential block. A blocking assignment uses the symbol = between the RHS and the LHS. Note that in parallel blocks (`fork–join` block) a blocking assignment does not block the next statement.
2. **Nonblocking assignment:** A nonblocking assignment allows assignment *scheduling* without blocking the execution of the following statement when it appears in a sequential block. We use nonblocking assignments when these assignments can be made within the same time step without dependence on each other. Thus, nonblocking assignment provides a mechanism to model *concurrent* or *parallel* data transfer within a block. A nonblocking assignment uses the symbol <= between the RHS and the LHS. Note that the symbol is the same as the *less than or equal to* operator. Tools recognize nonblocking assignments using the context of the symbol <= in the given code.

We illustrate the differences in the blocking and the nonblocking assignments in the following example.

Example 8.20 Consider the following piece of Verilog code.

```
module top();
    reg a, b, c, p, q, r;
    // blocking assignments
    initial begin
        a = #10 1'b1; //at time = 10
        b = #30 1'b1; //at time = 40
```

```
            c = #20 1'b1;  //at time = 60
    end
    // nonblocking assignments
    initial begin
            p <= #10 1'b1;  //at time = 10
            q <= #30 1'b1;  //at time = 30
            r <= #20 1'b1;  //at time = 20
    end
endmodule
```

First, note that we have used delay after the assignment operator (= and <=). This delay construct is known as *intra-assignment delay*. In this case, the RHS is evaluated before the delay, i.e., the RHS evaluation is not impacted by the delay. However, the assignment of the new value to the LHS is delayed by the specified amount. In contrast, if we specify delay at the beginning of a statement (e.g., #10 a=b;), the entire statement gets delayed by the specified amount.

In the above code, there are two initial blocks. Both of them start executing at the *simulation time* of zero.

In the first initial block, the assignments are blocking. Therefore, the first assignment completes at the simulation time of 10, the next one at 10+30=40, and the last at 40+20=60.

In the second initial block, the assignments are nonblocking. Thus, when the first nonblocking assignment is encountered, it schedules updating *p* to `1'b1` at the simulation time of 10 and moves on to the next statement. Subsequently, *q* and *r* are scheduled to be updated to `1'b1` at the simulation time of 30 and 20, respectively. Note that the evaluation of the RHS and the scheduling of the new value are done at the simulation time of zero for all three nonblocking assignments. However, the registers *p*, *q*, and *r* get updated to `1'b1` at the simulation time of 10, 30, and 20, respectively.

The above example suggests that the nonblocking assignments are processed in two steps by the simulators: *scheduling* and *updating*. Initially, the RHS is evaluated, and the LHS is *scheduled* to be updated with the new value. Note that the LHS is not yet updated. It is just scheduled to be updated by inserting that assignment in a queue known as the *nonblocking assignment update queue*. Finally, the nonblocking assignment update queue is processed in a first-in–first-out order, and the variables on the LHS are updated. The values of the RHS already determined in the scheduling steps are used in this step, i.e., the RHS is not re-evaluated.[3]

We illustrate the two-step processing of the nonblocking assignment in the following example.

Example 8.21 Consider the following piece of Verilog code.

```
module top ( );
    reg a, b, p, q, clock;
    initial begin
        a = 1'b1;
        b = 1'b0;
```

[3] The mechanism of scheduling and update will be discussed in Chapter 9 ("Simulation-based verification").

```
        p = 1'b1;
        q = 1'b0;
        clock = 1'b0;
    end
      always clock  =  #10 ~clock;  // clock with period 20 time units
      // nonblocking assignments
      always @(posedge clock) begin
            a <= b;  //a changes to 1'b0 at time=10, 1'b1 at time=30, and so on
            b <= a;  //b changes to 1'b1 at time=10, 1'b0 at time=30, and so on
      end
      // blocking assignments
      always @(posedge clock) begin
            p = q;  //p changes to 1'b0 at time=10 and remains there
            q = p;  //q changes to 1'b0 at time=10 and remains there
      end
endmodule
```

In the initial block, *a* gets assigned 1, while *b* gets assigned 0. Similarly, *p* gets assigned 1, while *q* gets assigned 0. Subsequently, the registers *a* and *b* are assigned in the nonblocking statements, while *p* and *q* are assigned in the blocking statements. Let us examine each of them.

Nonblocking assignments: The register *a* is scheduled to be updated to that of current value of *b* (i.e., `1'b0`) at simulation time of 10. Similarly, *b* is scheduled to be updated to that of the current value of *a* (i.e., `1'b1`) at simulation time 10. In the next step, the values of these registers are updated one after another (it does not matter, what is the sequence of this update, since the RHS value was already decided in the previous step). Thus, at the end of simulation time 10, the values are: *a*=0 and *b*=1. Similarly, on the occurrence of every positive edge of the *clock*, *a* and *b* will swap their contents.

Blocking assignments: The first blocking assignment updates the value of *p* to that of *q* (i.e., 0) at simulation time 10. The next blocking assignment updates the value of *q* to the current value of *p* (i.e., 0). Thus, both *p* and *q* attain a value of 0 and they remain at the same value in the subsequent clock edges.

8.2.9 System Tasks and Compiler Directives

Verilog language supports some in-built *system tasks* and *system functions* that help in debugging and verification hardware models. Their names start with `$`. Some system tasks and functions are shown in Table 8.3.

Table 8.3 System tasks and functions

Category	Example System Tasks/Functions
Dispaying	`$display, $monitor, $strobe, $write`
File manipulation	`$fopen, $fclose, $fgetc, $fwrite,` `$fdisplay, $fmonitor, $fstrobe, $fwrite`

Simulation time and control	$time, $realtime, $stime, $stop, $finish
Conversions	$rtoi, $itor, $bitstoreal, $realtobits $signed, $unsigned

Example 8.22 The following piece of Verilog code illustrates the behavior of some of the system tasks and functions.

```
reg [7:0] value;
reg active;
reg stableVal;
integer fileHandle;
// prints: Hello
$display("Hello");
// Assume simulation time is 104 and value=10101010
// prints: At time 104 value is 10101010
$display("At time %d value is %b", $time, value);
// monitor displays whenever its argument changes (other than $time, $realtime, $stime)
$monitor("At time %d stable value is %b", $time, stableVal);
// strobe displays values just before simulation time is advanced
// and after all other events at that time have been processed
// discussed in detail in the next chapter
$strobe("At time %d the active is %b", $time, active);
// opens a file
fileHandle = $fopen("result.dat");
// This goes into file result.dat
$fdisplay(fileHandle, "Hello again");
// closes the file
$fclose(fileHandle);
// simulation suspends, useful in debugging and examining variables during simulation
#100 $stop;
// simulator exits and passes control back to the host operating system
#1000 $finish;
```

Note that the above list of system tasks is not exhaustive. Readers should refer to [1] to know more about Verilog language supported system tasks and their options. Additionally, EDA tools (such as simulators) can support some system tasks that are not part of Verilog language standard. For example, `$reset`, `$save`, `$input`, `$key`, `$log`, and `$scope` [1]. Readers should refer to them in the manual of resepctive EDA tools.

Verilog also supports *compiler directives* that control the compilation by an EDA tool. They start with the symbol ` (accent grave). We can define compiler directives anywhere in a Verilog code, but we typically define them outside a module. The scope of compiler directives extends *across all files* from the processing point to the point where it is overridden.

Example 8.23 The following piece of Verilog code illustrates the usage of a few important compiler directives.

```
// inserts the entire contents of a file into another for compilation
`include "mydefinitions.v"
// entire content of mydefinitions.v appear here
// creates a macro for text substitution
`define DATA_BUS_SIZE 64
reg [`DATA_BUS_SIZE:1] mydatabus; // declared in module
// specifies the time unit and time precision of the modules
`timescale 1ns/1ps
// sets all compiler directives to their default values
`resetall
```

8.3 FEATURES OF SYSTEMVERILOG

With the increase in the design complexity, functional verification of a hardware model becomes complicated. To ease verification, we need additional features in HDLs that allow us to work at a higher level of abstraction. Therefore, a superset of Verilog, known as SystemVerilog, was proposed and standardized. It offers greater flexibility than Verilog and allows robust and easy design and verification.

In this section, we describe extra features of SystemVerilog compared to Verilog [9]. Many of these features are influenced by the programming language C++. Therefore, designers who are familiar with C++ and Verilog find it easy to learn SystemVerilog. Note that, in this book, we introduce readers to the basic features of SystemVerilog. For more understanding, we suggest that readers refer to dedicated resources on SystemVerilog such as [9–11].

8.3.1 Data Types

SystemVerilog has additional data types declared with keywords such as `logic`, `bit`, `byte`, `shortint`, `int`, and `longint`. The type `logic` can take four values similar to `reg` in Verilog. The types `bit`, `byte`, `shortint`, `int`, and `longint` are two-valued (`0/1`). By using two-valued signals in a hardware model, we can simplify the problem for the simulators and obtain quicker results. SystemVerilog also supports enumerated data types, structures, unions, and type definitions using keywords `enum`, `struct`, `union`, and `typedef`, respectively.

8.3.2 Procedural Blocks

SystemVerilog supports three additional types of always blocks. The always block with keywords `always_comb`, `always_latch`, and `always_ff` can model combinational logic, latches, and flip-flops, respectively. We can eliminate inadvertent design errors arising due to the generic always blocks by enforcing language semantics on these circuit elements. EDA tools such as simulators can infer sensitivity lists from the variables read in these blocks.

8.3.3 Classes

SystemVerilog supports *classes* to encapsulate data and functions. The data are referred to as *class properties* and functions are called *methods*.

Example 8.24 The following piece of SystemVerilog code illustrates the definition of a class [9].

```
class MyClass; // a simple class definition
    // class properties
    logic [3:0] mydata;
    integer status;
    // initialization (similar to constructor in C++)
    function new ();
        mydata = 4'b0;
        status = 0;
    endfunction
    // methods
    task setData(logic [3:0] d);
        mydata = d;
    endtask
    function integer getStatus();
        getStatus = status;
    endfunction
endclass
```

The instances of a class are referred to as *objects*. We can create an object of a class by declaring a variable of that class type and then assigning it to the result of the `function new`. We can then call various methods and access class properties on that object.

Example 8.25 The following piece of SystemVerilog code illustrates the creation of an object of the class *MyClass* (continued from Ex. 8.24).

```
MyClass myobj; // declaring a variable of class MyClass type
myobj = new; // initialize variable to a newly created object
logic [3:0] data = 4'b1010;
myobj.setData(data); // call method setData on myobj
int stat = myobj.getStatus(); // call method getStatus on myobj
```

SystemVerilog supports rich set of features for *object-oriented programming* (OOP), similar to C++. It allows *data hiding* (class properties not being publicly accessible) using keywords `local` and `protected`. It enables *inheritance* (deriving properties and methods from one class to another) using keyword `extends`. Additionally, classes can be made polymorphic (same method delivering different functionality or behavior) using `virtual` functions. Moreover, classes can be *parameterized* that allow generic programming. These OOP features make SystemVerilog very powerful in writing reusable, modular, easily maintainable, and robust code.

8.3.4 Interfaces

SystemVerilog supports *interfaces* that allow treating connections at a higher level of abstraction. At the lowest level, interfaces behave as a bundle of wires that can connect blocks or other instances. However, an interface can also encapsulate complex functionality using variables, parameters, tasks, and functions. We can instantiate an interface similar to a module in a design. Thus, the interface enables modeling communications between blocks at a higher level of abstraction. It enhances code reusability and reduces the chance of making wrong connections for complicated modules. We use the keyword `interface` to define an interface [9].

Example 8.26 The following piece of SystemVerilog code illustrates the definition of an interface and its use [9].

```
interface MyInterface; // interface definition, bundle of two wires
    logic request;
    logic grant;
endinterface
module MyMemory(MyInterface bMem, input logic clk);
    logic inuse;
    always @(posedge clk)
        bMem.grant <= bMem.request & ~inuse;
    ...
endmodule
module MyCpu(MyInterface bCpu, input logic clk);
    ...
endmodule
module Top(input logic clk);
    ...
    MyInterface intfc();
    MyMemory M1(intfc, clk); // connection by position
    MyCpu P1(.bCpu(intfc), .clk(clk)); // connection by name
    ...
endmodule
```

8.3.5 Assertions, Functional Coverage, and Constrained Random Testing

An *assertion* is a mechanism to capture the design behavior. We use EDA tools, such as simulators and model checkers to ensure the validity of a given assertion. If there are violations, tools report the input stimulus for which the assertion fails. Subsequently, we can debug and take corrective measures to resolve the failure.

SystemVerilog provides rich set of features to capture the *assertions* and *assumptions* of a hardware model. Moreover, SystemVerilog provides features to analyze *functional coverage* of a design. The functional coverage analysis helps us to identify untested portions of a design and meet verification goals. Additionally, SystemVerilog allows efficient testing by providing a mechanism for *constrainted random value generation*. We will discuss these aspects of SystemVerilog in Chapter 9 ("Simulation-based verification").

8.4 GOOD CODING PRACTICES

An RTL model goes through various stages of design and verification. Finally, we obtain a chip layout that a foundry can fabricate. The figures of merit of a chip, such as speed, power, cost, area, and reliability, depend strongly on the quality of the initial RTL. Additionally, the effort or cost incurred in designing, verifying, and testing a chip depends heavily on the RTL. Therefore, we need to code RTL models carefully [12–14].

8.4.1 Motivation

We can face problems in logic synthesis, physical design, and functional and timing verification for a badly written RTL code. Additionally, it can lead to testability issues in a circuit. Moreover, badly written code can be difficult to read, debug, and reuse in another design or toolset. To avoid these problems, we must adhere to some *good coding practices* or *coding rules* while creating RTL models.

8.4.2 Rules

The coding rules are derived by experience in creating RTL models and taking them through the entire design flow [13]. In the simplest case, rules can enforce some naming conventions for the design entities, such as nets, port, and registers, that improve the code readability. Note that these rules are *stricter* than what is allowed by the Verilog language. For example, rules can treat the mismatched width of a module port and its instantiation as a violation despite being allowed by the Verilog language.

To avoid potential issues or difficulty in verification, we can disallow some coding styles. For example, some usage of Verilog constructs can lead to nondeterministic behavior in simulation or synthesis. The rules can detect and flag such RTL codes. Thus, rule checking can help avoid some problems that can creep in through the lenient semantics of the HDLs.

Rules can enforce additional restrictions on the clock and reset signals to avoid nondeterministic behavior and testability issues in a circuit. Rules can also forbid using some hardware elements, such as latches, in a design because they increase the verification effort. Similarly, rules can detect and report circuit topologies, such as loops of combinational circuit elements, that can cause problems later in the design implementation and verification.

8.4.3 Rule Checking

In practice, we use automatic tools that check the RTL code and report violations of the given rules. These tools are known as *lint tools* or *rule checkers*. Lint tools check the RTL code *statically*, i.e., they do not use test stimulus or simulation for checking. Therefore, rule checkers are typically very fast. Many commercial linting tools permit customization based on the requirements of the projects. Some rule checkers can internally perform lightweight synthesis and report problems that are hard to debug later in the flow. Therefore, it is recommended to use rule checkers on the RTL model early in a design flow.

8.5 RECENT TRENDS

HDLs have evolved over several decades by fixing semantic problems and adopting features to support new design and verification methodologies. Currently, SystemVerilog is the dominant HDL for hardware modeling and functional verification. We expect HDLs to adopt more features to support new computing paradigms and design methodologies in the near future. As design flows become more automated, the transformation of HDLs ultimately to an internal format used by EDA tools, similar to the role of an assembly language in the software compilation, cannot be ruled out [15].

REVIEW QUESTIONS

8.1 How are hardware description languages different from traditional programming languages?

8.2 What are escaped identifiers in Verilog?

8.3 Why do we need four values in modeling hardware?

8.4 What are differences between following entities in Verilog:

- **(a)** Nets and variables,
- **(b)** Arrays and vectors,
- **(c)** The `casez` and `casex` statements,
- **(d)** Tasks and functions,
- **(e)** Blocking assignment and nonblocking assignment?

8.5 What features in Verilog are used to model concurrent operation of hardware elements?

8.6 What does the symbol question mark (`?`) indicate in Verilog?

8.7 How does parameters help in obtaining multiple implementations of the same module in Verilog?

8.8 There is an initial block and an always block in a module. Which of them will be executed first during simulation? Give reasons to support your answer.

8.9 Predict the output of the following piece of Verilog code:

```
module top ( ) ;
    reg a, b;
    initial begin
```

```
            $monitor("At time %d a=%b b=%b", $time, a, b);
            a = 1'b0;
            b = 1'b0;
            #10 a = 1'b1;
            #10 b = 1'b1;
            a <= #5 1'b0;
            b <= #5 1'b0;
      end
endmodule
```

8.10 Why are signals of type `wire` cannot be assigned in the procedural blocks?

8.11 Name a few distinct features of SystemVerilog that are particularly useful in verification.

8.12 How can object-oriented programming help in hardware modeling and verification?

8.13 What are the advantages of using interfaces of SystemVerilog instead of normal wire connections in hardware modeling?

8.14 Why should we follow "good" coding practices in hardware modeling?

8.15 Predict the output of the following piece of Verilog code:

```
module top();
      reg clk, en, a, b;
      initial begin
            clk = 1'b0;
            en = 1'b0;
            a = 1'b0;
            b = 1'b0;
            $monitor("At time %d clk=%b en=%b a=%b b=%b", $time, clk,
en, a, b);
            #3 en = 1'b1;
            #5 en = 1'b0;
            $finish;
      end
      always #2 clk =~clk;
      always @(en) a = clk;
      always begin
            wait (en) #1 b =~b;
      end
endmodule
```

TOOL-BASED ACTIVITY

In this activity, use any Verilog simulation tool, such as the open-source tool ICARUS Verilog [16]. Use waveform viewer, such as GTKWave [17].

1. Simulate the Verilog code shown in Questions 8.9 and 8.15.
2. Analyze the results. Take help of the waveform viewer, if required.

REFERENCES

[1] "IEEE standard Verilog hardware description language." *IEEE Std 1364-2001* (2001), pp. 1–792.

[2] D. Thomas and P. Moorby. *The Verilog® Hardware Description Language*. Springer Science & Business Media, 2008.

[3] S. Sutherland. *Verilog–2001: A Guide to the New Features of the Verilog® Hardware Description Language* 652. Springer Science & Business Media, 2012.

[4] P. Flake, P. Moorby, S. Golson, A. Salz, and S. Davidmann. "Verilog HDL and its ancestors and descendants." *Proceedings of the ACM on Programming Languages* 4, no. HOPL (2020), pp. 1–90.

[5] S. Golson (Interviewer), G. Hendrie (Videographer), and P. Moorby (Interviewee). *Oral History of Philip Raymond "Phil" Moorby*. No. 102746653, Computer History Museum, 2013.

[6] "IEEE standard for SystemVerilog—Unified hardware design, specification, and verification language." *IEEE STD 1800-2009* (2009), pp. 1–1285.

[7] "IEEE standard for VHDL language reference manual." *IEEE Std 1076-2019* (2019), pp. 1–673.

[8] "IEEE standard for binary floating-point arithmetic." *ANSI/IEEE Std 754-1985* (1985), pp. 1–20.

[9] "IEEE standard for SystemVerilog—Unified hardware design, specification, and verification language." *IEEE Std 1800-2017 (Revision of IEEE Std 1800-2012)* (2018), pp. 1–1315.

[10] S. Sutherland, S. Davidmann, and P. Flake. *SystemVerilog for Design Second Edition: A Guide to Using SystemVerilog for Hardware Design and Modeling*. Springer Science & Business Media, 2006.

[11] C. Spear. *SystemVerilog for Verification: A Guide to Learning the Testbench Language Features*. Springer Science & Business Media, 2008.

[12] L. Bening and H. Foster. *Principles of Verifiable RTL Design*. Springer, 2001.

[13] M. Keating and P. Bricaud. *Reuse Methodology Manual for System-on-a-chip Designs: For System-on-a-chip Designs*. Springer Science & Business Media, 2002.

[14] S. Churiwala and S. Garg. *Principles of VLSI RTL Design: A Practical Guide*. Springer Science & Business Media, 2011.

[15] S. A. Edwards. "Design and verification languages." *EDA for IC System Design, Verification, and Testing* (2018), pp. 373–399, CRC Press.

[16] S. Williams. "ICARUS Verilog." http://iverilog.icarus.com/. Last accessed on January 6, 2023.

[17] T. Bybell. "Welcome to GTKWave." http://gtkwave.sourceforge.net/. Last accessed on January 6, 2023.

9 Simulation-based Verification

...I am not bound to please thee with my answers...

—William Shakespeare, *The Merchant of Venice*, Act 4, Scene 1, 1596

We model the behavior of a digital circuit in hardware description languages (HDLs), such as Verilog and VHDL, as described in Chapter 8 ("Modeling hardware using Verilog"). Subsequently, we need to ensure that the *logical functionality* of the HDL model matches the given specification. This task is accomplished by *functional verification*. The goal of functional verification is to ensure that the *logic implementation* of the digital circuit is correct, i.e., it produces the right output bit sequence for a given input bit sequence. We can initially ignore the delay of combinational circuit elements during functional verification or make some simplistic assumptions about them. As more details are added to a design, we perform verification tasks related to delay, power dissipation, and correctness of layout later in the design flow. By segregating functional verification from other types of verification tasks, we are able to simplify the overall design verification process.

We can broadly categorize functional verification techniques into two classes: *simulation-based techniques* and *formal methods*. Simulating a hardware model is analogous to running a program written in a traditional programming language and ensuring the correctness of the program by observing its output. Therefore, simulation-based techniques are easy to use and can quickly discover bugs in a hardware model, especially in the early stages of design implementation. In practice, simulation-based techniques provide a foundation for functional verification. Subsequently, we augment and fill gaps in the simulation-based verification with more rigorous formal methods.

In this chapter, we will explain the simulation-based techniques for functional verification. In Chapter 11 ("Formal verification"), we will discuss techniques based on formal methods.

9.1 BASICS OF SIMULATION

A typical simulation framework is shown in Figure 9.1. It involves applying *stimulus* to the *design under verification* (DUV) and observing its *response* for correctness. We commonly refer to the verification environment created for applying stimuli and observing responses as a *testbench* [1]. A testbench interacts with a tool, known as *simulator*, to produce verification results and debugging information.

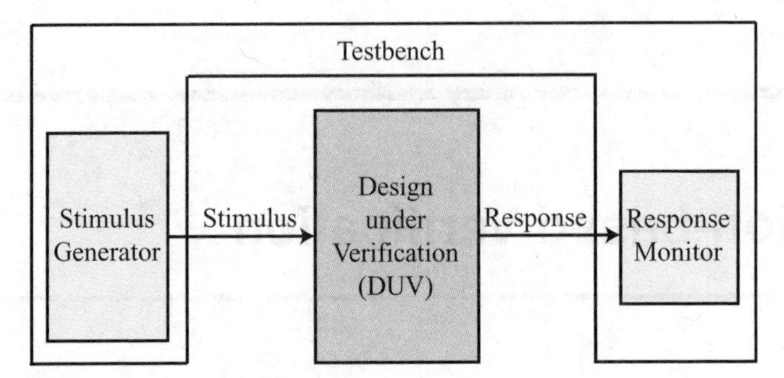

Figure 9.1 A typical simulation framework

A testbench can contain an internal stimulus generator, as shown in Figure 9.1. Alternatively, it can import a set of stimuli generated from an external source and apply them to a DUV. Similarly, a testbench can have an internal mechanism to observe responses and compare them with the *expected responses* to detect failures. Alternatively, a testbench can simply report the responses obtained from the DUV. Subsequently, we can compare these responses with the expected responses separately. The architecture of the simulation framework depends on the design type, the form of *golden model* (a reference model that can be considered correct), and the simulator.

The robustness of verification depends strongly on the testbench and the simulation framework. Moreover, the time and effort required in developing a testbench, performing simulations, and detecting bugs are comparable to the design implementation time and effort. Therefore, we should plan and build a verification framework concurrently with the design implementation.

In the following sections, we will explain the various components of the simulation framework in detail.

9.2 SIMULATOR

A simulator computes response of a circuit for a given input stimulus. The circuit is represented using an *abstract model* such as algorithmic description, register transfer level (RTL) model, gate-level netlist, or transistor-level model. A simulator attempts to mimic the behavior of a real-world circuit by running algorithms on a general-purpose computer. However, a real-world circuit produces a response using spontaneous physical phenomena (such as current flowing in a semiconductor) and millions of components (such as transistors) operating parallelly. Therefore, we cannot expect a simulator to be as fast as the real-world circuit that it simulates. Nevertheless, we can increase the speed of simulation by making the following observations.

1. **High-level abstraction of response:** We need to compute only an approximate response rather than the exact response of a circuit. For example, logic-level Verilog simulators need to produce only four possible responses `0/1/x/z` for a signal, despite a real mimicked circuit producing a continuous voltage level. Thus, simulators can exploit the model abstraction to reduce its computational complexity.

   In practice, simulators working at a higher level of abstraction are faster. For example, we often describe data transfer between buses (group of signals) or vectors in an RTL code. Hence, RTL simulation can work with a group of signals and attain greater speed than the gate-level simulation that operates on individual bits.

2. **Discretization of time:** Simulators need not produce the response for each time instant, in contrast to a real circuit that produces response continuously. Thus, simulators can compute the behavior of the simulated entity only at *discrete* instants of time. At other instants, nothing interesting happens to the system, and the simulators can avoid those computations. A simulator can choose the discrete-time instants for computing the response in two ways: *event-driven* and *cycle-based*. We describe them in the following paragraphs.

9.2.1 Event-driven Simulation

When an input changes, we consider it as an *event*. An event at a logic gate can change its output. This output change can further trigger new events for other logic gates. Thus, simulation involves computing a chain of events and their responses until no further events are available. This approach of simulation is known as *event-driven simulation*.

An event-driven simulation requires computation only at discrete-time instants corresponding to the events. However, it needs to consider the drivers of all signals in a circuit for evaluating and updating events. Furthermore, it needs to *schedule* and *process* events in a well-defined order governed by the HDL semantics. As a result, the computational complexity of the event-driven simulation is high. However, event-driven simulation can account for delays of combinational circuit elements and glitches. It makes event-driven simulation more accurate. We will discuss the mechanism of event-driven simulation in more detail later in this chapter.

9.2.2 Cycle-based Simulation

For a synchronous circuit, we can simplify the simulation problem by performing computation only at the *clock edges*. We assume that a simulator needs to compute only the evolution of the *states* in a circuit. We consider the output of the flip-flops and the primary output ports of a circuit as the *state elements*. A state is represented by a combination of values at the state elements. A cycle-based simulator simply evaluates the sequence of *next states* at successive *clock edges* as a function of the *present state* and the *present input*. Thus, it can avoid computing responses of the internal combinational logic gates. Consequently, it cannot model the delay of combinational circuit elements and glitches in the signals within a clock cycle. Nevertheless, it can run faster by avoiding the scheduling overheads of traditional event-driven simulation.[1]

Note that we can use a cycle-based simulator only in a synchronous circuit. A cycle-based simulator assumes that the data launched by a flip-flop propagates through the combinational circuit elements and reaches adjacent flip-flops within one clock cycle. This assumption must hold for a synchronous circuit. We ensure that the synchronous circuit maintains this behavior using *static timing analysis* (STA) tools.[2]

It is worth pointing out that if only a few state elements change their values in each clock cycle, the cycle-based simulation makes unnecessary next state computations and is inefficient. In practice, industry-standard simulators employ a combination of event-driven and cycle-based simulations. An event-driven simulation is performed for accuracy, while a cycle-based approach is used wherever possible to gain speed.

The following example illustrates the differences between an event-driven simulation and a cycle-based simulation. For all examples in this chapter, we have not mentioned time units for brevity. You can assume them to be nanoseconds or any other appropriate time unit.

---

[1] Traditional event-driven simulation can also be made faster by clock suppression and other techniques [2–4].

[2] We will discuss STA in detail in Chapter 14 ("Static timing analysis").

Example 9.1 Consider the circuit shown in Figure 9.2. Assume that all combinational gates have a delay of one time unit.

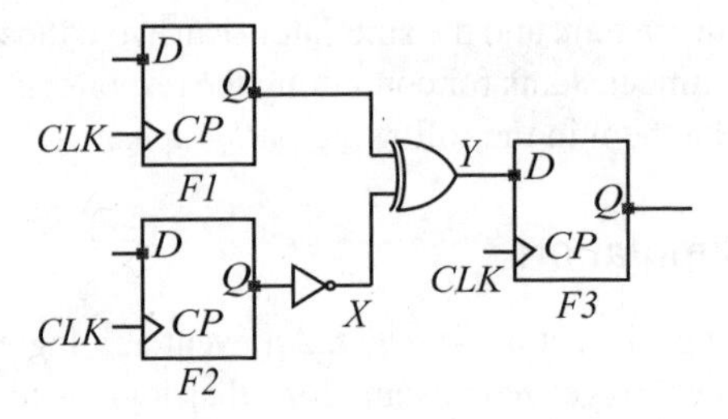

Figure 9.2 A given circuit for simulation

Assume that, at $t = 0$, *F1/D*=*F1/Q*=0, and *F2/D*=*F2/Q*=0. Hence, initially *X*=1, *F3/D*=*Y*=1. Assume that initially *F3/Q*=1 and the clock period is 10. Further, assume that *F1/D* and *F2/D* make a $0 \rightarrow 1$ transition at *t*=4, as shown in Figure 9.3.

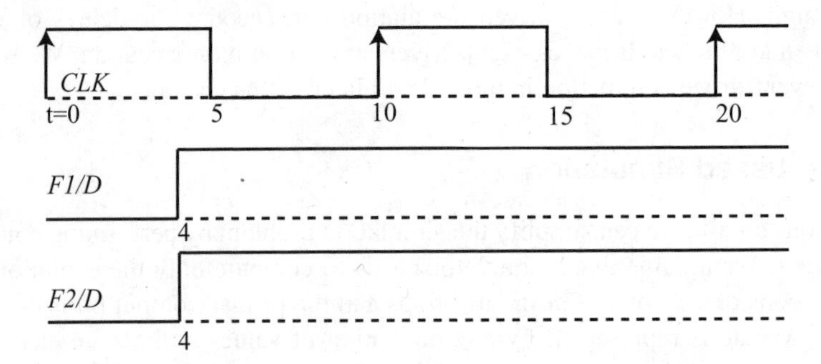

Figure 9.3 Given stimuli

Let us first examine the behavior of the circuit when it is computed using an event-driven simulator. When the next clock-edge arrives at *t*=10, *F1/Q* and *F2/Q* change to 1, as shown in Figure 9.4. Due to the change in *F1/Q*, the net *F3/D*=*Y* evaluates to 0 at *t*=10 ($Y = F1/Q \oplus X = 1 \oplus 1 = 0$). Due to the delay of 1 unit for the XOR gate, *F3/D*(=*Y*) changes to 0 at *t*=11. The net *X* changes its value to 0 at *t*=11 due to delay of 1 unit for the inverter. Hence, the value of the net *F3/D*(=*Y*) goes back to 1 at *t*=12. Thus, a glitch is observed at *F3/D*(=*Y*). However, by the time the next clock-edge arrives at *t*=20 the glitch disappears and the value at *F3/Q* remains unchanged.

Next, consider the cycle-based simulator. It computes values of the state elements *F1/Q*, *F2/Q*, and *F3/Q* at clock edges. At *t*=10, the value of *F1/D*=1, *F2/D*=1, and $F3/D = F1/Q \oplus (F2/Q)' = 0 \oplus 1 = 1$. Thus, at *t*=10, *F1/Q* and *F2/Q* change to 1 and *F3/Q* remains unchanged. At *t*=20, the values at D-pin remains unchanged (*F1/D*=1, *F2/D*=1, and $F3/D = F1/Q \oplus (F2/Q)' = 1 \oplus 0 = 1$). Thus, no change happens at *t*=20.

Remarks: Both simulators compute the same value at the state elements. However, the cycle-based simulator performs less computation. As a result, it cannot account for the delay in the combinational circuit elements and cannot determine glitches in the signals within a clock cycle.

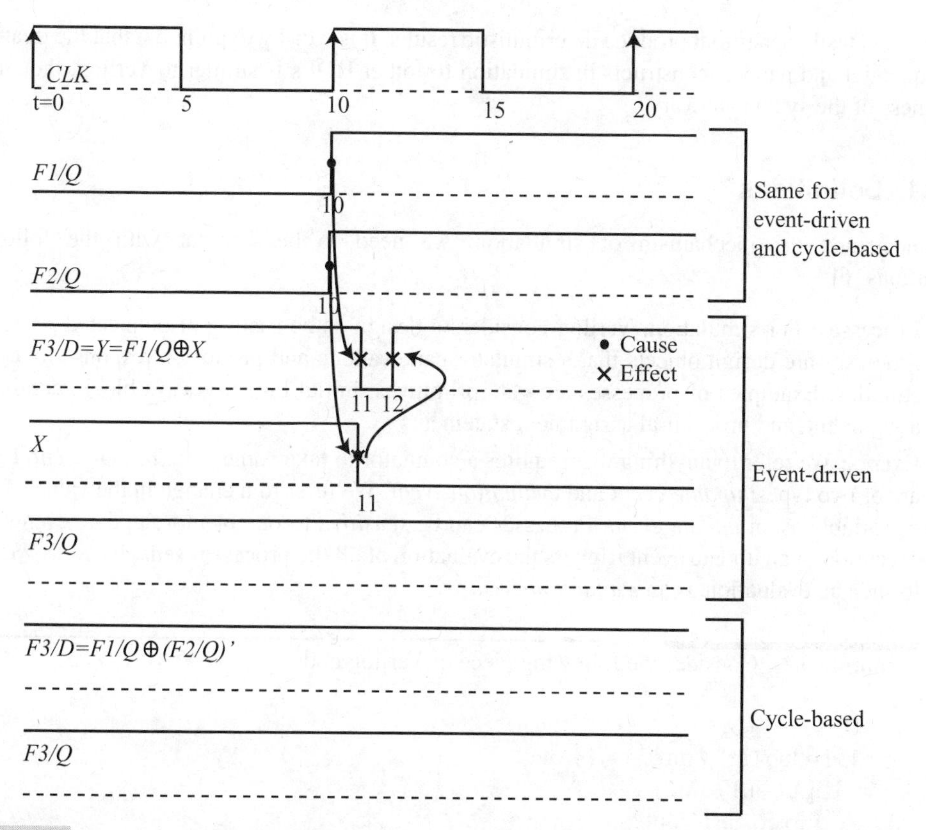

Figure 9.4 Comparison of response obtained using event-driven and cycle-based simulations

Another approach to make simulations more efficient is to use *compiled code* simulation [5]. A compiled code simulation runs in two phases. In the first phase, the event routines are compiled to the machine code (directly executable code) of the host machine (the machine on which the simulator runs). For example, a simulator can produce an intermediate C code for a given design and then compile it to an optimized machine code using a C compiler on the host machine. In the second phase, the machine code is executed on the host machine taking stimuli from the testbench. We can use the machine code in simulation because a design remains unchanged for multiple simulations. Though the compilation step consumes some extra time, it needs to be carried out only once, and the overall simulation time decreases considerably using this approach.

9.3 MECHANISM OF SIMULATION IN VERILOG

In this section, we describe how the simulation works and how various constructs of Verilog are treated in simulation [6]. Since the semantics of Verilog were defined with a focus on simulation, other usages of Verilog, such as synthesis, attempt to mimic the simulation behavior. Therefore, understanding the mechanism of Verilog simulation is essential for developing good hardware models and testbenches that produce deterministic results. It is worthy to point out that the treatment of sequential and parallel constructs in simulation for other HDLs is similar to Verilog, though the strictness of the syntax can vary.

9.3.1 Definitions

To understand the mechanism of simulation, we need to be familiar with the following definitions [6]:

1. **Processes:** For simulation, Verilog considers a design as consisting of connected *processes*. Processes are design objects that a simulator can evaluate and produce a response to a given stimulus. Examples of processes are gate primitives, initial block, always block, continuous assignment, and procedural assignment statements.
2. **Events:** We refer to anything that requires a simulator to take some action as an *event*. Events are of two types: *update event* and *evaluation event*. We refer to a change in the value of a net or variable as an *update event*. Processes can be *sensitive* to one or more update events. The execution of an update event triggers the evaluation of all the processes sensitive to it. We refer to such an evaluation as an *evaluation event*.

Example 9.2 Consider the following piece of Verilog code.

```
module top(in1, in2, out1, out2);
    input in1, in2;
    output out1, out2;
    reg out1, tt;
    and A1(out2, tt, in1); // gate primitive
    assign out1 = tt; // continuous assignment
    always @(in1 or in2) // always block
    begin
        tt = in1 & in2;
    end
endmodule
```

The primitive gate instance, continuous assignment statement, and the always block are the processes that a Verilog simulator will consider. The gate instance *A1* is sensitive to nets in its input, the continuous assignment statement is sensitive to all the signals on the right-hand side (RHS), and the sensitivity list is explicitly mentioned for the always block.

Consider that a $0 \rightarrow 1$ transition occurs at *in1*. It changes the value of net *in1*. Hence, an update event is created. The gate instance *A1* and the always block in the module are sensitive to *in1*. Hence, evaluation events get triggered at these processes. Note that a simulator can evaluate these processes in any order, despite the gate instance appearing before the always block in the Verilog code.

Similarly, if the signal *tt* changes its value, an update event is generated. It leads to an evaluation event in the gate instance *A1* and the continuous assignment statement in the module.

3. **Simulation time:** The time value maintained by a simulator to model the actual time in the simulated circuit is referred to as the *simulation time*. Note that the *simulation time* is different from the *time taken by the simulator to simulate*. The simulation time depends on the testbench and the given design. For a well-coded design and testbench, different simulators will yield the same simulation time for the occurrences of various events. However, the time taken by a simulator depends on the computer speed on which we run the simulation and the simulator's efficiency. Therefore, given a design and a testbench, the time taken by different simulators can vary.
4. **Event queue and scheduling:** Events have a unique simulation time associated with them. Simulators maintain a *queue* of events *ordered* by the simulation time to process events in the correct order. We refer to this queue as an *event queue*. When an event gets generated during simulation, it is put or scheduled on the event queue. We define a *time slot* for each simulation time on the event queue. We associate all the scheduled events for that simulation time with the corresponding time slot. As the simulation proceeds, all the events in the current time slot are executed. After executing an event, we remove it from the time slot. Once all events of a time slot are processed, the simulation proceeds to the next *nonempty* time slot. This procedure ensures that simulation never goes back in time.

   Simulators can choose any dynamic data structure, such as priority queues or circular linked lists, to implement event queues. One possible implementation of the event queue is shown in Figure 9.5. We refer to it as the *timing wheel*. It consists of an array of *M* slots that are indexed by the simulation time. The list of events corresponding to a simulation time is maintained as a linked list. If the simulation time is *T*, the corresponding index on the timing wheel is *T%M* (modulo division). Thus, we can find a slot for a given simulation time *T* in $O(1)$ time complexity.

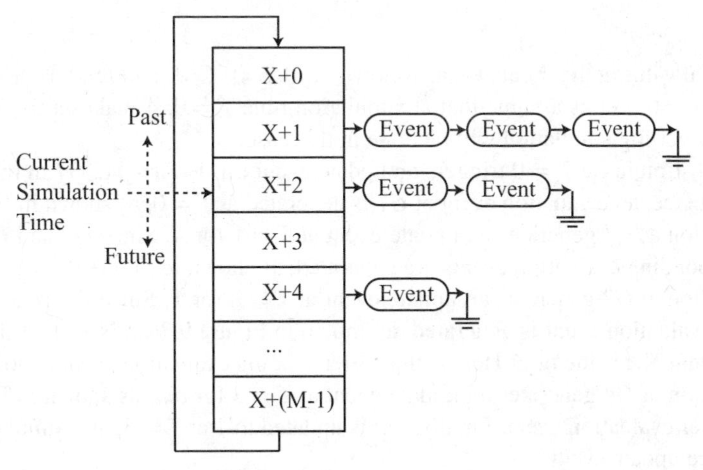

Figure 9.5 An *M*-sized timing wheel

The above implementation of the timing wheel works well in practice because most events are localized around the current simulation time. Additionally, after processing all events in a time slot, the time slot becomes empty, and we can reuse it. The simulation time moves forward, as shown by the arrow. When the end of the wheel is reached, it gets wrapped around. If a simulation time cannot fit in the timing wheel, we store it in another dynamic data structure that has no predefined size limit. As the simulation time advances and slots become available, we move the time slot from this dynamic data structure to the timing wheel for quicker processing.

9.3.2 Processing of Events

When a stimulus is applied to a design, it changes the value of some net or variable. As a result, an update event gets generated. An update event can trigger evaluation events. An evaluation event can change the value of another net or variable. Consequently, update events are generated for the affected net or variable, further triggering evaluation events. Thus, the sequence of update → evaluation → update → evaluation → ... goes on till the end of simulation time or till no more update events are generated. This sequence of operations is the basic mechanism of Verilog simulation. We illustrate this in the following example.

Example 9.3 Consider the circuit shown in Figure 9.6. Assume that the delay of each gate is one time unit.

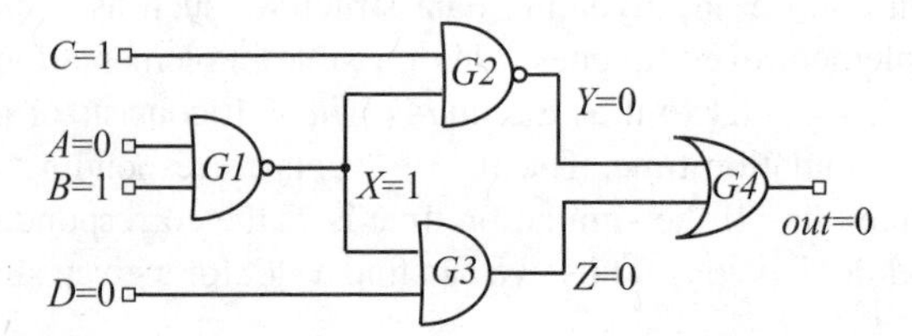

Figure 9.6 Given circuit for simulation

Let the initial value at the inputs are as follows: *A*=0, *B*=1, *C*=1, and *D*=0. Hence, initially *X*=1, *Y*=0, *Z*=0, and *out*=0. Let us assume that at simulation time $T = 0$, *A* makes a 0→1 transition. Let us examine how Verilog simulation will be done in this case.

The applied stimulus at $T = 0$ triggers an update event for *A*. The gate *G1* in the fanout of *A* is sensitive to it. Hence, an evaluation event at *G1* is generated at $T = 0$, as shown in Figure 9.7(a).

The evaluation at *G1* generates an update event at $T = 1$ for *X*. Since *G2* and *G3* are sensitive to *X*, the corresponding evaluation events are generated, as shown in Figure 9.7(b).

The evaluation at *G2* generates an update event at $T = 2$ for *Y*. Since *G4* is sensitive to *Y*, the corresponding evaluation event is generated, as shown in Figure 9.7(c). Note that the evaluation at *G3* does not change the value of *Z*. Hence, the corresponding evaluation event is not generated.

The evaluation at *G4* generates an update event at $T = 3$ for *out*, as shown in Figure 9.7(d). It triggers no further evaluation event. Finally, *out* is updated to 1 at $T = 3$, and simulation ends since there are no more update events.

Remarks: The advancement in simulation time and processing of other types of Verilog processes is done similar to the gate-level design of this example. A simulator considers all types of processes sensitive to an update event for evaluation. For example, the following piece of Verilog code will also be simulated as the given gate-level circuit.

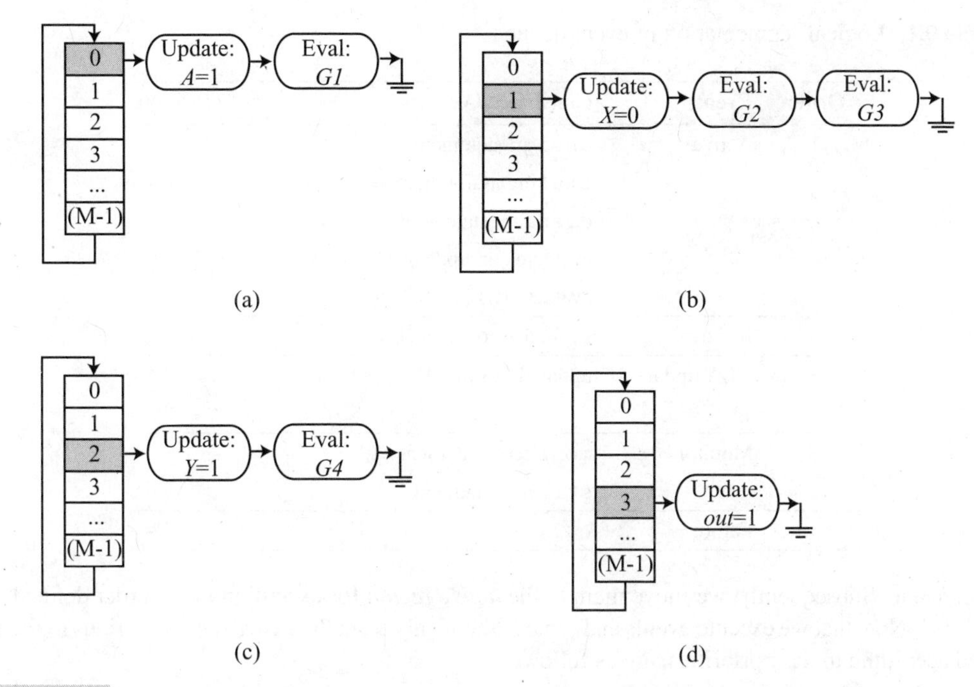

Figure 9.7 Timing wheel at (a) $T = 0$, (b) $T = 1$, (c) $T = 2$, and (d) $T = 3$

```
module top(A, B, C, D, out);
    input A, B, C, D;
    output out;
    assign #1 X = !(A & B);
    assign #1 Y = !(X & C);
    assign #1 Z = X & D;
    assign #1 out = Y | Z;
endmodule
```

However, when there are multiple events at the same time slot, the order of execution of different constructs and assignments can become ambiguous [7]. To resolve this, Verilog divides a time slot into *ordered regions* that provide predictable simulation behavior, as explained below.

9.3.3 Stratified Verilog Queue

Verilog defines the priority among various constructs scheduled for a given simulation time by logically dividing each *time slot* into five different regions, as shown in Table 9.1 [6]. We refer to it as Verilog *stratified event queue*. Initially, we add events to their respective regions in the stratified

Table 9.1 Logical segmentation of event queue

Order	Events	Constructs/Actions
1	Active	`assign` statement
		Blocking assignment (=)
		`display` statement
		evaluate and update IOs for instance/primitives
		evaluate RHS of NBA (=>)
2	Inactive	explicit zero delay (`#0`)
3	NBA update	update LHS of NBA (=>)
4	Monitor	`monitor` statement
		`strobe` statement
5	Future	

event queue. Subsequently, we move them to the *active region* for execution in the order defined in Table 9.1. Note that we execute events and remove them only from the *active region*. Various regions, listed according to their priorities, are as follows:

1. **Active region:** The *active region* initially consists of:
   (a) Continuous assignments (`assign` statements)
   (b) Blocking assignment (BA)
   (c) `display` statements
   (d) Evaluation of inputs followed by updates of their outputs for instances and primitive gates
   (e) Evaluation of the RHS of the nonblocking assignment (NBA): Recall from the previous chapter that BA separates the left-hand side (LHS) from the RHS using = symbol. In contrast, NBA uses <= symbol. Also, recall that NBA gets executed in two steps. In the first step, only the RHS of the NBA is evaluated. In the second step, its LHS is updated. The active event region initially has only the first part, i.e., the evaluation of the RHS of NBA.

   There is no priority among various events in the active region. Thus, a simulator can execute events in the active region in any arbitrary order. This is the primary source of *indeterminism* in the simulation of Verilog code, as we will explain in detail later in this section. For example, a simulator can execute two BAs of separate initial or always blocks in the active region in an arbitrary order. However, events defined within a sequential block (such as within `begin–end` block) must be executed in the order specified in the Verilog code. Thus, a simulator will execute two BAs within an initial or always block in the active region in the order *defined in the code*. This behavior of Verilog simulation is fully deterministic.
2. **Inactive region:** After executing all the events in the active region, the events in the *inactive region* are moved to the active region for execution. The inactive region consists of assignments with `#0` delay. Though assignments with `#0` delay are allowed in Verilog, we should avoid using them because its effect can be modeled using more efficient coding styles [7].
3. **NBA update region:** After executing all events in the active and inactive regions, the LHS *updates of NBAs* are moved to the active region for execution. Note that we have already computed the RHS of the NBAs in the earlier processing of the active region. Hence, we do not

re-evaluate the RHS while updating the LHS. Also, note that these updates can trigger other events, which are added to the respective regions and executed as described above.

4. **Monitor region:** The statements `monitor` and `strobe` are in the *monitor region*. After executing all events in the active, inactive, and NBA update regions, the events in the monitor region are moved to the active region for execution. Since all the updates for a signal are done before executing events of the monitor region, the commands `monitor` and `strobe` show its stable value for a given time slot. In contrast, the statement `display` is processed along with signal updates (as described in the active region). Hence, `display` can show intermediate values of a signal also.
5. **Future event region:** The events that occur at some future simulation time are in the *future event region*. After events of all other regions are executed, the simulation time is advanced, and the corresponding time slot is processed.

The following example illustrates the application of stratified event queue in simulation.

Example 9.4 Consider the following piece of Verilog code.

```
module top();
    reg a, b;
    initial begin
      a = 1'b0;
      a <= 1'b1;
      b = a;
      b <= a;
      $display("Display a=%b b=%b T=%d", a, b, $time);
      $monitor("Monitor a=%b b=%b T=%d", a, b, $time);
      #1 b = 1'b1;
    end
  endmodule
```

The code consists of an initial block. The statements within the initial block are executed sequentially. For the simulation time slot $T = 0$, the active region consists of the following events:

Event 1: blocking assignment `a = 1'b0;`
Event 2: RHS of NBA `a <= 1'b1;`
Event 3: blocking assignment `b = a;`
Event 4: RHS of NBA `b <= a;`
Event 5: `$display` statement

These events are executed in the above order and the consequences of each statement are as follows:

Consequence 1: value of *a* is updated to `1'b0`.
Consequence 2: RHS of NBA `a <= 1'b1;` is evaluated to `1'b1`. LHS is not yet updated.
Consequence 3: value of *b* is updated to `1'b0`.
Consequence 4: RHS of NBA `b <= a;` is evaluated to `1'b0` (because the current value of *a* is `1'b0`). LHS is not yet updated.
Consequence 5: `$display` statement reports $a = 0$ and $b = 0$.

There is no event in the inactive region. The following events are there in the NBA update region.

Event 1: LHS update of NBA `a <= 1'b1;`

Event 2: LHS update of NBA `b <= a;`

Next, these events are moved to the active region and executed and following are their consequences:

Consequence 1: value of *a* is updated to `1'b1`

Consequence 2: value of *b* is updated to `1'b0`. Note that the RHS was evaluated earlier and it is not re-evaluated after *a* has changed. Thus, *b* gets the previous value of *a*.

Next, the event from the monitor region (`$monitor` statement) is moved to the active region and executed. It reports $a = 1$ and $b = 0$.

Subsequently, simulator moves to the next simulation time slot ($T = 1$). It executes the blocking assignment `b = 1'b1;`. The value of *b* changes to `1'b1`, which is reported by the `$monitor` statement. Note that `monitor` and `strobe` statements are re-enabled in every successive time slots. Additionally, Verilog simulator sets up a mechanism such that when the value of any argument to `monitor` (other than `$time`) changes, it reports the new value. In summary, the following output will be produced by the simulator.

```
Display a=0 b=0 T= 0
Monitor a=1 b=0 T= 0
Monitor a=1 b=1 T= 1
```

9.3.4 Race in Simulation

We have seen that Verilog allows nonsequential statements in the active region to be executed in an arbitrary order. This flexibility of the language allows us to model inherently parallel operations of the hardware. However, it can also lead to a nondeterministic simulation behavior and mismatches in simulation and synthesis results, as explained below.

Sometimes, the simulation outcome for multiple events occurring concurrently (in the same time slot) depends on the order of execution of these events. We refer to this condition as a *simulation race*. In this case, a simulator is free to produce any one output. Therefore, we can obtain a mismatch in the results produced by different simulators. Moreover, there can be a mismatch between simulation and synthesis. It can manifest itself later in the design flow as a mismatch in the results of RTL simulation and gate-level netlist simulation. Therefore, simulation race conditions are undesirable in the hardware models and testbenches, and must be avoided.

Example 9.5 Consider the following Verilog code. Let us determine the expected simulation output.

```
module top();
    reg a;
    initial begin
        a = 1'b0;
    end
```

```
    initial begin
        a = 1'b1;
    end
    initial begin
        $display("Display a=%b", a);
    end
endmodule
```

There are three initial blocks in the module *top*. All of them execute concurrently. The two blocking assignments and the `$display` statement are added in the active region. A simulator is free to execute them in any order. Therefore, `$display` can show the value of *a* as `0`, `1`, or `x`. Thus, a simulation race exists in the above Verilog code. One of the ways to make the result deterministic is as follows:

```
module top();
    reg a;
    initial begin
        a = 1'b0; // or 1'b1
        $display("Display a=%b", a); // it will execute after the above statement
    end
endmodule
```

Remarks: The above example is trivial. Therefore, it is easy to discover and fix the race condition. However, race conditions can occur in many situations, and they often become difficult to debug in real designs. The problem exacerbates when we integrate IPs and VIPs from different vendors or sources [8].

We can avoid simulation race problems by following some coding guidelines [7]. A few guidelines are as follows [7]:

1. Use NBA for modeling sequential logic and BA for modeling combinational logic in an always block.
2. Do not mix BA and NBA in the same always block.
3. Do not write to a variable in more than one always block.

We can employ rule checkers to detect and flag violations of these rules or guidelines. It can help us avoid facing problems later in a design flow.

9.4 TESTBENCH

The creation of a testbench is a crucial task in simulation-based verification. All testbenches have some mechanism of generating and applying stimuli to a given DUV, observing their responses, and recording the result. Depending on the design and the verification requirements, we can create various types of testbenches. We demonstrate how to create a testbench using the following example.

Example 9.6 Consider a 4-bit synchronous counter shown in Figure 9.8(a). The Verilog description of the counter is shown below.

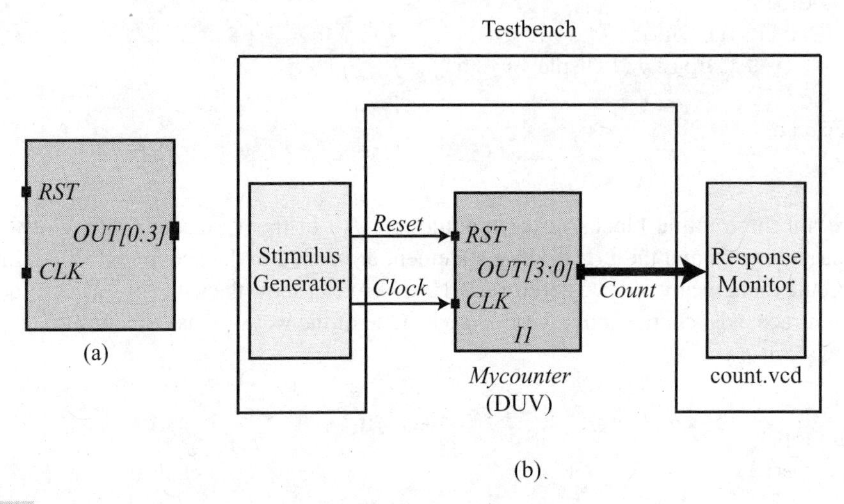

Figure 9.8 Creating testbench (a) given design and (b) testbench

```
module Mycounter (CLK, RST, OUT);
    input CLK, RST;
    output [3:0] OUT;
    reg [3:0] OUT;
    always @(posedgeCLK)
    begin
      if (RST == 1'b1)
          OUT <= 4'b0000;
      else
          OUT <= OUT+1;
    end
endmodule
```

We can create a testbench for the above design as in Figure 9.8(b). The Verilog implementation of the testbench is shown below.

```
module Testbench ();
    reg Clock, Reset;
    wire [3:0] Count;
    // instantiate the DUV and make connections
    Mycounter I1(.CLK(Clock), .RST(Reset), .OUT(Count));
```

```
    // initialize the testbench
    initial begin
      $display("Starting simulation ...");
      Clock = 1'b0;
      Reset = 1'b1; // reset the counter at t=0
      #100 Reset =1'b0; // remove reset at t=100
      #2000 $finish; // end the simulation after t=2000
    end
    // generate stimulus (in this case clock signal)
    always #50 Clock = ~Clock; // clock period = 100
    // monitor the response and save it in a file
    initial begin
      $dumpfile ("count.vcd"); // specifies the VCD file
      $dumpvars; // dumps all variables
      $monitor("%d,%b,%b,%d",$time, Clock, Reset, Count);
    end
endmodule
```

We carry out the following steps in the testbench:

1. **Instantiate and initialize:** We instantiate the DUV (*Mycounter*) in the testbench. We take it to a known (reset) state at $t = 0$. Subsequently, we remove the reset at $t = 100$. We also specify that the simulation finishes after 2000 time units.
2. **Generate stimulus:** We create a clock signal as a stimulus to the counter.
3. **Monitor signal:** We open a *value change dump* (VCD) file to record the response to the stimulus. A VCD file contains all the waveform-related information computed during simulation. We can load a VCD file in a waveform viewer to examine and debug simulation results. We also add commands in the testbench to dump and monitor all the variables.

Remarks: We can run the above testbench on a Verilog simulator and generate the VCD file. Since the design is simple, we can manually observe the output waveforms and check the functional correctness of the design.

Verification plan: In practice, designs are complicated, and performing simulation-based verification is challenging. We need to make a *verification plan* by identifying a prioritized list of features from the design specification that must be verified [1]. Some features can be hard-to-verify. For those features, we can modify the implementation to make it easily verifiable. Therefore, it is prudent to make a verification plan early in the design flow.

Directed testcase: We can group similar features that can share the verification effort into *testcases* [1]. Furthermore, we can group testcases into *testbenches* that can share design configurations, stimuli generators, and response monitoring framework. We can subsequently implement separate testbenches as per testcase specifications. This method of developing feature-based testcases is known as the *directed testcase approach*. However, it requires a high test development effort and is practical only when the number of testcases is small. Therefore, we employ the following alternate strategy.

Random testcase: We can also start with a *random stimuli generator* that is *constrained* to produce realistic input sequences. Moreover, we follow a *functional coverage model* to identify features that are not yet exercised. By observing functional coverage, we modify constraints applied to the random stimuli generator to exercise more features. This method of testing is known as *random testcase approach*. During the initial implementation of the verification plan, this approach can achieve high functional coverage with less effort. Finally, we can exercise still uncovered features by adding more constraints or using a directed testcase approach.

Self-checking: We can reduce the effort in detecting functional errors in a DUV by implementing self-checking strategies in a testbench. For example, we can use a reference model to automatically generate golden responses for comparison. Alternatively, we can determine a transfer function to model data transformation and consider its response as a golden response [1].

Raising abstraction: Implementing testbenches at a higher abstraction level and employing object-oriented programming can make testbenches maintainable and reusable. SystemVerilog provides many such features that allow easy testbench implementation. We will discuss them briefly in the following section.

9.5 FUNCTIONAL VERIFICATION USING SYSTEMVERILOG

In this section, we will briefly discuss SystemVerilog features for *constrained random stimuli generation*, *coverage-driven verification*, and *assertion-based verification*. Note that we will explain these features only with a few illustrative examples. A detailed discussion on SystemVerilog features is out of the scope of this book. To know more, readers can refer to [9–13].

9.5.1 Constrained Random Stimuli Generation

SystemVerilog provides the following basic features to generate constrained random stimuli:

1. **Random variable declaration:** The variables in SystemVerilog *class* can be declared *standard random variables* using the keyword `rand`.

Example 9.7 Consider the following piece of SystemVerilog code.

```
rand bit [3:0] myInput;
```

It declares a random variable *myInput* which can take a value in the range 0 to 15 with equal probability (if unconstrained).

2. **Constraint specification:** We can constrain the values of random variables using expressions declared in the *constraint blocks*. These blocks are class members and use the keyword `constraint`.

Example 9.8 Consider the following piece of SystemVerilog code.

```
class Generator;
    rand integer x;
    rand integer y;
    constraint c1 {
        x > 100;
        x <= 500;
    }
    constraint c2 {
        y inside {10, 20, 30, 40};
    }
endclass
```

We have defined two constraints above. The constraint *c1* defines the range for *x*. The constraint *c2* defines constraint as a set membership using the `inside` operator.

We can create complex constraints by defining the relationship among variables, implications, and iterations. A SystemVerilog constraint solver needs to find a solution for the given constraints if it exists.

3. **Randomization:** SystemVerilog provides an in-built class method `randomize()` that can generate random values for the *active random variables* in an object, subject to the *active constraints*. It returns 1 if the constraint solver could set all the random variables to valid values, and 0 otherwise. We can activate/deactivate randomization of variables using built-in method `rand_mode()`. Similarly, we can activate/deactivate constraints using built-in method `constraint_mode()`.

Example 9.9 Consider the following piece of SystemVerilog code.

```
Generator gen = new(); // object of class Generator of Ex. 9.8
for (int i = 0; i < 5; i++) begin
        gen.randomize(); // x, y random values, subject to c1 and c2
end
gen.x.rand_mode(0); // turns off randomization of variable x
gen.randomize(); // only y gets random value, subject to c1, c2 ...
gen.c2.constraint_mode(0); // turns off constraint c2
gen.randomize(); // constraint c2 not applied
...
```

It illustrates how to obtain random values for active random variables. It also shows how to switch off randomization and constraints.

9.5.2 Coverage-driven Verification

Developing testbench and performing verification requires significant effort, both manual and computational. Therefore, we should perform comprehensive design verification without wasting resources on dispensible tasks. Coverage-driven verification intends to achieve this.

If a testbench contains a sufficiently *diversified* set of stimuli, the coverage is high, and the verification becomes more robust. We can use coverage to measure the progress of the verification effort and fill the discovered *gaps*. Note that adding more test stimuli for already tested features may be redundant. Thus, coverage-driven verification can reduce wasted effort and help reach verification objectives quickly.

We can classify coverage broadly into two types: *code coverage* and *functional coverage*.

1. **Code coverage:** It identifies sections of DUV's source code that did not execute during verification. This technique is widely used in software engineering also. A coverage tool typically inserts some special tasks/instructions at strategic locations in the original source code for code coverage analysis. This process is known as *instrumentation*. Subsequently, the instrumented source code is simulated using the given testbench. The inserted instructions in the source code gather relevant information while simulating. Finally, a coverage tool collates all the information and reports coverage information to us. Typically, coverage tools report the following metrics [14]:
   (a) **Line or statement coverage:** It identifies the statements that were exercised during simulation. It is the simplest code coverage metrics, quite useful, and easy to compute.
   (b) **Branch or decision coverage:** It measures whether all branches of the code (as in `if-else` and `case` statements) were exercised in simulation. Note that it ensures that both the `true` and the `false` values are evaluated for an `if` statement, irrespective of whether the `else` part exists in the code. Therefore, it is different from the line coverage.
   (c) **Expression or condition coverage:** It measures whether all the combination of sub-expressions in a conditional expression have been exercised in simulation.

Example 9.10 Consider the following piece of Verilog code.

```
if (x && y) begin
    ...
end
```

We can achieve 100% branch coverage by executing two combinations: x=0, y=0 and x=1, y=1. However, despite 100% branch coverage, the following buggy code will produce identical result for the above combinations of x and y.

```
if (x || y) begin
    ...
end
```

Thus, the above code will pass simulation-based verification and the bug will escape. Hence, branch coverage is insufficient for the above example.

To catch the above bug, we also need to verify for x=1, y=0 or x=0, y=1. The expression coverage measures whether all the four combinations of x and y for the sub-expression are executed in the conditional expression. Therefore, if we do not have testcase for x=1, y=0 and x=0, y=1, it will report low expression coverage. Thus, expression coverage measure can help us catch the above bug.

(d) **State or transition coverage:** It measures whether all the states of a finite state machine have been activated and all the state transitions traversed.

(e) **Toggle coverage:** It measures whether all variables or bits in variables have risen and fallen.

Note that code coverage simply indicates the thoroughness of the testbench in exercising the given source code. It does not prove or verify the correctness of a DUV or its completeness. For example, a coverage tool will report an empty Verilog module as 100% covered while the complete functionality of a DUV is missing. Therefore, we should interpret the code coverage reports prudently.

2. **Functional coverage:** It measures whether the specified design features have been exercised during simulation. The design features are not automatically inferred by a coverage tool. We need to extract them from the *design specification* and provide them using appropriate *coverage model*. Note that a coverage model is independent of the design implementation. Thus, it can detect missing features also in the implementation. However, developing a coverage model requires an additional initial effort. The onus of developing a comprehensive coverage model is on the verification engineer/designer. The quality of functional coverage-driven verification strongly depends on the completeness of the coverage model. Typically, a combination of a good coverage model and random stimuli generation significantly reduces the overall verification effort.

SystemVerilog provides various constructs to create coverage models efficiently. The basic constructs employed in a coverage model are: `covergroup`, `coverpoint`, and `bins`.

A *covergroup* defines a coverage model. It is similar to a SystemVerilog class and can be instantiated multiple times using `new`. It can contain a set of *coverpoints*. It can have also an *event* that triggers sampling for the covergroup. We can also trigger sampling by a predefined function `sample()` for a covergroup.

A *coverpoint* defines an integral expression that needs to be covered. This expression gets evaluated when the covergroup is sampled. A coverpoint is divided into *bins*.

SystemVerilog provides predefined coverage methods `get_coverage()` and `get_inst_coverage()`. The method `get_coverage()` reports the coverage obtained for *all instances* of a particular coverage item. The method `get_inst_coverage()` is invoked on an instance of a coverage item and it reports the coverage for that instance.

We illustrate these coverage-related constructs in the following example.

Example 9.11 Consider the following SystemVerilog code.

```
module testbench;
    logic [3:0] mode;
    logic clk;
    always #10 clk = ~clk; // clock of time period 20
    covergroup MyCG @(posedge clk); // sampling triggered by event on clk
      coverpoint mode { // coverpoint on variable mode
        bins reset = {0};
        bins slow = { [1:4] };
        bins typical ={ [5:8] };
        bins fast ={ [12:15] }; // no bin defined for values 9-11
      }
    endgroup
    MyCG cg;
    initial begin
        cg = new;
        clk = 0; // initialize clock
        #30 mode = 0; // reset bin covered
        #30 mode = 1; // slow bin covered
        #30 mode = 12; // fast bin covered
        cg.sample(); // explicit call to sample
        $display("Coverage=%f", cg.get_inst_coverage());
        $finish;
    end
endmodule
```

The covergroup *MyCG* is defined. It samples data at the rising edge of *clk*. It contains a coverpoint for the variable *mode*. The range of *mode* is divided into four bins. Note that the range [9 – 11] does not lie in any bin. The samples in this range go into a *default bin*. However, default bins are not considered in the coverage calculation for a coverage point.

An instance *cg* of covergroup *MyCG* is created in the initial block. Values are assigned for *mode* and the bins are sampled explicitly using method `sample()`. Finally, the coverage is reported using method `get_inst_coverage()`. Since, three out of four bins are covered, it will report the coverage as 75%.

Remarks: We can combine random stimuli generation with the coverage computation. For example, we can set *mode* randomly and monitor the functional coverage. Subsequently, we can add more stimuli or constrain random stimuli generation until we reach the target verification goal. Thus, coverage-driven constrained random verification can help achieve our verification goal with less effort, especially for larger designs.

9.5.3 Assertion-based Verification

Assertions specify the behavior of a system using special language constructs. Note that these specifications are in addition to the functional description of a system. We use them for automatic validation during simulation or using formal verification techniques. SystemVerilog provides mainly four constructs to specify assertions:

1. `assert`: It is used to specify properties for a design that should be satisfied. Simulators and formal tools validate them.
2. `assume`: It is used to specify properties assumed for the environment. Simulators check that these properties hold. Formal tools consider them as constraints on the input.
3. `cover`: It is used to monitor coverage of the property evaluations.
4. `restrict` : It is used to specify constraints for the formal verification. Simulators do not validate these constraints.

The simplest form of assertion is the *immediate assertion* statement. It simply tests a nontemporal (that does not involve time) expression executed in a procedural code. If the expression evaluates to `x/z/0`, the assertion is said to *fail*, else it is said to `pass`. We can specify actions that the simulator should take for the passing and failing cases.

Example 9.12 Consider the following SystemVerilog assertion.

```
grant_check: assert(!(grant && !request)) $info("passed");
else $error("failed: grant without request");
```

The assertion *grant_check* checks that it should never happen that the signal *grant* is 1, and the signal *request* is still at 0. If the assertion passes, it gives an informational message that the assertion has "passed"; otherwise, it produces an error with the message "failed: grant without request." On failure, assertions can also produce warning or fatal error with system tasks `$warning` and `$fatal`, respectively.

Assertions can also use implication operator `->` as follows:

```
readycheck: assert(ready -> !reset) else $fatal("Ready signal high when reset
signal is still high");
```

The assertion *ready_check* checks that whenever the signal *ready* is 1, the signal *reset* must be 0. In general, a logical expression $A->B$ is equivalent to $[(!A)\,|\,(A\&B)]$, which can be simplified to $[(!A)\,|\,B]$.

There is another type of assertion in SystemVerilog that employs the semantics of *clocks* and uses the sampled values in the expressions. These assertions are known as *concurrent assertions* and follow cycle-based evaluation. We call them concurrent assertions because they execute in parallel or concurrently with the rest of the design. Concurrent assertions are powerful and particularly helpful in describing behavior for formal verification. They consist of *sequence* (defined using keywords `sequence–endsequence`), *property* definition (defined using keywords `property–endproperty`), and the *assertion* of a property (defined using keywords `assert property`). We explain them in the following example.

Example 9.13 Consider the following SystemVerilog concurrent assertion.

```
sequence myseq;
    x ##2 y;
endsequence
property myprop;
    @(posedge clk) myseq;
endproperty
cassrt1: assert property(myprop) $info ("passed");        else $error
("cassrt1 failed");
cassrt2: assert property( @(posedge clk)  a ##2 b ) $info ("passed");
else $error ("cassrt2 failed");
```

The property *myprop* is asserted using concurrent assertion *cassrt1*. This property checks for the sequence *myseq* at every rising edge of the clock signal *clk*. The sequence *myseq* checks that *x* is true and two clock cycles later *y* is true. For simple properties, we can omit explicit sequence and property definitions. An equivalent assertion *cassrt2* is shown that do not use any sequence explicitly.

Remarks: We represent clock cycle delays using symbol ##. We can also represent a range of sampling clock edges by using symbol `##[m:n]` construct. For example, `##[1:5]` represents first five clock edges.

In the above example, the sequence *myseq* (x ##2 y) is checked on every rising edge of the clock. If *x* is not high at any rising clock edge, `$error` is called. However, we often need to check a property conditionally. One of the constructs in SystemVerilog to check conditional properties is *implication*, as explained in the following example.

Example 9.14 Consider the following SystemVerilog concurrent assertion.

```
impassrt: assert property(@(posedge clk) request |-> request until grant);
```

The above assertion is checked at the rising edge of the clock signal *clk*. It uses the *implication* symbol (|->). An implication consists of the left-hand sequence known as *antecedent* (*request*) and the right-hand sequence known as consequent (*request* `until` *grant*). If there is no match to the antecedent (*request* is never 1), implication succeeds vacuously and returns true. For each antecedent sequence match (*request*=1 at some clock edge), the consequent sequence (*request* `until` *grant*) is separately evaluated starting from the antecedent-matching instant. The implication returns true if the result of all such consequent evaluations is true. Thus, the above assertion checks that whenever *request*=1, it should remain 1 until *grant*=1.

Remarks:

1. The keyword `until` specifies a temporal relationship. SystemVerilog supports other keywords to model temporal relationships, such as `always`, `nexttime`, and `s_eventually`.
2. The implication construct represented by the symbol `|->` is known as *overlapped* implication. For overlapped implication, the consequent is evaluated starting from the instant the corresponding antecedent matching completes. SystemVerilog supports another type of implication represented by the symbol `|=>`. It is known as *non-overlapped* implication. For a non-overlapped implication, the consequent is evaluated starting from the *next clock edge* at which the antecedent matching completes.

Similar to assertions, SystemVerilog supports *assumptions* using keyword `assume`. Both simulators and formal verification tools use these assumptions. If the assumption fails, simulators execute the statements specified for the failing scenario. Formal verification tools use assumptions to reduce the design space and simplify the problem they need to solve.[3] A few examples of assumptions are shown below.

```
assume_1: assume(func || test) $info("assumption holds"); else
$error("assumption violated");
assume_2: assume property (@(posedge clk) reset |-> !request);
```

SystemVerilog also supports specifying constraints using `restrict` statement for the formal verification tools. The constraints specified in the `restrict` statement are not used by the simulator. The semantics of the `restrict` construct is similar to the `assume` construct, except that it does not have an action block.

Example 9.15 Consider the following SystemVerilog `restrict` statement.

```
restrict1: restrict property(@(posedge clk) reset==1'b0);
```

The above restriction *reset*=0 is applied during formal verification. It can help in simplifying the problem for the formal verification tool. However, during simulation, this restriction does not apply. Therefore, testbench can have testcases with *reset*=1, and simulators will not treat it as an error. If *reset*=0 was an assumption, the simulators would have produced an error for *reset*=1. This differentiates an assumption from a restriction.

---

[3] Use of SystemVerilog assertions and assumptions in formal methods will be discussed in Chapter 11 ("Formal verification").

9.6 HARDWARE-ASSISTED VERIFICATION

The time taken in logic simulation becomes too high for multimillion gate designs and system-on-chips (SoCs). For such designs, we need to speed up the simulation. A popular technique to speed up simulation is to take help of some hardware to model the design. However, the following differences in logic simulation and hardware-assisted simulation are worth noting [15]:

1. Verilog logic simulation supports four-valued logic (`0/1/x/z`). However, evaluation of logic in hardware involves binary values (`0/1`) only.
2. Logic simulation typically processes RTL logic elements sequentially, while the hardware can execute them concurrently. Therefore, hardware-assisted verification can be orders of magnitude faster than the traditional logic simulation.
3. The creation of a setup of traditional logic simulation requires significantly less effort than the hardware-assisted simulation.
4. The traditional logic simulation is easy to use and offers more flexibility. We can observe and control any signal in a traditional logic simulation. It also allows easy debugging.

Due to the above differences, we prefer traditional logic simulation for functional verification of smaller blocks and during the early stages of design development. However, we need hardware-assisted simulation for large designs to achieve a reasonable turn-around time.

Depending on the design attributes, speed requirements, cost, and the availability of resources, we can employ various types of hardware-assisted simulation systems [16–18]. The following types of hardware-assisted verification systems or *emulators* are popular.

1. **Field programmable gate array (FPGA)-based system:** We can use FPGA to emulate the DUV. We need to synthesize a design and map it to the FPGA hardware. We often need multiple FPGAs to map a large design. Therefore, we use an array of FPGAs for emulation. It requires partitioning a design, routing signals across FPGAs, and ensuring timing correctness. These tasks require more initial effort in developing the emulation setup. However, FPGA-based emulation systems are fast and consume less power.
2. **Processor-based system:** It consists of a massive array of processors organized to share data among themselves. The hardware is custom-designed and forms a part of an emulation tool. The processors are scheduled and assigned tasks to implement the given design functionality. This task is accomplished by efficient compilers that map a design to the targetted processor-based emulator.

We can run emulation in the *in-circuit emulation* (ICE) mode. In this mode, we emulate a design in the environment where it will be deployed eventually. It allows a design to be verified with the real-world stimuli. Additionally, we can use an emulation setup to verify the software-hardware integration, low-level software testing, and system validation.

9.7 RECENT TRENDS

Simulation-based verification has been employed for digital circuits for the last several decades. Nevertheless, we have been making continuous advancement in it. With the increasing design size, the problem of simulation-based verification becomes more challenging. Therefore, we need to continuously tackle the problem of increased capacity, runtime, designer effort, and thoroughness of the simulation. Recently, the adoption of coverage-driven constrained random stimuli generation, assertion-based verification complemented by formal methods, development of verification IPs (VIPs), and advancement in emulation technology have provided increased scalability, reusability, and thoroughness to the simulation-based verification. Additionally, it is worth experimenting with machine learning and advanced data analytics for solving simulation-based verification problems [19–23].

REVIEW QUESTIONS

9.1 What is functional verification of a design? How is it different from timing verification?

9.2 How is event-driven simulation different from cycle-based simulation?

9.3 Why is cycle-based simulation a good functional verification technique for a synchronous design but not for an asynchronous design?

9.4 How is event-driven simulation able to capture glitches in a circuit?

9.5 For a synchronous design, how does static timing analysis complement the verification done by cycle-based simulation?

9.6 What is a timing wheel? What does a slot on a timing wheel contain?

9.7 How can we simulate a circuit for a simulation time significantly greater than the number of slots in the timing wheel?

9.8 What is a stratified Verilog queue?

9.9 How are blocking assignments and nonblocking assignments treated differently in Verilog simulation?

9.10 What causes indeterminism in Verilog simulation?

9.11 What is a race in Verilog simulation? Explain with the help of an example.

9.12 Consider the following three Verilog modules.

```
module top1 (clk, a, b, c);
    input clk, a;
    output b, c;
    reg b, c;
    always @(posedge clk) begin
      c = b;
      b = a;
    end
endmodule
```

```
module top2 (clk, a, b, c);
    input clk, a;
    output b, c;
    reg b, c;
    always @(posedge clk) c = b;
```

```
    always @(posedge clk) b = a;
endmodule
```

```
module top3(clk, a, b, c);
    input clk, a;
    output b, c;
    reg b, c;
    always @(posedge clk) c <= b;
    always @(posedge clk) b <= a;
endmodule
```

Analyze whether there is a simulation race condition in any of them.

9.13 How can we avoid a simulation race in a Verilog design?

9.14 What are the challenges in creating a testbench for functional verification?

9.15 What is a VCD file? How does it help in debugging a functional problem in a circuit?

9.16 How does constrained random stimuli generation and coverage analysis reduce the effort of testbench development?

9.17 Name a few features of SystemVerilog that ease testbench development.

9.18 Explain the behavior of the following SystemVerilog assertion.

```
casrt: assert property(@(posedge clk) request |-> ##[1:5] grant);
```

9.19 What are the differences between overlapping implication and nonoverlapping implication?

9.20 When should you prefer to use `restrict` over `assume` in defining environmental constraints in SystemVerilog?

9.21 What are the advantages and disadvantages of hardware-assisted simulation over traditional logic simulation?

9.22 When should you prefer traditional logic simulation over hardware-assisted simulation?

TOOL-BASED ACTIVITY

In this activity, use any Verilog simulation tool, such as the open-source tool ICARUS Verilog [24]. Use waveform viewer, such as GTKWave [25]. Use line coverage tools, such as Covered Verilog Code Coverage Analyzer [26].

1. Write a Verilog model for the finite state machine shown in Figure 9.9. The states are: *s0*, *s1*, *s2*, and *s3*. The input port *A* can take values {0, 1}. Output *Z* has two bits and can take values: {00, 01, 10, 11}. The initial state is *s0*.
2. Write a testbench to test the above Verilog model.

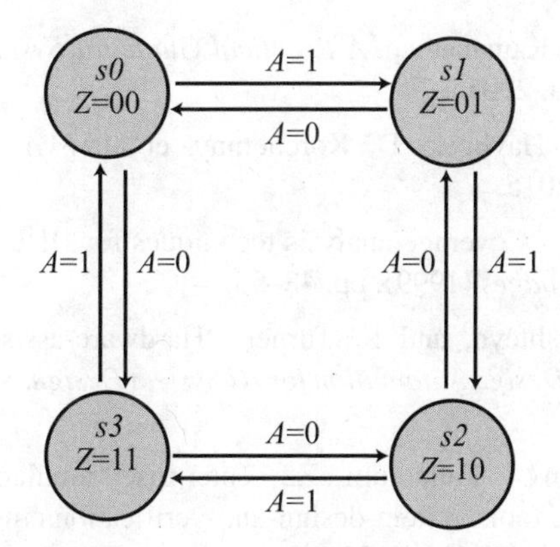

Figure 9.9 Given FSM

3. Observe the simulation output waveform and explain the correctness of the model.
4. Obtain line coverage report for the above testbench. Explain the code coverage report.

REFERENCES

[1] J. Bergeron. *Writing Testbenches: Functional Verification of HDL Models*. Springer Science & Business Media, 2012.

[2] H. Muhr and R. Holler. "Accelerating RTL simulation by several orders of magnitude using clock suppression." *2006 International Conference on Embedded Computer Systems: Architectures, Modeling and Simulation* (2006), pp. 123–128, IEEE.

[3] D. Kim, M. Ciesielski, and S. Yang. "A new distributed event-driven gate-level HDL simulation by accurate prediction." *2011 Design, Automation & Test in Europe* (2011), pp. 1–4, IEEE.

[4] S. Yang, J. Han, D. Kwak, N. Kim, D. Cha, J. Park, and J. Kim. "Predictive parallel event-driven HDL simulation with a new powerful prediction strategy." *2014 Design, Automation & Test in Europe Conference & Exhibition (DATE)* (2014), pp. 1–3, IEEE.

[5] Z. Barzilai, J. L. Carter, B. K. Rosen, and J. D. Rutledge. "HSS—A high-speed simulator." *IEEE Transactions on Computer-aided Design of Integrated Circuits and Systems* 6, no. 4 (1987), pp. 601–617.

[6] "IEEE standard Verilog hardware description language." *IEEE Std 1364-2001* (2001), pp. 1–792.

[7] C. E. Cummings et al. "Nonblocking assignments in verilog synthesis, coding styles that kill!" *SNUG (Synopsys Users Group) 2000 User Papers* (2000).

[8] P. Moorby, A. Salz, P. Flake, S. Dudani, and T. Fitzpatrick. "Achieving determinism in SystemVerilog 3.1 scheduling semantics." *Proceedings of the Design and Verification Conference*. 2003.

[9] "IEEE standard for SystemVerilog—Unified hardware design, specification, and verification language." *IEEE Std 1800-2017 (Revision of IEEE Std 1800-2012)* (2018), pp. 1–1315.

[10] J. Bergeron. *Writing Testbenches Using SystemVerilog*. Springer Science & Business Media, 2007.

[11] A. B. Mehta. *SystemVerilog Assertions and Functional Coverage*. Springer, 2020.

[12] S. Vijayaraghavan and M. Ramanathan. *A Practical Guide for SystemVerilog Assertions*. Springer Science & Business Media, 2005.

[13] E. Cerny, S. Dudani, J. Havlicek, D. Korchemny, et al. *SVA: The Power of Assertions in SystemVerilog*. Springer, 2015.

[14] J.-Y. Jou and C.-N. J. Liu. "Coverage analysis techniques for HDL design validation." *Proc. Asia Pacific CHip Design Languages* (1999), pp. 48–55.

[15] F. Schirrmeister, M. Bershteyn, and R. Turner. "Hardware-assisted verification and software development." *Electronic Design Automation for IC System Design, Verification, and Testing* (2017), p. 461.

[16] Cadence Design Systems. "Palladium Z2 enterprise emulation platform." https://www.cadence.com/en_US/home/tools/system-design-and-verification/emulation-and-prototyping/palladium.html. Last accessed on July 24, 2022.

[17] Siemens. "Veloce HW-assisted verification system." https://eda.sw.siemens.com/en-US/ic/veloce/. Last accessed on July 24, 2022.

[18] Synopsys. "Emulation." https://www.synopsys.com/verification/emulation.html. Last accessed on July 24, 2022.

[19] C. Ioannides and K. I. Eder. "Coverage-directed test generation automated by machine learning—A review." *ACM Transactions on Design Automation of Electronic Systems (TODAES)* 17, no. 1 (2012), pp. 1–21.

[20] F. Wang, H. Zhu, P. Popli, Y. Xiao, P. Bodgan, and S. Nazarian. "Accelerating coverage directed test generation for functional verification: A neural network-based framework." *Proceedings of the 2018 on Great Lakes Symposium on VLSI* (2018), pp. 207–212.

[21] S. Meraji and C. Tropper. "A machine learning approach for optimizing parallel logic simulation." *2010 39th International Conference on Parallel Processing* (2010), pp. 545–554, IEEE.

[22] S. Vasudevan, D. Sheridan, S. Patel, D. Tcheng, B. Tuohy, and D. Johnson. "Goldmine: Automatic assertion generation using data mining and static analysis." *2010 Design, Automation & Test in Europe Conference & Exhibition (DATE 2010)* (2010), pp. 626–629, IEEE.

[23] L.-C. Wang. "Experience of data analytics in EDA and test—Principles, promises, and challenges." *IEEE Transactions on Computer-aided Design of Integrated Circuits and Systems* 36, no. 6 (2016), pp. 885–898.

[24] S. Williams. "ICARUS Verilog." http://iverilog.icarus.com/. Last accessed on July 24, 2022.

[25] T. Bybell. "Welcome to GTKWave." http://gtkwave.sourceforge.net/. Last accessed on July 24, 2022.

[26] T. Williams. "Covered—Verilog Code Coverage Analyzer." http://covered.sourceforge.net/. Last accessed on July 24, 2022.

10 RTL Synthesis

...before attempting to translate our data into the rigorous language of symbols, it is above all things necessary to ascertain the ***intended*** *import of the words we are using. But this necessity cannot be regarded as an evil by those who value correctness of thought, and regard the right employment of language as both its instrument and its safeguard.*

—George Boole, *An Investigation of the Laws of Thought*, Chapter 4, 1854

We model hardware at the *register transfer level* (RTL) using hardware description languages (HDLs) such as Verilog and VHDL, as discussed in the preceding chapters. Subsequently, we *synthesize* the RTL model and obtain a netlist. During the initial phases of synthesis, we translate the RTL model to a netlist consisting of primitive logic gates, arithmetic blocks, and memory units, including registers. We refer to this step as *RTL synthesis*.

RTL synthesis involves analyzing the RTL model and instantiating appropriate circuit elements based on the semantics of the HDL. Therefore, it needs to comprehend various HDL constructs while translating them to circuit elements. In this chapter, we will illustrate the translation of a few essential Verilog constructs to hardware. These examples help us understand the correspondence between the Verilog constructs and the circuit elements. These concepts are often helpful while developing an RTL model, evaluating the impact of RTL code changes on the quality of result (QoR) measures, making manual RTL modifications, and interpreting the results of logic synthesis.

There are some optimization tasks that we can perform more efficiently at the level of RTL model. For example, we can efficiently carry out optimization related to resource allocation, arithmetic operators, multiplexer usage, and finite state machines (FSMs) at RTL because the HDL constructs allow easy identification of targets and make modifications at a higher abstraction level. We will discuss these optimizations also in this chapter.

10.1 LOGIC SYNTHESIS TASKS

We divide logic synthesis into a series of smaller tasks, as illustrated in Figure 10.1.

The initial portion of logic synthesis consists of *parsing* the RTL code, *elaboration*, RTL-specific optimization, and translation to primitive logic gates, arithmetic blocks, registers, memory units, and FSMs. We group these tasks as RTL synthesis and will discuss them in detail in this chapter.

After RTL synthesis, we carry out aggressive *logic optimization*, technology mapping, and technology-dependent logic optimizations. These tasks produce the final netlist that we can use for physical design. We will discuss these tasks in the subsequent chapters.

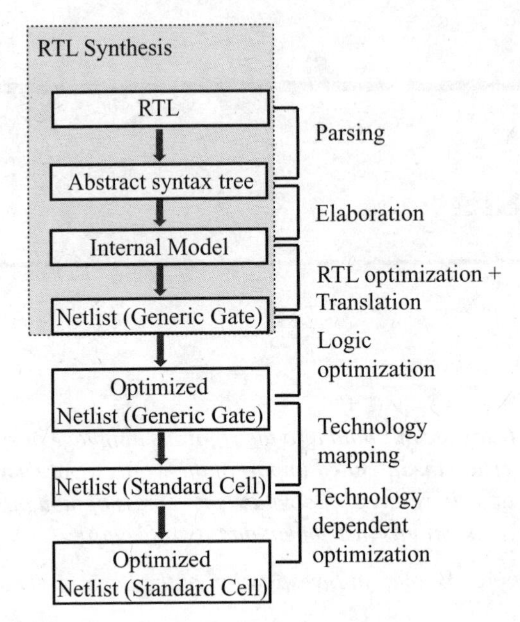

Figure 10.1 Major tasks in logic synthesis

10.2 PARSING

An RTL synthesis tool reads the given RTL files and populates a data structure for further processing. First, it breaks the sequence of characters in a file into a sequence of *tokens* or words that have special meaning for the given language. For example, the keywords in Verilog such as `always`, `module`, `endmodule`, `begin`, `end`, etc. can be the tokens [1]. This process is known as *lexical analysis*. Then, it analyzes whether the grammar of the given HDL is honored in the given RTL code. If there are any syntax errors, the tool reports them. If the RTL code is error-free, the tool populates a hierarchical data structure, typically in the form of a *syntax tree*. This process is known as *parsing*.

Example 10.1 Consider the following piece of Verilog code.

```
module top();
endmodule
module mid();
      ...
      always @ (*)
      begin
            a <= x+y;
            b <= 0;
            c = x;
      end
endmodule
```

The *abstract syntax tree* (AST) for the above code is shown in Figure 10.2. If a construct or design entity P contains another construct or design entity Q, we make P the parent of Q. Note that the AST of Figure 10.2 shows only the elements present in the above code.

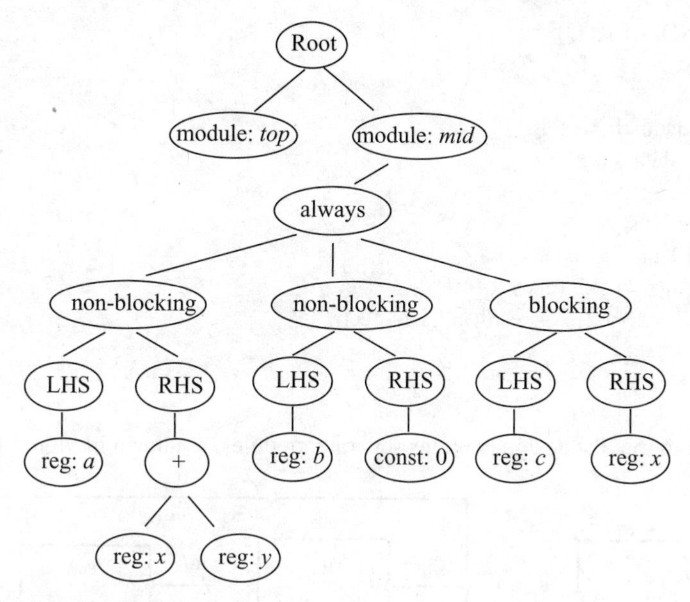

Figure 10.2 An abstract syntax tree

10.3 ELABORATION

After parsing, *elaboration* is carried out for a design. During elaboration, a tool checks whether the connections among RTL-specified components are legitimate. If they are legitimate, the tool will make the connections in the internally created design model, else it will report an error or warning depending on the severity of the violation. For example, if a module instantiates a sub-module with a port name A, the sub-module should contain a port named A. If the port named A does not exist, the tool will report an error. Similarly, if the port widths in the instantiation and its master module are inconsistent, the tool will report an error or warning.

Example 10.2 Consider the following piece of Verilog code.

```
module leaf(d, clk, q);
    input d, clk;
    output q;
    ...
endmodule
```

```
module middle(D, CLK, Q);
    input D, CLK;
    output Q;
    leaf F1(D, CLK, Q);
endmodule

module Top(data, clk, result);
    input data, clk;
    output result;
    wire w1, w2;
    leaf I1(data, clk, w1);
    leaf I2(.d(w1), .clk(clk), .q(w2));
    middle I3(.D(w2), .Q(result), .CLK(clk));
endmodule
```

Before elaboration, the design contains separate modules, as shown in Figure 10.3.

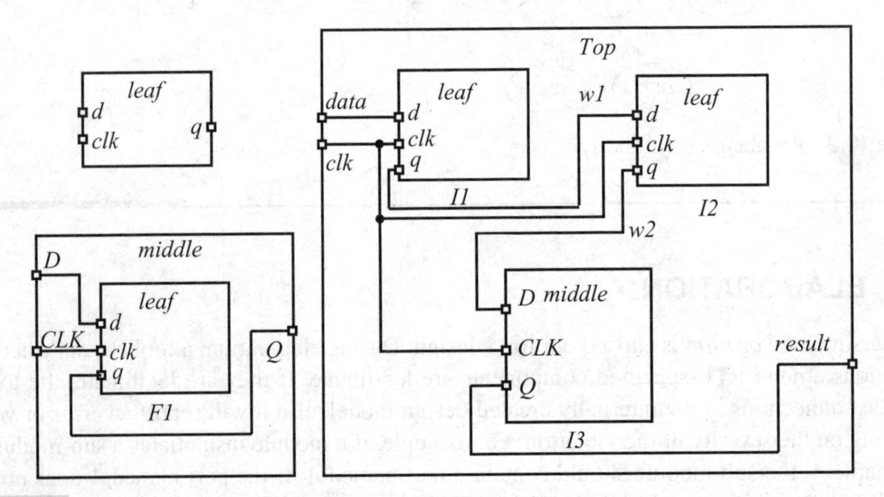

Figure 10.3 Given design before elaboration

The modules are not linked, and the hierarchy of modules is not yet created. Note that the information on the direction of the instance pin is initially only in the master module (e.g., the direction of pin *d* of instance *I1* is in the master module *leaf*). Hence, before elaboration, the tool can assume any direction for the instance pin. In this example, we have shown their direction as input.

After elaboration, the modules get linked, and the hierarchy of the modules gets created, as shown in Figure 10.4. Additionally, the direction of instance pins gets inferred, and appropriate connections are made. If there were any inconsistencies in the instance pin and master module's port (in name or width), the tool would have reported them.

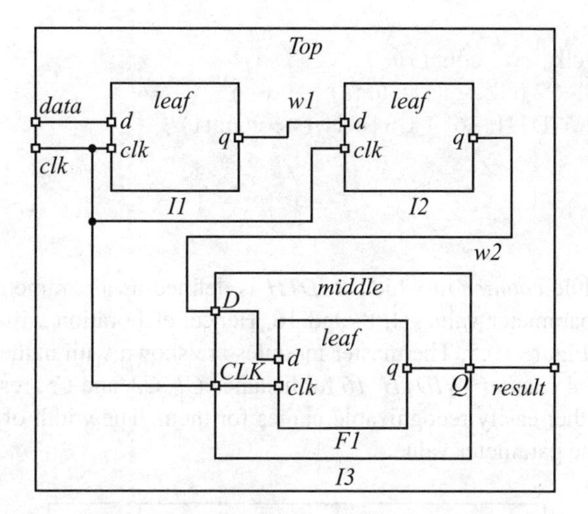

Figure 10.4 Given design after elaboration

If we instantiate a module that does not exist in the given RTL code, the tool will be unable to resolve its reference. Hence, a tool can report an error or treat that instance as a *black box* and proceed. However, treating instances as black boxes can often result in unexpected behavior, and we should try to avoid undefined references in a design hierarchy. Such problems typically occur when we have not provided the complete set of RTL files to the synthesis tool or provided the wrong module name during instantiation.

Elaboration needs to process *parameterized modules* differently because they can have different interfaces for varying parameters. Typically, elaboration creates *separate* modules with different interfaces for each distinct set of parameters. A tool can name such an internally created module with its own naming convention. We illustrate it in the following example.

Example 10.3 Consider the following piece of Verilog code.

```
module counter(clk, rst, count);
    parameter WIDTH=4;
    input clk, rst;
    output [WIDTH-1:0] count;
    reg [WIDTH-1:0] count;
    ...
endmodule
module top(clk, rst, count1, count2, count3);
    input clk, rst;
    output [15:0] count1;
    output [7:0] count2;
    output [3:0] count3;
    ...
```

```
      counter C1 (clk, rst, count3) ;
      counter # (8) C2 (clk, rst, count2) ;
      counter # ( . WIDTH (16) ) C3 (clk, rst, count1) ;
      ...
endmodule
```

It contains a module *counter* in which *WIDTH* is defined as a parameter. It is instantiated in the module *top* with parameter values 4, 8, and 16. Hence, elaboration will produce three master modules as shown in Figure 10.5. The master modules are shown with names *counter_WIDTH_4*, *counter_WIDTH_8*, and *counter_WIDTH_16* for instance *C1*, *C2*, and *C3*, respectively. A synthesis tool can choose any other easily recognizable names for them. The width of the output port *count* varies depending on the parameter value.

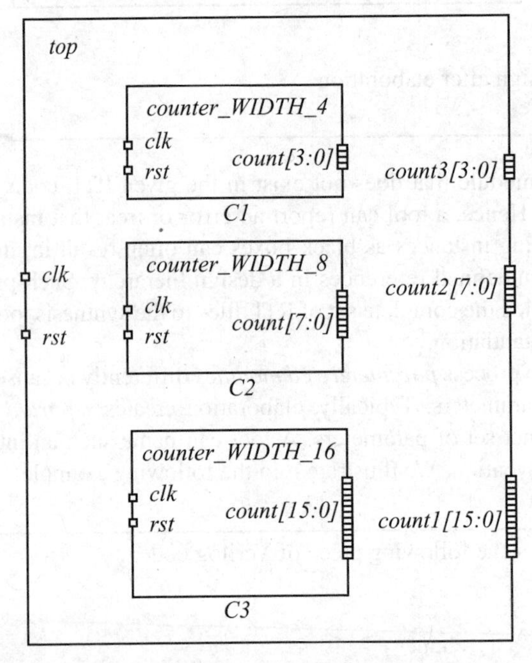

Figure 10.5 Given design after elaboration (connections not shown)

Sometimes we need to implement different instances of the same module differently, irrespective of whether they are parameterized. For example, one instance of a module can be on the critical path of a circuit, while two others instances of the same module are not timing critical. In such cases, we want different instances of the same module to be implemented differently. To achieve this, we can create separate copies of the module for different instances, despite having the same interface and the RTL code. This process is known as *uniquification*. Note that elaboration does not carry out uniquification by default. We need to instruct synthesis tools to perform uniquification, and industry-standard synthesis tools typically provide a mechanism for it.

10.4 VERILOG CONSTRUCTS TO HARDWARE

Once elaboration is done, the synthesis tool has created the design hierarchy and connected the RTL-specified instances. Thus, a general structure of the netlist is already built. However, the internal details of the modules are not yet determined. A module can contain various HDL constructs. An RTL synthesis tool needs to comprehend these constructs, transform them into circuit elements, and instantiate them. Though the transforming process can vary among RTL synthesis tools, the HDL semantics govern the translation of RTL constructs to hardware. In this section, we present the translation of some commonly used Verilog constructs to hardware.

Besides providing a mechanism for modeling hardware, Verilog has constructs that ease simulation and other design tasks. Therefore, some Verilog constructs cannot be synthesized into circuit elements, and they are helpful in other design tasks. Moreover, an RTL synthesis tool may not support all synthesizable Verilog constructs or modeling styles. Therefore, we need to be aware of the Verilog constructs supported by the synthesis tool that we are using for design implementation.

The behavior of an RTL synthesis tool for non-synthesizable HDL constructs is tool dependent. It can ignore constructs that it cannot synthesize and continue with the remaining RTL code. For example, # delay specification is not supported in RTL synthesis. When a synthesis tool encounters a Verilog statement such as:

```
out1 <= #12 a;
```

it will treat the statement simply as:

```
out1 <= a;
```

Note that the delay specification such as #12 is easy to model and consider during simulation. Therefore, we commonly use delay specification in testbenches for functional verification. However, designing hardware with an exact delay specification is not possible for various reasons, such as process-induced and environmental variations. Moreover, we do not need delay elements with an exact delay specification in a circuit. Typically, we need circuit elements that have delay within a given *range* or with some allowed variations.[1]

Some other constructs of Verilog that are typically not supported by RTL synthesis tools are `initial` block, `fork`, `join`, `force`, `release`, data types `real` and `time`, `$display`, `$monitor`, and other systems tasks.

In the following paragraphs, we describe the direct translation of some commonly used Verilog constructs to hardware. However, note that synthesis tools are free to implement them using other functionally equivalent hardware. Additionally, subsequent optimization can change the circuit structure obtained using direct translation.

10.4.1 Assign Statement

An `assign` statement assigns a Verilog expression to a wire or a vector of wires. Depending on the right-hand side (RHS) of the `assign` statement, various combinational circuit elements can be inferred. The following example illustrates it.

[1] We will discuss tackling delay of circuit elements in Chapter 14 ("Static timing analysis").

Example 10.4 Consider the following Verilog code.

```
module top (a, b, c, p, q, s, x, y, out1, out2, out3) ;
    input a, b, c, p, q, s;
    input [3:0]x, y;
    output out1;
    output [3:0]out2;
    output out3;
    assign out1 =  (a & b) | c; // logic network
    assign out2 =  (x & y) ; // bitwise logic network
    assign out3 =  (s)  ? q : p; // multiplexer
endmodule
```

One possible synthesis result is shown in Figure 10.6.

The first `assign` statement results in a logic network defined by the RHS expression.

The second `assign` statement has a 4-bit vector on the left-hand side (LHS). It results in four identical logic structures driving each bit of the vector.

The third `assign` statement results in a two-to-one multiplexer due to the conditional operator.

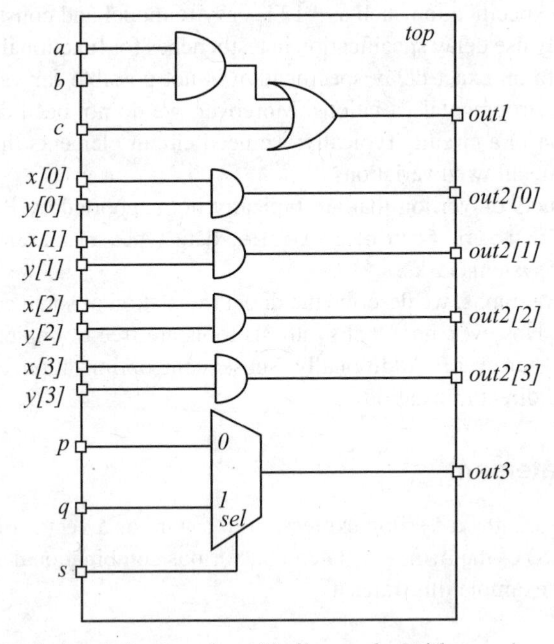

Figure 10.6 Synthesized circuit for the given Verilog code (with `assign` statements)

10.4.2 If–else Statement

An `if–else` statement translates to a multiplexer or selecting logic. If it results in a multiplexer, the signal appearing in the `if` clause goes to the select lines of the multiplexer.

Example 10.5 Consider the following Verilog code.

```
module top(a, b, s, out1);
    input a, b, s;
    output out1;
    reg out1;

    always @(*) begin
        if (s==1'b0)
            out1 = a;
        else
            out1 = b;
    end
endmodule
```

One possible synthesis result is shown in Figure 10.7.

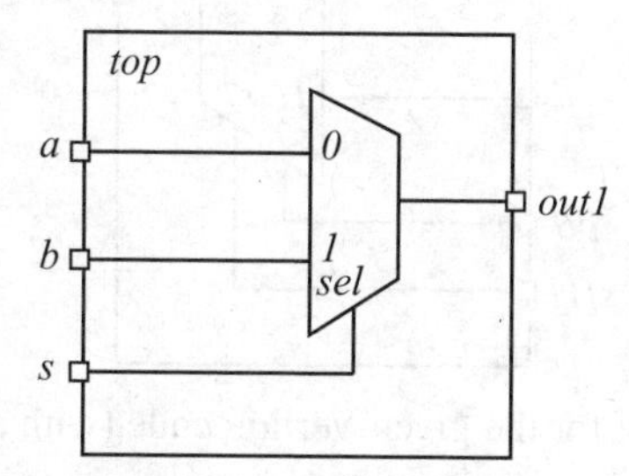

Figure 10.7 Synthesized circuit for the given Verilog code (with an `if–else` statement)

10.4.3 Case Statement

A case statement (specified using `case`, `casex`, or `casez`) also translate to multiplexers or select logic. The size of the multiplexer depends on the size of the control signal in the case statement.

Example 10.6 Consider the following Verilog code.

```
module top(a, b, c, d, s, out1);
    input a, b, c, d;
```

```
    input [1:0]s;
    output out1;
    reg out1;

    always @(*) begin
        case (s)
            2'b00: out1 = a;
            2'b01: out1 = b;
            2'b10: out1 = c;
            2'b11: out1 = d;
            default: out1=a
        endcase

    end
endmodule
```

One possible synthesis result is shown in Figure 10.8. It contains a four-to-one multiplexer.

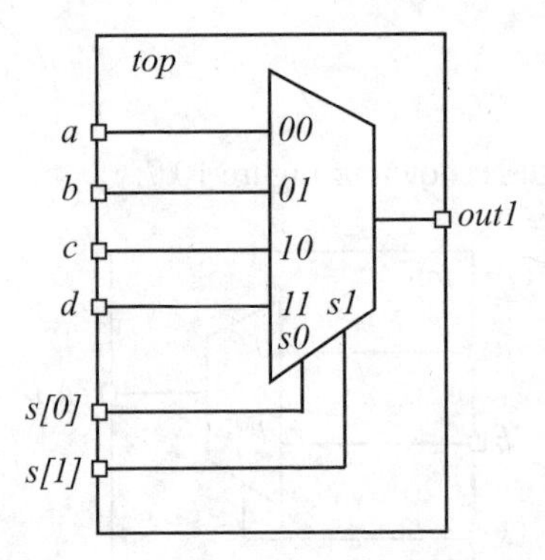

Figure 10.8 Synthesized circuit for the given Verilog code (with a case statement)

Verilog permits a case statement in which multiple items match. If there are multiple matches for a case statement, the *first* matched item gets the priority. Hence, priority multiplexers are inferred for overlapping matches in the case statement.

Example 10.7 Consider the following Verilog code.

```
module top(a, b, c, s, out1);
    input a, b, c;
    input [1:0]s;
    output out1;
```

```
    reg out1;
    always @(*) begin
        casez (s)
            2'b1?: out1 = a;
            2'b?1: out1 = b;
            default: out1 = c;
        endcase
    end
endmodule
```

One of the possible synthesis results is shown in Figure 10.9. Note that the multiplexer connections implement the priority specified in the case statement. The condition (*s[1]*=1) gets the highest priority, followed by (*s[0]*=1), and the default condition.

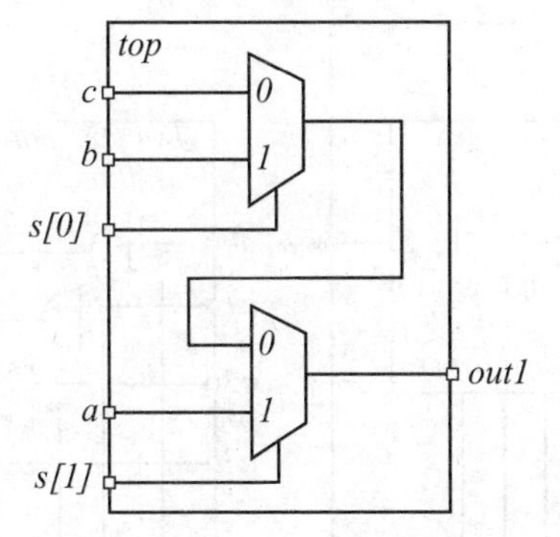

Figure 10.9 Synthesized circuit that retains the priority modeled by the case statement using multiplexers

A similar kind of priority can be defined using `if-else-if` construct also, as shown in the following code.

```
module top(a, b, c, s, out1);
    input a, b, c;
    input [1:0]s;
    output out1;
    reg out1;

    always @(*) begin
        if (s[1] == 1'b1)
            out1 = a;
        else if (s[0] == 1'b1)
            out1 = b;
        else
```

```
            out1 = c;
    end
endmodule
```

It will get synthesized similar to the circuit shown in Figure 10.9. However, note that an RTL synthesis tool can internally employ other kinds of multiplexers and produce a different but functionally equivalent circuit. For example, it can use a *one-hot multiplexer* for representing the circuit, as shown in Figure 10.10. A one-hot multiplexer has N data inputs and N select inputs. If ith select input is 1, it produces the ith data input at the output pin. Additionally, only one select input pin can have a value of 1 in a one-hot multiplexer (otherwise it can produce incorrect output). We can easily translate the control logic of `if–else–if` and case statements to one-hot multiplexer-based circuit. Additionally, it allows easier optimization of the control logic. Therefore, one-hot multiplexer-based representation of control logic has merits in RTL synthesis.

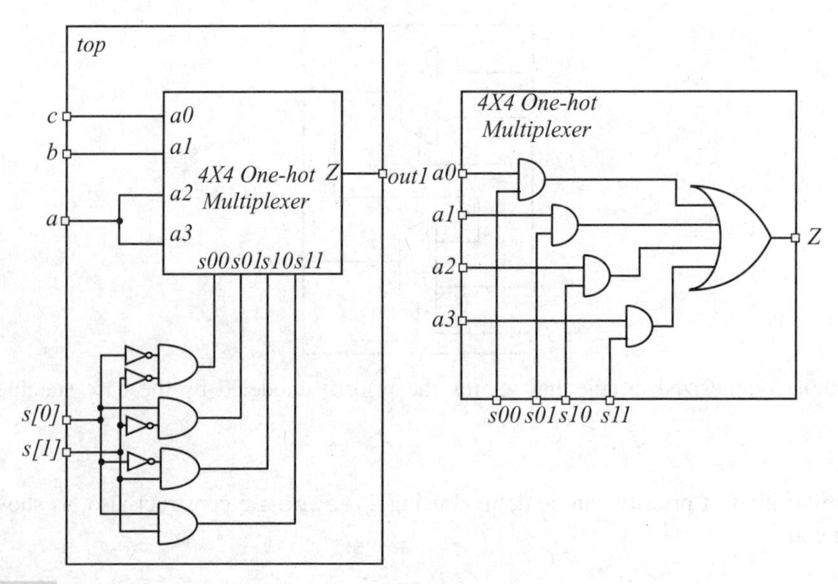

Figure 10.10 Synthesized circuit that uses one-hot multiplexer

For example, we can convert a 4×4 one-hot multiplexer of Figure 10.10 to a 3×3 multiplexer because two inputs are identical (port *a* feeds *a2* and *a3* pins of the multiplexer) [2]. The resultant circuit is shown in Figure 10.11. We have merged the data pin *a3* with *a2*. Additionally, the corresponding select pin *s10* is driven by the logical OR of the original control signals, i.e., $s[0]'.s[1]+s[0].s[1]=s[1]$. Further, we have eliminated pin *s11*. Thus, we obtain an optimized circuit shown in Figure 10.11. Subsequently, we can convert the optimized one-hot multiplexer-based circuit to binary multiplexer-based circuit or to other generic gates. For complicated control logic, the one-hot multiplexer-based optimizations can be more powerful and efficient.

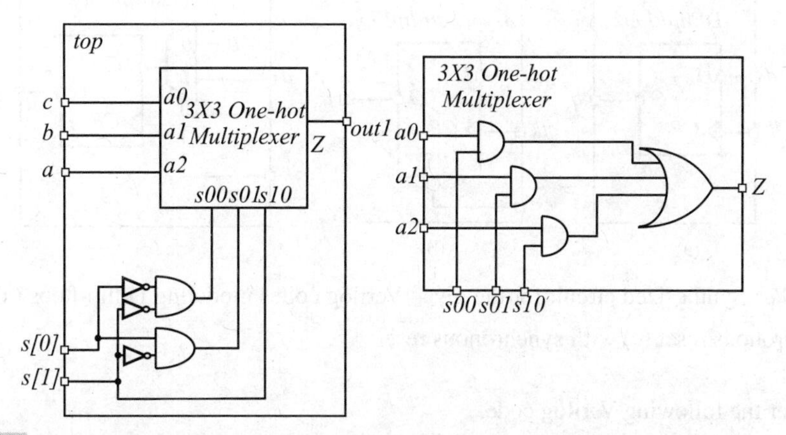

Figure 10.11 Synthesized circuit after optimizing one-hot multiplexer

10.4.4 Always Block

An always block can be inferred as a block of combinational logic or a block of sequential logic containing flip-flops and latches.

When the sensitivity list of an always block contains `posedge` or `negedge`, flip-flops are inferred.

Example 10.8 Consider the following Verilog code.

```
module DFlipFlop(d, clk, q);
    input d, clk;
    output q;
    reg q;

    always @(posedge clk)
        q <= d;
endmodule
```

The always block is sensitive to the rising edge of *clk*. When the rising edge of *clk* occurs, the value at *d* gets written to *q*. At other time instants, *q* holds its previous value. Hence, a positive-edge triggered D flip-flop gets inferred, as shown in Figure 10.12(a).[2]

[2] A negative-edge triggered flip-flop gets inferred when the `always` block is sensitive to the `negedge` of *clk*.

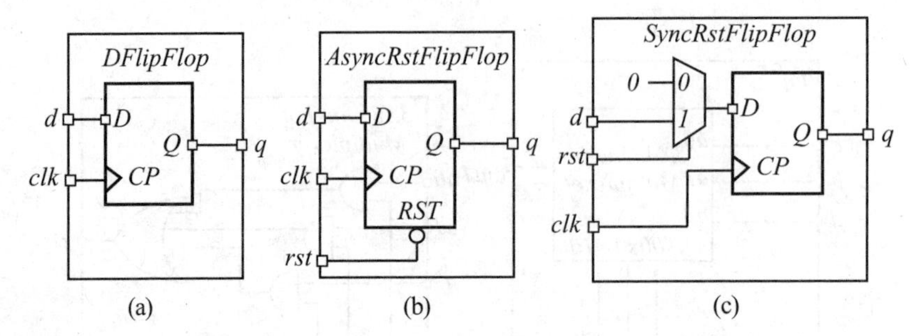

Figure 10.12 Synthesized circuit for the given Verilog codes modeling D flip-flops (a) simple (b) with asynchronous reset (c) with synchronous reset

Consider the following Verilog code.

```
module AsyncRstFlipFlop(d, clk, q, rst);
    input d, clk, rst;
    output q;
    reg q;

    always @(posedge clk or negedge rst)
        if (rst == 1'b0) begin
            q <= 1'b0;
        end else begin
            q <= d;
        end
endmodule
```

The always block is sensitive to rising edge of *clk* and falling edge of *rst*. When *rst*=0, *q* gets reset to 0. Hence, an asynchronously reset flip-flop gets inferred, as shown in Figure 10.12(b).

Consider the following Verilog code.

```
module SyncRstFlipFlop(d, clk, q, rst);
    input d, clk, rst;
    output q;
    reg q;

    always @(posedge clk)
        if (rst == 1'b0) begin
            q <= 1'b0;
        end else begin
            q <= d;
        end
endmodule
```

The always block is sensitive to only the rising edge of *clk*. When the rising edge of *clk* occurs, *q* becomes 0 if *rst*=0, else the value at *d* gets assigned to *q*. Hence, a flip-flop with synchronous reset gets inferred, as shown in Figure 10.12(c).

The statements in an always block are executed sequentially. Therefore, the connections of flip-flops inferred in an edge-sensitive always block depend on the assignments within that block. Since the mechanism of updating LHS is different for blocking and nonblocking assignments, the synthesis results can vary for these assignments. We illustrate it in the following example.

Example 10.9 Consider the following Verilog code.

```
module top1(in1, clk, out1);
    input in1, clk;
    output out1;
    reg reg1, reg2, reg3, out1;

    always @(posedge clk)
        begin
            reg1 = in1;
            reg2 = reg1;
            reg3 = reg2;
            out1 = reg3;
        end
endmodule
```

The assignments inside the always block are blocking. Therefore, whenever a rising clock edge occurs, the input *in1* will be assigned to register *reg1*, then to *reg2*, then to *reg3*, and finally to *out1*. Thus, *in1* gets transferred to *out1* at every clock edge. Hence, the above code can be synthesized to the circuit shown in Figure 10.13(a). The values of *reg1*, *reg2*, and *reg3* are redundant (not used in the circuit) and the flip-flops corresponding to them will be eliminated.

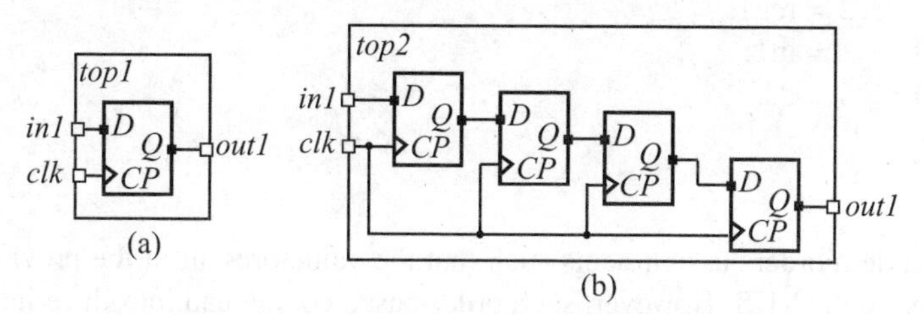

Figure 10.13 Synthesized circuit for the given Verilog codes

However, we often want a pipeline behavior (similar to a shift register) in the above Verilog code. We can achieve it using nonblocking assignments, as shown in the following code.

```
module top2(in1, clk, out1);
    input in1, clk;
    output out1;
    reg reg1, reg2, reg3, out1;

    always @(posedge clk)
        begin
            reg1 <= in1;
            reg2 <= reg1;
            reg3 <= reg2;
            out1 <= reg3;
        end
endmodule
```

Recall that a nonblocking assignment gets executed in two steps. In the first step, only the RHS is evaluated and the LHS is only scheduled to be updated with the RHS. In the second step, its LHS is updated. Hence, when the rising clock edge appears, the value *reg2* gets updated with the value of *reg1* held in the previous cycle. Similarly, *reg3* is updated with the value of *reg2* of the previous cycle, and *out1* is updated with the value of *reg3* in the previous cycle. Thus, we obtain the behavior of a shift register in the above code, and it will be synthesized as shown in Figure 10.13(b).

We can obtain a shift register by reordering blocking assignments also, as shown below.

```
module top1(in1, clk, out1);
    input in1, clk;
    output out1;
    reg reg1, reg2, reg3, out1;

    always @(posedge clk)
        begin
            out1 = reg3;
            reg3 = reg2;
            reg2 = reg1;
            reg1 = in1;
        end
endmodule
```

The above code reorders assignments such that the value present in the previous cycle on the LHS gets updated to the RHS. However, such order-based coding can introduce inadvertant errors. Therefore, we should prefer to use nonblocking assignments in always block that we expect to generate a sequential logic [3].

When the sensitivity list does not contain `posedge` or `negedge` constructs (i.e., it has only level-sensitive signals), the always block is inferred as a block consisting of purely combinational circuit elements or latches.

When the value of a variable is updated (refreshed) in every possible path (conditional branches in the code) within an always block, a purely combinational logic block gets inferred. However, if it retains its old value in some paths of the always block, a latch gets inferred. The following example illustrates it.

Example 10.10 Consider the following Verilog code.

```
module top(a, b, c, s, en, out1, out2);
    input a, b, c, s, en;
    output out1, out2;
    reg out1, out2;

    always @(*) begin
        if (s==1'b0)
            out1 = a;
        else
            out1 = b;
    end

    always @(*) begin
        if (en==1'b1)
            out2 = c;
    end
endmodule
```

In the first always block, the variable *out1* gets assigned. It gets updated to either *a* or *b*, whenever the always block is executed. Hence, the variable *out1* gets updated in all possible paths in the always block. Therefore, it gets synthesized to a purely combinational logic block. Since there are two possible cases for the `if` statement, a multiplexer is typically inferred for this always block, as shown in Figure 10.14(a).

In the second always block, the variable *out2* gets assigned. It gets updated to *c* only when *en*=1. It implies that it retains the old value when *en=0*. This always block is equivalent to the following Verilog code.

```
always @(*) begin
        if (en==1'b1)
            out2 = c;
        else
            out2 = out2;
end
```

The above portion of code can be synthesized as shown in Figure 10.14(b). It has a multiplexer in which the output recirculates if *en*=0. However, note that a multiplexer with a recirculating output is functionally equivalent to a latch.[3] Hence, the second always block gets synthesized to a latch, as shown in Figure 10.14(a).

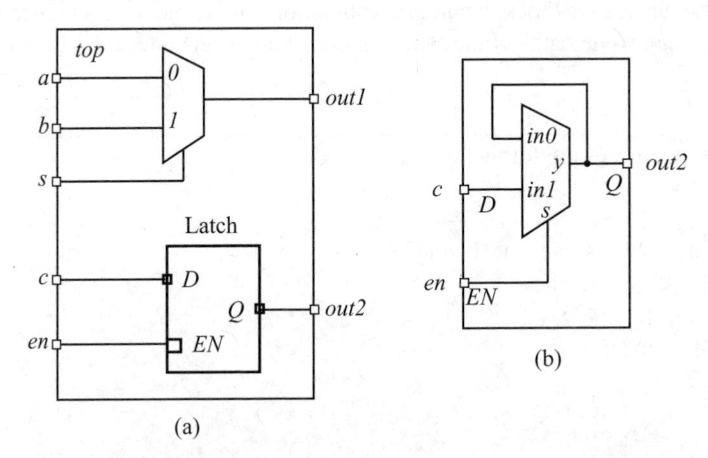

Figure 10.14 Synthesized circuit for the (a) given Verilog code (b) `if` statement in which the value gets retained when the condition is false

The above example illustrates that latches get inferred when a variable is not updated in all possible paths in an always block. Sometimes we need to model such a situation, and latches are indeed required in our design. However, latches often get inferred due to incorrect modeling of a combinational block. For example, if we inadvertently miss updating value in one of the branches of a `case` statement, a latch will be inferred. To avoid such problems, we should use `default` case to cover all possible paths in a `case` statement. Alternatively, we can set the output of a combinational logic to a default value at the beginning of the always block. It will cover the cases in which the output is not updated in some conditional branches within the always block, and we will obtain a combinational logic block.

Example 10.11 Consider the following Verilog code.

```
module top(in1, out1);
    input [0:1]in1;
    output out1;
    reg out1;

    always @* begin
        case(in1[0:1])
            2'b00: out1 = 1'b0;
```

[3] See Section 1.1.6 ("Sequential circuits") for the latch implementation.

```
            2'b01: out1 = 1'b1;
            2'b10: out1 = 1'b1;
        endcase
    end
endmodule
```

A latch will be inferred because there is no update on *out1* when *in1*=2'b11. Hence, *out1* would retain the previous value for this case. If we want that a purely combinational circuit is inferred, we need to define the behavior for the case *in1*=2'b11 also. An easy workaround for the `case` statement would be to add a `default` branch, as shown below.

```
module top(in1, out1);
    input [0:1]in1;
    output out1;
    reg out1;

    always @* begin
        case(in1[0:1])
            2'b00: out1 = 1'b0;
            2'b01: out1 = 1'b1;
            2'b10: out1 = 1'b1;
            default: out1 = 1'b0;
        endcase
    end
endmodule
```

Another mistake we can make is to omit some input signals of a combinational block from the sensitivity list of an always block. It will not change the synthesis output because the inference of inputs of combinational logic is not based on the sensitivity list but on the RHS of the assignments within that block [4]. However, it can lead to *mismatches in the simulation results* of the RTL model, and the post-synthesis netlist [4]. The problem can occur because the RTL simulation will not enter the always block for signal transitions missed in the sensitivity list. However, the gate-level simulation of the synthesized netlist will cause the combinational logic to respond to those transitions. Therefore, the simulation of the RTL model and gate-level netlist will be inconsistent. It can open up verification gaps and lead to design failure.

Example 10.12 Consider the following Verilog code.

```
module top (a, b, c, out1, out2) ;
    input a, b, c;
```

```
    output out1, out2;
    reg out1, out2;

    always @(a or b or c)
        out1 = a & b & c;

    always @(a or b)
        out2 = a & b & c;
endmodule
```

The first always block has all the signals on which *out1* depends in the sensitivity list. It will be synthesized into a three-input AND gate. The RTL model and gate-level netlist will give the same simulation results.

The second always block has *c* missing from the sensitivity list. However, the expression for *out2* contains *c* also. Nevertheless, it will be synthesized into a three-input AND gate based on the RHS expression. However, the RTL simulation of this always block will produce different results because a change to the variable *c* alone will not be observed at *out2*.

To avoid the above problem, we can use `@*` in the sensitivity list of the combinational logic.

Sometimes an always block can be non-synthesizable. For example, an always block in which the sensitivity list contains both edge- and level-triggered signals is non-synthesizable because, typically, we do not have hardware elements that match such behavior.

10.4.5 For Loops

A `for` loop can infer repeated logic structure. Typically, an RTL synthesis tool unrolls (expands) the loop and instantiates circuit elements for each iteration. Therefore, the synthesis tool must know the number of iterations in a `for` loop *during synthesis*. Hence, a `for` loop is synthesizable only when the number of iterations is *fixed*.

Example 10.13 Consider the following Verilog code.

```
module top(a, b, cin, sum, cout);
    input [3:0]a, b;
    input cin;
    output [3:0]sum;
    output cout;
    reg [3:0]sum;
    reg cout;
    reg carry;
    reg [2:0]idx;

    always @(*) begin
```

```
        carry = cin;
        for (idx=0; idx < 4; idx=idx+1)
        begin
            {carry, sum[idx]}=a[idx]+b[idx]+carry;
        end
        cout = carry;
    end
endmodule
```

The `for` loop iterates four times. Therefore, RTL synthesis tool will unroll the loop as follows:

```
{carry, sum[0]}=a[0]+b[0]+cin;
{carry, sum[1]}=a[1]+b[1]+carry;
{carry, sum[2]}=a[2]+b[2]+carry;
{cout, sum[3]}=a[3]+b[3]+carry;
```

Hence, it will use four 1-bit full adders in a cascaded configuration, as shown in Figure 10.15.

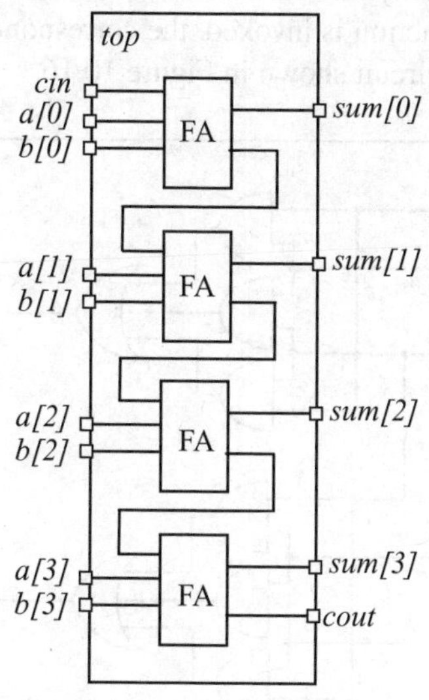

Figure 10.15 Synthesized circuit for the given Verilog code with a `for` loop

10.4.6 Functions

A Verilog `function` synthesizes to a combinational logic block with one output (scalar or vector).

Example 10.14 Consider the following Verilog code.

```
module top (a, b, c, d, e, out1, out2);
    input a, b, c, d, e;
```

```
    output out1, out2;
    reg out2;

    function MAJOR3;
        input A, B, C;
        begin
            MAJOR3 = (A&B) | (B&C) | (C&A);
        end
    endfunction
    assign out1 = MAJOR3(a, b, c);
    always @(*) begin
        out2 = MAJOR3(c, d, e);
    end
endmodule
```

The function *MAJOR3* will get synthesized to the combinational logic block based on the given Boolean expression. When the function is invoked, the corresponding combinational logic block is instantiated. Thus, we obtain the circuit shown in Figure 10.16.

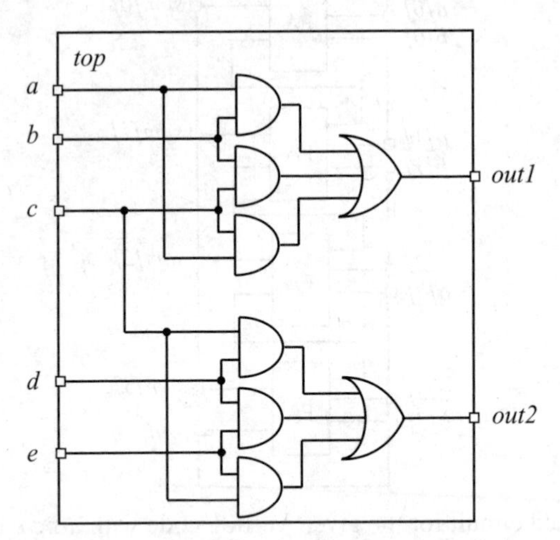

Figure 10.16 Synthesized circuit for the given Verilog code with `function`

10.4.7 Operators

An RTL synthesis tool can directly translate *bit-wise operators* and logical operators to their corresponding logic gates and optimize them subsequently. However, *arithmetic* and *relational* operators are handled differently. An RTL synthesis tool first translates them to some internal representation or data structure, such as a graph with the operator–operand relationship preserved. The data structure is chosen that facilitates *optimizations* at the *operator level*. We explain two such optimization techniques, *resource sharing* and *speculation*, in the following paragraphs.

1. **Resource sharing:** We can utilize the same computational element for multiple computations. This optimization strategy is known as *resource sharing*. By sharing resources we can save area. However, we need to ensure that its impact on the timing of a circuit is acceptable. We illustrate it in the following example.

Example 10.15 Consider the following piece of Verilog code.

```
if (sel == 1'b0)
    z = a*b;
else
    z = x*y;
```

Assume that *a*, *b*, *x*, and *y* are of 8 bits, and *z* is of 16 bits. We can represent the above code schematically using Figure 10.17(a).

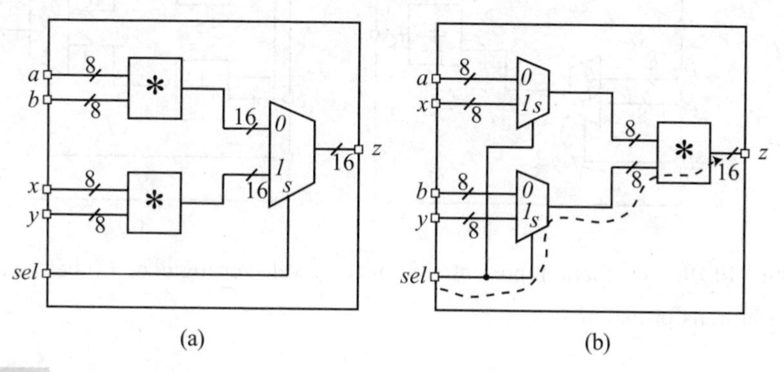

Figure 10.17 Schematic representation of the given Verilog code (a) before transformation, (b) after transformation

In this implementation, two multipliers are used. After multiplication, the result from one of the multipliers is selected. However, we can eliminate a multiplier by transforming the circuit, as shown in Figure 10.17(b). In this case, we select the operands that need to be multiplied before multiplying. Since multipliers are expensive resources in terms of area, this transformation helps us save area significantly. However, note that the multiplier now appears on the path through the select pins of the multiplexers (as indicated by the dashed line). Hence, the delay through the select pins of the multiplexers will increase due to this transformation, and we need to ensure that this path does not violate the timing constraint (in addition to other paths through the multiplier)..

2. **Speculation:** Sometimes, we can improve performance by adding extra resources or *unsharing resources*. We illustrate it in the following example.

Example 10.16 Consider the following piece of Verilog code. Assume that *a*, *b*, *c*, and *y* are of 8 bits.

```
if (sel == 1'b0)
    y = b;
else
    y = c;
z = a+y;
```

We can represent it schematically using Figure 10.18(a). Assume that the path through the *sel* port is critical, as indicated by the dashed line. Hence, the adder is on the critical path. It computes either (*a*+*b*) or (*a*+*c*) depending on the value of *sel*.

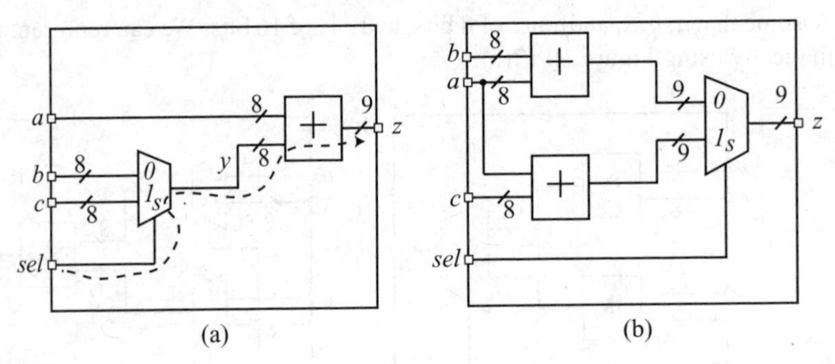

Figure 10.18 Schematic representation of the given Verilog code (a) before transformation, (b) after transformation

We can remove the adder from the critical path by performing addition before selection, as shown in Figure 10.18(b). In this implementation, two adders are used. After addition, the result from one of the adders is selected. As a result, the adders are not on the path through the port *sel*, and the critical path delay reduces. Note that an 8-bit adder typically exhibits a considerable delay. Therefore, the above transformation can reduce the critical path delay significantly. However, this transformation needs an extra adder and incurs an area penalty.

In the above transformation, we perform addition irrespective of the sum being needed (the result of only one adder will be finally selected). This transformation is an example of an optimization technique known as *speculation* or *eager computation*. It improves performance by employing more resources.

Mapping and architecture selection: After optimizing arithmetic and relational operators at the operator level, we map them to predefined modules implementing these operators. For example, an RTL synthesis tool can have a set of *internal parameterized modules* implementing arithmetic operators (+, −, *, /) and relational operators (==, >, >=, <, <=). It will instantiate these internal modules for arithmetic and relational operators, and choose the right set of parameters for the parameterized modules.

Example 10.17 Consider the following Verilog code.

```
module top(a, b, c, d, sum, carry, comp);
    input [7:0]a, b, c, d;
    output [7:0]sum;
    output carry, comp;

    assign {carry,sum}= a + b;
    assign comp = (c > d);
endmodule
```

The code contains operators + and >. An RTL synthesis tool will map the operator + to an internal module *ADD_8* and > to an internal module *GT_8*. As a result, we will obtain a circuit shown in Figure 10.19.

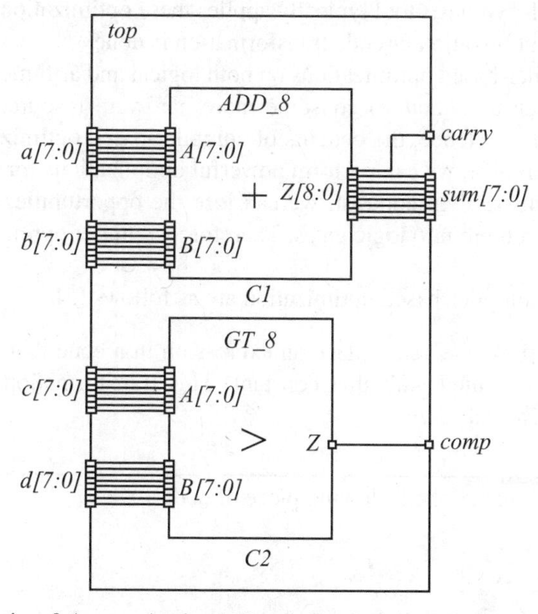

Figure 10.19 Schematic of the synthesized circuit for the given Verilog code (*ADD_8* and *GT_8* represent internal modules of an RTL synthesis tool). A tool can choose any name for these modules

The following points are noteworthy regarding the internal modules corresponding to the arithmetic operators:

1. The implementation of these modules is tool-specific. However, the logical functionality will be the same (as defined by the semantics of the language).

2. In subsequent stages of logic synthesis, the instances of these modules get *flattened* (module hierarchy gets dissolved), optimized, and mapped to the cells of the given technology libraries.
3. An RTL synthesis tool can have multiple implementations with varying architectures for the same operator. These architectures can represent different performance–area trade-offs. For example, it can have two implementations for the adder *ADD_8*: ripple carry adder (RCA) and carry-look ahead adder (CLA). An RCA has a smaller area, while CLA is faster. A synthesis tool can switch the architecture of the internal modules depending on whether the adder instance is timing critical or not. Such architectural switching can be done in the later stages of logic synthesis also. However, it should be done before flattening the internal modules because once they get flattened, it becomes challenging to reconstruct their original boundary and carry out the transformation.

10.5 COMPILER OPTIMIZATION

RTL synthesis tools employ various compiler optimization techniques by adapting them to the RTL code. These optimizations can be applied to the parse tree or the internal model of the RTL created after elaboration. An RTL synthesis tool typically applies these optimizations in passes, and in each pass, a specific type of optimization or code transformation is done.

We can apply compiler-based optimizations on both logical and arithmetic operators. However, arithmetic operators (such as + and *) consume more hardware resources compared to logical operators (such as & and |). Hence, the benefits of compiler-based optimizations are more for the arithmetic operators. Moreover, we can perform powerful optimizations for logical operators in the later stages of logic synthesis.[4] In contrast, we can lose the opportunities to optimize arithmetic operators once we convert them into logic gates. Therefore, compiler optimizations are targeted for arithmetic operators.

A few examples of compiler-based optimization are as follows [5].

1. **Constant propagation:** We can replace an expression in a code that evaluates to a *constant* every time it gets executed with that constant. This transformation is known as *constant propagation* or *constant folding*.

Example 10.18 Consider the following piece of code.

```
a = 8*8;
b = (a*1024)/32;
c = (b+32+b+32);
```

We can identify *a* as a constant and compute its value during compile time. Further, we can substitute its value in *b*. We can again identify *b* as a constant and propagate it to *c*. We can also compute *c* at the compile time. Thus, the above code transforms to the following code after constant propagation.

[4] We will describe logic optimization in Chapter 12 ("Logic optimization").

```
a = 64;
b = 2048;
c = 4160;
```

2. **Common subexpression elimination:** We can replace identical subexpression in multiple expressions with a single variable if it reduces the cost, such as area. A synthesis tool typically performs common subexpression elimination for the *arithmetic* operators during RTL synthesis, and leaves the task of common *logical* subexpression elimination to the subsequent logic optimization.

Example 10.19 Consider the following piece of code.

```
x = p+a*b;
y = q+a*b;
```

We can identify $a*b$ as a common subexpression and replace it with a variable c, as shown below.

```
c = a*b;
x = p+c;
y = q+c;
```

The above transformation allows us to use only one multiplier instead of two. Thus, it can help in saving area.

3. **Strength reduction:** Sometimes we can replace an expensive arithmetic operation with an equivalent less expensive operation. This transformation is known as *strength reduction.*

Example 10.20 Consider the following piece of code.

```
x = a*64;
y = b/4;
z = c*17;
```

It consists of expensive multiplication and division operators. We can carry out multiplication by 2 (or by powers of 2) by shifting left. Similarly, we can carry out division by 2 (or by powers of 2) by shifting right. The shifting operation is less expensive than multiplication and division. Therefore, we can transform the code as shown below and save area.

```
x = a<<6;
y = b>>2;
z = (c<<4)+c;
```

Note that we have transformed multiplication by 17 as shift and addition. Typically, such shifting and addition are less expensive than performing multiplication.

4. **Tree height reduction:** We can exploit the parallelism of hardware to improve the speed of computation for long arithmetic expressions.

   Assume that we represent a given arithmetic expression as a *binary tree* (*binary* because we typically implement arithmetic operators, such as adder and multiplier, as two-input arithmetic modules). The height of the tree represents the longest chain of operation and indicates the maximum delay in computation. We can employ the commutative and distributive properties of arithmetic operators to restructure the tree such that the height gets reduced. As a result, the maximum delay in computation gets reduced. Note that such restructuring is possible because separate hardware can work in parallel.

Example 10.21 Consider the expression $x=a+b+c+d$. We can perform the computation using two-input adders, as shown in Figure 10.20(a). Note that the computation is performed sequentially by three adders. Hence, the maximum delay in computing x is three adders' delay.

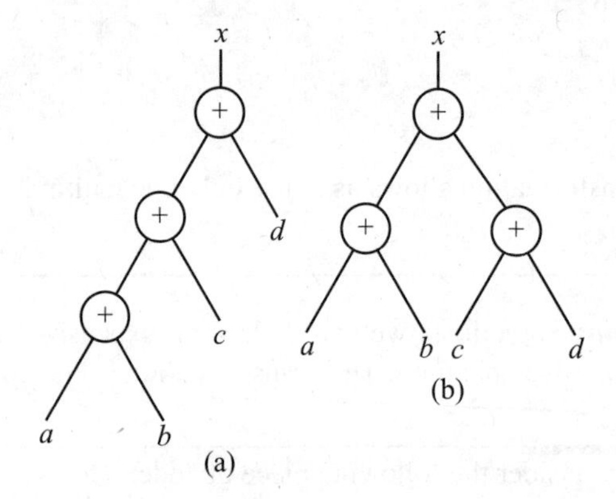

Figure 10.20 Binary tree for the given expression (a) before transformation, (b) after transformation

We can reduce the height of the expression tree as shown in Figure 10.20(b). Note that the number of adders required is the same in both implementations. However, the computations of $(a+b)$ and $(c+d)$ are being performed parallelly. Hence, the maximum delay in computing x reduces to two adders' delay.

5. **Dead code elimination:** We can remove the portion of a code that will never be executed without impacting the design functionality. Similarly, we can remove the portion of a code whose result is never used. We refer to these transformations as *dead code elimination*. Note that a designer never introduces a dead code intentionally. Nevertheless, dead code can appear in an RTL code due to previous transformations. By eliminating dead code, we can save hardware resources.

Example 10.22 Consider the following piece of code.

```
debug = 0;
if (debug == 1) begin // dead code
    x = a; // dead code
end // dead code
```

The RTL synthesis tool can deduce that the value of *debug* is 0 when it reaches the `if` statement. Hence, it can eliminate the `if` statement and everything within it as a dead code.

6. **Copy propagation:** We can sometimes use a copy assignment of the form $x = a$ for optimizing. We can replace the occurrences of the LHS (i.e., x) with the RHS (i.e., a) in the subsequent code, unless the value of a changes. This transformation is known as *copy propagation*.

Example 10.23 Consider the following piece of code.

```
x = a;
y = x+5;
z = x+9;
```

We can transform the code using copy propagation as follows:

```
x = a;
y = a+5;
z = a+9;
```

Now, the copy assignment becomes a dead code because its result is never used, and we can eliminate it as follows:

```
y = a+5;
z = a+9;
```

10.6 FSM SYNTHESIS

During RTL synthesis, FSMs need to be synthesized to hardware. The size and performance of the hardware depend on the number of states, the state encoding scheme, and the implementation of the combinational logic defining the state transitions and the output logic. Initially, an RTL synthesis tool can choose a simple encoding scheme such as one-hot encoding or what exists in the RTL code. In the later stages of logic synthesis, it can perform FSM optimization.[5]

Example 10.24 Consider the state diagram shown in Figure 10.21. There are three states {*S0*, *S1*, *S2*}. The state *S0* is the *initial state* or the *reset state*. The FSM takes 1 bit as input, and the state transitions are defined to detect a nonoverlapping sequence of 101. On detecting the 101 sequence, it produces 1 and goes to the reset state. For all other transitions, it produces 0. We can write the Verilog code for this FSM as shown below.[6] We have used asynchronous reset for the state elements.

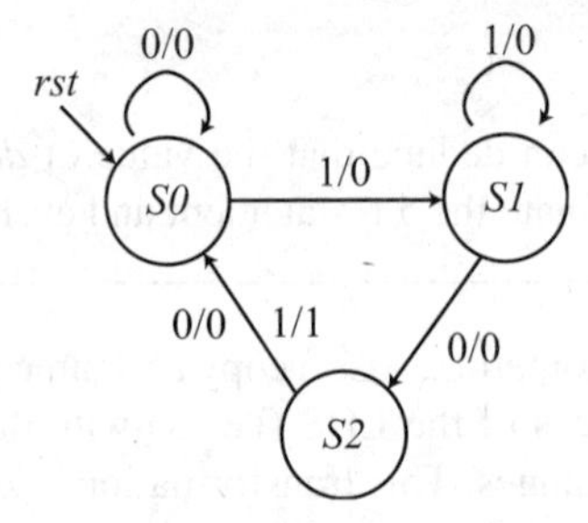

Figure 10.21 State diagram for a 101 nonoverlapping sequence detector

```
module MyFSM(in, clk, rst, out);
    input in, clk, rst;
    output out;
    reg out;

    reg  [2:0] cS;  // state elements of the FSM
    reg  [2:0] nS;  // holds the next state

    // state encoding (one-hot)
    parameter S0 = 3'b001, S1 = 3'b010, S2 = 3'b100;
```

[5] We will explain FSM optimization in Chapter 12 ("Logic optimization").

[6] The Verilog code consists of initialization (variable declaration and state encoding), combinational logic for state transition, output combinational logic, and state transition. It often improves the readability and debuggability of the Verilog code if we write these parts of an FSM in separate blocks.

```
    // combinational logic for state transition
    always @(*) begin
        case (cS)
            S0: if (in==1'b1) nS = S1; else nS = S0;
            S1: if (in==1'b0) nS = S2; else nS = S1;
            S2: nS = S0;
            default: nS = S0;
        endcase
    end

    // combinational logic for output
    always @(*) begin
        case (cS)
            S0: out = 1'b0;
            S1: out = 1'b0;
            S2: if (in==1'b1) out = 1'b1; else out = 1'b0;
            default: out = 1'b0;
        endcase
    end

    // state transition on clock edges
    always @(posedge clk or posedge rst) begin
        if (rst) cS <= S0;
        else cS <= nS;
    end
endmodule
```

An RTL synthesis tool can infer flip-flops for the state elements *cS[2:0]* using the edge-triggered `always` block. It will synthesize the code to three flip-flops (one for each bit of *cS*), as shown in Figure 10.22. Additionally, the RTL synthesis tool will infer the next state logic using assignments to the next state variable *nS[2:0]* in the first `case` statement. The logic derived using this `case` statement will drive the D-pins of the state element through the signals *nS[2:0]*. For example, we can observe from the statement "S1: if ..." that the FSM reaches the state *S2* (i.e., the next state bit *nS[2]* becomes 1), when the current state is *S1* and *in*=0.

The state *S1* is represented by 3'b010, or equivalently *cS[0]'.cS[1].cS[2]'*. Hence, the following logic will be inferred for *nS[2]*:

$$nS[2] = cS[0]'.cS[1].cS[2]'.in'$$

Similarly, an RTL synthesis tool can infer the output logic using the second `case` statement. We can observe that *out*=1 when the current state is *S2* (represented by 3'b100) and *in*=1. In all other cases, *out*=0. Hence, an RTL synthesis tool will infer the following logic for the output:

$$out = cS[0]'.cS[1]'.cS[2].in$$

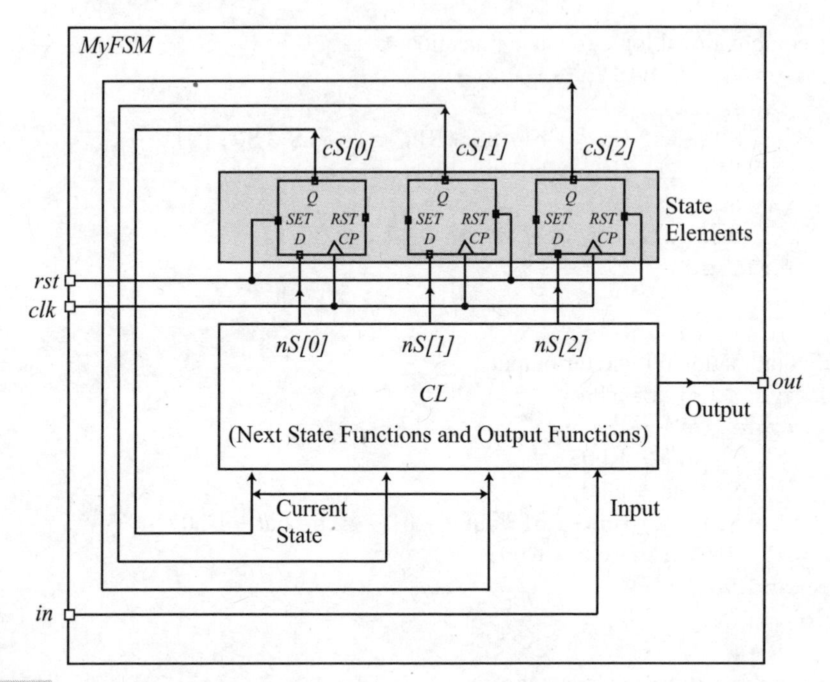

Figure 10.22 Schematic representation of the circuit for the given Verilog code (the next state and output logic functions are shown as combinational logic block *CL*)

An RTL synthesis tool can also possibly recognize the above code as an FSM and carry out FSM-based optimization such as state minimization and re-encoding. However, identifying an FSM in a Verilog code is a challenging problem because there are too many ways to represent an FSM. Therefore, tools use various heuristics to identify FSMs. Note that identifying an FSM is required not only in RTL synthesis, but also for other purposes such as functional verification of RTL using formal methods.

We know that the *next state* of an FSM depends on the *current state*. Additionally, the signals generated by the state elements propagate through the combinational logic and return to them, forming a *loop*. Hence, we can identify state elements by determining such loops in an RTL model [6]. However, such loops can occur even for sub-circuits that are not FSMs. Hence, we need to use heuristics to filter out such false matches.

RTL synthesis tools typically employ some pattern of RTL constructs or coding styles to recognize an FSM. For example, tools can search for Verilog `case` statements in a combinational `always` block and consider the variables compared in that statement as potential state elements [7]. This heuristic will work in the previous example. Additionally, a designer can provide some hints to the tool for recognizing an FSM. Moreover, we can follow a coding style that allows easy identification of an FSM [8].

10.7 RECENT TRENDS

RTL synthesis is a matured technology. Nevertheless, advancements are required to keep pace with the increased design complexity and tighter power, performance, and area (PPA) constraints.

We can take the help of SystemVerilog features to avoid simulation–synthesis mismatches, inadvertent modeling errors, and writing self-documenting codes [9]. For example, we can use explicit constructs such as `always_comb`, `always_latch`, and `always_ff` to avoid modeling errors. Additionally, we can define FSMs using `enum` that allows easy identification of the state elements.

The PPA of a design strongly depends on RTL synthesis. Optimizing data paths involving arithmetic operators is critical for meeting the PPA targets. The techniques such as resource sharing help us in minimizing circuit area. Nevertheless, we need to perform timing-aware data path optimization, especially for designs with long data paths. Typically, synthesis tools provide mechanisms and options to a designer for controlling and guiding their RTL optimizations.

After RTL synthesis, we need to verify the functional equivalence of the RTL and the generated netlist. It can be challenging when the boundaries of data path elements get dissolved or we optimize across data path elements. Also note that a synthesis tool is permitted to *refine* an `x`-valued (don't care) RTL signal to either 0 or 1 in the synthesized netlist. Such refinements also pose challenges because they make the problem more complicated for the equivalence checking tools.

REVIEW QUESTIONS

10.1 How are syntax errors detected in a Verilog code during RTL synthesis?

10.2 What do you understand by elaborating an RTL design? What kinds of errors can be detected by an RTL synthesis tool during elaboration?

10.3 When do we need to carry out uniquification of a design?

10.4 Why is delay specification in Verilog statements ignored by an RTL synthesis tool?

10.5 How are `assign` statements synthesized?

10.6 Why is it necessary that the number of iterations in a `for` loop in a Verilog code be fixed for RTL synthesis?

10.7 When can a latch be inferred in an `always` block?

10.8 How is the synthesis of Verilog `function` performed in an RTL model?

10.9 Why is it easier to carry out arithmetic operator-based optimizations at the RTL than at the gate level?

10.10 Comment on the impact of architecture chosen to implement an arithmetic operator on the performance–area trade-off.

10.11 Comment on the area and timing trade-off in the following optimization techniques:

- **(a)** Resource sharing
- **(b)** Speculation
- **(c)** Common subexpression elimination
- **(d)** Strength reduction
- **(e)** Tree height reduction

10.12 Why is it challenging to extract FSMs from a given Verilog code? How can we write a Verilog RTL model that makes this task easy for a tool?

10.13 What are merits of using `always_comb`, `always_latch`, and `always_ff` constructs over simple `always` in an RTL code?

10.14 Consider the following Verilog code.

```
module MyFSM(in, clk, rst, out);
    input in, clk, rst;
    output out;
    reg out;

    reg [1:0] cS;
    reg [1:0] nS;
    parameter S0 = 2'b00, S1 = 2'b01, S2 = 2'b10, S3 = 3'b11;

    always @(*) begin
        case (cS)
            S0: if (in==1'b1) nS = S1; else nS = S0;
            S1: if (in==1'b0) nS = S2; else nS = S1;
            S2: if (in==1'b1) nS = S3; else nS = S0;
            S3: nS = S0;
            default: nS = S0;
        endcase
    end

    always @(*) begin
        case (cS)
            S0: out = 1'b0;
            S1: out = 1'b0;
            S2: out = 1'b0;
            S3: if (in==1'b0) out = 1'b1; else out = 1'b0;
            default: out = 1'b0;
        endcase
    end

    always @(posedge clk or posedge rst) begin
        if (rst) cS <= S0;
        else cS <= nS;
    end

endmodule
```

Draw the state diagram for the FSM represented by it.

TOOL-BASED ACTIVITY

In this activity, use any logic synthesis tool, including open-source tools such as Yosys [10, 11]. You can use any library, including freely available technology libraries [12].

1. Write a Verilog module containing the code shown in Ex. 10.15. Synthesize it with a relaxed timing constraint. Examine the netlist and analyze how/whether the optimization is performed for the arithmetic operators.
2. Repeat the above activity for the code shown in Ex. 10.19.
3. Write a Verilog code shown in Ex. 10.24. Synthesize with a relaxed timing constraint. Observe the netlist and analyze how are the next state and the output functions implemented in the netlist.

REFERENCES

[1] "IEEE standard Verilog hardware description language." *IEEE Std 1364-2001* (2001), pp. 1–792.

[2] M. Budiu and S. C. Goldstein. "Pegasus: An efficient intermediate representation." Tech. rep., Carnegie-Mellon University Pittsburgh PA School of Computer Science, 2002.

[3] C. E. Cummings et al. "Nonblocking assignments in verilog synthesis, coding styles that kill!" *SNUG (Synopsys Users Group) 2000 User Papers*, 2000.

[4] D. Mills and C. E. Cummings. "RTL coding styles that yield simulation and synthesis mismatches." *SNUG (Synopsys Users Group) 1999 Proceedings* (1999).

[5] A. V. Aho, R. Sethi, and J. D. Ullman. "Compilers, principles, techniques." *Addison Wesley* 7, no. 8 (1986), p. 9.

[6] C.-N. J. Liu and J.-Y. Jou. "An automatic controller extractor for HDL descriptions at the RTL." *IEEE Design & Test of Computers* 17, no. 3 (2000), pp. 72–77.

[7] P. Jamieson and J. Rose. "A verilog RTL synthesis tool for heterogeneous FPGAs." *International Conference on Field Programmable Logic and Applications 2005* (2005), pp. 305–310, IEEE.

[8] T.-H. Wang and T. Edsall. "Practical FSM analysis for Verilog." *Proceedings International Verilog HDL Conference and VHDL International Users Forum* (1998), pp. 52–58, IEEE.

[9] "IEEE standard for SystemVerilog—Unified hardware design, specification, and verification language." *IEEE STD 1800-2009* (2009), pp. 1–1285.

[10] C. Wolf, J. Glaser, and J. Kepler. "Yosys—A free Verilog synthesis suite." *Proceedings of the 21st Austrian Workshop on Microelectronics (Austrochip)* (2013).

[11] T. Ajayi, V. A. Chhabria, M. Fogaça, S. Hashemi, A. Hosny, A. B. Kahng, M. Kim, J. Lee, U. Mallappa, M. Neseem, et al. "Toward an open-source digital flow: First learnings from the OpenROAD project." *Proceedings of the 56th Annual Design Automation Conference 2019* (2019), pp. 1–4.

[12] M. Martins, J. M. Matos, R. P. Ribas, A. Reis, G. Schlinker, L. Rech, and J. Michelsen. "Open cell library in 15nm FreePDK technology." *Proceedings of the 2015 Symposium on International Symposium on Physical Design* (2015), pp. 171–178.

11 Formal Verification

...program testing can be a very effective way to show the presence of bugs, but is hopelessly inadequate for showing their absence. The only effective way to raise the confidence level of a program significantly is to give a convincing proof of its correctness...

—Edsger W. Dijkstra, "The humble programmer," ACM Turing Lecture, 1972

A design undergoes several changes during a design flow. These changes are expected to meet or preserve the specified functionality. Nevertheless, sometimes the design functionality can deviate from the given specification due to various reasons.[1] We need to detect and fix these problems as soon as possible.

In general, verification takes considerable manual and computational effort [1–3]. Consequently, various types of verification techniques have evolved over the last few decades with different resource usage, manual intervention requirement, and rigor. Among them, *formal verification* is more rigorous, typically requires more computational resource, and is now routinely employed in a design flow. In this chapter, we will explain the basics of formal verification techniques.

11.1 LIMITATIONS OF SIMULATION-BASED VERIFICATION

The most commonly employed functional verification technique is *simulation*.[2] In this approach, we simulate a design for a set of *test vectors* and compare the output response with the expected response. If these two responses agree, then the design is *considered* to be functionally correct.

A simulation-based verification is fast and straightforward. It can efficiently find functional problems in a design and is especially useful for quickly detecting bugs and fixing functional problems in the early phases of design implementation. However, the biggest problem of a simulation-based verification is its *non-exhaustiveness*. A huge number of test vectors are possible for a given design, and we cannot simulate all of them.

Example 11.1 Consider a design consisting of a multiplier of 32-bit integers. The output Y is given as $Y = A \times B$, where A and B are 32-bit integers. Assume that each test vector takes 1 μs to simulate. Let us compute the time required to perform exhaustive simulation-based verification.

[1] Reasons for deviation from the specification can be human error, incorrect usage of tool, miscommunication, incorrect behavior of EDA tool, etc.

[2] For details of simulation, see Chapter 9 ("Simulation-based verification").

A and B can independently take one of the 2^{32} possible values. Consequently, the number of possible test vectors is $2^{32} \times 2^{32} = 2^{64}$. Simulation time required is $2^{64} \times 1 \times 10^{-6}$ seconds ≈ 0.5 million years.

Thus, simulation-based exhaustive verification is not feasible for real-world designs.

In practice, we simulate a design for a subset of all possible test vectors. Typically, we provide those test vectors that can discover some *anticipated bugs*. However, we often find bugs in those design portions where we did not pay full attention or anticipate problems. Pentium's FDIV bug is one of the most infamous bugs that crept in due to inadequate simulation-based testing [4, 5]. Due to the Pentium's FDIV bug, *sometimes* the microprocessor produced wrong results in the division of floating-point numbers. Intel had to replace the defective chips incurring losses in finances and reputation. Therefore, simulation-based verification is inadequate and can never prove the correctness of a given design. From this perspective, formal verification methodologies provide a feasible alternative.

11.2 SIMULATION VERSUS FORMAL VERIFICATION

In formal verification, we deduce mathematically or formally prove the correctness of a system. In contrast to simulation-based verification, formal verification does not explicitly rely on test vectors. However, its mathematical foundation guarantees that all the test vectors are implicitly covered while establishing system correctness. We illustrate the difference in these verification techniques in the following example.

Example 11.2 Consider a system in which the function is specified as $y = (x - 4)^2$, as shown in Figure 11.1(a). Here x is the input that can take any integer, and y is the output. Assume that we implement the given functionality as $y = x^2 - 8x + 16$, as shown in Figure 11.1(b). We need to verify that the implemented functionality is the same as the given specification. Let us examine how simulation-based verification and formal verification will tackle this problem.

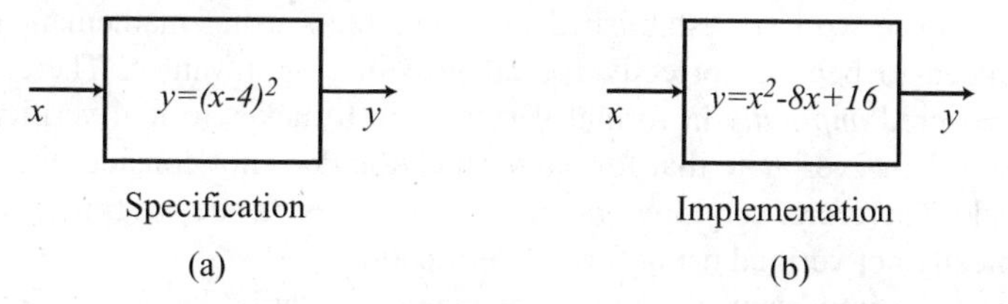

Figure 11.1 Given design (a) as specification and (b) as implementation

A simulation-based verification involves simulating the *specified functionality* and the *implementation* at different input values, as shown in Table 11.1. If their results agree at a given input value, the verification is said to "Pass" for that value. Although we have shown only five input values in Table 11.1, it is evident that x can take infinite possible values. However, simulating for all possible values is not feasible. Therefore, simulation-based verification cannot prove the correctness of the implementation.

Table 11.1 Illustration of simulation-based verification

Input (x)	Specification $(x-4)^2$	Implementation $x^2 - 8x + 16$	Pass/Fail?
0	16	16	Pass
2	4	4	Pass
4	0	0	Pass
−1	25	25	Pass
104	10000	10000	Pass

However, in formal verification, we do not enumerate different input values. We prove the correctness of the implementation by applying mathematical properties and tools. For example, we can start with the given specification and apply successive transformations (following the rules of arithmetic for integers) until we obtain the given implementation:

$(x-4)^2$
$= (x-4).(x-4)$ [By definition of square]
$= x.(x-4) - 4.(x-4)$ [By distributive property]
$= x.x - 4.x - 4.x + 16$ [By distributive/commutative property]
$= x.x - 8.x + 16$ [By distributive property]
$= x^2 - 8x + 16$ [By definition of square]

Since we obtain the implementation from the given specification by applying mathematical properties, we can conclude that the functionality of the implementation matches the given specification.

Note that we have not explicitly used input values during formal verification in the above example. However, once we have established the correctness using mathematical properties, the system is guaranteed to behave correctly for all possible input values. Therefore, all possible input values are covered *implicitly* in formal verification. It makes formal verification a *complete* verification method. However, note that formal verification does not consider the delay of a signal within a clock cycle. Therefore, *nonfunctional features* of a design such as timing safety[3] and signal integrity[4] are typically not verified using formal verification.

We summarize the important differences between simulation-based verification and formal verification in Table 11.2.

[3] Timing safety will be discussed in Chapter 14 ("Static timing analysis").

[4] Signal integrity will be discussed in Chapter 24 ("Basic concepts for physical design").

Table 11.2 Important differences in simulation-based verification and formal verification

Attributes	Simulation-based Verification	Formal Verification
Test vectors	Required	Not-required
Completeness	No	Yes
Mechanism	Simulated using test vectors and the output response expected response.	Mathematical proof of correctness developed.
Memory requirement	Comparatively low	Comparatively high
Scale of application in very large scale integration (VLSI)	Widely employed	Comparatively less employed

Note that proving the system's correctness by developing a mathematical proof is computationally challenging. However, since the early 1990s, efficient formal verification techniques have been developed for certain kinds of VLSI-specific verification problems. The development of these techniques was triggered and sustained by technological advancements in the following two areas:

1. Representation of Boolean functions using binary decision diagrams (BDDs) [6].
2. Efficient techniques to solve Boolean satisfiability (SAT) problems [7].

To understand the mechanism of formal verification, we should first understand the above techniques. Therefore, in the following sections, we will explain the fundamental aspects of BDDs and the SAT solvers (i.e., tools that solve the SAT problem). Subsequently, we will discuss the application of BDDs and SAT solvers in formal verification.

11.3 REPRESENTATION OF A BOOLEAN FUNCTION

Before understanding a BDD, we should be familiar with the following definitions related to Boolean functions and their representation.

1. **Boolean variable:** A Boolean *variable* can take one of the two values 0 or 1. We denote a Boolean variable by symbols such as x_1, a, x, etc.
2. **Boolean function:** A Boolean function is a function that takes a Boolean variable as an argument and evaluates to 0 or 1. We denote a Boolean function as $y = f\{x_1, x_2, ..., x_N\}$, where $x_1, x_2, ..., x_N$ are Boolean variables. We refer to y as the *output* of the function and the variables $x_1, x_2, ..., x_N$ as the *inputs* of the function.
3. **Literal:** A Boolean variable or its complement is known as a *literal*. Example: a, a', x, etc. We denote complement by the symbol $'$.
4. **Cube:** A product of literals is known as a *cube*. Example: $x_1x_2x_3$, x_1x_2', x_1, etc.

5. **Minterm:** Let a Boolean function y be dependent on N variables $y = f\{x_1, x_2, ..., x_N\}$. A *minterm* of the function y is the cube of N literals in which each variable or its complement appears exactly once.
6. **Maxterm:** Let a Boolean function y be dependent on N variables $y = f\{x_1, x_2, ..., x_N\}$. A *maxterm* of the function y is the sum of N literals in which each variable or its complement appears exactly once.

Example 11.3 Consider a function of three variables x_1, x_2, and x_3.
1. The minterms are: $x_1x_2x_3$, $x'_1x_2x_3$, $x'_1x'_2x'_3$, etc.
2. The cubes x_1x_2, $x_1x_2x'_2$, and x'_1x_2 are not its minterms because all variables or its complement do not appear exactly once in them.
3. The maxterms are: $(x_1 + x_2 + x_3)$, $(x'_1 + x_2 + x_3)$, $(x'_1 + x'_2 + x'_3)$, etc.
4. The sum terms $(x_1 + x_2)$, $(x_1 + x_2 + x_2')$, and $(x_1' + x_2)$ are not its maxterms because all variables or its complement do not appear exactly once in them.

7. **Truth table:** We can represent a function $y = f\{x_1, x_2, ..., x_N\}$ using a truth table. The row of a truth table shows the value (0 or 1) assigned to each input variable $\{x_1, x_2, ..., x_N\}$ and its corresponding output value. Since, 2^N possible combination of inputs is possible, a truth table has 2^N entries.
8. **Minterm representation of a function:** Each row of a truth table corresponds to a minterm. The minterm evaluates to 1 only for the assignment of variables corresponding to that row in the truth table. For all other assignments of variables, the minterm evaluates to 0. Using this observation, we can represent a Boolean function as a *sum of its minterms*. From the truth table, we take the sum of only those minterms for which the function evaluates to 1.

Example 11.4 The truth table of the function $y = x_1x_2 + x_2x_3 + x_3x_1$ is shown in Table 11.3. Since $N = 3$ in this case, there are $2^3 = 8$ rows in the truth table.

Table 11.3 Truth table for the function $y = x_1x_2 + x_2x_3 + x_3x_1$

x_1	x_2	x_3	y	Minterm
0	0	0	0	$x'_1x'_2x'_3$
0	0	1	0	$x'_1x'_2x_3$
0	1	0	0	$x'_1x_2x'_3$
0	1	1	1	$x'_1x_2x_3$
1	0	0	0	$x_1x'_2x'_3$
1	0	1	1	$x_1x'_2x_3$
1	1	0	1	$x_1x_2x'_3$
1	1	1	1	$x_1x_2x_3$

The last column in Table 11.3 shows the minterm for each row. We obtain a minterm for a given row by taking the complement of a variable assigned 0. For variables assigned 1, we take it without complementing. We can represent this function as sum of minterms as: $y = x_1'x_2x_3 + x_1x_2'x_3 + x_1x_2x_3' + x_1x_2x_3$. Note that for the row where $y = 1$, the corresponding minterm will be 1, and the *sum of minterms* will evaluate to 1. For the row where $y = 0$, all the minterms appearing in the *sum of minterms* will evaluate to 0. Thus, *sum of minterms* is an alternative representation of a truth table.

Similarly, we can represent a function and its truth table as *product of maxterms*.

We can represent a Boolean function in many ways. Some of the representations were shown above. We will discuss other representations of a Boolean function in the subsequent chapters. Note that, depending on the application, we can prefer one form of representation over the other. These representations differ in compactness, canonicity, and the ease with which we can manipulate them.

Compactness: We can quantify the compactness of a representation by examining the growth in its size with the increase in the number of Boolean variables. For example, for a Boolean function of n variables, there are 2^n rows in the truth table. Therefore, the truth table size increases exponentially with the number of Boolean variables. However, a Boolean function represented as a *logic formula* or a *logic network* can be more compact than a truth table.[5] In general, a Boolean function representation should be as small as possible. It allows the corresponding data structure to consume less memory in storage and can enable easy manipulation.

Canonicity: Another distinguishing characteristic of various representations of a Boolean function is their *canonicity*. If a representation is canonical, then the two functionally equivalent functions are represented identically. Conversely, if a representation is canonical, and if two functions have the same representation, they are functionally equivalent. For example, a truth table is a canonical representation of a Boolean function since two equivalent functions will have identical truth tables. Additionally, if the truth tables of two functions are identical, then the two functions are functionally equivalent. Similarly, *sum of minterms* and *product of maxterms* are canonical representation of a Boolean function. However, canonical representations such as the truth table and the sum of minterms grow exponentially with the number of variables. In contrast, a *logic formula* is not a canonical representation of a Boolean function. For example, the function in Table 11.3 can be written in the following five equivalent, yet different ways:

$$y = x_1x_2 + x_2x_3 + x_3x_1,$$
$$y = x_1x_2 + x_3(x_2 + x_1),$$
$$y = x_1x_2 + x_2x_3 + x_3x_1 + x_1x_2x_3,$$
$$y = (x_1x_2 + x_2x_3 + x_3x_1).(x_1 + x_1'),$$
$$y = x_1x_2 + x_2x_3 + x_3x_1 + x_1x_1'.$$

Many more functionally equivalent logic formulae are possible for the above function. In general, manipulating Boolean functions becomes easier in a canonical representation. For example, checking the equivalence of two functions is trivial with a canonical representation. Therefore, *compact canonical representation* of Boolean functions is of great interest for logic synthesis and verification.

11.4 BINARY DECISION DIAGRAMS

A BDD is a representation of a Boolean function that can be made canonical. Moreover, BDDs of several practically relevant Boolean functions are compact. Therefore, it can achieve both the desirable characteristics of a representation: canonicity and compactness. Furthermore, we can carry

[5] We will discuss logic formulae and logic networks in detail in Chapter 12 ("Logic optimization").

out some Boolean operations relevant to formal verification using BDDs very efficiently. Therefore, we often employ BDDs in formal verification. The invention of BDDs triggered the development of practically useful formal verification tools for VLSI.

11.4.1 Binary Decision Tree

A BDD is a modified form of *binary decision tree* [8, 9]. We can represent a Boolean function as a binary decision tree by recursively applying the *Shannon expansion theorem* [10]. First, we can represent a Boolean function as the sum of two sub-functions: positive and negative *cofactors* with respect to some variable in that function. The assignment of 1 to a variable in the function yields the *positive cofactor*, and 0 yields the *negative cofactor*. We can represent the given Boolean function as the root node of a binary decision tree and the negative/positive cofactors as the left/right child of the root node. Subsequently, the positive and negative cofactors are further expanded into cofactors and recursively represented as binary decision trees. The recursion continues until the cofactors reduce to constant Boolean values 0 or 1.

Example 11.5 Consider a Boolean function $f(x_1, x_2, x_3) = x_1x_2 + x_2'x_3 + x_1'x_3'$. Let us expand this function using Shannon expansion theorem.
We can write the negative and the positive cofactor of this function with respect to x_1 as:

$f_{x_1=0} = x_2'x_3 + x_3'$, and
$f_{x_1=1} = x_2 + x_2'x_3$.

The subscripts in the cofactors denote the variable assignment value corresponding to each cofactor. Thus, the function f can be written as: $f = x_1'.f_{x_1=0} + x_1.f_{x_1=1}$. We can further expand the cofactors $f_{x_1=0}$ and $f_{x_1=1}$ with respect to the second variable x_2 as follows:

$f_{x_1=0,x_2=0} = x_3 + x_3'$,
$f_{x_1=0,x_2=1} = x_3'$,
$f_{x_1=1,x_2=0} = x_3$,
$f_{x_1=1,x_2=1} = 1$.

Finally, we can expand these cofactors with respect to x_3, the last variable in the function, as follows:

$f_{x_1=0,x_2=0,x_3=0} = 1$,
$f_{x_1=0,x_2=0,x_3=1} = 1$,
$f_{x_1=0,x_2=1,x_3=0} = 1$,
$f_{x_1=0,x_2=1,x_3=1} = 0$,
$f_{x_1=1,x_2=0,x_3=0} = 0$,
$f_{x_1=1,x_2=0,x_3=1} = 1$,
$f_{x_1=1,x_2=1,x_3=0} = 1$,
$f_{x_1=1,x_2=1,x_3=1} = 1$.

Thus, we have expanded the function recursively to the lowest level.

We can represent the above function expansion as a binary decision tree shown in Figure 11.2. The root of the tree represents the given function f. The tree rooted at other nodes (rooted tree for a node v is a tree consisting of v and all its descendants) represents the cofactors. At any given node (other than the leaf nodes), there are two outgoing branches. The left one corresponds to the variable assignment of 0, and the right one corresponds to 1. The leaf nodes can take a value of 0 or 1.

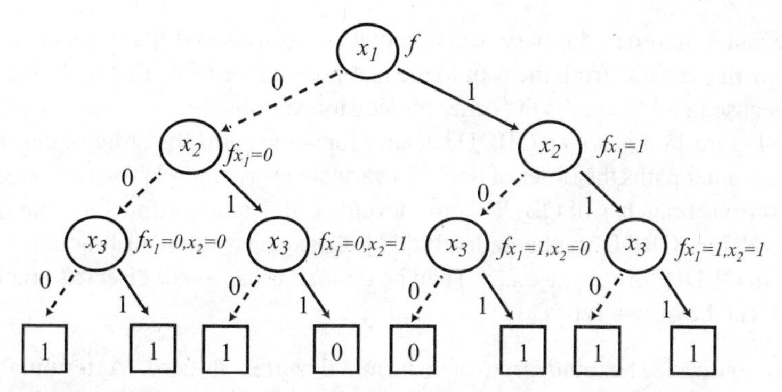

Figure 11.2 Representation of the function $f(x_1, x_2, x_3) = x_1x_2 + x_2'x_3 + x_1'x_3'$ in the form of a binary decision tree. The decision variables are marked inside the node. The function/cofactor represented by each node is marked alongside

For a function of n variables, there are 2^n leaf nodes in a binary decision tree. Each possible input combination gets explicitly represented by a unique *path* from the root to the leaf nodes. These paths correspond to 2^n rows in a truth table. Consequently, a binary decision tree is a kind of naïve BDD and is not compact. To make a BDD compact and canonical, we need to impose some constraints on the *decision variable ordering* and remove some *redundancies*, as described in the following paragraphs.

11.4.2 Ordered Binary Decision Diagram

In a binary decision tree, a rooted subtree at any given node v represents a Boolean function f_v. For a node v, we can choose any variable in f_v as a *decision variable*, as illustrated in the following example.

Example 11.6 For the binary decision tree shown in Ex. 11.5 (Figure 11.2), we can choose either x_2 or x_3 as the decision variable for the branch $x_1 = 1$. Earlier, we had chosen x_2 as the decision variable. However, if we choose x_3 as the decision variable, the binary decision tree would be as shown in Figure 11.3.

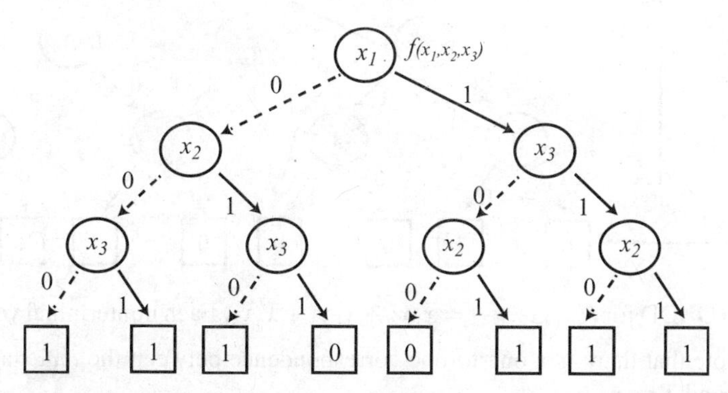

Figure 11.3 An unordered BDD for $f(x_1, x_2, x_3) = x_1x_2 + x_2'x_3 + x_1'x_3'$

Thus, we obtain different binary decision trees for different ordering of decision variables.

A BDD becomes an ordered binary decision diagram (OBDD) if the decision variables follow the same order in all the paths from the root to the leaf nodes. The BDD that is shown in Figure 11.2 is an OBDD because in all the paths the order of decision variable is $\{x_1, x_2, x_3\}$. However, the BDD that is shown in Figure 11.3 is not an OBDD because for some paths the order of decision variable is $\{x_1, x_2, x_3\}$ and for other paths the order of decision variable is $\{x_1, x_3, x_2\}$. The restriction on the order of variables in various branches of OBDD helps develop efficient algorithms for manipulating them.

Now, let us define OBDD mathematically [11]. Consider a Boolean function f of n variables $\{x_1, x_2, ..., x_n\}$. An OBDD for $f(x_1, x_2, ..., x_n)$ can be defined as a rooted, directed graph in which the set of vertices V can be of two types [11]:

1. **Terminal vertices:** The outdegree of a terminal vertex is zero. A terminal vertex v has an attribute *value*. The attribute $value(v) \in \{0, 1\}$ and represents the Boolean value of the function f.
2. **Nonterminal vertices:** A nonterminal vertex v has two children—$low(v) \in V$ and $high(v) \in V$. Additionally, a nonterminal vertex v has an attribute *index*. The attribute $index(v) \in \{1, 2, ..., n\}$ and refers to the corresponding variable in the set $\{x_1, x_2, ..., x_n\}$. The edge between a vertex v and $low(v)$ (or $high(v)$) represents the case when the coresponding variable assumes $x_{index(v)} = 0$ (or $x_{index(v)} = 1$). Therefore, a nonterminal vertex is also called a decision node.

Furthermore, to enforce ordering, following constraints are applied on an OBDD:

- For any nonterminal vertex v, if $low(v)$ is a nonterminal vertex, then $index(v) < index(low(v))$
- For any nonterminal vertex v, if $high(v)$ is a nonterminal vertex, then $index(v) < index(high(v))$.

Note that the above ordering constraints imply that along any path in the OBDD, the index of nonterminal vertices must be strictly increasing. Therefore, the same vertex cannot appear more than once along any path in an OBDD, implying that an OBDD is an acyclic graph. Based on the above definition, an OBDD for the function $f(x_1, x_2, x_3) = x_1x_2 + x_1x_3 + x_2'x_3$ is illustrated in Figure 11.4.

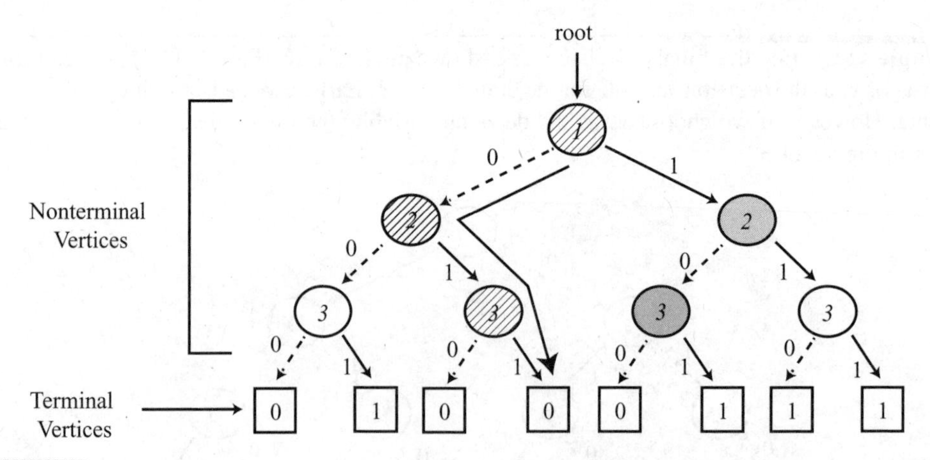

Figure 11.4 An OBDD for $f(x_1, x_2, x_3) = x_1x_2 + x_1x_3 + x_2'x_3$. Each nonterminal vertex v is marked with $index(v)$. Note that there is a one-to-one correspondence between the *index* and the *variable*: $1 \leftrightarrow x_1$, $2 \leftrightarrow x_2$, and $3 \leftrightarrow x_3$

Furthermore, an OBDD with a root vertex v denotes a Boolean function f_v, which is defined recursively as follows [11]:

- If v is a terminal vertex, then $f_v = value(v)$
- If v is a nonterminal vertex with $index(v) = i$, then $f_v = x'_i.f_{low(v)} + x_i.f_{high(v)}$

For example, the function at the fully shaded vertex 3 in Figure 11.4 is $f_3 = x'_3.0 + x_3.1 = x_3$. Similarly, the function at the fully shaded vertex 2 is $f_2 = x'_2(x'_3.0 + x_3.1) + x_2(x'_3.1 + x_3.1) = x'_2x_3 + x_2$.

A set of argument values for the variables $\{x_1, x_2, ..., x_n\}$ describes a path in an OBDD. For example, in the OBDD of Figure 11.4, let us assume that the variables are assigned as $\{x_1 = 0, x_2 = 1, x_3 = 1\}$. The corresponding path in OBDD is indicated by the thicker line with arrow. The path starts from the root of the graph. For a nonterminal vertex v, the edge v to $low(v)$ is traced if $x_{index(v)} = 0$, else the edge v to $high(v)$ is traced. For example, at the hatched vertex 2, the edge to the $high(2)$ is traced because $x_2 = 1$. The value of the function f for the given argument values for $\{x_1, x_2, ..., x_n\}$ is the *value* of the terminal vertex at the end of the traced path. For the path shown in Figure 11.4, the value of the function is 0, which agrees with the assignment of variables in the function $f(x_1, x_2, x_3) = x_1x_2 + x_1x_3 + x'_2x_3$. Additionally, it should be noted that the path traced by a set of argument values for $\{x_1, x_2, ..., x_n\}$ is unique in an OBDD.

From the above discussion, we can observe that an OBDD uniquely defines a Boolean function. However, a Boolean function is not defined uniquely by an OBDD. For example, another OBDD for the function $f(x_1, x_2, x_3) = x_1x_2 + x_1x_3 + x'_2x_3$ is shown in Figure 11.5. Therefore, an OBDD is not canonical. We need to remove redundancies in the representation to make OBDD canonical, as described in the following paragraphs.

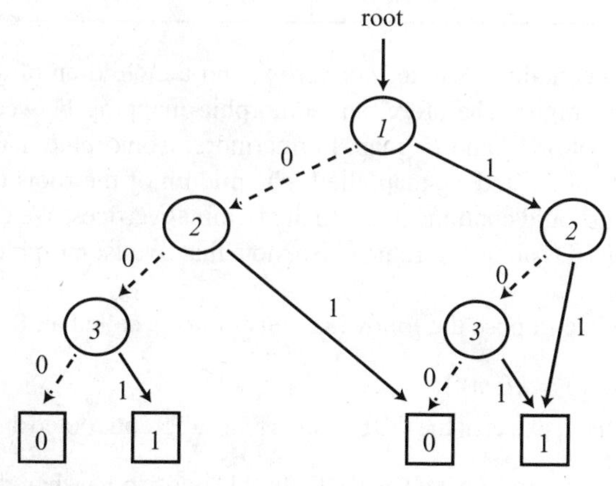

Figure 11.5 Another OBDD for $f(x_1, x_2, x_3) = x_1x_2 + x_1x_3 + x'_2x_3$

11.4.3 Reduced Ordered Binary Decision Diagram

To define *reduced ordered binary decision diagram* (ROBDD), we need to introduce the notion of *isomorphism*. Two OBDDs F_1 and F_2 are isomorphic if there exists a one-to-one mapping between their set of vertices such that the adjacency is preserved. Additionally, the correspondence of the *value* at the terminal vertices and *index* at the nonterminal vertices must exist. We illustrate the meaning of isomorphism in the following example.

Example 11.7 Consider OBDDs shown in Figure 11.6.

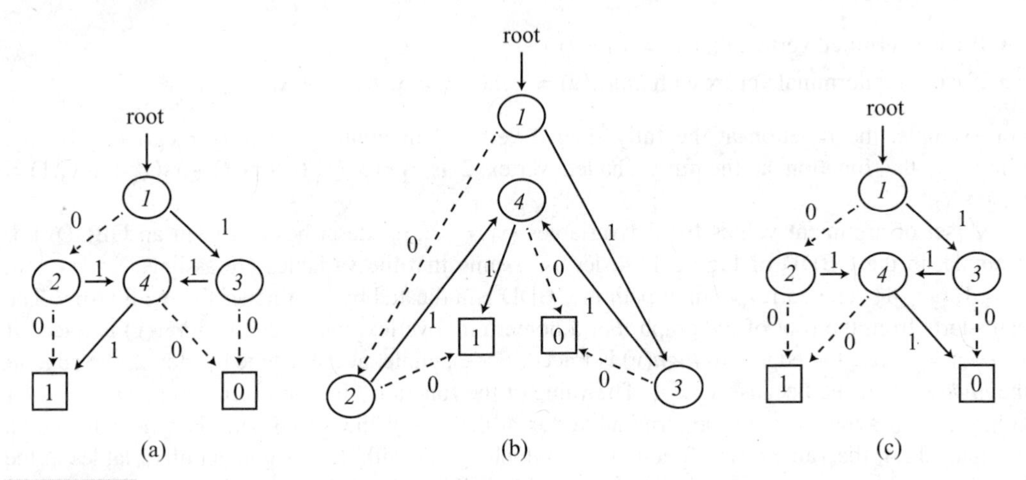

Figure 11.6 Illustration of isomorphism: OBDDs (a) and (b) are isomorphic. OBDD (c) is not isomorphic to (a) or (b)

We can establish a one-to-one mapping between the vertices of OBDDs shown in Figure 11.6(a) and (b). Therefore, these two OBDDs are isomorphic. However, if we compare the OBDDs in Figure 11.6(a) and (c), we observe that the *low*(4) and *high*(4) of these two OBDDs have different terminal vertices. Therefore, these two OBDDs are not isomorphic.

Note that an OBDD contains a single *root* vertex, and the children of any nonterminal vertex are distinguished (low or high). Therefore, an isomorphic mapping between OBDDs F_1 and F_2 firstly requires that the root in F_1 and F_2 map. Furthermore, isomorphic mapping requires that the low children of the roots of F_1 and F_2 map, the high children of the roots of F_1 and F_2 map, and recursively this mapping should continue down to the terminal vertices. We can use this observation to check whether two OBDDs are isomorphic. Also, note that two isomorphic OBDDs represent the same function.

An OBDD in which we impose the following constraints is called an ROBDD:

1. No vertex v has $low(v) = high(v)$
2. No pair of vertices $\{u, v\}$ exists in the OBDD such that subgraph rooted at u and v are isomorphic.

In widespread usage, we refer to ROBDD as BDD. In this book, other than this section, by the term BDD, we also refer to ROBDD. Bryant proved that an ROBDD is a canonical representation of a Boolean function and proposed an algorithm to obtain ROBDD by systematically removing vertices from the OBDD using the following steps [11]:

1. Any vertex with identical children is removed and replaced with any of its children.
2. Two vertices with identical OBDDs are merged into one.

These steps are carried out until we can remove vertices from the OBDD. We illustrate these steps in the following example.

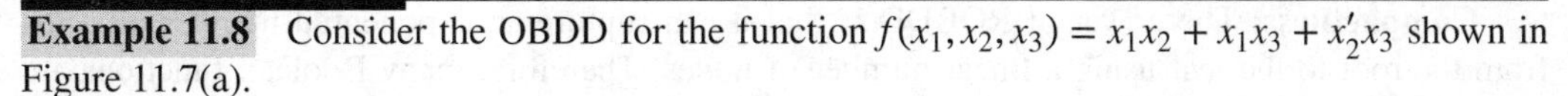

Example 11.8 Consider the OBDD for the function $f(x_1, x_2, x_3) = x_1x_2 + x_1x_3 + x_2'x_3$ shown in Figure 11.7(a).

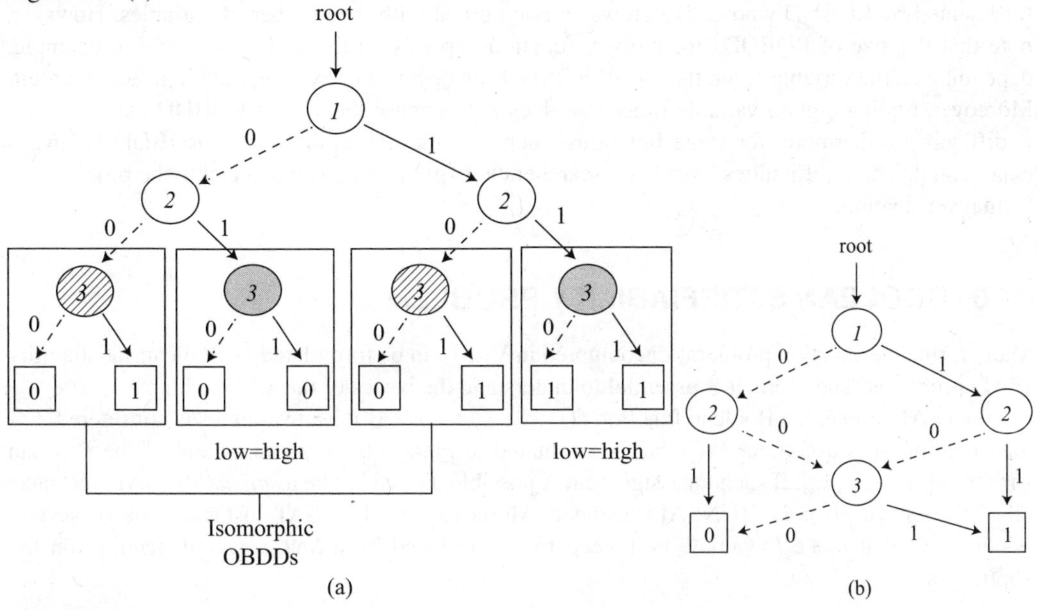

Figure 11.7 OBDD representing $f(x_1, x_2, x_3) = x_1x_2 + x_1x_3 + x_2'x_3$ (a) redundancies identified and (b) redundancies removed to obtain an ROBDD

The vertices with index 3 and fully shaded have the same low and high children. Moreover, the subgraph rooted at vertices with indices 3 and hashed are isomorphic. Therefore, the OBDD in Figure 11.7(a) is not reduced. For this OBDD, if we carry out the steps mentioned above (replace fully shaded vertices by terminal vertices and merge hashed vertices), we will obtain an ROBDD shown in Figure 11.7(b). It has no redundancy.

Practical implementations: The strategy of first creating an OBDD and then removing redundancy is generally not followed. The large size of intermediate OBDD can create memory overflow. Therefore, efficient algorithms have been developed to avoid the creation of redundant vertices while building an ROBDD. These algorithms cleverly use hash tables of functions to detect and prevent redundancies, keep ROBDD always in a reduced form, and reuse already created ROBDD for sub-functions or cofactors. Furthermore, efficient algorithms (complexity proportional to the size of the ROBDD) have been developed for manipulating and operating on BDDs of two Boolean functions.

Applications: After building ROBDDs of Boolean functions, some challenging tasks become easy. For example, checking *tautology* and *satisfiability* of a Boolean function becomes trivial. Checking for tautology involves finding whether a function produces 1 for all input combinations. Checking satisfiability involves finding whether a function produces 1 for at least one input combination. We can make these checks just by observing the root node of the corresponding ROBDD. Similarly, due to the canonicity of ROBDDs, testing the equivalence of two Boolean functions is trivial. These operations are crucial in implementing formal verification algorithms, as will become apparent in the following sections.

Compactness: The power of ROBDD is that it can capture an exponential number of paths from the root to the leaf using a linear number of nodes. Therefore, many Boolean functions are represented by ROBDD whose size grows as polynomial with the number of variables. However, note that the size of ROBDDs for a given function depends on the variable order. For example, depending on the variable order, the size of ROBDD can be linear or exponential for an adder circuit. Moreover, finding a good variable order that does not increase the size of ROBDD exponentially is difficult. Furthermore, for some functions, such as a multiplier, the size of ROBDD is always exponential. These difficulties have led researchers to explore other ways to tackle the problems of formal verification.

11.5 BOOLEAN SATISFIABILITY PROBLEM

Many formal verification problems encountered in VLSI can be formulated as Boolean satisfiability (SAT) problems. Therefore, it is essential to understand the basics of the SAT problem.

In a SAT problem, a Boolean function $f(x_1, x_2, ..., x_n)$ of n Boolean variables is considered. For this function, it is ascertained if f can be evaluated to logic 1 by any assignment of the Boolean variables $\{x_1, x_2, ..., x_n\}$. If such an assignment is possible, f is said to be a *satisfiable* (SAT) instance, otherwise an *unsatisfiable* (UNSAT) instance. Moreover, if f is a SAT instance, one or several assignments that make f evaluate to 1 need to be produced by a *SAT solver* depending on the application.

Example 11.9 Consider a function $f(x_1, x_2, x_3) = x_1x_2 + x_1x_3 + x_2'x_3$. The function f evaluates to 1 by the following assignments of $\{x_1, x_2, x_3\}$: $\{0, 0, 1\}$, $\{1, 0, 1\}$, $\{1, 1, 0\}$, and $\{1, 1, 1\}$. Therefore, f is a SAT instance.

Next, consider a function $g(x_1, x_2, x_3) = (x_1 + x_2)(x_1' + x_2')(x_1' + x_2)(x_1 + x_3)(x_1 + x_3')$. We can easily check by constructing the truth table that for no assignment of the variables $\{x_1, x_2, x_3\}$, the function g evaluates to 1. Therefore, g is an UNSAT instance.

11.5.1 Problem Formulation

When we give a SAT problem to a SAT solver, we represent the Boolean function in *conjunctive normal form (CNF)*. A Boolean function in CNF is an AND of *clauses*. A clause consists of OR of *literals*, and a literal can be a Boolean variable or its complement.

Example 11.10 Consider the Boolean function $f(x_1, x_2, x_3) = (x_1 + x_2)(x_1' + x_3)(x_2' + x_3)$.
The variables are: x_1, x_2, and x_3.
The literals used in this function are: x_1, x_2, x_3, x_1', and x_2'.
The clauses are: $(x_1 + x_2)$, $(x_1' + x_3)$, and $(x_2' + x_3)$.
The above function is built by taking AND of these clauses. Hence, it is in the CNF representation.
Some more examples of functions represented in CNF are: $f(x_1, x_2, x_3) = (x_1 + x_2 + x_3)(x_1 + x_3)(x_1' + x_2' + x_3)$ and $f(x_1, x_2, x_3, x_4) = (x_1' + x_2' + x_3')(x_1 + x_3 + x_4)(x_1' + x_2' + x_4)$.

A function represented in CNF is well-suited to be handled by a SAT solver. It reduces to 0 if any of the clauses is 0. Therefore, to make a function satisfiable, all its clauses must be simultaneously satisfied (made 1). A SAT solver can exploit this observation in detecting *conflicts* in the assignment of variables, apply *reasoning* in assigning values to variables, and *pruning* the search space. Furthermore, we can transform a given combinational logic circuit into a CNF representation in linear time and space [12, 13].

There are different forms of satisfiability problems that we typically encounter. If each clause in the CNF representation of a Boolean function is of *maximum k literals*, the corresponding SAT problem is called k-SAT problem. For example,

$f(x_1, x_2, x_3) = (x_1 + x_2)(x_1' + x_3)(x_2' + x_3)$

is a 2-SAT problem. Similarly,

$f(x_1, x_2, x_3) = (x_1 + x_2 + x_3)(x_1 + x_3)(x_1' + x_2' + x_3)$ and

$f(x_1, x_2, x_3, x_4) = (x_1' + x_2' + x_3')(x_1 + x_3 + x_4)(x_1' + x_2' + x_4)$ are 3-SAT problems.

11.5.2 Solving SAT Problem

The SAT problem belongs to the class of NP-complete problems [14]. Therefore, in the worst case, a SAT problem is expected to take exponential runtime to be solved. Despite theoretical hardness, several efficient techniques have been developed to solve SAT problems for *typical* cases that we encounter in the formal verification of VLSI circuits [15–17].

For a function of n variables, there are 2^n possible variable assignments. In the worst case, we need to try all of them. However, for many variables, say 1 million, it is not feasible to try all of them. Therefore, we typically tackle the SAT problem by performing a *systematic search* and *pruning* the search space where a satisfiable instance cannot exist.

DPLL algorithm: Most of the SAT solvers currently being used employ some form of Davis–Putnam–Logemann–Loveland (DPLL) algorithm for solving the CNF-formulated SAT problem [18, 19]. The DPLL algorithm is based on heuristically assigning a value 0/1 to an unassigned variable. Subsequently, it deduces the consequences of the assignments or determines forced assignments, as illustrated in the following example.

Example 11.11 Consider a function $f(x_1, x_2, x_3) = (x_1 + x_2)(x_1' + x_3)(x_2' + x_3')$.
Let us assign $x_1 = 1$. As a result, the clause $(x_1 + x_2)$ becomes 1. Therefore, the clause $(x_1 + x_2)$ can be removed from further considerations under the assignment $x_1 = 1$.
When $x_1 = 1$ is assigned, the second clause $(x_1' + x_3)$ reduces to x_3. Therefore, to make the second clause 1, the variable x_3 must be assigned 1. This is a forced assignment.

Deductions: A clause in which all but one literal takes a value 0 is called a *unit clause* and the corresponding forced assignment of the variable is called *implication*. Thus, assignment of variables can lead to unit clauses and implications. Note that an implication can further generate unit clauses and implications, as illustrated below.

Example 11.12 In the above example (Ex. 11.11), we have assigned $x_1 = 1$ in the function $f(x_1, x_2, x_3) = (x_1 + x_2)(x_1' + x_3)(x_2' + x_3')$.
As a result, in the clause $(x_1' + x_3)$, all but one variable takes a value 0. Hence, $(x_1' + x_3)$ becomes a unit clause. It leads to the implication $x_3 = 1$. The implication $x_3 = 1$, makes the third clause $(x_2' + x_3')$ a unit clause and lead to the implication $x_2 = 0$. This implication generates no further unit clause.

As illustrated above, given a variable assignment, we can deduce multiple implications iteratively until no more unit clause gets generated. This step is referred to as *Boolean constraint propagation* (BCP).

Conflict and backtracking: As a consequence of a variable assignment, a clause with all literals evaluating to 0 can appear. This scenario is known as a *conflict* and requires that some earlier decisions are *backtracked* by flipping the variable assignment.

Example 11.13 Consider a function $g(x_1,x_2,x_3) = (x_1 + x_2)(x_1' + x_3)(x_1' + x_3')$. Let us assign $x_1 = 1$.
As a result, the second clause becomes a unit clause and $x_3 = 1$ is an implication.
However, now, all the literals in the third clause $(x_1' + x_3')$ evaluate to 0. Therefore, under the current assignment $x_1 = 1$, the function g cannot evaluate to 1.
In this case, the variable assignment $(x_1 = 1)$ needs to be backtracked and flipped to $(x_1 = 0)$.

When is a function satisfiable: If no conflict is encountered, the solver goes on assigning variables until all variables get assigned. Finally, an assignment that satisfies a given function is obtained. In the previous example (Ex. 11.13), for the function $g(x_1,x_2,x_3) = (x_1 + x_2)(x_1' + x_3)(x_1' + x_3')$, after backtracking to $(x_1 = 0)$, the following assignment $(x_1 = 0),(x_2 = 1),(x_3 = 1)$ leads to no conflict even when all the variables have been assigned. Therefore, a satisfiable assignment is obtained.[6]

When is a function unsatisfiable: If we obtain a conflict and no more backtracking is possible, the function is unsatisfiable.

Example 11.14 Consider the function, $h(x_1,x_2) = (x_1 + x_2)(x_1' + x_3)(x_1' + x_3')(x_1 + x_2')$. For the assignment $(x_1 = 1)$, the second clause generates the implication $(x_3 = 1)$ and leads to a conflict in the third clause $(x_1' + x_3')$. Hence, we need to backtrack and assign $(x_1 = 0)$. However, the first clause now leads to an implication $(x_2 = 1)$ and a conflict is encountered in the fourth clause $(x_1 + x_2')$. Since, no more backtracking is possible for the variable x_1, we can conclude that the function h is an unsatisfiable function.

The pseudocode in Algo. 11.1, describes the general approach of solving a SAT problem formally. First, we initialize the variable *decision_level* to zero. Subsequently, we assign variables in DECIDE and increment the *decision_level* if DEDUCE finds no conflict. The function DEDUCE carries out BCP, generates implications, and detects a conflict. If conflict is found, DIAGNOSE identifies the variable assignment that needs to be *backtracked* and the corresponding *backtrack_level*. If all backtracking possibilities have been exhausted, we declare the given function *UNSAT*, otherwise backtracking is done by BACKTRACK. When all the variables get assigned, the function is declared *SAT*. A variable gets assigned to 0 or 1 in DECIDE, or by BCP in DEDUCE, or flipped in BACKTRACK.

Algo. 11.1 consists of DECIDE, DEDUCE, DIAGNOSE, and BACKTRACK. Different SAT solvers differ in the implementation of these functions. For example, a BACKTRACK implementation can pick the most *recently assigned not yet flipped* variable for backtracking. Another BACKTRACK implementation can pick a variable that led to *maximum conflict*.

[6] It could have reached other satisfiable solution $(x_1 = 0),(x_2 = 1),(x_3 = 0)$ too.

Algorithm 11.1 CHECK_SAT

Input:

- Given function *f* in CNF

Output:

- Returns *SAT* if *f* is satisfiable and *UNSAT* if *f* is unsatisfiable

Steps:

```
1: decision_level ← 0
2: while (DECIDE(f, decision_level) != ALL_ASSIGNED) do
3:    if (DEDUCE(f, decision_level) = CONFLICT) then
4:       backtrack_level ← DIAGNOSE(f, decision_level)
5:       if (backtrack_level = NOT_POSSIBLE) then
6:          return UNSAT
7:       else
8:          BACKTRACK(f, decision_level, _backtrack_level)
9:          decision_level ← backtrack_level
10:      end if
11:   else
12:      decision_level ← decision_level + 1
13:   end if
14: end while
15: return SAT
```

In recent years (since the late 1990s), many improvements have been made in the backtracking-based SAT solvers [15, 16, 20–23]. Some efficient heuristics have been developed to identify conflict-creating variable assignments. Preprocessing to simplify the SAT problem, employing efficient data structure for BCP, intelligent pruning of search spaces, random restarts, and multicore processing have enabled SAT solvers to work on large problem instances encountered in formal verification [17].

The rest of this chapter discusses how we can employ BDDs and SAT solvers in formal verification.

11.6 FORMAL VERIFICATION IN VLSI DESIGN FLOW

When we verify a design formally, in principle, we examine all possible behaviors. Therefore, using formal verification, we can obtain complete functional coverage. Thus, *in principle*, we can guarantee the functional correctness of a design for all input values. However, despite its power, we cannot apply formal verification to verify *all* the features of a given design. The high complexity of the problems (exponential in the worst case and NP-completeness) prohibits complete design verification using formal methods.

In practice, formal verification techniques supplement traditional simulation-based verification. We fill the verification gaps that remained after simulation-based verification using formal verification. We employ formal methods specifically in design subspaces or features that are critical for a design. However, in some situations, such as verification of microprocessors and in some problems, such as establishing equivalence of two circuits, formal verification techniques are used as the primary verification technique [24]. Additionally, we can employ formal verification methods to gather useful design information, identify unreachable regions or states of a model, debug, and analyze corner cases [25].

Formal verification techniques have been used in VLSI design verification since the early 1990s [25]. However, during the early days, formal verification tools could handle only small designs and had limited applications. However, with the development of new techniques (such as BDD and SAT) and novel heuristics, formal verification tools can now easily handle designs of multimillion size [25]. It has made formal verification techniques practically deployable on large VLSI design projects. We routinely apply two formal design verification techniques in a design flow: model checking and equivalence checking [26]. We will explain these two techniques in detail in the following sections.

11.7 MODEL CHECKING

A model checking tool or a model checker *formally* verifies whether a specified property holds for a given design. Thus, a model checker verifies by considering all input conditions and the complete evolution of the *states*. If the specified property is found to be violated, the model checker reports a *counter example* to enable further analysis and debugging. The inputs and the output for a model checker are shown in Figure 11.8 and described in the following paragraphs.

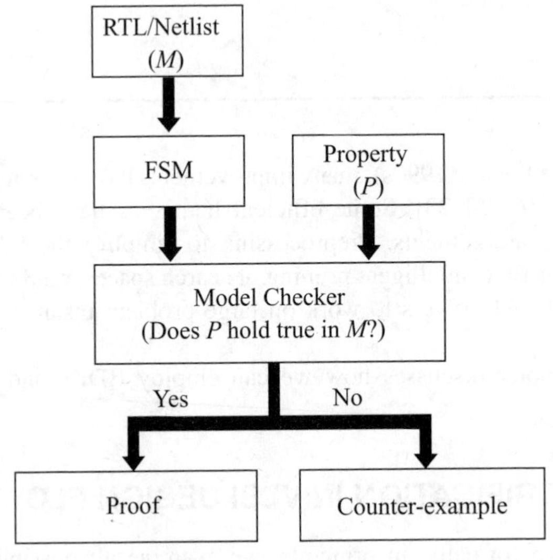

Figure 11.8 Model checking framework

11.7.1 Given Model

A model checker takes a design as an input, typically as a register transfer level (RTL) model or a netlist. A model checker internally represents a given design in an equivalent *finite state machine* (FSM). The states in an FSM correspond to the values at the Q-pin of the flip-flops. For example, consider a design that contains four flip-flops. Assume that the values at the Q-pin of these flip-flops are 1, 1, 0, and 0. Then, we can say that the state of the FSM is {1100}.

It is easy to note that if there are N flip-flops in a design, there can be 2^N possible states. Thus, the number of states in a system grows exponentially with the number of flip-flops. Even for a small design with 100 flip-flops, there can be an enormous number of possible states (2^{100}). A model checker needs to verify that the given property is valid as the system evolves through these enormously large number of possible states. This problem is known as the *state explosion problem* [27]. It is a nontrivial problem, and most research works on model checking have tried to tackle this fundamental problem.

In model checking, we often assume a design to be synchronous. The states of the system can change only at the edges of a clock signal. Additionally, we need to know the *initial state* of the FSM or the circuit for model checking. Typically, we have a *reset* signal in a circuit that initializes it or takes it to a known initial state. For example, the reset signal can produce zeros at the Q-pin of all the flip-flops. Thus, the initial state would be {000...000}. If we do not provide an initial state to the model checker, a model checking tool can assume some tool-specific default value.

11.7.2 Property Specification

We provide as an input a set of properties that a model checker needs to verify. To describe these properties, we need Boolean operators such as AND, OR, and NOT. Additionally, we often need to capture the notion of *time* in these properties. Some examples of properties are as follows (time-related description in italics):

1. *Whenever* a request for a resource is generated, the request is *eventually* granted.
2. At a traffic light post, one of the red, green, or yellow lights should *always* be ON.
3. The stack should *never* be popped *before* pushing.

In these properties, the ordering of events for time is implicit without explicitly specifying the time. We can conveniently represent these kinds of time-related properties by *temporal logic*, pioneered by Pnueli and Clarke/Emerson [28, 29].

The most popular technique to specify system properties is using *assertions* in SystemVerilog assertion (SVA) language. SVA is a part of the IEEE Std 1800-2005 SystemVerilog standard and is used to model linear-time temporal logic (LTL) [28, 30]. Modeling languages, such as Property Specification Language (PSL) and Open Verification Library (OVL), have also been used.

We can embed SVAs *within RTL* to define system properties. It allows SVAs to be checked not only by model checkers but also by simulators. The SVA language is rich enough to describe Boolean expressions, temporal relationships, and generalized regular expressions. Additionally, SystemVerilog allows specifying *assumptions* (defined using keyword `assume`) and *constraints* (defined using keyword `restrict`) about the environment and the internal signals of a design. They can simplify the verification task by providing hints to the checker and constraining the allowed behaviors. Model checkers treat them as axioms and employ them in proving assertions.[7]

Note that, in model checking, the responsibility of specifying the right set of properties is on the designer. For a nontrivial design, it is difficult for a designer to list out an exhaustive set of properties that must be satisfied by a system. Therefore, in practice, we specify a subset of the properties that

[7] For SVA, see Chapter 9 ("Simulation-based verification").

we consider critical. For example, we can define properties related to implementing handshaking protocols, avoiding deadlock conditions, and legitimate memory access for checking.

11.7.3 Techniques for Model Checking

The primary difficulty in model checking arises due to the *state explosion problem*. Earlier attempts to solve model checking by explicitly enumerating states and representing them as graphs could not scale to be practically useful [31]. A breakthrough came around 1990 when symbolic state-space exploration was proposed [7, 32, 33]. Initially, it employed BDDs and later SAT-based techniques. These techniques form the basis of the state-of-the-art model checkers, and we describe them in the following paragraphs.

BDD-based Model Checking

In BDD-based model checking, we traverse the state-space in a breadth-first manner and represent a *set of states* symbolically using BDDs as follows. Consider an FSM with a finite set of states Q. Then a subset of states $A \subset Q$ can be represented by a Boolean function f such that for any state $x \in Q, f(x) = 1$ if and only if $x \in A$.

Example 11.15 Consider an FSM with five states $Q = \{s_0, s_1, s_2, s_3, s_4\}$. Let the states be represented by 3 bits $\{x_2x_1x_0\}$. We refer to these bits as *state bits*. We can encode the states $Q = \{s_0, s_1, s_2, s_3, s_4\}$ as $\{000, 001, 010, 011, 100\}$, respectively.
In this representation, we can represent a subset of states such as $A = \{s_0, s_2, s_4\}$ as a Boolean function of state bits $f(x_2, x_1, x_0) = x_2'x_1'x_0' + x_2'x_1x_0 + x_2x_1'x_0$ (each minterm corresponds to one state). The Boolean function $f(x_2, x_1, x_0) = 1$ if and only if $\{x_2x_1x_0\} \in A$.

The Boolean function (such as f in Ex. 11.15) representing set A is called the *characteristic function* of the set A. We can use the characteristic function to represent even huge sets and utilize BDDs for compact and easy manipulation. Furthermore, we can efficiently represent the transition in an FSM using BDDs, as explained in the following paragraphs.

Consider an FSM with set of states Q. Let us denote the set of input values as I. Let us denote the *next-state function* as $x' = \delta(x, i)$. The next-state function $\delta(x, i)$ defines the transition from a given state $x \in Q$ to the next state $x' \in Q$ for a given input $i \in I$. The transition from a given state $x \in Q$ and a given input $i \in I$, to the next state $x' \in Q$ can be defined using a *transition relation* $T(x, i, x')$ such that $T(x, i, x') = 1$ if and only if $\delta(x, i) = x'$.

Example 11.16 Consider an FSM shown in Figure 11.9(a). There are two states $Q = \{sA, sB\}$ encoded as $\{\{0\}, \{1\}\}$. There are two inputs $I = \{A, B\}$ encoded as $\{\{0\}, \{1\}\}$.

The truth table of the transition relation is shown in Figure 11.9(b). For each edge that appears in Figure 11.9(a), the corresponding row in the transition relation $T(x, i, x')$ has a value of 1 and other rows have a value 0.

We can represent the transition relation of an FSM as a Boolean function using BDD.

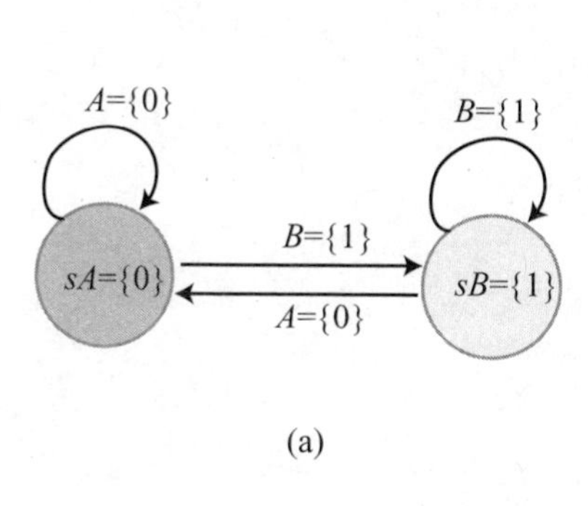

x	i	x'	$T(x,i,x')$
0	0	0	1
0	0	1	0
0	1	0	0
0	1	1	1
1	0	0	1
1	0	1	0
1	1	0	0
1	1	1	1

(b)

Figure 11.9 Illustration of transition relation (a) given FSM with states $Q = \{sA, sB\}$ encoded as $\{\{0\}, \{1\}\}$ and input set $I = \{A, B\}$ encoded as $\{\{0\}, \{1\}\}$ (the FSM is same as shown in Section 1.1.7 ("Finite state machines"), the outputs are not shown for clarity) (b) truth table of the transition relation $T(x, i, x')$ for given state x, input i, and next state x'

The BDD-based model checking relies on computing the next *set* of states from a given *set* of states by using transition relation. This process is known as *image computation*. Note that we do not consider each state *individually* during image computation. Instead, we evaluate and work on a *set of states*. Thus, for a given set of states S and transition relation $T(x, i, x')$, the image $S' = Image(S, T)$ is the set of states that we can reach in *one* step from S. Similarly, *preimage* of a set of states S' is a set of states S from which we can reach S' in *one* step. We denote preimage computation as $S = Preimage(S', T)$. For a given set of states, we can compute image and preimage very efficiently using BDDs. The efficient computation of image and preimage using BDDs makes BDD-based model checking feasible as explained below. Interested readers can look into references [34, 35] to understand the algorithmic details of image and preimage computation.

Finding reachable states: Given a starting set of states S_0, Algo. 11.2 shows how we can compute set of *all reachable states* S_{reach}.

For a given set of states, the next set of possible states is computed in each iteration by evaluating the image set and temporarily storing it in S_k. From the set of states S_k, the set of states that are already present in the reachable set of states S_{reach} is removed and stored in S_{new}.

As the iteration progresses, the new set of states S_{new} is added to the set of reachable states S_{reach}. Consequently, the set of reachable states S_{reach} grows as the iteration continues, as shown in Figure 11.10. Finally, when all the states that are reachable from S_0 have been added to S_{reach}, S_{new} reduces to an empty set. The algorithm is said to have attained a *fixed point* when S_{reach} does not change. When the fixed point is reached, the algorithm terminates.

Algorithm 11.2 COMPUTE_REACHABLE_STATES

Input:

- Given starting set of states S_0
- Transition relation $T(x, i, x')$

Output:

- Returns S_{reach} reachable set of states

Steps:

1: $S_{reach} \leftarrow S_0$
2: $S_{new} \leftarrow S_0$
3: $k = 0$
4: **while** ($S_{new} \neq \{\}$) **do**
5: $k \leftarrow k + 1$
6: $S_k \leftarrow Image(S_{new}, T)$
7: $S_{new} \leftarrow S_k - S_{reach}$
8: $S_{reach} \leftarrow S_{reach} \cup S_{new}$
9: **end while**
10: **return** S_{reach}

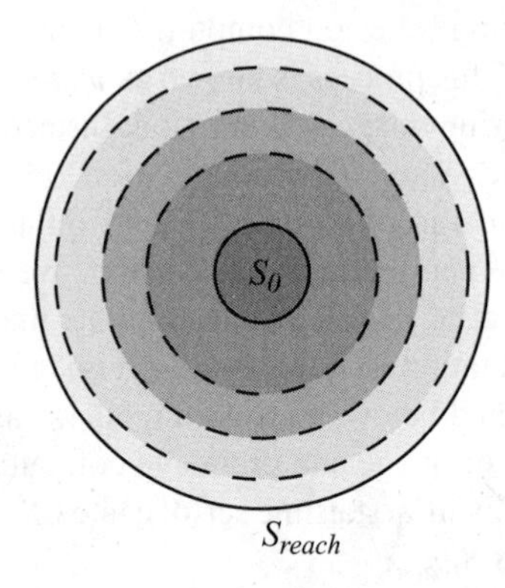

Figure 11.10 Computation of set of reachable states S_{reach} from a given set of starting states S_0. The set S_{reach} grows until the fix point is reached

Note that since the number of states is *finite* in an FSM, the algorithm is guaranteed to terminate at a fixed point. Furthermore, though the number of reachable states can be enormous, they can be represented compactly as characteristic functions using BDDs. Additionally, the canonicity of BDDs eases the task of testing the equality of sets and manipulating them.

Observe that instead of image computation, if we employ preimage computation iteratively in the above algorithm until fix-point, we obtain the set of all states from which S_0 can be reached. We denote this set of states as $S_{reach'}$ in subsequent discussions.

Model checking: BDD-based model checking employs strategies similar to finding reachable states to prove or refute a given property. Suppose it is required to check whether a state satisfying a Boolean function P is reachable from a given initial state s_0. Let both the states and the Boolean function P be represented in terms of state bits.

To prove this property, a model checker considers a set of states S_P for which P holds. Let us represent the set S_P using the characteristic function CF_{S_P}. However, we note that $CF_{S_P} = P$ (i.e., the characteristic function of the set of states for which P holds is nothing but P) because of the following argument:

Consider a state x for which P holds. Therefore, $P(x) = 1$ and x should belong to S_P. Hence, $CF_{S_P}(x) = 1$. Next, consider a state y for which $CF_{S_P}(y) = 1$. Therefore, y belongs to S_P and P should hold for it. Hence, $P(y) = 1$. Thus, $CF_{S_P} = P$.

Using preimage computation, we can determine the set of all states $S_{reach'}$ from which S_P can be reached. If $S_{reach'}$ includes the initial state s_0, we can conclude that the given property holds. On the other hand, if $S_{reach'}$ does not include the initial state s_0, we can conclude that the given property does not hold.

Limitations: As mentioned in the previous section, in the worst case, the size of BDD can be exponential in the number of inputs. Therefore, a BDD-based representation of transition relation can blow up with an increase in the number of state bits. We can avoid this in some cases by intelligently choosing the variable order. Nevertheless, a model checker can still run out of memory or cross a predefined runtime limit. In such cases, manual interventions, such as adding constraints, can help. They can simplify the problem for the model checker and make it solvable.

SAT-based Model Checking

Another technique for model checking is to formulate the property verification problem as a SAT problem and then use an efficient SAT solver to solve it.

Approach: The process involves obtaining a *counterexample* of a finite length n (n is the number of clock cycles from the initial state). We derive a Boolean function ϕ_n using the given circuit and the given property such that:

The function ϕ_n is satisfiable if and only if a counterexample of length n exists [7].

This type of model checking is known as *bounded model checking (BMC)* because the *maximum length* (n) or the bound of the counterexample is predefined. Typically, we carry out BMC iteratively by incrementing n. It continues until we have found a counterexample or the problem becomes too complicated to be handled by the SAT solver.

Mechanism: The Boolean function ϕ_n is formulated for each clock cycle n by the logical conjunction (ANDs) of clauses obtained from:

1. Given initial state.
2. The system behavior obtained from the next-state function.
3. A Boolean expression that evaluates to 1 for a counterexample (derived from the given property).

If ϕ_n is satisfiable at a given clock cycle n, the corresponding solution produces the counterexample. To derive ϕ_n, we unfold the behavior of the system one cycle at a time using the next-state function until it reaches nth clock cycle. We use different variables for the same signal in different clock cycles to represent their distinct values in each clock cycle. This is required to express the relationship between state bits and inputs in different cycles. We illustrate the mechanism of BMC in the following example.

Example 11.17 Consider a circuit consisting of a flip-flop with an output pin Q and an input port A, as shown in Figure 11.11(a). We define the circuit state by the value at the Q-pin. We denote a variable x in the nth clock cycle as x^n.

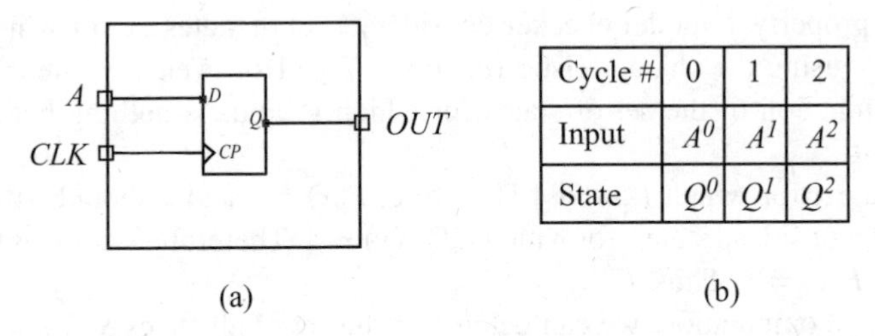

Cycle #	0	1	2
Input	A^0	A^1	A^2
State	Q^0	Q^1	Q^2

Figure 11.11 (a) Given circuit, (b) variables in different clock cycles

Suppose we need to check the following property:
P: When A is asserted (made 1), the state remains $\{1\}$ in the next *two* clock cycles.
Evidently, P does not hold because after making $A = 1$, when we make $A = 0$ in the next clock cycle, the state will change to $\{0\}$ in the second clock cycle. Let us understand how BMC can find this counterexample.

The input and the state bit Q are assigned different variable names in different clock cycles, as shown in Figure 11.11(b). Let us assume that the given initial state of the circuit is $Q^0 = 0$ and $A^0 = 1$. Thus, we are asserting 1 at the input A in the 0th clock cycle.

First clock cycle: Let us derive the Boolean expression ϕ_1 for the first clock cycle.

- The clause from the initial state (i.e., $Q^0 = 0$ and $A^0 = 1$) is $\bar{Q}^0 A^0$ (we use $\bar{\ }$ to represent complement).
- The clause from the next-state function is $(\bar{A}^0 \bar{Q}^1 + A^0 Q^1)$. It simply states that the values of A in the 0th clock cycle get assigned to Q in the first clock cycle, i.e., 0 at A^0 produces 0 at Q^1 and 1 at A^0 produces 1 at Q^1.
- The given property P gets refuted if we assert 1 on the input in the 0th clock cycle, and we obtain 0 at Q in the first clock cycle. Therefore, the refutation clause from the property is $A^0 \bar{Q}^1$.

Thus, we obtain the complete Boolean expression by the logical conjunction of the above three clauses: $\phi_1 = \bar{Q}^0 A^0.(\bar{A}^0 \bar{Q}^1 + A^0 Q^1).(A^0.\bar{Q}^1)$. By expanding the above expression, we can see that $\phi_1 = 0$ and is not satisfiable for any variable assignment. Therefore, we cannot refute P in the first clock cycle.

Second clock cycle: Let us derive the Boolean expression ϕ_2 for the second clock cycle.

- The clause from the initial state remains the same, i.e., $\bar{Q}^0 A^0$.
- By unfolding the behavior of the circuit for the second clock cycle, we express the system behavior as $(\bar{A}^1 \bar{Q}^2 + A^1 Q^2).(\bar{A}^0 \bar{Q}^1 + A^0 Q^1)$. Note that we obtain the behavior up to the second clock cycle by combining its behavior for the first and second clock cycles. Thus, as we unfold the circuit behavior for more clock cycles, the state evolution clauses become bigger.

- P gets refuted in the second clock cycle if we assert 1 on the input in 0th clock cycle, and we obtain 0 at Q in the first or the second clock cycle. This requirement can be represented by the clause $A^0.(\bar{Q}^1 + \bar{Q}^2)$. Another way of refuting P is by asserting 1 on the input in the first clock cycle and obtaining 0 at Q in the second clock cycle, i.e., $A^1.\bar{Q}^2$. Thus, the complete refutation clause from the property in the second clock cycle is $\{A^0.(\bar{Q}^1 + \bar{Q}^2) + A^1.\bar{Q}^2\}$.

The complete Boolean expression that must be satisfied to refute the given property in the second clock cycle is: $\phi_2 = \bar{Q}^0 A^0.(\bar{A}^1\bar{Q}^2 + A^1Q^2).(\bar{A}^0\bar{Q}^1 + A^0Q^1).\{A^0.(\bar{Q}^1 + \bar{Q}^2) + A^1.\bar{Q}^2\}$. By inspection, we can see that ϕ_2 is satisfied for $Q^0 = 0, A^0 = 1, A^1 = 0, Q^1 = 1, Q^2 = 0$ (A^2 can take any value). It proves that when $A^0 = 1$ (input is asserted), the output does not remain 1 (i.e., Q^2 becomes 0) in next two clock cycles. Thus, the expected counterexample is found by BMC when the behavior of the circuit is unfolded for two clock cycles.

Remark: In this example, BMC can find a counterexample in just two clock cycles. However, in general, we cannot guarantee or know whether we can find a counterexample within a given n clock cycle.

Merits: Note that BMC works on the *next-state function* represented as a compact Boolean expression. This expression grows linearly as the BMC traverses the next state in each cycle. Thus, it avoids the problem of memory blow-up in representing transition relations, typically encountered in the BDD-based model checking. However, with the introduction of more variables in each cycle, the solution space for the SAT solver rapidly grows. Therefore, the runtime of the SAT solver can become unacceptably high in BMC. Nevertheless, with the maturity attained by the SAT solvers over decades, SAT-based BMC can handle larger problem instances than BDD-based model checkers. The advantage of BMC is evident in cases where counterexamples exist, and BMC can find them quicker than BDD-based techniques [7].

Demerits: The disadvantage of BMC is that the method lacks *completeness*. In many cases, we cannot ascertain whether all possible transitions have been considered since BMC does not traverse up to the fixed point. Therefore, when a BMC cannot find a counterexample in n cycles, we cannot guarantee that the property holds for the model. Nevertheless, BMC is practically quite useful due to its ability to quickly find bugs in real circuits [25].

11.8 EQUIVALENCE CHECKING

In a design flow, we need to ensure that the *functionality* of a design remains intact even when we make some changes to it. To accomplish this, we verify that the two models of a design (one before the change and one after the change) exhibit the same logical functionality. This process is known as *equivalence checking*. The two models can be an RTL or a netlist, as demonstrated in Figure 11.12.

Sequential equivalence checking: In a generalized approach to equivalence checking, the given models are first converted to their corresponding FSMs. Then, given their initial states, it is checked whether the two FSMs produce matching output sequences for all input sequences [36]. This approach of equivalence checking is known as *sequential equivalence checking* (SEC). We need to explore all the reachable states from a given initial state in SEC. Thus, SEC needs to tackle the problem of state explosion. Therefore, SEC is a computationally difficult problem. Though multiple

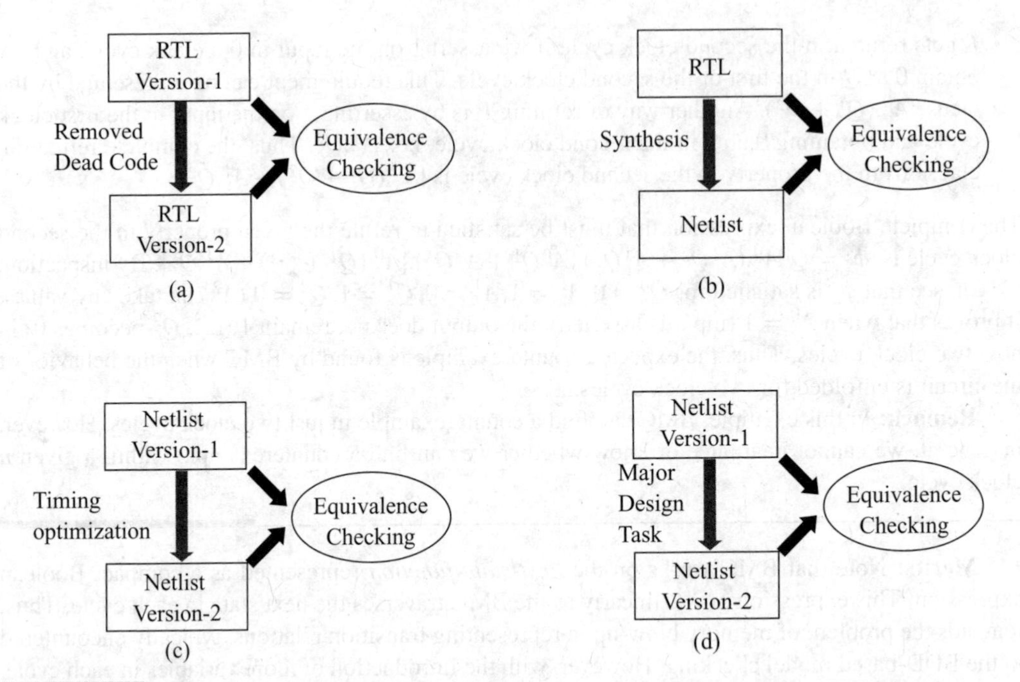

Figure 11.12 Examples of equivalence checking (a) two RTL models during RTL development, (b) RTL and netlist after synthesis, (c) two netlists after some optimization, and (d) after carrying out a major design tasks such as placement, buffering, routing, and clock tree insertion

techniques have been proposed to handle this problem, practically, we do not often use SEC in design flows [37]. Instead, we use a more straightforward equivalence checking method known as *combinational equivalence checking* (CEC). However, sometimes a design undergoes *sequential optimization* such as retiming and state minimization. In such cases, SEC becomes necessary. More often than not, sequential optimizations are avoided in a design flow to obviate verification difficulties.

Combinational equivalence checking: The CEC is based on the assumption that there is a *one-to-one correspondence* among memory elements or flip-flops and the ports of the two models. This assumption allows CEC to reduce the SEC problem to a more straightforward problem of establishing the equivalence of *pairs of combinational circuits*. Note that, due to the incremental nature of design flows, the above assumption is mostly valid. Therefore, CEC has now become an integral part of design flows. Furthermore, due to the recent advancements in CEC, it can easily handle billion-gate designs.

The steps carried out in CEC are shown in Figure 11.13. We provide the two models whose equivalence we are checking to the CEC tool. The input models are typically in the form of RTL or netlist. If the input is an RTL, then CEC internally translates the RTL to a netlist-like representation (without any significant optimization). It allows easy identification of registers (flip-flops and latches) and combinational logic elements. Subsequent steps in CEC for the RTL and the netlist representations are the same and explained in the following sections.

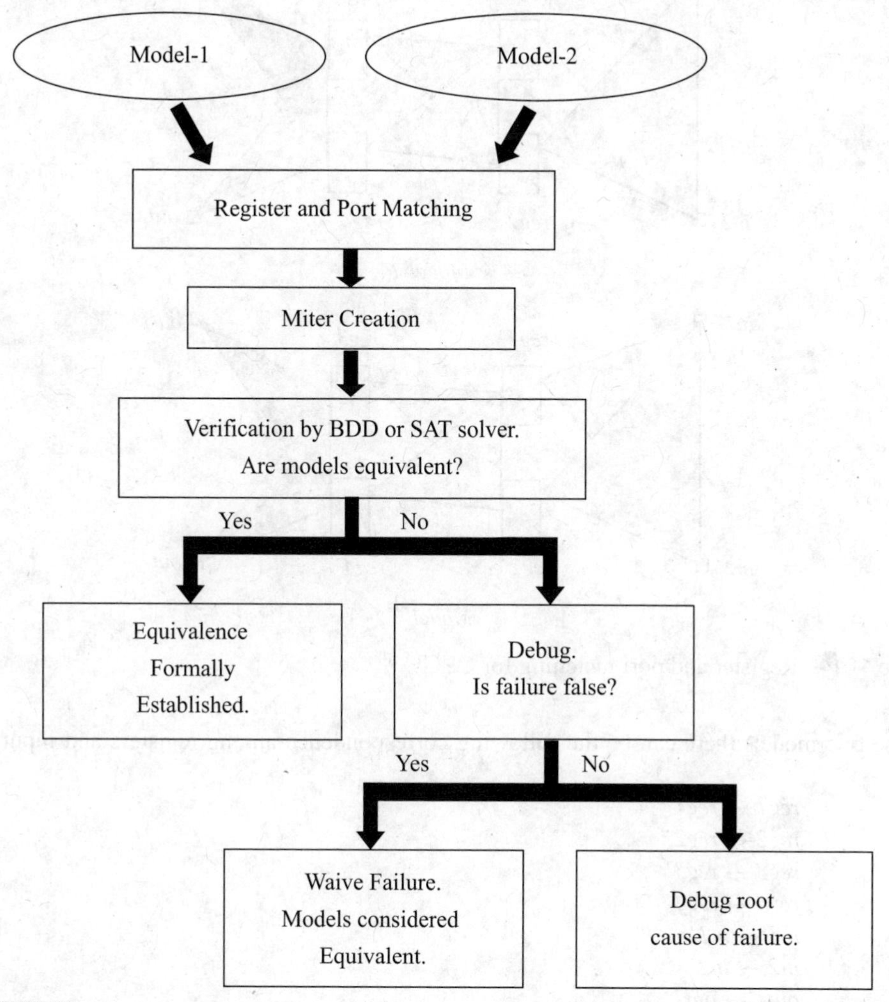

Figure 11.13 Steps in CEC

11.8.1 Register and Port Matching

Once two models are obtained in which registers have been identified, the process of *register and port matching* is carried out. This step involves establishing a correspondence between the registers and the input/output ports of the two models.

Example 11.18 Consider two models *Model-1* and *Model-2*, shown in Figure 11.14.

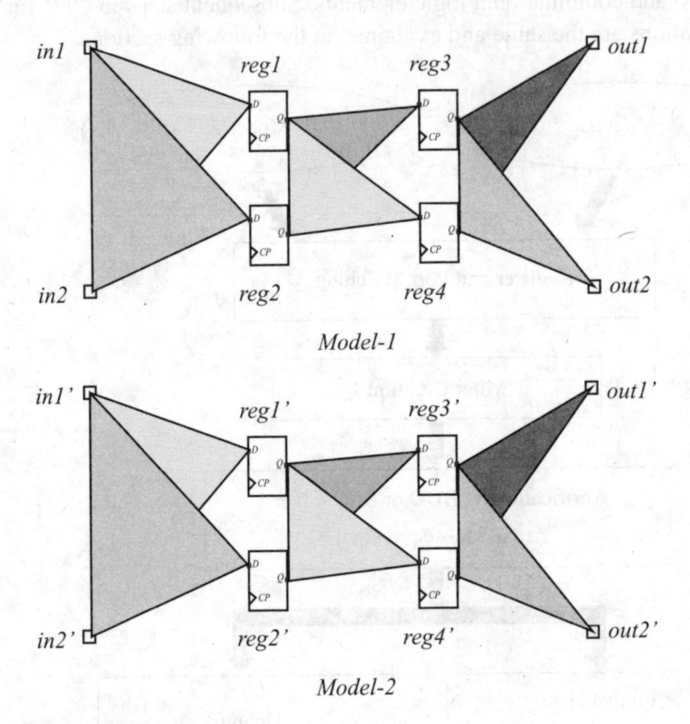

Figure 11.14 Register and port matching for CEC

In these two models there can be the following correspondence among registers and input/output ports:

reg1 ↔ *reg1'*
reg2 ↔ *reg2'*
reg3 ↔ *reg3'*
reg4 ↔ *reg4'*
in1 ↔ *in1'*
in2 ↔ *in2'*
out1 ↔ *out1'*
out2 ↔ *out2'*

These correspondences are established in the register and port matching step.

For register and port matching, there is no exact algorithm. Various heuristics are often employed by CEC tools. The most common heuristic is to establish correspondence based on names. It works very well for ports since their names typically do not change along the design flow. It also works quite well for flip-flops if the models have not undergone sequential optimization. Synthesis and implementation tools assist the name-based register matching by following some rules or conventions for register names. For example, consider the following Verilog code.

```
module flipflopD(d, clk, q);
    input d, clk;
    output q;
    reg q;
    always @(posedge clk) q <= d;
endmodule
```

The always block will be synthesized to a flip-flop that drives *q*. Though a synthesis tool can choose any arbitrary name for the synthesized flip-flop, they typically name the flip-flop by adding some suffix, such as *_reg*, to the corresponding entity in the RTL. Therefore, the name of the flip-flop for the above Verilog code can be *q_reg*. The register naming rules can guide a CEC tool in register matching. Additionally, a CEC tool can use sophisticated structural or functional analysis for register matching when name-based matching is inadequate [38]. As a designer, we can also guide or provide hints to a CEC tool for register matching.

11.8.2 Miter Creation and Checking

After determining the correspondence between the registers and the ports in the two models, we ensure that the corresponding *combinational circuits* lying between them are *equivalent*. To accomplish this, we derive multiple small combinational circuits, known as *miters*, from the given two models. A separate miter is derived for each register and output port of the two models. Using miter, we can establish the equivalence of two circuits, as we illustrate in the following example.

Example 11.19 Consider two matching registers *reg1* and *reg1'* in the models *Model-1* and *Model-2*, respectively. The primary purpose of CEC is to establish the equivalence of the combinational logic cones *CL* and *CL'* in the fanin of the D-pin of these registers, shown in Figure 11.15(a) and (b), respectively.

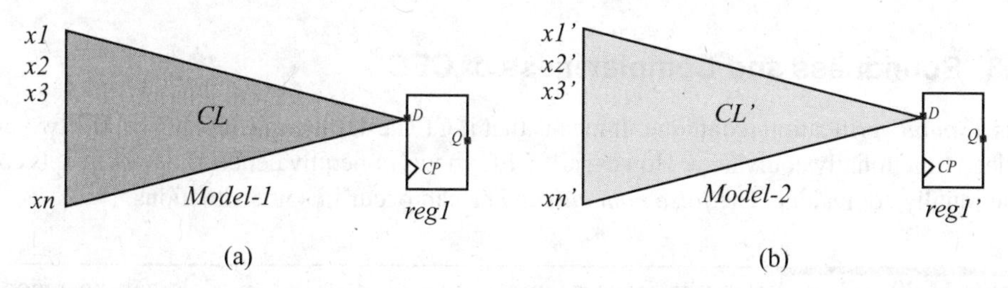

Figure 11.15 Given circuits for CEC (a) *Model-1* and (b) *Model-2*

Suppose *CL* is driven by signals $\{x1, x2, x3, ..., xn\}$, where these signals are directly connected to the Q-pin of some registers or to some input ports. Similarly, suppose *CL'* in *Model-2* is driven by signals $\{x1', x2', x3', ..., xn'\}$, where these signals are directly connected to the Q-pin of some registers or to some input ports. Since, register and port correspondences have already been established, typically, there will be a one-to-one correspondence between the signals $\{x1, x2, x3, ..., xn\}$ and $\{x1', x2', x3', ..., xn'\}$.

We obtain a miter circuit by extracting or copying the combinational logic cones *CL* and *CL'*, as shown in Figure 11.16. The inputs to the logic cones in the two models form the input ports of the miter. Note that the same input port of the miter drives the combinational logic cones of both models. The outputs of the logic cones are connected to an output port *Z* of the miter through an XOR gate.

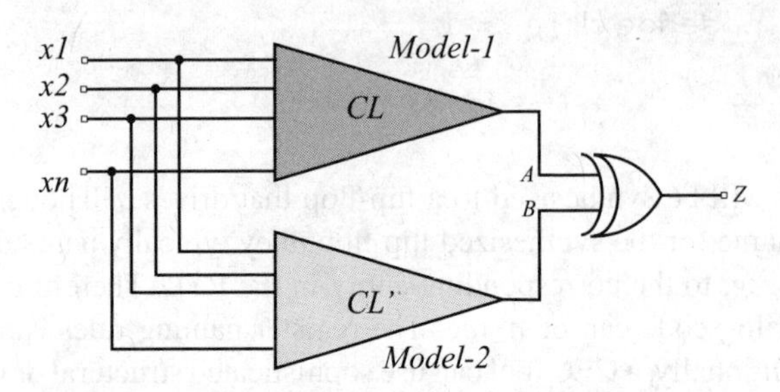

Figure 11.16 Miter for the circuits shown in Figure 11.15

If the combinational logic cones *CL* and *CL'* are equivalent, the behavior of *CL* and *CL'* would be exactly same for all combinations of input values at $\{x1, x2, x3, ..., xn\}$. Therefore, the miter output would always be 0 for equivalent logic cones. We can check this using BDD or a SAT solver. On the other hand, if the two logic cones are not equivalent, there will be some value at the inputs $\{x1, x2, x3, ..., xn\}$ for which the miter output is 1.

Establishing equivalence: If we build a BDD for a miter, it should reduce to a terminal vertex with the value of 0 for equivalent models, while for nonequivalent models, it should yield some other vertex. Similarly, we can invoke a SAT solver on the Boolean expression representing a miter. For equivalent models, the result would be UNSAT, while for nonequivalent models, it would be SAT. Thus, we can *formally* establish the equivalence of *CL* and *CL'*. Similarly, a CEC tool will create miter circuits for all other flip-flops and output ports. For equivalent models, all miter circuits should be 0 (or UNSAT).

11.8.3 Soundness and Completeness of CEC

CEC is a *sound* verification technique. It means that if CEC establishes equivalence, the two models are indeed functionally equivalent. However, if CEC shows nonequivalence, the two models can still be functionally equivalent. The *false nonequivalence* can occur in some situations.

Example 11.20 Consider the previous example (Ex. 11.19). Two models can be reported as nonequivalent when a miter can be proven SAT for some combination of input ($\{x1, x2, x3, ..., xn\}$ in Figure 11.16. Assume that CEC reports $\{x1 = 1, x2 = 1, x3 = 1, x4 = 0\}$ as the input combination for which the miter is satisfiable.

However, note that the input to a miter can be the output of registers (representing some state for an FSM). For example, $\{x1, x2, x3, x4\}$ can represent the states of an FSM encoded with four state bits. However, an FSM can prohibit the occurrence of some state bit combination. Assume that the valid states in this FSM are {0000, 0001, 0010, 00011, 0100, 0101, 0110, 0111, 1000, 1001}. Thus, the CEC reported combination {1110} would never occur in this FSM. Hence, the two models can still be equivalent despite CEC reporting nonequivalence.

Similarly, false nonequivalence can be reported by CEC if register correspondence is not correct. Therefore, CEC is considered an *incomplete* verification technique.

However, from a practical perspective, soundness and incompleteness imply that CEC is more restrictive, and employing CEC is still *safe*. Moreover, in real designs, the case of *false nonequivalence* is uncommon and can be waived off by a designer after analysis. Therefore, we apply CEC routinely in design flows.

11.9 RECENT TRENDS

Formal verification has now established itself as an integral part of VLSI design flow. Whenever a design changes, we carry out equivalence checking to ensure the sanity of the change. Furthermore, formal property verification has evolved over the last two decades. It has become a necessary complement to traditional functional verification methods and also helps in software development [39, 40]. Additionally, formal technologies are leveraged in design comprehension and *what-if analysis* [41]. The formal verification tools have matured over the years. Now, we need not be an expert in formal technologies to use these tools [25]. Therefore, we routinely employ formal verification tools in industrial design projects for various purposes.

One of the biggest challenges in property verification is the generation of properties for a given design. It is time-consuming to derive them manually. Additionally, ascertaining the quality and coverage of these properties is challenging. Therefore, automatic assertion synthesis to generate high-quality assertions for property verification is being explored and supported by many commercial tools [42, 43]. Furthermore, for standard interfaces and protocols, properties are captured in *verification intellectual properties* (VIPs) and reused in property verification [44]. Several vendors of formal verification tools also provide VIPs.

REVIEW QUESTIONS

11.1 Why is it not feasible to check all the functionality of a typical IC using simulation-based verification?

11.2 What are the advantages and disadvantages of formal verification compared to simulation-based verification?

11.3 Draw the ROBDD for the following functions with the variable order (a, b, c):

(a) $a + b + c$

(b) $a'.b'.c'$

11.4 Identify the Boolean function represented by the BDDs shown in Figure 11.17.

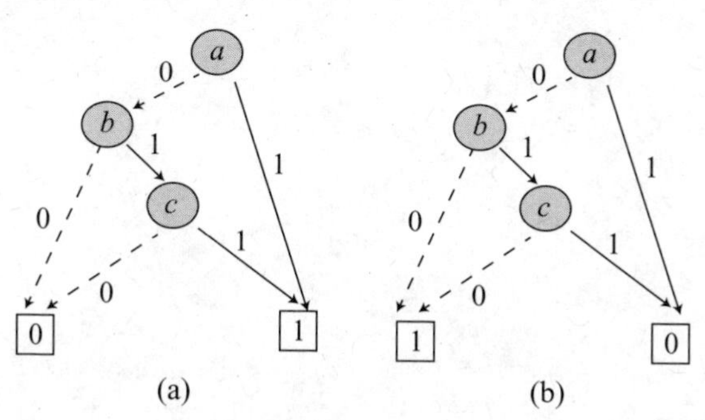

Figure 11.17 Given BDDs

11.5 What would be ROBDD for a Boolean function represented as the sum of all its minterms?

11.6 Why are SAT problems typically formulated in CNF representation?

11.7 How is Boolean constraint propagation applied in SAT solvers?

11.8 Why is the state explosion problem encountered in model checking?

11.9 How is the state explosion problem solved in BDD-based symbolic model checking?

11.10 Why is BMC not complete?

11.11 Why is SEC computationally difficult?

11.12 Why is sequential optimization such as retiming typically avoided in design flows?

11.13 Two designs are known to be equivalent (it produces the same output sequence for any given input sequence). Is it possible that CEC reports that these two designs are not equivalent? If yes, explain when it is possible?

11.14 Two models *Model-1* and *Model-2* are shown in Figure 11.18. The correspondence between ports is established using their names. The ports *A*, *B*, and *Z* in *Model-1* correspond to ports *A*, *B*, and *Z* in *Model-2*. Draw the miter circuit for checking the equivalence of *Model-1* and *Model-2*. Give two patterns of values that we can assign to ports *A* and *B* such that the output of the miter is 1 (the patterns for which the models would give different values at their output).

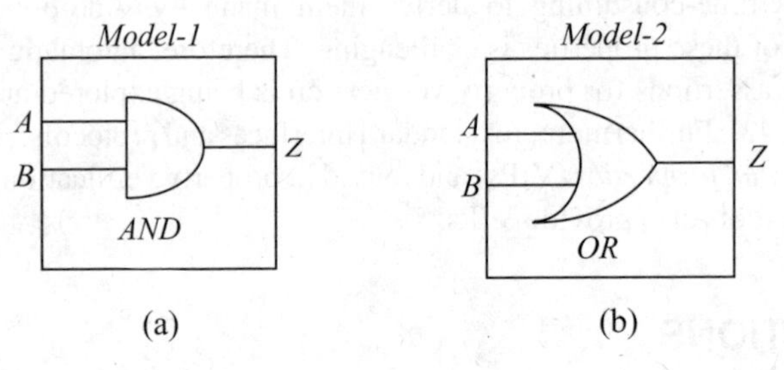

Figure 11.18 Given models for CEC

11.15 Two models *Model-1* and *Model-2* are shown in Figure 11.19. The correspondence between ports and flip-flops is established using their names. Draw the miter circuit for which CEC will report failure and the corresponding failing patterns. Are these failures reported by CEC correct (two models will give different responses), or are they false failures? Explain your answer.

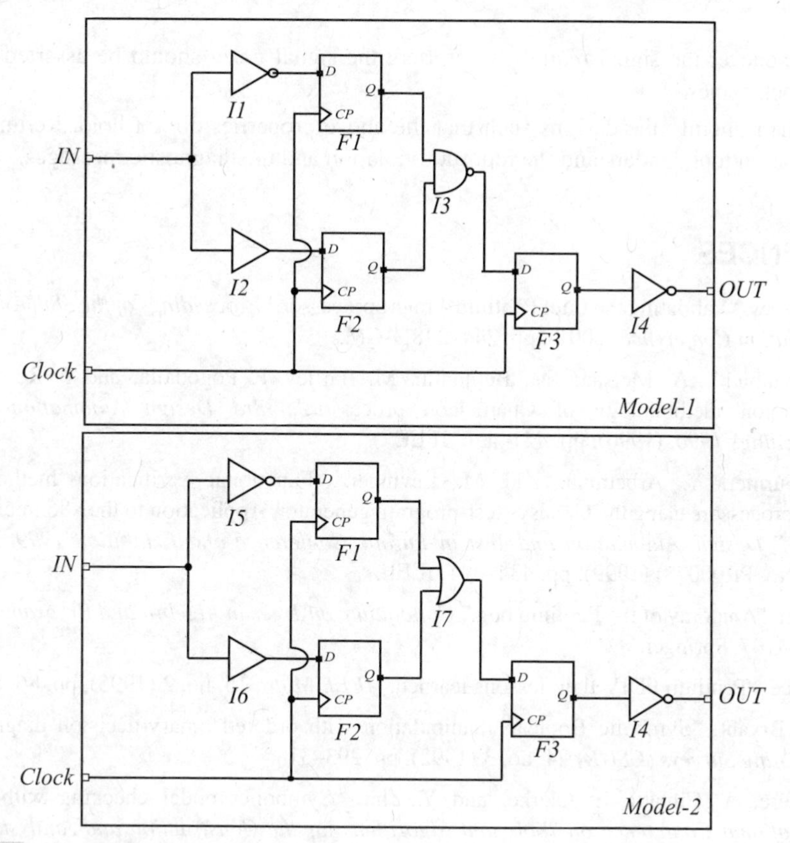

Figure 11.19 Given circuits for CEC

TOOL-BASED ACTIVITY

1. Take any **formal equivalence checking tool** that is available to you. You can take open-source tools such as Yosys that support equivalence checking [45, 46]. Use any library. For example, you can download freely available technology libraries for academic purposes [47].
   (a) Write Verilog netlist for the two models shown in 11.19(a) and (b). Choose matching net names and instance names in the two models. Run the CEC tool for the above models. Explain the result obtained.
   (b) Change the OR gate to NAND gate in *Model-2*. Rerun the CEC tool. Explain the result.
   (c) Use the netlists corresponding to 11.19(a) and (b). Run SEC if the tool supports it. Explain the result.
   (d) Change the buffers (*I2* and *I6*) with inverters in both the netlists above. Run SEC again. Explain the result.
2. Take any **formal property verification tool** that supports SVAs.
   Make SystemVerilog design(s) in which the following properties hold. Write the SVA corresponding to these properties in the same RTL. Run the formal verification tool to verify these properties. Ensure that these properties are verified formally rather than using simulation.
   (a) The signal *ready* should be high only when the signal *reset* is low.
   (b) The signal *grant* should never be high when the signal *request* is low.

(c) Whenever the signal *request* is asserted, the signal *grant* should be asserted within five clock cycles.

Then, modify the designs such that the above properties do not hold. Rerun the formal verification tool. Understand the report of violation and the diagnostic messages.

REFERENCES

[1] B. Bentley. "Validating the Intel Pentium 4 microprocessor." *Proceedings of the 38th Annual Design Automation Conference* (2001), pp. 244–248, ACM.

[2] F. Casaubieilh, A. McIsaac, M. Benjamin, M. Bartley, F. Pogodalla, and F. Ro. "Functional verification methodology of Chameleon processor." *33rd Design Automation Conference Proceedings 1996*, (1996), pp. 421–426, IEEE.

[3] L. Fournier, Y. Arbetman, and M. Levinger. "Functional verification methodology for microprocessors using the Genesys test-program generator. Application to the x86 microprocessors family." *Design, Automation and Test in Europe Conference and Exhibition 1999.* Proceedings (Cat. No. PR00078) (1999), pp. 434–441, IEEE.

[4] V. Pratt. "Anatomy of the Pentium bug." *Colloquium on Trees in Algebra and Programming* (1995), pp. 97–107, Springer.

[5] D. Price. "Pentium FDIV flaw-lessons learned." *IEEE Micro* 15, no. 2 (1995), pp. 86–88.

[6] R. E. Bryant. "Symbolic Boolean manipulation with ordered binary-decision diagrams." *ACM Computing Surveys (CSUR)* 24, no. 3 (1992), pp. 293–318.

[7] A. Biere, A. Cimatti, E. Clarke, and Y. Zhu. "Symbolic model checking without BDDs." *International Conference on Tools and Algorithms for the Construction and Analysis of Systems* (1999), pp. 193–207, Springer.

[8] C. Y. Lee. "Representation of switching circuits by binary-decision programs." *The Bell System Technical Journal* 38 (July 1959), pp. 985–999.

[9] S. B. Akers. "Binary decision diagrams." *IEEE Transactions on Computers*, no. 6 (1978), pp. 509–516.

[10] C. E. Shannon. "A symbolic analysis of relay and switching circuits." *Electrical Engineering* 57 (Dec. 1938), pp. 713–723.

[11] R. E. Bryant. "Graph-based algorithms for Boolean function manipulation." Tech. rep., Carnegie-Mellon University Pittsburgh PA School of Computer Science, 2001.

[12] G. S. Tseitin. "On the complexity of derivation in propositional calculus." *Automation of Reasoning* (1983), pp. 466–483, Springer.

[13] F. A. Aloul, I. L. Markov, and K. A. Sakallah. "MINCE: A static global variable-ordering heuristic for SAT search and BDD manipulation." *The Journal of Universal Computer Science* 10, no. 12 (2004), pp. 1562–1596.

[14] S. A. Cook. "The complexity of theorem-proving procedures." *Proceedings of the Third Annual ACM Symposium on Theory of Computing*, STOC '71, (New York, NY, USA) (1971), pp. 151–158, ACM.

[15] M. W. Moskewicz, C. F. Madigan, Y. Zhao, L. Zhang, and S. Malik. "Chaff: Engineering an efficient SAT solver." *Proceedings of the 38th Annual Design Automation Conference* (2001), pp. 530–535, ACM.

[16] N. Eén and A. Biere. "Effective preprocessing in SAT through variable and clause elimination." *International Conference on Theory and applications of Satisfiability Testing* (2005), pp. 61–75, Springer.

[17] S. Malik and L. Zhang. "Boolean satisfiability from theoretical hardness to practical success." *Communications of the ACM* 52 (Aug. 2009), pp. 76–82.

[18] M. Davis and H. Putnam. "A computing procedure for quantification theory." *Journals of the ACM* 7 (July 1960), pp. 201–215.

[19] M. Davis, G. Logemann, and D. Loveland. "A machine program for theorem-proving." *Communications of the ACM* 5, no. 7 (1962), pp. 394–397.

[20] J. P. Marques-Silva and K. A. Sakallah. "GRASP: A search algorithm for propositional satisfiability." *IEEE Transactions on Computers* 48, no. 5 (1999), pp. 506–521.

[21] L. Xu, F. Hutter, H. H. Hoos, and K. Leyton-Brown. "SATzilla: Portfolio-based algorithm selection for SAT." *Journal of Artificial Intelligence Research* 32 (2008), pp. 565–606.

[22] F. Hutter, D. Babic, H. H. Hoos, and A. J. Hu. "Boosting verification by automatic tuning of decision procedures." *Formal Methods in Computer Aided Design* (FMCAD'07) (2007), pp. 27–34, IEEE.

[23] Y. Hamadi, S. Jabbour, and L. Sais. "ManySAT: A parallel SAT solver." *Journal on Satisfiability, Boolean Modeling and Computation* 6 (2008), pp. 245–262.

[24] R. Kaivola, R. Ghughal, N. Narasimhan, A. Telfer, J. Whittemore, S. Pandav, A. Slobodová, C. Taylor, V. Frolov, E. Reeber, et al. "Replacing testing with formal verification in Intel Core i7 processor execution engine validation." *International Conference on Computer Aided Verification* (2009), pp. 414–429, Springer.

[25] E. Seligman, T. Schubert, and M. A. K. Kumar. *Formal Verification: An Essential Toolkit for Modern VLSI Design*. Morgan Kaufmann, 2015.

[26] L. Fix. "Fifteen years of formal property verification in Intel." *25 Years of Model Checking* (2008), pp. 139–144, Springer.

[27] E. M. Clarke, E. A. Emerson, and J. Sifakis. "Model checking: Algorithmic verification and debugging." *Communications of the ACM* 52, no. 11 (2009), pp. 74–84.

[28] A. Pnueli. "The temporal logic of programs." *18th Annual Symposium on Foundations of Computer Science (sfcs 1977)* (1977), pp. 46–57, IEEE.

[29] E. M. Clarke and E. A. Emerson. "Design and synthesis of synchronization skeletons using branching time temporal logic." *Workshop on Logic of Programs* (1981), pp. 52–71, Springer.

[30] "IEEE standard for SystemVerilog: Unified hardware design, specification and verification language." *IEEE Std1800-2005* (2005), pp. 1–648.

[31] E. M. Clarke and Q. Wang. "25 years of model checking." *International Andrei Ershov Memorial Conference on Perspectives of System Informatics* (2014), pp. 26–40, Springer.

[32] J. R. Burch, E. M. Clarke, K. L. McMillan, D. L. Dill, and L.-J. Hwang. "Symbolic model checking: 10^{20} states and beyond." *Information and Computation* 98 (1992), no. 2, pp. 142–170.

[33] O. Coudert, J. C. Madre, and C. Berthet. "Verifying temporal properties of sequential machines without building their state diagrams." *Proceedings of the 2nd International Workshop on Computer Aided Verification*, CAV '90, (Berlin, Heidelberg) (1991), pp. 23–32, Springer-Verlag.

[34] J. Herve, S. Hamid, L. Bill, K. B. Robert, and S.-V. Alberto. "Implicit state enumeration of finite state machines using BDD's." Computer-aided Design, 1990ICCAD-90. Digest of Technical Papers. *1990 IEEE International Conference on* (1990), pp. 130–133.

[35] I.-H. Moon, J. H. Kukula, K. Ravi, and F. Somenzi. "To split or to conjoin: The question in image computation." *Proceedings of the 37th Annual Design Automation Conference* (2000), pp. 23–28.

[36] M. N. Mneimneh and K. A. Sakallah. "Principles of sequential-equivalence verification." *IEEE Design & Test of Computers* 22, no. 3 (2005), pp. 248–257.

[37] C. Van Eijk. "Sequential equivalence checking without state space traversal." *Proceedings Design, Automation and Test in Europe* (1998), pp. 618–623, IEEE.

[38] J. Mohnke, P. Molitor, and S. Malik. "Establishing latch correspondence for sequential circuits using distinguishing signatures." *Proceedings of 40th Midwest Symposium on Circuits and Systems. Dedicated to the Memory of Professor Mac Van Valkenburg* 1 (Aug. 1997), pp. 472–476.

[39] E. Villani, R. P. Pontes, G. K. Coracini, and A. M. Ambrósio. "Integrating model checking and model based testing for industrial software development." *Computers in Industry* 104 (2019), pp. 88–102.

[40] A. Cimatti and A. Griggio. "Software model checking via IC3." *International Conference on Computer Aided Verification* (2012), pp. 277–293, Springer.

[41] R. K. Ranjan, C. Coelho, and S. Skalberg. "Beyond verification: Leveraging formal for debugging." *Proceedings of the 46th Annual Design Automation Conference* (2009), pp. 648–651, ACM.

[42] L.-C. Wang, M. S. Abadir, and N. Krishnamurthy. "Automatic generation of assertions for formal verification of PowerPC/sup TM/microprocessor arrays using symbolic trajectory evaluation." *Proceedings 1998 Design and Automation Conference.* 35th DAC. (Cat. No.98CH36175) (June 1998), pp. 534–537.

[43] S. Vasudevan, D. Sheridan, S. Patel, D. Tcheng, B. Tuohy, and D. Johnson. "Goldmine: Automatic assertion generation using data mining and static analysis." *2010 Design, Automation & Test in Europe Conference & Exhibition (DATE 2010)* (2010), pp. 626–629, IEEE.

[44] W. Chen, S. Ray, J. Bhadra, M. Abadir, and L.-C. Wang. "Challenges and trends in modern SoC design verification." *IEEE Design & Test* 34, no. 5 (2017), pp. 7–22.

[45] C. Wolf, J. Glaser, and J. Kepler. "Yosys—A free Verilog synthesis suite." *Proceedings of the 21st Austrian Workshop on Microelectronics (Austrochip)* (2013).

[46] T. Ajayi, V. A. Chhabria, M. Fogaça, S. Hashemi, A. Hosny, A. B. Kahng, M. Kim, J. Lee, U. Mallappa, M. Neseem, et al. "Toward an open-source digital flow: First learnings from the OpenROAD project." *Proceedings of the 56th Annual Design Automation Conference 2019* (2019), pp. 1–4.

[47] M. Martins, J. M. Matos, R. P. Ribas, A. Reis, G. Schlinker, L. Rech, and J. Michelsen. "Open cell library in 15nm FreePDK technology." *Proceedings of the 2015 Symposium on International Symposium on Physical Design* (2015), pp. 171–178.

12 Logic Optimization

We ascribe beauty to that which is simple; which has no superfluous parts; which exactly answers its end...

—R. W. Emerson, *The Conduct of Life*, on "Beauty," 1871

Logic optimization *transforms* a given function such that some quality of result (QoR) is improved, preserving the *functional equivalence* and meeting the given *constraints*. The QoR measure can be area, performance, power, testability, routability, and manufacturability.

We can employ logic optimization at various stages in the VLSI design flow. After register transfer level (RTL) synthesis, we obtain a netlist in terms of *generic gates*. For a generic gate, the function is well-defined. However, other attributes of a generic gate, such as timing, area, and power, are not well-defined. A logic optimization tool crudely estimates them and improves some QoR measures of the circuit. This task is known as *technology-independent logic optimization*.

Post technology-independent logic optimization, we map the netlist to the *cells* of the given *technology libraries*. This task is known as *technology mapping* or *mapping*. Post-mapping, we can apply logic optimization and improve QoR measures of the *cell-instantiated netlist*. This task is known as *technology-dependent logic optimization*.

In this chapter, we will discuss technology-independent logic optimizations. We will discuss technology-dependent logic optimizations in Chapter 17 ("Timing-driven optimizations") and Chapter 19 ("Power-driven optimizations").

There are two distinct types of technology-independent logic optimizations: *sequential logic optimization* and *combinational logic optimization*. In sequential logic optimization, we typically optimize the finite state machine (FSM) representation by *state minimization* and *state encoding*. In combinational logic optimization, we optimize the combinational logic elements in a circuit. We can divide combinational logic optimization techniques further into two categories: *two-level logic optimization* and *multilevel logic optimization* [1]. These two types of techniques differ in their Boolean function representation and have distinct merits and demerits. In practice, synthesis tools apply both these types of optimizations.

In this chapter, we describe logic optimization that works on a generic-gate netlist. However, estimating the cost of various implementations for a generic-gate netlist is challenging. We can use crude measures such as *gate count*, *literal count*, and number of *logic levels* to estimate the cost of an implementation. These measures typically work well for area estimation but are not good for timing or delay estimation. Therefore, often, the objective of technology-independent logic optimization is to *minimize the area* of a circuit. We defer timing optimization to technology mapping and subsequent design steps when the critical paths become trustworthy.

In this chapter, we will first discuss combinational logic optimization and then sequential logic optimization. The concepts of combinational logic optimizations help understand sequential logic optimization. Therefore, for pedagogical reasons, we have followed the above sequence. However, note that synthesis tools typically apply sequential optimizations such as FSM optimization before combinational logic optimization.

12.1 TWO-LEVEL LOGIC OPTIMIZATION

A *two-level logic circuit* consists of only two levels of logic gates from the input to the output. However, *multilevel logic circuit* consists of more than two levels of logic gates from the input to the output. A two-level logic can be in the *sum of products* (SOP) form or *product of sums* (POS) form. An SOP form involves taking the sum (logical OR) of the products (logical AND) of Boolean variables or their complements. A POS form involves taking the product (logical AND) of the sums (logical OR) of Boolean variables or their complements. Examples of two-level and multilevel logic networks are shown in Figure 12.1. In this section, we consider two-level logic optimization for Boolean functions represented in the SOP form.

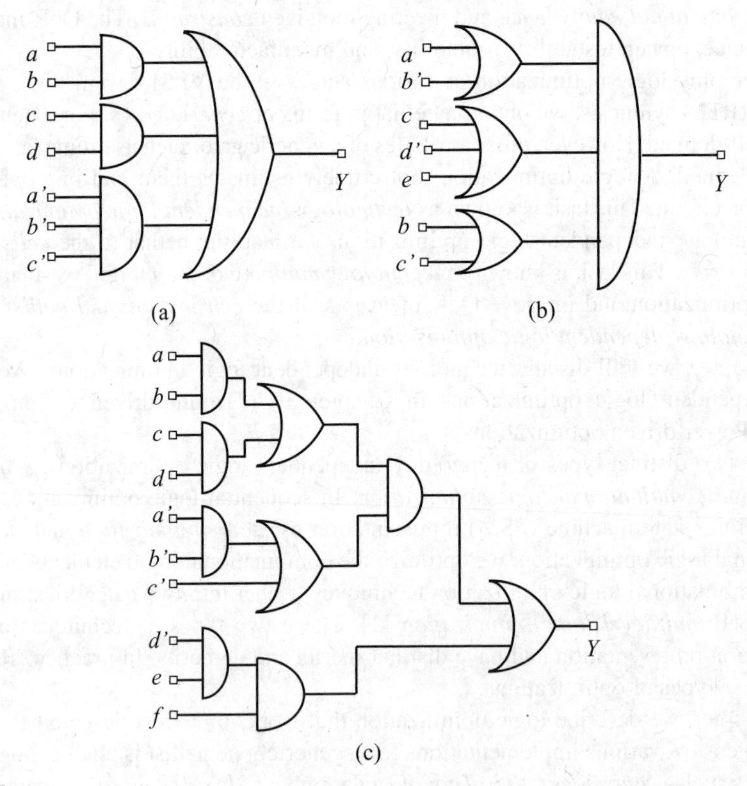

Figure 12.1 Examples of logic circuits (a) and (b) two-level logic (SOP and POS), and (c) multilevel logic

12.1.1 Terminologies

To understand the techniques employed in two-level logic optimization, we should be familiar with the definitions of *Boolean variable*, *Boolean function*, *literal*, *cube*, and *minterm* discussed in Section 11.3 ("Representation of a Boolean function"). Additionally, we should be familiar with the following terminologies and notations [2].

1. **Boolean space and hypercube:** A Boolean function of N-variables ($y = f(x_1, x_2, \dots, x_N)$) spans an N-dimensional *Boolean space*. An N-dimensional Boolean space can be represented and visualized using an N-dimensional *Boolean hypercube*. We associate each variable with one dimension of the hypercube. The corners of the hypercube represent binary-valued N-dimensional vectors, ith entry in the vector corresponds to the value of variable x_i. A Boolean hypercube has 2^N corners, representing 2^N input combinations.

Example 12.1 We show a three-dimensional Boolean space in Figure 12.2(a), with binary-valued inputs at each corner. We can associate a minterm with each corner of the hypercube, similar to the rows of a truth table, as shown in Figure 12.2(b).

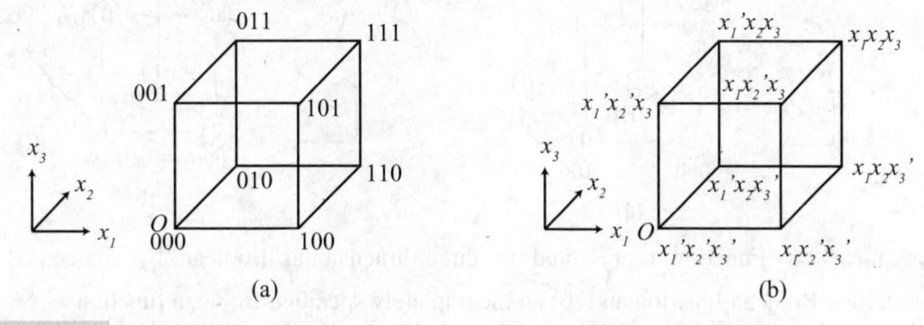

Figure 12.2 Three-dimensional Boolean hypercube for Boolean variables $\{x_1, x_2, x_3\}$ (a) binary-valued input vectors associated with each corner and (b) minterms associated with each corner

2. **Representation of a function in Boolean space:** To represent a Boolean function of N-variables, we mark the corners in the Boolean hypercube with the value of the function for the associated input combination.
3. **Incompletely specified Boolean function:** For some input combinations, the function $y = f(x_1, x_2, \dots, x_N)$ may not be specified. These input combinations are known as *don't care* (DC) conditions. The DC conditions are related to the input combinations that can never occur. They can also be those input combinations for which the output is not observed. For example, when a function receives binary-coded decimal digits 0–9 (coded as 0000, 0001, ..., 1001), it cannot receive input combinations {1010, 1011, ..., 1111}. Therefore, {1010, 1011, ..., 1111} are DC conditions. We denote the DC condition as X.

A function that has a DC condition is known as *incompletely specified Boolean function*. We can represent an incompletely specified Boolean function by three sets of input vectors: (a) ON-set (input combinations for which the output is 1, (b) OFF-set (input combinations for which the output is 0), and (c) DC-set (input combinations for which the output is X).

Example 12.2 (a) An example of a completely specified Boolean function is $y = x'_1x'_2x'_3 + x'_1x_2x_3 + x_1x_2x_3$. We can represent it in a three-dimensional Boolean hypercube as in Figure 12.3(a).
(b) An example of an incompletely specified Boolean function of three variables x_1, x_2, and x_3 is:

ON-set={010, 011},
OFF-set={000, 001, 100, 101}, and
DC-set={110, 111}.

We can represent this function in three-dimensional Boolean hypercube as in Figure 12.3(b).

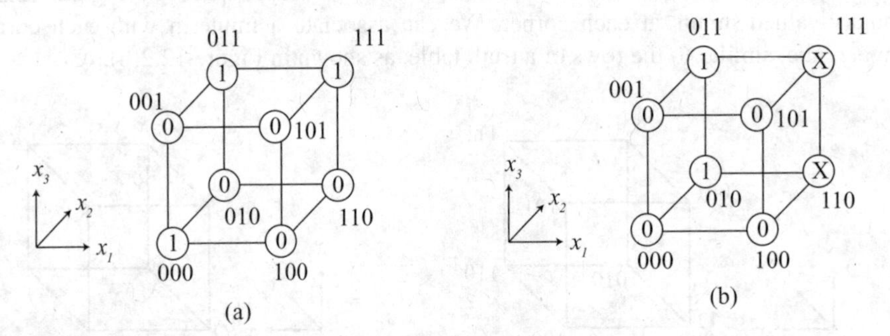

Figure 12.3 Function represented in three-dimensional Boolean hypercube (a) a fully specified Boolean function and (b) an incompletely specified Boolean function

Remark: The hypercube representation is similar to a truth table. However, its graphical arrangement of zeroes and ones helps us intuitively understand Boolean function minimization that is discussed subsequently.

---

4. **Implicant:** An *implicant* of a Boolean function is a cube whose corners are all in the ON-set or DC-set of that function. Some examples of implicants for the function shown in Figure 12.3(b) are $x'_1x_2x'_3$, x_1x_2, $x_2x'_3$, x'_1x_2, and x_2. The cube $x'_1x'_3$ is not an implicant because it contains (000) which is in the OFF-set.
5. **Cover:** A *cover* of a Boolean function is a set of implicants that includes all its minterms. The number of implicants in a cover is known as the *size* of the cover. We can reduce it by utilizing elements of the DC-set for incompletely specified Boolean functions. The cover with the minimum size is known as the *minimum cover*.

Example 12.3 For the function shown in Figure 12.3(b), we show some of the covers in Figure 12.4. Evidently, the minimum cover is $\{x_2\}$ with a size of 1.

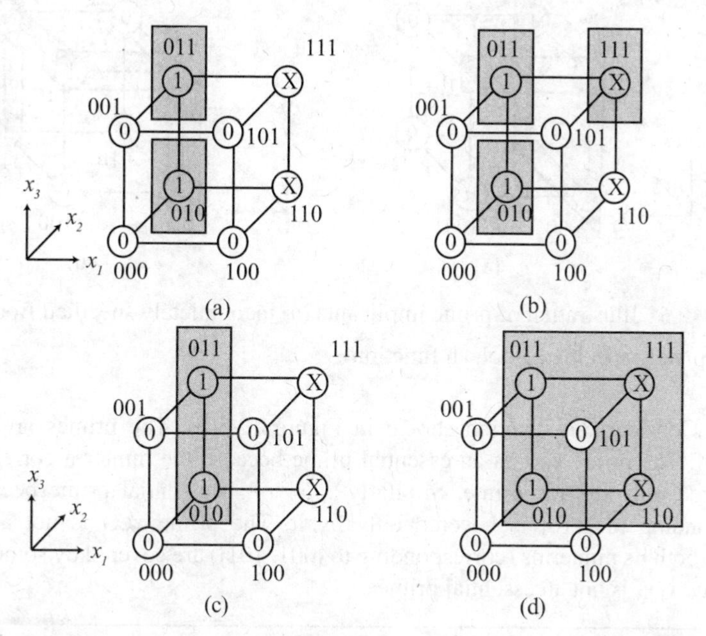

Figure 12.4 Different covers of a function (cubes of a cover are shown as shaded rectangles) (a) $\{x_1'x_2x_3', x_1'x_2x_3\}$, (b) $\{x_1'x_2x_3', x_1'x_2x_3, x_1x_2x_3\}$, (c) $\{x_1'x_2\}$, and (d) $\{x_2\}$

6. **Prime implicant:** An implicant of a function not covered by any other implicant of that function is known as a *prime implicant*. For the function shown in Figure 12.4, x_2 is a prime implicant because it is not covered by any other implicant. We refer to prime implicants as *primes* in short. A cover that consists only of primes is known as a *prime cover*. For the function shown in Figure 12.4, the prime cover is $\{x_2\}$.

   If we drop any literal from a prime, it will intersect corners in the OFF-set. We can use this attribute of the primes to check whether an implicant is a prime implicant, as explained in the following example.

   A prime implicant is an *essential prime implicant* if there is at least one minterm that is covered by only that prime implicant. We refer to an essential prime implicant as *essential prime* in short.

Example 12.4 (a) Consider the function shown in Figure 12.5(a). The primes are $x_2'x_3$ and x_1'. Note that, from the prime $x_2'x_3$, if we drop x_2' then the OFF-set is intersected at $\{111\}$, and if we drop x_3 then the OFF-set is intersected at $\{100\}$.

Additionally, we can observe that both the primes, $x_2'x_3$ and x_1', are essential primes. The minterm corresponding to $\{101\}$ is covered only by $x_2'x_3$. The minterms corresponding to $\{011, 010\}$ are covered only by x_1'.

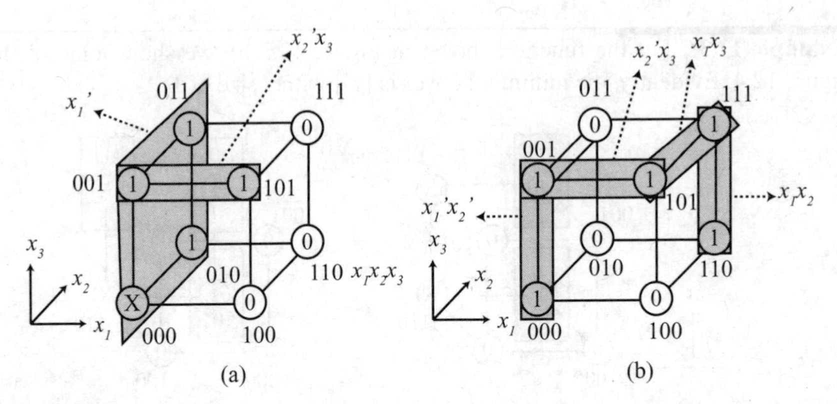

Figure 12.5 Illustration of prime implicants (a) incompletely specified Boolean function and (b) completely specified Boolean function

(b) Consider the function shown in Figure 12.5(b). The primes are $x_1'x_2'$, $x_2'x_3$, x_1x_3, and x_1x_2. The prime $x_1'x_2'$ is an essential prime because the minterm corresponding to {000} is covered only by this prime. Similarly, x_1x_2 is an essential prime because the minterm corresponding to {110} is covered only by it. The prime $x_2'x_3$ is not an essential prime because both its minterms (corresponding to {001, 101}) are covered by some other prime also. Similarly, x_1x_3 is not an essential prime.

12.1.2 Exact Logic Minimization

We illustrated in Ex. 12.3 that we can cover a Boolean function in several ways. The exact two-level logic minimization aims to find the *minimum cover*. We are interested in determining the minimum cover because it correlates with the circuit's reduced hardware or reduced area. We can perform two-level logic minimization using conventional techniques such as Boolean algebra-based manipulations and Karnaugh maps. However, these techniques become too complicated when the number of variables is high (e.g., more than 20). Therefore, a logic synthesis tool typically employs alternative approaches that can handle large-sized problems. We explain some of these techniques in the following paragraphs.

Quine's theorem: In finding the minimum cover of a given Boolean function, we can reduce the search space by employing Quine's theorem. We can state Quine's theorem as follows:

There is a *minimum cover* that is a *prime cover*.

This theorem is based on the following observation. Consider a minimum cover that is not a prime cover. It implies that it contains some non-prime implicants. We can replace each non-prime implicant with a prime implicant that *contains* it. Thus, we obtain a new cover that consists of primes only and is of the same size. Hence, there is a minimum cover that is a prime cover.

Prime implicant table: Quine's theorem limits the search space for a minimum cover to prime covers. We can find minimum cover among prime covers by finding the set of prime implicants and building the *prime implicant table*.

We build a prime implicant table as follows. First, we arrange the prime implicants in separate columns and the minterms in individual rows of the prime implicant table. Then, we fill in the entries in the table. If a given prime (column) covers a given minterm (row), then the corresponding entry is made 1, else it is made 0.

Example 12.5 Let us build the prime implicant table for the function shown in Figure 12.5(b). We arrange the primes $x'_1x'_2$, x'_2x_3, x_1x_3, and x_1x_2 along the column, as shown in Table 12.1. We list minterms in separate rows. Then, we make an entry as 1 if a given prime (column) covers a given minterm (row), else 0. Note that we have not shown the minterms that evaluate to 0 for the function.

Table 12.1 Prime implicant table for the function shown in Figure 12.5(b)

	Prime Implicants			
Minterms ↓	$x'_1x'_2$	x'_2x_3	x_1x_3	x_1x_2
$x'_1x'_2x'_3$ (000)	1	0	0	0
$x_1x_2x'_3$ (110)	0	0	0	1
$x'_1x'_2x_3$ (001)	1	1	0	0
$x_1x'_2x_3$ (101)	0	1	1	0
$x_1x_2x_3$ (111)	0	0	1	1

Set covering problem: The task of finding the minimum cover boils down to finding the minimum set of columns that covers all the rows in the prime implicant table. It involves solving the *set covering problem* which is known to be NP-complete. For a function of N variables, the number of minterms and the number of primes grow exponentially [3]. Nevertheless, we can solve the set covering problem exactly if the number of variables in the given Boolean function is small.

Reducing problem size: We can reduce the problem size by identifying essential primes. Note that we need to include all essential primes in the prime cover. For the function shown in Figure 12.5(b), we can start forming the prime cover by including the essential primes $\{x'_1x'_2, x_1x_2\}$. Subsequently, we need to cover only one minterm that is still not covered (corresponding to 101). We can cover it using x'_2x_3 or x_1x_3. Thus, we obtain the minimum cover $\{x'_1x'_2, x_1x_2, x'_2x_3\}$ or $\{x'_1x'_2, x_1x_2, x_1x_3\}$.

We can also examine row and column *dominance* to reduce the table size. A row x dominates over another row x' if all the columns that cover x also cover x' [4]. In this case, when x is covered, x' will be automatically covered. Hence, we can remove the dominated row x'.

We can reduce the prime implicant table by extracting essential primes and applying the dominancy rules until no more reduction is possible. We refer to the reduced table as a *cyclic core*. After deriving cyclic core, there are many efficient techniques to obtain a minimum cover. These techniques prune the search space cleverly and employ efficient data structures to find an optimal solution quickly [5–8].

12.1.3 Heuristic Minimizer

For large problems we prefer *heuristic minimizer* over exact minimization [5, 9, 10]. Heuristic minimizers are faster for large problem sizes. Moreover, for many applications, we do not need the exact *minimum cover*. For them, any solution that can be found quickly and is near-optimal is acceptable.

Minimal cover: A heuristic minimizer finds *minimal cover* instead of the *minimum cover*. A minimal cover satisfies certain *local* minimum cover property rather than the *global* minimum property. For example, a heuristic minimizer can find a cover in which no implicant is contained in any other implicant of the cover. The resultant cover is minimal with respect to *single-implicant containment*. The size of a minimal cover can be more than the size of the minimum cover. Nevertheless, in practice, the size of minimal cover obtained by a heuristic minimizer is often close to the minimum cover.

Approach: A heuristic minimizer does not list all the primes. It starts with an initial cover and iteratively improves the solution by applying some *operators* on it.The iteration terminates when the algorithm can no longer improve the solution.

A heuristic minimizer can apply the following operators on the implicants of a given cover [1]:

1. **Expand:** It expands a non-prime implicant to make it prime. If the expanded implicant covers some other implicant, it removes the covered implicant. Thus, the *expand* operator can decrease the cover size. After the *expand* operation, no implicant remains covered by any other implicant of that cover.
2. **Reduce:** It replaces an implicant with a reduced implicant (covering fewer minterms) such that the function is still covered. Thus, the *reduce* operator makes a prime cover to a non-prime cover, and the cover size remains the same.
3. **Reshape:** Reshape operates on a pair of implicants. It expands one implicant and reduces others such that the function is still covered. The size of the cover remains the same.
4. **Irredundant:** It operates on a cover to make it an *irredundant cover*. A cover is said to be irredundant if removing any implicant from it leaves the function uncovered. In an irredundant cover, no implicant gets covered by a subset of implicants of that cover. An irredundant cover is also a *minimal cover*. The *irredundant operation* removes the implicants covered by a subset of implicants in that cover. Thus, it decreases the cover size. The removed implicants are known as *redundant implicants*.

Example 12.6 Consider the function shown in Figure 12.6(a). Let us minimize it using the above operators.

Initially, the cover consists of only the minterms, as shown in Figure 12.6(a). We apply the following operators:

1. **Expand:** On applying the *expand* operator, we obtain a prime cover of size 4, as shown in Figure 12.6(b). To further decrease the cover size, we apply the *reduce*, *reshape*, and *expand* operators to get out of the local minima.[1]
2. **Reduce:** On applying *reduce* operator we obtain the cover shown in Figure 12.6(c). The implicant $x_2'x_3$ gets reduced to $x_1'x_2'x_3$ covering only {001}. Note that the cover size remains 4, and the cover is no more prime. We carry out *reduce* in the hope that subsequent operations can find a better solution. Thus, *reduce* can possibly take the solution out of local minima.
3. **Reshape:** We apply *reshape* operator for the implicants $x_1'x_2'x_3$ and x_1x_3. The implicant $x_1'x_2'x_3$ is expanded to cover {001, 101}. The implicant x_1x_3 is reduced to cover only {111}. Thus, we still obtain a cover of size 4, as shown in Figure 12.6(d).

[1] A local minima refers to a minimum within some neighborhood of a given function. It may not be the global minimum for the function.

4. **Expand:** We apply the expand operator to remove the implicant $x_1x_2x_3$. Thus, we obtain the cover of size 3, as shown in Figure 12.6(e).

Remark: The size of the cover obtained above is minimum. Nevertheless, we cannot guarantee that the cover size obtained using a heuristic minimizer always will be minimum.

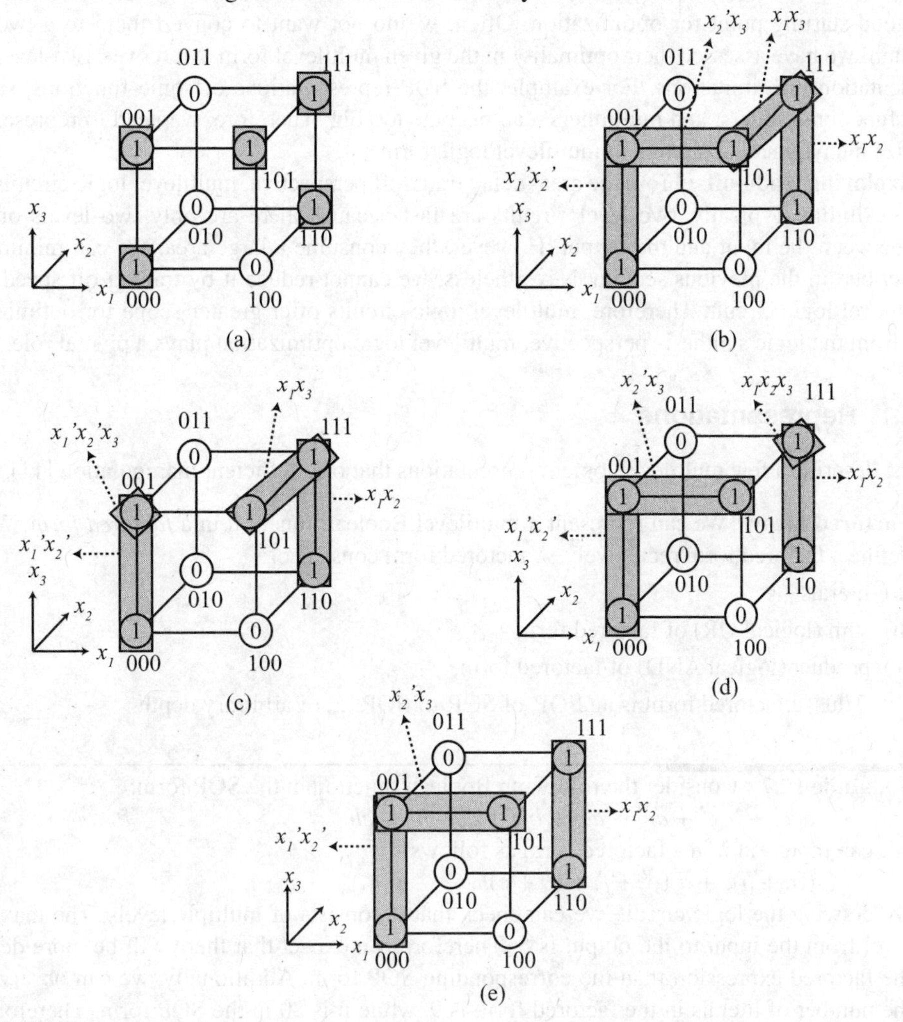

Figure 12.6 Illustration of heuristic minimization (a) start with cover as sum of minterms, (b) after *expand*, (c) after *reduce*, (d) after *reshape*, and (e) after *expand*

The cover obtained in the above example is *irredundant*. However, in general, the cover obtained after expand operator may not be irredundant. Therefore, after the iteration of *reduce*, *reshape*, and *expand* operators yield no improvement, we carry out the *irredundant* operation once.

The iteration of *reduce*, *reshape*, and *expand* operators attempts to move the solution toward the minimum cover. The runtime and result quality are strongly dependent on the implementation detail of the above operators. In practice, efficient two-level logic minimizers, such as ESPRESSO, can minimize functions of hundreds of variables and thousands of cubes in much less than a second of the CPU time [10].

12.2 MULTILEVEL LOGIC OPTIMIZATION

In practice, we often encounter multilevel logic circuits. During RTL synthesis, they appear naturally because of the way we describe the circuit behavior. The RTL-specified multilevel logic circuits may not be optimal, and we might need to optimize them. The RTL-specified multilevel logic can be a good starting point for optimization. Often, we do not want to convert them to a two-level form until we have assessed their optimality in the given multilevel form. Moreover, two-level logic representation has limitations. For example, the SOP representations of some functions, such as parity functions, adders, and multipliers, can become too big. Therefore, we need to represent and optimize many practical circuits in multilevel logic form.

Exploring trade-offs: From the area–delay trade-off perspective, multilevel logic circuits offer more flexibility. Typically, two-level circuits are fast because there are only two levels of logic gates between the input and the output. However, they consume a large area. We can minimize it, as described in the previous section. Nevertheless, we cannot reduce it by trading off speed, as in a multilevel logic circuit. Therefore, multilevel logic circuits offer greater scope for optimization. Thus, from the logic synthesis perspective, multilevel logic optimization plays a pivotal role.

12.2.1 Representations

We first describe a few multilevel logic representations that offer efficient manipulation [11].

1. **Factored forms:** We can represent a multilevel Boolean function in a *factored form.* We can define a factored form recursively. A factored form consists of:
   (a) literals
   (b) sum (logical OR) of factored form
   (c) product (logical AND) of factored form.
   Thus, a factored form is an SOP, of SOP, of SOP, ..., of arbitrary depth.

Example 12.7 Consider the following Boolean function in the SOP form:

$$ace' + bce' + de' + acf + bcf + df + bh + ch$$

We can represent it in a factored form as follows:

$$((a + b)c + d)(e' + f) + (b + c)h.$$

By drawing the logic circuit, we can check that it consists of multiple levels. The maximum level from the input to the output is 5. Therefore, we expect that there will be more delay in the factored expression than the corresponding SOP form. Additionally, we can observe that the number of literals in the factored form is 9, while it is 20 in the SOP form. Therefore, we expect that the area of the factored form logic circuit would be smaller than the corresponding SOP form.

Factoring: Given an SOP, we can represent it in the factored form using a process known as *factoring*. Thus, factoring can convert a two-level representation to a multilevel representation. However, given an SOP, its factored form is not unique.

Example 12.8 Consider the following Boolean function in the SOP form:

$$ac + ad + bc + bd + ce$$

We can represent it as:

$$(a + b)(c + d) + ce$$

Or we can represent it as:

$$(a + b)d + (a + b + e)c$$

Moreover, stand-alone SOP and POS are also Boolean functions in factored forms. Thus, we can represent and implement a Boolean function in a factored form in multiple ways.

Different implementations of a Boolean function can have different costs (in terms of the number of literals and delays). During logic optimization, we try to find the implementation that *minimizes* the cost.

In complementary metal–oxide–semiconductor (CMOS) technology, we can implement a Boolean function given in the factored form with N literals using $2N$ transistors (excluding the output inverter). We illustrate this in the following example.

Example 12.9 Consider the Boolean function $F = (ab + c)d + e$. It has five literals. We can implement it in static CMOS circuit, as shown in Figure 12.7. It has 10 transistors (excluding the output inverter). We obtain the implementation by constructing the pull-down network with a series connection for AND operation and a parallel connection for OR operation. We construct the pull-up network by taking the dual of the given function.

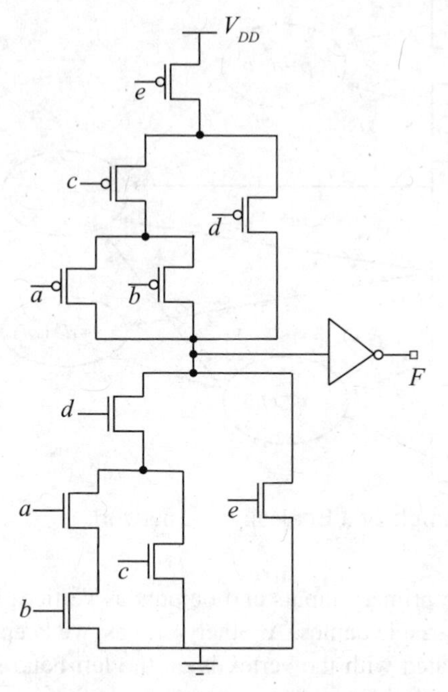

Figure 12.7 CMOS implementation of $F = (ab + c)d + e$

Remark: As illustrated above, the number of literals in the factored form correlates with the number of transistors (or the circuit area). Hence, we attempt to minimize the number of literals for Boolean functions in the factored form.

2. **Boolean logic network:** We often need to optimize a multilevel logic circuit that can have more than one output. It is generally convenient to model those circuits as a *Boolean logic network*. It is a *directed acyclic graph* in which we annotate each vertex with a *single-output local Boolean function*. The local function on a vertex describes the behavior of that vertex. The edges between vertices describe the structure of the logic implementation.

Example 12.10 Consider the following set of equations:

$$p = a + b$$
$$q = ef$$
$$r = p + c'd + q$$
$$s = d' + q'$$
$$x = r$$
$$y = s$$

Assume that the primary inputs or the inputs that go into the model are *a*, *b*, *c*, *d*, *e*, and *f*. The primary outputs (i.e., the outputs that are observed externally) are *x* and *y*. The internal inputs and outputs are *p*, *q*, *r*, and *s*. We can represent the above set of equations as a Boolean logic network shown in Figure 12.8.

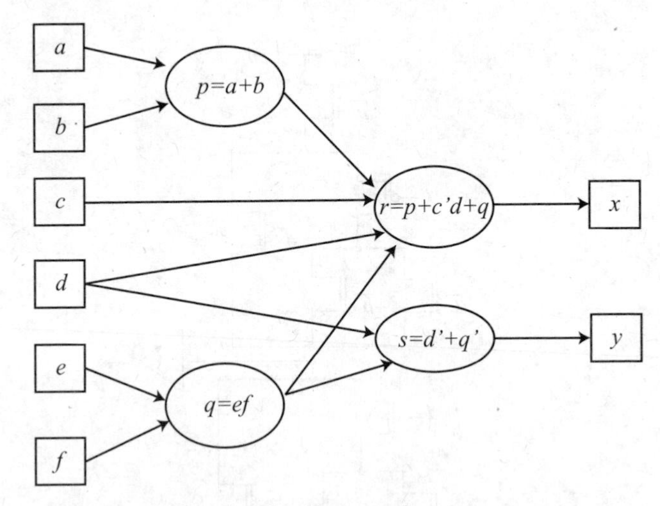

Figure 12.8 An example of a Boolean logic network

We represent the primary inputs and outputs as vertices in the graph and annotate them with the associated variable names. At other vertices, we keep two pieces of information. The variable name associated with the vertex is on the left-hand side (LHS) of the equality sign and the local Boolean function produced by that vertex is on the right-hand side (RHS). The incoming edges to a vertex denote the variables on which the local function at that vertex depends.[2] For example, the local function at vertex s depends on d and q. Hence, edges exist between $d \rightarrow s$ and $q \rightarrow s$. Similarly, an outgoing edge from a vertex denotes the local functions that depend on its associated variable. For example, the local functions $r = p + c'd + q$ and $s = d' + q'$ depend on q. Hence, edges $q \rightarrow r$ and $q \rightarrow s$ exist.

Flexibility of Boolean logic network: Unlike conventional logic circuits consisting of logic gates, the local functions in a Boolean logic network can be arbitrarily complicated. During optimization, we can manipulate both its underlying graph and the local functions at the vertices. Thus, we can explore both the behavioral and the structural features of the implementation using a Boolean logic network.

Estimating QoR: For easy manipulation, we restrict the local functions to be in an SOP form. Furthermore, we keep them as *minimal* with respect to *single-implicant containment.* It allows us to quantify area just by counting the total number of literals in the SOP expression because they are correlated. To estimate the area of a Boolean logic network, we take the sum of all the literal counts of the local functions. For the Boolean logic network in Ex. 12.10, the total literal count is the sum of literal count for the vertices p, q, r, and s, which evaluates to 10. Additionally, restricting local functions to the SOP form allows us to relate the *stages of vertices* in the logic network with the logic level or logic depth. Note that, during intermediate steps of optimization, we can represent a local function in factored form also. Nevertheless, the subsequent conversion of a factored form to a multiple-stage Boolean logic network is straightforward.

12.2.2 Transformations of Boolean Logic Network

Multilevel logic optimization is a challenging problem. Similar to heuristic-based two-level logic optimization, we perform multilevel logic optimization by applying transformations. We can view these transformations as *operators* for the Boolean logic network. We apply these operators iteratively until no more improvement in some QoR measures is possible. However, the final QoR depends on the order of operation and is hard to predict. A multilevel logic optimization tool can provide a mechanism through which we can experiment with different orders of applying operators. A logic synthesis tool can also use a default order for these operators that produces reasonably good QoR on benchmark designs.

Let us look at some of the operators that transform a Boolean logic network.

1. **Eliminate:** It removes a vertex from the graph and replaces all its occurrences in the network with the corresponding local function. It only removes some structural information from the network. Therefore, it is easy to implement in a tool.

---

[2] While considering dependency on a variable, we consider both the variable and its complement.

Example 12.11 Consider the Boolean logic network shown in Figure 12.9(a). We eliminate the vertex p and move its functionality to q and r, as shown in Figure 12.9(b). We carry out *eliminate* in the hope that subsequent operators can reduce the cost.

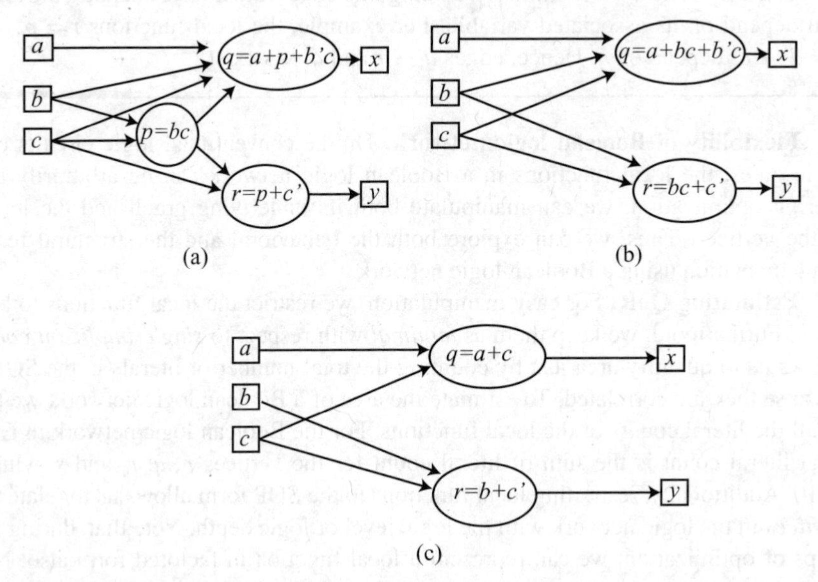

Figure 12.9 Illustration of transformations (a) given Boolean logic network, (b) after *eliminate*, and (c) after *simplify*

2. **Simplify:** It works on the vertices of the network individually. It simplifies the associated local SOP expression to reduce the literal count. We can view it as two-level logic optimization carried out on each local function individually. We can use the techniques discussed in the previous section for simplifying.

Example 12.12 Consider the logic network in Figure 12.9(b). We *simplify* the vertices q and r to obtain the network shown in Figure 12.9(c). Thus, by applying *eliminate* and *simplify* in sequence, we have reduced the literal count from 8 to 4. Since *simplify* works locally on a vertex, it is easy to implement in a tool.

3. **Substitute:** It replaces the local function with a simpler SOP by creating new dependencies and possibly removing other dependencies. The *substitute* operator creates dependencies by searching for an appropriate match. It adds more structural information to the network. Therefore, implementing *substitute* is more challenging than *eliminate*. Nevertheless, we can carry out substitution efficiently, as discussed in the following paragraphs.

Example 12.13 Consider the Boolean logic network shown in Figure 12.10(a).

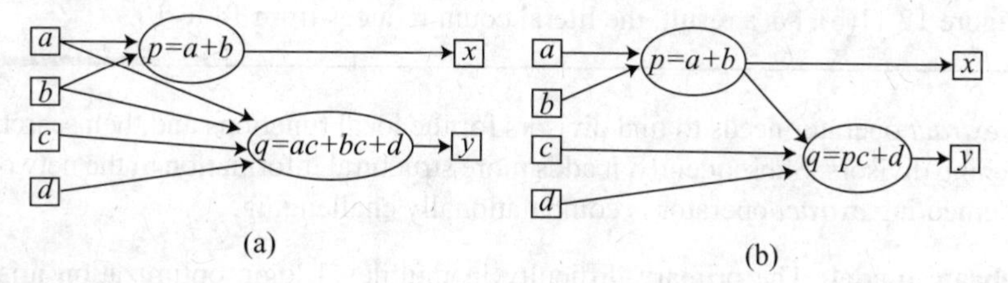

Figure 12.10 Illustration of *substitute* (a) given Boolean logic network and (b) after *substitute*

We can replace the function at q with a simpler function by creating a dependency over p. We also drop dependency over a and b, as shown in Figure 12.10(b). As a result, the literal count reduces from 7 to 5.

4. **Extract:** It finds a *common subexpression* for functions associated with two or more vertices. Subsequently, it creates a new vertex associated with the subexpression. Then it replaces the common subexpressions in the original functions with the variable of the new vertex.

Example 12.14 Consider the Boolean logic network shown in Figure 12.11(a).

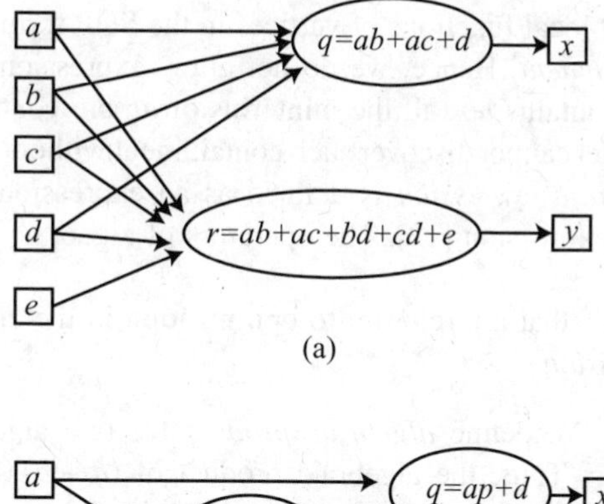

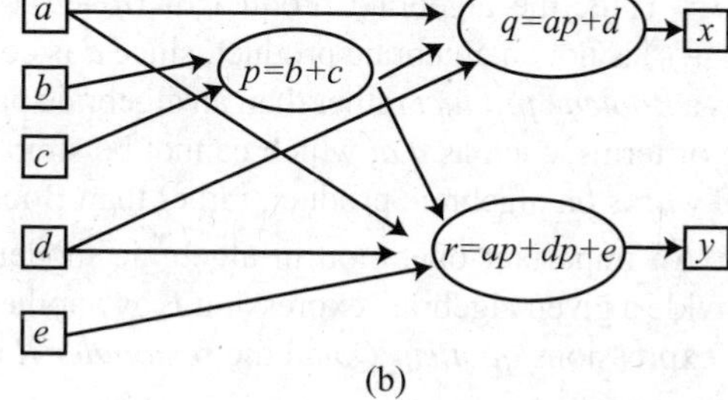

Figure 12.11 Illustration of *extract* (a) given Boolean logic network and (b) after *extract*

We find that the subexpression $(b + c)$ is common in the functions corresponding to vertices q and r. We create a new vertex p and replace $(b + c)$ with p in the original functions, as shown in Figure 12.11(b). As a result, the literal count reduces from 14 to 10.

The *extract* operator needs to find divisors for the local functions and then search vertices with matching divisors. Subsequently, it adds more structural information to the network. Therefore, implementing *extract* operator is computationally challenging.

Algebraic model: The primary difficulty in multilevel logic optimization arises due to the huge search space. We can simplify the problem by neglecting some Boolean properties of the local functions during transformations. The simplified model treats the local Boolean functions as polynomials and is known as *algebraic model*. We can quickly find effective transformations, such as common subexpressions extraction, in the algebraic model.

An algebraic model is weaker than a Boolean model for optimization. It cannot fully optimize a Boolean logic network. Therefore, post algebraic model-based optimization, we apply transforms utilizing the power of the Boolean model. Thus, we derive the benefits of both these models in multilevel logic optimization. We will briefly discuss these transformations in the following paragraphs.

12.2.3 Optimizing Using Algebraic Models

In an algebraic model, we use the polynomial algebra rules neglecting the specific features of Boolean algebra. Consequently, we treat a variable and its complement as separate variables. Additionally, we represent local functions of vertices in the SOP form that is minimal with respect to *single-implicant containment*. Hence, we do not allow expressions like $a + ab$ to exist in the algebraic model since a contains ab (all the minterms of ab are contained in the minterms of a). Note that an algebraic model cannot discover such containment without invoking features of Boolean algebra. Hence, an *algebraic expression* is defined as an expression in which no cube contains some other cubes of that expression [12]. The operators of algebraic model work on and produce algebraic expressions.

Two critical operations that are relevant to optimization in the algebraic model are *algebraic product* and *algebraic division*.

1. **Algebraic product:** We define *algebraic product* for two algebraic expressions that have *disjoint support sets.*[3] Thus, the algebraic product of $(a + b).(c + d)$ is $ac + ad + bc + bd$. However, $(a + b).(a + c)$ is not an algebraic product, since a is common in the support set. We say $(a + b).(a + c)$ is a *Boolean product* rather than an algebraic product. The algebraic product prohibits generation of terms, such as $a.a$, which cannot be simplified in the algebraic model. The algebraic model works on algebraic product, rather than Boolean product.
2. **Algebraic division:** An important operation in algebraic model is the *division* of algebraic expression. Let us divide a given algebraic expression F by another algebraic expression D. We obtain the algebraic expressions *quotient* Q and the *remainder* R if the following holds:

$$F = D.Q + R \tag{12.1}$$

[3] A support set is a set of input variables in an expression.

We say that D is an *algebraic divisor* of F, when the following conditions are satisfied:

(a) $D.Q \neq 0$

(b) the operator . in $D.Q$ is an algebraic product (i.e., the support of D and Q are disjoint).

(c) R has as few cubes as possible.

When $R = 0$, we say that the divisor D is a *factor*.

Example 12.15 (a) Consider the following algebraic expression:

$$F = ac + ad + bc + bd + e$$

We can write it as $F = (a + b).(c + d) + e$. Hence, for a given divisor $D = (a + b)$, $Q = (c + d)$ is the quotient and $R = e$ is the remainder. We can also treat $(c + d)$ as the divisor and $(a + b)$ as the quotient.

(b) The divisor ab is a factor for the function $G = abc + abd$ with a quotient of $(c + d)$ because $R = 0$.

Comparison with Boolean division: The algebraic division is "weaker" than the Boolean division. It does not consider all the possibilities for the quotient and remainder. It restricts the space for algebraic divisors. However, the *algebraic* division ensures that for a given function F and a given divisor D, the quotient and remainder are unique. It allows us to formulate efficient algorithms to carry out division in the *algebraic model*, rather than in the Boolean model. We can carry out algebraic division in *linear* or *linear–logarithmic* time complexity (in terms of number of product terms in F and D) [1, 12].

Applications of algebraic division: During multilevel logic optimization, we need to carry out division too many times. For example, consider the substitute operation discussed in the previous section. For substitution, we need to check whether the local function f_i divides another local function f_j. If it divides, we can substitute f_i in f_j as follows: $f_j = f_i.Q + R$, where Q is the quotient and R is the remainder of the division. Practical circuits have thousands of vertices in the Boolean logic network. Therefore, we might need to divide two functions $O(n^2)$ times, where n is the number of vertices in the Boolean logic network. Other optimization steps also use algebraic division in their inner loop. Thus, the efficiency of algebraic division is critical for multilevel logic optimization.

To optimize a multilevel Boolean logic network, we need to find good divisors (one that can reduce cost) for Boolean expressions. Finding a *good* set of divisors for a given Boolean expression is nontrivial. Nevertheless, in the algebraic model, the problem becomes easier by deriving *kernels* and *co-kernels* [13]. We discuss kernel-based optimization in the following paragraphs.

Kernels and Co-kernels

To define a kernel, we need to introduce the notion of a *cube-free expression.* A cube-free expression is an expression that does not leave remainder 0 when divided by a cube. For example, $a + bc$ and $ab + ac + bd$ are cube-free expressions because no cube can divide them leaving remainder 0. However, $ab + bc$ is not a cube-free expression because it can be divided by b leaving remainder 0. Any stand-alone cube, such as $abcd$, is also not a cube-free expression. Note that a cube-free expression will have always more than one cubes.

A kernel of an expression F is a cube-free quotient k obtained by algebraically dividing F by a single cube c. Thus, $F = c.k + R$, where c is a cube, k is a cube-free expression, and R is the remaining terms. The cube c corresponding to the kernel k is called the co-kernel of F. We denote the set of kernels k for a function F by $K(f)$.

Example 12.16 Consider the function $F = abc + abde$.
(a) Let us divide F with the cube ab. Thus, $F = ab.(c + de)$. The quotient $c + de$ is cube-free. Hence, $c + de$ is a kernel of F and ab is the co-kernel.
(b) The cube a also divides F with quotient $bc + bde$. However, $bc + bde$ is not cube-free. Hence, $bc + bde$ is not a kernel.
(c) The trivial cube 1 is always a co-kernel of a cube-free function.

Hierarchy of kernels: Consider the kernel k of a function F, such that $F = c.k + R$. The kernels of a function are also algebraic expressions. Therefore, we can find the kernels of k also. Let us denote it by k', such that $k = c'k' + R'$, where c' is a cube, k' is a cube-free expression, and R' is the remaining terms. We can substitute the expression $k = c'k' + R'$ back into F. Hence, $F = c.(c'k' + R') + R = cc'k' + cR' + R$. Note that cc' divides F and has quotient k'. Since, k' is cube-free and cc' is a cube, we can consider k' as a kernel and cc' as its co-kernel. Hence, we can infer the following important property of the kernels:

The kernels of a kernel of a function F are also the kernels of F. We illustrate this property in the following example.

Example 12.17 Consider the function $F = abde + cde + df + g$.
We can write $F = d.(abe + ce + f) + g$. Thus, $abe + ce + f$ is a kernel and d is the co-kernel.
Next, we can write $abe + ce + f = e.(ab + c) + f$. Thus, $ab + c$ is a kernel and e is its co-kernel.
Note that, we can also write $F = abde + cde + df + g = de.(ab + c) + df + g$. Thus, $ab + c$ is also the kernel of F. It demonstrates that the kernel of a kernel of a function is also a kernel of that function.

The above description suggests that kernels of a function have an *hierarchical structure*. A level-0 kernel contains no other kernels except itself. A level-1 kernel contains only level-0 kernels inside it. A level-2 kernel contains only level-0 and level-1 kernels inside it. The hierarchy of kernels continues similarly. In the above example, $ab + c$ is the level-0 kernel, $abe + ce + f$ is the level-1 kernel, and $abde + cde + df + g$ (with 1 as co-kernel) is the level-2 kernel.

Using the hierarchical nature of kernels, we can devise a recursive algorithm to find all the kernels of a function. We can make it more efficient by restricting the search space for the co-kernels and recording the already searched solution space. Interested readers can refer to [12] for details on the algorithm to find kernels.

Extraction

A Boolean logic network in an industrial scale design can have thousands of vertices. The kernels and co-kernels help us to *extract* good common divisors from multiple vertices in a Boolean logic network due to the following theorem.[4]

[4] A *good* common divisor helps reduce the cost, such as the number of literals.

Brayton and McMullen's theorem: Consider two algebraic expressions F and G. Let k_f denote a kernel of F and k_g denote a kernel of G. Then, F and G have a common multi-cube divisor D *if and only if* there exists k_f and k_g such that $k_f \cap k_g$ has two or more terms. The intersection refers to taking the common terms in k_f and k_g.

Thus, to find multi-cube common divisors for two expressions, we can look for the kernel intersection of these expressions. If there is no intersection or the intersection contains only a single cube, those expressions can have *no multi-cube common divisors*. This observation allows us to avoid finding multi-cube common divisors for many vertices just by noting their kernels. Thus, the above theorem enables us to prune the search space for finding good common multi-cube divisors.

Example 12.18 Consider the functions $q = ab + ac + d$ and $r = ab + ac + bd + cd + e$ (shown in Figure 12.11(a)).
The kernels of q are $(b + c)$ and $(ab + ac + d)$.
The kernels of r are $(b + c)$, $(a + d)$, $(b + c)$, and $ab + ac + bd + cd + e$.
We obtain $(b + c)$ as their kernel intersection.
Hence, we can extract $(b + c)$ from q and r, demonstrated earlier in Ex. 12.14 (Figure 12.11).

The kernels and co-kernels are powerful aids in extracting good divisors or common subexpressions in a complex Boolean logic network. We can start by finding the kernel set or a subset of kernels. Then, we can compute their intersection to determine good common divisors. We can represent the kernel and co-kernels in a matrix form [14]. It allows us to find a favorable kernel set intersection by solving the *rectangle covering problem*. There are efficient algorithms to solve the rectangle covering problem. We can also use a similar technique to find good single-cube divisors in a complex Boolean logic network. The mathematical foundation of kernels and efficient heuristics to find their intersection form the basis of fast multilevel optimization in the contemporary logic optimization tools.

12.2.4 Optimizing Using Boolean Models

The algebraic models are fast for logic optimization. However, some optimizations that can exploit the specific properties of Boolean algebra are out of their scope. Therefore, after optimizing a Boolean logic network based on the algebraic model, we perform optimizations based on Boolean models.

DC conditions: While optimizing a Boolean logic circuit manually, we often assume that DC conditions are given to us. However, in a Boolean logic network, DCs arise naturally due to the graph structure and dependencies among local functions. These DCs are a rich source of optimization in multilevel logic synthesis. However, a logic synthesis tool needs to discover them using Boolean algebra (in contrast to given DCs). Subsequently, it can exploit the discovered DCs in simplifying local functions and improving the circuit's overall QoR.

There are two types of DC that are useful in simplifying local functions in a Boolean logic network: *controllability don't cares* (CDCs) and *observability don't cares* (ODCs). We describe them in the following paragraphs.

Controllability Don't Cares

The combination of input variables that can never occur at a given vertex in a Boolean logic network produces CDCs. We can simplify local functions by accounting for CDCs and invoking efficient two-level logic minimizers, such as ESPRESSO. We illustrate this technique in the following example.

Example 12.19 Consider the Boolean logic network in Figure 12.12(a). The local function corresponding to the vertex $q = pb + bc$ cannot be simplified further using two-level logic minimizer. Figure 12.12(b) shows the cover for $q = pb + bc$.

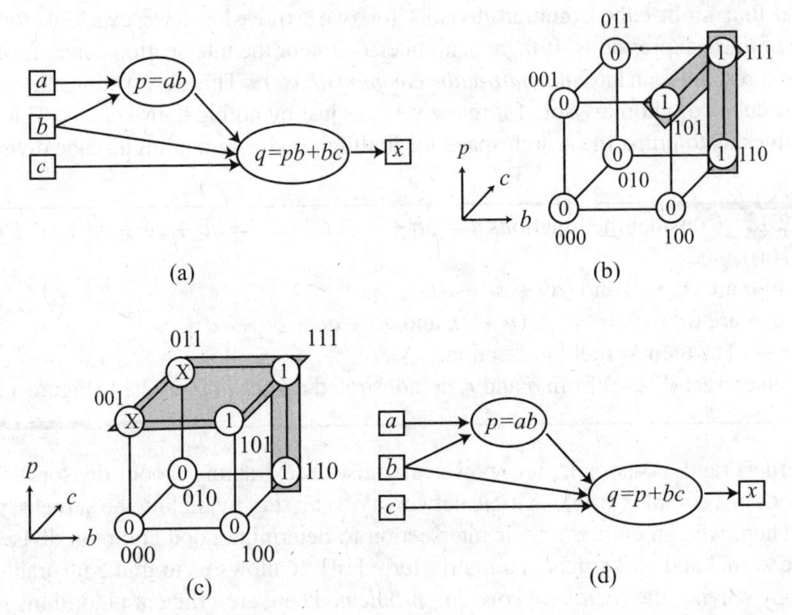

Figure 12.12 Illustration of CDCs (a) given Boolean logic network, (b) cover without considering CDC, (c) cover when CDC is considered, and (d) simplified Boolean logic network

Next, consider the vertex $p = ab$. It will produce $p = 1$ only when $b = 1$. Hence, $p = 1$ and $b = 0$ can never occur at the input of q. Therefore, we can consider pb' as a DC for $q = pb + bc$. Note that this DC appears at q because of the incoming edge from the vertex $p = ab$. Such type of DCs is known as CDCs.

Figure 12.12(c) shows the cover for $q = pb + bc$ when we consider pb' as a CDC. Hence, we obtain $q = p + bc$. The resultant simplified Boolean logic network is shown in Figure 12.12(d). We have reduced the literal count by 1.

In the above example, we could guess the CDC because it has only two vertices and the local functions are simple. However, practical circuits have hundreds of vertices in the Boolean logic network. We can compute CDCs in them using efficient algorithms that exploit *satisfiability don't cares* (SDCs). SDCs get enforced by the local functions associated with a vertex at its output. We explain it in the following example.

Example 12.20 Consider the vertex $p = ab$ for the logic network shown in Figure 12.12(a) (same as in Ex. 12.19). Thus, p should always be equal to ab, and we cannot satisfy the following function:

$$p \oplus ab = pa' + pb' + p'ab$$

Hence, the following patterns that satisfy the above function can never occur in the network:
$p = 1, a = 0,$
$p = 1, b = 0,$
$p = 0, a = 1, b = 1$
We can treat the above combination of variables as DCs for the network. Note that the DC pb' of the previous example is included above. These DCs appear due to the relationship between the local functions and its associated variable. We refer to them as SDCs. We can easily obtain them for a given Boolean logic network.

Logic synthesis tools typically derive CDCs algorithmically using SDCs. Subsequently, the CDCs get utilized in simplifying local functions. Interested readers can refer to [1] for the algorithmic details of the CDC-based optimization.

Observability Don't Cares

We derive ODCs induced by vertices in the fanout of a given vertex. The ODCs are input variable combinations that obstruct the *vertex output* from being observed at the *network output*. By network output, we mean the primary output of the Boolean logic network.

Example 12.21 Consider the Boolean logic network shown in Figure 12.13(a). The network output is x and the corresponding function is $q = pcd$. We consider the local function $p = ab + bc + ac$ for ODC. If we do not consider any DC, we cannot simplify p, as it is evident from the cover shown in Figure 12.13(b).

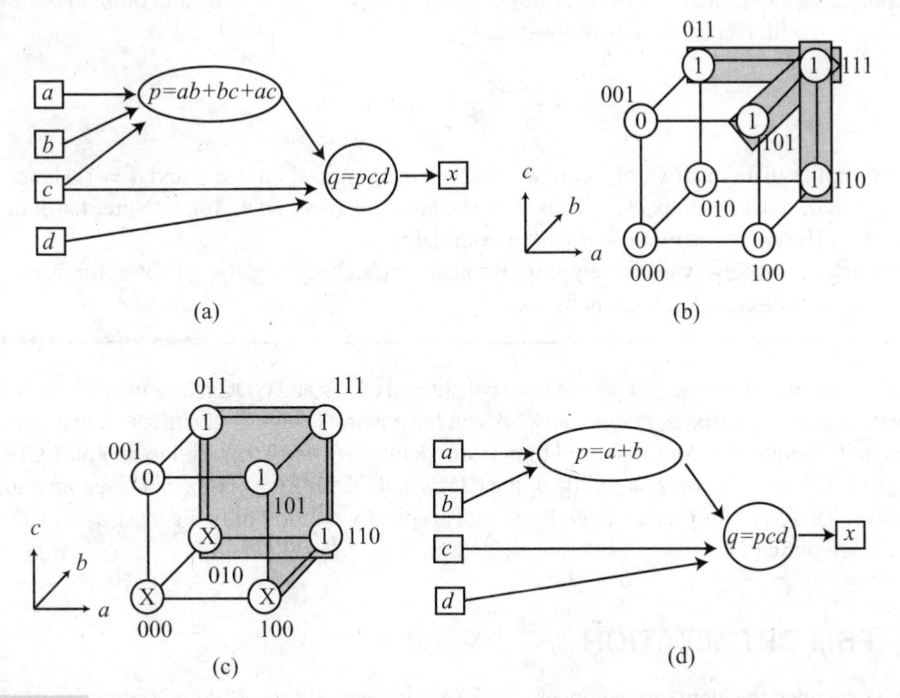

Figure 12.13 Illustration of ODCs (a) given Boolean logic network, (b) cover without considering ODC, (c) cover when ODC is considered, and (d) simplified Boolean logic network

Next, let us consider the input variable combination that will obstruct p from having any impact on x. We can note that if $c = 0$, then $x = 0$, irrespective of the value of p. Therefore, p is free to produce any output when $c = 0$. Hence, we can use $c = 0$ (or the corresponding DC function c') as DC condition for simplifying p. It is illustrated in Figure 12.13(c). We obtain the minimal cover $p = a + b$ by including the DC minterms corresponding to 010 and 100. The literal count decreases from 9 to 5, as shown in Figure 12.13(d).

In the above example, we could guess the ODC because of its simplicity. In a Boolean logic network with hundreds of vertices, we need to compute ODCs algorithmically. Typically, we employ *Boolean difference* to determine ODCs, as illustrated in the following paragraphs.

Consider a Boolean function $F = f(x_1, x_2, x_3, ..., x_n)$ that depends on n Boolean variables x_1, x_2, x_3, ..., x_n. We denote Boolean difference of F with respect to variable x_i as $\frac{\partial F}{\partial x_i}$, and compute it as follows:

$$\frac{\partial F}{\partial x_i} = F_{x_i=0} \oplus F_{x_i=1} \tag{12.2}$$

where $F_{x_i=0}$ refers to the function obtained on setting $x_i = 0$ (called negative cofactor) and $F_{x_i=1}$ refers to the function obtained on setting $x_i = 1$ (called positive cofactor). The Boolean difference $\frac{\partial F}{\partial x_i}$ is 0 when flipping x_i has no impact on F. Intuitively, when $\frac{\partial F}{\partial x_i} = 0$, we can assign arbitrary value to x_i without impacting F.

Example 12.22 Consider the Boolean logic network of Ex. 12.21. We can compute the Boolean difference of q with respect to p as follows:

$$\frac{\partial q}{\partial p} = \frac{\partial (pcd)}{\partial p} = 0.cd \oplus 1.cd = cd$$

This implies that the effect of p can be observed on q only when $c = 1$ and $d = 1$. Hence, when $c = 0$ or $d = 0$, p cannot impact q. Thus, $c = 0$ can be taken as ODC for p. Note that p does not depend on d. Hence, we cannot use $d = 0$ in optimizing p.

Remark: Similarly, we can employ Boolean difference to derive ODCs for each vertex algorithmically in a complex logic network.

In the examples discussed so far, we have considered CDCs and ODCs for only one level of logic. However, in general, a Boolean logic network can have many levels. A given vertex can have many vertices in its fanin and fanout. We need to perform network (graph) traversal to compute CDCs and ODCs in such cases. However, deriving all the CDCs and ODCs for a vertex can be computationally prohibitive. Therefore, in practice, synthesis tools typically employ only a subset of all CDCs and ODCs for simplification.

12.3 FSM OPTIMIZATION

We have explained the elementary concepts of FSMs in Section 1.1.7 ("Finite state machines"). We have discussed FSM synthesis and its hardware implementation in Section 10.6 ("FSM synthesis"). In this section, we will discuss FSM optimization.

12.3.1 State Minimization

We typically encode states of an FSM using bit patterns. To represent an FSM with n_s states, we need at least $\lceil log_2 n_s \rceil$ bits. We represent each bit by a storage element, such as flip-flop. Therefore, if we reduce n_s, the number of bits and the number of flip-flops representing them are expected to decrease. Additionally, by reducing n_s, we can reduce the complexity of *state transitions* in the FSM. Consequently, the number of combinational circuit elements required for implementing the state transition can also reduce. Thus, the overall size of the circuit implementing an FSM can reduce by reducing n_s.

Equivalent states: *State minimization* involves deriving an FSM that has the *minimum* number of states and exhibits the same behavior as the original FSM [15]. We can minimize the number of states in an FSM by determining sets of *equivalent states*. Two states in an FSM are said to be equivalent if we initialize the FSM with either of the two states and obtain the *same output sequence* for any input sequence. When two states of an FSM are equivalent, the following conditions hold for any input:

1. They produce identical outputs, and
2. The corresponding next states are the same or equivalent.

We can use these properties to determine sets of equivalent states for an FSM. After identifying a set of equivalent states E, we can retain any one of the *equivalent states* $r \in E$ and remove all other equivalent states $(E - r)$ from the FSM. We update the transition function to maintain the same behavior as the original FSM.

Example 12.23 Consider an FSM represented with the state diagram shown in Figure 12.14(a). There are two possible inputs $I = \{A, B\}$ and two possible outputs $O = \{0, 1\}$.

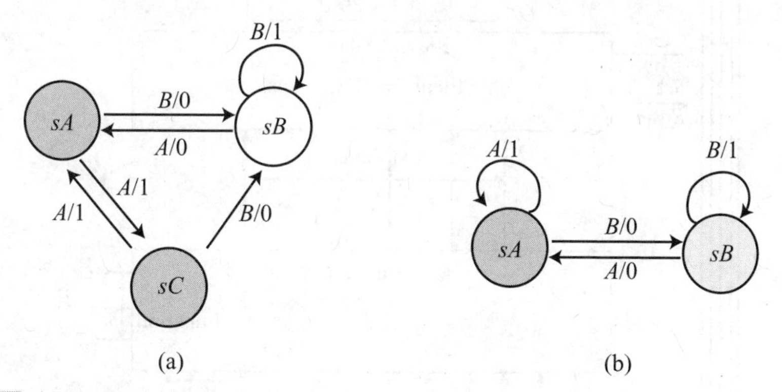

Figure 12.14 Illustration of state minimization (a) given FSM and (b) minimized FSM

We can infer that the states sA and sC are equivalent by making the following observations:

1. When the input is A, both states produce 1, and when input is B, both states produce 0.
2. When the input is A, sA transitions to sC (equivalent state) and sC transitions to sA (equivalent state). When the input is B, both sA and sC transitions to sB.

After identifying equivalent states, we can remove the equivalent state *sC* and retain *sA* (we could have removed *sA* and retained *sC* also). As a result, we obtain a state-minimized FSM shown in Figure 12.14(b).

Approach: We can carry out state minimization algorithmically as follows. We first partition the states according to their outputs (we put states producing identical outputs in the same partition). Subsequently, we refine these partitions by iteratively examining the successor state on a given input. We keep the states whose successor states are in different partitions for a given input into separate partitions. When no further refinement is possible, all the states in a partition are equivalent. We can retain one of the *equivalent states* in each partition. The above method has a complexity of $O(n_s^2)$, where n_s is the number of states. Using Hopcroft's approach, we can achieve the worst-case complexity of $O(n_s \log n_s)$ [16].

12.3.2 State Encoding

We implement an FSM using flip-flops and combinational circuit elements, as shown in Figure 12.15. We represent a state by the combination of the Q-pin value of flip-flops. Since the Q-pin value can take only Boolean values (0 or 1), we need to encode states by a string of zeroes and ones. The state encoding process assigns a *binary* representation to each state of an FSM. Note that as we change the state encoding in an FSM, its next state function and the output function can change. Consequently, the combinational circuit element *CL* and the associated QoR will also change. Depending on the attributes of the logic gates, one of the implementations can be more favorable than the others. We illustrate it in the following example.

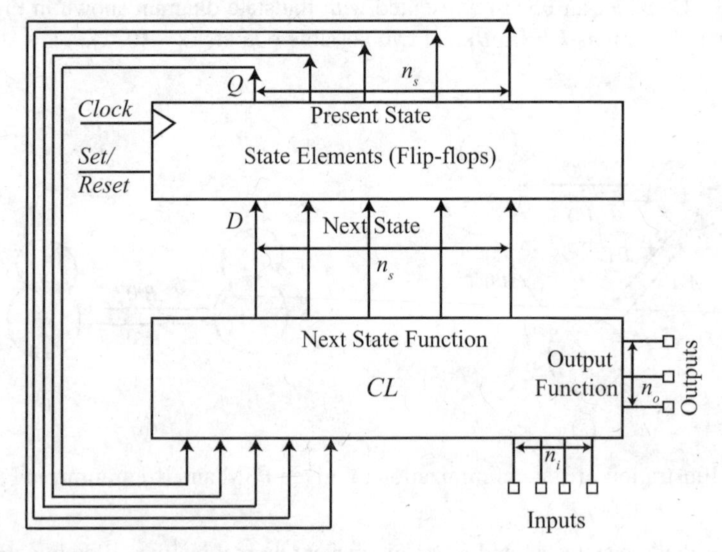

Figure 12.15 An FSM implementation

Example 12.24 Consider the FSM shown in Figure 12.16(a). We can also represent it as state transition table shown in Figure 12.16(b). We denote the input as *a* and the output as *z*. It has two states *s*1 and *s*2. We can encode them using 1 bit.

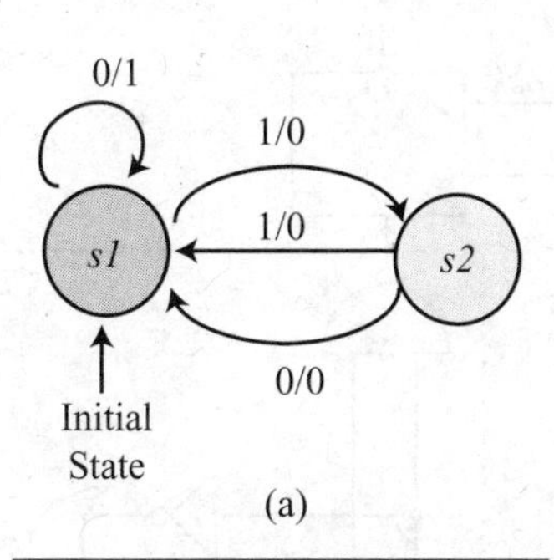

(a)

a (input)	Present State (PS)	Next State (NS)	z (output)
0	$s1$	$s1$	1
1	$s1$	$s2$	0
0	$s2$	$s1$	0
1	$s2$	$s1$	0

(b)

a (input)	p (PS)	x (NS)	z (output)
0	0	0	1
1	0	1	0
0	1	0	0
1	1	0	0

(c)

a (input)	p (PS)	x (NS)	z (output)
0	1	1	1
1	1	0	0
0	0	1	0
1	0	1	0

(d)

Figure 12.16 Illustration of different FSM encodings: (a) state transition diagram, (b) state transition table, (c) $s1 = 0$, $s2 = 1$, and (d) $s1 = 1$, $s2 = 0$

Let us first encode states $s1 = 0$ and $s2 = 1$. In this case, we can represent the transition table of the FSM as shown in Figure 12.16(c). We represent the present state with variable p and the next state as variable x. From the transition table, we can derive the next state function and the output functions as:

$$x = ap' = (a' + p)'$$
$$z = a'p' = (a + p)'.$$

Next, let us encode states $s1 = 1$ and $s2 = 0$. In this case, we can represent the transition table of the FSM as shown in Figure 12.16(d). From the transition table:

$$x = a'p + a'p' + ap' = a' + p' = (ap)'$$
$$z = a'p = (a + p')'.$$

Thus, we obtain different next state functions x and output functions z for two different encodings. We can implement the FSM using these encodings as shown in Figure 12.17.

The $s1 = 0$, $s2 = 1$ encoding requires 2 NOR gates and 1 inverter. The $s1 = 1$, $s2 = 0$ encoding requires 1 NAND gate, 1 NOR gate, and 1 inverter. Depending on the implementation technology, one of these encodings can be more favorable than the other.

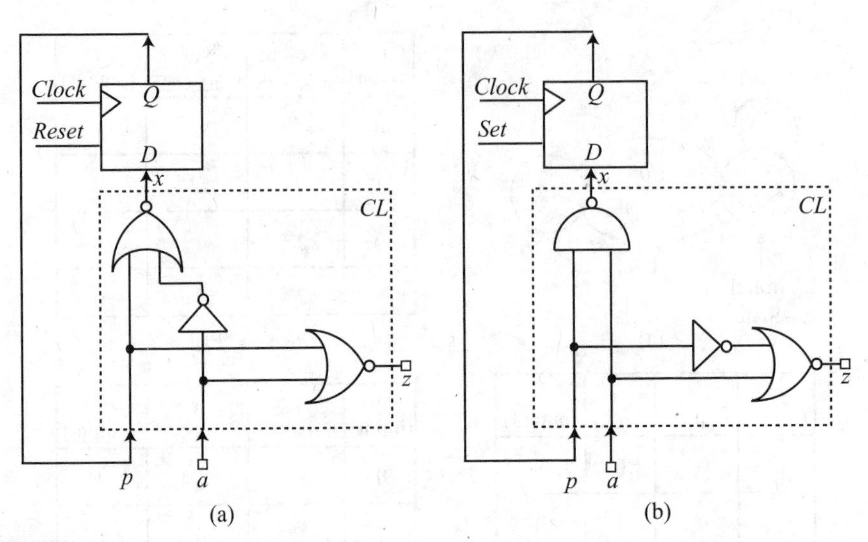

Figure 12.17 Implementations of FSM with encoding (a) $s1 = 0, s2 = 1$ and (b) $s1 = 1, s2 = 0$

Encoding length: We can also optimize the encoding length for an FSM. For n_s states, we need at least $\lceil log_2(n_s) \rceil$ encoding length. Nevertheless, we can choose longer encoding lengths. For example, we can use *one-hot encoding* by reserving a bit for each state. Thus, the encoding length becomes n_s, and we need n_s flip-flops in the FSM. We illustrate one-hot encoding in the following example.

Example 12.25 Consider the FSM described in Ex. 12.24 (shown in Figure 12.16(a) and (b)).

We can encode states with 2 bits: $s1 = \{01\}$ and $s2 = \{10\}$. The corresponding, state transition table is shown in Figure 12.18(a). We denote the present state using variables $\{p, q\}$, and the next state using variables $\{x, y\}$. Note that the states $\{00\}$ and $\{11\}$ are not allowed in one-hot encoding. Therefore, we have not shown the rows corresponding to those states in the truth table. We can treat them as DC conditions and optimize *CL*. We can derive the following from the truth table (and taking help of the DC conditions):

$x = y'$
$y = (aq)'$
$z = (a + p')$.

The corresponding circuit implementation is shown in Figure 12.18(b). We initialize the FSM to one of the valid states ($p = 0, q = 1$). Note that we prefer using NAND, NOR, and inverter because these gates can be efficiently implemented in the CMOS technology.

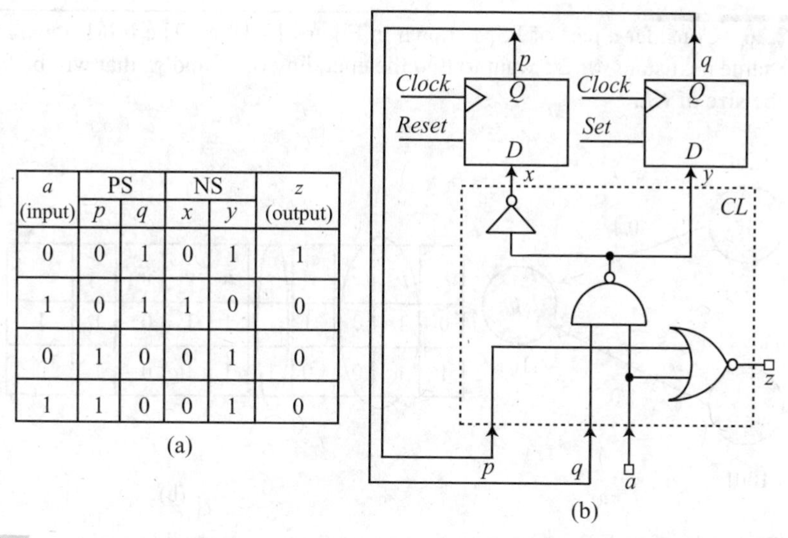

a	PS		NS		z
(input)	p	q	x	y	(output)
0	0	1	0	1	1
1	0	1	1	0	0
0	1	0	0	1	0
1	1	0	0	1	0

(a)

Figure 12.18 Illustration of one-hot encoding (a) state transition table and (b) circuit implementation

The encoding length determines the number of flip-flops in the FSM. Therefore, we need to keep the encoding length smaller. Moreover, as we increase the encoding length, the number of inputs $(n_i + n_s)$ and outputs $(n_o + n_s)$ of the combinational logic *CL* increases (see Figure 12.15). Consequently, the function *CL* will change. Sometimes we can simplify *CL* due to these changes. Intuitively, to identify a state, we need to examine just *one state* bit in one-hot encoding, while we might need a *minterm* in other encodings. Consequently, one-hot encoding can have fewer logic levels between flip-flops and be faster, though at the expense of more number of flip-flops and the associated area increase.

Challenges: The problem of state encoding is challenging because of an exponential number of possibilities. We can have FSMs with hundreds of states. Therefore, we cannot apply brute force to examine all solutions. Additionally, during encoding, we cannot implement and optimize *CL* to compare different solutions. After FSM encoding, we optimize *CL* using techniques discussed in the previous sections. Thus, we need to assess the encoding quality based on the *expected* FSM implementation. It makes the state encoding problem more difficult.

Various heuristic-based algorithms have been proposed for the FSM encoding problem [17–19]. We can formulate heuristics based on the following observation. The combinational block *CL* needs to generate n_s next state functions and n_o output functions. If we can generate these functions such that they *share* logic elements, we can reduce the area of *CL*. Note that sharing is not in direct control of the algorithm that generates encoding. Nevertheless, while encoding, we can create opportunities for optimization by generating codes that have more common cubes and common sub-expressions in the next state and output functions. We expect multilevel logic optimizer to exploit the common cubes and sub-expressions and generate compact *CL* later in the synthesis flow.

We present two observations that can be used as guidelines in manual state encoding or generating optimal state encoding [17].

Example 12.26 Consider a part of FSM shown in Figure 12.19(a). The FSM transitions from $s1$ and $s2$ to the same next state $s0$. We want to find the encoding of $s1$ and $s2$ that will be favorable for minimizing the size of *CL*.

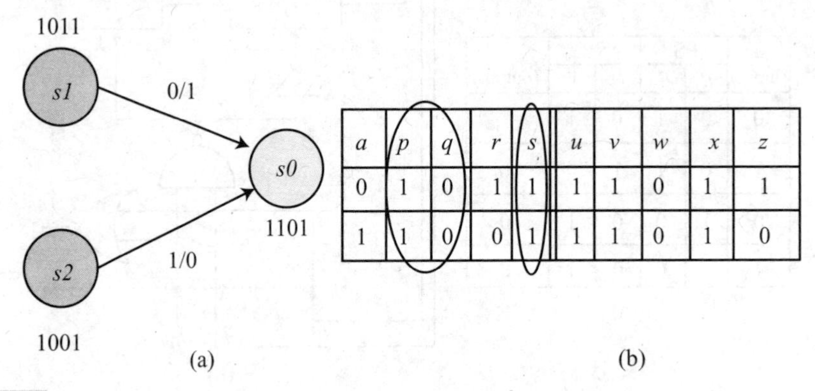

Figure 12.19 A part of an FSM (a) state diagram and (b) transition table

Let us represent the input bit as a and the output bit as z. Let us represent the present state with variables $\{p, q, r, s\}$ and the next state with $\{u, v, w, x\}$. Assume that we encode $s0 = \{1101\}$, $s1 = \{1011\}$, and $s2 = \{1001\}$. In this case, *CL* would contain the following functions (... represents terms related to other transitions in the FSM):

$u = a'pq'rs + apq'r's + ...$
$v = a'pq'rs + apq'r's + ...$
$w = ...$
$x = a'pq'rs + apq'r's + ...$
$z = a'pq'rs + ...$

We can extract the most frequently used common cube $M = pq's$ as follows:

$M = pq's$
$u = a'rM + ar'M + ...$
$v = a'rM + ar'M + ...$
$w = ...$
$x = a'rM + ar'M + ...$
$z = a'rM + ...$

Note that the size of the shared cube M is three. It is the consequence of same bit occurring in $s1$ and $s2$ at three positions in the encoding, as highlighted in Figure 12.19(b).

The above example motivates the following guideline:

Guideline 1: We should encode states that lead to the *same* next state with minimally distant codes in the Boolean space.

It will enable logic optimizer to share *maximum-sized cubes* in *CL*. We can measure the distance between two codes by counting the bits at which they differ. This measure of distance in the Boolean space is known as *Hamming distance*.

Example 12.27 Consider a part of FSM shown in Figure 12.20(a). The FSM transitions from $s0$ to the next states $s1$ and $s2$. We want to find encoding of $s1$ and $s2$ that will be favorable for minimizing CL. The variable names in the state transition table have the same meaning as in the previous example.

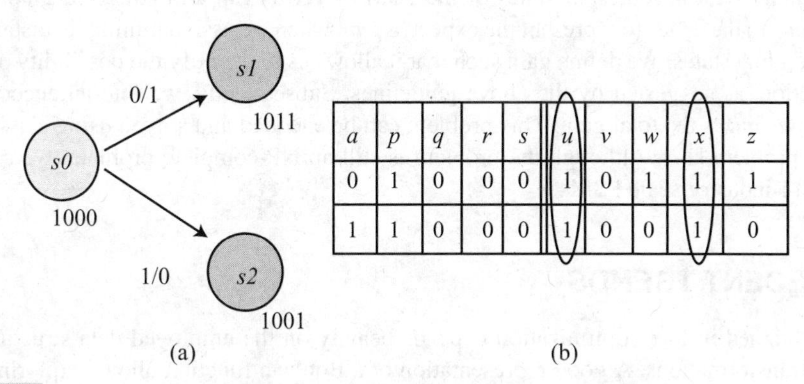

a	p	q	r	s	u	v	w	x	z
0	1	0	0	0	1	0	1	1	1
1	1	0	0	0	1	0	0	1	0

Figure 12.20 A part of an FSM (a) state diagram and (b) transition table

Assume that we encode $s0 = \{1000\}$, $s1 = \{1011\}$, and $s2 = \{1001\}$. The CL would contain the following terms:

$u = a'pq'r's' + apq'r's' + ...$
$v = ...$
$w = a'pq'r's' + ...$
$x = a'pq'r's' + apq'r's' + ...$
$z = a'pq'r's' + ...$

We can extract the most frequently used common cubes $M = pq'r's'$ as follows:

$M = pq'r's'$
$u = a'M + aM + ... = M(a' + a) + ...$
$v = ...$
$w = a'M + ...$
$x = a'M + aM + ... = M(a' + a) + ...$
$z = a'M + ...$

The size of M is four literals because there are 4 bits in $s0$. It is independent of our choice of encoding of $s1$ and $s2$. The common cube M appears for each 1 entry in the next state and output functions in the transition table. However, when both $s1$ and $s2$ have 1's in their encoding (as highlighted in Figure 12.20(b)), we can extract common cube in the associated next state function. It allows u and x to be minimized further. Note that if both $s1$ and $s2$ have 0's at some location, it would lead to no minimization because M will not appear in the corresponding bit (example v).

The above example motivates the following guideline:

Guideline 2: We should encode the next states that transition from the *same* current state with minimally distant codes in the Boolean space for 1's.

It will *maximize the number of occurrences* of the largest common cubes in the next state function.

Note that we have shown only two transitions in the above examples. In a practical FSM, multiple transitions will pull the encoding of a given state closer to many other states. Ensuring such requirements simultaneously for all states is difficult, if not impossible, to achieve manually. Nevertheless, we can automate the state encoding as follows [17].

Approach: We can represent states of the FSM as vertices in an undirected graph [17]. We assign *weights* to the edges to represent the expected *gain* when we assign minimally distant codes to the corresponding states. We define gain such that it allows us to quantify the possibility of common cube extraction, as suggested by the above guidelines. Subsequently, we find an encoding of the states that *maximizes* the total gain. This problem can be encoded and solved using classical *graph embedding* problem [17]. Although the problem is still an NP-complete problem, we can solve it efficiently for industry-scale FSMs.

12.4 RECENT TRENDS

The QoR obtained in logic optimization depends heavily on the employed data structure and the associated transformations. A good representation of a Boolean function allows exploring solution space more efficiently. There has been continuous progress in the Boolean function representations in the last three decades.

Binary decision diagrams (BDDs) have been used for logic optimization since the early 1990s [20, 21]. Homogeneous networks, such as *and-inverter graphs* (AIGs) (a network of two-input ANDs and inverters), have been attractive since the early 2000s [22]. The homogenous representations are simple and easy to manipulate. On large designs, they enable efficient optimization by consuming less memory and runtimes. AIG-based optimization is implemented in a public-domain tool ABC [23].

More recently, *majority inverter graph* (MIG) and associated transformations have been proposed for Boolean function representation and optimization [24–26]. An MIG consists of three-input *majority nodes* and regular/complemented edges. The majority function of three Boolean variables evaluates to 1 if and only if at least two of them are 1. An extension of MIG, called *XOR-majority graph* (XMG), has also been proposed for logic optimization. The algebraic and Boolean optimization techniques implemented on MIG have shown better QoR than AIG-based ABC [24].

Recently, machine learning techniques are also being explored to automatically tune synthesis parameters and choose the best optimizer for the given portion of a circuit [27, 28].

REVIEW QUESTIONS

12.1 What is the difference between a two-level logic circuit and a multilevel logic circuit?

12.2 A Boolean function in SOP form is not canonical, while it is canonical in the sum of minterm form. Explain.

12.3 What is an incompletely specified Boolean function?

12.4 How can we exploit DC conditions to reduce the size of a circuit?

12.5 How does Quine's theorem help develop an efficient algorithm for two-level logic optimization?

12.6 What is the difference between the minimum cover and an irredundant cover?

12.7 What is a minimal cover with respect to single-implicant containment?

12.8 Why do we need to use a heuristic two-level logic minimizer instead of an exact two-level logic minimizer?

12.9 Why do we typically minimize the number of literals in a multilevel Boolean logic network?

12.10 How is an algebraic model different from a Boolean model?

12.11 What are the advantages and disadvantages of optimizing using an algebraic model compared to the Boolean model?

12.12 What are kernels and co-kernels of an algebraic expression?

12.13 How does Brayton and McMullen's theorem help develop an efficient algorithm for finding multi-cube common divisors?

12.14 Determine all the kernels of the algebraic expression $ae + be + acd + bcd$.

12.15 What are SDCs?

12.16 How can we use CDCs and ODCs to optimize a Boolean logic network?

12.17 Define Boolean difference of a Boolean function with respect to a given variable. What can we say about the function if it evaluates to zero?

12.18 Derive the ODC for F in the Boolean logic network shown in Figure 12.21. Subsequently, optimize the logic network using the derived ODC.

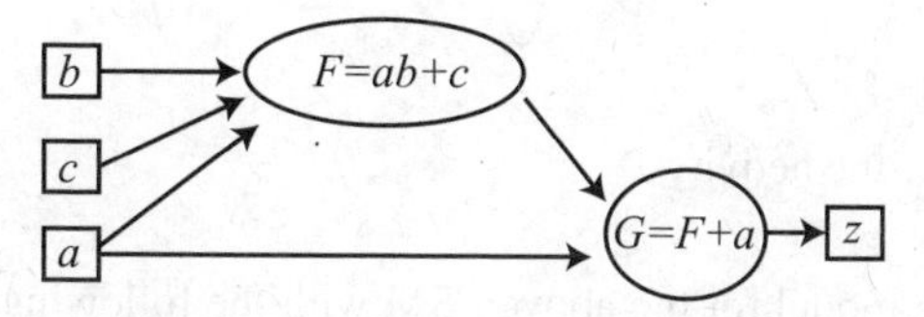

Figure 12.21 A given logic network

12.19 What is the motivation of state minimization?

12.20 When are two states in an FSM said to be equivalent?

12.21 How does a good state encoding of an FSM help minimize its circuit implementation?

12.22 List a few guidelines we should follow for encoding states in an FSM.

TOOL-BASED ACTIVITY

1. In this activity, use any logic synthesis tool, including open-source tools such as Yosys [29, 30]. For each Boolean logic network shown in Figures 12.9(a), 12.10(a), 12.11(a), 12.12(a), and 12.13(a) carry out the following tasks individually:
   - **(a)** Write Verilog codes as equations for individual vertices (you can use `assign` statements).
   - **(b)** Synthesize with a relaxed timing constraint (objective is to optimize for the minimum area). You can use any library, including freely available technology libraries [31]. Temporarily remove all cells from the technology library except inverters, AND gates, and OR gates.
   - **(c)** Observe the optimized netlist and explain the optimizations done for each circuit.

2. In this activity, use tools and libraries as in the previous activity (however, retain all cells in the library).

Consider the state diagram of an FSM shown in Figure 12.22. It has five states $\{s1, s2, s3, s4, s5\}$, one input bit, and one output bit. The output is {0, 1, 0, 1, 0} for the states $\{s1, s2, s3, s4, s5\}$, respectively. The reset state is $s1$. Carry out the following tasks:

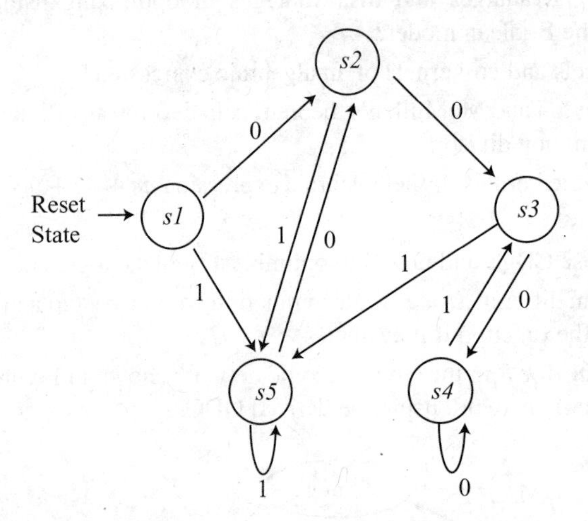

Figure 12.22 An FSM for the activity

(a) Develop a Verilog model for the above FSM with the following encodings:

(i) $\{s1 = 000, s2 = 001, s3 = 010, s4 = 101, s5 = 111\}$

(ii) $\{s1 = 001, s2 = 010, s3 = 011, s4 = 111, s5 = 100\}$

(iii) $\{s1 = 0001, s2 = 0010, s3 = 0100, s4 = 1100, s5 = 1000\}$

(iv) $\{s1 = 00001, s2 = 00010, s3 = 00100, s4 = 01000, s5 = 10000\}$

(b) Synthesize and report the total area of the circuit for each case for relaxed timing constraints.

(c) Analyze the result from the perspective of which encoding helps reduce area.

REFERENCES

[1] G. D. Micheli. *Synthesis and Optimization of Digital Circuits*. McGraw-Hill Higher Education, 1994.

[2] W. V. Quine. "On cores and prime implicants of truth functions." *The American Mathematical Monthly* 66, no. 9 (1959), pp. 755–760.

[3] W. V. Quine. "The problem of simplifying truth functions." *The American Mathematical Monthly* 59, no. 8 (1952), pp. 521–531.

[4] O. Coudert and T. Sasao. "Two-level logic minimization." In *Logic Synthesis and Verification* (2002), pp. 1–27, Springer.

[5] R. L. Rudell and A. Sangiovanni-Vincentelli. "Multiple-valued minimization for PLA optimization." *IEEE Transactions on Computer-aided Design of Integrated Circuits and Systems* 6, no. 5 (1987), pp. 727–750.

[6] P. C. McGeer, J. V. Sanghavi, R. K. Brayton, and A. Sangiovanni-Vicentelli. "ESPRESSO-SIGNATURE: A new exact minimizer for logic functions." *IEEE Transactions on Very Large Scale Integration (VLSI) Systems* 1, no. 4 (1993), pp. 432–440.

[7] M. R. Dagenais, V. K. Agarwal, and N. C. Rumin. "McBOOLE: A new procedure for exact logic minimization." *IEEE Transactions on Computer-aided Design of Integrated Circuits and Systems* 5, no. 1 (1986), pp. 229–238.

[8] O. Coudert and J. C. Madre. "Implicit and incremental computation of primes and essential primes of Boolean functions." *DAC* 92 (1992), pp. 36–39.

[9] S. J. Hong, R. G. Cain, and D. L. Ostapko. "MINI: A heuristic approach for logic minimization." *IBM Journal of Research and Development* 18, no. 5 (1974), pp. 443–458.

[10] R. K. Brayton, G. D. Hachtel, C. McMullen, and A. Sangiovanni-Vincentelli. *Logic Minimization Algorithms for VLSI Synthesis,* vol. 2. Springer Science & Business Media, 1984.

[11] R. K. Brayton, R. Rudell, A. Sangiovanni-Vincentelli, and A. R. Wang. "MIS: A multiple-level logic optimization system." *IEEE Transactions on Computer-aided Design of Integrated Circuits and Systems* 6, no. 6 (1987), pp. 1062–1081.

[12] R. K. Brayton, G. D. Hachtel, and A. L. Sangiovanni-Vincentelli. "Multilevel logic synthesis." *Proceedings of the IEEE* 78, no. 2 (1990), pp. 264–300.

[13] R. K. Brayton. "The decomposition and factorization of Boolean expressions." *Proceedings International Symposium on Circuits and Systems (ISCAS 82) Rome* (1982).

[14] R. Rudell. *Logic Synthesis for VLSI Design*. PhD thesis, University of California, Berkeley, 1989.

[15] H. Bhatnagar. *Advanced ASIC Chip Synthesis*. Springer, 2002.

[16] J. Hopcroft. "An n log n algorithm for minimizing states in a finite automaton." In *Theory of Machines and Computations* (1971), pp. 189–196, Elsevier.

[17] S. Devadas, H.-K. Ma, A. R. Newton, and A. Sangiovanni-Vincentelli. "MUSTANG: State assignment of finite state machines targeting multilevel logic implementations." *IEEE Transactions on Computer-aided Design of Integrated Circuits and Systems* 7, no. 12 (1988), pp. 1290–1300.

[18] G. De Micheli, R. K. Brayton, and A. Sangiovanni-Vincentelli. "Optimal state assignment for finite state machines." *IEEE Transactions on Computer-aided Design of Integrated Circuits and Systems* 4, no. 3 (1985), pp. 269–285.

[19] T. Villa and A. Sangiovanni-Vincentelli. "NOVA: State assignment of finite state machines for optimal two-level logic implementation." *IEEE Transactions on Computer-aided Design of Integrated Circuits and Systems* 9, no. 9 (1990), pp. 905–924.

[20] Y. Matsunaga. "Multi-level logic optimization using binary decision diagrams." *Proceedings IEEE International Conference on Computer-aided Design (ICCAD-89)* (1989), pp. 556–559.

[21] B. Lin, H. J. Touati, and A. R. Newton. "Don't care minimization of multi-level sequential logic networks." *1990 IEEE International Conference on Computer-aided Design. Digest of Technical Papers* (1990), pp. 414–417.

[22] A. Mishchenko, S. Chatterjee, and R. Brayton. "DAG-aware AIG rewriting: A fresh look at combinational logic synthesis." *2006 43rd ACM/IEEE Design Automation Conference* (2006), pp. 532–535, IEEE.

[23] R. Brayton and A. Mishchenko. "ABC: An academic industrial-strength verification tool." In *International Conference on Computer Aided Verification* (2010), pp. 24–40, Springer.

[24] L. Amaru, P.-E. Gaillardon, and G. De Micheli. "Majority-inverter graph: A new paradigm for logic optimization." *IEEE Transactions on Computer-aided Design of Integrated Circuits and Systems* 35, no. 5 (2015), pp. 806–819.

[25] L. Amarù, E. Testa, M. Couceiro, O. Zografos, G. De Micheli, and M. Soeken. "Majority logic synthesis." *Proceedings of the International Conference on Computer-aided Design* (2018), pp. 1–6.

[26] Z. Chu, M. Soeken, Y. Xia, L. Wang, and G. De Micheli. "Structural rewriting in XOR-majority graphs." *Proceedings of the 24th Asia and South Pacific Design Automation Conference* ASPDAC '19 (2019), pp. 663–668, Association for Computing Machinery.

[27] M. M. Ziegler, H.-Y. Liu, G. Gristede, B. Owens, R. Nigaglioni, and L. P. Carloni. "A synthesis-parameter tuning system for autonomous design-space exploration." *2016 Design, Automation & Test in Europe Conference & Exhibition (DATE)* (2016), pp. 1148–1151, IEEE.

[28] W. L. Neto, M. Austin, S. Temple, L. Amaru, X. Tang, and P.-E. Gaillardon. "LSOracle: A logic synthesis framework driven by artificial intelligence." *2019 IEEE/ACM International Conference on Computer-aided Design (ICCAD)* (2019), pp. 1–6, IEEE.

[29] C. Wolf, J. Glaser, and J. Kepler. "Yosys—A free Verilog synthesis suite." *Proceedings of the 21st Austrian Workshop on Microelectronics (Austrochip)* (2013).

[30] T. Ajayi, V. A. Chhabria, M. Fogaça, S. Hashemi, A. Hosny, A. B. Kahng, M. Kim, J. Lee, U. Mallappa, et al. "Toward an open-source digital flow: First learnings from the OpenROAD project." *Proceedings of the 56th Annual Design Automation Conference 2019* (2019), pp. 1–4.

[31] M. Martins, J. M. Matos, R. P. Ribas, A. Reis, G. Schlinker, L. Rech, and J. Michelsen. "Open cell library in 15nm FreePDK technology." *Proceedings of the 2015 Symposium on International Symposium on Physical Design* (2015), pp. 171–178.

13 Technology Library

...A library implies an act of faith ...

—Victor Hugo, "À qui la faute?" in *L'Année Terrible* (translated from French), 1872

We have earlier seen that a cell is the basic building block of application-specific integrated circuits (ASICs) and cell-based designs. A cell is a circuit that delivers a specified functionality, such as logic gates, flip-flops, arithmetic logical units (ALUs), and memory blocks. We carefully design and verify the function of a cell at the transistor level. We also measure characteristics such as delay, power dissipation, area, and the impact of process-induced variations for each cell. Subsequently, we organize these cell-specific attributes in a widely accepted database known as *library*. We instantiate and connect these cells in our design by utilizing the information contained in the library. We also need a library to verify whether a design delivers the desired functionality.

In VLSI design flow, we employ primarily two types of libraries: *technology library* and *physical library*.

1. **Technology library:** Historically, technology libraries were introduced for *logic synthesis* for defining the functionality of each cell. However, during the last three decades, technology libraries have evolved to support various design tasks such as physical implementation, timing verification, and test activities such as scan chain insertion. Nevertheless, a large portion of a technology library defines the timing attributes of the cells. Therefore, we also refer to a technology library as *timing library*. We represent a technology library in *Liberty format* [1]. The library files in the Liberty format are American Standard Code for Information Interchange (ASCII) files, typically with an extension of .lib. However, since library files are voluminous, sometimes we compress and store these files. Moreover, some tools can convert and store the library information in their proprietary format.
2. **Physical library:** A physical library contains information about the geometry or layout of the cells, including design rules, in some abstract form. Physical libraries are popularly represented in *library exchange format* (LEF) [2].

In this chapter, we will discuss technology libraries in detail. We will discuss physical libraries in Chapter 24 ("Basic concepts for physical design").

13.1 MOTIVATION FOR LIBRARY-BASED DESIGN

By employing technology libraries in a design flow, we decompose the overall design process into two steps:

1. **Creating library:** We design each cell at the transistor level and determine its optimal layout. We extract essential information about the cells and write them in the library. While creating libraries, we can optimally design and rigorously verify each cell. Though creating a library is time-consuming and costly, many designs can employ the same library. Therefore, the cost of developing a high-quality library gets distributed over multiple designs. Thus, the library-based design flows become profitable in the long run.
2. **Using library:** To achieve the desired functionality, we instantiate numerous cells (often more than a million) from a library. The library cells allow us to focus on their instantiations rather than on their internal details. Thus, the design time and effort decrease. Additionally, we also reduce the chances of errors by eliminating problems within the cells. Note that for multi million gate designs, it is infeasible to design and verify individual cells at the transistor level. Therefore, a library-based design methodology raises the *abstraction* from the *transistor* level to the *cell* level, and makes complex synthesis, static timing analysis (STA), and physical design tasks feasible.

Thus, a library-based design flow simplifies the overall design problem. It allows quicker design space exploration, easier verification due to the reduced number of components, and more efficient design closure.

13.2 LIBRARY CHARACTERIZATION

The process of creating a technology library is known as *library characterization*. We design each cell at the transistor level, create their layouts, and determine their electrical behavior using simulations, as shown in Figure 13.1. We use a *process design kit* (PDK) for obtaining fabrication-dependent parameters and transistor simulation models. We perform circuit simulations for a cell using *simulation program with integrated circuit emphasis* (SPICE). A SPICE simulation produces output voltage and current levels for the given input stimuli at a given operating condition. First, we verify the functional correctness of a cell for the given input stimuli. Then, we measure or extract physical quantities such as delay, slew, capacitance, and power dissipation. Using these results, we build an *abstract* model for each cell that various electronic design automation (EDA) tools can use in the design flow. Finally, we write these models in a widely accepted format in technology libraries.

We can obtain a library from various sources. A foundry can develop libraries and share them in foundry-provided PDKs. Alternatively, a design house can characterize its library using foundry-provided PDK, as illustrated in Figure 13.1. For academic and research purposes, we can use freely available libraries also [3].

Computational effort: During the SPICE simulation of a cell, differential equations are formulated and typically solved using iterative techniques. Therefore, SPICE simulation of a cell can be time taking. Moreover, simulating for various operating conditions and multiple input stimuli further increases the runtime of library characterization. Note that we need to perform numerous simulations for each cell to ensure that the library models can produce accurate results for varying instantiating conditions.

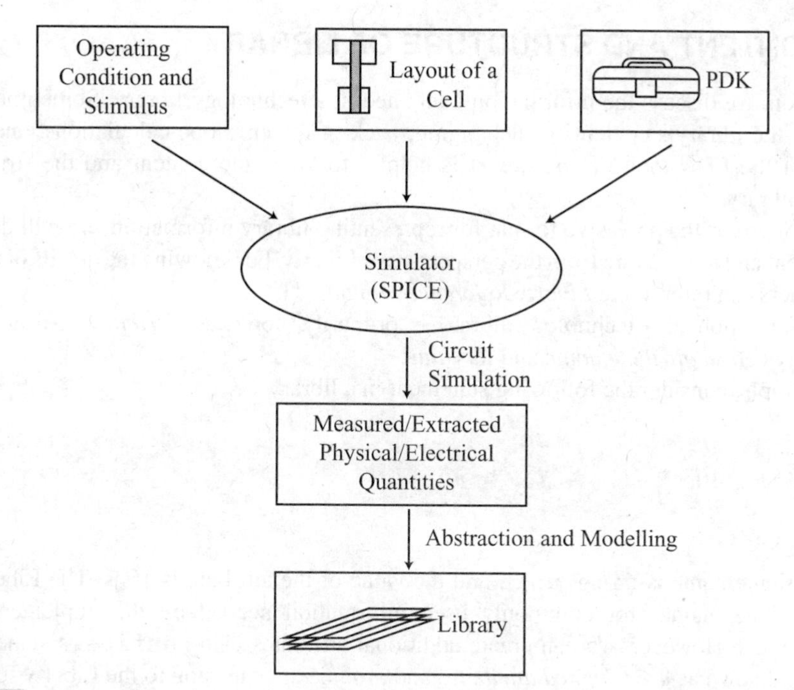

Figure 13.1 Library characterization framework

Library models: Once we have created a library, we can avoid time-taking SPICE simulations in the design flow. Various EDA tools can simply employ the abstract library models to compute quantities such as delay, slew, power dissipation, and voltage fluctuations. By avoiding SPICE simulations, EDA tools can compute these quantities several orders of magnitude faster and with reasonable accuracy. A critical element of this methodology is to develop appropriate library models that can compute these quantities significantly faster than SPICE yet be accurate enough that we can rely on their results. We will look at some of these models later in this chapter.

Variety and uniformity: For a given technology, we can develop multiple libraries that differ in optimization criteria. For example, we typically develop separate libraries for *high performance*, *low power*, and *high density* [4]. We can also develop libraries with cells implemented using transistors of different threshold voltages (V_T). These libraries are known as *multi-V_T libraries*. We will discuss the application of these libraries in the subsequent chapters.

In a given library, we typically design multiple cells with the same functionality. For example, we can implement and model ten different inverters in a library with varying areas, speeds, drive strengths, and power dissipation. Nevertheless, to ease physical design, we keep the height of the cells uniform in a technology library. Although it can sacrifice some figures of merit for the cells, it allows placing cells compactly in multiple fixed-height rows and eases the physical design tasks. To increase the speed or drive strength of a cell, we increase the *width* of the cell, keeping its *height* constant. To achieve this, we can employ multiple transistors in parallel such that the effective width of the pull-up and the pull-down transistors increase without changing the height of the cell.

13.3 CONTENT AND STRUCTURE OF LIBRARY

In this section, we discuss the information contained in a technology library. Sometimes we need to look into the library's content for debugging, back-of-the-envelope calculations, and checking the sanity of the EDA tools. Therefore, it is helpful to know the content and the structure of a technology library.

Since Liberty is the pervasive format for representing library information, we will describe the library content and its structure from the perspective of Liberty. For knowing the details of the Liberty format, readers can refer to the *Liberty Reference Manual* [1].

The information in a technology library is primarily stored as *attributes*. An attribute is a mapping between an *attribute name* and its *value*.

For example, consider the following statement in a library.

```
time_unit : "10ps";
```

The attribute name is `time_unit` and the value of the attribute is 10ps. The Liberty format provides attribute names for commonly used information needed by the implementation and verification tools. However, we can create additional attributes using `define` statements. These attributes are known as *user-defined attributes* and provide an extension to the Liberty format.

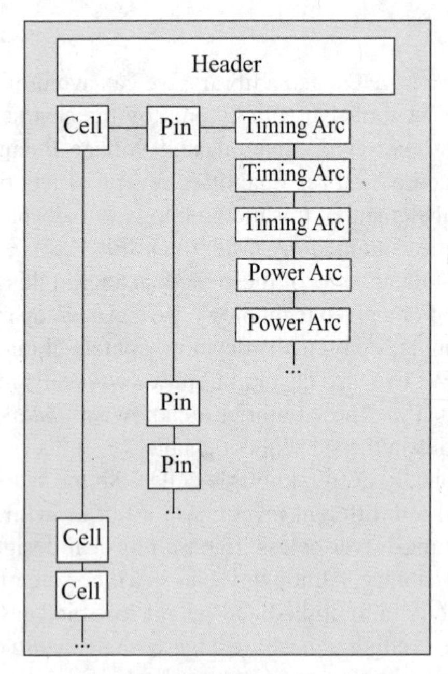

Figure 13.2 Organization of a technology library in Liberty format

For organizing data in a technology library, we define *groups* at various levels. Some of the groups existing in the Liberty format are shown in Figure 13.2. A *library* contains a group of *cells*, a cell contains a group of *pins*, and a pin contains a group of *timing arcs* and *power arcs*. The information contained in a technology library is organized at the following levels:

1. **Global data:** The global data is valid for all entities of a technology library and is specified in its *header*. The header is located at the top of a technology library, as shown in Figure 13.2. The header can contain the following information:
   (a) Library name
   (b) Units for voltage, time, capacitance, resistance, current, power, etc.
   (c) *Operating conditions* of characterization: The operating condition is a combination of process parameters, supply voltage, and temperature (PVT).
   (d) Threshold for the delay and slew computation (details are in the following section).
   (e) Templates for *look-up tables* (LUTs) that are used in the library.
   (f) The header can also contain *wire load* model. An implementation tool can use a wire load model to estimate net capacitances as a function of the number of pins and design size.
   (g) The header can also contain the derating factors to compute delays for the PVT conditions different from the characterized PVT conditions.

Example 13.1 Consider the following fragment of a technology library.

```
library (mylib) { ...
    time_unit : "10ps";
    nom_process : 1.00;
    nom_temperature : 100.00;
    nom_voltage : 0.80;
    slew_lower_threshold_pct_rise : 10.00;
    slew_upper_threshold_pct_rise : 90.00; ...
}
```

It defines a library named *mylib*. It also defines the unit for time, PVT conditions, and threshold for slew computation.[1]

2. **Cell-data:** A library can contain hundreds of cells with different functions and drive strengths. A cell can contain cell-specific information and a list of pins. The cell-specific information can be *name*, `area`, `cell_leakage_power`, `dont_touch` flag, and `pad_cell` flag.

[1] The examples of libraries in this chapter show only a part of a library for the sake of easy explanation. An actual technology library will contain additional information.

Example 13.2 Consider the following fragment of a technology library.

```
library (mylib) { ...
  cell (INVX1) {
    area : 0.15013;
    cell_leakage_power : 276959.1; ...
  }
  cell (AND2X1) {
    area : 0.28015;
    cell_leakage_power : 5676958.3; ...
  } ...
}
```

It defines two cells named *INVX1* and *AND2X1*. It also defines simple attributes `area` and `cell_leakage_power` for them.

3. **Pin-data:** A cell contains a list of pins. We define pin-specific information, such as its *name*, `direction`, and `function`, at the pin level. We can also define pin-specific design rules, such as `max_capacitance`, at the pin level.

   Some attributes of a cell are related to two pins of that cell. For example, the delay corresponds to an input pin and an output pin of a cell. We model such attributes using *arcs*. An arc is a pair of two pins: a *source pin* and a *destination pin*. We store the attribute of the arc on the destination pin and specify the corresponding source pin by an attribute named `related_pin`. An arc that corresponds to the timing is known as a *timing arc*. Similarly, an arc that corresponds to the power is known as a *power arc*. We employ timing arcs and power arcs to model delay and power dissipation in a cell.

Example 13.3 Consider the following fragment of a technology library.

```
library (mylib) { ...
  cell (INVX1) { ...
    pin (I) {
      direction : input;
      capacitance : 0.761564; ...
    }
    pin (ZN) {
      direction : output;
      max_capacitance : 150.000000;
      function : "!I";
      timing () {
        related_pin : "I";
```

```
        intrinsic_rise:0.2;
        intrinsic_fall:0.3;
      }...
    }...
  }...
}
```

It defines two pins *I* and *ZN* for the cell *INVX1* alongwith the pin-level attributes `direction`, `capacitance`, `max_capacitance`, and `function`. It also defines a timing arc between the source pin *I* and the destination pin *ZN*.

13.4 TIMING ATTRIBUTES AND DEFINITIONS

To understand library timing models, we should be familiar with the following definitions and concepts.

13.4.1 Slew

We measure the steepness or abruptness of $0 \rightarrow 1$ and $1 \rightarrow 0$ transitions using a quantity known as *slew*. We assume that the waveform is linear and quantify the sharpness of the transition by its average slope.

We define slew for the rise transition as the time taken for a signal to reach from a *lower threshold percentage* (LTP) to a *upper threshold percentage* (UTP) (with respect to the supply voltage V_{DD}), as illustrated in Figure 13.3(a). The slew for the rise transition is also called *rise transition time* or *rise time*. Similarly, we define slew for the fall transition as the time taken for a signal to reach from a UTP to an LTP, as illustrated in Figure 13.3(b). The slew for the fall transition is also called *fall transition time* or *fall time*.

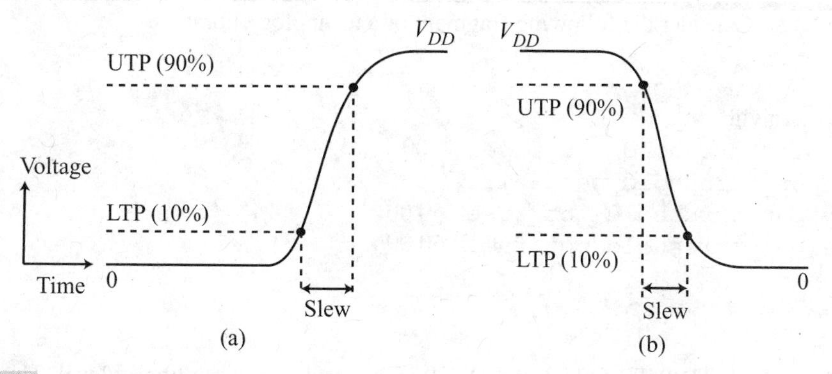

Figure 13.3 Definition of slew of a signal (a) rise transition and (b) fall transition

While characterizing a library, we choose an LTP and a UTP for each type of transition (rise and fall). We specify them as attributes at the library level, as shown in the following example.

Example 13.4 Consider the following fragment of a technology library.

```
library (mylib) { ...
    slew_lower_threshold_pct_rise : 10.00;
    slew_lower_threshold_pct_fall : 10.00;
    slew_upper_threshold_pct_rise : 90.00;
    slew_upper_threshold_pct_fall : 90.00; ...
}
```

It specifies the LTP and UTP for the rise and fall transitions as attributes at the library level. Typically, we take LTP as 10% and UTP as 90%, as shown in Figure 13.3. We refer to it as a 10–90 slew threshold. Some libraries can have 20–80 or 30–70 slew thresholds.

13.4.2 Delay

To measure the time taken for the change in the input to propagate to the output, we use a quantity known as the *propagation delay* or *delay* in short. The delay can be different for the rise transition and the fall transition at the output. We call the delay corresponding to the rise transition at the *output pin* as the *rise delay* and the delay corresponding to the fall transition at the *output pin* as the *fall delay*.

Similar to the slew, we quantify delay using some predefined threshold points on the input waveform and the output waveform. Before library characterization, we choose these threshold points. For the input signal for an arc, we choose thresholds called *input rise threshold percentage* (IRTP) and *input fall threshold percentage* (IFTP). Similarly, for the output signal for an arc, we choose thresholds called *output rise threshold percentage* (ORTP) and *output fall threshold percentage* (OFTP). We define these thresholds as attributes at the library level.

Example 13.5 Consider the following fragment of a technology library.

```
library (mylib) { ...
    input_threshold_pct_rise : 50.00;
    input_threshold_pct_fall : 50.00;
    output_threshold_pct_rise : 50.00;
    output_threshold_pct_fall : 50.00; ...
}
```

It defines the IRTP, IFTP, ORTP, and OFTP using attributes at the library level. The names of these attributes are self-explanatory. Typically, we take these thresholds as 50% of the supply voltage V_{DD}, as shown in the example.

We use *appropriate* pair of thresholds while measuring delay during library characterization. We need to use the same pair of thresholds in delay calculation during timing analysis. The following example explains how to choose an appropriate pair of thresholds for delay computation.

Example 13.6 Consider the timing arc shown in Figure 13.4(a).

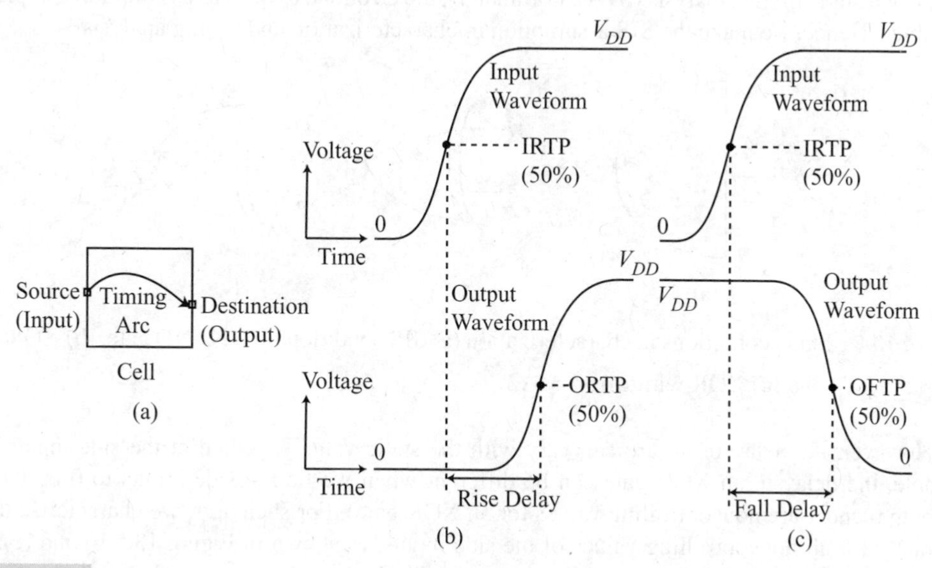

Figure 13.4 Definition of delay of a signal (a) timing arc, (b) rise delay, and (c) fall delay

Assume that the rise transition at the source results in a rise transition at the destination (as in a buffer). In this case, we define the rise delay as the time elapsed between the input signal crossing IRTP and the output signal crossing ORTP, as shown in Figure 13.4(b).

Similarly, assume that the rise transition at the source results in a fall transition at the destination (as in an inverter). In this case, we define the fall delay at the destination as the time elapsed between the input signal crossing IRTP and the output signal crossing OFTP, as shown in Figure 13.4(c).

Similarly, we employ IFTP for the fall transition at the source pin.

13.4.3 Controlling and Noncontrolling Values

We need to know the definition of *controlling* and *noncontrolling* values for understanding the delay characterization method for logic gates. The *controlling value* of a gate is the input value that can *solely* decide the value of its output. For example, 0 is the controlling value for AND/NAND gates, and 1 is for OR/NOR gates. The values other than the controlling value for a gate are known as the *noncontrolling values*. For example, 1 is the noncontrolling value for AND/NAND gates, 0 is the noncontrolling value for OR/NOR gates, and both 1 and 0 are noncontrolling values for XOR/XNOR gates.

13.4.4 Conditional Attributes

While characterizing the delay of multi-input gates, we tie the side-inputs (input pins not belonging to that arc) to *noncontrolling values*. For example, when we characterize the arc $A \rightarrow Z$ in the NAND gate of Figure 13.5(a), we tie the side-inputs B and C to 1. This assumption is called *single input switching* (SIS) assumption. It greatly simplifies the characterization problem, modeling of delays, and static timing analysis (STA). Fortunately, the error due to the SIS assumption is typically tolerable.[2] Hence, we make the SIS assumption in characterization and timing analysis.

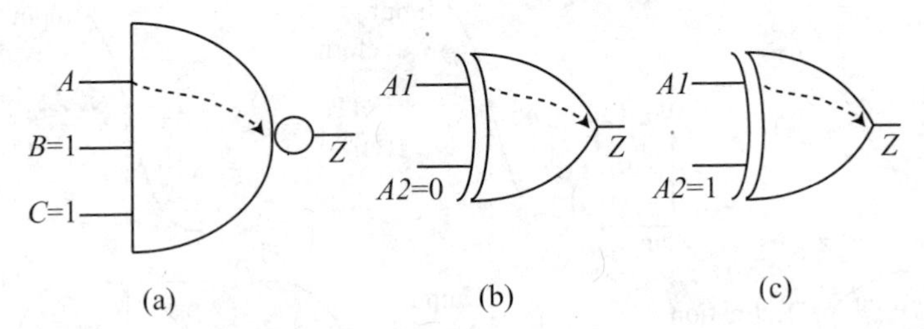

Figure 13.5 Static conditions in characterization (a) SIS condition for a NAND gate, (b) XOR with `when : "!A2"`, and (c) XOR with `when : "A2"`

However, the delay of an arc can vary with the static value specified at the side inputs. For example, the delay in an XOR gate can be different when we tie the side inputs to 0 or 1 (note that both 0 and 1 are noncontrolling values for an XOR gate). For such arcs, we characterize delay separately for all noncontrolling values of the side inputs, as shown in Figure 13.5(b) and (c). We can differentiate these timing delay arcs using an attribute `when` in the technology library defined at the arc-level.

Example 13.7 Consider the following fragment of a technology library for an XOR gate.

```
pin (Z) {
      function : "(A1 & !A2) |(!A1 & A2)";
      timing () {
        related_pin : "A1";
        when : "!A2"; ...
      }
      timing () {
        related_pin : "A1";
        when : "A2"; ...
      } ...
} ...
```

[2] The SIS assumption can lead to significant errors at advanced process nodes in some cases. Hence, *multi input switching* (MIS) models are now becoming of interest [5].

It defines `when` condition for the delay arc $A1 \rightarrow Z$. For computing delay, the delay calculator should consider the first timing arc when $A2=0$, and the second timing arc when $A2=1$. Similarly, when the leakage power dissipation, current, resistance, and capacitance are dependent on the input pin values, we can specify them using `when` conditions in the library.

13.4.5 Unateness

The *unateness* of an arc defines the relationship between the source transition and the destination transition.

Consider a cell arc in which the rise transition at the source results in a rise transition at the destination. Additionally, assume that the fall transition at the source results in a fall transition at the destination. Such arcs of a cell are known as *positive unate*. For example, timing arcs in buffers, AND gate, OR gate, two-input XOR gate with the side-input at 0, and a two-input XNOR gate with the side-input at 1 are positive unate.

Consider a cell arc where the rise transition at the source results in a fall transition at the destination. Additionally, assume that the fall transition at the source results in a rise transition at the destination. Such arcs of a cell are known as *negative unate*. For example, timing arcs in inverters, NAND gate, NOR gate, two-input XOR gate with the side-input at 1, and two-input XNOR gate with the side-input at 0 are negative unate.

For some arcs, there exists no relationship between the source transition and the destination transition. For example, consider a timing arc between the clock-pin (*CP*) and the Q-pin of a flip-flop. The transition type at the Q-pin depends on the current value at the D-pin and the Q-pin. It is not related to the transition type at the *CP* of the flip-flop. Such types of arcs are known as *non-unate*.

In a technology library, we specify the unateness of an arc using the attribute `timing_sense`, with valid values being `positive_unate`, `negative_unate`, and `non_unate`.

13.4.6 Cells and Interconnects

In a cell-based circuit, instances of library cells are interconnected by wires. The signal generated by a driving cell propagates through the interconnect and reaches the receiving cells. A receiving cell can further produce a signal that propagates through the interconnects to the next receiving cell. Thus, signal generation and propagation can occur sequentially throughout a circuit whenever it is active.

Cell delay and wire delay: A signal experiences delay and distortion (change in the shape of waveform) as it is generated and propagated through the circuit elements. We refer to the delay introduced by a *cell* in producing a signal at its output pin as the *cell delay* or *gate delay*. We refer to the delay in the signal propagation through an *interconnect* as the *wire delay* or *interconnect delay*.

Stage for delay calculation: Tools employed for STA, noise analysis, synthesis, and physical design need to internally compute delay and output waveforms. These tools compute delay and output waveforms by considering a small sub-circuit at a time, known as *stage* [6]. A stage consists of a driving cell and the receiving pins connected directly through interconnects, as shown in Figure 13.6(a). The cell driving the interconnect is referred to as the *driving cell*, and the receiving pins of a given stage are known as *sinks*. Note that there can be multiple sinks for a stage, and each sink can have a distinct output waveform. We decompose a given circuit into several such stages, and waveform propagation is performed from one stage to another. We will explain this aspect of delay calculation for a circuit in Chapter 14 ("Static timing analysis").

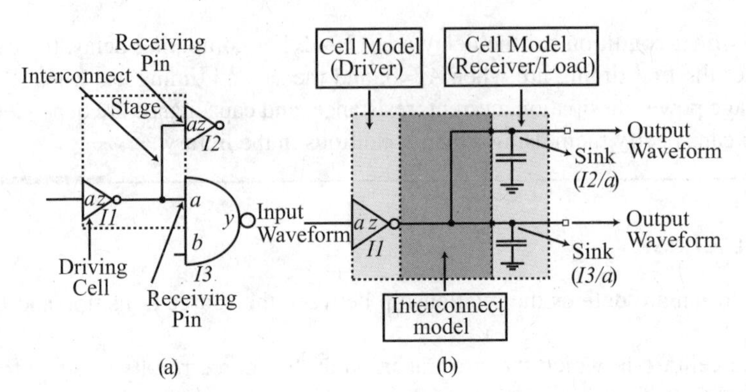

Figure 13.6 Delay computation (a) a stage (higlighted by a rectangle), (b) cell model and interconnect model in a stage

Driver and receiver models: The library cell models allow delay calculators to compute output waveforms for a given stage and given input waveform. The same cell instance can act as a driver in one stage and as a receiver in the preceding stage. Thus, we need to develop two types of models for a cell for delay calculation purposes: a *driver model* and a *receiver model*, as shown in Figure 13.6(b).

Interconnect models: Interconnects between cells play a crucial role in determining the output waveform for a given stage. With the advancement in technology, the interconnect delay dominates over the cell delay. Therefore, while developing cell models, we should consider how to account for the effect of the interconnects. This problem is challenging because a interconnect's electrical behavior depends strongly on its layout, the neighboring interconnects, and the fabrication technology. We employ various types of *interconnect models* to capture the electrical properties of an interconnect with varying degrees of accuracies.[3] We also refer to an interconnect model as a wire model.

The interconnect models complicate developing cell models also. A cell model is expected to work with any realistic interconnect model and produce a reasonably accurate delay. In practice, it is challenging to ensure this. Therefore, we make some simplifying *assumptions* for the interconnect model while developing cell delay models. During delay calculation for timing analysis, we make appropriate adjustments to the interconnect model such that the assumptions of the cell model remain valid.

Receiver model: The cell models for *receivers* are based on the capacitances of the input pins. The following example illustrates the capacitance-based receiver model.

Example 13.8 Consider the following fragment of a technology library.

```
pin (I) {
    direction : input;
    capacitance : 0.7;
    rise_capacitance : 0.6;
    fall_capacitance : 0.8;
    rise_capacitance_range (0.3,0.9);
    fall_capacitance_range (0.5,1.1); ...
}
```

[3] We will discuss interconnect models in Chapter 24 ("Basic concepts for physical design").

It models single lumped capacitance using attribute `capacitance`. For greater accuracy, we define separate capacitance for the rise and fall transitions using attributes `rise_capacitance` and `fall_capacitance`. It also defines `rise_capacitance_range` and `fall_capacitance_range` for computing the maximum and minimum capacitances for timing analysis.

Note that the input capacitance depends on various circuit conditions. We should capture these dependencies in the receiver capacitance models for a greater accuracy, as discussed in the following section.

13.5 DELAY MODELS

Some of the widely used library delay models are as follows.

13.5.1 Nonlinear Delay Model

The delay of a timing arc primarily depends on the input waveform and the output load for a given operating condition and given static values at the side inputs.

Consider the delay D of the timing arc shown in Figure 13.7(a). Typically, D is a nonlinear function of the input slew (S_I) and the capacitive load (C_L). However, it is difficult to model the nonlinear delay function $D(S_I, C_L)$ exactly. As an approximation, we represent D as a two-dimensional table T with $M \times N$ entries, as shown in Figure 13.7(b). We define delays for M discrete input slew values $S_{I,1}, S_{I,2}, ..., S_{I,M}$ and N discrete capacitive load values $C_{L,1}, C_{L,2}, ..., C_{L,N}$. Each entry at the ith row and the jth column in the table T represents the delay for the slew $S_{I,i}$ and load $C_{L,j}$. This type of model is known as a *nonlinear delay model* (NLDM).

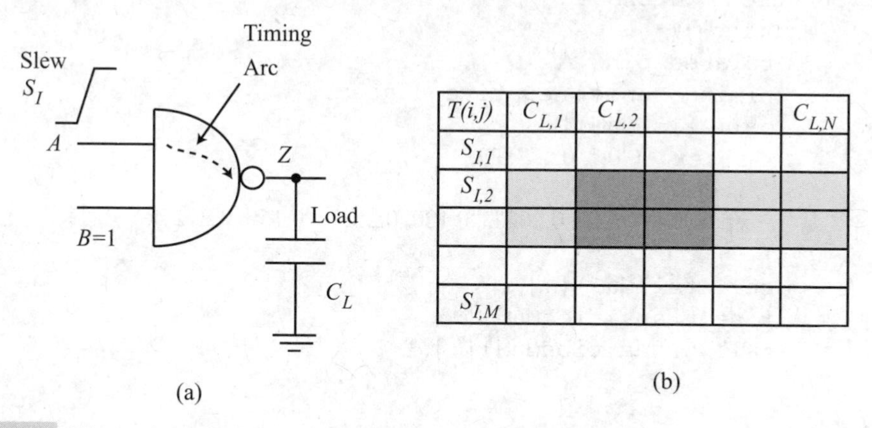

Figure 13.7 Illustration of NLDM (a) dependency of delay on slew and load, (b) two-dimensional NLDM table T

We explain the use of the NLDM table in the following example.

Example 13.9 Suppose we need to determine the delay for a slew S_I and load C_L using a given NLDM table. We proceed as follows.

First, among M slew points, we determine the two closest matches within which S_I lies. Then, we select two rows corresponding to these slews in T, as shown by darkened rows in Figure 13.7(b). Then, among N loads, we determine the two closest matches within which C_L lies. Then, we select two columns corresponding to these loads in the selected rows. Thus, we obtain the four closest matches within which the given slew and load lie. Using these four entries in the NLDM table, we can approximate $D(S_I, C_L)$ by *interpolation*. We can use various interpolation techniques such as bilinear, quadratic, and cubic splines [6].

We define NLDM tables in technology libraries using timing arcs. The following example illustrates it.

Example 13.10 Consider the following fragment of a technology library.

```
library (mylib) {...
  lu_table_template (lut1) {
    variable_1 : input_net_transition;
    variable_2 : total_output_net_capacitance;
    ...
  } ...
    pin (Z) {
      function : "!(A & B)"; ...
        timing () {
          related_pin : "A";
          timing_sense : negative_unate;
          cell_rise(lut1) {
            index_1("0.1, 0.5, 2.0");
            index_2("0.1, 1.0");
            values("0.110, 0.202", "0.120, 0.252", "0.150, 0.272");
          }
          cell_fall(lut1) {...}
          rise_transition(lut1) {...}
          fall_transition(lut1) {...}
        }...
    } ...
}
```

It defines the rise delay table using an attribute `cell_rise`. The slew and load points are specified by `lu_table_template` at the library level and is overridden while specifying the table entries. The `variable_1` and `index_1` correspond to the input net transition or slew and `variable_2` and `index_2` correspond to output net capacitance. The values in the table can be viewed as two nested loops: the first variable varies in the outer loop and the second variable varies in the inner loop. Thus, $T(0.1,0.1)$=0.110, $T(0.1,1.0)$=0.202, $T(0.5,0.1)$=0.120, etc.

Similarly, it defines the fall delay using the attribute `cell_fall`.

The output slew of a timing arc also varies nonlinearly with the input slew and the load capacitance. Similar to the delay, we model the output slew using two-dimensional NLDM tables. In the technology library, we define the output slew using attributes `rise_transition` and `fall_transition`, as shown in the example.

For NLDM to be usable in a design flow, we must ensure that the M slew points and N load points are closely spaced such that the interpolation gives sufficiently accurate results compared to the SPICE simulations. The problem gets simplified because delay and output slew are typically *nondecreasing functions* of S_I and C_L. Therefore, even with fewer points (say $M, N = 8$), we can ensure reasonable accuracy of the interpolation.

While using NLDM in delay calculation, we should ensure that the input slew S_I and the output load C_L are within the NLDM table range, i.e., $S_{I,1} \leq S_I \leq S_{I,M}$ and $C_{L,1} \leq C_L \leq C_{L,N}$. If this condition is not satisfied, we need to *extrapolate*, and it can produce large errors in delay calculation. Hence, we should avoid extrapolation and recheck our design when a tool reports extrapolation in delay calculation or timing analysis. Additionally, we should be aware of the following limitations of the NLDM:

1. **Assumptions on the waveforms:** NLDM assumes that the delay and the output slew of an arc depend on the *slew* of the input signal. Therefore, NLDM implicitly assumes that the input waveform is linear because a slew accurately captures a waveform shape for a linear signal. Therefore, when a cell receives an input that we cannot approximate as linear, the delay and the output slew computed using NLDM can have significant errors.

   At advanced process nodes, the waveforms can become nonlinear due to back-Miller effects, high impedance of the wires, and nonlinear pin capacitance [7]. For *noise analysis* or *signal integrity* (SI), we cannot ignore the nonlinearity of waveform and non-monotonic transitions [6].

2. **Assumptions on the receiver:** NLDM assumes that the delay and the output slew of an arc depend on the *load capacitance*. Therefore, in a strict sense, NLDM tables should be employed only for capacitive loads. For wires with negligible resistance, the accuracy of NLDM is acceptable.

   For wires with significant resistance, delay calculators typically employ *effective capacitance* (C_{eff}) to compute cell delays [6, 8, 9]. The C_{eff} for a wire is determined such that the average current (or total charge transfer) through C_{eff} and the actual wire are the same [8]. This value of C_{eff} ensures that the cell delay (50% crossing point) computed using C_{eff} matches the cell delay computed using the actual wire despite the corresponding waveforms being vastly different.

   By looking up C_{eff} in the NLDM table, we can compute the cell delay and the output slew. Using the output slew and the given wire model, we can compute the waveforms at the receiver pins of a given stage. However, modeling complex interconnects using a single linear C_{eff} is difficult (mathematically impossible), and the NLDM-computed delay can exhibit significant errors at advanced process nodes [6].

Though NLDM is compact and easy to compute, due to the above problems, we employ alternative delay models when greater accuracy is required (such as during signoff) and when waveforms can be highly nonlinear (such as during SI analysis). We describe two such models in the following paragraphs.

13.5.2 Composite Current Source Model

The traditional driving cell model consists of a voltage source V_D in series with a driving resistance of R_D. When the external impedance is Z, the traditional driving cell model will produce cell output voltage V_O as:

$$V_O = \frac{Z}{Z + R_D} \times V_D$$

If Z dominates R_D, ($V_O \approx V_D$). Hence, V_O becomes essentially independent of the load and can be erroneous. We remove this deficiency in the current source-based models by considering the driving cell as a time-dependent voltage-dependent current source [6, 10, 11]. In effect, this makes the drive resistance infinite and can model the response of a complex load more accurately.

Note that the current source-based models do not attempt to model the transistor behavior accurately. They achieve better accuracy in delay calculation by keeping a more flexible correspondence between the lumped load response and the response to any arbitrary interconnect model and circuit conditions [12].

Driver and receiver model: In *composite current source* (CCS) model, the *output current* is modeled as a function of input slew, output capacitance and time instants [1]. We define output current and time instant pairs obtained for discrete values of input slew and output load. We keep this information at the arc-level (i.e., source–destination pin combination). We can define the CCS model for a receiver at the arc level by specifying capacitances at discrete values of input slew and output load. We can specify capacitances separately for the rise/fall transitions using some attributes.

Using CCS model: We can compute the output waveform with the help of the CCS model and the given input waveform. The driver, interconnect, and receiver models are combined to obtain a stage, as illustrated in Figure 13.6(b). The driver and receiver are nonlinear current source and capacitance, respectively. The output waveform is computed by formulating the circuit equations and solving them numerically [6]. From the output waveform, we can extract delay and other information. Note that, in contrast to NLDM, the CCS model does not capture the delay information explicitly.

13.5.3 Effective Current Source Model

In *effective current source model* (ECSM), the *output voltage* is modeled at the arc-level as a function of input slew, output capacitance and time instants [11]. We define output waveform as a three-dimensional table at discrete values of input slew, output capacitance, and time instants.[4]

Though we model voltage waveform in ECSM, we can easily derive a current source model from it by differentiation because the current I through a capacitor C and voltage V across it are related as follows [6]:

$$I = C \times \frac{dV}{dt}$$

[4] The ECSM is supported by Liberty extensions. Readers can refer to [11] for more details on ECSM.

Nevertheless, numerical differentiation can introduce additional errors in a model. Also, note that the characterization of CCS and ECSM is straightforward. However, both these models are bulky due to three-dimensional tables and require more computational effort to produce delay and waveforms [6, 7, 12].

13.6 MODELING CONSTRAINTS

To properly operate some circuit elements such as a flip-flop, we need to ensure that certain timing constraints are satisfied between signals arriving at their inputs. These constraints originate due to the internal structure of library cells and are cell-specific. Therefore, we define these constraints in technology libraries. We explain some of these constraints in this section.

Consider the positive edge-triggered D-flip-flop shown in Figure 13.8(a). We have named the signal received at the D-pin (*D*) as *DATA* and the signal received at the *CP* as *CLOCK*. When the positive clock edge is received at the *CP*, we expect that the data present at the D-pin gets latched, and the latched value becomes available at the Q-pin after some delay. However, for deterministic data latching, the signal *DATA* is forbidden to change close to the clock edge. While characterizing a flip-flop, we define a *forbidden window* around a clock-edge where *DATA* is not allowed to change.

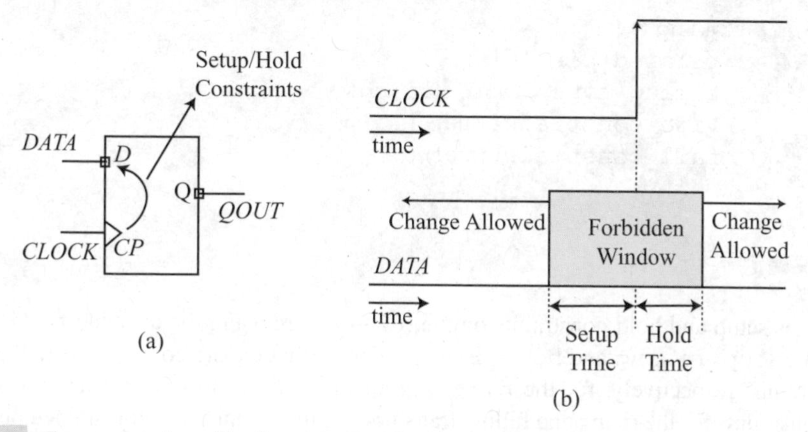

Figure 13.8 Constraints in a flip-flop (a) setup/hold arcs and (b) forbidden timing window

The minimum amount of time *before* the *CLOCK* edge that the *DATA* should be held steady is known as the *setup time* of the flip-flop [13, 14]. Similarly, the minimum amount of time *after* the *CLOCK* edge that the *DATA* should be held steady is known as the *hold time* of the flip-flop. Thus, setup time and hold time define the forbidden window around the clock edge where *DATA* is not allowed to change, as shown in Figure 13.8(b). We verify that the *DATA* does not change in the forbidden window in a circuit using STA tools. We will explain this aspect of verification in detail in Chapter 14 ("Static timing analysis") .

In general, the setup and hold time of a sequential element depend on the *data slew* (slew at the D-pin) and the *clock slew* (slew at the *CP*). Since the relationships between the setup/hold time and the data/clock slews are nonlinear, we model these constraints as two-dimensional discrete point tables (similar to NLDM), as explained in the following example.

Example 13.11 Consider the following fragment of a technology library.

```
library (mylib) {
  lu_table_template (lutff) {
    variable_1 : constrained_pin_transition;
    variable_2 : related_pin_transition; ...
  } ...
      pin (D) { ...
        timing () {
          related_pin : "CP";
          timing_type : setup_rising;
          rise_constraint(lutff) {
            index_1("0.1, 0.5, 2.0");
            index_2("0.1, 1.0");
            values("0.07, 0.12", "0.08, 0.13", "0.12, 0.20");
          }
          fall_constraint(lutff) { ... }
        }
        timing () {
          related_pin : "CP";
          timing_type : hold_rising;
          rise_constraint(lutff) { ... }
          fall_constraint(lutff) { ... }
        }...
      }...
}
```

It defines setup and hold constraints on the ($CP \rightarrow D$) arc using the attribute `timing_type`. The values `setup_rising` and `hold_rising` denote that the arc corresponds to the setup and hold constraints, respectively, for the *rising edge* of the related pin (*CP* in this case). It defines separate constraints for the rising and falling transitions of the signal at the constrained pin (*D* in this case) using the attributes `rise_constraint` and `fall_constraint`, respectively.

In addition to the setup and hold constraints, some other constraints must be honored to ensure that the circuit performs deterministically. These constraints are cell-specific, and we model them in a technology library. Most of these constraints are modeled on arcs as one- or two-dimensional table. We can identify these constraints in a technology library by the attribute name `timing_type`. We ensure that these constraints are honored in a design using STA tools. Some of these constraints are as follows:

1. **Minimum pulse width and clock period:** The minimum pulse width constraint defines the minimum value of the pulse width to produce the expected change in the data. For example, the clock signal may not latch the data if the clock pulse width is smaller than the specified minimum. We specify minimum pulse width constraint at a *self-arc* ($CP \rightarrow CP$) using the attribute–value pair `timing_type : min_pulse_width`. The minimum

pulse width is typically considered a function of input slew. Similarly, we can define the minimum period constraint on a self-arc using the attribute–value pair `timing_type : minimum_period`.

2. **Recovery and removal constraints:** Some flip-flops can have asynchronous pins such as *reset* and *preset*. When an asynchronous pin becomes active, the output can change, irrespective of the clock signal. However, when the asynchronous pin turns inactive, the clock signal regains control, and it latches data at the active clock edges. To ensure that the synchronous behavior gets restored deterministically, certain timing constraints must be satisfied between the asynchronous pin (such as *reset* and *preset*) and the clock-pin. We model these constraints using *recovery* and *removal* constraints.

   The *recovery time* is the minimum time that the signal at an asynchronous pin should remain stable after being de-asserted (made inactive from active) *before* the next clock edge. It is similar to the setup check for the D-pin of a flip-flop. We model recovery time on an arc using the attribute `timing_type` with values `recovery_rising` and `recovery_falling` for rising and falling clock-edges, respectively. Similarly, the *removal time* is the minimum time *after* the clock edge that the signal at the asynchronous pin should remain active before it can be de-asserted. It is similar to the hold check for the D-pin of a flip-flop. We model removal time on an arc using the attribute `timing_type` with values `removal_rising` and `removal_falling` for rising and falling clock-edges, respectively. Note that the recovery and removal time create a forbidden window around the clock edge where we are not allowed to de-assert the signal at the asynchronous pins of a flip-flop.

13.7 OTHER INFORMATION IN LIBRARY

Similar to the timing model, there are power models in a technology library. Power models capture various types of power dissipation inside a cell. We will discuss library power models in Chapter 18 ("Power analysis").

A technology library also contains crosstalk and noise analysis models, models for accounting process-induced variations, models defining design rules, and models supporting low-power implementations. Interested readers can refer to them in the *Liberty Reference Manual* [1]. Note that a technology library can contain tool-specific attributes and information in addition to the attributes natively supported by Liberty. Moreover, as new effects and phenomena become dominant with the advancement in process nodes, we need to model additional cell-specific information in a technology library.

13.8 RECENT TRENDS

When we adopt newer process nodes in a design flow, technology libraries are the first components to be impacted. At advanced process nodes, we often need to capture new phenomena and new dependency of some earlier ignored parameters. However, traditional table-based approaches do not scale with the increased number of parameters and make technology libraries voluminous, often more than tens of GBs. Consequently, the runtime required in loading data contained in a set of technology libraries becomes unacceptably high [15]. Therefore, techniques that can represent library information compactly become important. In this regard, we can utilize machine learning-based models since they can exploit the intrinsic degrees of variation in the data more efficiently compared to the table-based approaches [5]. With the increasing popularity of machine learning techniques in VLSI design flow, in the future, completely replacing table-based models of technology libraries with machine-learned models with varying abstraction levels, accuracy requirements, environmental conditions, and a richer set of input features is possible.

REVIEW QUESTIONS

13.1 Explain the motivation of using technology libraries in a design flow.

13.2 How is the characterization of cells done for creating a technology library?

13.3 What are the characteristics of a "good" technology library?

13.4 Assume that a signal can be approximated as shown in Figure 13.9. Compute the slew of the signal using:

(a) 10–90 slew threshold

(b) 30–70 slew threshold.

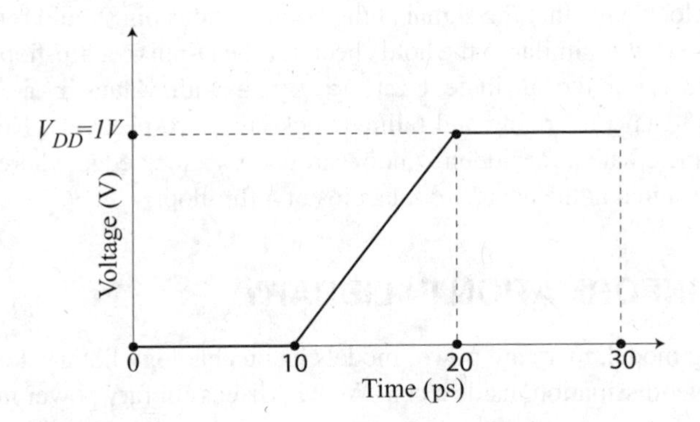

Figure 13.9 Question 13.4

13.5 Why do we typically employ the SIS model for delay computation for combinational cells? What are the limitations of the SIS model?

13.6 What are the advantages and disadvantages of employing NLDM for delay computation?

13.7 Using NLDM, how do you compute cell delay for a cell driving an interconnect that has an appreciable resistance?

13.8 What is the motivation of employing the current source-based model for delay computation?

13.9 What is the difference between CCS and ECSM? Are they interchangeable? If yes, how?

13.10 What are the following constraints used for: setup time, hold time, recovery time, removal time, minimum pulse width, and minimum period?

TOOL-BASED ACTIVITY

Take any **technology library** (you can also download freely available technology libraries [3]) and carry out the following activities:

1. Edit the content of the technology library to model the information shown in Table 13.1 (Note that, typically, we do not need to edit technology libraries. The objective of this activity is to understand the timing information modeling in a technology library).

Table 13.1

Cell Name	Attribute	Value (ps)
DFF	Setup time (all arcs)	50
	Hold time (all arcs)	10
	Clock to Q delay (all transitions)	110
	Clock to Q slew (all transitions)	5
AND2	Delay (all arcs)	37
	Output slew (all arcs)	23
INV1	Delay (rise)	27
	Delay (fall)	17
	Output slew (rise)	19
	Output slew (fall)	12
XOR2	Delay (all arcs)	43
	Output slew (all arcs)	25

2. Using delay calculation (or STA), check that the above information is modeled correctly. For checking you can use the circuit shown in Figure 13.10. You can use the Synopsys Design Constraint (SDC) shown in the following box (assuming time is in ps and load is in fF):

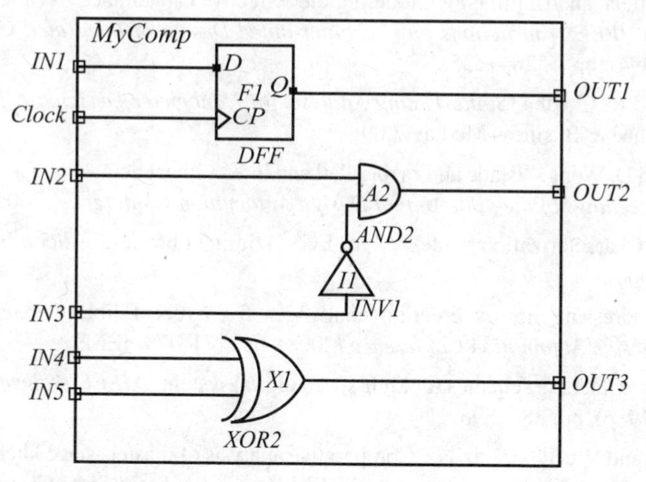

Figure 13.10 Given circuit for the activity

```
create_clock -name CLK -period 2000 [get_ports clock]
set_clock_transition 50 [get_clocks CLK]
set_input_transition 20 [all_inputs]
set_input_delay 10 -clock [get_clocks CLK] [all_inputs]
set_load 25 [all_outputs]
```

To perform delay calculation (or STA), you can use any tool, including open-source tools such as OpenSTA [16].

REFERENCES

[1] Synopsys Inc. "Liberty." https://www.synopsys.com/community/interoperability-programs/tap-in.html. Last accessed on 7 January, 2023.

[2] Cadence. "LEF/DEF Language Reference, Product Version 5.7, November 2009." http://www.ispd.cc/contests/18/lefdefref.pdf. Last accessed on 7 January, 2023.

[3] M. Martins, J. M. Matos, R. P. Ribas, A. Reis, G. Schlinker, L. Rech, and J. Michelsen. "Open cell library in 15nm FreePDK technology." *Proceedings of the 2015 Symposium on International Symposium on Physical Design* (2015), pp. 171–178.

[4] W. Roethig. "Library characterization and modeling for 130 nm and 90 nm SOC design." *IEEE International [Systems-on-chip] SOC Conference, 2003. Proceedings* (2003), pp. 383–386, IEEE.

[5] S. R. OVS and S. Saurabh. "Modeling multiple input switching in timing analysis using machine learning." *IEEE Transactions on Computer-aided Design of Integrated Circuits and Systems* (2020).

[6] I. Keller, K. H. Tam, and V. Kariat. "Challenges in gate level modeling for delay and SI at 65nm and below." *2008 45th ACM/IEEE Design Automation Conference* (2008), pp. 468–473, IEEE.

[7] S. Saurabh, H. Shah, and S. Singh. "Timing closure problem: Review of challenges at advanced process nodes and solutions." *IETE Technical Review* (2018).

[8] J. Qian, S. Pullela, and L. Pillage. "Modeling the 'effective capacitance' for the RC interconnect of CMOS gates." *IEEE Transactions on Computer-aided Design of Integrated Circuits and Systems* 13, no. 12 (1994), pp. 1526–1535.

[9] J. Bhasker and R. Chadha. *Static Timing Analysis for Nanometer Designs: A Practical Approach.* Springer Science & Business Media, 2009.

[10] J. F. Croix and D. Wong. "Blade and razor: Cell and interconnect delay analysis using current-based models." *Proceedings of the 40th Annual Design Automation Conference* (2003), pp. 386–389.

[11] R. C. Kezer. "Characterization guidelines for ECSM timing libraries." *Silicon Integration Initiative, Inc.*, 9111 (2006).

[12] R. Trihy. "Addressing library creation challenges from recent liberty extensions." *2008 45th ACM/IEEE Design Automation Conference* (2008), pp. 474–479, IEEE.

[13] C.-J. Tan et al. "Clocking schemes for high-speed digital systems." *IEEE Transactions on Computers* 100, no. 10 (1986), pp. 880–895.

[14] V. Stojanovic and V. G. Oklobdzija. "Comparative analysis of master–slave latches and flip-flops for high-performance and low-power systems." *IEEE Journal of Solid-State Circuits* 34, no. 4 (1999), pp. 536–548.

[15] S. Saurabh and P. Mittal. "A practical methodology to compress technology libraries using recursive polynomial representation." *2018 31st International Conference on VLSI Design and 2018 17th International Conference on Embedded Systems (VLSID)* (2018), pp. 301–306, IEEE.

[16] T. Ajayi, V. A. Chhabria, M. Fogaça, S. Hashemi, A. Hosny, A. B. Kahng, M. Kim, J. Lee, U. Mallappa, M. Neseem, et al. "Toward an open-source digital flow: First learnings from the OpenROAD project." *Proceedings of the 56th Annual Design Automation Conference 2019* (2019), pp. 1–4.

14 Static Timing Analysis

...any real body must have extension in four directions: it must have Length, Breadth, Thickness, and—Duration ... It is only another way of looking at Time ... For instance, here is a portrait of a man at eight years old, another at fifteen, another at seventeen ... All these are evidently sections, as it were, Three-Dimensional representations of his Four-Dimensioned being...

—H. G. Wells, *The Time Machine*, Introduction, 1895

In a synchronous circuit, the clock signal synchronizes the operation of various circuit elements. For deterministic circuit operation, certain timing constraints must be satisfied between the data signal relative to the clock signal. If these constraints are violated then the circuit can go into a *metastable state* or an *invalid state*. Therefore, we need to verify that these constraints are indeed satisfied in a synchronous circuit [1–4]. We use *static timing analysis* (STA) for this purpose. The simplicity and computational efficiency of STA have made synchronous design methodology the de facto standard for complicated digital circuits.

An STA tool verifies that a given synchronous circuit operates deterministically and remains in a valid state for a given frequency even in *the worst-case* scenario. Note that the purpose of STA is *not* to find a frequency at which a circuit can operate. Its purpose is just to ascertain whether a given circuit can *safely* operate at a given frequency and operating conditions. Therefore, STA needs to examine only the worst-case behavior for the circuit. It greatly simplifies the problem for an STA tool. It need not evaluate the circuit response to *all possible input stimuli* because most of them do not contribute to the *worst-case* behavior. Hence, STA employs methods that do not require applying stimuli and observing their dynamic responses. In this sense, STA is a *static* verification technique. It employs efficient stimuli-independent techniques to examine the worst-case behavior of a circuit.

In this chapter, we first describe the expected behavior of a synchronous circuit. Then, we derive the timing constraints that must be met to achieve this synchronous behavior. Next, we explain the techniques employed by the STA tools to ensure that these constraints are satisfied. We also highlight the assumptions used by the STA tools and point out the merits and demerits of these assumptions. Finally, we explain some of the popular techniques to account for variations in STA.

It is worthy to point out that a sequential circuit can contain various kinds of sequential elements. For simplicity, we assume in this chapter that the circuit consists of only *positive edge-triggered D flip-flops* as sequential circuit elements. We can easily extend the concept discussed in this chapter to other types of sequential elements.

14.1 BEHAVIOR OF SYNCHRONOUS CIRCUIT

First, let us understand the expected behavior of a synchronous circuit. We illustrate this using a simple circuit shown in Figure 14.1(a).

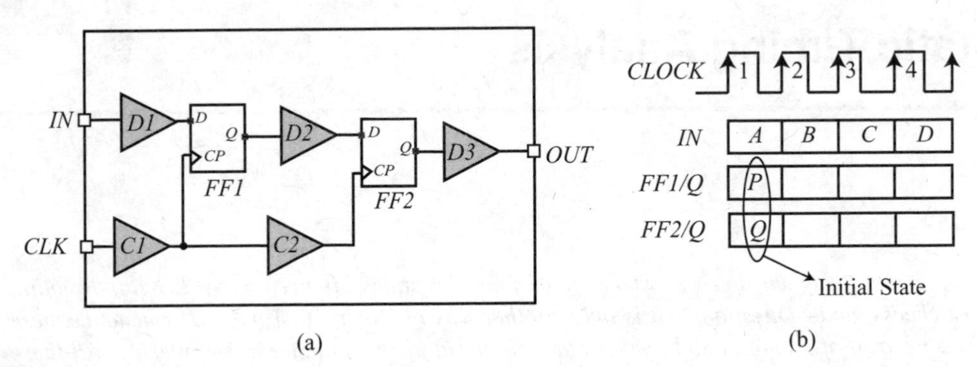

Figure 14.1 Examining synchronous behavior (a) given circuit and (b) given initial state and stimuli

It consists of flip-flops *FF1* and *FF2*. For easy analysis, we first assume that D flip-flops are *ideal*. Therefore, the setup time, hold time, and clock-to-Q delay are zero for all flip-flops. It contains three buffers *D1*, *D2*, and *D3* on the data side of flip-flops. They exhibit some delay. There are buffers *C1* and *C2* on the clock-pin side of the flip-flops. They also exhibit some delay.

We illustrate the behavior of the above circuit for various combination of delays of *D1*, *D2*, *D3*, *C1*, and *C2* in the following paragraphs. We assume the initial state and the stimuli are as shown in Figure 14.1(b). We can represent the state of the circuit as the combination of Q-pin value of the flip-flops *FF1* and *FF2*. Thus, the initial state of the circuit is $\{PQ\}$. Further, we have assumed that the data arriving at the input port *IN* is *A*, *B*, *C*, and *D* in the first, second, third, and fourth clock cycles, respectively. Note that the valid values of nets in a digital circuit is always Boolean (0 or 1). However, for the sake of clarity, we have given unique identifiers (*A*, *B*, *C*, etc.) to them.

14.1.1 Synchronous Behavior

We illustrate the synchronous behavior in the following example.

Example 14.1 Let the delay of *D1*, *D2*, and *D3* be some finite value less than the clock period, and the delays of *C1* and *C2* are negligible. Therefore, the circuit in Figure 14.1(a) reduces to as shown in Figure 14.2(a). Let us examine the state of the circuit in subsequent cycles.

The value *A* appearing at the port *IN* in the first clock cycle would appear at *FF1/Q* in the second cycle and *FF2/Q* in the third clock cycle. Similarly, *B*, *C*, *D*, and *P* will propagate through *FF1* and *FF2* in subsequent clock cycles, as shown in Figure 14.2(b). Thus, the states in the first, second, third, and fourth clock cycles are as shown in Figure 14.2(c). We obtain these states when the circuit behaves synchronously. We refer to them as the *valid states*.

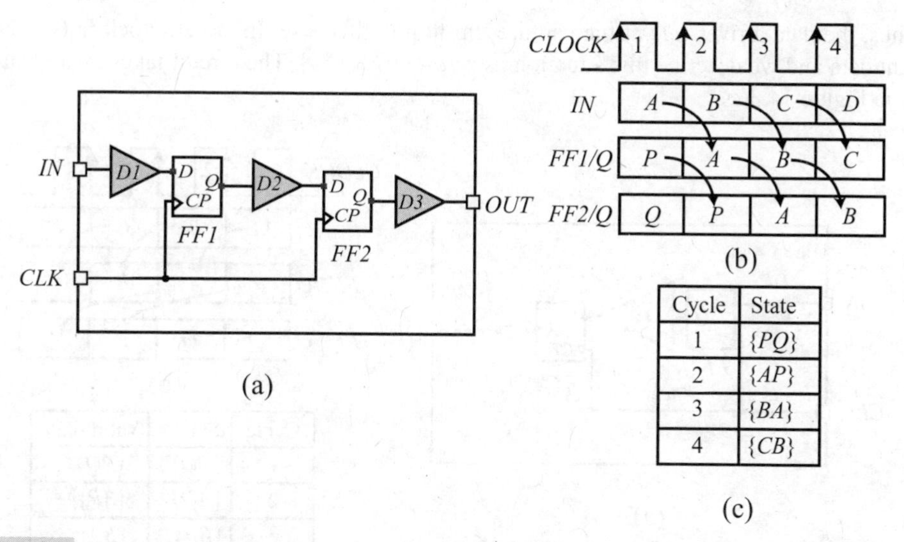

Cycle	State
1	{PQ}
2	{AP}
3	{BA}
4	{CB}

Figure 14.2 Synchronous behavior (a) a simplified circuit for Figure 14.1(a), (b) synchronous state transitions, and (c) valid states

14.1.2 Zero-clocking

To ensure that the circuit remains only in the valid states at all clock cycles, the value at the Q-pin of the launch flip-flop (*FF1*) must reach the D-pin of the capture flip-flop (*FF2*) by the beginning of the *next clock cycle*. If this constraint is violated, the circuit can go into an invalid state, as illustrated in the following example.

Example 14.2 Assume that the delay of *C1* and *C2* are negligible. Let us assume that the delay of the circuit element *D2* is more than one clock period, but less than two clock periods. For simplicity, let us assume that the delay of *D1* and *D3* are minimal. Thus, we can represent the circuit as shown in Figure 14.3(a).

The value of *FF1/Q* will be the same as in the previous example. However, the value of *FF2/Q* will be as shown in Figure 14.3(b). We can understand this behavior as follows.

Ideally, by the time the second clock edge arrives at *FF2*, the value of *FF1/Q* (=*P*) should have propagated to *FF2/D*. However, since the delay of *D2* is more than one clock cycle, the value *P* does not arrive at *FF2/D*. Consequently, the previous value of *FF2/Q* (=*Q*) repeats in the second clock cycle. Somewhere, between the second clock edge and the third clock edge, *P* will arrive at *FF2/D*.

When the third clock edge arrives at *FF2*, ideally *A* (the same as the value of *FF1/Q* in the second clock cycle) should have arrived at *FF2/D*. However, due to the delay of *D2*, *FF2/D* is still *P*. Therefore, in the third clock cycle, *P* is latched at *FF2/Q*. A similar problem exists when the fourth clock edge arrives at *FF2*.

Thus, the data arrives *late* at the capture flip-flop in this case. In effect, clock fails to capture the right data and we describe this situation as *zero-clocking* [5]. The circuit takes invalid states as shown in Figure 14.3(c).

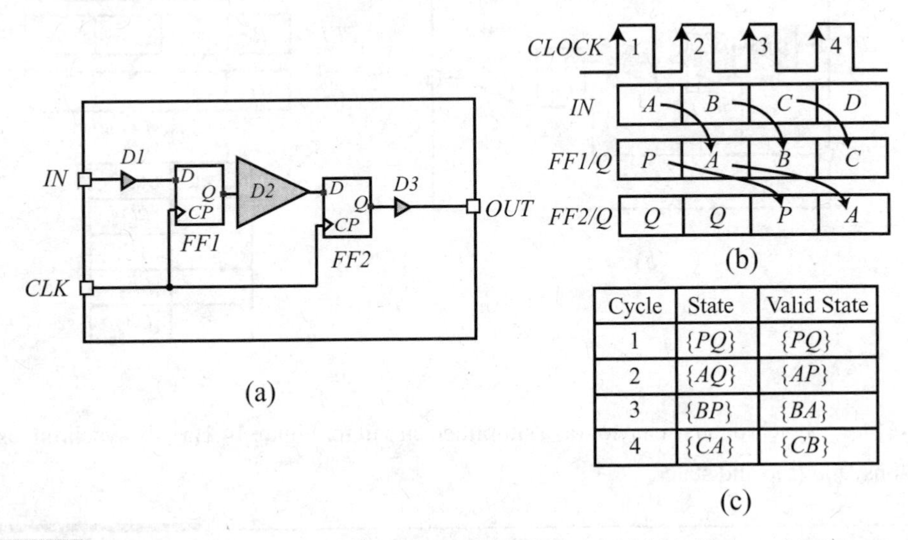

Cycle	State	Valid State
1	{PQ}	{PQ}
2	{AQ}	{AP}
3	{BP}	{BA}
4	{CA}	{CB}

Figure 14.3 Zero-clocking (a) a simplified circuit for Figure 14.1(a) (delay of *D2* is more than one clock period and less than two clock periods), (b) state transitions for zero-clocking, and (c) comparison of states in zero-clocking with the valid states in the synchronous circuit

The above example shows that zero-clocking disturbs the synchronous behavior of a circuit. It takes a circuit into *invalid states*. For a synchronous circuit, we must ensure that this situation never occurs. STA tools detect this problem using *setup analysis*, which we will describe in the following sections.

14.1.3 Double Clocking

Another problem can occur in a synchronous circuit when the *delay between two flip-flops is too small*. In this case, the launched data reaches the capture flip-flop too early. As a result, the data will be captured by *two flip-flops* by the *same* clock edge, as illustrated in the following example.

Example 14.3 Assume that the delay of circuit elements *D1*, *D2*, *D3*, and *C1* are insignificant. Let us assume that the delay of the circuit element *C2* is Δ. Assume that Δ is large, but less than one clock period. Thus, we can represent the circuit as shown in Figure 14.4(a).

The value of *FF1/Q* will be the same as in Ex. 14.1. However, the clock signal reaches *FF2/CP* delayed by time duration Δ because of *C2*. As a result, the value of *FF2/Q* will be as shown in Figure 14.4(b). We can understand this behavior as follows.

When the first clock edge arrives at *FF1*, *P* is latched at *FF1/Q* and starts propagating toward *FF2/D*. Since the delay between *FF1/Q* and *FF2/D* is negligibly small, *P* reaches *FF2/D* quickly. Note that the second clock edge arrives Δ time later at *FF2/CP* than at *FF1/CP*. By the time the clock reaches *FF2/CP*, the value at *FF2/D* gets overwritten with *P*. Thus, in the first clock edge, *P* will be captured by *FF2* also.

Similarly, when the second clock edge arrives at *FF1*, *A* is latched at *FF1/Q* and starts propagating toward *FF2/D*. By the time the second clock edge reaches *FF2/CP*, the original value (*P* propagated by *FF1/Q*) gets overwritten with *A*. Thus, in the second clock edge, *A* will be captured by *FF2* instead of *P*.

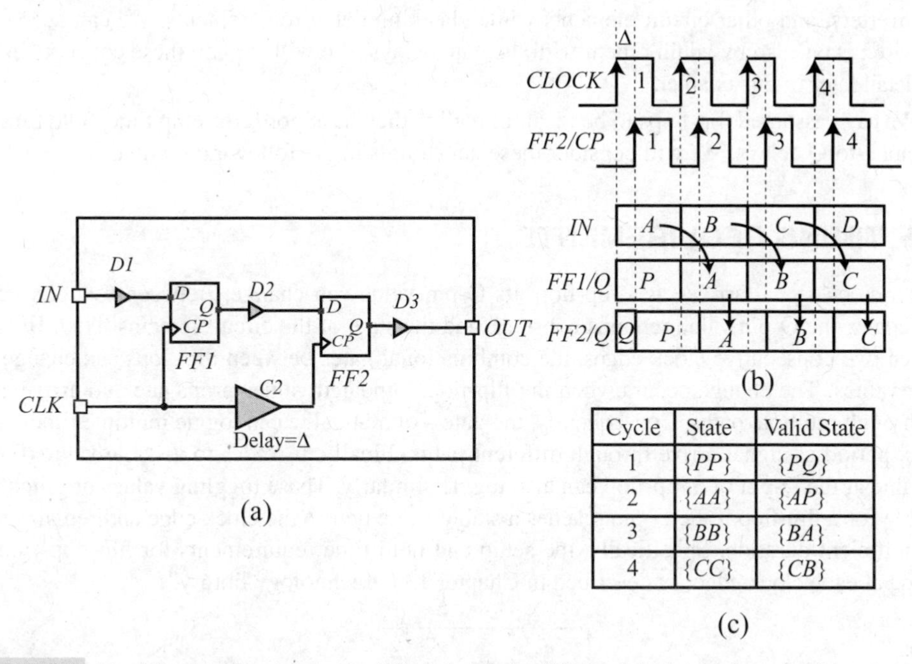

Figure 14.4 Double clocking (a) a simplified circuit for Figure 14.1(a) (delay of *C2* is appreciable and less than one clock period), (b) state transitions for double clocking, and (c) comparison of states in double clocking with the valid states in the synchronous circuit

A similar problem occurs at the third and the fourth clock edges, as illustrated in Figure 14.4(b). Thus, in this case, the data *races* through a fast path ahead of the clock. As a result, it overwrites the existing data at the capture flip-flop by reaching too *early*. In effect, the data gets captured by two flop-flops by the same clock edge. This behavior is referred to as *double clocking* [3]. The circuit goes into the *invalid states* shown in Figure 14.4(c).

Thus, double clocking disturbs the synchronous behavior of a circuit. For a synchronous circuit, we must ensure that this situation never occurs. STA tools detect this problem using *hold analysis*, which we will describe in the following sections.

Examples 14.1–14.3 show that for proper synchronous circuit operation, the delay of *D1*, *D2*, and *D3* on the data side, delay of *C1* and *C2* on the clock side, and the clock period must satisfy some constraints. If these constraints are violated, then the circuit goes into an invalid state. However, we have made several simplifying assumptions in the above examples. Let us relook at them:

1. We have considered only two flip-flops. In practical circuits, the data can propagate sequentially through a *pipeline* of many flip-flops before reaching the output. In such cases, we can *examine each pair* of launch and capture flip-flops separately to ensure the synchronous operation of the circuit.

2. We have used only buffers on the data path side of flip-flops. In practical circuits, there can be multiple gates on those paths realizing some complex Boolean functions. In such cases, we can add the delay of all the combinational circuit elements in the path and check for delay requirements. On the clock path, too, we can add the delay of all the intervening buffers, inverters, and other circuit elements while checking delay requirements. We can account for wire delays also by adding them with the gate delays. We will explain these concepts in more detail later in this chapter.
3. We have assumed flip-flops to be ideal. In reality, they have nonzero setup time, hold time, and clock-to-Q delays. We will consider these constraints in the following section.

14.2 TIMING REQUIREMENTS

When a clock signal arrives at a flip-flop, its Q-pin value can change. Between two consecutive clock edges, the Q-pin value remains constant, and the state of the circuit remains fixed. However, between two consecutive clock edges, the combinational gates between flip-flops can change their output values. The change occurs when the flip-flops' updated values propagate and arrive at that gate through multiple paths. Consequently, the gate's output value can toggle multiple times within a clock period as signal arrive through different paths. Finally, it settles to a stable value (0 or 1). The value at the D-pin of a flip-flop can also toggle similarly. These toggling values or glitches are tolerable for a flip-flop if the D-pin reaches a stable value before the clock edge and remains stable for a sufficient duration. Specifically, the setup and hold time requirements for flip-flops must be honored at each clock edge, as described in Chapter 13 ("Technology library").

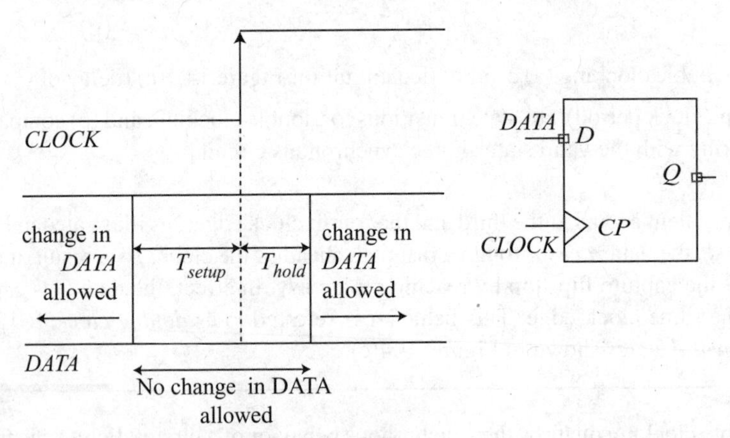

Figure 14.5 Setup and hold requirements for a flip-flop

To recap, the signal at the D-pin of a flip-flop must be stable for at least T_{setup} time *before* the active clock edge arrives at the flip-flop's clock-pin, as illustrated in Figure 14.5. This is the *setup requirement* for a flip-flop and T_{setup} is called the *setup time* of the flip-flop. Additionally, the signal at the D-pin of a flip-flop must be stable for at least T_{hold} time *after* the active clock edge arrives at the flip-flop's clock-pin. This is the *hold requirement* for a flip-flop and T_{hold} is called the *hold time* of the flip-flop.

Violations: If the setup or hold requirement of a flip-flop gets violated, it can go into a *metastable state.*[1] When a circuit goes into a metastable state, its behavior becomes unpredictable. We should

[1] See Section 1.1.6 for metastability.

avoid the circuit going into this state. Therefore, an STA tool verifies if the setup and hold requirements for flip-flops are honored besides ensuring the synchronous behavior of the circuit.

In the following paragraphs, we explain how are the above requirements modeled mathematically.

14.2.1 Setup Requirement

Consider a portion of a circuit shown in Figure 14.6. We represent a block of combinational circuit elements as ovals.

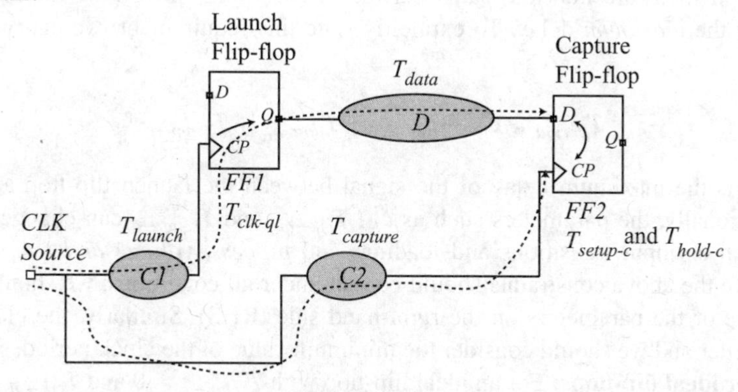

Figure 14.6 A portion of circuit to compute setup requirement

Clock source: The clock signal is generated at the *CLK Source*. Let us assume that the clock period is T_{period}. The flip-flops *FF1* and *FF2* receive this clock signal through the delay elements *C1* and *C2*, respectively. In practice, *C1* and *C2* consist of buffers (or pairs of inverters).
Arrival time: Lets assume that the positive clock edge is generated at the *CLK Source* at time $t = 0$. This clock edge will be received at *FF1/CP* at time $t = T_{launch}$, where T_{launch} is the delay of *C1*. On receiving the clock edge, *FF1* will latch the data and it becomes available at *FF1/Q* at time $t = T_{launch} + T_{clk-ql}$, where T_{clk-ql} is the delay between the clock-pin and the Q-pin of the launch flip-flop *FF1*. The data at *FF1/Q* propagates through the combinational block *D* and *arrives* at *FF2/D* at time $t_{arrival} = T_{launch} + T_{clk-ql} + T_{data}$, where T_{data} is the delay of the combinational block *D*.
Required time: The data at *FF2/D* must be captured by *FF2* when the *next* clock edge arrives at *FF2/CP*. The next edge of the clock is generated at the clock source at time $t = T_{period}$. This edge arrives at *FF2/CP* at time $t = T_{period} + T_{capture}$, where $T_{capture}$ is the delay of *C2*. The data at *FF2/D* must settle at least $T_{setup-c}$ time before the arrival of the next clock edge, where $T_{setup-c}$ is the setup time of the capture flip-flop *FF2*. Thus, it is *required* that the launched data must settle at *FF2/D* by the *required time* $t_{req,set}$, given as:

$$t_{req,set} = T_{period} + T_{capture} - T_{setup-c}. \tag{14.1}$$

Setup requirement: To ensure that zero-clocking does not occur and the flip-flop's setup time requirement is fulfilled, the latest arrival time $t_{arrival}$ of data at *FF2/D* must be less than the setup required time $t_{req,set}$. Hence,

$$t_{req,set} > t_{arrival}$$

$$\Rightarrow T_{period} + T_{capture} - T_{setup-c} > T_{launch} + T_{clk-ql} + T_{data}$$

$$\Rightarrow T_{period} > \{T_{launch} - T_{capture}\} + T_{clk-ql} + T_{data} + T_{setup-c} \tag{14.2}$$

Equation 14.2 describes the setup requirement for a synchronous circuit. The difference between the arrival time of the clock edge at the launch flip-flop and the capture flip-flop is known as the *clock skew* ($\delta_{lc} = T_{launch} - T_{capture}$). Thus, Eq. 14.2 can be written as:

$$T_{period} > \delta_{lc} + T_{clk-ql} + T_{data} + T_{setup-c} \quad (14.3)$$

The most restrictive constraint: For avoiding zero-clocking and meeting the setup requirement of the capture flip-flop, Eq. 14.3 should be valid for all paths between the launch and the capture flip-flop. Thus, if there are multiple paths between *FF1/Q* and *FF2/D*, we should consider the path exhibiting the *maximum* delay. To explicitly state this requirement, we can write the above constraint as:

$$T_{period} > \delta_{lc} + T_{clk-ql} + T_{data_{max}} + T_{setup-c} \quad (14.4)$$

where $T_{data_{max}}$ is the maximum delay of the signal between the launch flip-flop and the capture flip-flop. Additionally, the parameters such as δ_{lc}, T_{clk-ql}, and $T_{setup-c}$ can change due to circuit conditions (such as input transitions and loading) and *process, voltage and temperature* (PVT) variations. Since the above constraints should be met under all conditions, we should consider the maximum value of the parameters on the right-hand side (RHS). Similarly, the clock period can reduce due to jitter and we should consider the minimum value of the clock period.
Constraints for ideal flip-flops: For an ideal flip-flop with $T_{setup-c} = 0$ and $T_{clk-ql} = 0$ and for an ideal clock distribution network with $\delta_{lc} = 0$, Eq. 14.4 boils down to:

$$T_{period} > T_{data_{max}} \quad (14.5)$$

The above constraint is to avoid zero-clocking. Thus, even for a circuit with ideal flip-flops and an ideal clock network, the setup requirement still needs to be checked. The non-idealities in the flip-flops and the clock network tighten this constraint. Practically, the setup requirement is primarily governed by $T_{data_{max}}$. Therefore, to increase the maximum operable frequency ($f = 1/T_{period}$) of a circuit, *the maximum combinational delay* between *pairs* of launch–capture flip-flops must be decreased.

14.2.2 Hold Requirement

Consider the circuit of Figure 14.7 (same as Figure 14.6, shown again for convenience). We formulate the hold requirement to avoid double clocking, i.e., avoiding the launched data from being captured by another flip-flop in the *same* clock cycle.
Arrival time: Assume that at time $t = 0$, a postive edge of clock is generated at the *CLK Source*. As computed earlier, the data launched by *FF1 arrives* at the capture flip-flop *FF2* at time $t_{arrival} = T_{launch} + T_{clk-ql} + T_{data}$.
Required time: The same edge of the clock signal reaches capture flip-flop *FF2* at time $t = T_{capture}$. To fulfill the hold constraint of the capture flip-flop, the data should arrive *FF2/D* at least T_{hold-c} time after this clock edge, where T_{hold-c} is the hold time of the capture flip-flop. Thus, it is *required* that data should arrive *FF2/D* after the required time $t_{req,hold}$, given as:

$$t_{req,hold} = T_{capture} + T_{hold-c}. \quad (14.6)$$

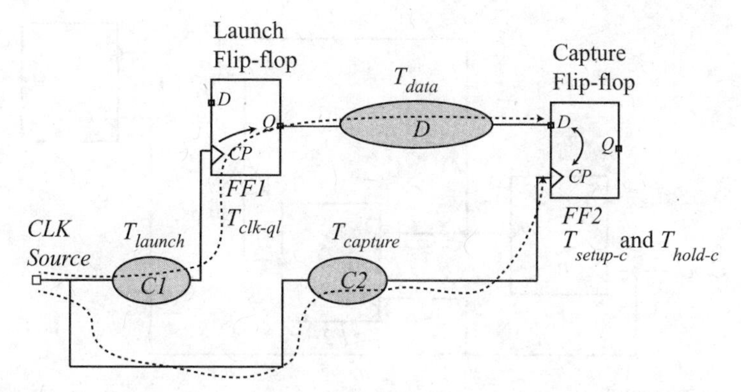

Figure 14.7 A portion of circuit to compute hold requirement

Hold requirement: To avoid double clocking and to meet hold requirement, the earliest arrival time $t_{arrival}$ must be more than the hold required time $t_{req,hold}$. Hence,

$$\begin{aligned} t_{arrival} &> t_{req,hold} \\ \Rightarrow T_{launch} + T_{clk-ql} + T_{data} &> T_{capture} + T_{hold-c} \\ \Rightarrow \delta_{lc} + T_{clk-ql} + T_{data} &> T_{hold-c} \end{aligned} \tag{14.7}$$

The most restrictive constraints: Eq. 14.7 should be valid for all the paths between the launch and the capture flip-flop. Thus, if there are multiple paths between *FF1/Q* and *FF2/D*, we should consider the path exhibiting the *minimum* delay (because the above constraint is most likely to be violated for the minimum delay path). To explicitly state this requirement, we can write the above constraint as:

$$\delta_{lc} + T_{clk-ql} + T_{data_{min}} > T_{hold-c} \tag{14.8}$$

where $T_{data_{min}}$ is the minimum delay of a signal between the launch flip-flop and the capture flip-flop.

The parameters in Eq. 14.8 can change due to circuit conditions and PVT variations. We need to ensure that the hold constraint is always valid. Hence, we should consider the minimum value of δ_{lc} and T_{clk-ql}, and the maximum value of T_{hold-c}.

We illustrate checking setup and hold requirements in the following example.

Example 14.4 Consider a portion of a sequential synchronous circuit shown in Figure 14.8.

The following attributes are valid for all the flip-flops *F1*, *F2*, and *F3*: $T_{setup-c}$=30 ps, T_{hold-c}=20 ps, and T_{clk-ql} = 10 ps.

The delay of the NAND gate *N1* is 50 ps and inverter *I1* is 30 ps. The frequency of the *Clock* is 1 GHz.

Ignore the delay of all the wires. Find whether there is any timing problem in the given circuit.

The clock frequency *f* is 1 GHz. Hence, time period T_{period}=1/*f*=1000 ps.

There are no circuit elements on the clock path. Hence, T_{launch} and $T_{capture}$ are zero. The circuit has three pairs of launch–capture flip-flops. We need to consider each of them for checking setup and hold requirements in Eqs. 14.2 and 14.7, respectively. The results are shown in Table 14.1.

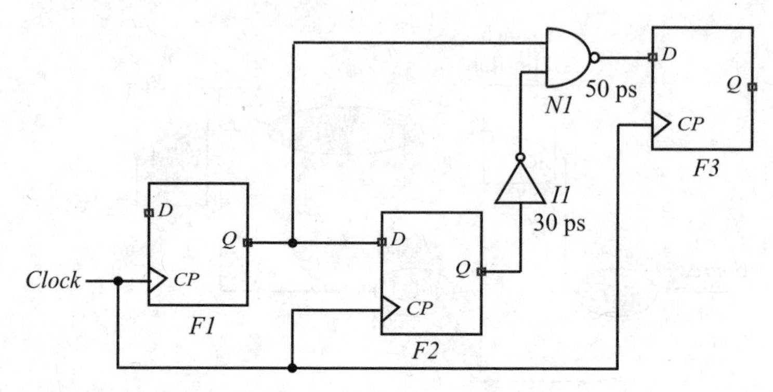

Figure 14.8 Given circuit

Table 14.1 Checking timing requirements

Path	Arrival Time (ps)	Required Time (ps)	Requirement	Met?
$F1 \rightarrow F2$	T_{clk-ql} $= 10$	$T_{period} - T_{setup-c}$ $= 1000 - 30 = 970$	Setup ($t_{arrival} < t_{req,set}$)	Yes
$F1 \rightarrow F3$	$T_{clk-ql} + T_{data}$ $= 10 + 50 = 60$	$T_{period} - T_{setup-c}$ $= 1000 - 30 = 970$	Setup ($t_{arrival} < t_{req,set}$)	Yes
$F2 \rightarrow F3$,	$T_{clk-ql} + T_{data}$ $= 10 + 50 + 30 = 90$	$T_{period} - T_{setup-c}$ $= 1000 - 30 = 970$	Setup ($t_{arrival} < t_{req,set}$)	Yes
$F1 \rightarrow F2$	T_{clk-ql} $= 10$	T_{hold-c} $= 20$	Hold ($t_{arrival} > t_{req,hold}$)	No
$F1 \rightarrow F3$	$T_{clk-ql} + T_{data}$ $= 10 + 50 = 60$	T_{hold-c} $= 20$	Hold ($t_{arrival} > t_{req,hold}$)	Yes
$F2 \rightarrow F3$	$T_{clk-ql} + T_{data}$ $= 10 + 50 + 30 = 90$	T_{hold-c} $= 20$	Hold ($t_{arrival} > t_{req,hold}$)	Yes

The setup requirements are fulfilled for all paths. However, there is a hold violation in the path from $F1 \rightarrow F2$. The data launched by $F1$ will reach $F2$ so quickly that the hold time requirement of $F2$ gets violated.

Constraints for ideal flip-flops: We can make an important observation from Eq. 14.8. For an ideal flip-flop with $T_{hold-c} = 0$ and $T_{clk-ql} = 0$, the hold requirement boils down to:

$$\delta_{lc} + T_{data_{min}} > 0 \tag{14.9}$$

The above constraint is to avoid double clocking, irrespective of the attributes of the flip-flop. Thus, even when the flip-flops are ideal, hold requirements must still be verified for a circuit to maintain its synchronicity. Moreover, note that the clock skews δ_{lc} can be both positive and negative, while $T_{data_{min}}$ will always be positive. When $\delta_{lc} > 0$, the clock edge reaches late at the launch flip-flop compared to the capture flip-flop. In this case, the hold constraint becomes lenient and easier to meet. It is in contrast to the setup requirement that becomes more difficult to satisfy when $\delta_{lc} > 0$ (see Eq. 14.3). Similarly, when $T_{data_{min}}$ is large, then the hold constraint is easy to satisfy, while the setup constraint becomes difficult to satisfy.

Hold requirement in pre-CTS stages: In the pre-CTS stages (i.e., before clock tree synthesis), the clock network does not exist. Hence, the clock skews are unknown. Since hold requirements depend heavily on clock skews, verifying hold requirements in the pre-CTS stages is inaccurate. Typically, we assume ideal clock networks with $\delta_{lc} = 0$ before CTS, and the hold requirement of Eq. 14.8 becomes:

$$T_{clk-ql} + T_{data_{min}} > T_{hold-c} \tag{14.10}$$

The above requirement is easily satisfied when there are combinational logic elements between two flip-flops. However, when two flip-flops are directly connected (as in a shift register), it becomes necessary to verify the hold requirement. Some of the hold requirements can be met automatically by the wire delays in the data path. Therefore, we typically analyze and fix the hold requirements only after CTS.

Differences between setup and hold requirements: We summarize the key differences between the setup and hold requirements in Table 14.2. It is worthy to note that the setup requirement depends on the clock frequency, in contrast to the hold requirement. Hence, we can fix a setup violation by lowering the clock frequency, while we cannot fix a hold violation using this method.

Table 14.2 Difference in setup and hold check in STA

Attribute	Setup Requirement	Hold Requirement
Clock period	Becomes stricter when the clock period is decreased.	Independent of the clock period.
Data path	Worst case is when the data path delay is maximum.	Worst case is when the data path delay is minimum.
Clock path	Worst case is when the skew is a large positive value.	Worst case is when the skew is large negative value.

14.3 DATA PATHS AND CLOCK PATHS

STA tools consider two types of path while verifying setup and hold requirements: *data path* and *clock path*.

Data path: A data path starts at an input port or a clock pin of a flip-flop. We also refer to these entities as *timing startpoints*. A data path ends at a D-pin of a flip-flop or an output port. We also refer to these entities as *timing endpoints*. The setup checks and hold checks are performed at timing endpoints. The previous section shows the need to perform setup and hold checks at the D-pin of a flip-flop. The setup and hold checks are required at the output ports also because the circuit's external environment enforces required time constraints on them. We will discuss this aspect of STA in Chapter 15 ("Constraints").

A path starting from a timing startpoint, passing through combinational logic gates, and ending on a timing endpoint is referred to as a *data path* for STA purposes.

Clock path: A clock path starts at the *source* of the clock signal. A clock source can be an internal pin, such as an output pin of a clock generator. Alternatively, it can be an input port that receives a clock signal from an external clock generator. An STA tool infers a clock source and the clock frequency from the *constraints* that we provide to the tool, typically in Synopsys Design Constraints (SDC) format. We will explain how to provide these constraints in Chapter 15 ("Constraints").

A clock path ends at the clock-pin of a flip-flop. A path starting from a clock source, passing through combinational circuit elements (typically buffers and inverters), and ending on a clock-pin of a flip-flop is referred to as a *clock path* for STA purposes.

Example 14.5 Consider the circuit shown in Figure 14.9. Assume that the clock source is at the input port *CLK*. Let us identify various types of path in it.

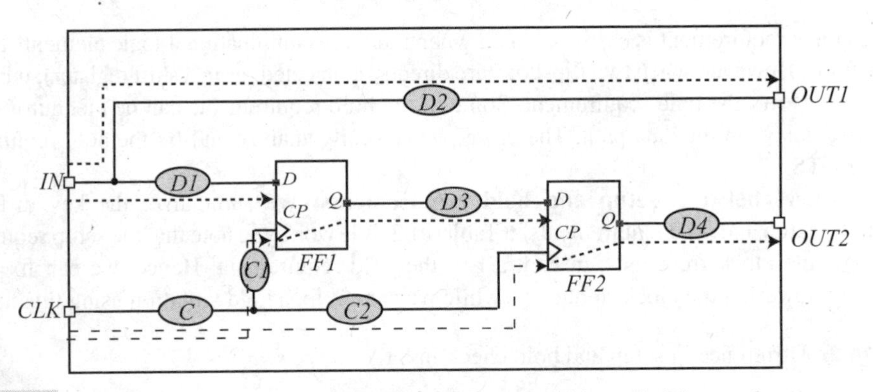

Figure 14.9 Different kinds of paths considered by an STA tool. The ovals represent paths comprising combinational circuit elements. Data paths shown as short dashes and clock paths shown as long dashes

There are four types of data paths:

1. The path starting at the input port *IN*, going through *D1*, and ending at the D-pin *FF1/D*.
2. The path starting at the input port *IN*, going through *D2*, and ending at the output port *OUT1*.
3. The path starting at the clock-pin *FF1/CP*, going through *D3*, and ending at the D-pin *FF2/D*.
4. The path starting at the clock-pin *FF2/CP*, going through *D4*, and ending at the output port *OUT2*.

There are two clock paths:

1. The path starting from *CLK*, going through *C* and *C1*, and ending at *FF1/CP*.
2. The path starting from *CLK*, going through *C* and *C2*, and ending at *FF2/CP*.

14.4 TIMING GRAPH

An STA tool performs setup and hold checks by building *timing graphs* for a given circuit. Therefore, we need to understand timing graphs before looking into the mechanism of STA.

A timing graph is a *directed acyclic graph* $G=(V, E)$, where V is the vertex set and E is the edge set. A vertex $v \in V$ corresponds to a pin or a port in the circuit. An edge $e \in E$ represents a timing arc in the circuit, with the direction being the same as that of the timing arc. An edge $e_{i,j} = (v_i, v_j)$ exists in E if and only if there exists a timing arc between the corresponding pins or ports in the circuit. We refer to the vertex v_i as the *tail* of the edge and v_j as the *head* of the edge. We refer to the vertex v_i as the *input* of the vertex v_j, and the vertex v_j as the *output* of the vertex v_i.

A timing arc can be two types:

1. **Cell arc:** A timing arc between two pins of the *same cell* is known as a *cell arc*. It is inferred from the timing model of a cell in the given technology library.
2. **Net arc:** A timing arc between two pins of different cells connected directly by a *net* is known as a *net arc*. A net arc is inferred from the interconnection of the cells as described in the netlist.

Example 14.6 Consider a portion of circuit shown in Figure 14.10(a).

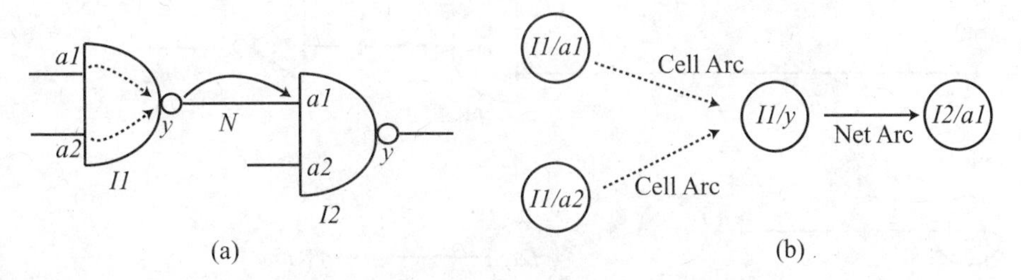

Figure 14.10 Types of timing arcs (a) given portion of a circuit and (b) corresponding timing graph

The corresponding timing graph is shown in Figure 14.10(b). The vertices corresponds to the pins *I1/a1*, *I1/a2*, *I1/y*, and *I2/a1* (vertices corresponding to other pins are not shown in this example). We refer to the timing arcs within a cell (*I1/a1*→*I1/y* and *I1/a2*→*I1/y*) as cell arcs and the timing arc between pins of different cells (*I1/y*→*I2/a1*) as a net arc.

Annotations on edges and vertices: Each edge $e_{ij} = (v_i, v_j)$ has an annotated delay D_{ij} in a timing graph. If an edge corresponds to a cell arc, then D_{ij} is the *cell delay* of the timing arc between pins corresponding to the vertices v_i and v_j. If an edge corresponds to a net arc, then D_{ij} is the *interconnect delay* between pins corresponding to the vertices v_i and v_j. As STA computation progresses, we store or annotate various other timing attributes on the edges and vertices of the timing graph. Typically, STA tools provide some mechanism to the designers to retrieve these attributes from the timing graph.

Source: The vertices in the timing graph corresponding to the input ports have no incoming edges. Similarly, ports or pins corresponding to the clock sources do not have incoming edges. We refer to a vertex with no incoming edges as a *source* of the timing graph.

Sink: The vertices in the timing graph corresponding to the output ports have no outgoing edges. Since the D-pins of flip-flops are considered timing endpoints, we treat them like an output port in the timing graph. Thus, no outgoing edge emanates from the D-pin of a flip-flop. We refer to a vertex with no outgoing edges as a *sink* of the timing graph.

Acyclic nature: The timing graph G is *acyclic*. It means that it does not contain any loop or a path in which a vertex repeats. If a circuit has loops consisting of combinational circuit elements, it is generally impossible to ensure its timing safety using STA. Therefore, an STA tool needs to break combinational loops (in case they exist in a circuit) by disabling some timing arcs before building a timing graph.

Example 14.7 Consider the combinational circuit shown in Figure 14.11(a).

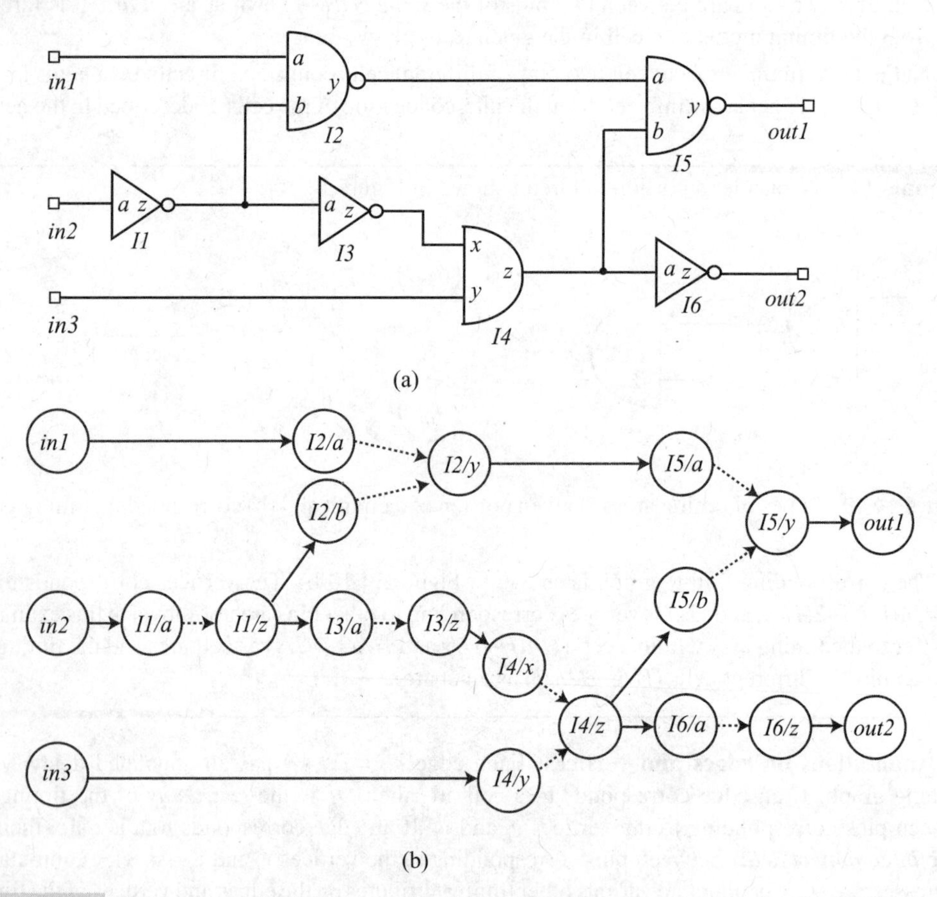

Figure 14.11 Illustration of a timing graph (a) given circuit and (b) corresponding timing graph (dashed edges represent cell arcs and solid edges represent net arcs)

Creating a timing graph by creating vertices for ports and pins, and the corresponding edges for the timing arcs is straightforward, as illustrated in Figure 14.11(b).

Similarly, we can build a timing graph for any *sequential circuit* by considering relevant library-based timing arcs for the sequential circuit elements. For example, we need to create edges for the clock-to-Q timing arc for flip-flops.

14.5 DELAY CALCULATION

An STA tool needs to compute delays for the timing arcs existing in the timing graph. Typically, it has an inbuilt *delay calculator* or is coupled with some *delay calculator* to retrieve delay for the timing arcs. An STA tool decomposes a given circuit into separate *stages* and invokes delay calculation on each stage. A stage is composed of a driving cell and its driven pins connected through wires.

Example 14.8 Consider a part of circuit shown in Figure 14.12(a).

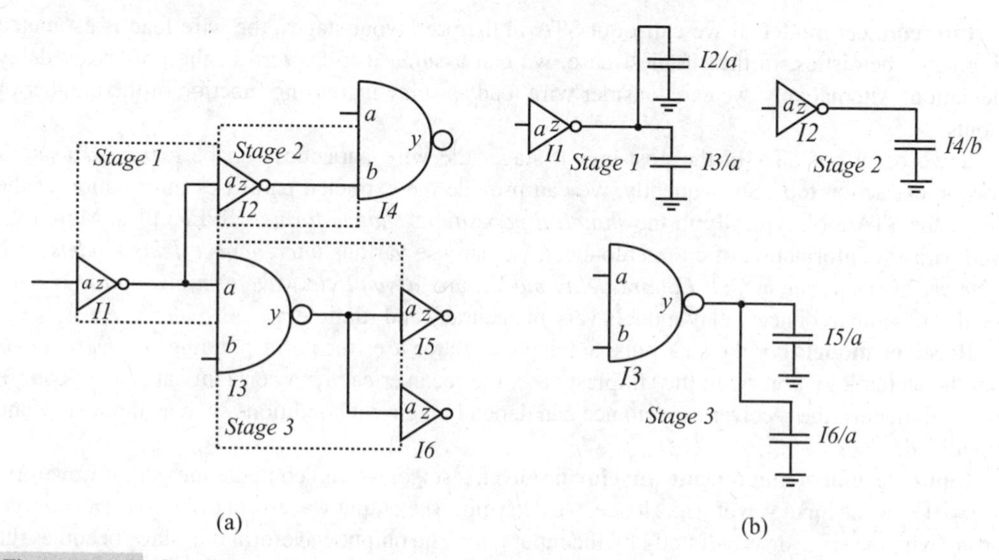

Figure 14.12 Creating stages (a) portion of a circuit and (b) corresponding stages

The stages corresponding to the driving cells *I1*, *I2*, and *I3* are shown in Figure 14.12(b). Each stage consists of a driving cell, interconnect, and the driven pins (shown as capacitive loads, though we can use a more complicated receiver model if available in the library).

The framework for delay computation in STA is shown in Figure 14.13. The essential components are:

Driver model: The delay calculator employs the *timing models* of the driving cell as defined in the technology library. These timing models can be the traditional *nonlinear delay model* (NLDM) or the advanced delay models such as *composite current source* (CCS) model and *effective current source model* (ECSM).

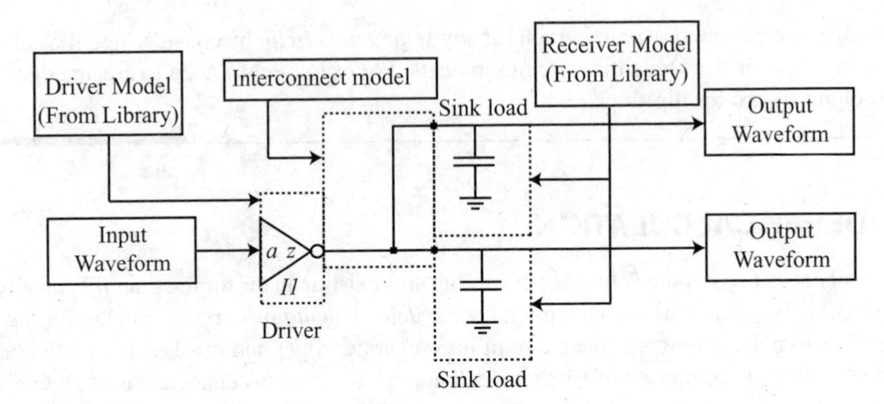

Figure 14.13 Framework for delay calculation in STA

Interconnect model: If we carry out STA in the pre-layout stages, the wire load is estimated using some heuristics. In the simplest case, we can assume it to be zero in the pre-layout delay calculation. Alternatively, we can consider wire load as some increasing function of the number of fanouts.

If we carry out STA in the post-layout stage, the wire parasitics can be *extracted* using a parasitic extraction tool. Subsequently, we can provide the extracted parasitics information of the wire to the STA tool, typically in the *standard parasitic exchange format* (SPEF) files. Moreover, given parasitic information of an interconnect, we can use various *interconnect delay models* such as *lumped capacitance model*, *Elmore delay model*, and *asymptotic waveform evaluation* (AWE) model. These interconnect delay models vary in accuracy and runtime.[2]

Receiver model: For the sink pins, a delay calculator uses the corresponding *receiver models* from the technology library. In the simplest case, the receiver can be a constant capacitive load. In advanced models, the receiver capacitance can depend on circuit conditions such as input slew and output load.

Input waveform and output waveform: Given a stage, we can compute the output waveform once we know the input waveform. Hence, we determine the output waveform in a *topological order*, starting with the stage driven directly by the input port. The output waveform of a stage becomes the input waveform for the following stage. Thus, the waveform propagates in a circuit from an input port to all stages in its fanout.

The incoming waveform at the input ports depends on the external environment, which we can specify in the *constraints file* (also referred to as SDC file, since it is typically in the SDC format). Alternatively, an STA tool can assume some default waveform for the input ports if we do not specify it in the constraints file. Typically, we take the input waveform as a voltage ramp with a given slope or *slew* (also called *transition time* or *rise/fall time*).

A delay calculator computes the output waveforms at all the sinks for a given stage and given input waveform. Using the computed output waveforms, it estimates the cell arc delay and the net arc delay and annotates them on the associated edges in the timing graph as D_{ij}. It also estimates the output slews S_{ij} using the output waveforms and employs them in the delay computation of the next stage.

In the subsequent discussion of this chapter, we will assume that D_{ij} and S_{ij} can either be computed by a delay calculator or be retrieved from the timing graph whenever needed.

[2] We will explain various interconnect delay models in Chapter 24 ("Basic concepts for physical design").

14.6 ARRIVAL TIME COMPUTATION

An STA tool employs Eqs. 14.2 and 14.7 to verify whether the setup and hold requirements are met. For evaluating them, an STA tool computes the *arrival time* and the *required time* at each vertex in the timing graph. First, let us look at the arrival time computation. Then, we will look at the required time computation.

The arrival time is defined as the time taken for a signal to reach that vertex. When a vertex v_j has a single incoming edge e_{ij} from another vertex v_i, the arrival time A_j at that vertex can be computed simply as:

$$A_j = A_i + D_{ij}$$

where A_i is the arrival time at vertex v_i, and D_{ij} is the delay for the edge e_{ij}. However, when there are multiple incoming edges for a vertex some complications arise, as explained below.

14.6.1 Minimum and Maximum Bounds

When a vertex has multiple incoming edges, every incoming edge can impact the signal at that vertex. We can associate different arrival times with each incoming edge. As edges drive different values (0 or 1) at the vertex, the signal can toggle multiple times before settling to a final value.

Example 14.9 Consider a 2-to-1 multiplexer shown in Figure 14.14(a). Assume that the inputs A, B, and S have waveforms as shown in Figure 14.14(b). Thus, the arrival times at the inputs are: A_A=1 ns, A_B=3 ns, and A_S=2 ns. Assume that delay of all arcs from the inputs to the output Z is 1 ns. Thus, we obtain the timing graph as shown in Figure 14.14(c). Let us examine the waveform at the output.

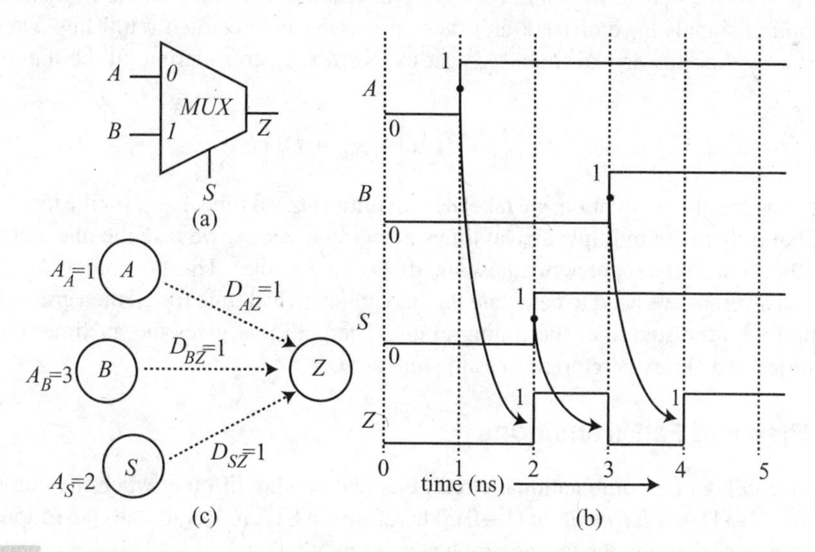

Figure 14.14 Path-dependent arrival times (in nanosecond) (a) a 2-to-1 multiplexer, (b) timing diagram, and (c) corresponding timing graph

At 0 ns, Z will be 0 (if the inputs were 0 since long).

The change in A arrives at 1 ns. Z changes to 1 after a delay of 1 ns. Hence, the arrival time at Z due to A is $A_A + D_{AZ}$=2 ns.

The change in S arrives at 2 ns. Z changes to 0 after a delay of 1 ns. Hence, the arrival time at Z due to S is $A_S + D_{SZ}$=3 ns.

The change in B arrives at 3 ns. Z changes to 1 after a delay of 1 ns. Hence, the arrival time at Z due to B is $A_B + D_{BZ}$=4 ns.

The above example illustrated that we can associate arrival time with each incoming edge in the timer graph. Hence, we can define the minimum and the maximum arrival time for a vertex. In the above example, the minimum arrival time for Z is $A_{Z,min}$=2 ns and the maximum arrival time for Z is $A_{Z,max}$=4 ns.

Assume that a signal is already stable at a vertex v_j. The signal at v_j can change *earliest* when among all the arriving signals the first one shows its effect. Thus, we can define a *minimum bound* on the arrival time $A_{j,min}$ at that vertex by considering all the input vertices v_i as follows:

$$A_{j,min} = Min\left[A_{i,min} + D_{ij}\right] \tag{14.11}$$

Note that in the above equation, we take the minimum arrival time $A_{i,min}$ for the input vertices. It ensures that we take the earliest one if there are multiple arrival times at an input vertex. Thus, to compute the minimum arrival time at a vertex, we need to calculate the minimum arrival time for all vertices in its fanin cone, starting with the source of the timing graph. The path that gives the minimum arrival time from the source to a vertex is referred to as the *early path*.

After the first signal has shown its effect on the vertex v_j, the value at v_j can toggle many times as signals from various other incoming edges arrive. Finally, it would settle to a stable value when all the incoming signals have arrived, and their effects on v_j have died out. Thus, we can define a *maximum bound* on the arrival time $A_{j,max}$ at that vertex by considering all the input vertices v_i as follows:

$$A_{j,max} = Max\left[A_{i,max} + D_{ij}\right] \tag{14.12}$$

Note that in the above equation, we take the maximum arrival time $A_{i,max}$ for the incoming edges. It ensures that if there are multiple arrival times at an input vertex, we take the one that arrives last (i.e., when the value at the input vertex has already become stable). Thus, to compute the maximum arrival time at a vertex, we need to calculate the maximum arrival time for all the vertices in its fanin cone, starting with the source of the timing graph. The path that gives the maximum arrival time from the source to a vertex is referred to as the *late path*.

14.6.2 Rise and Fall Transitions

In general, the delay of a combinational circuit element can be different when the output makes a *rise transition* (0→1) or *fall transition* (1→0). Therefore, an STA tool computes the following arrival times and annotates them on the timing graph at a vertex v_j:

1. $A_{j,min,rise}$: the minimum arrival time for the rise transition.
2. $A_{j,min,fall}$: the minimum arrival time for the fall transition.
3. $A_{j,max,rise}$: the maximum arrival time for the rise transition.
4. $A_{j,max,fall}$: the maximum arrival time for the fall transition.

Note that, for the *cell arcs*, we should consider the *unateness* of a gate while using Eqs. 14.11 and 14.12, as explained in the following example.[3]

Example 14.10 Consider a positive unate gate (such as buffer, AND gate, and OR gate). For computing the arrival time for the rise transitions ($A_{j,min,rise}$ and $A_{j,max,rise}$) at a vertex v_j using Eqs. 14.11 and 14.12, we should consider the rise transitions at the input vertices v_i (i.e., $A_{i,min,rise}$ and $A_{i,max,rise}$). Similarly, for computing arrival time for the fall transitions we should consider the fall transitions at the input vertices.

Next, consider a negative unate gate (such as inverter, NAND gate, and NOR gate). For computing the arrival time for the rise transitions ($A_{j,min,rise}$ and $A_{j,max,rise}$) at a vertex v_j, we should consider the fall transitions at the input vertices v_i (i.e., $A_{i,min,fall}$ and $A_{i,max,fall}$). Similarly, for computing arrival time for the fall transitions we should consider the rise transitions at the input vertices.

For non-unate gates (such as an XOR gate), we should consider both rise and fall transitions at the input and choose the one that yields a more conservative value.

14.6.3 Arrival Time Propagation

The arrival time computation is done by the forward traversal of the timing graph, starting with the source vertices.

Arrival time at the source: We consider the input ports and the startpoints of the clock signal as the source vertices. The arrival time at an input port depends on the external environment of the circuit. We can convey this information to an STA tool using SDC file. We can also specify a nonzero clock arrival time at the clock source in the SDC file. If we have not specified them, an STA tool assumes them as zero.

Process: The arrival time computation starts from the source of the timing graph and progresses in the fanout in the topological order.

During arrival time computation, an STA tool maintains a *ready queue* Q_r. The vertices are processed by dequeuing them from Q_r in the *first-in first-out sequence* until it becomes empty. A vertex is enqueued into Q_r if *all its input vertices have already been processed.* This criterion for enqueuing into Q_r is maintained throughout arrival time computation.

Initially, all the *sources* of the timing graph are enqueued into Q_r. Subsequently, after processing a vertex, all its output vertices that satisfy the above enqueuing criterion are enqueued into Q_r. While processing a vertex, typically, all the previously described four bounds on the arrival time are computed using Eqs. 14.11 and 14.12 and annotated on the vertices. Note that the arrival time is dependent on the delay D_{ij} of the edges. If D_{ij} is not already computed, the tool invokes the *delay calculator* for its calculation. Subsequently, it stores the obtained results for delays and slews on the corresponding edges for reusing. After a vertex is processed, it is dequeued from Q_r.

The above enqueuing, processing, and dequeuing of vertices in Q_r are carried out until Q_r becomes empty. Finally, arrival times get annotated on all the vertices connected to the source vertices in the timing graph. We illustrate the above process in the following example.

[3] For the definition of unateness, see Chapter 13 ("Technology library").

Example 14.11 Consider the timing graph shown in Figure 14.15 (same as shown in Figure 14.11). We have labeled the vertices with a number indicating the processing order. For simplicity, let us assume that the delay of all the timing arc is 1 ns. Therefore, we can ignore the impact of rising and falling transitions in this example. Let us assume that the arrival times at the source vertices are zero.

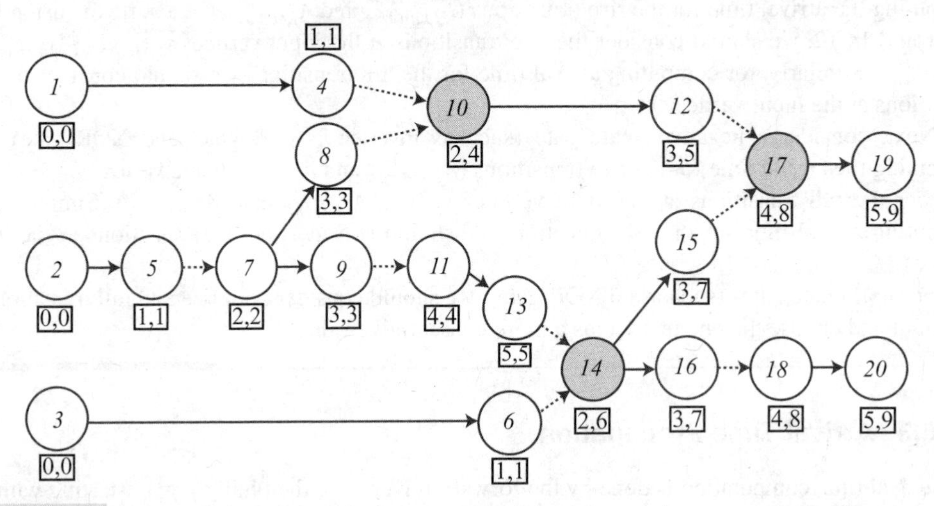

Figure 14.15 Illustration of arrival timing computation (time unit is nanosecond). The computed minimum and maximum arrival times are shown within the rectangles

The computed minimum and the maximum arrival times are shown within the rectangles. The processing of vertices is done as follows.
First, the source vertices *1*, *2*, and *3* are enqueued into Q_r. Then, *1* is processed (its arrival times are computed using Eqs. 14.11 and 14.12). After *1* is processed, *4* is enqueued to Q_r, since all the input vertices of *4* are already processed. Then, *1* is dequeued from Q_r. Similarly, *2* is processed, *5* is enqueued, and *2* is dequeued. Then, *3* is processed, *6* is enqueued, and *3* is dequeued. Then, *4* is processed. After processing *4*, *10* cannot be enqueued since the arrival time at its other input (vertex *8*) is not yet computed. The vertex *10* is enqueued after *8* has been processed. Similarly, the rest vertices are processed in the order of increasing vertex label. Note that the minimum and the maximum arrival times differ when the input vertices have different arrival times (e.g., for the vertices *10*, *14*, and *17* shown as shaded circles).

Complexity: The arrival time computation, as described above, can be done in one traversal of the vertices and the edges of the timing graph. Therefore, the runtime of arrival time computation is $O(|V|, |E|)$. Additionally, only fixed size of information (minimum and maximum arrival times in this case) is stored at each vertex. Thus, the memory consumption is $O(|V|)$ in the above computation.

Reason for efficiency: Note that we have reduced the runtime and the memory consumption in the arrival time computation by computing and storing only the bounds (maximum and minimum) on the arrival times. If we store the exact arrival times for different paths, then memory consumption can grow drastically. In the above example, consider the vertex labeled as *17*. The arrival time of the signal from *2* through *5→7→8→10→12→17* is 6 ns. However, we have avoided computing and storing this information. Instead, we have just stored the bounds [4, 8] on the arrival time at *17*. Note that for ensuring the *safety* of a circuit, the bounds on the arrival times are sufficient. It is the key concept that makes the arrival time computation and STA efficient.

14.6.4 Slew Propagation

There is one difficulty in the above approach of arrival time computation. For computing delays for a given stage, the slews at the inputs must be known. We can specify slews at the source of the timing graph using constraints, or an STA tool can assume some reasonable *minimum* and *maximum* slews. However, slews at other vertices must also be known for delay calculation. Therefore, an STA tool must also propagate the slews in the timing graph, starting from the source vertex. The propagated slews through different combinational paths can be different. However, computing, storing, and propagating all the possible slews at a vertex through different paths is infeasible because of too many possibilities.

Bounds on delay: An important property of complementary metal–oxide–semiconductor (CMOS) logic gates simplifies the above problem. The delay and the output slew are typically *monotonically nondecreasing functions* of the input slew. Therefore, if we choose two input slews (S_a and S_b) such that ($S_a < S_b$), then the corresponding delays (D_a and D_b) and the output slews (OS_a and OS_b) hold the following relationship: $D_a \leq D_b$ and $OS_a \leq OS_b$. It allows computing, storing, and propagating only the *minimum* and the *maximum* output slews at the vertices. We can obtain the *bounds* on the delay and the output slews using the *bounds* on the input slews. Thus, we can guarantee that arrival times are within the computed bounds and the circuit is timing safe.

Bounds on the arrival time and slews: To compute safe bounds on the arrival time, an STA tool also stores and propagates the minimum and the maximum slews at each vertex. Consider a cell arc (edge) between two vertices v_i and v_j shown in Figure 14.16. Assume that we know the bounds on the input vertex v_i, i.e., we know $A_{i,min}$, $A_{i,max}$, $S_{i,min}$, and $S_{i,max}$. Using the load capacitance associated with the vertex v_j and the minimum/maximum input slew $S_{i,min}/S_{i,max}$, we can compute the minimum/maximum delay $D_{ij,min}/D_{ij,max}$. Similarly, we can compute the minimum/maximum output slew $OS_{ij,min}/OS_{ij,max}$.

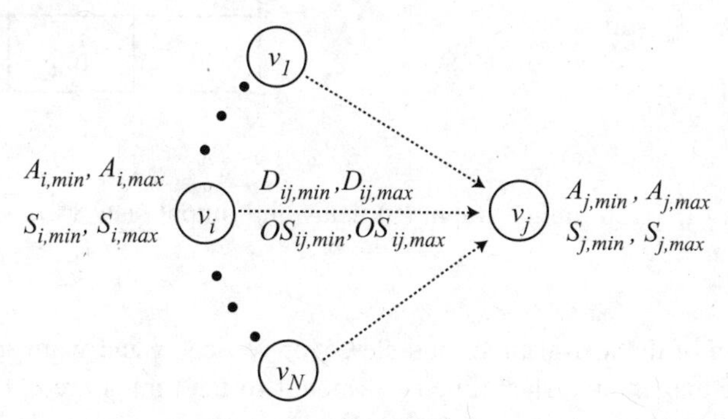

Figure 14.16 Impact of slew in the arrival time computation

When v_j has multiple incoming edges, we obtain the bounds on the arrival time *and the output slew* by considering all the incoming edges:

$$
\begin{aligned}
A_{j,min} &= Min\left[A_{i,min} + D_{ij,min}\right] \\
A_{j,max} &= Max\left[A_{i,max} + D_{ij,max}\right] \\
S_{j,min} &= Min\left[OS_{ij,min}\right] \\
S_{j,max} &= Min\left[OS_{ij,max}\right]
\end{aligned}
\tag{14.13}
$$

Graph-based analysis: The bounds on the arrival time and the slews are propagated together in a timing graph using the method described earlier. Note that, for simplicity, we have not considered transition types (i.e., rise and fall) in the above description. We need to compute and store values separately for each transition to account for the distinct rise and fall delays. Thus, eight values will be calculated using Eq. 14.13 (four each for rising and falling transitions).

The above method of slew propagation is known as *worst-case slew propagation* or *graph-based analysis* (GBA). It is fast and the most popular methodology of STA. It ensures the timing safety of a circuit but can lead to pessimism. We illustrate it in the following example.

Example 14.12 Consider the timing graph shown in Figure 14.17(a). Let us determine the maximum arrival time and the maximum slews using GBA method for the vertices x and y. The dependency of the delay and output slew for the arc $y \rightarrow x$ on the input slew for the given load is shown in Figure 14.17(b).

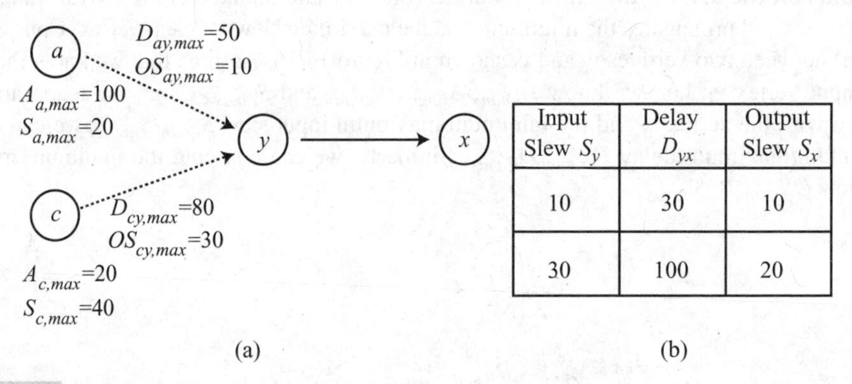

Input Slew S_y	Delay D_{yx}	Output Slew S_x
10	30	10
30	100	20

Figure 14.17 GBA (a) given timing graph, (b) delay and output slew vs input slew (time unit is picosecond)

The computation of the arrival times and slews for vertices x and y are shown in Table 14.3. For the vertex y, the maximum arrival time corresponds to the timing arc $a \rightarrow y$ (because it yields a higher arrival time compared to the other arc). However, the maximum slew corresponds to the timing arc $c \rightarrow y$ (because it produces a greater output slew compared to the other arc). However, this is *unrealistic* because only one of the two arcs will exhibit the maximum arrival time, and we should propagate slew through that path.

Table 14.3 Slew propagation using GBA

	y	x
A_{max} (ps)	Max [100 + 50, 20 + 80] = 150	150 + 100 = 250
S_{max} (ps)	Max [10, 30] = 30	20

Nevertheless, if we propagate the slew corresponding to the same arc that exhibited the maximum arrival time, we can miss the real upper bound on the arrival time. For example, let us propagate the arrival time and the slew through the same timing arc $a \rightarrow y$ (one that exhibited maximum arrival time). The computation is shown in Table 14.4. The maximum arrival time is found to be 180 ps. Next, let us propagate arrival time and slew through the other arc $c \rightarrow y$. The computation is shown in Table 14.5. The arrival time found in this case is 200 ps, which is more than what we obtained if both the arrival time and slews were propagated through $a \rightarrow y$. Thus, the correct maximum arrival time for the circuit is 200 ps. It demonstrates that if we propagate both the maximum arrival time and slew through the same path, we can get an optimistic arrival time which is unsafe.

Table 14.4 Slew propagation using path $a \rightarrow y$

	y	x
A_{max} (ps)	100 + 50 = 150	150 + 30 = 180
S_{max} (ps)	10	10

Table 14.5 Slew propagation using path $c \rightarrow y$

	y	x
A_{max} (ps)	20 + 80 = 100	100 + 100 = 200
S_{max} (ps)	30	20

Note that the arrival time computed by GBA is 250 ps. It is more than the actual bound (200 ps) but is still safe. To make arrival time computation in GBA tighter, we need to propagate arrival times and slews through multiple paths and store them on vertices. For real circuits, due to too many possibilities, it is not feasible. Therefore, GBA trades off some accuracy to gain a dramatic reduction in runtime and memory consumption.

Path-based analysis: Industry-standard STA tools are equipped with another STA methodology known as *path-based analysis* (PBA). In PBA, tools analyze a path by strictly using the timing arcs lying on that path. In the above example, PBA for the path $a \rightarrow y \rightarrow x$ would produce the arrival time as 180 ps. Similarly, PBA for the path $c \rightarrow y \rightarrow x$ would compute the arrival time as 200 ps. Thus, PBA gives a realistic arrival time and removes the pessimism of GBA. Therefore, we can carry out PBA for some top thousands of paths that exhibit the worst behavior to remove pessimism associated with GBA. Note that we cannot carry out PBA for the complete design because practical designs contain too many paths. Therefore, traditional GBA-based STA is still the preferred methodology of STA. At later stages of design flows, we can use PBA to remove some pessimism and ease achieving timing closure.

14.7 REQUIRED TIME COMPUTATION

After arrival time computation, an STA tool computes *required time* on the timing graph. The required times at a vertex put constraints on the corresponding arrival times such that we avoid the setup and the hold violations.

Setup slack: For setup check, the required time provides the *maximum time* by which a signal should arrive to avoid violation. For a given vertex v in a timing graph, the difference between the setup required time $R_{v,set}$ and the maximum arrival time $A_{v,max}$ is known as the setup slack $Sl_{v,set}$:

$$Sl_{v,set} = R_{v,set} - A_{v,max} \tag{14.14}$$

A $Sl_{v,set} > 0$ denotes that the setup constraint is met. The value of slack $Sl_{v,set}$ is the time by which we can further delay a data signal without violating the setup requirement. If $Sl_{v,set} < 0$, then the signal is arriving at v too late and there is a setup violation. In such cases, we should carefully analyze the timing violation and fix the circuit if required and possible. Alternatively, we can operate a circuit at a lower clock frequency to obviate setup timing violations.

Hold slack: For hold check, the required time provides the *minimum* time after which a signal should arrive to avoid timing violation. For a given vertex v in a timing graph, the difference between the minimum arrival time $A_{v,min}$ and the hold required time $R_{v,hold}$ is known as hold slack $Sl_{v,hold}$:

$$Sl_{v,hold} = A_{v,min} - R_{v,hold} \tag{14.15}$$

A $Sl_{v,hold} > 0$ denotes that the hold constraint is met. The value of slack $Sl_{v,hold}$ is the amount of time by which a data signal can be made faster without violating the hold requirement. If $Sl_{v,hold} < 0$, then the signal is arriving at v too early and there is a hold violation. In such cases, we should carefully analyze the timing violation and fix the circuit if required. If a fabricated circuit has a hold violation, then the circuit will malfunction. We cannot fix hold violations after fabrication, in contrast to setup violations which can be avoided by operating a circuit at a lower clock frequency. Therefore, though often easier to fix during designing, hold violations are more dangerous if they escape verification.

Required time at the sink: The required time is computed in a timing graph, starting from the sinks. The D-pin of the flip-flops and the output ports are considered as the sinks in a timing graph. At the D-pin of a flip-flop, we can compute the setup required time $R_{v,set}$ using Eq. 14.1 (difference of the arrival time of the *next* edge of the clock and the maximum setup time of the flip-flop) and the hold required time $R_{v,hold}$ using Eq. 14.6 (sum of the arrival time of the *same* edge of clock and the maximum hold time of the flip-flop).

At the output port, the required time depends on the environment of the design. We convey this information to an STA tool using SDC file.[4] We assume that the setup and required times are available at the sinks for the rest of this section.

Bounds on required times: The required times are computed by the *backward traversal* of the timing graph starting from the sinks. Given a vertex v_i, the hold required time $R_{i,hold}$ can be computed by considering all the *output vertices* v_j as follows:

$$R_{i,hold} = Max\left[A_{j,hold} - D_{ij}\right] \tag{14.16}$$

Similarly, the setup required time $R_{i,set}$ can be computed by considering all the *output vertices* v_j as follows:

$$R_{i,set} = Min\left[A_{j,set} - D_{ij}\right] \tag{14.17}$$

Among all the output vertices, taking the *maximum* for the hold required time and the *minimum* for the setup required time ensures that the most conservative constraints get imposed at a vertex. Note that the delay, and also the required time, can be different for the rise transition and the fall transition. Therefore, at each vertex v_i four different required times are computed and stored:

1. $R_{i,hold,rise}$: the hold required time for rise transition.
2. $R_{i,hold,fall}$: the hold required time for fall transition.
3. $R_{i,set,rise}$: the setup required time for rise transition.
4. $R_{i,set,fall}$: the setup required time for fall transition.

Process: Similar to the arrival time computation, an STA tool maintains a *ready queue* Q_r for the required time computation. The vertices are processed by dequeuing vertices from Q_r in the *first-in first-out sequence* until it becomes empty. A vertex is enqueued into Q_r if *all its output vertices have already been processed.* This criterion for enqueuing a vertex into Q_r is always maintained during required time computation.

Initially, all the *sinks* of the timing graph are enqueued into Q_r. Subsequently, after processing a vertex, all its input vertices that satisfy the above enqueuing criterion are enqueued. While processing a vertex, typically, all the above four required times are computed using Eqs. 14.16 and 14.17 and annotated on the timing graph. Note that the required time is dependent on the delay D_{ij} annotated at the edges. Since we compute the required time after arrival time, D_{ij} is already available at every edge from the arrival time computation. After a vertex is processed, it is dequeued from Q_r. We illustrate this process in the following example.

[4] We will discuss the required time details for the output port in Chapter 15 ("Constraints").

Example 14.13 Consider the timing graph shown in Figure 14.18 (same as shown in Figure 14.15). We assume that the delay of each edge is 1 ns. We ignore the impact of transition type in this example. We assume that for the sink vertex *19* $R_{19,hold}$=4 and $R_{19,set}$=8, and for the sink vertex *20* $R_{20,hold}$=5 and $R_{20,set}$=12.

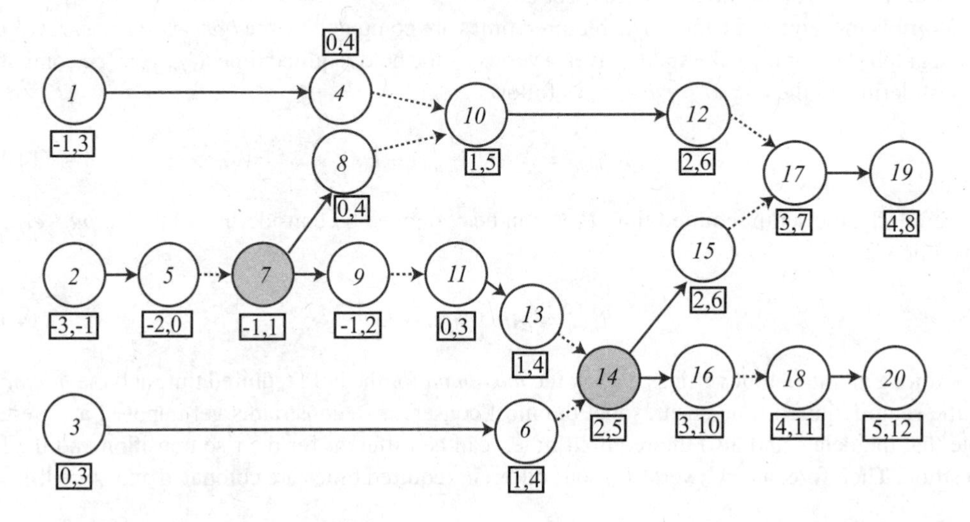

Figure 14.18 Illustration of required time computation (time unit is nanosecond). The computed minimum and maximum required times are shown within the rectangles

The computed hold and setup required time for other vertices are shown within the rectangles. Note the bounds are obtained using the *maximum* and *minimum* operations on the required times at the output vertices for *7* and *14*. Other vertices have only one output. Hence propagation does not require the maximum and minimum operations.

Complexity: The required time computation, as described above, can be done in one traversal of the vertices and the edges of the timing graph. Therefore, its runtime is $O(|V|, |E|)$. Additionally, only a fixed size of information (four values) is stored at each vertex. Thus, the memory consumption is $O(|V|)$ in the above computation. Like the arrival time computation, we have avoided computing and storing nonessential information by considering only the most conservative setup and hold requirements. Note that for ensuring the *safety* of a circuit, the bounds on the arrival times and required times are sufficient.

14.8 SLACK COMPUTATION

After required time computation, the computation of slack is straightforward using Eqs. 14.14 and 14.15.

Example 14.14 Consider Figures 14.15 and 14.18, in which we have already computed the arrival times and required times, respectively. We compute the slack using Eqs. 14.14 and 14.15. The slacks for hold and setup requirements are shown in Figure 14.19.

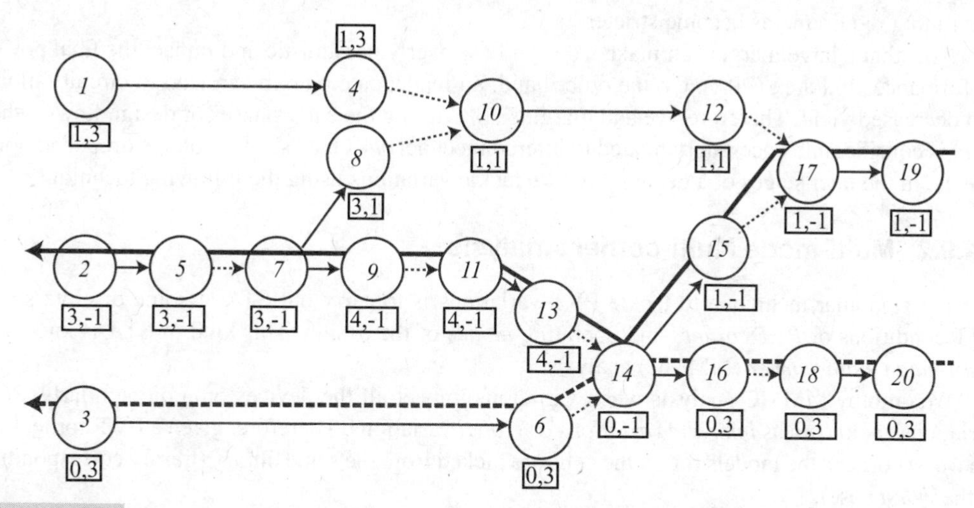

Figure 14.19 Illustration of slack computation (time unit is nanosecond). The computed hold and setup slacks are shown within the rectangles

Among the sinks, the vertex *20* has the worst hold slack of 0, and vertex *19* has the worst setup slack of −1. The *path* exhibiting the worst hold slack can be retraced by starting with the sink with the worst hold slack and traversing in the fanin along the vertices exhibiting the same hold slack. In this case, the path traced is *20←18←16←14←6←3*.

Similarly, we can trace the path exhibiting the worst setup slack starting from the sink that exhibits the worst setup slack. In this case, the path traced is *19←17←15←14←13←11←9←7←5←2*.

The path that has the *minimum slack* in the circuit is called the *critical path* of the circuit. In this case, the worst setup slack is negative. Hence, there is a setup constraint violation. We will need to fix this violation in the circuit.

14.9 ACCOUNTING FOR VARIATIONS

In practical circuits, the behavior can differ from the *nominal behavior* due to process-induced variations and fluctuations in temperature and voltage. These variations can impact the delay of the timing arcs and the timing constraints (such as setup and hold time of flip-flops). Therefore, STA tools must account for variations while ascertaining the safety of a design.

To tackle variations, different techniques are employed. These techniques differ in accuracy, modeling effort, computational resource requirement, and design effort [7]. We discuss some of the popular ones in the following paragraphs.

14.9.1 Safety Margins

The easiest technique to tackle variations is by adding suitable *safety margins* [8]. We can convey margins to an STA tool using appropriate *constraints*. These margins adjust the required time such that timing requirements become stricter.

Note that a large margin can make a design flow overly pessimistic and impact the final power, performance, and area (PPA). On the other hand, an inadequate margin can lead to circuit failures and decreased yield. Therefore, we add margins only during the early phases of design flows, when there is considerable uncertainty related to interconnect delays, clock skews, voltage drop, and noise effects. In the later stages of a design flow, we tackle variations using the following techniques.

14.9.2 Multi-mode Multi-corner Analysis

The most popular technique to tackle PVT variations is to carry out STA at some discrete set of PVT conditions or *PVT corners* and operating *modes* of the design. This kind of STA is known as *multi-mode multi-corner* (MMMC) analysis.

We employ MMMC analysis when variations impact all the devices on a die similarly. Such variations are known as *inter-die variations* or *global* variations. Therefore, given a PVT corner (say the *worst* corner), the models for *all* the cells are picked from the same library (library corresponding to the *worst* case).

The MMMC-based STA methodology becomes tedious at advanced process nodes due to many possible scenarios obtained by a combination of multiple corners and numerous modes [6]. Typically, MMMC analysis is made faster by parallel processing.[5]

14.9.3 On-chip Variation

There can be *local variations* also in the properties of devices and interconnects on the same die. These variations are known as *intra-die variations* and can be tackled by *on-chip variation* (OCV) *derating factors*. An STA tool multiplies the nominal delay obtained using a delay calculator with the OCV derating factor to determine the timing arc's effective delay. Thus, the OCV derating factor provides a mechanism to adjust the upper and lower delay bounds based on the expected intra-die variations.

We can define different OCV derating factors based on the following considerations [6, 9]:

1. Path bounds (early or late)
2. Path type (data or clock)
3. Delay type (gate delay or interconnect delay)
4. Corners (best, worst, typical, etc.).

Example 14.15 Consider the following OCV derate factors:

Late path derating factor = 1.1.

Early path derating factor = 0.9.

These derating factors will have the following impact:

1. The delay of *all* the timing arcs in the late paths (data paths and clock launch paths for the setup checks and capture clock paths for the hold checks) get scaled by a factor of 1.1.

---

[5] We will discuss MMMC methodology in detail in Chapter 24 ("Basic concepts for physical design").

2. The delay of *all* the timing arcs in the early paths (capture clock paths for the setup checks, and data paths and clock launch paths for the hold checks) get scaled by a factor of 0.9.

Note that, we increase delays in the late path and reduce them in the early path. Hence, the given derate factors will make STA more pessimistic.

The drawback of OCV derating is that the delays of *all* the timing arcs are scaled by the same factor for a given path type. Therefore, we implicitly assume that a perfect positive correlation exists among timing arcs of a given group.

Similarly, different groups of timing arcs get scaled differently. In the above example, we have scaled the timing arcs of the launch and the capture clock paths differently. Therefore, we implicitly assume that a perfect negative correlation exists between the timing arcs of these groups. However, such an ideal correlation (positive or negative) is unrealistic [10]. It makes OCV derating technique overly pessimistic.

14.9.4 Statistical Static Timing Analysis

Another technique to tackle process-induced variations is to treat process-sensitive device and interconnect parameters as *Gaussian random variables* [11]. We also treat the delay and the arrival time as random variables. We obtain their statistical distribution by defining the *sum* and the *maximum* operations statistically [12]. This technique of STA is known as *statistical static timing analysis* (SSTA).

Though we have researched SSTA intensively since the early 2000s, it is not widely employed. Despite its accuracy, the unfavorable cost–benefit of characterizing and developing statistical library models and the complexities of deploying SSTA in design flows have limited its wide-scale usage. We prefer simpler techniques that account for variations, as explained below.

14.9.5 Advanced On-chip Variation

In *advanced on-chip variation* (AOCV) technique, a *derating table* is defined for each cell in a technology library [8]. The derating factor of a cell depends on the following attributes of the path in which it lies on the layout [8]:

1. **Spatial extent:** We observe that the nearby cells exhibit lesser systematic variations. Therefore, we use larger derating factors for cells located in a path spanning over a larger area.
2. **Logic depth:** We observe that the effects of random variations get nullified among cells when they are part of a path with more levels of logic gates. Consequently, on an average, smaller random variation is contributed by each cell in that path. Therefore, we use a smaller derating factor for instances in a path with more logic depth.

A demerit of the AOCV method is that we ignore the impact of input slew and capacitive loading on the variations.

14.9.6 Parametric On-chip Variation

The *parametric on-chip variation* (POCV) is a simplified form of statistical timing analysis [6, 13]. It avoids the tedious and costly statistical library characterization and statistical parasitic extraction required for SSTA. Nevertheless, it requires information on the relative variation in the timing arc delays due to variation in a given parameter that we expect to vary. This information is easier to gather, e.g., by Monte Carlo simulations. Using this information and the expected variation in a given parameter, we can compute the statistical variations in delay and arrival time similar to SSTA.

In the POCV methodology, we can also account for the dependency of delay variations on the input slew and the capacitive load by coupling with the *Liberty variation format* (LVF) files [10, 14]. An LVF file models the variation in delay, output slew, and constraints (setup/check values) as a function of input slew and output load. Thus, POCV can model variations at the timing arc-level and is more accurate than AOCV [10].

14.10 RECENT TRENDS

We employ STA technology extensively in VLSI design flow. We use it as a stand-alone verification tool, and many design optimization tools internally employ STA to guide them and evaluate various solutions. Therefore, the speed of STA implementations is a primary concern. One of the popular techniques to speed up STA tools is to exploit multithreading, and distributed computing [15, 16].

With the advancement in technology, we face many design challenges and difficulties in achieving timing closure. The impact of process-induced variations becomes more severe, the nonlinearity in the voltage waveforms increases, and the signal integrity issues aggravate. Addressing these issues requires changes in many tools, including the STA tool.

At advanced process nodes, meeting the required frequency target is challenging. Therefore, approaches such as PBA that can remove pessimism in STA become critical. Recently, efforts are being made to make PBA feasible and more efficient for large designs [17]. Additionally, pessimism introduced by safety margins is avoided as much as possible. For example, to account for dynamic voltage fluctuations of the supply rails and their impact on the timing, we can use path-specific margins rather than a blanket margin [18]. Recently, the application of machine learning in addressing some of the challenges of STA is also being explored [19–21].

REVIEW QUESTIONS

14.1 What is a zero-clocking? Why should we avoid it in a synchronous circuit?

14.2 What is a double clocking? Why should we avoid it in a synchronous circuit?

14.3 The critical portion of a circuit is shown in Figure 14.20. For flip-flop *FF1*, $CP \rightarrow Q$ delay is 20 ps and setup time is 12 ps. For flip-flop *FF2*, $CP \rightarrow Q$ delay is 26 ps and setup time is 4 ps. Find the maximum value of the delay D of the inverter *I1* that will allow the circuit to operate at 1 GHz clock frequency by honoring setup requirements. Ignore the delays of wires.

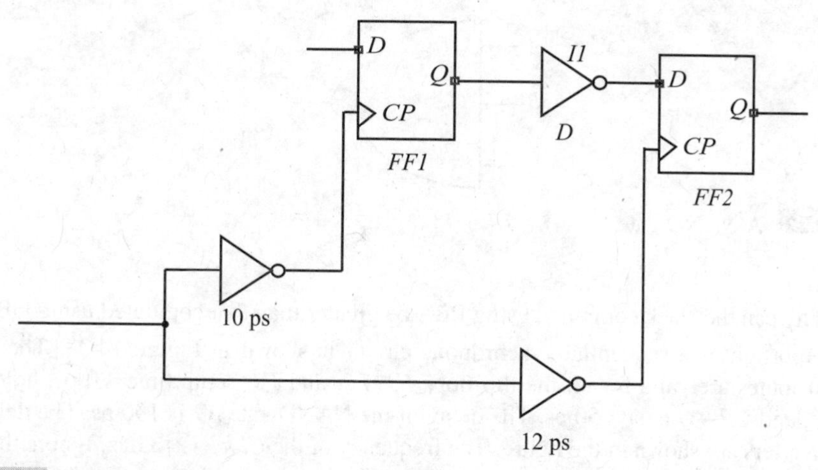

Figure 14.20 Given circuit

14.4 Draw the timing graph for the circuit shown in Figure 14.21.

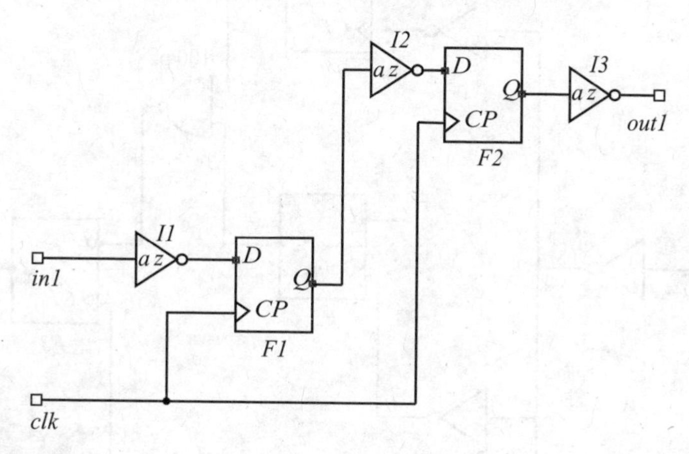

Figure 14.21 Given circuit

14.5 A fabricated chip has some setup time violation. What can we do to make it functional (we can tolerate some PPA sacrifice in the system)?

14.6 A fabricated chip has some hold time violation. Can anything be done to make it functional (functionally deterministic and correct)?

14.7 For the NAND gate shown in Figure 14.22, the arrival times at the inputs are: A_A=5 ps, A_B=6 ps, and A_C=12 ps. The delay of the cell arcs is: $D_{A,Z}$=10 ps, $D_{B,Z}$=20 ps, and $D_{C,Z}$=30 ps. The output slew of the cell arcs is: $S_{A,Z}$=20 ps, $S_{B,Z}$=5 ps, and $S_{C,Z}$=10 ps.

What will be the arrival time and the output slew at the pin Z under the following modes of analysis:

(a) GBA for setup

(b) PBA for setup through arc $B \rightarrow Z$

(c) GBA for hold

(d) PBA for hold through arc $B \rightarrow Z$?

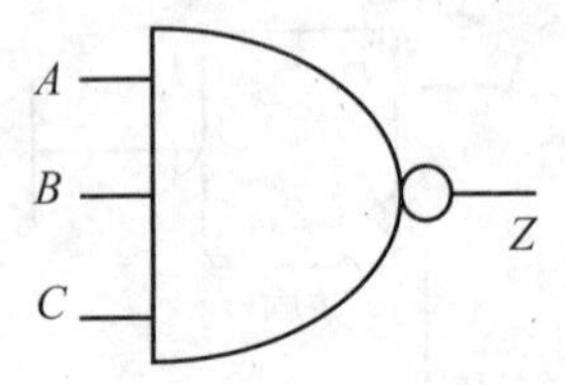

Figure 14.22 A NAND gate

14.8 Why can the slack computed using PBA be greater than that computed using GBA?

14.9 A portion of a sequential synchronous circuit is shown in Figure 14.23. The following attributes are valid for all the flip-flops *F1*, *F2*, and *F3*: setup time=30 ps, hold time=10 ps, and *CP*→*Q* delay=50 ps. The delay of the NAND gate *N1* is 100 ps. The delays of the inverters are shown in the figure. The frequency of the *Clock* is 1 GHz. Ignore the delay of all the wires. Find the worst setup slack and the worst hold slack of the circuit.

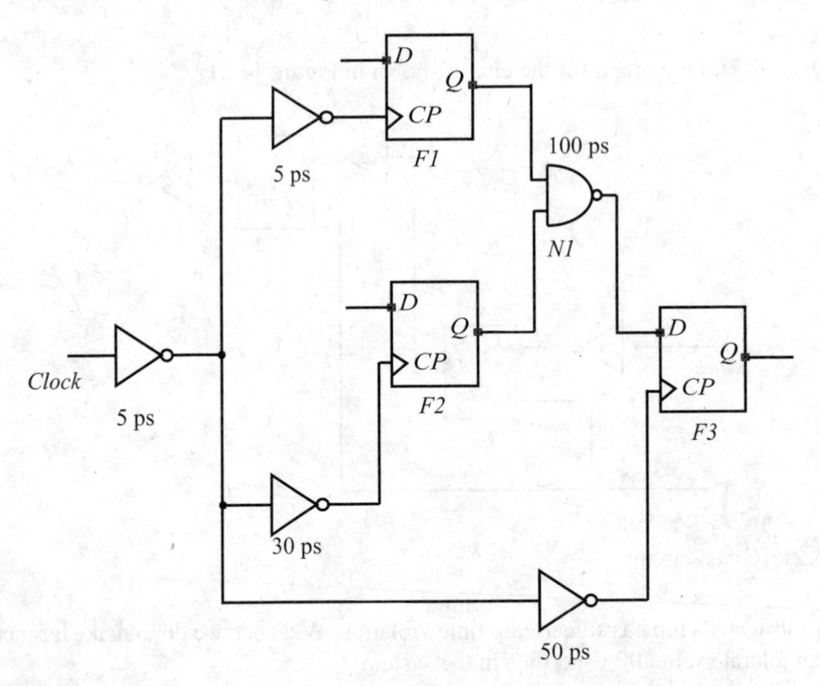

Figure 14.23 Given circuit

14.10 In Problem 14.9, assume that we add a safety margin of 50 ps to the design for STA. Find the worst setup slack and the worst hold slack of the circuit.

14.11 In Problem 14.9, assume we add an OCV derate factor of 1.1 for the late paths and 0.9 for the early paths. Find the worst setup slack and the worst hold slack of the circuit.

TOOL-BASED ACTIVITY

1. Take any STA tool that is available to you and carry out the following activities:
   (a) Load a design in the form of a netlist of approximately 100–1000 logic gates, along with a technology library that is available to you and some realistic timing constraints using the SDC file.
   (b) Perform STA and report the path with the worst setup slack and the worst hold slack.
   (c) Change the clock period and observe the impact on the setup slack and the hold slack.
   (d) Add safety margin (as constraints) and observe the impact on the setup slack and the hold slack.
   (e) If your STA tool supports OCV analysis, add different derate on the early and the late paths. Observe the impact on the setup slack and the hold slack.
2. Take any STA tool that supports both PBA and GBA modes (most industry-standard STA supports them) and carry out the following activities:
   (a) Write a Verilog netlist manually such that the critical path has many side inputs.
   (b) Use SDC (for input delay and input transition) such that the GBA slack is worse than the PBA slack.
   (c) By reporting various cell attributes, examine the difference between the GBA and the PBA modes of slew propagation.

REFERENCES

[1] L. W. Cotton. "Circuit implementation of high-speed pipeline systems." *Proceedings of the November 30–December 1, 1965, Fall Joint Computer Conference, Part I* (1965), pp. 489–504.

[2] R. A. Harrison and D. J. Olson. "Race analysis of digital systems without logic simulation." *Proceedings of the 8th Design Automation Workshop* (1971), pp. 82–94.

[3] M. A. Wold. "Design verification and performance analysis." *15th Design Automation Conference* (1978), pp. 264–270, IEEE.

[4] T. M. McWilliams. "Verification of timing constraints on large digital systems." *Proceedings of the 17th Design Automation Conference* (1980), pp. 139–147.

[5] J. P. Fishburn. "Clock skew optimization." *IEEE Transactions on Computers* 39, no. 7 (1990), pp. 945–951.

[6] S. Saurabh, H. Shah, and S. Singh. "Timing closure problem: Review of challenges at advanced process nodes and solutions." *IETE Technical Review* (2018).

[7] R. Chen, L. Zhang, V. Zolotov, C. Visweswariah, and J. Xiong. "Static timing: Back to our roots." *2008 Asia and South Pacific Design Automation Conference* (2008), pp. 310–315, IEEE.

[8] H. Bhatnagar. *Advanced ASIC Chip Synthesis*. Springer, 2002.

[9] J. Bhasker and R. Chadha. *Static Timing Analysis for Nanometer Designs: A Practical Approach*. Springer Science & Business Media, 2009.

[10] A. Mutlu, J. Le, R. Molina, and M. Celik. "A parametric approach for handling local variation effects in timing analysis." *Proceedings of the 46th Annual Design Automation Conference* (2009), pp. 126–129.

[11] D. Blaauw, K. Chopra, A. Srivastava, and L. Scheffer. "Statistical timing analysis: From basic principles to state of the art." *IEEE Transactions on Computer-aided Design of Integrated Circuits and Systems* 27, no. 4 (2008), pp. 89–607.

[12] C. Visweswariah, K. Ravindran, K. Kalafala, S. G. Walker, S. Narayan, D. K. Beece, J. Piaget, N. Venkateswaran, and J. G. Hemmett. "First-order incremental block-based statistical timing analysis." *IEEE Transactions on Computer-aided Design of Integrated Circuits and Systems* 25, no. 10 (2006), pp. 2170–2180.

[13] X. Peng, H. Wang, S. Wang, and J. Du. "A new generation of static timing analysis technology based on N7+ process—POCV." *2019 IEEE 4th International Conference on Integrated Circuits and Microsystems (ICICM)* (2019), pp. 199–203, IEEE.

[14] B. Bautz and S. Lokanadham. "A slew/load-dependent approach to single-variable statistical delay modeling." *Proceedings Tau Workshop* (2014), pp. 1–18.

[15] P. Ghanta, A. Goel, F. P. Taraporevala, M. Ovchinnikov, J. Liu, and K. Kucukcakar. "Simultaneous multi-corner static timing analysis using samples-based static timing infrastructure." Dec. 24, 2013. US Patent 8, pp. 615, 727.

[16] K. E. Murray and V. Betz. "Tatum: Parallel timing analysis for faster design cycles and improved optimization." *2018 International Conference on Field-Programmable Technology (FPT)* (2018), pp. 110–117, IEEE.

[17] A. B. Kahng, U. Mallappa, and L. Saul. "Using machine learning to predict path-based slack from graph-based timing analysis." *2018 IEEE 36th International Conference on Computer Design (ICCD)* (2018), pp. 603–612.

[18] A. Vakil, H. Homayoun, and A. Sasan. "IR-ATA: IR annotated timing analysis, a flow for closing the loop between PDN design, IR analysis & timing closure." *Proceedings of the 24th Asia and South Pacific Design Automation Conference* (2019), pp. 152–159.

[19] S.-S. Han, A. B. Kahng, S. Nath, and A. S. Vydyanathan. "A deep learning methodology to proliferate golden signoff timing." *2014 Design, Automation & Test in Europe Conference & Exhibition (DATE)* (2014), pp. 1–6, IEEE.

[20] S. Bian, M. Hiromoto, M. Shintani, and T. Sato. "LSTA: Learning-based static timing analysis for high-dimensional correlated on-chip variations." *2017 54th ACM/EDAC/IEEE Design Automation Conference (DAC)* (2017), pp. 1–6, IEEE.

[21] S. R. OVS and S. Saurabh. "Modeling multiple input switching in timing analysis using machine learning." *IEEE Transactions on Computer-aided Design of Integrated Circuits and Systems* (2020).

15 Constraints

To define is to limit.

—Oscar Wilde, *The Picture of Dorian Gray*, Chapter 17, 1890

We can implement a given functionality in various ways. For example, to add two binary numbers, we can implement it using different architectures, employ standard cells of different sizes, place them on the layout in many legitimate ways, and connect them using several alternative routes. These implementations can differ in power, performance, and area (PPA). For example, a carry-lookahead adder (CLA) can exhibit better speed than a ripple carry adder, though at the expense of increased area. However, as a designer, we know the target application and the required attributes of the circuit. We define and convey these requirements to electronic design automation (EDA) tools using constructs known as *constraints*.

During implementation steps, such as logic and physical synthesis, EDA tools attempt to ensure that the *constraints* are honored. Similarly, verification tools such as the static timing analysis (STA) tool check whether the specified *constraints* are met, independent of the implementation tool. Thus, constraints act as a common denominator for the implementation and verification tools.

We specify requirements or design constraints using *Synopsys Design Constraint* (SDC) commands in a set of ASCII files [1]. These files are known as *constraint files* or *SDC files*. Note that there can be different constraint files for different *modes of operation*, such as functional mode, test mode, and scan mode, due to mode-dependent timing requirements.

In this chapter, we will discuss various design constraints and their relevance. We will also demonstrate, using examples, how we can define them in the SDC format. However, note that this chapter does not exhaustively describe the SDC commands and their arguments (options). For an exhaustive list of SDC commands, readers should refer to the SDC user guide or tool-specific manuals. Additionally, it is worth pointing out that EDA tools can also provide some *proprietary* commands and mechanisms to gather additional information about the design. In this book, we will not discuss tool-specific commands. Readers should refer to them in tool-specific user guides and reference manuals.

Most of the design constraints specified in a constraint file are related to the timing and used during STA. Therefore, a *constraint file* is also referred to as a *timing constraint file*. In this chapter, we will explain STA-specific design constraints in detail. Therefore, we suggest that readers understand the basic concepts of STA, which are discussed in Chapter 14 ("Static timing analysis"), before reading this chapter.

The SDC format is based on *Tool Command Language* (Tcl). We can use Tcl features such as variables, data structures, and built-in commands in the SDC file. Therefore, we suggest that readers should become familiar with the basic commands of Tcl. Readers can refer to some excellent online materials on Tcl for this purpose.

Many industry-standard EDA tools accept Tcl script as input. We can combine the features of SDC, Tcl, and tool-specific commands to develop powerful scripts that can automate complicated design tasks and extract helpful design information for us. Therefore, proficiency in Tcl often helps in the very large scale integration (VLSI) design process.

15.1 TYPES OF CONSTRAINTS

The constraints in an SDC file can relate to the following aspects of a design:

1. **Clock signal:** The attributes of a clock signal are crucial for a synchronous design. However, the clock signal can be generated outside a chip or outside the block that we are designing. Hence, the information about the clock signal generation and its characteristics may not exist within the design under consideration. Sometimes, we can derive some characteristics of a clock signal, such as the start point and the relationship between two synchronous clock signals, by circuit analysis. Nevertheless, deriving them can be computationally tricky and be inaccurate. Moreover, some attributes of the clock signal, such as frequency, duty cycle, skew, jitter, and delay, can be impossible to derive using digital circuit analysis. Therefore, we define the attributes of the clock signal using SDCs. These constraints are used by the implementation tools in guiding their optimization and by the timing and power analysis tools in verification.
2. **Environment:** To implement a design correctly, we must know its environment. For example, implementation tools must instantiate suitable cells at the interfaces considering the external signals and loads. Furthermore, we need to verify that our design can work in a given environment. However, design implementation and verification tools cannot infer the environmental constraints on their own. Therefore, we convey this information to various EDA tools using SDC commands.
3. **Functionality:** Some functional attributes of a circuit are difficult to derive but are conveniently known to a circuit designer. For example, there can be a *timing path* through which a signal cannot propagate due to the circuit functionality. Moreover, there can be a timing path for which we allow a signal to propagate differently than the traditional synchronous data transfer. Such timing paths are known as *timing exceptions*. STA tools need to model setup and hold requirements differently for those paths. We can ease the task for STA and other EDA tools by informing them of these timing exceptions using SDC commands.
4. **Design rules and optimization constraints:** Some design rules, such as the maximum pin capacitance and maximum transition time at a port, must be obeyed in a design. These rules can be conveyed to the implementation and verification tools using constraints. These are *hard constraints* because if they get violated, a design can fail. Additionally, there are some desirable attributes or optimization constraints. These are *soft constraints*. We can inform EDA tools about soft constraints also using SDC commands.

We will discuss in detail the above constraints in the following sections.

15.2 CLOCK SOURCES

The definition of clock constraints begins by defining the clock sources. A clock source can be of two types: *primary clock source* and *derived clock source*. The waveform generated by a primary clock source is independent of other clock sources in that design. However, the waveform generated by a derived clock source depends on other clock sources.

Example 15.1 Consider the clock-divider circuit as shown in Figure 15.1(a).

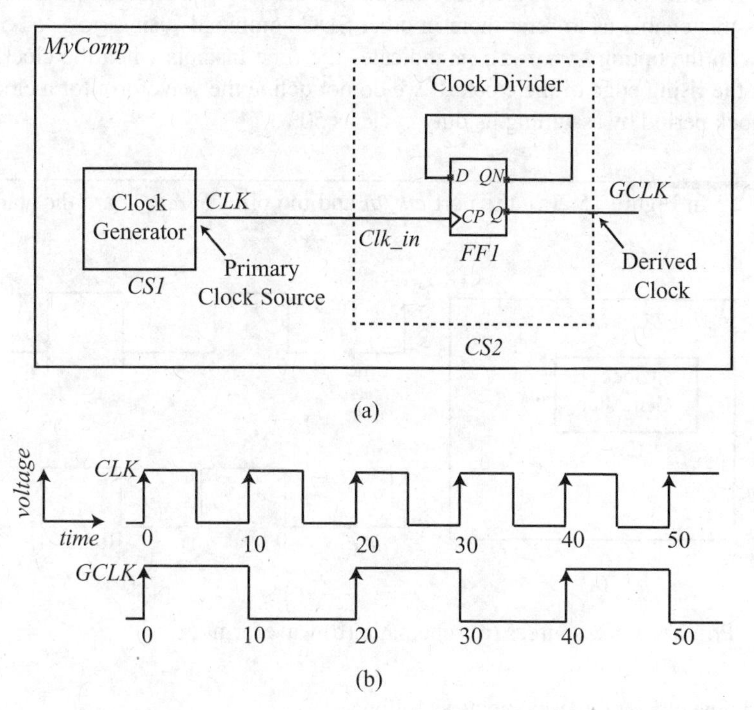

Figure 15.1 (a) Circuit with a primary and generated clock and (b) clock waveforms

The output *GCLK* of the clock-divider circuit depends on its input clock signal *Clk_in*. Therefore, the clock-divider circuit *CS2* is a *derived clock source*. The input to the clock-divider circuit comes from the primary clock source *CS1*. The clock from which we derive another clock is known as the *master clock* of the derived clock. Thus, in the above example, *CS1* is the master clock source of *CS2*. One of the possible waveforms for the *CLK* (same as *Clk_in*) and *GCLK* is shown in Figure 15.1(b).

A primary clock source can be an *internal clock generator* that generates a clock signal within a design using special circuitry. Alternatively, there can be an external clock generator. A design receives the external clock signal at some primary input. Thus, the starting point of a primary clock source can be an internal pin or an input port of a design.

Primary clock source: The SDC command `create_clock` is used to define the attributes of the *primary clocks* in a design. Though EDA tools can sometimes automatically infer the design name, we can explicitly mention it in the SDC file using command `current_design`. Note that, we apply each SDC to some design object or a group of design objects. A design object or a group of design objects can be referred to in an SDC file using command `get_*`, where `*` can be `pins`, `ports`, `cells`, etc. We can apply the SDC `create_clock` on a pin or an input port corresponding to the start point of the clock source.

We can define other attributes of the clock signal using command options `-name`, `-waveform`, `-period`, etc. (command options are preceded by `-` in SDC). We assign unique names to different clock sources that enable us to refer them in other SDC commands using `get_clocks`. The list of numbers with the option `-waveform` indicates the time instants when the clock edges occur, starting from the rising edge of the clock. If we do not define the waveform for a clock, tools infer it from the clock period by assuming its duty cycle as 50%.

Example 15.2 In Figure 15.2(a), the port *clk_in* and the pin *CS1/clk_g* are the startpoints of the clock sources.

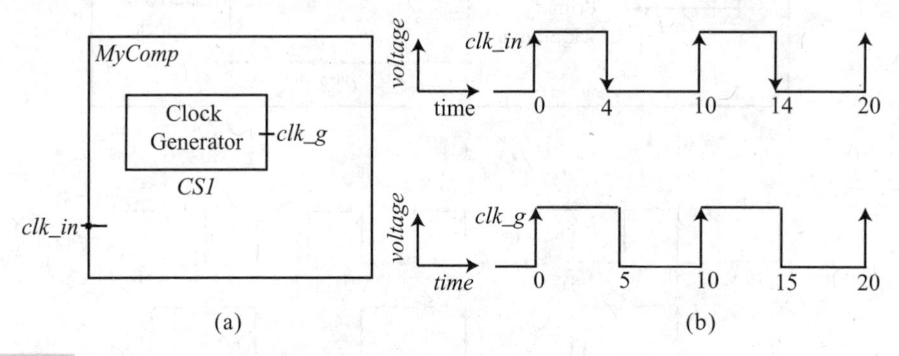

Figure 15.2 Primary clock sources (a) schematic (b) waveforms

We can define clock on these objects as follows.

```
current_design MyComp
create_clock -name EXT_CLK -period 10 -waveform {0 4}
[get_ports clk_in]
create_clock -name INT_CLK -period 10 [get_pins CS1/clk_g]
```

We refer to the internal pin by its hierarchical name *CS1/clk_g* in the command `get_pins`. The waveform inferred for the above commands is shown in Figure 15.2(b).

Note that when we invoke an SDC command, it only provides some definition or constraints to an *existing* design object. Thus, the command `create_clock` will provide a clock definition to an *existing* port or pin in a design. We do not expect the SDC command to make changes to the design. For example, assume that a design does not have a clock generator. If we invoke the command create_clock on a *non-existing* clock generator pin, we should not expect that an implementation tool will create a clock generator for us. Typically, a tool will show an error or warning in this case.

Another point to note is that, in an SDC file, we refer to various physical quantities such as time, capacitance, resistance, voltage, current, and power by simple numbers. The units for these quantities are typically inferred from the technology libraries. Alternatively, we can define them by the SDC command set_units.

Derived clock source: We define a derived clock on a port or a pin using the SDC command create_generated_clock. Therefore, we refer to a derived clock also as a *generated clock*.

Example 15.3 To define the clocks in Figure 15.1(a), we can use the SDC commands, as shown in the following box.

```
create_clock -name CLK -period 10 [get_pins CS1/CLK]
create_generated_clock -name GCLK -divide_by 2 -source
[get_pins CS1/CLK] [get_pins CS2/GCLK]
```

First, we define the primary clock or the master clock *CLK*. Then, we define a generated clock at the output pin *CS2/GCLK* of the clock divider. We also define the source of the generated clock as *CS1/GCLK* using the option -source. It refers to the *master clock* source pin and acts as a reference from which the attributes of the generated clock can be derived. The relationship between the master clock and the generated clock is defined by the option -divide_by. Thus, the waveform shown in Figure 15.1(b) is obtained for the generated clock.

15.3 ATTRIBUTES OF A CLOCK SIGNAL

For STA, we need to define some more attributes of a clock signal. In this section, we describe how we can specify them in the SDC file.

15.3.1 Latency

When a clock signal is launched from the clock source, it can get delayed before reaching the clock pin of a flip-flop due to wire capacitance and cell delays in its path. Therefore, we need to account for these delays during timing analysis.

We construct the clock distribution network in physical design during the *clock tree synthesis* (CTS) step. Therefore, in the pre-CTS stage, we cannot compute the clock signal delay. However, in the pre-CTS stages, we can make a rough estimate of the clock delay and define it in the SDC file for the STA tools to consider. The estimated delay is known as *clock latency*. We define clock latency using SDC command set_clock_latency. We can define two types of latencies in an SDC file:

1. **Source latency:** The source latency is the time taken by the clock signal to propagate from the clock source to the design object (port or pin) where the clock is *defined* in the command create_clock or create_generated_clock. We can attribute the source latency to the external delay or delay within a clock generator.

2. **Network latency:** The network latency is the time taken by the clock signal to propagate from the design object (port or pin) where the clock is *defined* using `create_clock` or `create_generated_clock` to the clock-pin of the flip-flop. We can attribute the network latency to the delay due to the clock distribution network within a design.

Example 15.4 Consider the system shown in Figure 15.3. The clock generator lies outside the circuit. The clock signal is received at the input port *clk_port* after a delay of 5 time units. The flip-flops receive the clock signal after a delay of 10 time units.

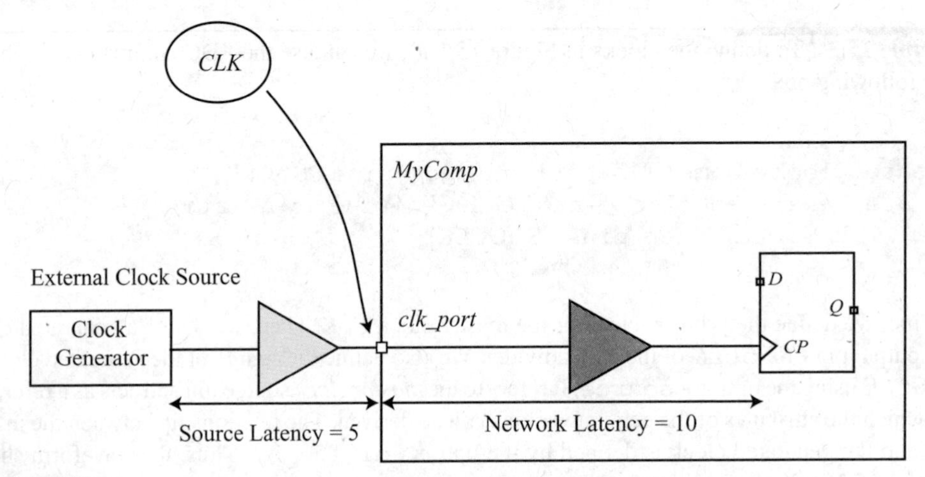

Figure 15.3 Different types of clock latencies

We can define these constraints as shown below.

```
create_clock -name CLK -period 200 [get_ports clk_port]
set_clock_latency 5 -source [get_clocks CLK]
set_clock_latency 10 [get_clocks CLK]
```

Note that, for defining source latency we add the option `-source`. If we do not specify `-source` in `set_clock_latency` then it is assumed that the constraint refers to the network latency.

Post-CTS, the clock distribution network exists in a design. Consequently, STA tools can compute the delay in the clock network using its layout. Therefore, post-CTS, we need not specify the estimated clock network latency using `set_clock_latency`. Instead, we add the SDC command `set_propagated_clock` as shown below.

```
create_clock -name CLK -period 200 [get_ports clk_port]
set_propagated_clock [get_clocks CLK]
```

15.3.2 Uncertainty

As discussed in the previous section, we specify the clock period, duty cycle, and the clock edges in the clock constraints. Ideally, these attributes of the clock signal should not change over *time* and *space* (layout). Thus, *all* the clocked elements (flip-flops, latches, and memories) should receive the clock edges *simultaneously*. However, the clock signal attributes deviate from the *ideal* value in real circuits. We model unpredictable deviation of the clock edges from the ideal value by a quantity known as *clock uncertainty*.

We use the SDC command `set_clock_uncertainty` to specify the clock uncertainty. We use the options `-hold` and `-setup` to specify the type of analysis in which the given clock uncertainty is applicable. An STA tool accounts for the clock uncertainty by adjusting the corresponding required time so that the analysis becomes more pessimistic for a greater value of uncertainty.

Example 15.5 The following SDCs model clock uncertainty.

```
create_clock -name CLK -period 200 [get_ports clk_port]
set_clock_uncertainty 15 -hold [get_clocks CLK]
set_clock_uncertainty 20 -setup [get_clocks CLK]
```

An STA tool will increase the required time by 15 time units in the hold analysis and reduce it by 20 time units in the setup analysis.

We can model the following nonideal effects using clock uncertainty:

1. **Jitter:** Jitter is the *temporal* variations in the clock edges compared to the ideal clock signal. The temporal variations can be caused by the power supply noise, thermal noise, and crosstalk [2]. The jitter in the clock signal gets manifested as the variation in the instantaneous clock period. Therefore, we must consider jitters in STA.

   Recall that, for a given launch–capture flip-flop pair, the hold checks (to avoid double-clocking) are done for the *same* clock edges, while the setup checks (to avoid zero-clocking) are done for the *next* clock edges. Since jitter in the clock signal changes the instantaneous clock period, the uncertainty can differ for the hold and setup timing analysis.
2. **Skew:** Skew originates due to the *spatial* variations in the clock signal across the layout of a circuit. It is caused by an unbalanced delay in the clock paths for a given launch–capture flip-flop pair. It gets manifested as the difference in the clock arrival time at the flip-flops. During the pre-CTS stage, we can model the estimated skew as clock uncertainty. Post-CTS, STA tools can compute the clock skew by employing the layout along with the wire delays.
3. **Safety margins:** The clock arrival time at a flip-flop can deviate from the ideal value due to process-induced variations. We can account for such unpredictable deviations by maintaining some *safety margins* using clock uncertainty.

   Before routing, the wires do not exist in a design. Hence, we cannot compute wire delays. Pre-routing implementation tools can do an overly *optimistic* timing analysis by ignoring wire delays. Consequently, these tools can forego some timing optimizations by wrongly assuming positive slacks. However, some of these optimizations could be essential for meeting post-route timing requirements. Thus, an overly optimistic view of a design can cause problems in attaining *timing closure* [3]. Therefore, to compensate for the optimism in the timing analysis, we often add some margins by including them in the clock uncertainty in the early phases of the design flow. As the flow progresses, we reduce and finally eliminate this component of clock uncertainty. However, the component of clock uncertainty due to jitter remains throughout the design flow.

15.3.3 Transition

Ideally, a clock signal should rise and fall instantaneously. However, as it propagates through the clock distribution network, the transitions become *nonideal*. Since the flip-flop's timing attributes (such as delay, setup time, and hold time) depend on the clock transition, an STA tool should consider nonideal clock transitions for better accuracy.

During the pre-CTS stage, a clock is treated as *ideal* by an STA tool. Nevertheless, we can make STA more realistic by specifying an estimated clock transition time. We use the SDC command `set_clock_transition` for this purpose.

Example 15.6 The following SDC commands model clock transition of 10 time units.

```
create_clock -name CLK -period 200 [get_ports clk_port]
set_clock_transition 10 [get_clocks CLK]
```

Post-CTS, an STA tool can compute the clock transition with the help of clock network layout and the associated parasitics. However, we might need to specify transition in the post-CTS stage also to model nonideal transition at the *starting point* of the clock network. For example, if a clock signal received at the input port has a nonideal transition, we need to specify that transition in the SDC file since a tool cannot automatically infer it.

15.4 SYNCHRONOUS DATA TRANSFER

The circuit that we design shares signals with other components at the system level or the board level. A popular technique is to share data *synchronously* among them. The synchronous communication offers high performance since the clock signal provides easy time reference. It avoids separate communication overhead [2].

Synchronicity with external components: For a system to function correctly, we must ensure the synchronicity of data transfer. To accomplish this, we define certain *timing constraints* at the input/output (I/O) ports of each component. Subsequently, an STA tool verifies that these I/O constraints are honored. Hence, by ensuring that we do not have any STA violations in our design, we confirm that the data transfer with the external components will be synchronous. Note that the *consistency* among constraints for various components is critical for this methodology to work properly [4]. Moreover, not only our design but external components should also honor their respective I/O constraints. We will discuss this aspect of constraints later in this chapter.

We can understand the requirements of synchronous data transfer across components using Figure 15.4. Assume that we are designing the component *MyComp*. It shares data with the external world using ports *IN*, *OUT*, and *CLK*.

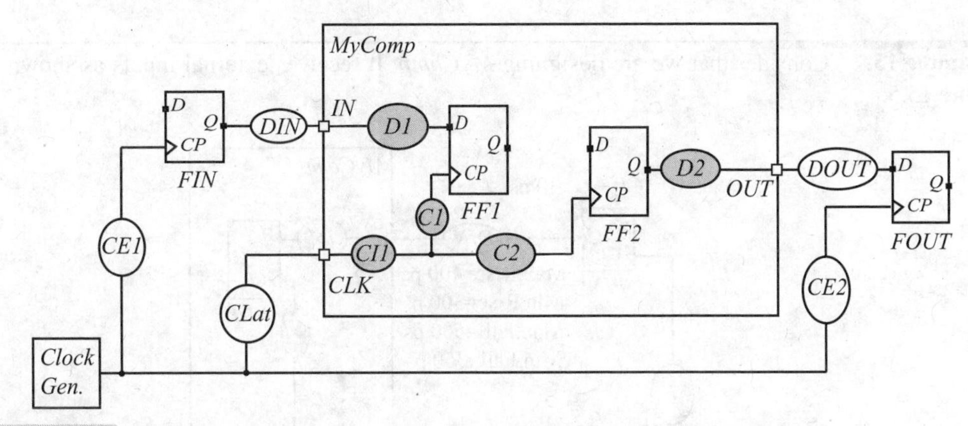

Figure 15.4 Environment in which a circuit operates in a synchronous system

Input considerations: The data received at *IN* is launched by an *external* flip-flop *FIN* and is captured by an *internal* flip-flop *FF1*. To receive data synchronously, *FIN* and *FF1* should receive clock signals from the same or synchronous clock sources. The delay between the external flip-flop *FIN* and the port *IN* is represented by the combinational block *DIN*. It includes the wires that physically connect *IN* to the external world. We should account for the delay due to the combinational block *DIN* while checking the setup and hold requirements at *FF1*. Additionally, the clock signal is delayed by the external blocks *CE1* on the launch clock path and *CLat* on the capture clock path. These delays must also be accounted for while checking the setup and hold requirements at the flip-flop *FF1*. We can account for the delay due to *CLat* by defining clock constraints (using `create_clock` and `set_clock_latency -source` as described earlier). We can account for the other external circuit elements connected to *IN* using the SDC command `set_input_delay`. We will explain this command in the following section.

Output considerations: The data leaving *OUT* is launched by an internal flip-flop *FF2* and captured by an external flip-flop *FOUT*. To receive the data synchronously, *FF2* and *FOUT* should receive clock signals from the same or synchronous clock sources. Additionally, the signal produced at *OUT* experiences delays due to the combinational block *DOUT* and should meet the setup and hold requirements of *FOUT*. Therefore, timing analysis of the internal path *FF2*→*D2*→*OUT* should consider the timing requirements of the capture flip-flop *FOUT*. We can ensure this by enforcing appropriate setup and hold requirements at the output port *OUT*. In these requirements, we should consider the delays of the external blocks *CE2* and *DOUT*. We should also consider the setup and hold time of *FOUT*. We can enforce these requirements by the SDC command `set_output_delay`, as explained in the following section.

15.5 I/O CONSTRAINTS

In this section, we explain how to define constraints at the I/O ports.

15.5.1 Input Constraints

We need to model two attributes of the signal applied at an input port: delay and waveform. First, we illustrate the delay modeling in the following example.

Example 15.7 Consider that we are designing *MyComp*. It receives external inputs as shown in Figure 15.5.

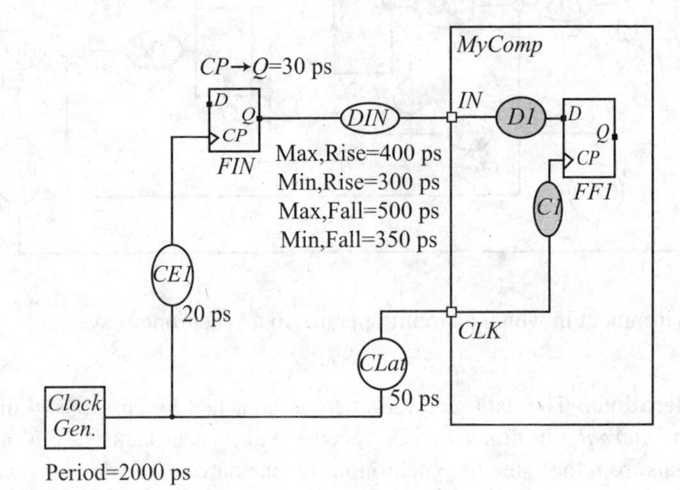

Figure 15.5 Given environment for determining `set_input_delay` constraints

Further, assume the following:

1. The clock generator generates a clock signal of time period 2000 ps.
2. The external circuit elements *CE1* has a delay of 20 ps.
3. The *CP*→*Q* delay of *FIN* is 30 ps.
4. The external circuit element *CLat* has a delay of 50 ps.
5. The combinational element *DIN* has the following delays:
   (a) The maximum rising delay is 400 ps,
   (b) The minimum rising delay is 300 ps,
   (c) The maximum falling delay is 500 ps, and
   (d) The minimum falling delay is 350 ps.

The data signal reaching the input port *IN* experiences delay due to *CE1*, *CP*→*Q* arc of *FIN*, and *DIN*. Therefore, the maximum rising delay at *IN* is $\tau_{IN} = 20 + 30 + 400 = 450$ ps. Similarly, we can compute other delay attributes.

We use the SDC command `set_input_delay` to model the above attributes as shown below.

```
current_design MyComp
create_clock -name SYS_CLOCK -period 2000 [get_ports CLK]
set_clock_latency 50 -source [get_clocks SYS_CLK]
```

```
set_input_delay  450  -max -rise -clock  [get_clocks  SYS_CLK]
[get_ports IN]
set_input_delay  350  -min -rise -clock  [get_clocks  SYS_CLK]
[get_ports IN]
set_input_delay  550  -max -fall -clock  [get_clocks  SYS_CLK]
[get_ports IN]
set_input_delay  400  -min -fall -clock  [get_clocks  SYS_CLK]
[get_ports IN]
```

Before defining the delay constraints, we must define the clock constraints. The clock constraints provide the reference time for delay specification. We can account for the external clock path delay *CLat* using source latency. Note that in the command `set_input_delay` we use the option `-clock` to provide the clock reference. Additionally, we can use the flags `-min/max` and `-rise/fall` to indicate the specific analysis type and the transition type, respectively.

During STA, the arrival time at the input port will be taken as the specified input delay for the given analysis type and transition. Therefore, when the setup (max) requirement is computed for the internal flip-flop *FF1*, then the arrival time at *IN* will be taken as 450 ps and 550 ps, for the rise and fall transitions, respectively. Similarly, for the hold (min) requirement, the arrival time at *IN* will be taken as 350 ps and 400 ps for the rise and fall transitions, respectively.

While creating constraints for a design, we may not know the details of the external paths. However, we may know the expected data–clock timing relationship at the input ports. In such cases, we can define the data–clock relationship as constraints by assuming a *virtual flip-flop* at the input ports. We illustrate it in the following example.

Example 15.8 Consider that we are designing *MyComp*. It receives external inputs at the port *IN* and clock at the input port *CLK*, as shown in Figure 15.6(a).

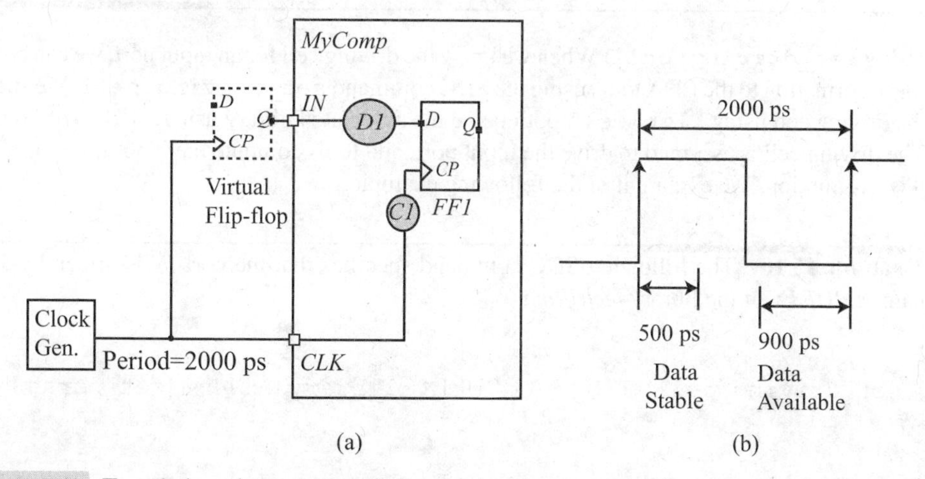

Figure 15.6 Translating data–clock relationship to input constraints (a) circuit with a virtual flip-flop and (b) given data–clock relationship

Assume that the clock period is 2000 ps. Additionally, assume that the data at *IN* is available at least 900 ps before the clock edge and is stable for at least 500 ps after the clock edge, as shown in Figure 15.6(b). It implies that the maximum delay through the virtual flip-flop shown in the figure will be 2000 − 900 = 1100 ps. Moreover, the minimum delay through the virtual flip-flop should be 500 ps (otherwise, after the clock edge, the data cannot be guaranteed to remain unchanged for 500 ps). Therefore, we can define the constraint as shown below.

```
create_clock -name SYS_CLOCK -period 2000 [get_ports CLK]
set_input_delay 1100 -max -clock [get_clocks SYS_CLK] [get_ports IN]
set_input_delay 500 -min -clock [get_clocks SYS_CLK] [get_ports IN]
```

Another constraint that we must define for an input port is the waveform of the incoming signal. Depending on the waveform type, availablity of the information about the environment, and the STA accuracy requirements, we can define waveform in one of the following ways.

1. **Using `set_input_transition`:** When the input signal can be approximated by a linear input, then we can define the waveform by its slope using SDC command `set_input_transition`. We can use the flags `-min/max` and `-rise/fall` to indicate the specific analysis type and the transition type, respectively.

Example 15.9 The following SDC command defines a rise transition time of 10 time units at the port *IN* for setup (maximum) analysis.

```
set_input_transition 10 -max -rise [get_ports IN]
```

2. **Using `set_driving_cell`:** When we know the driving cell for an input port, we can convey this information to the EDA tools using the SDC command `set_driving_cell`. We specify the driving cell using `-lib_cell` option and the associated library using `-library` option. The driving cell is assumed to drive the input port, and tools compute the input waveform with this assumption. We explain it in the following example.

Example 15.10 The following SDC command specifies that the port *IN* is driven by a cell named *BUF1X* of the library *tech14nm*.

```
set_driving_cell -lib_cell BUF1X -library tech14nm [get_ports IN]
```

A delay calculator will consider a *stage* similar to shown in Figure 15.7 to determine the waveform *Y*.

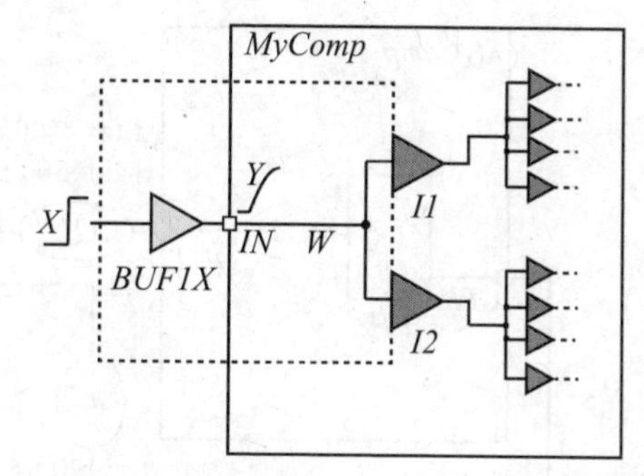

Figure 15.7 Input waveform computation using driving cell *BUF1X* (stage considered by delay calculator is shown by dashed rectangle)

It accounts for the parasitics of the internal wire *W* and the load of driven cells (*I1* and *I2*) in its computation. Furthermore, the stimulus *X* to the driving cell can be assumed to be ideal or to have a tool-specific default transition.

Note that the computed waveform *Y* can be nonlinear depending on the attributes of the wire *W* and the driven loads. Moreover, we can derive the waveforms using advanced delay models such as the composite current source (CCS) model and effective current source model (ECSM) and propagate them further. Thus, we can model the signal more realistically using a driving cell.

Implementation tools can also use driving cell information to determine the buffering requirements at the input. For example, if the driving cell is insufficient to drive internal loads, buffers can be inserted at the inputs to reduce the loading effects on the driving cell.

15.5.2 Output Constraints

We model timing requirements of the signals produced at the output ports using output constraints. Additionally, output constraints are used to model the external loading effects on the output ports.

We explain the timing requirements on the signal produced at the output port in the following example.

Example 15.11 Consider that we are designing *MyComp* shown in Figure 15.8. It produces an output at the port *OUT* using the clock signal received at *CLK*. The output gets captured externally using the same clock signal.

Further, assume the following:

1. The clock generator generates a clock signal of period 2000 ps.
2. The external circuit elements *CE2* has a delay of 20 ps.
3. The setup and hold time of the external flip-flop *FOUT* are 30 ps and 10 ps, respectively.
4. The external circuit element *DOUT* has the maximum delay of 400 ps and the minimum delay of 300 ps.

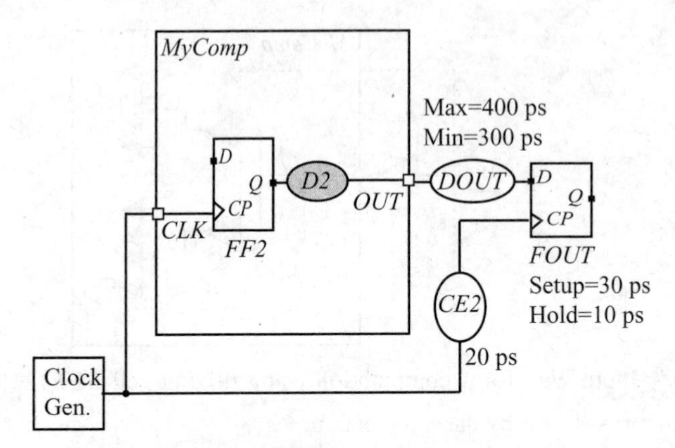

Figure 15.8 Given circuit *MyComp* with output signal being captured by an external flip-flop

The data produced by the output port *OUT* must satisfy the setup and hold requirements of *FOUT*. To ensure this, we enforce setup and hold checks at the output port *OUT*. However, since the attributes of the external flip-flop are unknown during STA of *MyComp*, an STA tool checks these requirements by considering an *equivalent* situation shown in Figure 15.9.

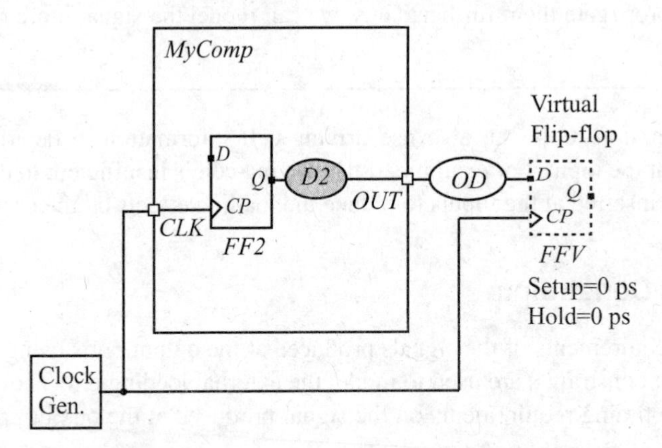

Figure 15.9 Evironment for determining `set_output_delay` constraints

A *virtual flip-flop FFV* with zero setup and hold times is connected to the output port through a combinational element *OD*. The delay of *OD* (τ_{OD}) needs to be chosen such that the setup and hold requirements in the actual circuit (Figure 15.8) and the STA-inferred circuit (Figure 15.9) are the same. We specify τ_{OD} in the SDC command `set_output_delay` and compute it separately for the setup and hold requirements, as shown below.

The setup requirement at *FOUT* in Figure 15.8 can be stated as follows:

$$T_{clk-q,FF2} + T_{D2,max} + 400 + 30 - 20 < T_{period}$$
$$\Rightarrow T_{clk-q,FF2} + T_{D2,max} + 410 < T_{period}$$

where $T_{clk-q,FF2}$ is the $CP \rightarrow Q$ delay of *FF2* and $T_{D2,max}$ is the maximum delay through *D2*. The setup requirement at *FFV* in Figure 15.9 can be stated as:

$$T_{clk-q,FF2} + T_{D2,max} + \tau_{OD} < T_{period}$$

The inequations for both circuits should represent the same constraints. Hence, by comparing the inequations, we can infer that τ_{OD} should be 410 ps for the setup analysis.

The hold requirement at *FOUT* in Figure 15.8 can be stated as:

$$T_{clk-q,FF2} + T_{D2,min} + 300 > 10$$

$$\Rightarrow T_{clk-q,FF2} + T_{D2,min} + 290 > 0$$

where $T_{D2,min}$ is the minimum delay through *D2*. The hold requirement at *FFV* in Figure 15.9 can be stated as:

$$T_{clk-q,FF2} + T_{D2,min} + \tau_{OD} > 0$$

The above two inequations should represent the same constraints. Hence, by comparing the inequations, we can infer that τ_{OD} should be 290 ps for the hold analysis.

We specify the setup constraint using `-max` option and the hold constraint using `-min` option, as shown below.

```
create_clock -name SYS_CLOCK -period 2000 [get_ports CLK]
set_output_delay 410 -max -clock [get_clocks SYS_CLK] [get_ports
OUT]
set_output_delay 290 -min -clock [get_clocks SYS_CLK] [get_ports
OUT]
```

Note that sometimes we know the constraints of an output port as the expected data–clock relationship. We can model these constraints using `set_output_delay` as shown in the following example.

Example 15.12 Assume that the period of a clock signal is 2000 ps. Further, assume that we need to model the following requirements for an ouptut port *OUT*:
(a) the data signal should be available by 900 ps before the clock edge, and
(b) data should be stable for 600 ps after the clock edge.
The requirement is similar if we consider a flip-flop at the output port with a setup and hold times of 900 ps and 600 ps, respectively. However, an STA tool considers a virtual flip-flop with zero setup and hold times. Therefore, we need to derive τ_{OD} as in the previous example that allows to model the given constraint using virtual flip-flop of zero setup and hold times. We leave this as an exercise for the readers. Finally, we need to define `set_output_delay` as shown below.

```
create_clock -name SYS_CLOCK -period 2000 [get_ports CLK]
set_output_delay 900 -max -clock [get_clocks SYS_CLK] [get_ports
OUT]
set_output_delay -600 -min -clock [get_clocks SYS_CLK] [get_ports
OUT]
```

Note that τ_{OD} is negative for the hold check.

Another constraint that is important for an output port is the external load capacitance. We can specify the load capacitance being driven by an output port using SDC command `set_load`. The delay calculator employs this constraint for computing the delay of the cell that drives the associated output port.

Example 15.13 The following SDC command defines a load of 0.039 capacitance unit on the output port *OUT*.

```
set_load 0.039 [get_ports OUT]
```

15.6 VIRTUAL CLOCKS

A *virtual clock* is a clock definition in the SDC file for which there is no associated design port or pin. Nevertheless, we define its name, period, waveform, and other attributes similar to other clocks. We can refer to a virtual clock in other commands of SDC, such as `get_clocks`, `set_clock_uncertainty`, `set_clock_transition`, `set_input_delay`, and `set_output_delay`, by its name.

Example 15.14 Consider the following clock definition.

```
create_clock -name V_CLK -period 10
```

It defines a clock with the name *V_CLK* and a clock period of 10 time units. However, we have not applied it at any design port or pin. Therefore, it is a virtual clock.

We can use virtual clocks to define timing constraints in several situations [5]. We illustrate one such application of a virtual clock in the following example.

Example 15.15 Consider the circuit shown in Figure 15.10(a). Assume that we are designing *MyComp*. The clock *M_CLK* captures data in an internal flip-flop *FF1*. Additionally, it receives input at the port *IN* from the external flip-flop *FIN*. The external flip-flop launches data using another external clock *P_CLK*. Assume that *M_CLK* and *P_CLK* have the same clock period (10 time units), but different duty cycle and phase, as shown in Figure 15.10(b).

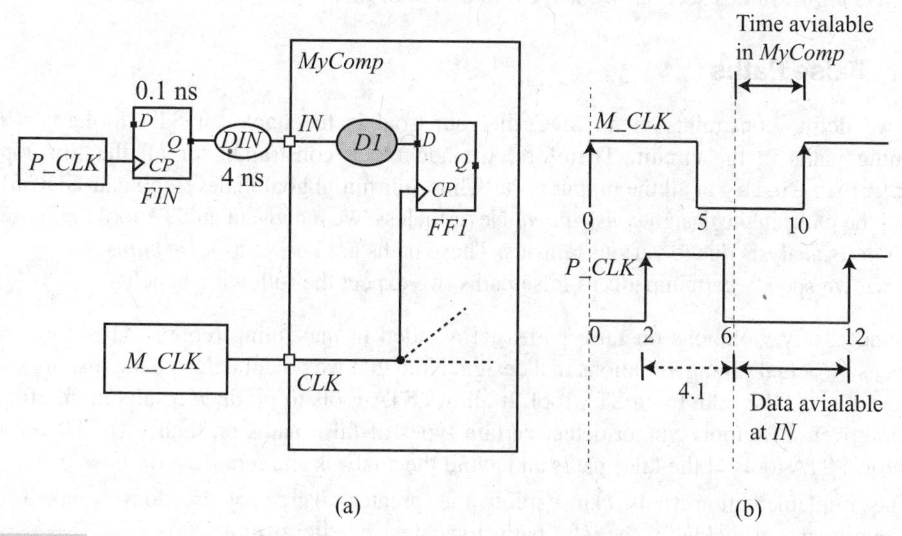

Figure 15.10 Defining `set_input_delay` using virtual clock (a) schematic (b) waveform

The data received at *IN* undergoes a delay of 4 + 0.1 = 4.1 ns with respect to clock edge of *P_CLK*. Therefore, we need to define the clock *P_CLK* in *MyComp* to constrain the port *IN*. However, there is no clock pin/port associated with *P_CLK*. Hence, we define it as a virtual clock and refer to it in the command `set_input_delay`, as shown below.

```
#time unit is nanosecond
create_clock -name M_CLK -period 10 [get_ports CLK]
create_clock -name P_CLK -period 10 -waveform {2 6}
set_input_delay 4.1 -max -clock [get_clocks P_CLK] [get_ports IN]
```

Note that the available time window for capturing the data inside *MyComp* gets reduced due to the phase difference in *M_CLK* and *P_CLK*, as illustrated in Figure 15.10(b). The clock definitions capture this information.

Similarly, we can use virtual clocks to define constraints at the output ports. Moreover, we can use a virtual clock to constrain a *port-to-port combinational path* in a design. An externally launched data propagates through such port-to-port paths in a given design and finally gets captured externally. To constrain such paths, we proceed as follows. We first define a virtual clock, and then define `set_input_delay` and `set_output_delay` with respect to this virtual clock at the input and the output ports, respectively. If we choose the appropriate values for these constraints, it will constrain the port-to-port combinational path delay to the desired range.

15.7 TIMING EXCEPTIONS

Sometimes, we want STA to treat a timing path or a group of timing paths differently from the traditional timing paths in a synchronous circuit. We convey this information to the EDA tools using constraints known as *timing exceptions*. There are two types of timing exceptions: *false paths* and *multi-cycle paths*. In this section, we will explain both of them.

15.7.1 False Paths

When we define constraints in an SDC file, our goal is to enable an STA tool to check *all* the timing paths in the circuit. Therefore, we add clock constraints for all the flip-flops and `set_output_delay` at all the output ports. Thus, our initial goal in developing an SDC file is to make all the paths in the design *constrained*. Nevertheless, we may want an STA tool to ignore some of them in its analysis due to various reasons. These paths are known as *false paths*.

When we specify certain paths as false paths, we expect the following benefits:

1. Unnecessary violations on false paths get avoided in the timing reports. Therefore, we can focus on actual timing violations in a design. Note that we do not delegate the responsibility of detecting a false path to an STA tool. It allows STA tools to produce results more efficiently. Moreover, STA tools cannot detect certain types of false paths on their own. Therefore, we inform STA tools of the false paths and avoid the analysis and reporting of those paths.
2. The implementation tools can exploit the greater degree of freedom (relaxed timing requirement) available for the false paths to achieve a better overall PPA.

How to specify: We use SDC command `set_false_path` to specify false paths in a design. We can use design objects such as ports, pins, or SDC clocks (specified by `create_clock` or `create_generated_clock`) to define false paths. We specify the startpoint of paths using the option `-from`, the intermediate points using the option `-through`, and the endpoint using the option `-to`. Note that if multiple `-through` options are specified for a path, then that path must pass through *all* those `-through` objects in the *sequence* specified.

Example 15.16 A few examples of `set_false_path` commands are shown below.

```
set_false_path -from [get_ports RST]
set_false_path -to [get_pins FF0/SE]
set_false_path -from [get_pins F1/CP] -through [get_pins add/a0]
-through [get_pins sub/a0] -to [get_pins F2/CP]
```

The first false path includes all timing paths starting from *RST*. The second false path includes all timing paths ending on *FF0/SE*. The third false path includes all timing paths starting from *F1/CP*, then passing through *add/a0*, then passing through *sub/a0*, and ending on *F2/CP*.

Note that if we do not specify false paths in a design, it cannot directly lead to a timing problem. However, defining them in several situations is desirable because an STA tool can produce many unnecessary violations. We describe below a couple of such situations.

Non-sensitizable paths: We can define a timing path through which a transition cannot propagate as a false path. However, we must analyze such timing paths carefully before specifying them in the SDC file. Note that just checking whether a path is statically *sensitizable* (observing the impact of start point on the endpoint) by *logical analysis* is *not sufficient* to detect a false path. A path can be statically not sensitizable, yet glitches can propagate through that path. If flip-flops are at the end of such paths, flip-flops can capture a glitch, and the circuit can fail. Therefore, we cannot always treat a statically *non-sensitizable* path as a false path.

Example 15.17 Consider a portion of circuit shown in Figure 15.11(a). The path from *FF1/Q* to *Z* is not statically sensitizable, because *Z* remains zero irrespective of whether *FF1/Q*=0 or *FF1/Q*=1.

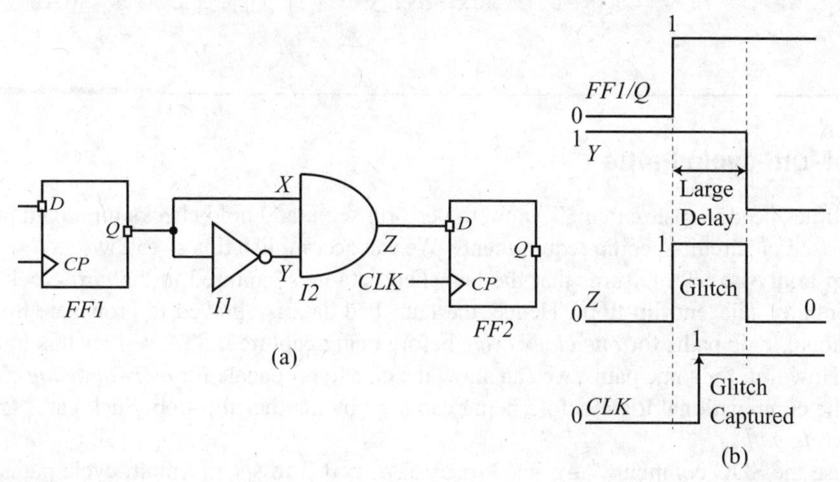

Figure 15.11 Path which is statically not sensitizable, yet transition can propagate through it (a) given portion of a circuit and (b) waveform showing possible glitch capture by *FF2*

Consider that *FF1/Q* makes a 0→1 transition, as shown in Figure 15.11(b). Assume that the delay from *FF1/Q* to *Y* is large, while the delay from *FF1/Q* to *X* and the delay through the AND gate are negligible. As a result, a glitch will be generated at *Z*. If the clock signal arrives at *FF2* when *Z*=1, *FF2* can reach a wrong state (*FF2/Q*=1), as shown in the above figure. Therefore, we cannot declare the path between *FF1* and *FF2* as a false path in the SDC file despite being statically non-sensitizable (if we declare, STA will not perform setup and hold checks at *FF2*). However, the normal setup and hold checks at *FF2* will ensure that only a stable value of *Z* (i.e., 0) is captured by *FF2*, and we should allow STA to perform them in this situation.

The previous example is trivial. Nevertheless, in practical circuits, there exist many statically non-sensitizable paths that we can find using Boolean logic analysis. However, these paths can allow glitches to propagate. Hence, we should declare them as false paths after ensuring they cannot cause flip-flops to fail under the given delay model.

Data transfer between clock domains: We may require to specify false paths between two different *clock domains*. A clock domain consists of flip-flops that are clocked by the same clock source. However, when data propagates from one clock domain to another, we need to be more careful. We need to add *synchronizers* at the interface of the clock domains. A synchronizer is a circuit element that tackles metastability issues in flip-flops.[1] However, an STA tool can produce many unnecessary violations at the clock domain crossings even when synchronizers exist in our circuit. Therefore, we may need to define false paths between clock domains.

Example 15.18 Consider that we have two clock domains in a circuit *MAIN_CLK* and *INT_CLK*. We can define false path between these clock domains as shown below.

```
create_clock -name MAIN_CLK -period 10 [get_ports CLK]
create_clock -name INT_CLK -period 37 [get_pins CLK_GEN/CLKOUT]
set_false_path -from [get_clocks MAIN_CLK] -to [get_clocks INT_CLK]
set_false_path -from [get_clocks INT_CLK] -to [get_clocks MAIN_CLK]
```

15.7.2 Multi-cycle Paths

We sometimes need to instruct an STA tool to perform setup and hold checks for a path or a group of paths based on circuit-specific requirements. We can accomplish this as follows.

Setup analysis: STA ensures that the launched data gets captured in the *next* clock cycle by the sequentially adjacent flip-flops. Hence, the launched data is allowed to propagate through the combinational logic paths for *one clock cycle* before being captured. STA verifies this in the *setup* analysis. However, for some paths, we can allow the data to propagate for *more than one clock cycle* through the combinational logic before being captured by another flip-flop. Such paths are known as *multi-cycle paths*.

We use the SDC command `set_multicycle_path` to specify multi-cycle paths. We use `-setup`/`-hold` flags to specify the analysis type. We define paths using `-from`, `-through`, and `-to` options (similar to the `set_false_path` command). Additionally, we specify the path multiplier to indicate the number of cycles allowed for the setup/hold check.

Example 15.19 Consider a flip-flop *FF1* which changes its value once every fourth clock cycle. We can allow the data launched by *FF1* to propagate for four clock cycles before being captured by the adjacent flip-flop *FF2*. We can use the following SDC command to specify this.

```
set_multicycle_path 4 -setup -from [get_pins FF1/CP] -to [get_pins
FF2/D]
```

The above `set_multicycle_path` specifies that we allow four clock cycles between flip-flops *FF1* and *FF2* for setup analysis. Without this command, an STA tool will allow only one clock cycle for this. Note that we do not want an STA tool to ignore the above path. We want to verify that the launched data gets captured by the sequentially adjacent flip-flops by the fourth clock cycle. In this sense, a multi-cycle path is different from a false path.

[1] We will discuss clock domains and synchronizers in Chapter 19 ("Power-driven optimizations").

Hold analysis: In addition to the setup analysis, an STA tool performs *hold* analysis. As explained in Chapter 14 ("Static timing analysis"), hold analysis verifies whether the data is propagating too fast and reaching the capture flip-flop in the *same* clock cycle, instead of the *next* clock cycle. Thus, we make a hold check at the clock edge *one cycle before* the corresponding setup check edge, as shown in Figure 15.12(a).

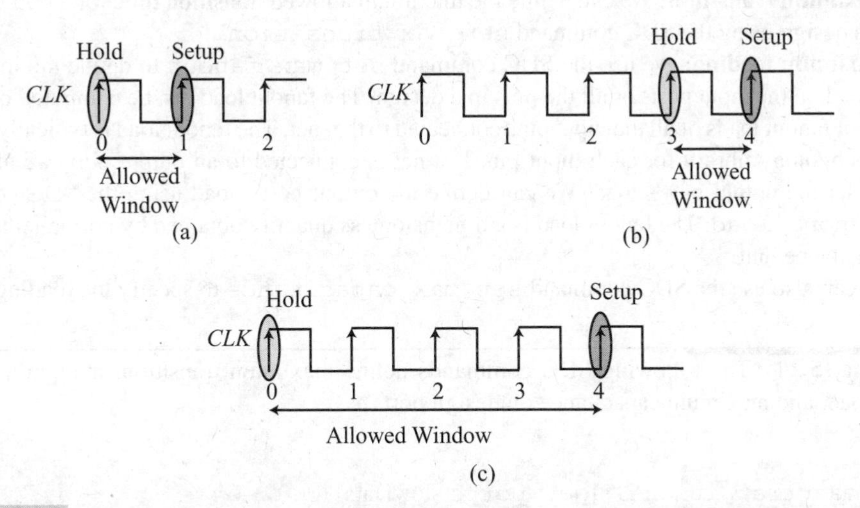

Figure 15.12 Clock edges where setup and hold checks are made and the allowed timing windows for data propagation (a) no multi-cycle specification, (b) multi-cycle with setup multiplier=4 and hold not specified, and (c) multi-cycle with setup multiplier=4 and hold multiplier=3

When we specify a multi-cycle path for setup analysis, an STA tool typically mimics the above behavior by default. It performs hold check *one cycle before* the setup check [5]. Therefore, when we specify a multi-cycle setup check of four clock cycles, the hold check is performed at the third clock cycle ($4 - 1 = 3$). As a result, an STA tool will report a violation when the data reaches the capture flip-flop anytime before the third clock edge. Thus, when we specify multi-cycle of four for the setup check *only*, the allowed time window for data propagation is between the third and the fourth clock edges, as illustrated in Figure 15.12(b). In general, when we specify an N-cycle multi-cycle path, by default, the hold check is performed at the $(N-1)$th clock edge.

However, the above behavior of hold check is too restrictive. Often, we want that the STA tool makes a hold check at the *0*th clock edge (data being launched and captured in the same cycle), as shown in Figure 15.12(c). It allows the time window for data propagation to widen to N clock cycles, where N is the path multiplier for the setup check. The relaxed hold timing constraint will help implementation tools in avoiding the insertion of unnecessary delay elements. To achieve this, we use `set_multicycle_path` with `-hold` option and path multiplier of $(N-1)$, as shown below.

```
set_multicycle_path 3 -hold -from [get_pins FF1/CP] -to [get_pins FF2/D]
```

To summarize, when we specify a path multiplier of N for the setup analysis, the setup check moves to the Nth edge and the hold analysis moves to the $(N-1)$th edge. Additionally, when we specify a path multiplier of M for hold analysis, the hold analysis shifts back to the $(N-1-M)$th edge.

15.8 DESIGN RULES AND OPTIMIZATION CONSTRAINTS

The technology libraries typically specify design rules such as the maximum allowed transition time, fanout, and capacitance for design objects. If we want to make design-specific *more restrictive design rules*, we can use SDC commands.

Maximum transition: We can define the maximum allowed transition time for a port or for all pins in a design using the SDC command `set_max_transition`.

Maximum loading: We use the SDC command `set_max_fanout` to define the maximum fanout load on the input ports or all the pins in a design. The fanout load can be computed by taking the sum of fanout loads of all the input pins connected to that net. The fanout load is typically defined in the technology library for each input pin. If a net is connected to an output port, we also need to consider the output port's load. We can define the output port's load using the SDC command `set_fanout_load`. The fanout load is a dimensionless quantity obtained by normalizing it with the capacitance unit.

We can also use the SDC command `set_max_capacitance` to specify the loading limit.

Example 15.20 The following SDC commands define maximum transition, maximum fanout, fanout load, and maximum capacitance on design ports.

```
set_max_transition 2.5 [get_ports slowData]
set_max_fanout 20 [get_ports control]
set_fanout_load 12 [get_ports pop]
set_max_capacitance 3.5 [get_ports RST]
```

The constraints described above are *design rule constraints*. If these constraints get violated, a circuit can fail. Therefore, implementation tools can give higher priority to design rule constraints over *optmization constraints*. The optimization constraints convey the desirable attributes of a design. Some SDC commands that specify optimization constraints are `set_max_delay`, `set_min_delay`, `set_max_area`, `set_max_leakage_power`, and `set_max_dynamic_power`.

It is worth pointing out that the priority among conflicting optimization constraints is tool-dependent [6, 7]. Hence, readers should refer to it in the tool's documentation.

15.9 EVOLUTION OF CONSTRAINTS IN A DESIGN FLOW

We use SDC files throughout a design flow. The implementation tools use them to compute design-specific attributes and guide their optimization. Various timing and power verification tools use them to know the targets, requirements, and rules. However, as a design evolves during implementation, the SDC files also need to be modified or updated. We need to change or update an SDC file because of the following reasons:

1. **Change in design entities:** SDC commands refer to design objects such as ports, pins, nets, and cells. As a design evolves, design objects are created, modified, and destroyed. For example, after register transfer level (RTL) synthesis, new cells, pins, and nets get created. Moreover,

some RTL objects can become unavailable. Therefore, after RTL synthesis, we may need to modify the SDC file. Nevertheless, as far as practicable, we should refer to invariant design objects such as I/O ports in the SDC commands. It helps avoid making error-prone changes to the SDC files.

2. **Change in design attributes:** Design attributes evolve with design flow. Therefore, the details of the timing analysis technique can change as the design flow progresses. We inform the implementation and verification tools of the changes in design attributes by updating or modifying SDC files.

   For example, after scan chain insertion, a design can have extra ports, timing paths, and analysis *modes*. Typically, we employ separate SDC files for performing STA for different modes.

Example 15.21 Assume that we have two modes in a design after scan chain insertion: *functional mode* and *scan mode*. Assume that we can take a circuit into the functional mode by making the port *SCAN_ENABLE*=0 and into the scan mode using *SCAN_ENABLE*=1. We can assign a constant value to a port or an internal pin in the SDC file by the command `set_case_analysis`.

In the functional mode SDC file, we will add the following command:

```
set_case_analysis 0 [get_ports SCAN_ENABLE]
```

In the scan mode SDC file we will add the following command:

```
set_case_analysis 1 [get_ports SCAN_ENABLE]
```

Another example of changing or updating design attributes in the SDC file is after CTS. The clock distribution network becomes available after CTS. Hence, STA tools can compute clock signal delay using the clock distribution network after CTS. We add the command `set_propagated_clock` in the post-CTS SDC file to specify that the delay calculator should use the clock distribution network to compute the clock signal delays.

15.10 CONSTRAINTS FOR HIERARCHICAL DESIGN FLOW

For large designs, hierarchical design flow is a popular method of implementation and verification. In a hierarchical design flow, we first *partition* a big circuit into smaller sub-circuits. These smaller sub-circuits are called *partitions* or *blocks*. Subsequently, we implement and verify each block individually. Finally, after each block has passed verification individually, we integrate them at the top level and verify that the integrated design works properly.

Block-level constraints: In a hierarchical design flow, we need to create block-specific SDC files to perform STA for each block separately. Given a top-level design and constraints, we can develop block-level constraints. Constraining flip-flop to flip-flop paths within a block is often straightforward. We just need to specify clock constraints at the block level for clocks propagating from the top level. However, defining constraints for the block I/O ports is more challenging because they share signals with the top-level design components and other blocks. Nevertheless, we can observe that the synchronous data transfer from/to a block is similar to the system level synchronous data transfer discussed earlier (from/to *MyComp* in Figure 15.4). Therefore, we can derive block-level I/O constraints as we have discussed in Section 15.5 ("I/O constraints"). It will allocate sufficient margins for the external components in the block-level I/O constraints.

Budgeting: The process of allocating some *fraction of a clock cycle* to different blocks and the top-level design for signals crossing block boundaries is known as *budgeting*. We need to allocate a larger fraction of the clock cycle to a block through which a signal experiences more delay. We illustrate it in the following example.

Example 15.22 Consider the circuit shown in Figure 15.13. The data is launched from *B1/FF1* and it passes through combinational logic *D1→D2→D3→D4*, and is captured by *B3/FF2*. For proper operation, the launched data must reach the capture flip-flop in one clock cycle. To ensure this different blocks *B1*, *B2*, and *B3* are alloted a fraction of the clock period based on the delay of *D1*, *D2*, *D3*, and *D4*.

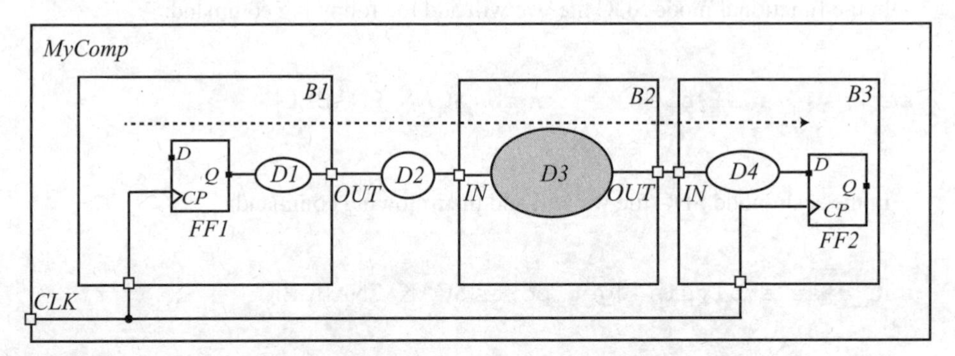

Figure 15.13 Illustration of time budgeting for three blocks *B1*, *B2*, and *B3*

Assume that a majority of computation occurs in the block *B2* for the above path. We can allocate a large portion of the clock period, say 70% to block *B2*, and 10% each to *B1*, *B3*, and the top-level. Note that even when there is no logic element at the top level, we should allocate some fraction of the clock cycle for wires running at top-level between the blocks.

We can use the SDC command `set_input_delay`, `set_output_delay` and clock constraints (real or virtual clock definitions) to model block-level timing constraints or timing budget requirements.

Challenges in budgeting: Determining the timing budget for various blocks is challenging. We need to decide budgets, *before* we have implemented the blocks. The implementation tools employ the given budgets to synthesize interface logic. If the budget is too relaxed for a block, that block can be implemented sub-optimally, while the budget can become too tight for the other blocks incurring an area penalty in them. Consequently, the overall PPA of a design can suffer.

In the worst case, if the budget is too tight, the implementation tool can fail to find a feasible solution for a block. In such cases, we need to revise timing budgets and re-implement various blocks. However, modifying timing budgets can have a cascading effect on multiple blocks. It can disrupt the complete design schedule. Thus, proper budgeting is critical for efficient hierarchical design flows.

15.11 SANITY AND CONSISTENCY OF CONSTRAINTS

We give constraints as input to various EDA tools. Many information contained in the constraints file is impossible to derive automatically. For example, a tool cannot derive attributes of clock signals, environmental requirements, and design rules. Additionally, some information, such as timing exceptions, is typically not verified for correctness by STA tools. Therefore, there is a greater responsibility on a designer to ensure the correctness of constraints. Some thumb rules and commercially available tools can avoid critical constraint-related bugs [4].

Another issue is the consistency of various constraints: both within a single SDC file and across multiple SDC files. For example, the definitions of `set_input_delay/set_output_delay` should be consistent with the corresponding capture/launch clock signals. Moreover, in the hierarchical design flows, the I/O constraints of various blocks should be consistent. For example, the total budgeted delay for a path should not exceed the clock period. At the system level also, the I/O constraints for various components should be consistent. Some tools and automation can help in maintaining the consistency of constraints. Nevertheless, constraints should be developed, updated, and managed carefully in VLSI design flow [4].

15.12 RECENT TRENDS

The methodology of defining design constraints has evolved over the last few decades. Still, fully automatic constraints generation and validation is a challenge [8–12]. The existence of numerous clock domains and their crossings, increasing number of circuit modes, and complex timing exceptions make this task more challenging. Additionally, deriving optimal timing budgets and avoiding iterations in the hierarchical design flows remain critical problems for a designer.

REVIEW QUESTIONS

15.1 Write an SDC command to define a clock named *Clock* at an input port *clk* in a design named *MYDES* with the waveform shown in Figure 15.14.

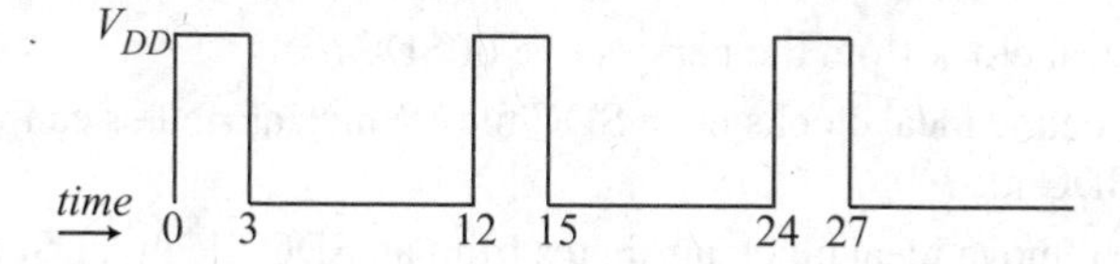

Figure 15.14 Given waveform

15.2 The following commands appear in an SDC file. The unit of time is nanosecond. What is the frequency of the clocks *MCLK* and *GCLK*?

```
create_clock -name MCLK -period 10 -waveform {0 4} [get_pins U1/CLK]
create_generated_clock -name GCLK -divide_by 3 -source [get_pins U1/CLK] [get_pins FF1/Q]
```

15.3 Which SDC commands can you use to model the following attributes of a clock signal:

(a) Delays in the clock network

(b) Skew in the clock network

(c) Jitter in the clock network

15.4 What is the purpose of the following SDC commands:

(a) `set_input_transition`

(b) `set_driving_cell`

(c) `set_load`

(d) `set_case_analysis`

15.5 How will you model the setup requirement of the external flip-flop *FF* for the circuit shown in Figure 15.15. The name of the design is *MYDES*. The frequency of the clock signal generated from the clock source is 100 MHz. Ignore wire delays. Take the setup time of the flip-flop *FF* as 50 ps.

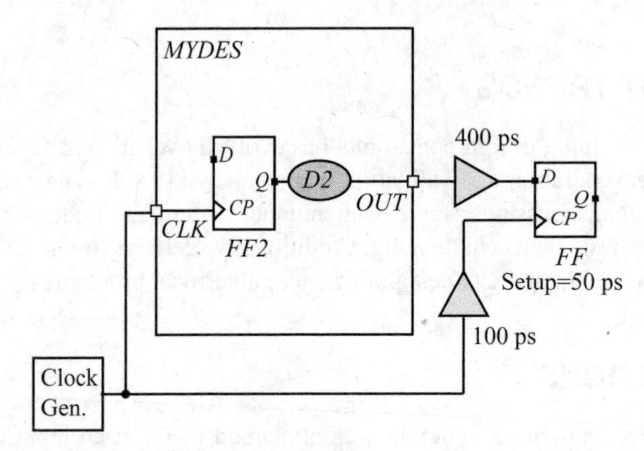

Figure 15.15 A given circuit and its environment

15.6 What is an ideal clock from the perspective of SDC?

15.7 When do we define ideal clocks in an SDC file? Which attributes can we define for an ideal clock in an SDC file?

15.8 When do we remove ideal clock attributes from an SDC file in VLSI design flow?

15.9 What are timing exceptions? How can timing exceptions remove unnecessary violations from the timing reports?

15.10 Why do we need to update an SDC file as the design flow progresses?

TOOL-BASED ACTIVITY

In this activity, you can use any tool that support STA and SDC. You can use any technology library. For example, you can use open-source tools such as OpenSTA [13], and you can also download freely available technology libraries [14].

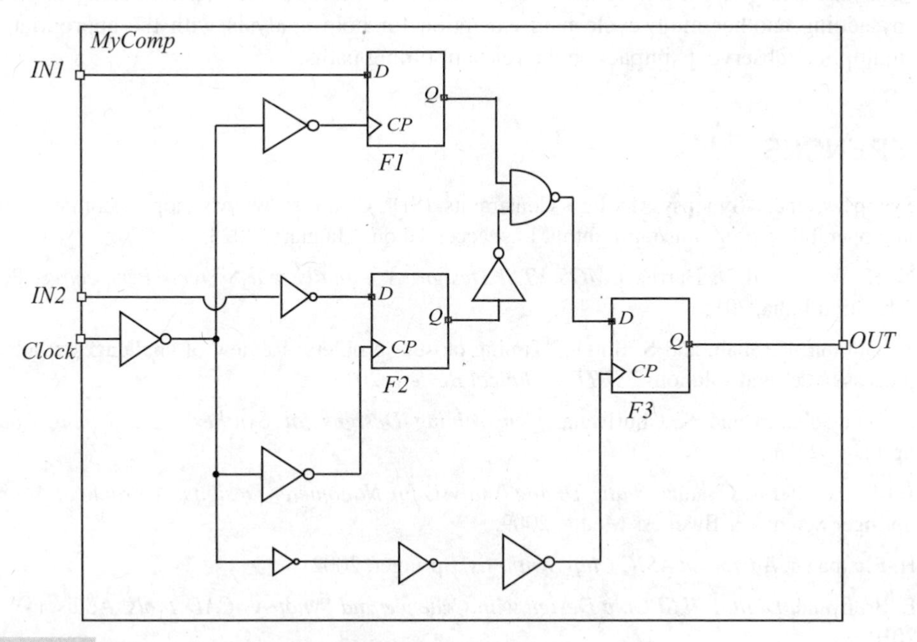

Figure 15.16 Given circuit for the activity

1. Write a Verilog netlist for the circuit shown in Figure 15.16. Choose net names and instance names as you wish.
2. Write an SDC file such that all the flip-flops and the output ports are constrained. Define a clock at the port *Clock*. Choose the clock period that makes the slack positive. Perform STA and report the timing details of the path with the worst setup slack and the worst hold slack.
3. Add clock latency, transition, and uncertainty for the clock signal. Perform STA. Explain the effect of these commands on the timing of the worst path.
4. Double the clock frequency and explain the impact on the setup slack and the hold slack.
5. Make the input constraints tighter for the setup check only and explain the change in setup slack through the input ports.
6. Make the input constraints tighter for the hold check only and explain the change in hold slack through the input ports.
7. Make the output constraints tighter for the setup check only and explain the change in slack through the output port.
8. Make the output constraints tighter for the hold check only and explain the change in slack through the output port.
9. Increase the transition time at the input ports significantly. Explain the effect on the setup and hold slacks of the timing paths through the input ports.

10. Increase the load at the output ports significantly. Explain the effect on the setup and hold slacks of the timing path through the output port.
11. Add multi-cycle path exception for the setup analysis with a path multiplier of 4 on the path showing the worst setup slack. Observe the timing report for both the setup and the hold analysis. Did you get a hold analysis violation? If yes, remove the hold analysis violation by adding another multi-cycle path exception for hold analysis with the appropriate path multiplier. Observe its impact on the relevant timing paths.

REFERENCES

[1] Synopsys, Inc. "Synopsys Design Constraints (SDC)." https://www.synopsys.com/community/interoperability-programs/tap-in.html. Last accessed on 7 January, 2023.

[2] N. H. Weste and D. Harris. *CMOS VLSI Design: A Circuits and Systems Perspective*. Pearson Education India, 2015.

[3] S. Saurabh, H. Shah, and S. Singh. "Timing closure problem: Review of challenges at advanced process nodes and solutions." *IETE Technical Review*, 2018.

[4] S. Gangadharan and S. Churiwala. *Constraining Designs for Synthesis and Timing Analysis*. Springer, 2013.

[5] J. Bhasker and R. Chadha. *Static Timing Analysis for Nanometer Designs: A Practical Approach*. Springer Science & Business Media, 2009.

[6] H. Bhatnagar. *Advanced ASIC Chip Synthesis*. Springer, 2002.

[7] E. Brunvand. *Digital VLSI Chip Design with Cadence and Synopsys CAD Tools*. Addison-Wesley, 2010.

[8] O. Coudert. "An efficient algorithm to verify generalized false paths." *Design Automation Conference*, pp. 188–193, IEEE, 2010.

[9] H. Bhatnagar and P. Petrov. "Method to automatically generate and promote timing constraints in a Synopsys Design Constraint format." August 27, 2019. US Patent 10,394,983.

[10] Y. Xia, D. R. Amirtharaj, A. Vahidsafa, A. Smith, S. Diraviam, and M. J. Mohd. "Unified tool for automatic design constraints generation and verification." May 31, 2016. US Patent 9,355,211.

[11] R. K. M. Sadhu, G. Maheshwari, and S. U. Bhirud. "Methods and systems for timing constraint generation in IP/SoC design." August 16, 2018. US Patent App. 15/475,376.

[12] A. Sequeira, S. Sripada, and S. N. M. Palla. "Simplifying modes of an electronic circuit by reducing constraints." November 8, 2016. US Patent 9,489,478.

[13] T. Ajayi, V. A. Chhabria, M. Fogaça, S. Hashemi, A. Hosny, A. B. Kahng, M. Kim, J. Lee, U. Mallappa, M. Neseem, et al. "Toward an open-source digital flow: First learnings from the OpenROAD project." *Proceedings of the 56th Annual Design Automation Conference 2019* (2019), pp. 1–4.

[14] M. Martins, J. M. Matos, R. P. Ribas, A. Reis, G. Schlinker, L. Rech, and J. Michelsen. "Open cell library in 15nm FreePDK technology." *Proceedings of the 2015 Symposium on International Symposium on Physical Design* (2015), pp. 171–178.

16 Technology Mapping

... a symbolical system of position, in which the figures ... expressed by ... objects suggesting the particular numbers in question ... Thus, for 1 were used the words moon, Brahma, Creator, or form ... such notations made it possible to represent a number in several different ways. This greatly facilitated the framing of verses containing arithmetical rules or scientific constants, which could thus be more easily remembered.

—Florian Cajori, *A History of Mathematics*, Macmillan & Company, 1893

As described in Chapter 4 ("RTL to GDS implementation flow"), we divide logic synthesis into two distinct phases: *technology-independent phase* and *technology-dependent phase*. The first task in technology-dependent phase is to *map* a netlist of generic gates to the cells of the given set of *technology libraries*. This task is known as *technology mapping*, or *mapping* in short. The tool or the software that carries out technology mapping is referred to as *technology mapper*, or *mapper*. Post-mapping we perform further *technology-dependent* logic optimizations [1, 2].

The attributes of timing, area, and power for cells are defined in a technology library. Therefore, mapping brings a significant transformation to a design by describing the timing, area, and power characteristics of each *instance* in a design. Consequently, the final quality of result (QoR) of a design depends heavily on the mapping. Moreover, post-mapping, the level of abstraction in a design decreases, and making further changes to the *instances* of a design becomes difficult. Thus, mapping is one of the most critical tasks of a design flow. We will discuss various aspects of mapping in this chapter.

16.1 BASICS OF MAPPING

The application framework of a technology mapper is shown in Figure 16.1. Let us examine its various components.

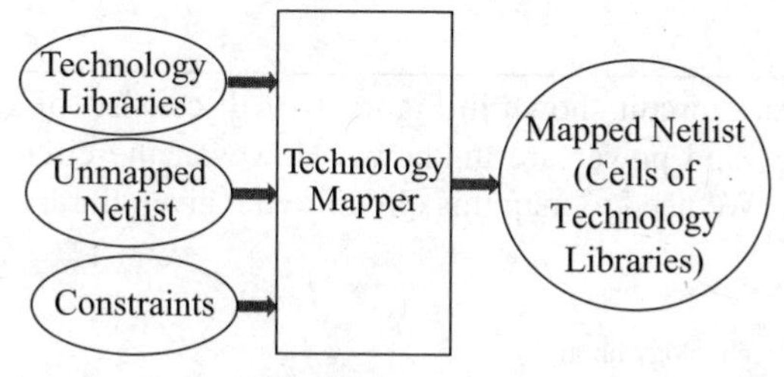

Figure 16.1 Inputs and outputs for a mapper

Inputs: A mapper takes a set of technology libraries, a given design, and constraints as inputs.

1. **Technology libraries:** A technology library consists of cells with different functions. For example, a library can contain cells implementing logic functions such as AND, OR, NOT, multiplexer, flip-flops, and latches. These cells can have a different number of inputs. For example, there can be cells implementing two-input AND gate, three-input AND gate, four-input AND gate, etc. A mapper can use a combination of smaller gates or an equivalent big gate during mapping.

   A technology library can contain multiple cells with the same function but different sizes and drive strengths.[1] For example, a library can contain cells named *AND2X1*, *AND2X2*, and *AND2X4*. All these cells can be performing the same logic function, i.e., two-input AND gate. However, the implementation of these cells can contain transistors of different sizes. As a result, though the functionality of these cells is the same, the attributes such as area, timing, and power can be different. For example, the area of *AND2X2* can be 2× greater than *AND2X1* and the area of *AND2X4* can be 4× greater than *AND2X1*. On the other hand, *AND2X2* can be faster than *AND2X1* and *AND2X4* can be faster than *AND2X2*. A mapper can instantiate any of these functionally equivalent cells during mapping to achieve a given design objective.

   We can give multiple libraries as an input to a mapper. For example, we can have two libraries: a high-performance (HP) library having cells with higher speed and higher power dissipation and a low-power (LP) library having cells with lower speed and lower power dissipation. We can give both these libraries to a mapper. The mapper can pick cells from either library for optimization depending on the speed–power trade-offs.
2. **Design:** The input to a mapper is a netlist of generic gates. It consists of primary ports, generic gate instances, and nets connecting them. We refer to this netlist as the *unmapped netlist*.
3. **Constraints:** Based on our design intent, we can specify various objectives and constraints to a mapper. A few examples are:
   (a) Minimize area under a given delay constraint
   (b) Minimize delay under a given area constraint
   (c) Minimize power under a given delay constraint

   We can provide timing constraints to the mapper using a Synopsys Design Constraint (SDC) file.[2] We indicate our optimization priorities to the mapper, typically using tool-specific mechanisms. For example, a mapper can provide some tool option to determine whether it should minimize area, delay, or power.

Output: Given all the above inputs, a mapper produces a *mapped netlist* consisting of library cell instances. The unmapped netlist and the mapped netlist must be *functionally equivalent*. A mapper can produce different mapped netlists depending on the specified objective and design constraints. We illustrate it in the following example.

Example 16.1 Consider a circuit shown in Figure 16.2. It consists of generic gates *G1* and *G2*. Therefore, its area, delay, and power are unknown. However, the circuit's functionality is well-defined $Y = ((A.B) + C)'$. We need to map this circuit to the given library.

[1] For details, see Chapter 13 ("Technology library").

[2] For details, see Chapter 15 ("Constraints").

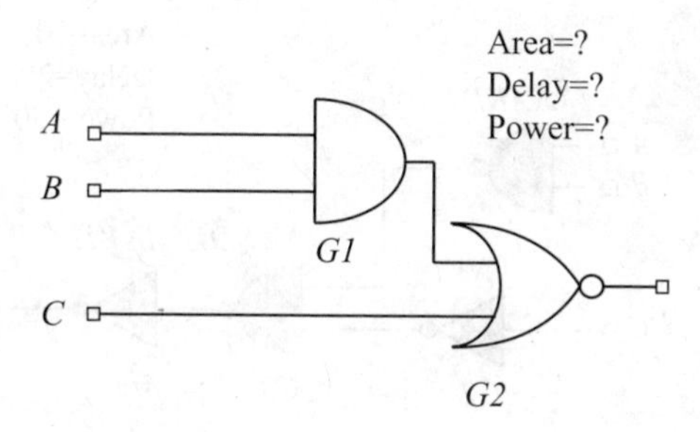

Figure 16.2 Given netlist for mapping

For simplicity, let us assume that the given library contains only four cells: *INV1*, *NAND1*, *NAND1_LP*, and *AND1*, as shown in Figure 16.3.

Cell Name	Symbol	Function	Area	Delay	Power
INV1	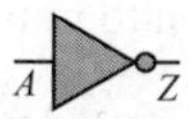	*Z=A'*	1	4	5
NAND1	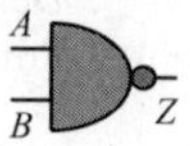	*Z=(A.B)'*	4	8	20
NAND1_LP	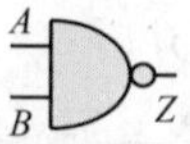	*Z=(A.B)'*	5	12	6
AND1	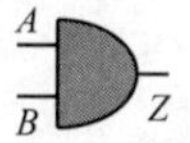	*Z=A.B*	8	9	30

Figure 16.3 Given library cells for mapping

The attributes of these cells are also listed in the figure. Note that we have two cells for the NAND logic: *NAND1* and *NAND1_LP*. The cell *NAND1* has smaller area and delay compared to *NAND1_LP*, while *NAND1_LP* dissipates less power. Therefore, a mapper can choose either of these NAND cells based on the optimization criteria. Normally, technology libraries contain multiple cells with the same function and different electrical characteristics.

Let us examine the solutions a mapper will produce under the following optimization criteria (1) minimum area, (2) minimum delay, and (3) minimum power.

1. **Minimum area:** We can implement the function $Y = ((A.B) + C)'$ as shown in Figure 16.4. It uses only NAND gates and inverters. Using DeMorgan's theorem, we can verify that the implementation $Y = (((A.B)'.C')')'$ is equivalent to the given function $Y = ((A.B) + C)'$. To minimize area, we have used smaller cells of the library. The total area is 10 units. The longest path delay in the circuit is 20 units corresponding to the path $G1 \rightarrow G3 \rightarrow G4 \rightarrow Y$. The total power dissipation is 50 units.

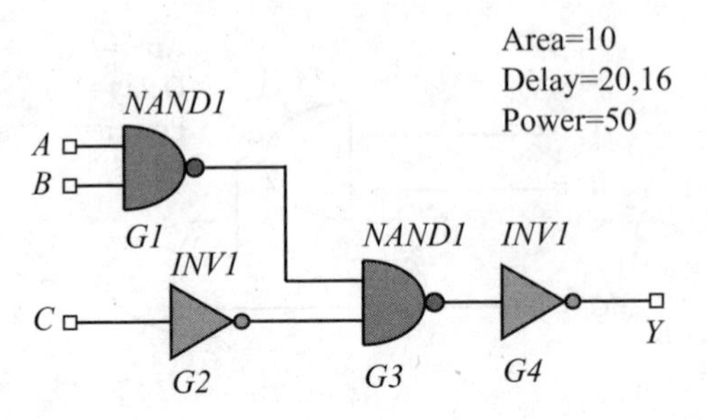

Figure 16.4 Mapped circuit for the minimum area

2. **Minimum delay:** We can observe from the library that the *AND1* cell delay is smaller than the combined delay of *NAND1* and *INV1*. Therefore, we can replace the combination of *NAND1* and *INV1* in Figure 16.4 with *AND1* cell to reduce delay. Thus, we obtain the circuit shown in Figure 16.5. The maximum delay of the circuit reduces from 20 to 17 units. However, the area increases from 10 to 13 units, and the power dissipation increases from 50 to 55 units. Thus, we have traded off the area and the power dissipation to minimize the maximum delay.

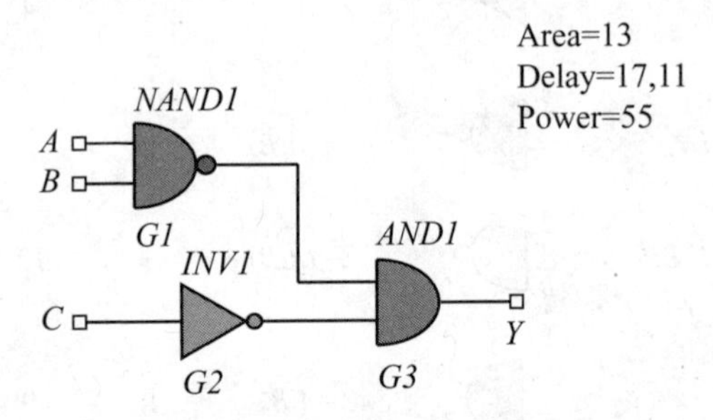

Figure 16.5 Mapped circuit for the minimum delay

3. **Minimum power:** The cell *NAND1_LP* dissipates less power than *NAND1* and *AND1* cells. Hence, a mapper can produce the circuit shown in Figure 16.6 to minimize power dissipation. The power dissipation reduces to 22 units. However, the area increases to 12 units, and the maximum delay increases to 28 units. Thus, we have traded off the area and the delay to minimize the power dissipation in the circuit.

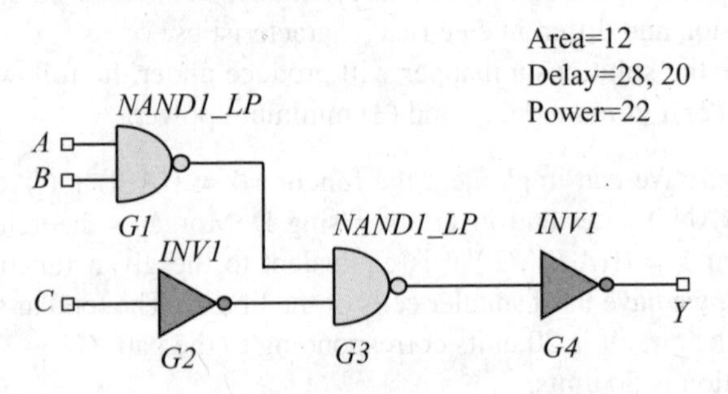

Figure 16.6 Mapped circuit for the minimum power

The above example illustrates that a mapper can choose a solution out of many possibilities based on the optimization criteria. For practical designs consisting of millions of logic gates, the solution space becomes enormous. Hence, we need to devise efficient mapping techniques, which we describe in the following section.

16.2 MAPPING TECHNIQUES

A generic netlist consists of combinational and sequential circuit elements. The mapping of generic sequential elements such as flip-flops and latches to the corresponding library cells is typically straightforward [3]. The most challenging problem for a mapper is to map generic combinational circuit elements. Therefore, we will discuss mapping combinational logic circuits in the rest of this chapter.

Boolean logic network graph: We can model a combinational circuit as a *Boolean logic network.*[3] We usually represent a Boolean logic network as a directed acyclic graph $G = (V, E)$, where V is the vertex set, and E is the edge set. A vertex $v \in V$ has a one-to-one correspondence with the given circuit's input ports, output ports, and logic gates. We annotate each vertex with the local Boolean function associated with the corresponding gate. We create an edge $(u, v) \in E$ if the gate or the output port corresponding to v receives an input from the gate or the input port corresponding to u. Thus, edges model connections or nets in the given combinational circuit. We can decompose multiterminal nets into two-terminal nets and then derive edges for the Boolean logic network.

If a combinational circuit has only one output port, its logic network will contain only one zero-fanout vertex (corresponding to that output port). We designate this vertex as the *root* of the network.

For a given Boolean logic network $G = (V, E)$, we can obtain a subnetwork by taking a subset of vertices and edges. If the subnetwork has only one vertex v with a zero fanout, we refer to that subnetwork as the *subnetwork rooted at v*. We illustrate this concept in the following examples.

Example 16.2 Consider the circuit shown in Figure 16.7.

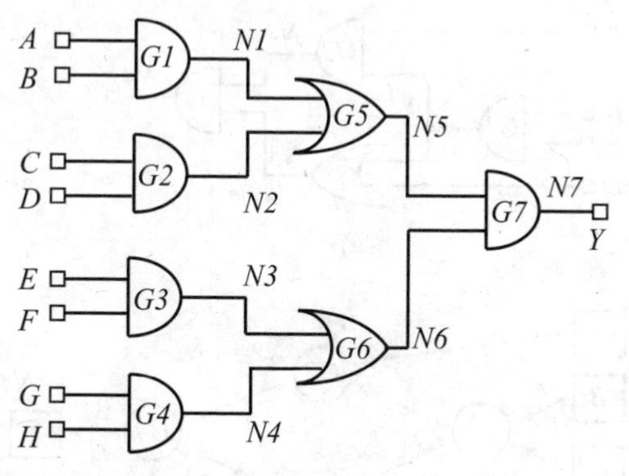

Figure 16.7 Given combinational circuit

[3] For details, see Chapter 12 ("Logic optimization").

We can represent it as a Boolean logic network as shown in Figure 16.8.

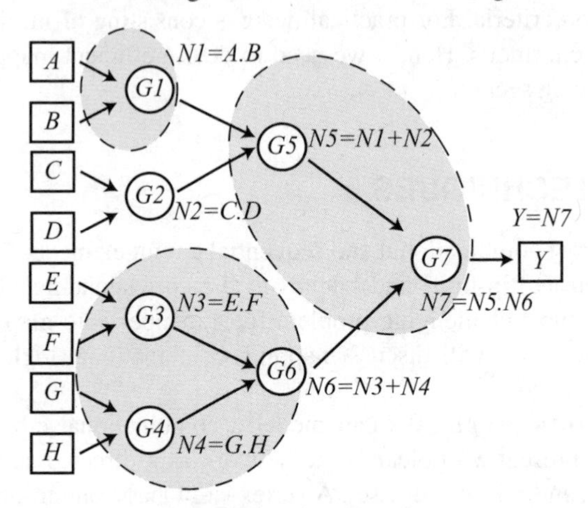

Figure 16.8 Boolean logic network for the circuit in Figure 16.7

The vertices *V={G1, G2, ..., G7}* correspond to the gates in the circuit. The edges are formed by examining the nets. For example, there is an edge *(G1,G5)* because *G1* drives *G5* through the net *N1*. Similarly, we infer other edges of the graph. We have annotated the local functions on the vertices. Since there is only one vertex with zero fanout (*Y*), we say that the logic network is rooted at *Y*. We have highlighted three subnetworks rooted at *G1*, *G6*, and *G7*. Note that there can be many such subnetworks.

We take another example of the circuit shown in Figure 16.9(a). It contains a three-terminal net *N1*. We create two edges (*(G1, G2)*, *(G1, G3)*) for this net by decomposing into two two-terminal nets, as shown in Figure 16.9(b). We have highlighted three subnetworks, all rooted at *G4*. It illustrates that we can obtain multiple subnetworks rooted at the same vertex by taking different subsets of vertices and edges.

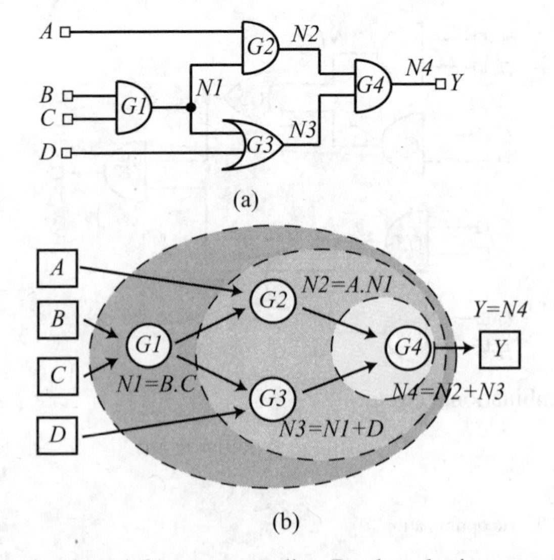

Figure 16.9 (a) Given circuit and (b) corresponding Boolean logic network

Matching set and cover: Consider a vertex v of a logic network graph. We say that a library cell matches a subnetwork rooted at v if the cell and the subnetwork are *functionally* equivalent. By functional equivalence, we mean that both the cell and the subnetwork implement the same Boolean function for some one-to-one correspondence of their inputs. As explained earlier, there can be multiple rooted subgraphs at v. Therefore, we can have a set of library cells that matches subgraphs rooted at v. We denote that set as M_v. When matching occurs, a cell in M_v is said to *cover* the vertex v and other vertices in the subgraph rooted at v [2]. We illustrate these concepts in the following example.

Example 16.3 Consider that a library consists of the cells *C1*, *C2*, *C3*, and *C4*, as shown in Figure 16.10.

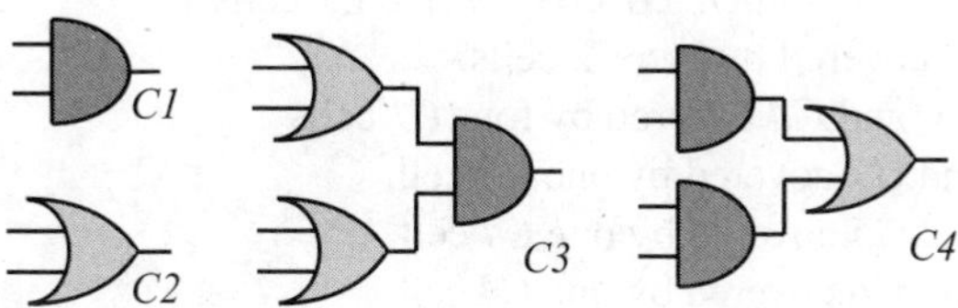

Figure 16.10 An example library

For the logic network in Figure 16.8, we can identify the following matches:
$M_{G1} = \{C1\}$
$M_{G2} = \{C1\}$
$M_{G3} = \{C1\}$
$M_{G4} = \{C1\}$
$M_{G5} = \{C2, C4\}$
$M_{G6} = \{C2, C4\}$
$M_{G7} = \{C1, C3\}$
These matches are shown in Figure 16.11. The corresponding subnetworks are also highlighted.

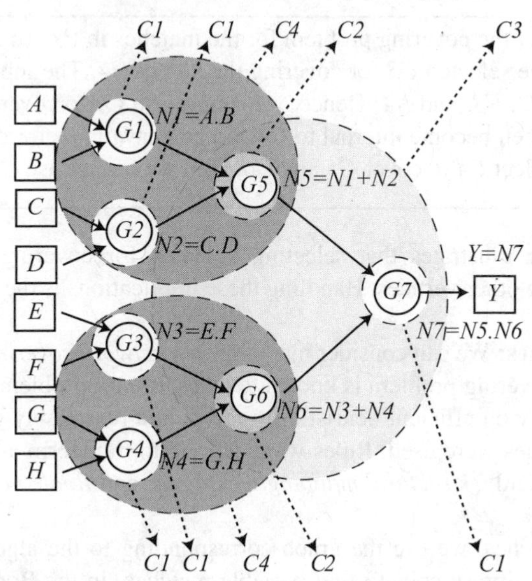

Figure 16.11 Matches for the logic subnetworks at each vertex

Matching problem: The task of finding possible matches M_v for each vertex in a graph is known as the *matching* problem [3]. The set of all possible matches M for a graph G can be obtained by taking the union of matches for all vertices $v \in V$, such that $M = \cup M_v$. In the above example $M = \{C1, C2, C3, C4\}$.

Covering problem: A mapper finds enough matches in M such that *all* vertices $v \in V$ get covered according to a given optimization criteria. This task is known as the *covering* problem [3].

Example 16.4

Consider the matches in Ex. 16.3 (Figure 16.11). The possible covers are:

Cover 1: *G1*, *G2*, *G3*, *G4*, and *G7* covered by five *C1* cells
G5 and *G6* covered by two *C2* cells.
Cover 2: *G1*, *G2*, *G3*, and *G4* covered by four *C1* cells
G5, *G6*, and *G7* covered by one *C3* cell.
Cover 3: *G1*, *G2*, and *G5* covered by one *C4* cell
G3, *G4*, and *G6* covered by one *C4* cell
G7 covered by one *C1* cell.

A mapper will produce one of the covers as the final solution based on the attributes of cells in the library and the given optimization criteria.

A sufficient condition for mapping is that each vertex v has at least one match, i.e., $|M_v| > 0$, $\forall v \in V$.

Connectivity constraints: For a given vertex v the selection of a particular match from the set M_v precludes some matches in the adjacent vertices. It happens because the inputs of the selected match need to be connected to the outputs of the adjacent matches. We explain it in the following example.

Example 16.5 Consider the covering problem for the matches in Ex. 16.3 (Figure 16.11).

Assume that we have selected *C3* for covering the vertex *G7*. The input pins of *C3* need to be driven by the nets *N1*, *N2*, *N3*, and *N4*. Hence, we cannot use *C4* to cover the vertices *G5* and *G6* because these nets will then become internal to *C4* and be unavailable for driving other cells.

Conversely, if we select *C4* to cover *G5* and *G6*, then we cannot use *C3* to cover *G7*.

The above example illustrates that selecting a match for covering a vertex can result in *implications* for covering other vertices. Handling these implications is the primary difficulty of the covering problem [2].

Mapping approaches: We can consider mapping as a *network covering problem*, as explained above. Since network covering problem is known to be a difficult problem, rather than looking for an exact solution, we rely on efficient heuristics to arrive at a reasonably good solution [2]. In the past, rule-based techniques were used. Rules were defined to transform a sub-circuit to a network of library cells [4]. Currently, *structural mapping techniques* and *Boolean mapping techniques* are popular.

In structural approaches, we use the graph corresponding to the algebraic decomposition of the given circuit and the library cells to find possible matches. In the Boolean method, we check the Boolean functional equivalence of the local functions for the given circuit with the functions implemented by the library cells. We discuss these techniques in the following sections.

16.3 STRUCTURAL MAPPING

In structural mapping, we represent the given circuit as a graph, known as the *subject graph*. We can use the Boolean logic network discussed earlier for the graph representation. We also represent each cell of the library as a graph in the same format or the *algebraic decomposition*. These graphs of library cells are known as the *pattern graphs*. Next, we find the *matches* of the pattern graphs in the subject graph. We say that a pattern graph matches a subject subgraph if they are *isomorphic*.[4] When the graphs of two functions represented in the same format are isomorphic, we can say that these functions are functionally equivalent. Therefore, in general, we can view the structural mapping problem as a *subgraph isomorphism problem* [3, 5].

The subgraph isomorphism problem is known to be difficult or *intractable*. Hence, we employ some heuristics in structural mapping. In this section, we explain a rudimentary structural mapping technique [6]. The details of the structural mapping algorithm can vary in different implementations.

16.3.1 Decomposition and Partitioning

In structural mapping, we simplify the task of the covering problem encountered later in the algorithm using *decomposition* and *partitioning*.

Decomposition: We first decompose the given circuit and the library cells in terms of some *base functions*. The base functions should be *functionally complete*, i.e., they should be able to represent any Boolean function. Additionally, the library should contain cells that implement the base functions to ensure that the mapping solution exists for any chosen cover. Some of the popular base functions are: (two-input NAND gate+inverter), (two-input NOR gate+inverter), and (AND/OR gate+inverter) [1, 7]. A finer granularity of the base functions allows more flexibility in the cover.

To decompose a given unmapped netlist, we follow two steps. First, we decompose the generic combinational gates, such as multi-input AND gates, OR gates, multiplexers, and demultiplexers, in terms of base functions. Next, we directly replace generic gates in the unmapped netlist with their decompositions found above. In general, the decomposition of the unmapped netlist will contain more gates than the original netlist due to the restriction of using only the base functions.

Example 16.6 Consider an unmapped circuit shown in Figure 16.12.

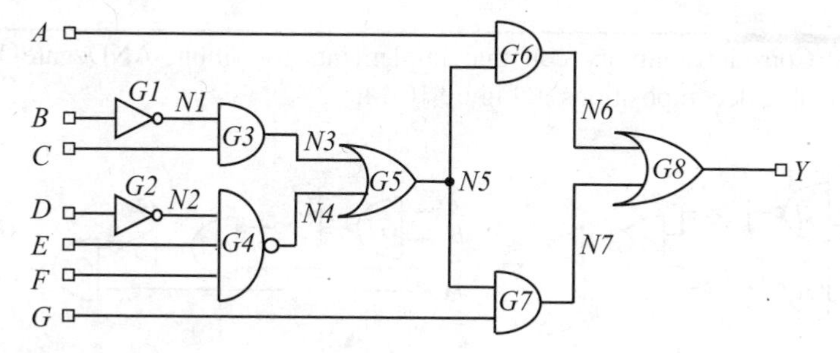

Figure 16.12 A given unmapped circuit

[4] For isomorphism, see Chapter 11 ("Formal verification").

Let us decompose it in terms of two-input NAND gates and inverters. There are four types of generic gates: inverter, two-input AND gate, two-input OR gate, and three-input NAND gate. We decompose them in terms of two-input NAND gates and inverter. Then, we replace the decomposed generic gates directly in the unmapped circuit. Thus, we obtain the circuit shown in Figure 16.13.

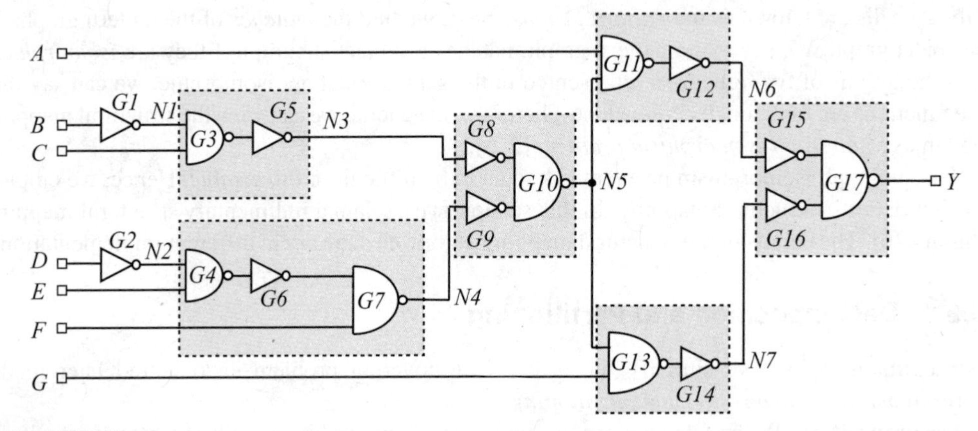

Figure 16.13 The decomposition of the circuit shown in Figure 16.12 in terms of two-input NAND gates and inverters

Remark: There is some redundancy in the decomposed circuit. For example, a series of two inverters is redundant and can be replaced by a wire (*G5-G8*, *G12-G15*, and *G14-G16*). However, we can choose to retain such inverter-induced redundancies in the decomposed circuit. Sometimes, they help in discovering covers that cannot be inferred without them [1].

We also decompose library cells using the same base functions that we used in decomposing the unmapped netlist. For some cells, the decomposition will not be unique. In those cases, we try to obtain all possible decompositions of the cells and store them. While finding possible matches for a vertex in the subject graph, we check all the decompositions of the pattern graph.

Example 16.7 Consider a library cell that implements four-input AND gate $Y = ABCD$. We illustrate two possible decompositions in Figure 16.14.

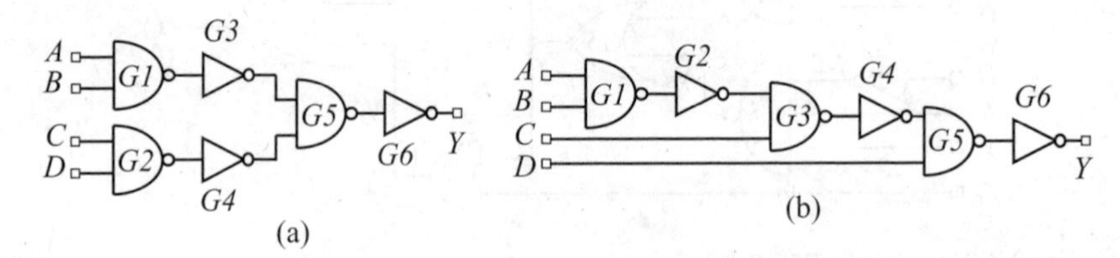

Figure 16.14 Two different decompositions of the function $Y = ABCD$

If the base function is highly granular, it can result in more decompositions and increase the matching effort. It is worthy to note that enumerating all possible structural representations for some complex cells may not be feasible. Hence, we store a subset of decompositions that are more likely to occur in the subject graph for those cells. However, this approach can lead to some QoR loss.

Partitioning: We further simplify the structural matching problem by *partitioning* the decomposed circuit. One of the most straightforward partitioning schemes is to partition the circuit at the multiple-fanout nets (i.e., nets having fanout more than one) [6]. We treat the multiple-fanout net as the primary output of one sub-circuit and the primary input of the next sub-circuit(s). Thus, within each sub-circuit, there is no multiple-fanout net. It results in transforming a big subject graph into many small *subject trees* rooted at the multiple-fanout net. We illustrate this partitioning scheme in the following example.

Example 16.8 Consider the circuit shown in Figure 16.13 (Ex. 16.6). We partition it at the multiple-fanout net *N5* (fanout is two). As a result, we obtain two sub-circuits, as shown in Figure 16.15.

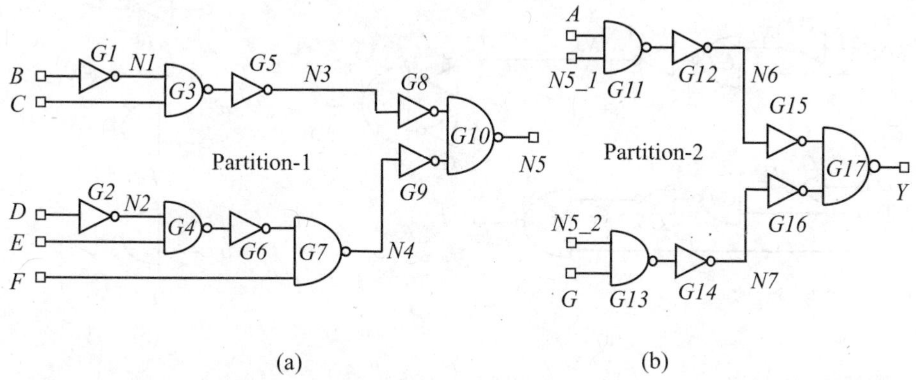

Figure 16.15 The partitioning of the circuit shown in Figure 16.13

We can represent these sub-circuits as trees. We refer to these trees as subject trees. One subject tree is rooted at *N5*, and the other is rooted at *Y*.

We can represent the library cells also in the form of rooted trees, called *pattern trees*. However, some cells, such as XOR and XNOR, do not have a direct tree representation. We can still derive a tree-like representation for them by splitting at the multiple-fanout input variable and using some special notation for those variables [2].

Thus, we can transform the subject and pattern *graphs* to subject and pattern *trees* by decomposing and partitioning. We cover each subject tree separately by a set of pattern trees, minimizing costs such as area. Finally, we combine these solutions to obtain the overall mapping for the circuit. Note that the above transformations simplify the difficult *subgraph isomorphism* and *graph covering* problems to *tree matching* and *tree covering* problems. Next, we discuss how to solve tree matching and tree covering problems.

16.3.2 Tree Matching

Given a vertex v of a subject tree and a root of a pattern tree P, we can determine whether the rooted trees at v and P correctly match using several efficient algorithms. If we decompose them in terms of two-input NAND gates and inverters, the *comparison* of two vertices during matching involves just comparing the fanin count of the corresponding vertices. It is possible because the fanin count of two-input NAND gate, inverter, and leaf vertices is two, one, and zero, respectively, and are distinct. Therefore, we can determine the match just by comparing the root of the pattern tree P with the vertex v of the subject tree and visiting their children recursively. If all vertices of P match the corresponding vertices of the subject tree rooted at v, then P is said to be a valid match of the rooted tree at v. We illustrate this method in the following example.

Example 16.9 Consider the circuits shown in Figure 16.16. The subject tree is same as in Figure 16.15(a).

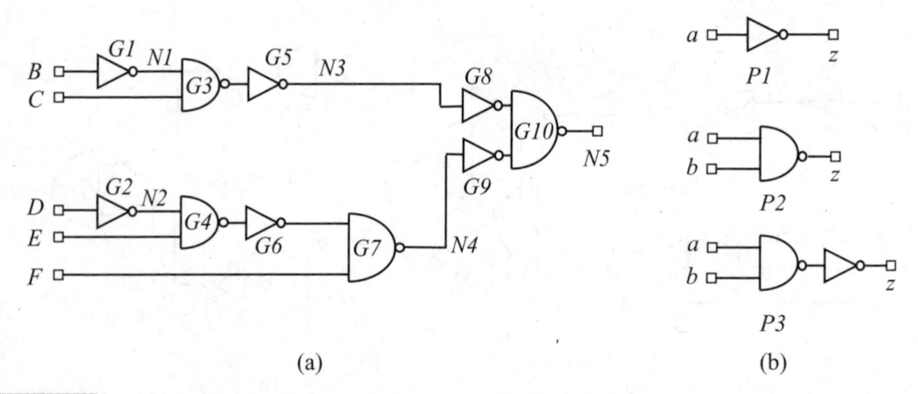

Figure 16.16 (a) A given subject tree and (b) pattern trees

By viewing ports and instances as vertices (nodes) of a tree and their connections as edges (branches), we can easily infer the subject and pattern trees from respective circuits. These circuits have been obtained by decomposing them in terms of inverters and two-input NAND gates. Since an inverter (*P1*) and a two-input NAND gate (*P2*) are also the members of the pattern tree set, each vertex in the subject tree is guaranteed to match either to *P1* or *P2*.

Let us examine how matching of some of the vertices of the subject tree is done.

1. *G1*: It has a fanin of one. It matches *P1* and does not match *P2*. It matches *P3* at the first level. However, it does not match the next level (vertex of the pattern tree has a fanin of two, while the subject tree terminates). Thus, *G1* matches with only *P1*.
2. *G3*: It has a fanin of two. Hence, it does not match *P1* and *P3* (since they have fanin of 1). It matches with *P2* only.
3. *G5*: It has a fanin of one. It matches *P1* and does not match *P2*. It matches *P3* at the first level. Then, it compares the next level and matches (both have fanin of two). Thus, *G5* matches with *P1* and *P3*.
4. *G8*: It has a fanin of one. It matches *P1* and does not match *P2*. It matches *P3* at the first level. However, it does not match in the next level (vertex of the subject tree has fanin of one, while the pattern tree has fanin of two). Thus, it matches with *P1* only.

Similarly, we can find matches for the other vertices. The summary is:
G1, *G2*, and *G8* match with *P1* only.
G3, *G4*, *G7*, and *G10* match with *P2* only.
G5, *G6*, and *G9* match with both *P1* and *P3*.

16.3.3 Tree Covering

We determine *minimum cost cover* for a subject tree by selecting a subset of matches found at each vertex. We can solve this problem efficiently using *dynamic programming* if the objective is to minimize area or minimize the delay (assuming constant delay model for gates).

In the following discussion, we assume that the objective of mapping is to *minimize area*. However, we can take a similar approach for other optimization criteria [8–13]. For example, if we require to minimize area under given *delay constraint*, we can first find a minimum area solution (ignoring delay constraints). Then, we can iteratively remap the *critical paths* with faster cells till the delay constraints are met. An alternative method could be considering a *area–delay trade-off curves* during selection of matches [11].

We can fetch the area of each library cell from the technology library. Hence, we can associate the area with each pattern tree. Assume that a set of *P* pattern trees matches a vertex *v* in the subject tree. Among them, let us assume that we have selected the pattern tree *M* for covering *v*. We can compute the area associated with the match *M* at vertex *v* as follows:

$$Area(v) = Area(M) + \sum_{u} Area(u) \tag{16.1}$$

where *Area*(*M*) is the area associated with the pattern tree *M* and *u* is a vertex in the subject tree that would form leaf of *M* when *M* is replaced in the subject tree at *v*. Additionally, we consider the area of the vertices associated with the input ports as zero. We illustrate this computation in the following example.

Example 16.10 Consider the subject tree and pattern trees in Figure 16.17 (same as Ex. 16.9). Assume that the area associated with the pattern trees *P1*, *P2*, and *P3* are 2, 4, and 5, respectively.

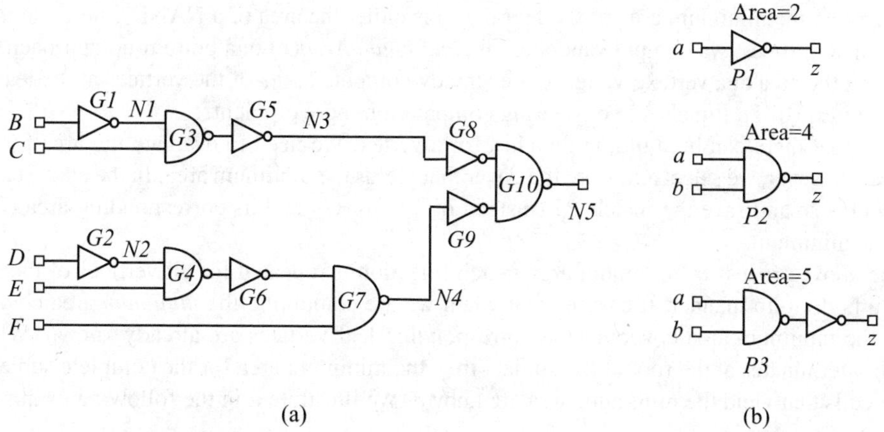

Figure 16.17 (a) A given subject tree and (b) pattern trees

Let us compute the area for some of the vertices for a given cover.

1. *G1*: It has *P1* as the only match. When *P1* covers *G1*, the leaf vertex is the input port *B*. We consider the area of the input port as zero. Hence, the area for this cover is:

$$Area(G1) = Area(P1) + Area(B)$$
$$= 2 + 0 = 2$$

2. *G3*: It has *P2* as the only match. When *P2* covers *G3*, the leaf vertices are *G1* and the input port *C*. Hence, the area for this cover is:

$$Area(G3) = Area(P2) + Area(G1) + Area(C)$$
$$= 4 + 2 + 0 = 6$$

3. *G5*: It has *P1* and *P3* as the matches. First, let us cover it using *P1*. When *P1* covers *G3*, the leaf vertex is *G3*. Hence, the area is:

$$Area(G5) = Area(P1) + Area(G3)$$
$$= 2 + 6 = 8$$

   Next, let us cover it using *P3*. When *P3* covers *G5*, the leaf vertices are *G1* and the input port *C*. Hence, the area is:

$$Area(G5) = Area(P3) + Area(G1) + Area(C)$$
$$= 5 + 2 + 0 = 7$$

   Thus, we obtain different areas for different covers.

Using the above definition of $Area(v)$, we can devise an efficient dynamic programming-based algorithm to obtain the minimum area cover. The algorithm runs in a bottom-up approach (from primary input toward the root of the tree). Since a leaf vertex can be either a two-input NAND gate or an inverter, the minimum area of the leaf nodes is either the area of a NAND gate or an inverter. For all other vertices, we compute and store the *minimum* $Area(v)$ in a bottom-up approach. While computing the area at a vertex, we reuse the already computed area of the vertices at the leaf of the cover (u in Eq. 16.1). It makes tree covering computation more efficient.

Note that there can be multiple matches for a vertex. We need to evaluate the area for all the matches. However, we select and store the cover that yields the minimum area. In the above example, we will choose and save the match *P3* (instead of *P1*) for *G5* and its corresponding area (7 units) since it is minimum.

The above process of minimum area match selection is repeated for all vertices of the subject tree in a bottom-up manner. It ensures that when we are computing the *minimum* area cover for a vertex, the minimum area covers of the corresponding leaf vertices are already known. When the algorithm terminates at the root of the subject tree, the minimum area for the complete subject tree, and the cells that yield the minimum area are known. We illustrate it in the following example.

Example 16.11 Assume that a library contains following cells: *NOT* (inverter), *NAND2* (two-input NAND gate), *AND2* (two-input AND gate), and *OR2* (two-input OR gate). We decompose these cells in terms of inverters and two-input NAND gates, as shown in Figure 16.18. Note that the mapper internally keeps a pair of inverters *P5* with zero areas to model a wire or direct connection. Whenever the mapper finds that using a direct connection reduces area, it uses *P5*.

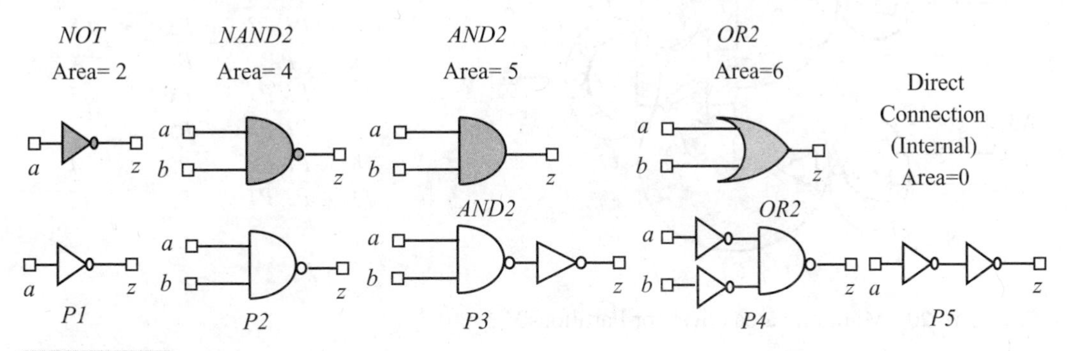

Figure 16.18 Pattern trees

Next, we use these pattern trees to match and cover each partition of Figure 16.15 (Ex. 16.8).

Figure 16.19 shows minimum area cover for each vertex for Partition-1. The table alongside lists out all valid matches and the selected match that results in the minimum area. The minimum area for this partition is 22 units.

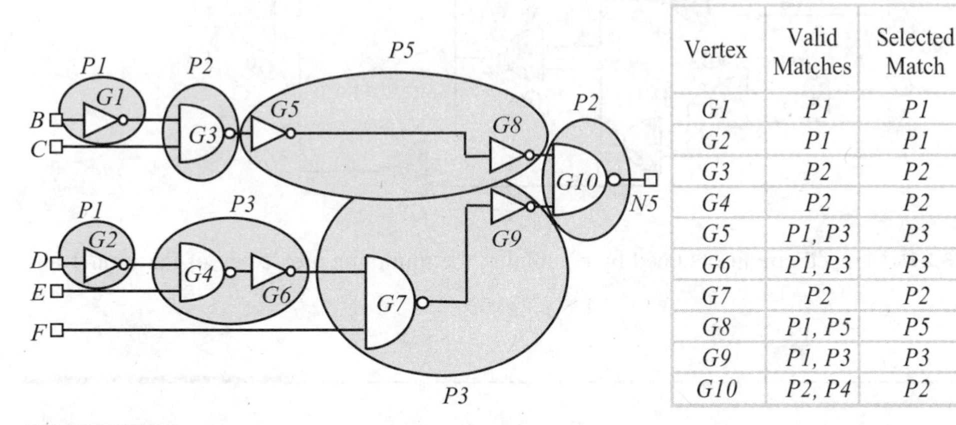

Vertex	Valid Matches	Selected Match	Area
G1	*P1*	*P1*	2
G2	*P1*	*P1*	2
G3	*P2*	*P2*	6
G4	*P2*	*P2*	6
G5	*P1, P3*	*P3*	7
G6	*P1, P3*	*P3*	7
G7	*P2*	*P2*	11
G8	*P1, P5*	*P5*	6
G9	*P1, P3*	*P3*	12
G10	*P2, P4*	*P2*	22

Figure 16.19 Minimum area cover for Partition-1

Figure 16.20 shows minimum area cover for each vertex for Partition-2. The minimum area for this partition is 12 units.

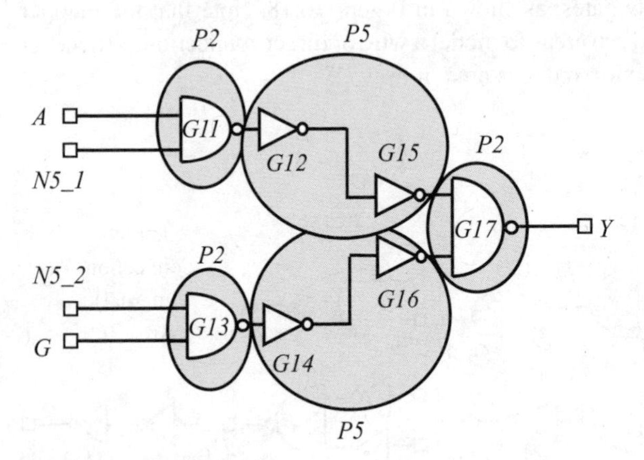

Vertex	Valid Matches	Selected Match	Area
G11	P2	P2	4
G12	P1, P3	P3	5
G13	P2	P2	4
G14	P1, P3	P3	5
G15	P1, P5	P5	4
G16	P1, P5	P5	4
G17	P2, P4	P2	12

Figure 16.20 Minimum area cover for Partition-2

Finally, we combine the minimum area solutions of both the partitions. We obtain the mapped circuit shown in Figure 16.21. It achieves the minimum area for the mapped circuit.

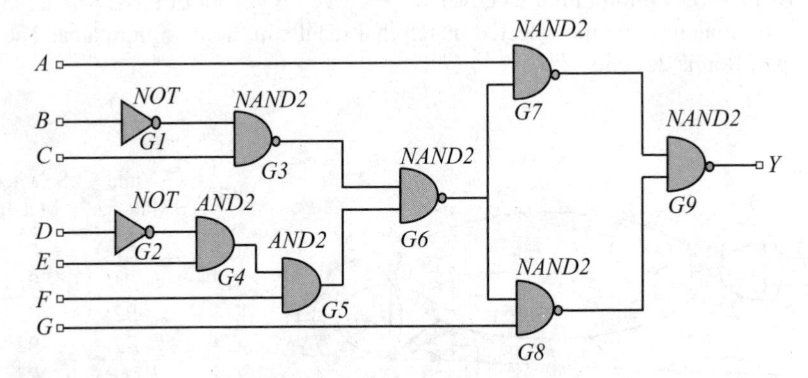

Figure 16.21 Mapped circuit obtained by combining the minimum area cover of Partition-1 and Partition-2

16.4 BOOLEAN MAPPING

The structural mapping described in the previous section explained the challenges in mapping. Some of these challenges can be tackled more efficiently using Boolean mapping techniques. Therefore, in recent times, Boolean mapping techniques have become more popular [14–16].

The Boolean mapping techniques require establishing *functional equivalence* of a portion of a given circuit with the library cells. By definition, if a part of a circuit is functionally equivalent to a library cell, that part of the circuit and the library cell will produce the same output for any given input. Therefore, we can replace that part of the circuit with the library cell.

To establish functional equivalence efficiently, we can use representations of a Boolean function such as *binary decision diagram* (BDD) and *functionally reduced AND-Invert graph* (FRAIG) [15–20]. Note that two trees can be structurally different yet be functionally equivalent (because we can represent a function *structurally* in many different ways). Therefore, establishing the functional equivalence using Boolean methods is more versatile than structural techniques. Moreover, Boolean methods can exploit don't care conditions also. Thus, Boolean methods can produce more matches for a vertex in the subject graph and help explore the solution space more extensively. We can make Boolean methods more efficient by pruning the search space intelligently. Therefore, Boolean mapping techniques typically deliver better results despite the computational difficulty and are now more popular.

16.5 RECENT TRENDS

Recent works on technology mapping propose to exploit satisfiability (SAT) solver in finding minimum-sized logic circuit for a given Boolean function and given delay constraints [21]. With a trend toward tighter coupling of logic synthesis and physical design, mapping techniques based on iterative improvement and predictive analysis are becoming popular [22, 23]. Advanced design techniques, such as the use of 3D standard cells, novel devices and interconnects, and approximate computing, pose newer challenges to technology mapping [24–27].

REVIEW QUESTIONS

16.1 What is the purpose of technology mapping in VLSI design flow?

16.2 Why does the QoR of a design depends heavily on the technology mapping?

16.3 An unmapped circuit and the cells in a library are shown in Figure 16.22.
Find the mapping solution separately for the following requirements:

- **(a)** Worst case delay be minimized
- **(b)** Power dissipated be minimized
- **(c)** Area be minimized.

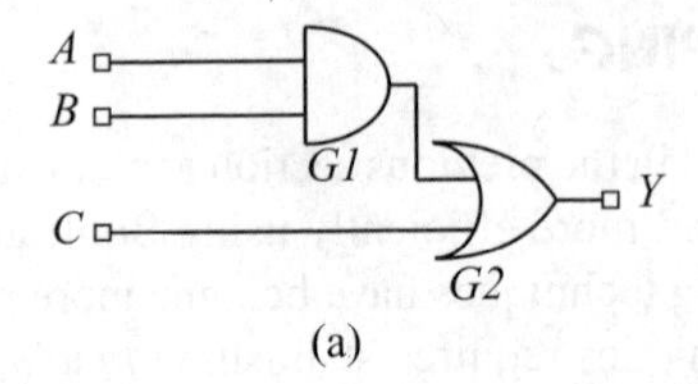

(a)

Cell Name	Symbol	Function	Area	Delay	Power
INV1		*Z=A'*	1	4	5
NAND8		*Z=(A.B)'*	8	4	40
NOR1		*Z=(A+B)'*	3	12	10
AND1		*Z=A.B*	4	16	50

(b)

Figure 16.22 (a) An unmapped circuit and (b) cells in the library and their attributes

16.4 (a) Decompose the circuit shown in Figure 16.22(a) in terms of two-input NOR gate and inverter.

(b) Decompose all the cells in the library shown in Figure 16.22(b) in terms of two-input NOR gate and inverter.

(c) Find all possible matches of cells at each node in the circuit derived above. Find the cover with minimum area.

TOOL-BASED ACTIVITY

In this activity, use any logic synthesis tool, including open-source tools such as Yosys that supports technology mapping [28, 29]. You can use any library, including freely available technology libraries [30].

Write a Verilog code for a 1-bit full adder (a combinational circuit with three inputs and two outputs). Synthesize and examine the netlist in the following cases (observe the cells used and total area). You can remove cells from the library by editing.

1. Library consists of only an inverter and a NAND gate
2. Library consists of only an inverter and a NOR gate
3. Library consists of only an inverter and an OR gate
4. Library consists of only an inverter and an AND gate
5. Library consists of only an inverter, AND gate, and an OR gate
6. Library consists of all cells in the library.

Explain the observed trend in the total area and delay in the above cases.

REFERENCES

[1] R. K. Brayton, G. D. Hachtel, and A. L. Sangiovanni-Vincentelli. "Multilevel logic synthesis." *Proceedings of the IEEE* 78, no. 2 (1990), pp. 264–300.

[2] G. D. Micheli. *Synthesis and Optimization of Digital Circuits*. McGraw-Hill Higher Education, 1994.

[3] G. De Micheli. "Technology mapping of digital circuits." *[1991] Proceedings, Advanced Computer Technology, Reliable Systems and Applications* (1991), pp. 580–586, IEEE.

[4] J. Ishikawa, H. Sato, M. Hiramine, K. Ishida, S. Oguri, Y. Kazuma, and S. Murai. "A rule based logic reorganization system lores/ex." *Proceedings 1988 IEEE International Conference on Computer Design: VLSI* (1988), pp. 262–266, IEEE.

[5] M. Ohlrich, C. Ebeling, E. Ginting, and L. Sather. "SubGemini: Identifying subcircuits using a fast subgraph isomorphism algorithm." *30th ACM/IEEE Design Automation Conference* (1993), pp. 31–37, IEEE.

[6] K. Keutzer. "DAGON: Technology binding and local optimization by DAG matching." *Proceedings of the 24th ACM/IEEE Design Automation Conference* (1987), pp. 341–347, ACM.

[7] E. Lehman, Y. Watanabe, J. Grodstein, and H. Harkness. "Logic decomposition during technology mapping." *IEEE Transactions on Computer-aided Design of Integrated Circuits and Systems* 16, no. 8 (1997), pp. 813–834.

[8] V. Tiwari, P. Ashar, and S. Malik. "Technology mapping for low power." *30th ACM/IEEE Design Automation Conference* (1993), pp. 74–79, IEEE.

[9] C.-Y. Tsui, M. Pedram, and A. M. Despain. "Technology decomposition and mapping targeting low power dissipation." *30th ACM/IEEE Design Automation Conference* (1993), pp. 68–73, IEEE.

[10] Y. Kukimoto, R. K. Brayton, and P. Sawkar. "Delay-optimal technology mapping by DAG covering." *Proceedings 1998 Design and Automation Conference. 35th DAC. (Cat. No. 98CH36175)* (1998), pp. 348–351, IEEE.

[11] K. Chaudhary and M. Pedram. "A near optimal algorithm for technology mapping minimizing area under delay constraints." *[1992] Proceedings 29th ACM/IEEE Design Automation Conference* (1992), pp. 492–498, IEEE.

[12] B. Lin and H. De Man. "Low-power driven technology mapping under timing constraints." *Proceedings of 1993 IEEE International Conference on Computer Design ICCD'93* (Oct. 1993), pp. 421–427.

[13] D. Pandini, L. T. Pileggi, and A. J. Strojwas. "Global and local congestion optimization in technology mapping." *IEEE Transactions on Computer-aided Design of Integrated Circuits and Systems* 22, no. 4 (2003), pp. 498–505.

[14] F. Mailhot and G. De Micheli. "Technology mapping using Boolean matching and don't care sets." *Proceedings of the Conference on European Design Automation* (1990), pp. 212–216, IEEE Computer Society Press.

[15] F. Mailhot and G. Di Micheli. "Algorithms for technology mapping based on binary decision diagrams and on Boolean operations." *IEEE Transactions on Computer-aided Design of Integrated Circuits and Systems* 12, no. 5 (1993), pp. 599–620.

[16] R. Brayton and A. Mishchenko. "ABC: An academic industrial-strength verification tool." *International Conference on Computer Aided Verification*, pp. 24–40. Springer, 2010.

[17] J. Cortadella, M. Kishinevsky, A. Kondratyev, L. Lavagno, E. Pastor, and A. Yakovlev. "Decomposition and technology mapping of speed-independent circuits using Boolean relations." *IEEE Transactions on Computer-aided Design of Integrated Circuits and Systems* 18, no. 9 (1999), pp. 1221–1236.

[18] A. Mishchenko, S. Chatterjee, R. Brayton, X. Wang, and T. Kam. "Technology mapping with Boolean matching, supergates and choices." 2005. https://people.eecs.berkeley.edu/~alanmi/publications/2005/tech05_map.pdf. Last accessed on 7 January, 2023.

[19] A. Mishchenko, S. Chatterjee, R. Jiang, and R. K. Brayton. "FRAIGs: A unifying representation for logic synthesis and verification." ELRI Technical Report, 2005.

[20] J. M. Matos, J. Carrabina, and A. Reis. "Efficiently mapping VLSI circuits with simple cells." *IEEE Transactions on Computer-aided Design of Integrated Circuits and Systems* 38, no. 4 (2019), pp. 692–704.

[21] M. Soeken, G. De Micheli, and A. Mishchenko. "Busy man's synthesis: Combinational delay optimization with SAT." *Design, Automation & Test in Europe Conference & Exhibition (DATE), 2017* (2017), pp. 830–835, IEEE.

[22] L. Machado, M. G. A. Martins, V. Callegaro, R. P. Ribas, and A. I. Reis. "Iterative remapping respecting timing constraints." *2013 IEEE Computer Society Annual Symposium on VLSI (ISVLSI)* (Aug. 2013), pp. 236–241.

[23] A. Reis. "Towards a VLSI design flow based on logic computation and signal distribution." *Proceedings of the 2018 International Symposium on Physical Design*, ISPD '18, (New York, NY, USA), pp. 58–59, ACM, 2018.

[24] S. Bobba, A. Chakraborty, O. Thomas, P. Batude, T. Ernst, O. Faynot, D. Z. Pan, and G. De Micheli. "CELONCEL: Effective design technique for 3-D monolithic integration targeting high performance integrated circuits." *16th Asia and South Pacific Design Automation Conference (ASP-DAC 2011)* (2011), pp. 336–343, IEEE.

[25] L. Amarú, P.-E. Gaillardon, S. Mitra, and G. De Micheli. "New logic synthesis as nanotechnology enabler." *Proceedings of the IEEE* 103, no. 11 (2015), pp. 2168–2195.

[26] H. Jiang, C. Liu, L. Liu, F. Lombardi, and J. Han. "A review, classification, and comparative evaluation of approximate arithmetic circuits." *Journal on Emerging Technologies in Computing Systems* 13 (Aug. 2017), pp. 60:1–60:34.

[27] E. Testa, M. Soeken, L. G. Amar, and G. De Micheli. "Logic synthesis for established and emerging computing." *Proceedings of the IEEE* 107, no. 1 (2019), pp. 165–184.

[28] C. Wolf, J. Glaser, and J. Kepler. "Yosys—A free Verilog synthesis suite." *Proceedings of the 21st Austrian Workshop on Microelectronics (Austrochip)* 2013.

[29] T. Ajayi, V. A. Chhabria, M. Fogaça, S. Hashemi, A. Hosny, A. B. Kahng, M. Kim, J. Lee, U. Mallappa, et al., "Toward an open-source digital flow: First learnings from the OpenROAD project." *Proceedings of the 56th Annual Design Automation Conference 2019* (2019), pp. 1–4.

[30] M. Martins, J. M. Matos, R. P. Ribas, A. Reis, G. Schlinker, L. Rech, and J. Michelsen. "Open cell library in 15nm FreePDK technology." *Proceedings of the 2015 Symposium on International Symposium on Physical Design* (2015), pp. 171–178.

17 Timing-driven Optimizations

... I recommend to you to take care of the minutes; for hours will take care of themselves ... Never think any portion of time whatsoever too short to be employed ...

—Philip Dormer Stanhope, 4th Earl of Chesterfield, in *Letters to His Son*, November 6, 1747

Before technology mapping, the netlist consists of *generic gates*. We refer to this netlist as *unmapped netlist*. After technology mapping, the netlist consists of *cells* from the technology libraries. We refer to this netlist as *mapped netlist*.

We have discussed logic optimization of unmapped netlist in Chapter 12 ("Logic optimization"). It primarily minimizes the circuit *area*. The delay estimates in an unmapped netlist are typically based on logic depth. These delay estimates can differ widely from the delay estimates of the mapped netlist. Therefore, we do not carry out timing-specific logic optimizations on an unmapped netlist.

We can perform static timing analysis (STA) on a mapped netlist using the associated technology libraries and the Synopsys Design Constraint (SDC) files. The timing models of library cells are trustworthy. Therefore, post-mapping STA fairly exposes potential timing issues in a design. Consequently, we can perform *timing-driven optimizations* on a mapped netlist. By timing-driven optimizations, we refer to design transformations that increase the maximum operable frequency of a circuit or fix timing issues in it.

In this chapter, we will discuss timing-driven optimizations that are performed on a mapped netlist.

17.1 MOTIVATION

To improve timing of a *mapped netlist*, one of the approaches can be to include the *gate delays* in the cost function within the *mapping algorithm* and allow the mapper to perform timing-driven optimizations [1–3]. However, this task is challenging, as explained below, and we can often find several opportunities for timing optimizations on the mapped netlist.

Challenges of timing-driven technology mapping: Technology mapping proceeds in two steps: *finding matches* for an unmapped gate (or vertex in an equivalent graph) and *covering* it.[1] While finding suitable *matches* for an unmapped gate, we still have unmapped logic gates in its fanout. We do not know the input pin capacitance of the unmapped logic gates. Hence, it is challenging to compute delays of the possible matches because the gate delay depends on its output *load*. Therefore, we cannot select optimum cells that will be suitable for the future loading condition.

[1] For details, see Chapter 16 ("Technology mapping").

We can employ some heuristics to overcome the above difficulty [1–3]. For example, we can determine the optimum matches for each possible loading condition. Subsequently, when we traverse in the fanin for *covering*, we pick an optimum match corresponding to the actual load. Nevertheless, such approaches are complicated and require huge memory and runtime. For a realistic nonlinear delay model (NLDM), timing-driven mapping leaves ample scope for further timing optimizations. Additionally, mapping cannot perform some timing optimizations that require *global* view or that require a significant change in the logic structure. Therefore, despite timing-driven technology mapping, we need to perform timing-driven optimizations on a mapped netlist [4].

17.2 TIMING OPTIMIZATION PROCESS

A typical timing optimization flow is shown in Figure 17.1 [5]. First, we carry out STA on the mapped netlist using technology libraries and SDC files. Then, using the timing information (delay, arrival time (AT), and slack), we find the target for the transformations, such as the critical path. Subsequently, we apply the transformations such as *resizing*, *restructuring*, *buffering*, *replicating*, and *retiming*. We will explain these transformations in the following sections. We carry out these transformations iteratively until we have satisfied timing requirements or no further timing improvement is possible.

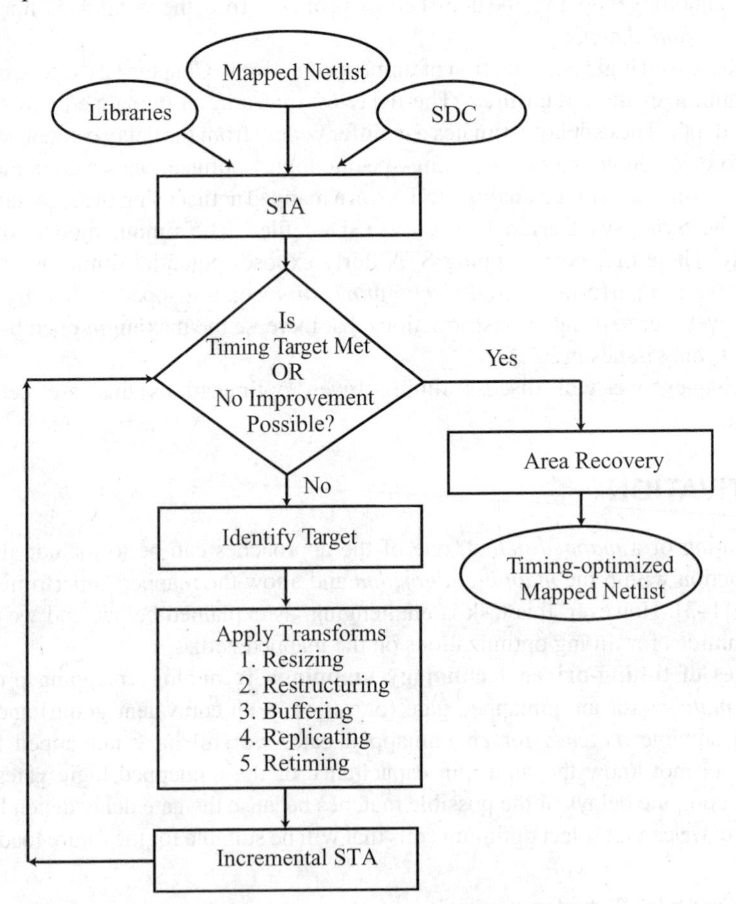

Figure 17.1 Timing optimization flow

Post-timing-driven transformations, the timing profile of the circuit can change drastically. The timing constraints for some parts of a circuit can become significantly relaxed due to the transformations applied in other regions. Therefore, we can re-implement parts of a circuit having considerable positive slack with less area. This transformation is known as the *area recovery*. A timing optimization tool typically produces the final timing-optimized netlist after area recovery.

A challenge in post-mapping timing-driven optimization is that we have not yet performed physical design. Therefore, we cannot estimate *wire delays* accurately enough. Thus, at this stage, we focus on timing issues arising solely due to the *logic* implementation. However, note that we also apply some of these transformations during physical design when the estimates of the wire delays become more accurate.

17.2.1 Identifying Targets

The delay-reducing transformations often incur area and power penalties. Therefore, we cannot apply them over the entire circuit. Instead, we should apply them only where it is *necessary* by identifying pertinent targets. The strategy to identify targets varies with the transformations. For example, a transformation can attempt to reduce the *worst negative slack* (WNS) of a design. It can identify signals in a circuit that have slacks within ϵ of the WNS (typically ϵ is a small number, say a few picoseconds to tens of picoseconds). The network or the sub-circuit consisting of such signals and their drivers is known as an ϵ-network [6]. Many timing-driven optimizations work on a sub-circuit derived from the ϵ-network. As we increase ϵ, the target of transformations moves to the noncritical part of the circuit. Consequently, more transformations can be applied, and the timing optimization process can take more runtime.

17.2.2 Incremental Static Timing Analysis

As we apply timing-driven transformations, the slack can change for many paths in a circuit. We need to evaluate the timing impact of a transformation and accept the favorable ones. To assess the effect of a transformation, we need to carry out an *incremental static timing analysis* (iSTA) [7–10]. It is *incremental* in the sense that it updates the timing of only a small part of a circuit where the changes are noticeable. We typically need to perform iSTA more than a million times during timing-driven optimizations. Therefore, iSTA needs to be extremely fast and reasonably accurate.

We can make iSTA efficient by updating the delays, the output slews, and the ATs only when necessary.

1. **Delay updates:** When we make a local transformation in a circuit, it is sufficient to update the delays of the cells only in the neighborhood of that transformation. Thus, we can avoid delay updates of the major portion of a circuit, as explained in the following example.

Example 17.1 Consider a circuit consisting of CMOS inverters *A*, *B*, ... *E*, as shown in Figure 17.2(a).

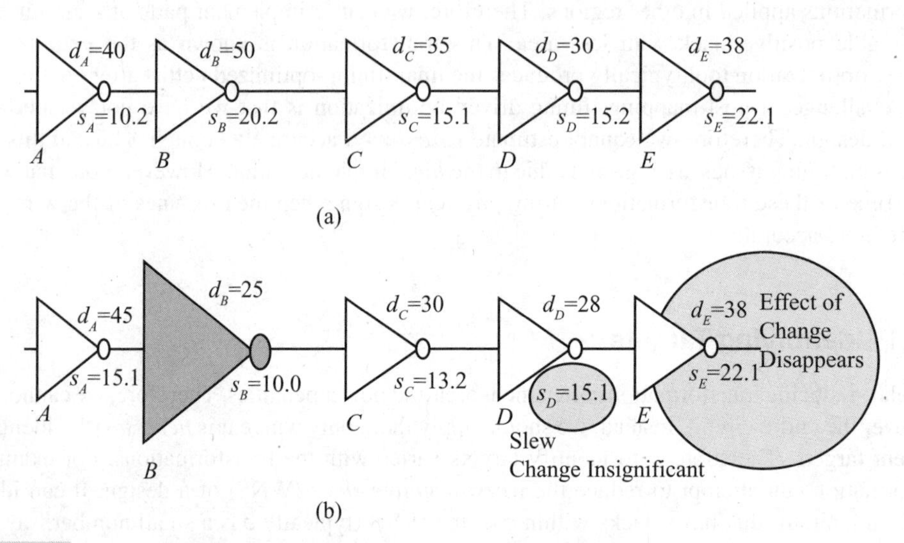

Figure 17.2 Chain of inverters (a) original and (b) *B* is resized (For an instance *X*, its delay is denoted as d_X and slew as s_X. Delays and slews are in ps.)

Suppose we increase the size of the inverter *B*, as shown in Figure 17.2(b). As a result, the delay and the output slew of all the instances in the circuit can change. Let us examine these changes (we have taken some illustrative delay and slew values for the CMOS inverters in the following discussion).

(a) Inverter *A*: Due to the increased load, the delay increases from 40 ps to 45 ps, while the output slew increases from 10.2 ps to 15.1 ps. We cannot ignore these changes.

(b) Inverter *B*: Its delay and the output slew change due to the change in the input slew and the increased size of *B*. Let us assume that the delay decreases from 50 ps to 25 ps due to increased size of *B*, while the output slew decreases from 20.2 ps to 10 ps. The change in its output slew is expected to trigger further timing changes in its fanout.

(c) Inverter *C*: Due to the decrease in its input slew, its delay and the output slew are expected to decrease. Let us assume that the delay decreases from 35 ps to 30 ps, while the output slew decreases from 15.1 ps to 13.2 ps. Note that the change in output slew will be typically less than in the inverter *B*.

(d) Inverter *D*: Due to the decrease in the input slew, let us assume that its delay reduces from 30 ps to 28 ps, while the output slew decreases from 15.2 ps to 15.1 ps. Note that the change in the output slew is expected to reduce further.

(e) Inverter *E*: Since the input slew remains pretty unchanged, its delay and the output slew remain unchanged. Thus, the effect of transformation is expected to die down in the fanout after a few stages.

Thus, when we resize *B*, we need to update the delays, and the output slews in its fanout. However, the changes in the output slew decay after a few stages because the output slew of a CMOS inverter depends weakly on the input slew [11, 12]. Therefore, we need not update the delay and output slew for the entire circuit when making a local change. It is sufficient to re-compute delays and output slews for a few stages in the neighborhood of the local transformation. We can use this observation to reduce the runtime of iSTA.

Note that *A* (driver of upsized inverter *B*) has only one fanout in the above example. If there were multiple fanouts of *A*, we need to re-compute delays for other gates in its fanout also.

2. **AT updates:** When the delay of a cell changes, the AT of signals propagating through it can change. Therefore, after making local transformations, we need to update the ATs. Nevertheless, depending on the location of *the critical paths*, updating the AT of only a small part of a circuit can be sufficient.

Example 17.2 Consider the circuit shown in Figure 17.3. The ciritical path is *P1*→ *B/x*→ *B/z*→ *C/x*→ *C/z*. Assume that we are performing setup analysis. Thus, we need to store only the maximum AT at each pin.

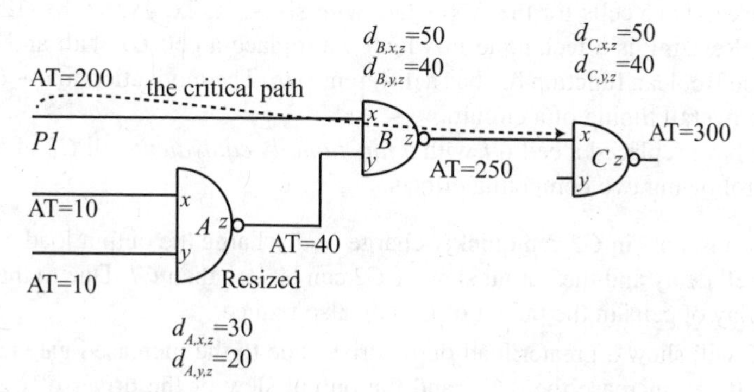

Figure 17.3 Given portion of a circuit (for an instance X and timing arc $x \rightarrow z$, its delay is denoted as $d_{X,x,z}$ and arrival time as AT. Delays and arrival times are in ps.)

Assume that we reduce the size of the instance *A* that is not on the critical path.

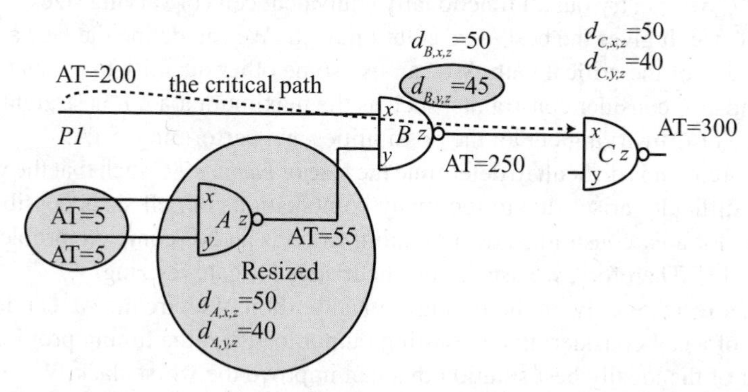

Figure 17.4 AT updates after resizing *A*. The updated timing values after resizing are higlighted inside ovals

Consequently, we expect the delay and the AT of signals at *A* and its fanout to change. However, only the values highlighted in Figure 17.4 changes. Note that the AT at *B/z* does not change because the worst-case AT is still through *B/x*. Therefore, we need not update the AT at *B/z* and further in its fanout. Thus, we can avoid the AT updates in large part of a circuit by considering the worst-case behavior.

17.3 TRANSFORMATIONS

In this section, we discuss various transformations that can be employed in timing-driven optimizations.

17.3.1 Resizing

Technology libraries contain cells that produce the same Boolean function but are of different sizes. For example, we can have cells for the AND gate with sizes 1x, 2x, 4x, 6x, 8x, 12x, and 16x (x is some area unit). Resizing is a technique in which we replace a cell *C1* with another cell *C2* that produces the same Boolean function but has a different size. The motivation to use *C2* instead of *C1* is to improve the overall timing of a circuit.

Suppose we have replaced a cell *C1* with a *functionally equivalent* cell *C2* of a larger size. We can observe the following two competing effects:

1. The larger transistors in *C2* can quickly charge or discharge the output load capacitance. As a result, the cell delay and the output slew of *C2* can be less than *C1*. Due to the reduced output slew, the delay of cells in the fanout of *C2* can also reduce.
2. The cell *C2* will show a greater load on its driver due to the increased gate capacitance of its transistors. It can increase the delay and the output slew of the driver of *C2*. The increased output slew of the driver can further increase the delay of the cells in its fanout, including that of *C2*.

The overall impact of resizing *C1* on the timing depends on which of the above two effects dominate.

Given a cell *C*, we can try out all functionally equivalent cells of varying sizes in the library and pick the *best* cell size. It gives the best *local* solution for *C*. We can define the *best* as the one which maximizes the slack of the critical path. We can use some other quantity to evaluate the merit of a tranformation and also consider constraints, such as the maximum load. It is straightforward to find the local best size of *C* by trying out all the possibilities and performing iSTA.

For a given circuit, it is difficult to determine the size of *each cell C* such that the worst slack gets minimized. The difficulty arises due to too many combinations of cell sizes possible for a circuit. The gate resizing for area-constrained delay minimization is an NP-complete problem for practical delay models [4, 13]. Therefore, we use various heuristics for gate resizing.

We can target resizing only on the timing-critical portion of a circuit. We can initially find the *locally* best size of a cell considering the existing output load and the timing profile. Subsequently, we pick a subset of the locally best solution that can improve the worst slack. We repeat the above transformations until no more improvement is possible, or we violate area or other constraints.

We can increase the slack of the critical path by resizing cells on the *adjacent paths* also. We illustrate it in the following example.

Example 17.3 Consider the circuit shown in Figure 17.5(a).

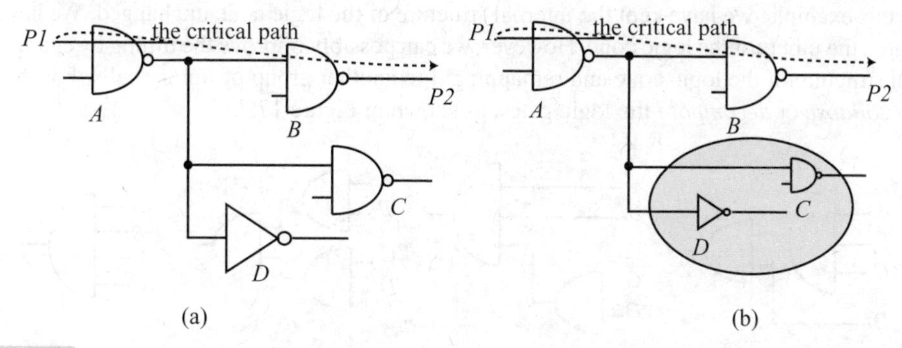

Figure 17.5 Resizing on the path adjacent to the critical path (a) given circuit and (b) resized circuit

The critical path is $P1 \rightarrow A \rightarrow B \rightarrow P2$. We can reduce the size of the cells C and D on the adjacent path to reduce the output load on A, as shown in Figure 17.5(b). As a result, the cell delay of A can reduce, and we obtain speed up on the critical path.

We can also apply gate resizing in the physical design once we have a better estimate of the wire delay. Since local resizing does not significantly perturb a design's placement or routing, we can also apply this transformation post-placement and post-routing to fix timing issues.

17.3.2 Restructring

We can carry out various kinds of logic *restructuring* to improve the overall timing of a circuit. We can iteratively identify timing-critical logic cones of a circuit and restructure them. Typically, the goal of restructuring is to minimize the AT at the output of a logic cone. We illustrate a few types of restructuring in the following examples.

Example 17.4 Consider the logic cone shown in Figure 17.6(a). The AT is in the order $B > A > D > C$. The maximum AT at the output is 220 ps.

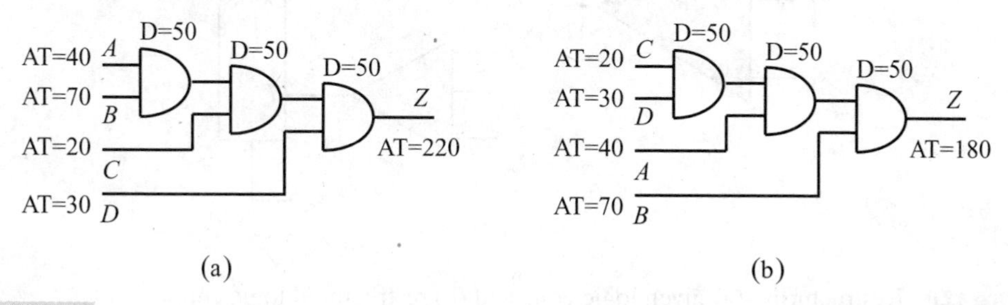

Figure 17.6 (a) Given logic cone and (b) inputs re-ordered (arrival time (AT) and delay (D) are in ps)

By re-ordering the inputs, as shown in Figure 17.6(b), we can reduce the maximum AT at the output to 180 ps.

In this example, we have kept the internal structure of the logic cone unchanged. We have only re-ordered the inputs of the logic cone. However, we can possibly improve the timing by changing the internal structure of the logic cone and remapping it to another group of library cells. For example, we can *collapse* or *decompose* the logic gates, as shown in Figure 17.7.

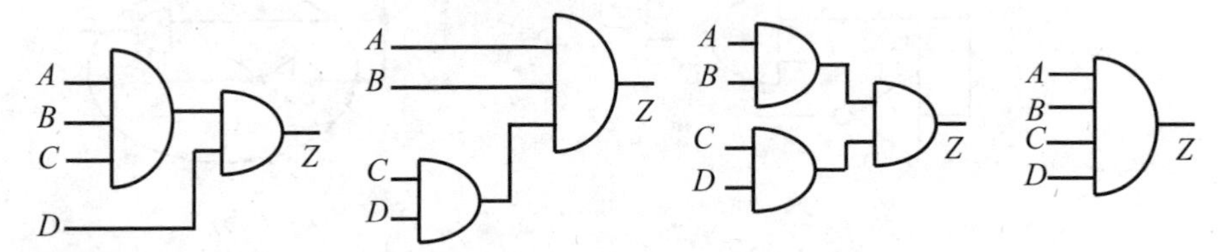

Figure 17.7 Possible transformations of the logic cone shown in Figure 17.6

After evaluating these transformations, we can finally choose the one which minimizes the AT at the output and meets the area constraint.

In the above example, the Boolean function is symmetric, and it is easy to enumerate the possible transformations. When the Boolean function of a logic cone is complex, we can employ Shannon's expansion of Boolean function to determine favorable functional decomposition. We illustrate it in the following example.

Example 17.5 Consider the logic cone shown in Figure 17.8(a). It implements the Boolean function $Y = F(x_0, x_1, x_2, ..., x_N)$. Assume that x_0 lies on the critical path. Furthermore, assume that the signal through x_0 goes through several logic levels (stages) before reaching Y. We can reduce the maximum AT of the signal at Y through x_0 by moving it closer to the output Y as explained below.

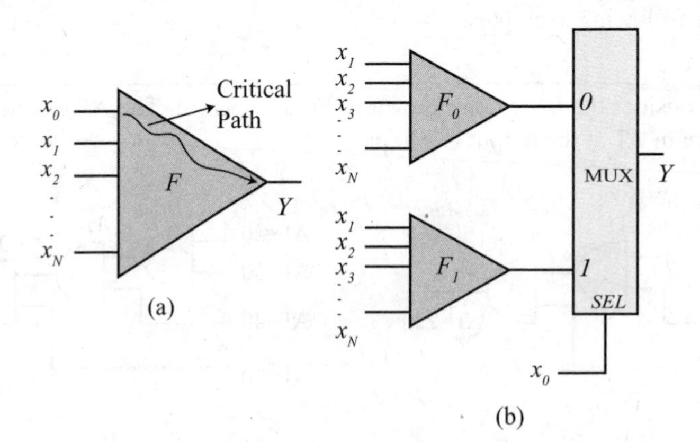

Figure 17.8 Restructuring (a) given logic cone and (b) restructured logic cone

We decompose the function F using Shannon's expansion with respect to x_0. Thus, we obtain the negative cofactor $F_0 = F(0, x_1, x_2, ..., x_N)$ and the positive cofactor $F_1 = F(1, x_1, x_2, ..., x_N)$.

Subsequently, we implement F as shown in Figure 17.8(b). The signal x_0 reaches Y through the select input of the multiplexer. Thus, it experiences smaller delay than in the original logic cone and the maximum AT to Y can improve.

Note that the AT of other signals to Y can increase. As a result, the overall maximum AT at Y can sometimes increase. Therefore, a restructuring tool needs to recheck AT after restructuring before permanently applying the transform to the design. We can carry out the above process recursively by choosing the next late-arriving signal at the input of F. However, we need to be aware of the area increase associated with this transformation.

17.3.3 Fanout Optimization

We can improve the timing of a circuit by inserting buffers or replicating the drivers of a high-fanout net. We illustrate these techniques in the following examples.

Example 17.6 Consider an output pin A connected to a high-fanout net N, as shown in Figure 17.9(a). There are k input pins $P1$, $P2$, $P3$, ..., Pk in the fanout of N. The driver pin A is also referred to as the *source* and the driven input pins are referred to as *sinks*.

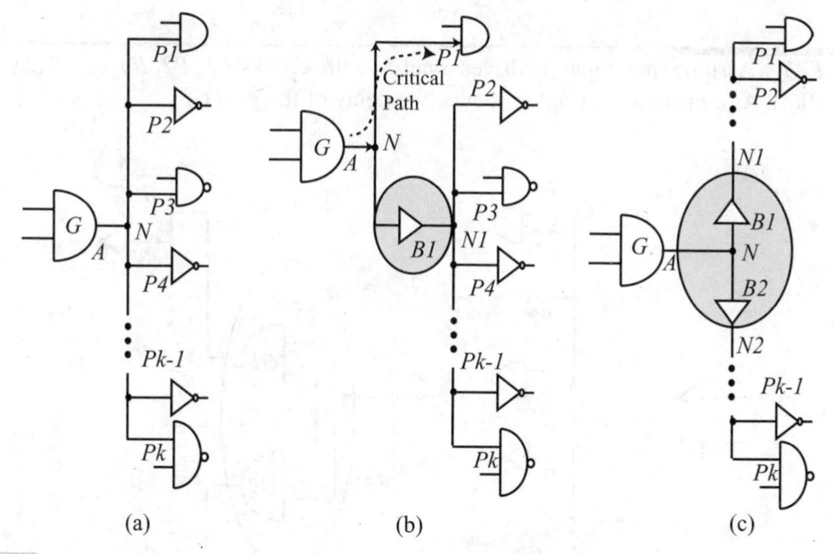

Figure 17.9 Fanout optimizations (a) a high-fanout net N, (b) inserted buffer $B1$, and (c) inserted buffers $B1$ and $B2$

We can insert buffers for the following benefits:

1. **Reduce delay of the critical path:** Assume that the pins A and $P1$ are on the critical path. We can reduce the delay of the gate G by adding a buffer $B1$, as shown in Figure 17.9(b). It reduces the load on A and makes the signal reach quickly to $P1$. Nevertheless, the signal can take longer time to reach other sinks $P2$, $P3$, ..., Pk. The insertion of $B1$ is acceptable if other paths do not become critical. If other pins are also near-critical, we need to perform buffer insertion by considering the required times (RT) at all the input pins.

2. **Fix load and slew constraints violations:** Assume that the net *N* has an overload or a large slew problem. We can distribute the fanout load of *N* to two buffers *B1* and *B2*, as shown in Figure 17.9(c). Since each buffer *B1* and *B2* drives a smaller load, they can possibly meet the load and slew constraints. If we are not able to fix violations even after distributing the loads to two buffers, we can carry out further load distribution for the nets *N1* and *N2*. However, note that as we add buffers to the net, the total delay from the source to the sinks can increase. Therefore, we need to optimize the number and size of the buffers during fanout optimization.

We can apply the above fanout optimization techniques in physical design also. However, adding buffers to a net increases the circuit area and perturbs the layout. Therefore, we need to choose the size of the buffers optimally [14]. During physical design, we also need to determine an optimal location for the inserted buffers [15].

The above examples treat the problem of fanout optimization *locally* for a given net. However, for a given circuit, we can carry out buffer insertion globally to meet the given timing constraint and minimize the circuit area [16–19]. One of the approaches could be to traverse the circuit and carry out local fanout optimization for relevant nets.

Another approach to improve the timing of a high-fanout net is to replicate the driver of the net. We illustrate it in the following example.

Example 17.7 Assume that a gate *G* drives a net *N* with *k* sinks *P1*, *P2*, *P3*, ..., *Pk*, as shown in Figure 17.10(a). Assume that we want to reduce the delay of the gate *G*.

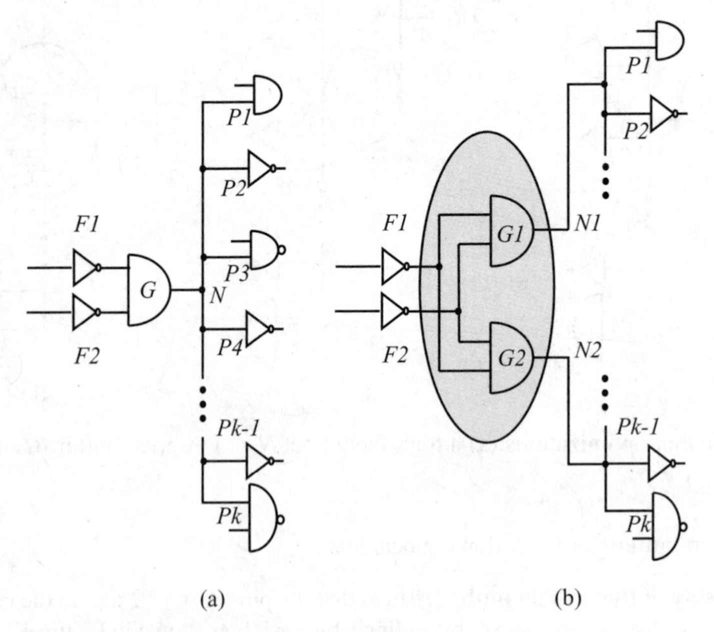

Figure 17.10 Gate replication (a) gate *G* driving high-fanout net *N* and (b) *G* replicated

We can create two copies of *G* (i.e., *G1* and *G2*) and re-distribute the load, as shown in Figure 17.10(b). As a result, the load on each gate (*G1* and *G2*) is less than what was on *G*. Thus, the delay of *G1* and *G2* is expected to be less than *G*. However, the load on the drivers of *G* (*F1* and *F2*) will increase. Hence, the delay of *F1* and *F2* can increase. Nevertheless, we can sometimes improve the worst slack of the circuit by intelligently re-distributing the fanout as shown in this example.

Note that both gate replication and buffering reduce the load on the driver to obtain speedup. However, there are some differences between these approaches [4]. The gate replication can reduce the delay to all the sinks. However, buffering can increase the delay to the sinks where the buffer delay adds up and decrease it where there is no extra buffer on the path. Thus, gate replication is favorable when all sinks are similarly timing-critical, while we would prefer buffering when the slack varies widely among sinks.

17.3.4 Pin Permutation

The delay of timing arcs of a symmetric multi-input gate can vary for different input pins. We can connect the late-arriving signal to the input that exhibits a smaller delay. Thus, for symmetric gates, we can improve the AT at the output by considering various permutations of the input pins.

17.3.5 Retiming

In the previous sections, we have considered transforming combinational logic to improve timing. Registers or flip-flops separate the combinational local blocks. Thus, the combinational optimization techniques get restricted by the register boundaries. However, we can allow registers to move across the combinational logic elements and increase the maximum operable frequency or other figures of merit. This is the motivation for *retiming* [20, 21].

Retiming is a *sequential* timing optimization technique. It involves moving registers or flip-flops across combinational logic elements on the D-pin side or the Q-pin side. By moving registers, the surplus slack on one side of the register gets transferred to the deficit side. Thus, we can increase the maximum operable frequency of a circuit.

Example 17.8 Consider the timing-critical portion of a synchronous circuit shown in Figure 17.11. The total delay of the combinational circuit elements between flip-flops is shown inside ovals. Assume that the flip-flops are ideal and the clock skew is zero. Thus, the maximum operable frequency of the circuit is determined by the largest data path delay.

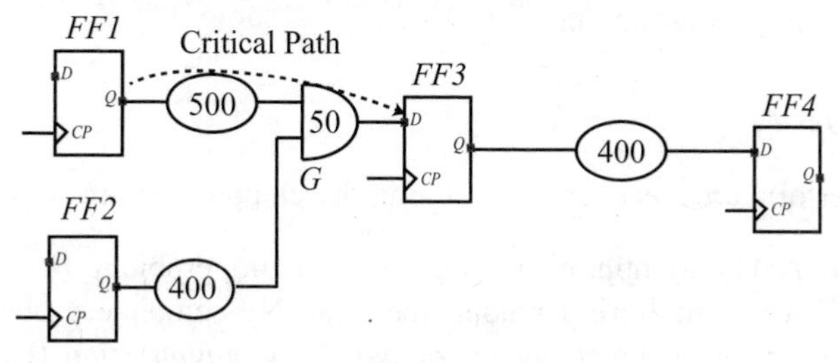

Figure 17.11 Given the critical portion of a circuit (delay in ps)

The largest delay between flip-flops is 550 ps (between *FF1* and *FF3*). Thus, the time period (T) of the clock should satisfy $T > 550\ ps$. We can reduce T by moving the flip-flop *FF3* to the other side of the AND gate G, as shown in Figure 17.12.

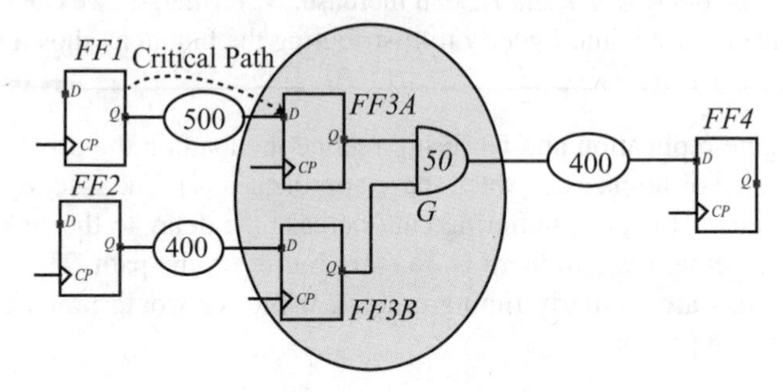

Figure 17.12 The circuit of Figure 17.11 retimed

Since the AND gate has two inputs, we need two flip-flops *FF3A* and *FF3B* to drive it. The constraint gets relaxed due to the AND gate moving to the noncritical path (we utilize the surplus slack in the path between *FF3* and *FF4*). Now, the clock should satisfy $T > 500\ ps$, and we can operate the circuit at a greater frequency. We can move *FF3A* and *FF3B* further in the fanin to balance the slacks. Note that we have introduced an extra flip-flop in the retimed circuit. Therefore, retiming can increase the area of the circuit.

Remark: Retiming can also be employed to reduce the area of a circuit. For example, we can view the transformation from Figure 17.12 to Figure 17.11 as reducing area by one flip-flop. Note that it could improve the timing also if the path *FF3*→ *FF4* was more critical.

Retiming changes the internal sequential behavior of a circuit. The number of state elements and the state transition function can change after retiming. However, the input/output behavior of the circuit does not change due to retiming. Moreover, its *latency* or the number of cycles required to produce the output does not change. Thus, retiming preserves the external cycle-level behavior of a circuit, despite its internal sequential behavior changing.

The Boolean function between two flip-flops (such as between *FF1* and *FF3*) can change due to retiming. Therefore, traditional *combinational equivalence checking* (CEC) tools can show equivalence failure after retiming. Typically, we need to inform a CEC tool about the retiming transformations. Alternatively, we can use *sequential equivalence checking* to verify the cycle-level equivalence of the retiming transformations. To avoid challenges in verifying a retimed circuit, often, we forego improvement due to retiming.

17.4 Area Recovery

The motivation to perform area recovery after timing-driven optimizations is as follows:

1. The area-constrained delay minimization is a challenging problem. For example, creating an area-constrained delay-minimized fanout tree is an NP-complete problem [14]. Therefore, sometimes we carry out simpler *unconstrained delay minimization* (UDM) during timing-driven optimizations [14]. After UDM, we can find many opportunities for reducing the

area further. Even when we solve area-constrained delay minimization, we leave the scope for further area reduction. Therefore, after area-constrained delay minimization, we carry out *constrained area minimization* to recover area from the noncritical portions of a circuit [4].

2. After timing-driven optimization is performed for some portion of a circuit, the timing constraints can relax for some other portion of the circuit. We can reduce area in the portion of the circuit that has now a relaxed timing constraint.

To recover area, we often decrease the cell size in the noncritical portion of the circuit [5, 22]. Additionally, we can remove unnecessary buffers from the circuit to recover the area. However, we can get a reasonable estimate of the wire delays and buffering requirements only after routing. Therefore, post-global routing, we can carry out resizing and buffer removal to recover area, ensuring that it does not violate timing or overload constraints [23].

17.5 Physical Design Aware Logic Optimizations

Traditional logic synthesis framework that is oblivious of the physical design suffers from two major drawbacks:

1. It cannot account for the *wire delay* accurately. The wire delay is the time taken by a signal in propagating through the interconnects. Since wires are laid out only during physical design, it is challenging to estimate wire delays. The wire delays contribute significantly to the total path delay. Therefore, timing-driven optimizations based on inaccurate wire delays can identify wrong targets for optimization.
2. Logic synthesis has no information about the floorplan, routing resources, congestion, and signal integrity. Hence, it can produce a netlist that will pose problems to the physical design tool. In the worst case, a physical design tool can fail to generate a legal layout for the given netlist. We might need to perform logic synthesis again based on the feedback from the physical design. It can introduce costly iterations between logic synthesis and physical design in the very large scale integration (VLSI) design flow.

To tackle these problems, we can take the following approaches:

1. **Accounting for wire delays:** We can use *wire load models* in logic synthesis. Statistical models based on fanout, chip size, and technology are found to be inadequate. The wire length estimates based on quick placement or routing can improve the accuracy [24]. We can also add margins to the slack to account for wire delays.
2. **Physical design aware logic synthesis:** We can make logic synthesis aware of physical design. We provide physical libraries and a rough floorplan to the logic synthesis tool. Internally, it creates an *abstract physical model* using quick placement, clock network design, and global routing. Using this model, it can perform physical design aware technology mapping, restructuring, resizing, fanout optimization, and clock gating [25–27]. It updates the abstract physical model incrementally as the logic optimization proceeds. Thus, it can target real critical paths, reduce overall wire length, alleviate congestion, and improve routability.
3. **Performing logic optimizations in physical design:** Despite logic synthesis being physically aware, we might need to perform some logic optimizations in physical design. Performing logic optimization in physical design is challenging because it can perturb the layout significantly.

Nevertheless, the state-of-the-art physical design tools can perform local incremental logic optimizations such as restructuring, resizing, pin permutations, fanout optimizations, and area recovery.

In practice, we use a combination of the above approaches in a design flow.

17.6 RECENT TRENDS

With the advancement in technology, the impact of variations on a design becomes dominant. Traditionally, we apply margins in logic synthesis to deal with the process, voltage, and temperature (PVT) variations and inaccuracies of delay estimates. We need to reduce these margins due to the shrinking clock period. Thus, in recent times, research has focused on considering temperature, voltage, and reliability during timing-driven optimizations [28–30]. It allows timing-driven optimizations to work on real critical paths and achieve higher clock frequencies.

At advanced process nodes, the timing closure problem also exacerbates. We can alleviate this problem by making logic synthesis aware of the physical design. We can possibly use machine learning (ML) in logic synthesis to predict and prevent problems from occurring in the physical design [31, 32].

REVIEW QUESTIONS

17.1 What are the challenges in timing-driven technology mapping?

17.2 Why do we need to carry out timing-driven optimization even after timing-driven technology mapping?

17.3 How do we identify the part of a design where we need to carry out timing-driven optimizations?

17.4 How does an incremental STA differ from the traditional STA?

17.5 Why is it sufficient to update the timing attributes only in the neighborhood of perturbation in an incremental STA?

17.6 How can increasing the size of a cell in the critical path sometimes improve the worst setup slack?

17.7 How can the worst setup slack of a circuit sometimes improve by decreasing the size of a cell in a noncritical path?

17.8 Can decomposing a logic gate with many inputs into multiple two-input logic gates impact the physical design? If yes, how?

17.9 Consider the circuit shown in Figure 17.13. The arrival time (*AT*) at the input ports and the delay (*D*) of OR, AND, and multiplexer are shown. The time unit is ps. Consider the setup analysis.

(a) Compute the worst *AT* at the output *Z*.

(b) Restructure the logic to minimize the worst *AT* at *Z*. Compute the worst *AT* in the restructured circuit.

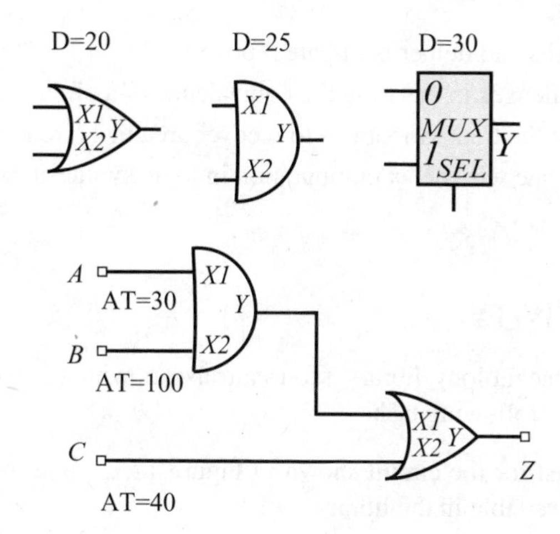

Figure 17.13 Problem 17.9

17.10 How does inserting buffers in a high-fanout net help in the following cases:

(a) increase maximum operable frequency

(b) fix overload violation?

17.11 Consider a CMOS buffer of 0.2 unit area. The load capacitance of its input pin is 1 pF. Its load-dependent delay model is as shown in Table 17.1 (for simplicity, we assume that the delay is independent of the input slew). Consider the fanout trees shown in Figure 17.14. The load on the terminating buffers (*B1*, *B2*, *B3*, ..., *B8*) is 1 pF. The AT at input *IN* is 0 ps. Find the AT at the output of the fanout tree in each case. Which implementation has the minimum AT and also less area?

Table 17.1 Given delay model

Capacitance (pF)	1	2	4	8	16
Delay (ps)	10	18	34	65	120

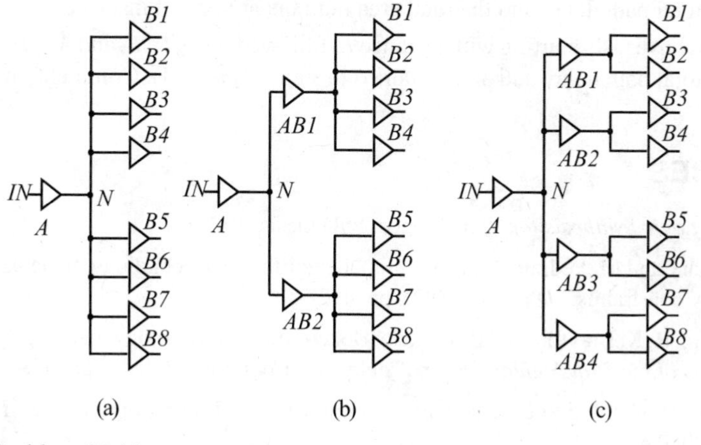

Figure 17.14 Problem 17.11

17.12 What are the merits and demerits of gate replication?

17.13 What are the challenges in ensuring the equivalence of a circuit after retiming?

17.14 What are some of the transformations to recover area in a circuit?

17.15 How does inaccurate wire delay computation in logic synthesis pose a problem in physical design?

TOOL-BASED ACTIVITY

In this activity, use any technology library (you can also download freely available technology libraries [33]). Perform the following tasks:

1. Write a Verilog netlist for the circuit shown in Figure 17.15. Use instances of the minimum-sized (1X) inverter available in the library.

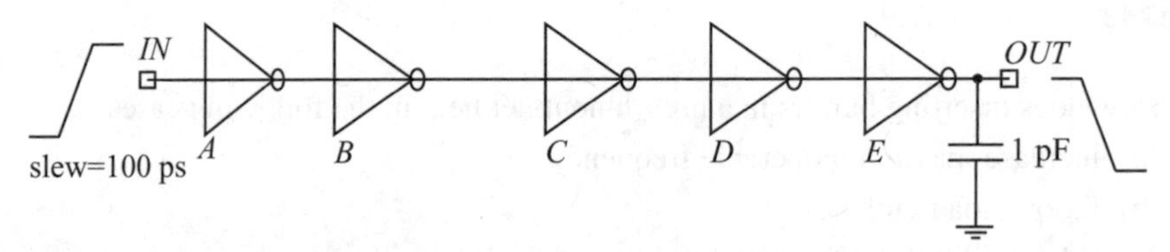

Figure 17.15 Circuit for the activity

2. Specify the following constraints: input slew at port *IN* is 100 ps and the output load at port *OUT* is 1 pF. Use any STA tool (you can use open-source tools such as OpenSTA [34]) to measure the delay between *IN* and *OUT*.
3. Manually change the instance *A* with cells of higher sizes (e.g., 2X, 4X, 8X, ...) in the technology library. Observe the change in the delay of each instance and the total path delay. Tabulate the information. Comment on the result. For the instance *A*, fix the cell that produced the smallest total path delay. Next, repeat the above process for the instance *B* and find the cell in the technology library that minimizes the total path delay. Now, fix the instance B also to the cell that minimizes the delay. Repeat the same for instances *C*, *D*, and *E* in succession. Report the minimum total path delay and the total area (add the area of all instances).
4. Repeat the above task, starting with instance *E*, followed by *D*, *C*, *B*, and *A* in steps. Compare the minimum total path delay and area obtained in steps (3) and (4). Comment on the observation.

REFERENCES

[1] R. Rudell. *Logic Synthesis for VLSI Design*. PhD thesis, 1989.

[2] K. Chaudhary and M. Pedram. "A near optimal algorithm for technology mapping minimizing area under delay constraints." *DAC* 92 (1992), pp. 492–498.

[3] Y. Kukimoto, R. K. Brayton, and P. Sawkar. "Delay-optimal technology mapping by DAG covering." *Proceedings of the 35th Annual Design Automation Conference* (1998), pp. 348–351.

[4] R. Murgai. "Technology-dependent logic optimization." *Proceedings of the IEEE* 103 (2015), pp. 2004–2020.

[5] K. Yoshikawa, H. Ichiryu, H. Tanishita, S. Suzuki, N. Nomizu, and A. Kondoh. "Timing optimization on mapped circuits." *Proceedings of the 28th ACM/IEEE Design Automation Conference* (1991), pp. 112–117.

[6] K. J. Singh, A. R. Wang, R. K. Brayton, and A. L. Sangiovanni-Vincentelli. "Timing optimization of combinational logic." *ICCAD* 88 (1988), 282–285.

[7] A. B. Kahng, S. Kang, H. Lee, S. Nath, and J. Wadhwani. "Learning-based approximation of interconnect delay and slew in signoff timing tools." *2013 ACM/IEEE International Workshop on System Level Interconnect Prediction (SLIP)* (2013), pp. 1–8, IEEE.

[8] J. Hu, G. Schaeffer, and V. Garg. "TAU 2015 contest on incremental timing analysis." *2015 IEEE/ACM International Conference on Computer-aided Design (ICCAD)* (2015), pp. 882–889, IEEE.

[9] M.-C. Kim, J. Hu, J. Li, and N. Viswanathan. "ICCAD-2015 CAD contest in incremental timing-driven placement and benchmark suite." *2015 IEEE/ACM International Conference on Computer-aided Design (ICCAD)* (2015), pp. 921–926, IEEE.

[10] T.-W. Huang, G. Guo, C.-X. Lin, and M. D. Wong. "Opentimer v2: A new parallel incremental timing analysis engine." *IEEE Transactions on Computer-aided Design of Integrated Circuits and Systems* (2020).

[11] S. Dutta, S. M. Shetti, and S. L. Lusky. "A comprehensive delay model for CMOS inverters." *IEEE Journal of Solid-State Circuits* 30, no. 8 (1995), pp. 864–871.

[12] L. W. Massengill and P. W. Tuinenga. "Single-event transient pulse propagation in digital CMOS." *IEEE Transactions on Nuclear Science* 55, no. 6 (2008), pp. 2861–2871.

[13] W.-N. Li, A. Lim, P. Agrawal, and S. Sahni. "On the circuit implementation problem." *IEEE Transactions on Computer-aided Design of Integrated Circuits and Systems* 12, no. 8 (1993), pp. 1147–1156.

[14] K. J. Singh and A. Sangiovanni-Vincentelli. "A heuristic algorithm for the fanout problem." *27th ACM/IEEE Design Automation Conference* (1990), pp. 357–360, IEEE.

[15] L. P. Van Ginneken. "Buffer placement in distributed RC-tree networks for minimal Elmore delay." *IEEE International Symposium on Circuits and Systems* (1990), pp. 865–868, IEEE.

[16] H. J. Hoover, M. M. Klawe, and N. J. Pippenger. "Bounding fan-out in logical networks." *Journal of the ACM (JACM)* 31, no. 1 (1984), pp. 13–18.

[17] K. Kodandapani, J. Grodstein, A. Domic, and H. Touati. "A simple algorithm for fanout optimization using high-performance buffer libraries." *Proceedings of 1993 International Conference on Computer Aided Design (ICCAD)* (1993), pp. 466–471, IEEE.

[18] D. S. Kung. "A fast fanout optimization algorithm for near-continuous buffer libraries." *Proceedings 1998 Design and Automation Conference. 35th DAC. (Cat. No. 98CH36175)* (1998), pp. 352–355, IEEE.

[19] R. Murgai. "On the global fanout optimization problem." *Proceedings of the 1999 IEEE/ACM International Conference on Computer-aided Design, ICCAD '99* (1999), pp. 511–515, IEEE Press.

[20] S. Malik, E. M. Sentovich, R. K. Brayton, and A. Sangiovanni-Vincentelli. "Retiming and resynthesis: Optimizing sequential networks with combinational techniques." *IEEE Transactions on Computer-aided Design of Integrated Circuits and Systems* 10, no. 1 (1991), pp. 74–84.

[21] C. E. Leiserson and J. B. Saxe. "Retiming synchronous circuitry." *Algorithmica* 6, no. 1 (1991), pp. 5–35.

[22] J. P. Fishburn. "LATTIS: An iterative speedup heuristic for mapped logic." *Proceedings 29th ACM/IEEE Design Automation Conference* (1992), pp. 488–491, IEEE.

[23] R. Murgai. "Delay-constrained area recovery via layout-driven buffer optimization." *VLSI Design 2000.* Wireless and Digital Imaging in the Millennium. *Proceedings of 13th International Conference on VLSI Design* (2000), pp. 240–245, IEEE.

[24] N. Shenoy, M. Iyer, R. Damiano, K. Harer, H.-K. Ma, and P. Thilking. "A robust solution to the timing convergence problem in high-performance design." *Proceedings 1999 IEEE International Conference on Computer Design: VLSI in Computers and Processors (Cat. No. 99CB37040)* (1999), pp. 250–257, IEEE.

[25] T. Kutzschebauch and L. Stok. "Congestion aware layout driven logic synthesis." *IEEE/ACM International Conference on Computer Aided Design. ICCAD 2001. IEEE/ACM Digest of Technical Papers (Cat. No. 01CH37281)* (2001), pp. 216–223, IEEE.

[26] D. Pandini, L. T. Pileggi, and A. J. Strojwas. "Congestion-aware logic synthesis." *Proceedings 2002 Design, Automation and Test in Europe Conference and Exhibition* (2002), pp. 664–671, IEEE.

[27] L. Trevillyan, D. Kung, R. Puri, L. N. Reddy, and M. A. Kazda. "An integrated environment for technology closure of deep-submicron IC designs." *IEEE Design & Test of Computers* 21, no. 1 (2004), pp. 14–22.

[28] S. G. Ramasubramanian, S. Venkataramani, A. Parandhaman, and A. Raghunathan. "Relax-and-retime: A methodology for energy-efficient recovery based design." *2013 50th ACM/EDAC/IEEE Design Automation Conference (DAC)* (2013), pp. 1–6, IEEE.

[29] H. Amrouch, B. Khaleghi, and J. Henkel. "Optimizing temperature guardbands." *Design, Automation & Test in Europe Conference & Exhibition (DATE), 2017* (2017), pp. 175–180, IEEE.

[30] H. Amrouch, B. Khaleghi, A. Gerstlauer, and J. Henkel. "Reliability-aware design to suppress aging." *2016 53nd ACM/EDAC/IEEE Design Automation Conference (DAC)* (2016), pp. 1–6, IEEE.

[31] R. Kirby, S. Godil, R. Roy, and B. Catanzaro. "CongestionNet: Routing congestion prediction using deep graph neural networks." *2019 IFIP/IEEE 27th International Conference on Very Large Scale Integration (VLSI-SoC)* (2019), pp. 217–222, IEEE.

[32] D. Hyun, Y. Fan, and Y. Shin. "Accurate wirelength prediction for placement-aware synthesis through machine learning." *2019 Design, Automation & Test in Europe Conference & Exhibition (DATE)* (2019), pp. 324–327, IEEE.

[33] M. Martins, J. M. Matos, R. P. Ribas, A. Reis, G. Schlinker, L. Rech, and J. Michelsen. "Open cell library in 15nm FreePDK technology." *Proceedings of the 2015 Symposium on International Symposium on Physical Design* (2015), pp. 171–178.

[34] T. Ajayi, V. A. Chhabria, M. Fogaça, S. Hashemi, A. Hosny, A. B. Kahng, M. Kim, J. Lee, U. Mallappa, M. Neseem, et al. "Toward an open-source digital flow: First learnings from the OpenROAD project." *Proceedings of the 56th Annual Design Automation Conference 2019* (2019), pp. 1–4.

18 Power Analysis

There are few facts in science more interesting than those which establish a connexion between heat and electricity.

—James Prescott Joule, "On the heat evolved by metallic conductors of electricity, and in the cells of a battery during electrolysis," *The London, Edinburgh, and Dublin Philosophical Magazine and Journal of Science*, 1841

The power–performance trade-off has now become a key ingredient in VLSI design flows. The increased power dissipation in integrated circuits (ICs) and the ubiquitous use of battery-powered devices have made incorporating power-saving techniques essential to the design flows. Since power-saving strategies are more effective early in the flow, we adopt low-power design methodologies right from the pre-RTL stages. Subsequently, the power-related tasks permeate throughout the design flow.

We can broadly classify the power-related tasks as: *power analysis* and *power-driven optimizations*. In this chapter, we will explain power analysis methods. In the next chapter ("Power-driven optimization"), we will discuss power optimization techniques.

18.1 COMPONENTS OF POWER DISSIPATION

There are two components of power dissipation in a CMOS circuit: *dynamic power dissipation* and *static power dissipation*. The dynamic power dissipation occurs when a circuit performs computation actively, i.e., a signal or the output of a logic gate changes its value. The static power dissipation occurs when the circuit is powered on (supply voltages are applied), but it does not perform active computation. Let us understand these components of power dissipation in more detail.

18.1.1 Dynamic Power Dissipation

Consider a CMOS inverter *I1*, shown in Figure 18.1(a). The output pin drives the input of other logic gates through wires. The total load due to these M input pins is $C_I = \sum_{i=1}^{M} C_i$, where C_i is the capacitance of the ith input pin. The wires offer load capacitance C_w. Additionally, the driving pin

Z has parasitic capacitances C_d due to the drain diffusion regions of the transistors. Thus, an output pin of a CMOS logic gate has total load capacitance:

$$C_L = C_I + C_w + C_d \tag{18.1}$$

We can use the circuit shown in Figure 18.1(b) for understanding dynamic power dissipation. Assume that the inverter undergoes a transition from 0→1 or 1→0. These transitions lead to power dissipation due to switching of capacitors and due to drawing short-circuit current from the power supply. We discuss these components of power dissipation in the following paragraphs.

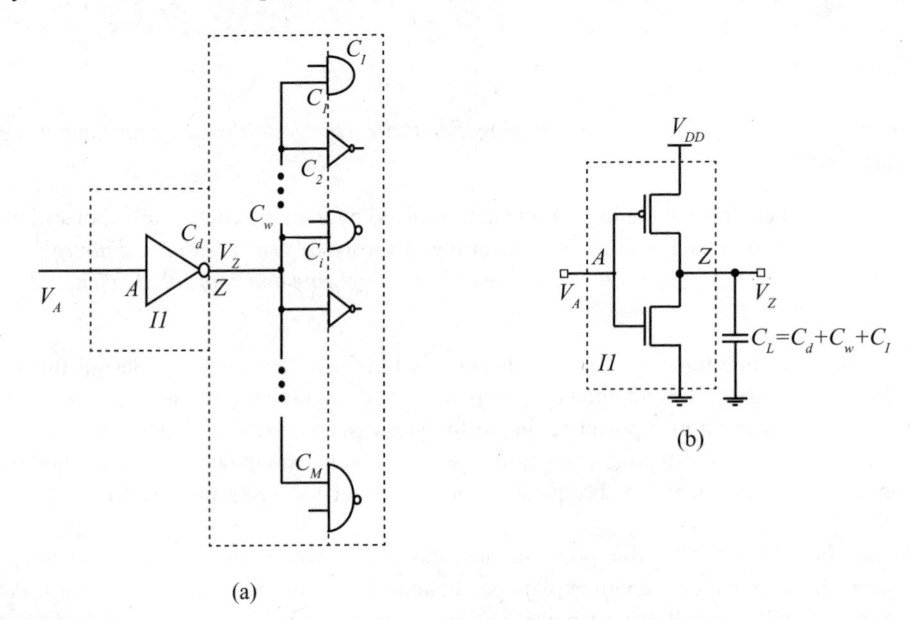

Figure 18.1 Illustration of dynamic power dissipation (a) given circuit and (b) equivalent circuit

Switching of Capacitors

When the input pin A of the inverter transitions from 1→0, its output pin Z transitions from 0→1. When A=0, the PMOS is switched ON, and the NMOS is switched OFF. Thus, we can represent the inverter as in Figure 18.2(a). The output load capacitance C_L gets charged to V_{DD} by drawing charge from the power source. The energy drawn from the power source is given as:

$$E_{batt} = \int_0^{\infty} I(t)V_{DD}dt = V_{DD}\int_0^{\infty} C_L\frac{dV_Z}{dt}dt$$

$$= C_L V_{DD}\int_0^{V_{DD}} dV_Z = C_L V_{DD}^2 \tag{18.2}$$

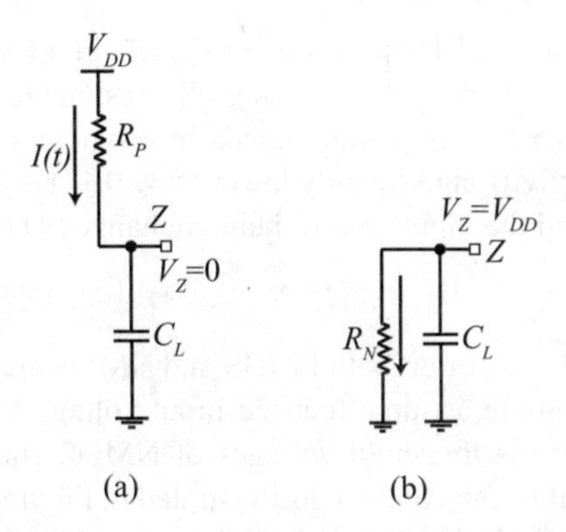

Figure 18.2 Dynamic power dissipation (a) charging of capacitor and (b) discharging of capacitor

Out of the total energy drawn from the power source, half of it ($\frac{1}{2}C_L V_{DD}^2$) gets stored in C_L, while the other half ($\frac{1}{2}C_L V_{DD}^2$) gets dissipated as heat in the resistance R_P.

When the output transitions from 1→0, PMOS is switched OFF, and the NMOS is switched ON, as shown in Figure 18.2(b). The stored energy in C_L (= $\frac{1}{2}C_L V_{DD}^2$) gets dissipated as heat in the resistance R_N of the NMOS. Thus, in one cycle of 0→1→0 transition, $C_L V_{DD}^2$ energy gets dissipated as heat. This component of power dissipation is also known as *switching power* dissipation. Note that the switching power does not directly depend on the size or the speed of the inverter.

Let us assume that the output pin Z makes 0→1→0 transition N times in the time interval t_0. Thus, the power P_{sw} dissipated in switching is:

$$P_{sw} = NC_L V_{DD}^2 / t_0 \tag{18.3}$$

Let us assume that the inverter is a part of a synchronous circuit in which the clock has a period T_{clk}. Let us assume that the clock toggles K times in a time interval t_0. Thus, $t_0 = KT_{clk}$ and the power dissipated by the inverter is $P_{sw} = NC_L V_{DD}^2 / KT_{clk}$. The frequency of the clock $f_{clk} = 1/T_{clk}$. We can write the power dissipated as:

$$P_{sw} = (N/K) C_L V_{DD}^2 f_{clk} \tag{18.4}$$

The ratio N/K denotes the times that a signal transitions relative to the clock signal. It is known as the *activity* of the signal Z and is denoted as α. Thus, the power dissipated in charging and discharging capacitors at the output pin Z is:

$$P_{sw} = \alpha C_L V_{DD}^2 f_{clk} \tag{18.5}$$

In other complex CMOS logic gates, we replace the top PMOS with a pull-up network and the bottom NMOS with a pulldown network. We can explain the switching power dissipation in them similarly.

In the above equations, we define $\alpha = 1$ when the output completes one cycle of transitions (1→0→1) in one clock period. Thus, when there are two transitions in a clock cycle, we say $\alpha = 1$.

Therefore, a clock signal has $\alpha = 1$.[1] However, other signals in a synchronous circuit typically do not change more than once in a clock cycle (ignoring glitches due to arrival time differences at the inputs). Thus, typically $\alpha < 0.5$ for non-clock signals in a synchronous circuit. In practice, static CMOS logic gates can have activity significantly lower ($\alpha \approx 0.1$) because outputs of many gates are skewed toward logic 0 or 1, and the inputs can remain unchanged [1].

Short-circuit Current

During the transition of a CMOS inverter, both PMOS and NMOS are partially switched ON during some time interval t_s. For example, assume that the input voltage V_A is between V_{TN} and $V_{DD} - |V_{TP}|$, where V_{TN} and V_{TP} are the threshold voltages of NMOS and PMOS, respectively. In this condition both PMOS and NMOS can conduct, as illustrated in Figure 18.3. We observe that a direct path exists between V_{DD} and GND through the transistors, and we have a short-circuit condition. For both 0→1 and 1→0 transitions, the short-circuit current I_{sc} flows in addition to the capacitor charging/discharging current. The power P_{sc} dissipated due to the short-circuit current is given as:

$$P_{sc} = V_{DD} I_{sc} \tag{18.6}$$

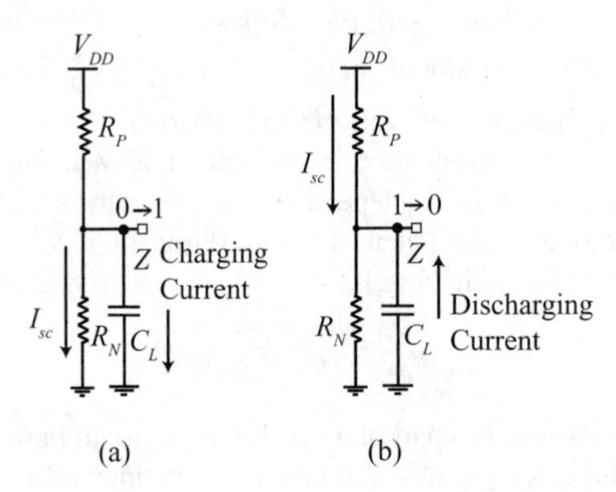

Figure 18.3 Short-circuit current for (a) 0→1 transition and (b) 1→0 transition

This component of power dissipation is known as *short-circuit power* dissipation. Fortunately, the short-circuit time duration t_s is small in practical circuits. Therefore, the energy consumed per transition due to short-circuit is small. We can reduce the short-circuit power dissipation by decreasing the transition time or the slew of the input signal. It allows a CMOS inverter to quickly pass through the short-circuit operating region and reduces the associated power dissipation.

Other complex CMOS logic gates also consume P_{sc}. However, we can observe it only when the pertinent transition occurs at the input pins that creates a short-circuit condition.

The dynamic power dissipated P_{dyn} is the sum of the above two components. Thus,

$$P_{dyn} = P_{sw} + P_{sc} \tag{18.7}$$

[1] Some tools and literature define α as number of transitions per clock cycle. This makes $\alpha = 2$ for the clock signal, and the formula for P_{sw} becomes $P_{sw} = \frac{\alpha}{2} C_L V_{DD}^2 f_{clk}$.

18.1.2 Static Power Dissipation

The static power dissipation (P_{stat}) occurs in a circuit when it is not actively computing and its transistors are not switching while the power supply is connected. It occurs due to the leakage current (I_{leak}) in the transistors of the circuit. Therefore, static power dissipation is also called *leakage power*. The leakage current flows in a MOSFET due to the following reasons:

1. **Subthreshold current:** When the gate voltage V_{GS} of an NMOS is below its threshold voltage V_{TN}, and we apply positive voltage V_{DS} between the source and the drain, the current I_{DS} between them should be ideally zero. However, some high-energy electrons of the source can surmount the gate-induced potential barrier and reach the drain terminal. This flow of electrons constitutes the *subthreshold current*. Similarly, subthreshold current flows in a PMOS also. A simple model for the subthreshold current I_{sub} in a MOSFET is [2, 3]:

$$I_{sub} = I_0 e^{\frac{V_{GS}-V_{TN}}{\eta V_{therm}}} (1 - e^{\frac{-V_{DS}}{V_{therm}}}) \tag{18.8}$$

   where I_0 and η are parameters dependent on the fabrication technology and device dimensions, V_{GS} is the gate–source voltage, V_{TN} is the threshold voltage of the transistor, V_{therm} is the thermal voltage, and V_{DS} is the drain–source voltage. We compute the thermal voltage as $V_{therm} = kT/q$, where q is the electron charge, k is the Boltzmann constant, and T is the absolute temperature. We can infer from Eq. 18.8 that a smaller V_{TN} increases I_{sub}. With the advancement in technology, the threshold voltage of the transistors typically decreases. Therefore, subthreshold leakage current becomes significantly high at advanced process nodes. Additionally, the gate control over the channel potential diminishes at reduced channel lengths. Consequently, it becomes easier for the electrons from the source to reach the drain terminal. As a result, I_{sub} increases at smaller channel lengths. Moreover, due to *drain-induced barrier lowering* (DIBL) I_{sub} increases with the increasing drain voltage.
2. **Gate leakage:** The gate capacitance of a MOSFET can be given as follows:

$$C_g = \frac{\epsilon_r \epsilon_0 A}{t_{ox}} \tag{18.9}$$

   where ϵ_r is the relative permittivity of the gate oxide material, ϵ_0 is the absolute permittivity of vacuum, A is the gate area, and t_{ox} is the thickness of the gate oxide. At advanced process nodes, device dimensions reduce. Thus, for a reduced area, we need to reduce t_{ox} to obtain a given gate capacitance. However, when the gate oxide is thin, some current flows between the channel and the gate terminal *through the gate oxide* due to the tunneling of carriers. It constitutes gate leakage and contributes to the static power dissipation.

   To reduce gate leakage current, we use a *high-k dielectric material*, such as HfO_2 ($\epsilon_r \approx 16$), instead of the traditional SiO_2 ($\epsilon_r = 3.9$) as gate oxide. Thus, we can obtain a given C_g for a higher t_{ox}. A thicker gate oxide suppresses the gate leakage current considerably, and the contribution of gate leakage to the total static power dissipation becomes insignificant.
3. **Junction leakage:** We keep the p–n junction between the drain/source and the substrate in the reverse-biased mode. Nevertheless, some current can flow in the reverse-biased p–n junction. It contributes to the leakage current of the MOSFET.

Among the three components of I_{leak}, the subthreshold current typically dominates. Thus, we can write:

$$I_{leak} \approx I_{sub} \tag{18.10}$$

We can compute the static power dissipation due to the leakage current as:

$$P_{stat} = V_{DD} I_{leak} \tag{18.11}$$

The total power dissipation in a CMOS circuit can be given as:

$$P_{tot} = P_{dyn} + P_{stat} \tag{18.12}$$

18.2 POWER MODELS IN LIBRARIES

Consider a CMOS circuit in which library cell instances are interconnected using physical wires (layers of interconnects). In this circuit, some power gets dissipated due to the switching of capacitors associated with the wires and inputs of library cell instances, while some power gets dissipated *internally* in each library cell instance. The internal power dissipation of a cell instance is due to the switching of internal capacitors, short-circuit current, and leakage current.

The power dissipated within a cell depends on the implementation of that cell. We can characterize each library cell using SPICE simulations and create *power models* for them separately. Moreover, similar to the timing models,[2] we can organize power models of the cells in a technology library. The power analysis tools use the library power models to compute the power dissipated *within* each cell instance. By reusing the library power model for various instantiations and in multiple designs, we can reduce the *effective cost* of developing a power model.

In this section, we explain the library power models. The *dynamic power disspation* and *static power dissipation* are modeled separately in a library.

18.2.1 Dynamic Power Dissipation

Consider the circuit shown in Figure 18.1. The inverter internally consumes the switching power dissipated due to the drain diffusion capacitance C_d and the short-circuit current I_{sc}. Other complex CMOS logic gates can consume power in charging/discharging of the internal node capacitances also. These components of power dissipation depend on the internal implementation of the cell. We model these internally dissipated power in a technology library. We do not model the external switching power dissipated due to C_w and C_I in the library. A power analysis tool can account for them separately with the help of computed C_w and C_I for a given net in a circuit. In practice, we measure internal power dissipated by a cell by subtracting the power dissipated due to the external load from the total power dissipated in the characterization circuit [4].

[2] For details of timing models, see Chapter 13 ("Technology library").

For a complex CMOS logic gate, the output can transition when an input makes a transition, depending on the values at its other input pins. In some cases, an input transition may not lead to an output transition. However, there can be internal power dissipation due to the charging and discharging of internal capacitors. Consequently, even when we do not observe an output transition, a cell can dissipate power internally. We explain it in the following example.

Example 18.1 Consider the NAND gate shown in Figure 18.4(a). Let us examine the power dissipation in it due to the internal node capacitors.

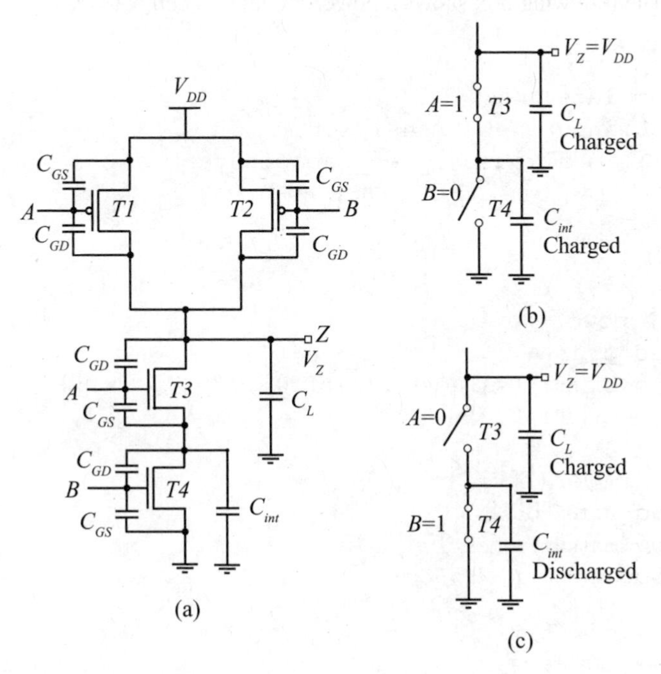

Figure 18.4 Illustration of internal power dissipation (a) two-input NAND gate, (b) internal capacitor C_{int} charged, and (c) internal capacitor C_{int} discharged

There are parasitic capacitors C_{GS} and C_{GD} that couple the input signals to the internal nodes and the output signals. Additionally, there are parasitic capacitors, such as C_{int}, associated with the drain/source junctions. These parasitic capacitors get charged and discharged even when the output signal does not transition [5, 6]. When A=1, B=0, as shown in Figure 18.4(b), C_{int} gets charged. When inputs change to A=0, B=1, as shown in Figure 18.4(c), C_{int} gets discharged, despite no change in the output value of the NAND gate. The charging and discharging of parasitic capacitances inside a CMOS library cell leads to internal power dissipation. Additionally, the feedback due to C_{GS} and C_{GD} can lead to *overshoots* (instantaneous voltage being above V_{DD}) and *undershoots* (instantaneous voltage going below *GND*). They can contribute to additional internal power dissipation.

We model the power dissipated internally by a cell by defining *power arcs*. Similar to the timing arcs, a power arc consists of a pair of source and destination pins. We model power arcs in a technology library using the attribute `internal_power`. We define a power arc on the output pin and specify the corresponding input pin as the `related_pin`.

The internal power dissipation depends on the input slew and the output load, similar to nonlinear delay model (NLDM).[3] Therefore, we model a power arc as two-dimensional tables known as *nonlinear power model* (NLPM). The values in the table represent the *energy per transition* consumed internally by the cell, and can be separately specified for the rise and fall transitions. We illustrate the modeling of power arcs in the following example.

Example 18.2 The following box shows a power model of a cell *NAND2*.

```
lu_table_template (lut1) {
  variable_1 : input_net_transition;
  variable_2 : total_output_net_capacitance;
  index_1 ("0.1, 0.5, 2.0"); index_2 ("0.1, 1.0");
}
cell (NAND2) {
  pin (Z) { ...
    internal_power () {
      related_pin : "A";
      rise_power(lut1) {values("0.4, 0.6","0.7, 0.8","0.8, 0.9"); }
      fall_power(lut1) { ... }
    }
    internal_power () {
      related_pin : "B";
      rise_power(lut1) { ... }
      fall_power(lut1) { ... }
    } ...
```

It shows two power arcs $A \rightarrow Z$ and $B \rightarrow Z$. Each arc has the rise and the fall internal power dissipation modeled separately as a two-dimensional transition-load table.

For digital signals, the number of $0 \rightarrow 1$ transitions is almost equal to the number of $1 \rightarrow 0$ transitions when the number of transitions is large [5]. Hence, we can use the average value of power dissipated in the rise and fall transitions in power estimation. We can also define *when conditions* (using attribute named `when`) with the tables to specify the values at the other pins of the cell. For some transitions at the input pin, the output pin may not transition, as explained earlier. We model internal power dissipation in such cases as a one-dimensional table.

[3] For details of NLDM, see Chapter 13 ("Technology library").

Example 18.3 The following box shows some power arcs of a cell *NAND2*.

```
lu_table_template (hidden) {
  variable_1 : input_net_transition; index_1 ("0.1, 0.5, 2.0");
}
cell (NAND2) {
  pin (A) {
    internal_power () {
      when : "!B & Z";
      fall_power(hidden) {values("0.4,0.6,1.0"); }
      rise_power(hidden) { ... }
    } ...
  pin (B) {
    internal_power () {
      when : "!A & Z";
      fall_power(hidden) {values("0.3,0.7,1.1"); }
      rise_power(hidden) { ... }
    } ...
```

It shows power arcs on the input pins *A* and *B*. The value of other pins or the conditions for which these arcs should be considered is specified using the `when` attribute. The first `when` condition "!B & Z" means that the corresponding arc is for the case when B=0 and Z=1. Additionally, note that there is no output pin related to these arcs because changes in the input pin associated with these arcs do not cause a change in any output pin for the given when condition. Hence, the internal power dissipation for these arcs depends only on the input transition, and the internal pin load is fixed. Hence, these power arcs are modeled simply as a one-dimensional array.

Remark: Similarly, in flip-flops, latches, and multilevel cells, input transitions can lead to internal pin transitions without output pin transition. For example, clock signal transitions in flip-flops can lead to significant internal power dissipation in a flip-flop, while the output pin value remains unchanged. We model such internal power dissipation also as one-dimensional tables at the clock-pin (*CP*).

Note that, we can define `internal_power` both at the pin-level and the arc-level for some input pins. For example, we can define `internal_power` at the *CP* of a flip-flop and also for the $CP \rightarrow Q$ power arcs. Therefore, to avoid double counting, we do not include the `internal_power` defined at the pin-level in the corresponding `internal_power` defined at the arc-level [7].

18.2.2 Static Power Dissipation

The static power dissipated inside a CMOS logic gate depends on the value (0 or 1) at its input pin. We illustrate it using the following example.

Example 18.4 Consider a two-input NAND gate shown in Figure 18.5(a). Let us examine the leakage power dissipation in the four possible combinations of inputs.

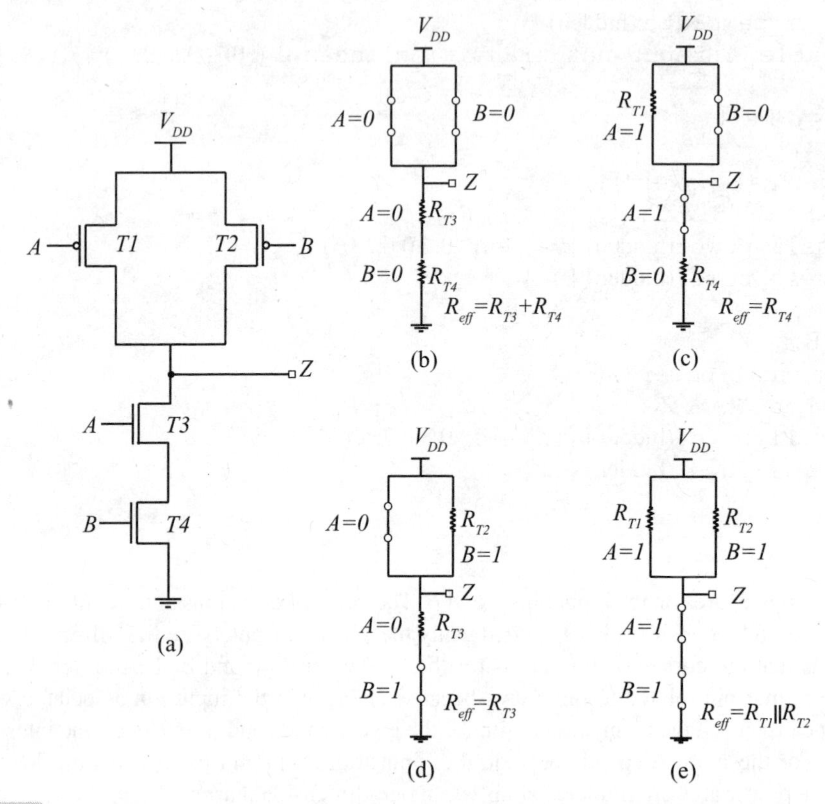

Figure 18.5 Static power dissipation in different states (a) a two-input NAND gate and (b–e) equivalent circuit in different states

Assume that *A*=*B*=0, as shown in Figure 18.5(b). Both the PMOS (*T1* and *T2*) are switched ON. They exhibit a small resistance between the source and the drain terminals in this state. Therefore, we can assume that these terminals are short-circuited. Due to *A*=*B*=0, both the NMOS (*T3* and *T4*) are switched OFF. If these transistors were ideal, we could have considered that their drain and source terminals are open-circuited. However, due to the subthreshold conduction, some current will flow, and we can assume high resistances (R_{T3} and R_{T4}) exist between them. The amount of current flowing from V_{DD} to ground is determined by the effective resistance $R_{eff} = R_{T3} + R_{T4}$. Similarly, we can compute R_{eff} for the other states, as illustrated in Figure 18.5(c–e). Note that the R_{eff} is expected to be maximum in the *A*=*B*=0 state due to series resistances adding up. It is expected to be minimum in the state *A*=*B*=1 due to parallel paths. Thus, the leakage current is expected to be the minimum in the *A*=*B*=0 state and the maximum in the *A*=*B*=1 state.

We model static power dissipation in libraries using the attribute `leakage_power` at the cell level. Since the leakage power of a cell depends on the static value at its pin, we can associate leakage power `value` with its corresponding `when` condition. We can also define the default leakage power of a cell by the attribute `cell_leakage_power` without specifying any `when` condition. The power analysis tools that ignore the states of a cell will report the default `cell_leakage_power`.

Example 18.5 A typical leakage power specification for the two-input NAND gate is shown in the following box.

```
cell (NAND2) { ...
  cell_leakage_power : 125;
  leakage_power () {
    when : "!A & !B"; value : 20; }
  leakage_power () {
    when : "A & !B"; value : 150; }
  leakage_power () {
    when : "!A & B"; value : 200; }
  leakage_power () {
    when : "A & B"; value : 300; } ...
```

It has a default cell leakage of 125 power units. It has also leakage power separately specified for all four possible combinations of the inputs.

18.3 COMPUTING POWER DISSIPATION

We need to compute power dissipation in a circuit for two purposes:

1. **Guide power optimization tools:** Optimization tools need to compare different solutions and pick the more power-efficient ones. To make these comparisons, they need to estimate the power dissipation in a circuit. Often, we can tolerate accuracy loss in the *absolute power* estimates in these applications, though we need to preserve the fidelity of comparison of different solutions. By sacrificing some accuracy in the absolute power estimates, we can reduce runtime and computational effort.
2. **Verification:** We need to compute the absolute level of power dissipated in a circuit to ensure that it meets the allocated power budget. Additionally, we need to compute power dissipation for estimating the voltage drop on the power supply grids and their impact on the circuit timing. For these applications, the accuracy of power computation is critical.

Computing power dissipation is a challenging problem. It involves accounting for the *activity* of signals, in addition to the other circuit details. A signal's activity depends on the application being run on an IC, logical structure, and the circuit topology and is difficult to estimate. Moreover, the power computation varies for different levels of abstraction. When there are fewer details, such as in RTL, the power estimation is crude. As we add more information to a design, we can compute power using extracted capacitance and actual signal waveforms. Though the computational complexity increases with more details being added to a design, the accuracy of power analysis improves.

18.3.1 Activity Measures

Let us represent the signal associated with a net x as a time-dependent Boolean function $x(t)$. As the signal transitions, the total power dissipated for that net depends on the following measures [8]:

1. **Static probabilities:** We denote the fraction of time that the signal $x(t)$ is in the 1 state as P_1^x. For example, $P_1^x = 0.3$ for a clock signal with 30% duty cycle. Similarly, we denote the fraction of time that the signal $x(t)$ is in the 0 state as P_0^x. Evidently, $P_0^x = 1 - P_1^x$. We refer to P_1^x and P_0^x as *static probabilities*. P_1^x is the average value of the signal over a time duration. Thus,

$$P_1^x = \lim_{T\to\infty} \frac{1}{T} \int_{-T/2}^{+T/2} x(t)dt \tag{18.13}$$

2. **Transition rate:** Let us assume that the signal makes $n_x(T)$ transitions in the time interval $(-T/2, +T/2]$. Thus, we can compute the average number of transitions made per unit time as follows [8]:

$$Tr^x = \lim_{T\to\infty} \frac{n_x(T)}{T} \tag{18.14}$$

   We refer to Tr^x as the *transition rate* or toggle rate of the signal associated with the net x. For example, a 1 MHz clock signal has $Tr^x = 2 \times 10^6/s$. Note that the transition rate is twice the frequency because a signal transitions twice in a clock cycle.

We define P_1^x and Tr^x for both *periodic* and *nonperiodic* signals by taking the time duration T to be sufficiently large. Note that both P_1^x and Tr^x are statistical measures of a signal. They abstract out the *time instant* and the slope of the transitions. Still, we can estimate power dissipation using P_1^x and Tr^x, as explained in the following sections.

18.3.2 Estimation of Activity

The key problem in computing power dissipation is estimating the activity measures P_1^x and Tr^x for each net in a design. We can categorize the activity estimation techniques into two types: *simulation-based techniques* and *probabilistic techniques*. The simulation-based techniques are also referred to as *vector-based techniques*, while probabilistic techniques are also referred to as *vector-less techniques*.

Simulation-based Techniques

We can estimate P_1^x and Tr^x using simulation. Depending on the design description, we can perform RTL simulation, gate-level simulation, or transistor-level simulation. We provide the test stimulus or test bench to the simulator. Often, we use the same test bench created for the functional verification for the power computation also.[4] The simulator generates the output response for all the nets in a design. The output of the simulation is typically obtained in *value change dump* (VCD) files. We can convert a VCD file into a format from which the activity measures can be easily extracted, such as *switching activity interchange format* (SAIF) [9–11]. We can provide the SAIF file to the power analysis tool. Sometimes we do not have activity information that we can provide to a power analysis

[4] For details of functional verification, see Chapter 9 ("Simulation-based verification").

tool. In such cases, typically, a power analysis tool will assume some default activity measures for the signals (such as activity=0.2 and static probabilities=0.5).

The delay model plays a critical role in computing the activity of a signal. If we consider zero delays for gates and wires, a net will transition not more than once in each clock cycle. However, if we consider realistic delay models, glitches can occur as signals arrive at a converging point through multiple paths at different time instants. These glitches increase the activity of a signal.

Example 18.6 Consider the circuit shown in Figure 18.6(a). Assume that, initially A=1 and B=0. Hence, X=1 and Z=0. Assume that B transitions from 0→1 at $t = 0$. Let us examine the behavior of the circuit when (a) delay of gates is zero (b) delay of gates is 1 time unit.

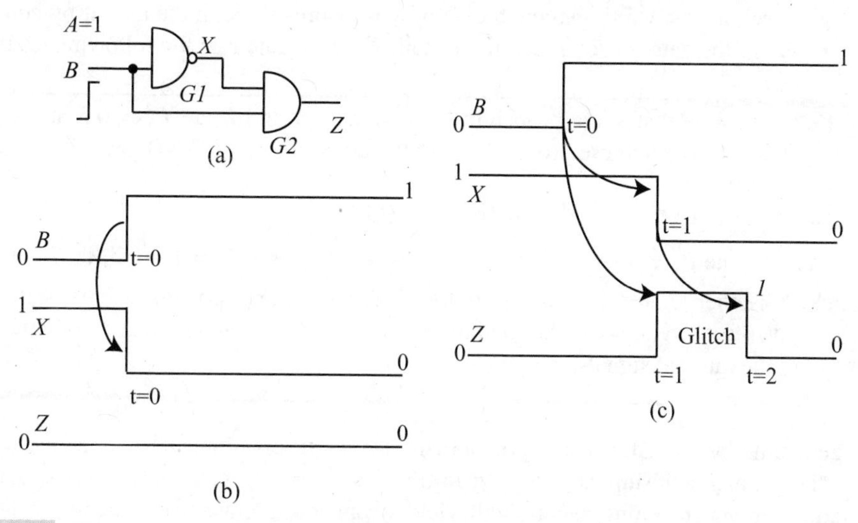

Figure 18.6 Glitches due to delay (a) given circuit, (b) waveform for zero delay model, and (c) waveform when the delay of each gate is 1 time unit

(a) When the gates $G1$ and $G2$ have zero delay, X changes to 0 at $t = 0$ and Z remains unchanged. No glitch is observed, as shown in Figure 18.6(b).

(b) When $G1$ and $G2$ have delay of 1 time unit, we obtain the waveform shown in Figure 18.6(c). The output Z changes to 1 at $t = 1$, as B arrives quickly at the input of $G2$. However, when the effect of B arrives through $G1$, Z restores to 0. Thus, we observe a glitch when we account for gate delays.

In practical circuits, glitches often occur, especially in circuits, such as multipliers, that have re-convergent fanouts. A glitch increases the transition rate of the associated signal and contributes to the switching power dissipation. We need to consider realistic delay models for gates and wires to account for glitches. Therefore, power analysis tools typically consider timing models of technology libraries and wire delays to determine transition rate.

The accuracy of the activity estimates depends on the quality of the test bench. The accuracy typically increases with the increasing number of test stimuli ($T \rightarrow \infty$ in Eqs. 18.13 and 18.14). Practically, a minimal number of test stimuli (thousands) are sufficient to estimate average switching activity for typical combinational logic circuits with reasonable accuracy [12]. To correctly estimate the activity of low-switching nets, we might need to provide many test stimuli. Nevertheless, we can tolerate low accuracy for those nets since they contribute less to the total switching power dissipation.

Estimating activity measures for a sequential circuit is more challenging than that for a combinational circuit. The test bench should ensure that the state space is traversed realistically. Sometimes, it can require a very long sequence of test inputs.

An alternative to the designer-specified testbench is Monte Carlo-based technique [13]. It consists of applying random test stimuli and computing the total power dissipated in the circuit with the help of simulation [13]. The process is continued until we obtain the desired accuracy in the total average power dissipation at a specified confidence level.

Probabilistic Techniques

In probabilistic techniques, we *propagate* the activity measures through the circuit by considering the *logic function* of the gates encountered in the path. We illustrate it in the following example.

Example 18.7 Assume that static probabilities of signals A and B are $P_1^A = 0.5$ and $P_1^B = 0.3$, respectively. Let us propagate these probabilities to the output of (a) an AND gate ($Z = A.B$) (b) an OR gate ($Z = A + B$)

(a) For an AND gate $P_1^Z = P_1^A.P_1^B = 0.5 \times 0.3 = 0.15$.

(b) For an OR gate $P_1^Z = 1 - (P_0^A.P_0^B) = 1 - (1 - 0.5) \times (1 - 0.3) = 1 - 0.5 \times 0.7 = 0.65$.

Remark: For other logic gates, we can frame similar rules and propagate signal statistics in a circuit. Thus, given activity measures at the inputs of a circuit, we can efficiently compute activity measures for all the internal signals.

Though estimating activities using probabilistic techniques is efficient, we face the following difficulties. The signals at the input of a logic gate are not always independent. In such cases, ignoring the correlation among incoming signals will yield wrong results, as illustrated in the following example.

Example 18.8 Consider the circuit shown in Figure 18.7(a). Assume that the static probabilities for the inputs are: $P_1^A = P_1^B = P_1^C = 0.5$. Let us compute static probability at its output by (a) ignoring correlation of signals (b) accounting for correlation of signals.

Computing static probabilities for signals X and Y is straight forward:

$$P_1^X = P_1^A.P_1^B = 0.25$$

$$P_1^Y = P_1^B.P_1^C = 0.25.$$

(a) If we treat X and Y as independent:

$$P_1^Z = 1 - P_0^X.P_0^Y = 1 - 0.75 \times 0.75 = 0.4375.$$

However, we cannot treat the signals X and Y as independent because both depend on B and are *spatially correlated*. Therefore, the above computation is incorrect.

(b) We can derive P_1^Z as follows.

For, an OR gate the output is 1 if any of its input is 1. Hence, we can write:

$$P_1^Z = P(X = 1 \cup Y = 1)$$

The expression on the right-hand side (RHS) represents the probability of $X = 1$ *OR* $Y = 1$. Further, by the rules of probability, we can write:

$$P(X = 1 \cup Y = 1) = P(X = 1) + P(Y = 1) - P(X = 1 \cap Y = 1)$$

We can compute the terms on the RHS as follows:

$$P(X = 1) = P_1^A.P_1^B = 0.5 \times 0.5 = 0.25$$
$$P(Y = 1) = P_1^B.P_1^C = 0.5 \times 0.5 = 0.25$$

The term $P(X = 1 \cap Y = 1)$ denotes the probability of $X = 1$ *AND* $Y = 1$. Hence,

$$P(X = 1 \cap Y = 1) = P_1^A.P_1^B.P_1^C = 0.5 \times 0.5 \times 0.5 = 0.125$$

Therefore,

$$P_1^Z = 0.25 + 0.25 - 0.125 = 0.375$$

We can verify that the above computation is correct using the truth table of the function $Z = AB + BC$, shown in Figure 18.7(b). The truth table has eight rows. Since each input has an equal probability of being 0 and 1 (because static probabilities at inputs are given as 0.5), each truth table row is likely to occur with the probability of $1/8$. The truth table has three rows with Z=1. Therefore, $P_1^Z = 3/8 = 0.375$.

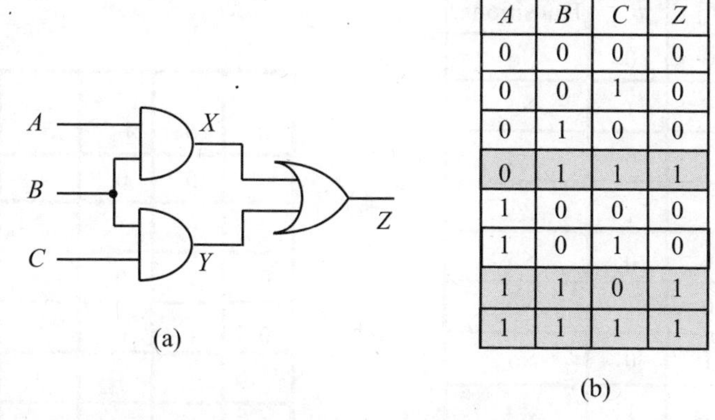

A	B	C	Z
0	0	0	0
0	0	1	0
0	1	0	0
0	1	1	1
1	0	0	0
1	0	1	0
1	1	0	1
1	1	1	1

(b)

Figure 18.7 Spatial correlation in signals (a) given circuit and (b) truth table

Another challenge in computing transition probabilities is accounting for the *temporal correlations* [14]. Such correlations are possible due to the implemented circuit function. For example, flip-flops can be part of a finite state machine (FSM) whose state bits are temporally correlated. Consequently, the outputs of these flip-flops have a temporal correlation. Therefore, logic gates in the fanout of these flip-flops can receive temporally correlated signals. Temporal correlation in signals impacts transition probabilities, and we need to account for them in the computation. We illustrate it in the following example.

Example 18.9 Consider an AND gate $Z = A.B$, shown in Figure 18.8(a). Assume that the input signals are temporally correlated as follows: (a) 1 at A is *immediately* followed by a 0 in the next cycle (b) 0 at B is *immediately* followed by a 1 in the next cycle. Let us compute the transition probabilities for the following cases:
(a) when temporal correlations are ignored
(b) when temporal correlations are considered.

(a) If we ignore temporal correlation, from a given combination of (A, B), the circuit can transition to any combination (A_{next}, B_{next}) in the next cycle. Thus, 16 possible combinations of (A, B) and (A_{next}, B_{next}) are possible, as shown in Figure 18.8(b). Out of 16 possible combinations, 6 lead to transition in Z. Hence, the transition probability of Z is $6/16 = 0.375$.

(b) When we consider temporal correlations, some transitions do not occur. For example, from (A, B)=(0,0) the circuit cannot go to (0,0) or (1,0) in the next cycle because B=1 follows immediately B=0. Therefore, instead of 16 transitions, 9 transitions are possible as shown in Figure 18.8(c). Out of these nine possible transitions, four transitions lead to a transition in Z. Hence, the transition probability of Z is $4/9 = 0.44$. Thus, temporal correlations impact transition probability, and ignoring them can result in the wrong computation of activity measures.

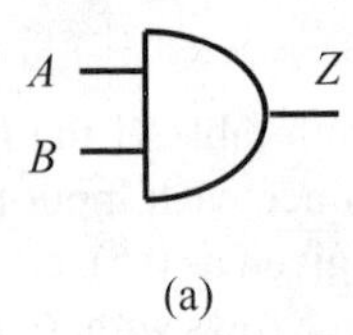

(a)

A	B	A_{next}	B_{next}	Z Transitions?
0	0	0	0	No
0	0	0	1	No
0	0	1	0	No
0	0	1	1	Yes
0	1	0	0	No
0	1	0	1	No
0	1	1	0	No
0	1	1	1	Yes
1	0	0	0	No
1	0	0	1	No
1	0	1	0	No
1	0	1	1	Yes
1	1	0	0	Yes
1	1	0	1	Yes
1	1	1	0	Yes
1	1	1	1	No

(b)

A	B	A_{next}	B_{next}	Z Transitions?
0	0	0	1	No
0	0	1	1	Yes
0	1	0	0	No
0	1	0	1	No
0	1	1	0	No
0	1	1	1	Yes
1	0	0	1	No
1	1	0	0	Yes
1	1	0	1	Yes

(c)

Figure 18.8 Temporal correlation in signals (a) AND gate, (b) transitions with no temporal correlations, and (c) transitions with temporal correlations (A, B are the present values and A_{next}, B_{next} are the next values)

In practice, the probability-based estimation of activity measures is faster than the simulation-based estimation. However, the simulation-based activity estimation is more accurate.

18.3.3 Using Library Power Models

After estimating activity measures, we can compute total power dissipation for a cell using technology libraries and the given circuit conditions. We can obtain total power dissipated by a circuit by adding the power dissipated for each cell. We illustrate computing power dissipation for a cell in the following example.

Example 18.10 Let us consider a NAND gate $Z = (A.B)'$ shown in Figure 18.9. It drives a load of $C_L = 25\,fF$. Assume that the estimated static probabilities and transition rates at the pins are as shown in the figure. Assume that $V_{DD} = 1.0\,V$. Let us compute dynamic and static power dissipation for the NAND gate.

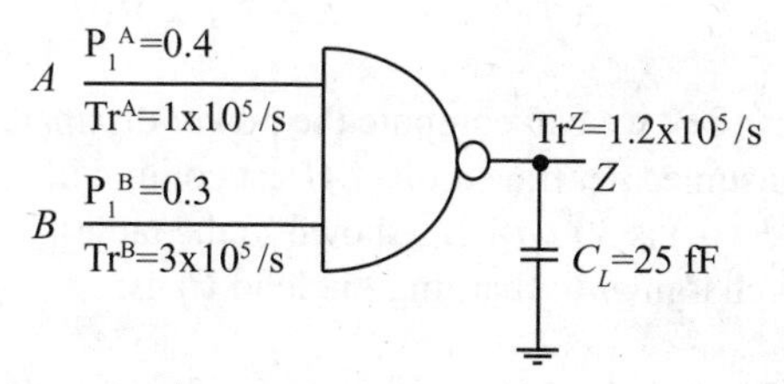

Figure 18.9 A NAND gate with given activity measures

First, we extract the power arc information from the technology library using the input slews at A and B and output load C_L. Assume that the values found by table lookup for these arcs are as shown in Table 18.1 (first three columns). We take the average of internal power dissipation values extracted for the rise and fall transitions.

Table 18.1 Internal dynamic power computation

Pin/Arc	When Condition	Average Rise/Fall Energy (E) (fJ)	Transition Rate (Tr) (/s)	Power ($P = E \times Tr$) (pW)
$A \to Z$	-	2.0	0.3×10^5	60
$B \to Z$	-	2.5	0.9×10^5	225
A	$!B\&Z$	0.3	0.7×10^5	21
B	$!A\&Z$	0.4	2.1×10^5	84
	Total internal dynamic power ($P_{d,int}$)			390

The transition rate at the output Tr^Z is $1.2 \times 10^5/s$. The transition rate at the inputs is $Tr^A = 1 \times 10^5/s$ and $Tr^B = 3 \times 10^5/s$. We can attribute the output transition to the input transitions in proportion to their transition rates. Thus, we can write:

$$Tr^{A\to Z} = Tr^Z \times \frac{Tr^A}{Tr^A + Tr^B} = 1.2 \times 10^5 \times \frac{1}{1+3} = 0.3 \times 10^5/s$$

$$Tr^{B\to Z} = Tr^Z \times \frac{Tr^B}{Tr^A + Tr^B} = 1.2 \times 10^5 \times \frac{3}{1+3} = 0.9 \times 10^5/s$$

Further, we can compute the input transition rates that resulted in no output transition as follows:

$$Tr^{A\nrightarrow Z} = Tr^A - Tr^{A\to Z} = 0.7 \times 10^5/s$$

$$Tr^{B\nrightarrow Z} = Tr^B - Tr^{B\to Z} = 2.1 \times 10^5/s$$

Using the transition rates (*Tr*), we can compute the power dissipated (*P*) by multiplying it with the corresponding energy consumed per transition (*E*) (last column in Table 18.1). The total internal dynamic power dissipation ($P_{d,int}$) is 390 pW (as shown in the table).

The power dissipated in charging/discharging the load C_L is:

$$P_{C_L} = \frac{1}{2} \times C_L V_{DD}^2 Tr^Z = \frac{1}{2} \times 25 \times 10^{-15} \times (1.0)^2 \times 1.2 \times 10^5\ W = 1500\, pW$$

Note that the factor $\frac{1}{2}$ appears in the above equation because we count rise and fall transitions separately in our definition of the transition rate *Tr*. The total dynamic power dissipation is:

$$P_{dyn} = P_{d,int} + P_{C_L} = 390 + 1500 = 1890\, pW$$

We can compute the static power dissipation using the values specified in the technology library for each state (defined by the `when` conditions). Assume that the leakage power values extracted from the technology library for the NAND gate are as shown in Table 18.2 (first two columns).

Table 18.2 Static power computation

State (When)	State Leakage Power ($P_{L,S}$) (pW)	Probability (P_1)	Leakage Power ($P_L = P_{L,S} \times P_1$) (pW)
$A = 0\ B = 0$	20	0.7×0.6=0.42	8.4
$A = 0\ B = 1$	200	0.7×0.4=0.28	56
$A = 1\ B = 0$	150	0.3×0.6=0.18	27
$A = 1\ B = 1$	300	0.3×0.4=0.12	36
		Total static power (P_{stat})	127.4

Using the static probabilities $P_1^A = 0.3$ and $P_1^B = 0.4$, we can compute the static probabilities of each state, as shown in column 3. By multiplying the state-dependent leakage power ($P_{L,S}$) with its corresponding probability (P_1), we obtain the leakage power (P_L) for each state (last column). The total leakage power is $P_{stat} = 127.4\,pW$.

The total power dissipated by the NAND cell is $P_{tot} = P_{dyn} + P_{stat} = 1890 + 127.4 = 2017.4\,pW$.

18.4 RECENT TRENDS

A power analysis tool faces different challenges at different abstraction levels. For RTL power analysis, the ability to perform what-if analysis, evaluate power-driven optimizations, and correlate with the physical design metrics are critical. At the gate level comprehending advanced library power models, accomodating glitches in the analysis, and exploring power–delay interactions are crucial. Therefore, we need continuous advancements in power analysis tools to tackle these challenges. Moreover, with the increasing design complexity, power analysis tools need to be scalable and produce results on larger designs within a reasonable time. Electronic design automation (EDA) tools typically tackle these problems using multithreading. In the future, advanced data mining techniques can help improve the scalability and debuggability of power analysis tools.

REVIEW QUESTIONS

18.1 What are different components of power dissipation in a CMOS circuit?

18.2 Why is the supply voltage a dominant factor in deciding the power dissipation in a CMOS circuit?

18.3 Why is the contribution of static power dissipation to the total power dissipation increasing with the advancement in technology nodes?

18.4 For a technology node, there are two types of n-type MOSFETS. One type of transistor operates at a threshold voltage of 0.2 V. The other works at a threshold voltage of 0.4 V. Assuming that both transistors operate at the same supply voltage, compute the ratio of static power dissipation in the two transistors. Assume that $\eta = 1.2$, $V_{therm} = 26$ mV and $I_0 = 1$ nA are the same for both types of transistors (symbols have same meaning as in Eq. 18.8). Assume that the subthreshold leakage current dominates static power dissipation.

18.5 Name the factors on which the short-circuit power dissipation in a CMOS circuit depends.

18.6 What does the internal power of a CMOS logic gate in a technology library represent?

18.7 Where does internal power dissipation occur in a CMOS logic gate even when there is no transition at its output?

18.8 Why does static power dissipation in a CMOS logic gate depend on its state?

18.9 Why does activity depend on the delay of logic gates in a circuit?

18.10 A circuit operates at a 1 GHz clock frequency and 1 V supply voltage. The output of an inverter switches as shown below. It drives a load of 1 fF. Compute the switching power dissipation of the inverter (Figure 18.10).

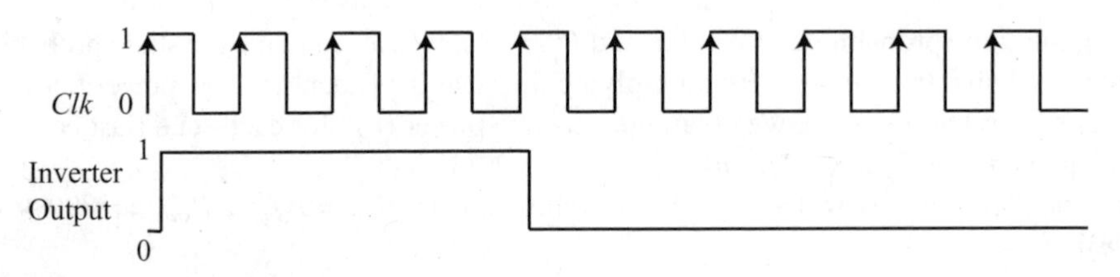

Figure 18.10 Waveform for Q 18.10

TOOL-BASED ACTIVITY

In this activity, use any gate-level power analysis tool, including open-source tools such as OpenSTA [15]. You can use any library, including freely available technology libraries meant for the academic purposes [16].

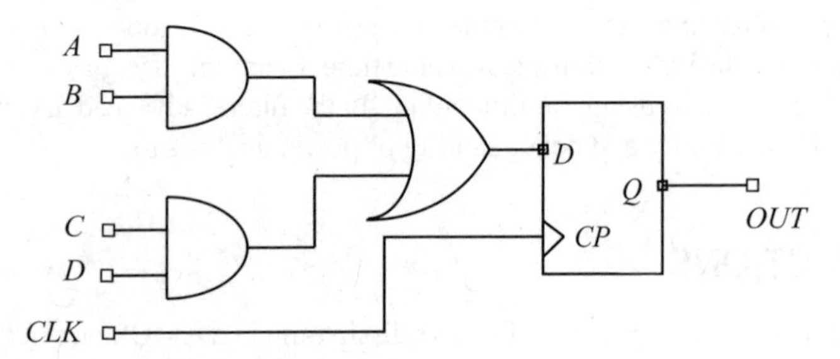

Figure 18.11 The circuit for the power analysis

Consider the circuit shown in Figure 18.11. Perform the following tasks:

1. Write a Verilog netlist for the above circuit. Use cells from the chosen technology library.
2. After specifying suitable clock frequency and global activity, perform power calculation for the complete circuit. Examine the static and dynamic power dissipated by each circuit element.
3. Compute the static power dissipated by each circuit element manually. Use the information of the technology library (power tables), circuit conditions (load, frequency, and voltage), and the specified activity.
4. Compare the results obtained by the tool with the manual computation.

REFERENCES

[1] N. H. Weste and D. Harris. *CMOS VLSI Design: A Circuits and Systems Perspective*. Pearson Education India, 2015.

[2] Y. Taur and T. H. Ning. *Fundamentals of Modern VLSI Devices*. Cambridge University Press, 2013.

[3] K. Roy, S. Mukhopadhyay, and H. Mahmoodi-Meimand. "Leakage current mechanisms and leakage reduction techniques in deep-submicrometer CMOS circuits." *Proceedings of the IEEE* 91, no. 2 (2003), pp. 305–327.

[4] M. Martins, J. M. Matos, R. P. Ribas, A. Reis, G. Schlinker, L. Rech, and J. Michelsen. "Open cell library in 15nm FreePDK technology." *Proceedings of the 2015 Symposium on International Symposium on Physical Design* (2015), pp. 171–178.

[5] J.-Y. Lin, T.-C. Liu, and W.-Z. Shen. "A cell-based power estimation in CMOS combinational circuits." *Proceedings of the 1994 IEEE/ACM International conference on Computer-aided Design* (1994), pp. 304–309.

[6] W.-Z. Shen, J.-Y. Lin, and F.-W. Wang. "Transistor reordering rules for power reduction in CMOS gates." *Proceedings of ASP-DAC'95/CHDL'95/VLSI'95 with EDA Technofair* (1995), pp. 1–6, IEEE.

[7] T. Cui, Q. Xie, Y. Wang, S. Nazarian, and M. Pedram. "7nm FinFET standard cell layout characterization and power density prediction in near-and super-threshold voltage regimes." *International Green Computing Conference* (2014), pp. 1–7, IEEE.

[8] F. N. Najm. "Transition density: A new measure of activity in digital circuits." *IEEE Transactions on Computer-aided Design of Integrated Circuits and Systems* 12, no. 2 (1993), pp. 310–323.

[9] "IEEE standard for design and verification of low-power, energy-aware electronic systems." *IEEE Std 1801-2018* (2019), pp. 1–548.

[10] R. Chadha and J. Bhasker. *An ASIC Low Power Primer: Analysis, Techniques and Specification.* Springer Science & Business Media, 2012.

[11] M. Srivastav, S. Rao, and H. Bhatnagar. "Power reduction technique using multi-Vt libraries." *Fifth International Workshop on System-on-chip for Real-time Applications (IWSOC'05)* (2005), pp. 363–367, IEEE.

[12] J. Monteiro, R. Patel, and V. Tiwari. "Power analysis and optimization from circuit to register-transfer levels." *Electronic Design Automation for IC Implementation, Circuit Design, and Process Technology: Circuit Design, and Process Technology* 3 (2016), p. 57.

[13] R. Burch, F. N. Najm, P. Yang, and T. N. Trick. "A Monte Carlo approach for power estimation." *IEEE Transactions on Very Large Scale Integration (VLSI) Systems* 1, no. 1 (1993), pp. 63–71.

[14] M. Pedram. "Power minimization in IC design: Principles and applications." *ACM Transactions on Design Automation of Electronic Systems (TODAES)* 1, no. 1 (1996), pp. 3–56.

[15] T. Ajayi, V. A. Chhabria, M. Fogaça, S. Hashemi, A. Hosny, A. B. Kahng, M. Kim, J. Lee, U. Mallappa, M. Neseem, et al. "Toward an open-source digital flow: First learnings from the OpenROAD project." *Proceedings of the 56th Annual Design Automation Conference 2019* (2019), pp. 1–4.

[16] M. Martins, J. M. Matos, R. P. Ribas, A. Reis, G. Schlinker, L. Rech, and J. Michelsen. "Open cell library in 15nm FreePDK technology." *Proceedings of the 2015 Symposium on International Symposium on Physical Design* (2015), pp. 171–178.

19

Power-driven Optimizations

It is impossible to enjoy idling thoroughly unless one has plenty of work to do ... Idleness ... to be sweet must be stolen.

—Jerome K. Jerome, "On being idle," *Idle Thoughts of an Idle Fellow*, 1886

Power-driven optimization is an integral part of a design flow. Various design steps, such as system-level design, logic synthesis, and physical design, consider reducing power dissipation as one of their objectives. We carry out power-driven optimization and include power reduction techniques throughout the flow. Nevertheless, for easy readability, we present a consolidated view of these techniques in this chapter.

19.1 MOTIVATION

We need to reduce the power dissipated by an integrated circuit (IC) due to the following reasons:

1. The energy that a circuit draws from the power source gets stored internally or gets dissipated to the environment through *packaging* and *heat sinks* [1]. We can relax the cooling requirement of an IC by reducing its power dissipation. Thus, it allows us to use simpler and less costly packaging and heat sinks.
2. An IC draws power from a battery, especially in portable devices such as mobiles and laptops. For a given battery, we can reduce the frequency of recharges by reducing its energy consumption. To an approximation, we can view a fully charged battery as delivering a fixed amount of energy. Therefore, we need to reduce the average power or total energy dissipated by the circuit to reduce the frequency of recharges [2]. Alternatively, we can reduce the battery weight by reducing the average power dissipation for a given recharging frequency. From the environmental perspective too, consuming less power is desirable.
3. The power dissipated in an IC gets manifested as an increase in its temperature. When the temperature of an IC increases, some device failure mechanisms exacerbate. By reducing power dissipation in an IC, we can avoid significant temperature increases and the associated reliability issues.

When we reduce the power dissipation in an IC, we often sacrifice other figures of merit, such as performance and area. Therefore, as a designer, power reduction is never our sole objective. We intelligently trade-off other metrics such as performance while achieving power reduction.

Depending on the applications, such as mobiles, embedded systems, laptops, and high-end servers, we can give more preference to power reduction over other figures of merit [3].

19.2 STRATEGIES

We can devise power reduction techniques by observing the parameters that impact power dissipation. In Chapter 18 ("Power analysis"), we have discussed that the total power dissipated P_{tot} is given as follows[1]:

$$P_{tot} = \alpha C_L V_{DD}^2 f_{clk} + V_{DD} I_{sc} + V_{DD} I_{leak} \tag{19.1}$$

Evidently, supply voltage (V_{DD}) is the most dominant parameter that impacts power dissipation. Therefore, we can reduce power dissipation by lowering V_{DD}. In the limiting case, we can reduce V_{DD} to zero, or we can switch OFF the power supply. Indeed, the most effective strategy to reduce power dissipation is to keep a circuit powered OFF whenever possible. It eliminates both dynamic and static power dissipations.

We can employ switching OFF strategies in the pre-RTL, RTL, logic, and circuit levels. However, they are more effective at a higher level of abstraction. Consequently, we can obtain more power savings by devising power-optimizing strategies at a higher level of abstraction. The percentage of power-saving reduces down the design flow. Therefore, it is prudent to plan and implement power-saving strategies in a top-down manner in VLSI design flow [4, 5]. Other power reduction techniques rely on reducing the load capacitance C_L, switching activity α, frequency of operation f, short-circuit current I_{sc}, and the leakage current I_{leak}.

We have summarized power reduction techniques in Table 19.1 and will describe them in the following sections. However, note that Table 19.1 is not an exhaustive list of power reduction techniques, and we can devise specialized power reduction techniques based on the application.

Table 19.1 Summary of power reduction techniques

Level	Tasks	Example Techniques
pre-RTL	HW/SW partitioning	1. energy-efficient HW for computational kernel
	Behavior synthesis	1. schedule that allows switching off FUs 2. resource allocation to reduce capacitance and switching
	Memory design	1. partition address space to exploit sleep mode
	Communication link design	1. reduce voltage swings of signals 2. data encoding to reduce power 3. power-efficient newtork architecture
	Operating system design	1. dynamic power management 2. dynamic voltage/frequency scaling
RTL	Coding, simulation, synthesis, and verification	1. partition into power domains and clock domains 2. power gating and clock gating 3. finite state machine (FSM) encoding such as Gray encoding

[1] For an explanation of the formula, see Section 18.1 ("Components of power dissipation").

Logic	Logic synthesis and verification	1. path equalization 2. precomputing output logic
Circuit	Physical design and verification	1. resizing cells 2. multi-V_T libraries

19.3 PRE-RTL TECHNIQUES

Some examples of power reduction techniques employed in the pre-RTL stages are as follows [2, 6].

1. **Hardware–software (HW–SW) partitioning:** We can find a cluster of instructions that consume most runtime for a system. We refer to such a cluster of instructions as the *computational kernel*. We map the computational kernel to dedicated energy-efficient hardware, i.e., hardware that consumes less energy for the given task to complete. During system operation, when the computational kernel is active, we switch on only the dedicated energy-efficient hardware and switch off other inactive components in the system [7]. Thus, we can save power significantly.
2. **Behavior synthesis:** Two key tasks in behavior synthesis are: *scheduling* (mapping operations to clock cycles) and *resource allocation* (mapping operations to hardware). We can make these tasks power-aware [8, 9]. For example, behavior synthesis can produce a schedule that allows maximizing the power-off time for the functional units [9]. Moreover, it can allocate resources such that the capacitance and the transition rate reduce on the power-consuming data paths [8].
3. **Memories:** We can reduce the energy consumed by memories by partitioning the address space such that the *sleep* mode can be fully exploited [10]. In the sleep mode, we can reduce energy consumption by de-activating the memory refresh, switching off the power supply, or disabling the clock signals.
4. **Communication links:** We can reduce power dissipation by lowering the voltage swing of the interface signals [11]. We can also encode the transmitted data such that it reduces transitions, especially for data transfers that involve charging/discharging of large capacitors [12, 13].

Example 19.1 Assume that we have transmitted 00000000 in the current cycle, and we need to send 10111111 in the next cycle. It will require 7 transitions in the bus.

Instead, we can transmit 01000000 and assert a line *isInverted* to indicate that the receiver needs to invert the received value. Thus, it will require only two transitions (one in the encoded data and one in the *isInverted* line).

We can also use power-efficient network architectures, such as structured wireless *networks-on-chip* (NoC), to reduce power dissipation in the on-chip communications [14–16].

5. **Operating system (OS):** At the system-level we can implement *dynamic power management* that shuts down or slows down the components lying idle [6, 17]. Depending on the workload and some predefined *policy*, the OS can trigger an appropriate power-saving scheme in the hardware. For example, it can shut down some components in a laptop after it meets a policy-defined inactivity measure. Delegating the power management task to the OS simplifies reconfiguring and implementing power-saving policies [6]. Nevertheless, the hardware needs to provide the required hooks to the OS because it is only the hardware that dissipates power.

A popular implementation of dynamic power management is *dynamic voltage/frequency scaling* (DVFS) for processors [18–21]. It exploits workload variations to save energy. Typically, the full speed of a processor is utilized by only a few tasks or for a small time duration. For the remaining period, the deadlines can be met at low speed and consuming significantly less energy.

In the DVFS scheme, the OS varies the supply voltage and the processor's clock frequency by monitoring the workload. When instructed by the OS, the processor provides the highest speed. During the remaining period, it consumes minimum energy by operating at reduced voltage and frequency.

Example 19.2 Consider a processor that can perform a task in 10 *ms* at 1.2 *GHz* and 1.2 *V*. When we reduce the clock frequency and the supply voltage to half (600 *MHz*, 0.6 *V*), the task will complete in double the original time, i.e., 20 *ms*. If we can afford the task completion in 20 *ms*, we can make the above change to the system. It will reduce the switching power dissipation by 1/8 ($P_{sw} \propto V_{DD}^2 f_{clk}$) and the energy consumption by 1/4 (less than power dissipation saving because the task continues for a longer duration).

The block diagram for a hardware implementation of the DVFS scheme is shown in Figure 19.1. The core logic provides the workload information to the *DVFS controller*. The DVFS controller decides the minimum frequency f_{min} at which the core logic can meet the deadline. It provides the control signal to the *clock generator*. The clock generator applies a clock signal with frequency f_{min} to the core logic. For a given frequency f_{min}, the DVFS controller also determines the safe voltage V_{DD} for the core logic. One of the most straightforward techniques to obtain V_{DD} is by looking it up from a pre-characterized f_{min}-V_{DD} table. The DVFS controller applies a control signal to the *switching voltage regulator* that steps down the external battery voltage to V_{DD}. We can improve the DVFS implementation by providing a *feedback* of variations in temperature, process, and supply voltage from the core logic to the DVFS controller.

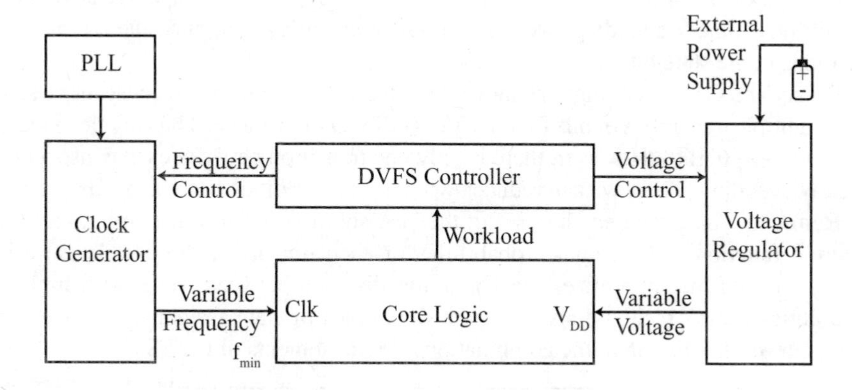

Figure 19.1 Hardware implementation of DVFS scheme

19.4 REGISTER TRANSFER LEVEL TECHNIQUES

The power-specific decisions made at the system and architectural levels need to be implemented and supported in the RTL. For example, if we have designed a DVFS scheme at the system level, it needs to be implemented in the corresponding hardware. Moreover, we need to verify these features in the RTL using simulation, static timing analysis, power analysis, and formal verification. Thus, we need to consider low-power design features throughout a design flow. Some of the other power-saving techniques employed in RTL are as follows:

1. **Power domains and clock domains:** We can save power in the non critical functional units by employing lower supply voltage. It requires partitioning a design into multiple *power domains* and mapping each functional unit to an appropriate power domain. Another technique to reduce power dissipation is to run the non critical functional units with a slower clock. It lowers the transition rate of signals and saves dynamic power dissipation. It requires implementing multiple *clock domains* in a circuit. We will explain these techniques in detail later in this chapter.
2. **Power gating and clock gating:** We can reduce static power dissipation by inserting a *power switch* between the logic circuit and the ground. The power switches are implemented using low leakage transistors. In the *sleep* mode, we turn off the power switches and save power dissipation. This technique of reducing power is known as *power gating*. Similarly, we can selectively stop the clock propagation when the associated flip-flops do not perform any useful computation. This technique is referred to as *clock gating*. It can result in substantial power savings because clock distribution networks consume high switching power in a synchronous circuit. We will discuss these techniques in more detail later in this chapter.
3. **FSM encoding:** If an FSM often transitions from a state p to a state q, we should assign codes with minimum *Hamming distance*[2] between them [22]. It minimizes the switching activity at the output of the flip-flops.

Example 19.3 Consider an FSM with sixteen states and represented using four state bits. Assume that the FSM transitions from $s_0 \rightarrow s_1 \rightarrow s_2 ... s_{14} \rightarrow s_{15} \rightarrow s_0 ...$ like a counter. Instead of traditional binary encoding, we can use *Gray encoding* to reduce the overall number of transitions in the state bits.

Binary and Gray encodings are shown in Table 19.2. Between two consecutive states, there can be multiple transitions of bits for binary encoding. For instance, 3 bits toggle in the transition s_3 (0011) $\rightarrow$ s_4 (0100). However, there is only one transition between each consecutive state in the Gray encoding. Thus, we can reduce switching power dissipation using Gray encoding.

Remark: The Gray encoding is not the best solution for an FSM with an arbitrary state transition function and unequal probabilities of each transition. We can devise a heuristic-based algorithm that minimizes the Hamming distance between states with high transition probabilities [22]. We should also consider the impact of state assignment on the complexity and the power dissipation in the combinational logic elements of the FSM.

---

[2] For definition of Hamming distance, see Section 12.3.2 ("State encoding").

Table 19.2 Binary and Gray encodings for state bits

State	Binary Encoding	Gray Encoding
s_0	0000	0000
s_1	0001	0001
s_2	0010	0011
s_3	0011	0010
s_4	0100	0110
s_5	0101	0111
s_6	0110	0101
s_7	0111	0100
s_8	1000	1100
s_9	1001	1101
s_{10}	1010	1111
s_{11}	1011	1110
s_{12}	1100	1010
s_{13}	1101	1011
s_{14}	1110	1001
s_{15}	1111	1000

19.5 LOGIC-LEVEL TECHNIQUES

We can apply power minimization techniques during various logic synthesis steps. For example, we can include some measure of power dissipation in the cost function in the common subexpression extraction, don't care optimizations, state assignment, and technology mapping [23–26]. It will enable logic synthesis to produce a netlist that draws less power. Additionally, we can apply the following power reduction techniques at the logic level.

1. **Path equalization:** We can reduce glitches in a circuit by equalizing the delay of paths converging at the inputs of a gate. It allows the input signals to arrive at the gate almost simultaneously. Thus, the output value becomes stable by making only one transition. We can restructure logic to balance delays through multiple paths. We can also add buffers to make the signals arrive simultaneously.

Example 19.4 Consider the circuit shown in Figure 19.2(a). Assume that, initially A=1 and B=0. Hence, X=1 and Z=0. Assume that B transitions from 0→1 at $t = 0$. We have seen in Ex. 18.6 that for unit delay of gates we obtain glitches in the output, as shown in 19.2(b).

However, we can equalize delay from the inputs to the output by adding a buffer $B1$ with unit delay, as shown in Figure 19.2(c). As a result, the arrival time of the signal at X and Y becomes equal and the glitch at Z gets eliminated, as shown in Figure 19.2(d). Note that the critical delay of the circuit does not change by path equalization. However, we should account for the additional capacitance of the inserted buffer *B1* that can offset the gains. This technique is effective, especially in arithmetic circuits such as multipliers [17].

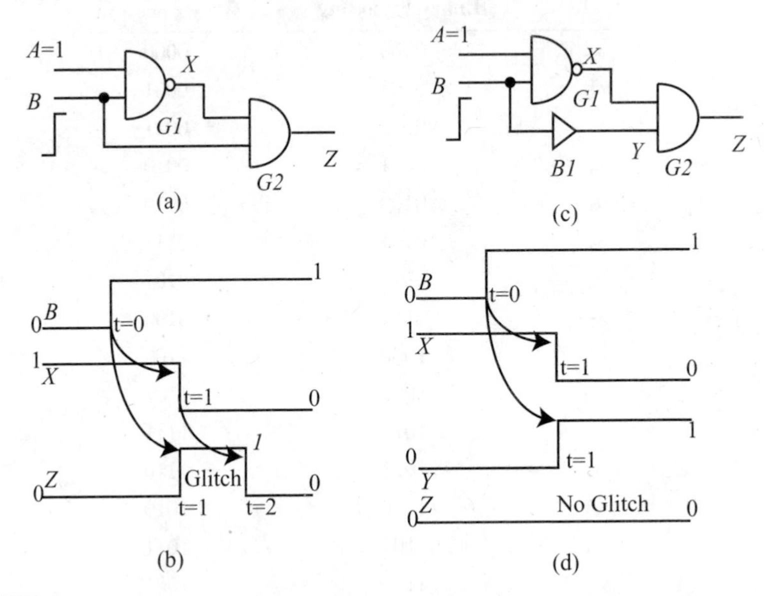

Figure 19.2 Delay equalization (a) initial circuit, (b) waveform with glitch, (c) buffer *B1* with unit delay is inserted, and (d) waveform with no glitch

2. **Precomputing output logic:** We can reduce activity in a circuit by precomputing the output values for a subset of input values one clock cycle before they are required [27]. If the output values can be precomputed using the subset of input values, we can turn off the original circuit in the next clock cycle and reduce switching activity.

Example 19.5 Consider the circuit shown in Figure 19.3(a). It computes $A > B$ for two n-bit numbers $A_{n-1}...A_1A_0$ and $B_{n-1}...B_1B_0$. On receiving the clock signal, the bits at the input of register banks *R1* and *R2* is presented to the combinational block $A > B$ and the result becomes available at *R3* in the next clock cycle. Note that the most significant bits A_{n-1} and B_{n-1} are sufficient to compute Z in the following cases:

$$Z = 1 \text{ if } A_{n-1} = 1 \text{ and } B_{n-1} = 0$$

$$Z = 0 \text{ if } A_{n-1} = 0 \text{ and } B_{n-1} = 1$$

Therefore, computation and associated switching power dissipation for lower-order bits in the $A > B$ block are unnecessary in these cases.

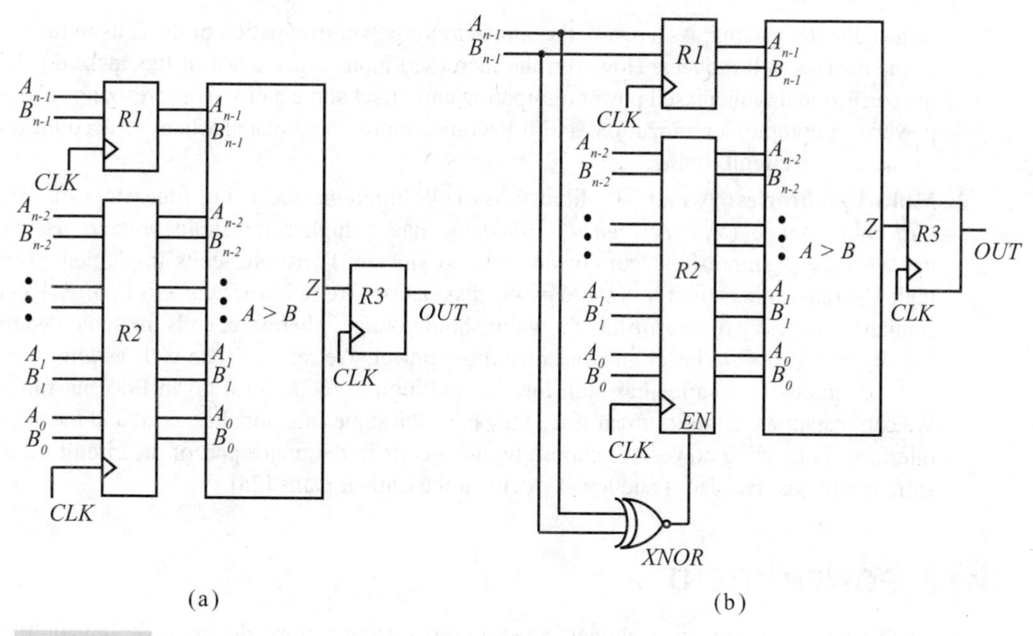

Figure 19.3 Precomputation (a) original circuit and (b) transformed circuit [27]

We can eliminate the unnecessary switching in the $A > B$ block by precomputing the output Z for the above cases. In other cases (A_{n-1} XNOR $B_{n-1} = 1$), we allow the lower-order bits to reach the $A > B$ block, as shown in Figure 19.3(b). Note that the *EN* pin of the register bank *R2* becomes 1 only when A_{n-1} XNOR $B_{n-1} = 1$. In other cases, *EN*=0 avoids switching of the lower-order bits in the $A > B$ block. Assuming that the inputs have a uniform probability, the XNOR gate evaluates to 1 with a probability of 0.5. Hence, we can reduce switching power dissipation by 50% (ignoring the power dissipation in the XNOR gate).

19.6 CIRCUIT-LEVEL TECHNIQUES

We can employ power reduction techniques during physical design to reduce power dissipation at the circuit level. The key idea is to reduce the capacitance of nets having high activity. During partitioning, we avoid cuts for nets having high activities because it can result in increased capacitance. During placement and routing, we assign greater weights or priorities to nets having high activity. It allows us to reduce the capacitance of highly active nets. For example, the placer can keep the cells connected to a highly active net nearby and reduce the net capacitance and the associated power dissipation. Some other power reduction techniques employed at the circuit level are as follows:

1. **Resizing:** We can *resize* cells to reduce power dissipation. For example, we can use smaller cells in the noncritical path of a circuit. It will reduce the power dissipation due to the reduced load capacitances. Sometimes we can reduce overall power dissipation by increasing the size of a heavily loaded cell. The output *slew* of the heavily loaded cell reduces (transition becomes

faster) due to upsizing. As a result, the short-circuit power dissipation of the cells in the fanout of the upsized cell reduces. However, the increased input capacitance of the upsized cell and the corresponding increased power dissipation can offset some gains. Thus, resizing for saving power is a complex optimization, and it becomes more complicated when we account for its impact on the circuit timing.

2. **Multi-V_T libraries:** A multi-V_T library has cells implemented using transistors of varying threshold voltage (V_T). A *high-V_T transistor* has a higher threshold voltage. Hence, it has smaller subthreshold current and is also slower. Therefore, cells implemented using high-V_T transistors exhibit less static power dissipation. We refer to such cells as *high-V_T cells*. Similarly, a *low-V_T transistor* has a low threshold voltage. Therefore, cells implemented using low-V_T transistors are faster and dissipate more power. We refer to such cells as *low-V_T cells*. A set of multi-V_T libraries has both low-V_T and high-V_T cells for a given Boolean function. We can instantiate either of them since they have the same functionality, area, and the external interface. For saving power, we choose high-V_T cells in the major part of the circuit (that has sufficiently positive slack) and low-V_T cells in the critical paths [28].

19.7 POWER INTENT

We implement low-power methodologies in design flows using various electronic design automation (EDA) tools. We need to provide these tools with power-related information such as the power domains in the design, supply voltage levels, mapping of nets with voltage levels, power modes, and specially designed circuit elements for power management. We refer to these power-related design specifications as *power intent*.

Design implementation tools produce features in a design for the given power intent. For example, a cell should be picked from a given set of technology libraries according to its power domain. After implementation, we need to verify whether the design honors the given power intent. Thus, the EDA tools must use the power intent consistently in system design, synthesis, verification, physical design, and testing. We can accomplish this by capturing the power intent in a portable *unified power format* (UPF) (IEEE Std 1801) files [29]. Most of the power-aware EDA tools support UPF. We explain UPF and how it captures the power intent in the following paragraphs.

19.7.1 Unified Power Format

Similar to Synopsys Design Constraint (SDC), UPF is based on Tool Command Language (Tcl). It provides *commands* to capture the power intent in a syntax consistent with the Tcl. Note that hardware description languages (HDLs), such as Verilog and VHDL, are not well equipped to carry the power intent. Moreover, UPF allows us to segregate the power intent from the logical intent captured in HDL. Thus, for the same logical design described in HDL, we can have multiple power management strategies and power intent captured by their corresponding UPF file sets. Hence, UPF provides modularity, reusability, and maintainability to the low-power design methodologies.

UPF file sets form part of the complete design description. We provide UPF files along with the logic description (RTL or netlist) to the implementation and verification tool, as shown in Figure 19.4(a). However, note that the UPF files can be successively refined in the design flow, as shown in Figure 19.4(b).

During initial stages of a design flow, we define power intent abstractly and do not refer to the implementation and process technology. We refer to the corresponding UPF fileset as the *constraint UPF*. In system-on-chip (SoC) design methodology, intellectual property (IP) vendors can provide the constraint UPF. A constraint UPF describes the power domains, power states, retention, and isolation requirements of an IP. We use commands such as `create_power_domain`, `add_power_state`, `set_retention_elements`, and `set_port_attributes` to define constraint UPF.

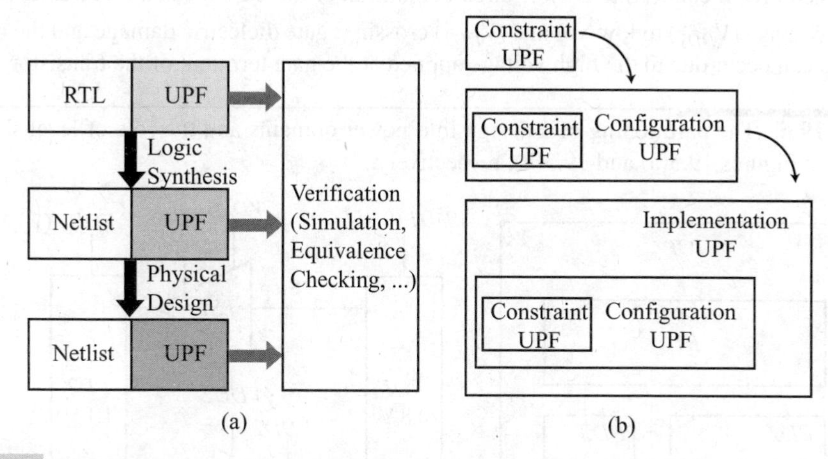

Figure 19.4 UPF-based design flows (a) sharing of information and (b) successive refinement [29]

When we integrate IPs at the system level, we define the *configuration UPF*. It creates design ports for power management control and defines instance-specific retention and isolation *strategies*. Note that while defining configuration UPF, we cannot change the corresponding constraint UPF, but need to honor it. We use commands such as `create_logic_port`, `create_logic_net`, `connect_logic_net`, `set_isolation`, `set_retention`, and `add_power_state -update` to define configuration UPF.

Finally, the *implementation UPF* defines the low level details such as power distribution network, control logic for power management, and technology-specific information. We use commands such as `create_supply_port`, `create_supply_net`, `create_supply_set`, and `create_power_switch` to define implementation UPF. We use the command `add_power_state` with appropriate options to define voltage values for each supply sets in the implementation UPF.

The above three types of UPF files taken together constitute the power intent. We provide them to various implementation and verification tools in a design flow. These tools can update the implementation UPF with the new features added to a design that fulfill the given power intent. Thus, the power intent and associated UPF files get successively refined along the design flow [29].

We have listed some of the UPF commands above. To understand the usage of the UPF commands and for illustrative examples, readers can refer to [29].

19.7.2 Power Domain

Power domain is a collection of instances that are treated as a group for power management. Typically, all instances of a power domain operate at the same power supply voltage.

When a signal propagates from one power domain to another that works at a different voltage level, we need to insert a *level shifter* between them. A level shifter translates signal values from an input voltage swing to a different output voltage swing. If a signal crosses two power domains operating at different voltage levels without a level shifter, the following problems can occur:

1. Low-voltage (V_{DD1}) to high-voltage (V_{DD2}) crossing: pull-up transistors in the high-voltage domain can be fully/partially ON even when the input is at V_{DD1} (ideally they should be switched off). It can lead to a short-circuit condition or increase the static power dissipation.
2. High-voltage (V_{DD1}) to low-voltage (V_{DD2}) crossing: gate dielectric damage and the reliability issues can occur due to the high voltage applied at the gate terminal of the transistor.

Example 19.6 The partitioning of a design into power domains and the role of level shifters are illustrated in Figures 19.5(a) and 19.5(b), respectively.

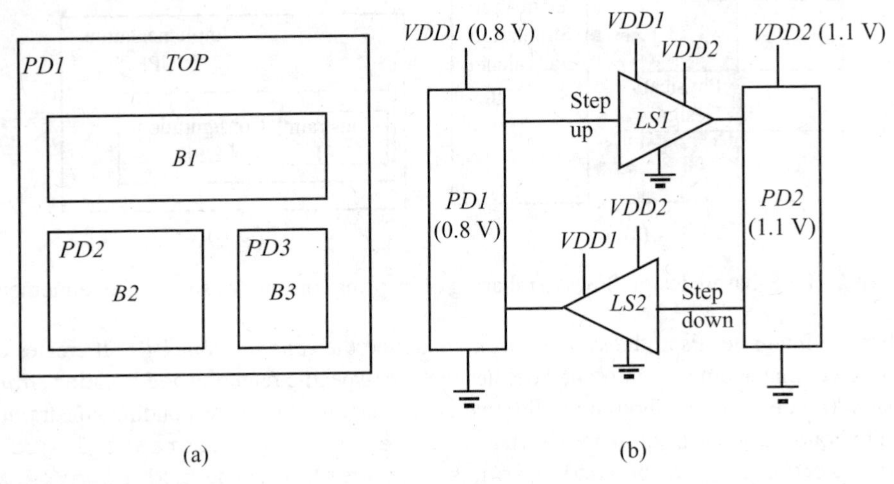

Figure 19.5 Power domains (a) an illustration and (b) level shifters

The *TOP* contains three blocks *B1*, *B2*, and *B3*. The power domain of the TOP is *PD1*. The power domain of *B1* is also *PD1*, while the power domains of *B2* and *B3* are *PD2* and *PD3*, respectively. A power domain can operate at a single voltage, be powered off, or operate at different voltages based on the state of the design. Thus, partitioning a design into multiple power domains enables efficient implementation of power management strategies, such as selectively powering off and reducing voltage levels.

The insertion of level shifters between *PD1* and *PD2* is shown in Figure 19.5(b). Since *PD1* works at a lower voltage than *PD2*, we need to insert a step-up level shifter for signals propagating from *PD1* to *PD2* and a step-down level shifter for signals propagating from *PD2* to *PD1*.

The definition of power domains puts additional constraints on the implementation tools. A logic synthesis tool should pick cells from a given set of technology libraries, considering their suitability for a given power domain. Additionally, it should insert level shifters at the interface of the power domains. Level shifters are designed, characterized, and made available in the technology libraries for instantiation. The synthesis tool should ensure that the chosen level shifter has proper drive strength and meets the timing constraints.

During chip planning, the implementation tool should place blocks by considering their power domains and creating a power delivery network accordingly. The placement and other optimization tasks in physical design should comprehend power domain boundaries and avoid moving cells from one power domain to another. Thus, power domain definition has a sweeping impact on the design flow.

19.7.3 Power Gating

Switching off the power supply for a block or *power gating* is an effective technique to tackle both static and dynamic components of power dissipation. However, it requires a careful circuit design and inserting specially designed circuit elements. We illustrate it using the following example.

Example 19.7 The schematic of a circuit that contains power gating features is shown in Figure 19.6(a). The power gating function is carried out as follows:

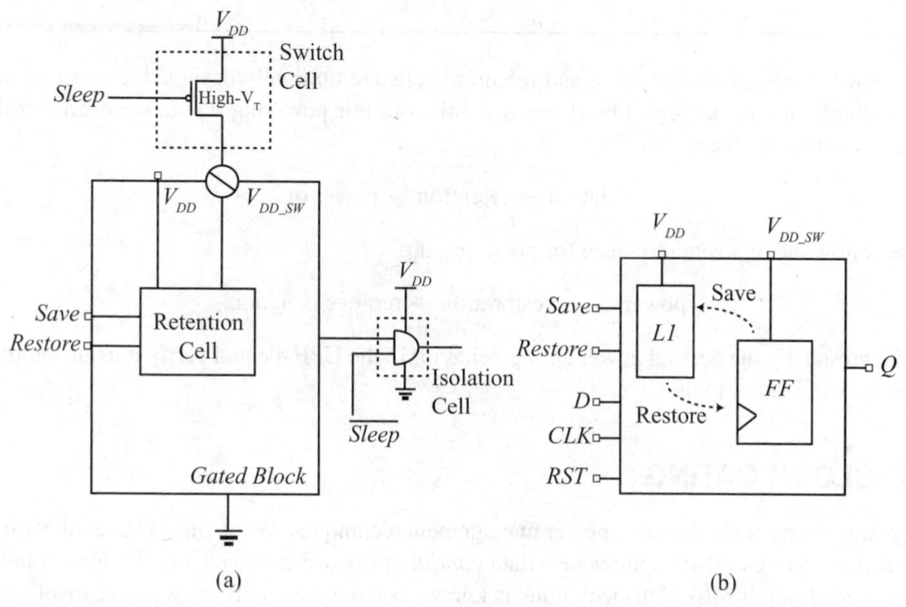

Figure 19.6 Power gating (a) salient features in the circuit and (b) a retention cell example

1. We switch off the power supply using a *switch cell* and asserting the signal *Sleep*=1. It switches off the PMOS in the switch cell and cuts off V_{DD} from the power-gated block. The switch cell consists of high-V_T transistors so that the leakage current is low. When *Sleep*=0, the switch cell connects V_{DD} to V_{DD_SW} and supplies power to the power-gated block. The width of the PMOS in the switch cell should be sufficiently high such that the ON-state resistance is low.
2. When *Sleep*=1, the output from the gated block can float between V_{DD} and 0. As a result, a large current can flow through the receiving block (because transistors can be made partially on in the receiving block by an output signal that has a floating voltage). Additionally, a floating output can cause a functional problem in the downstream logic. To avoid these problems, we connect the output of the power-gated block to an *isolation cell*. The isolation cell receives $\overline{Sleep}$ (complement of *Sleep*). When *Sleep*=0, it simply passes the output value. When *Sleep*=1, it clamps its output to a specified logic value (0 for the isolation cell shown in Figure 19.6(a)).

3. When power goes down for a block, the values stored in the flip-flops get lost. Hence, when the power gets restored, the circuit will come back in an *unknown state*. However, we might need to restore the values of some of the critical flip-flops after power is restored. To enable this, we use a particular type of cell for critical flip-flops known as *retention cell*.

   The internal working of a typical retention cell is shown in Figure 19.6(b). It has two power sources: the gated power source V_{DD_SW} and the always-ON power source V_{DD}. It contains a latch *L1* powered by always-ON power source V_{DD} and a flip-flop *FF* powered by gated power source V_{DD_SW}. Before powering off, we save the current value of the retention cell in the latch *L1* using the signal *Save*. After power is restored, the saved value in the latch is transferred back to the flip-flop *FF* using the signal *Restore*. When *Sleep*=0, the retention cell provides the function of a normal flip-flop with the help of the internal flip-flop *FF*. We typically implement *L1* using high-V_T transistors to reduce the static power dissipation in the sleep mode.

The switch cells, isolation cells, and retention cells are optimally designed, characterized, and made available in a technology library for instantiation. For powering down, we need to follow a specific sequence, such as:

$$\text{isolation} \rightarrow \text{retention} \rightarrow \text{power off.}$$

We follow the opposite sequence for powering up:

$$\text{power on} \rightarrow \text{restoration} \rightarrow \text{remove isolation.}$$

We can specify the desired power gating behavior in the UPF file and verify it using simulations and formal verification tools.

19.8 CLOCK GATING

Clock gating is one of the dynamic power management techniques. When there is a set of *N* flip-flops or a portion of a circuit that captures new data *conditionally*, we can *dynamically* shut off the clock when that condition is *false*. This technique is known as *clock gating*. It saves power by eliminating unnecessary charging and discharging of the capacitors in the clock network, including the flip-flops' internal capacitors on the clock path.

19.8.1 Transformation

A logic synthesis tool can extract the enabling condition for writing to a flip-flop using functional analysis. Subsequently, if it finds an opportunity of saving power by clock gating, it can transform the logic appropriately and insert clock gating elements. We explain it in the following example.

Example 19.8 Consider the following RTL code. The input *Data* gets captured only when *EN*=1.

```
always @(posedge CLK) begin
    if (EN) Q <= Data;
end
```

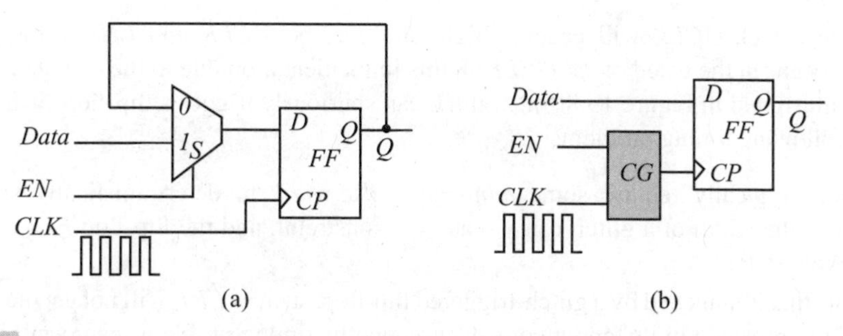

Figure 19.7 Implementation of enable condition (a) on the data path and (b) on the clock path

A synthesis tool can implement the above RTL as shown in Figure 19.7(a). The output *Q* recirculates through the multiplexer when *EN*=0, and the *Data* is captured when *EN*=1. Note that the clock signal causes transitions in the internal capacitors of the flip-flop even when the value at its D-pin is unchanged for *EN*=0. A power-aware synthesis tool can avoid such transitions by implementing the circuit as shown in Figure 19.7(b). A clock gater *CG* is inserted on the clock path. It stops the clock propagation when *EN*=0 and allows it when *EN*=1. Hence, *Data* gets latched when *EN*=1. In this implementation, the switching inside the flip-flop is disabled when *EN*=0. However, the switching power dissipation in the clock gater *CG* gets introduced. Thus, if the same clock gater can stop the clock propagation for *many* flip-flops and other circuit elements in the clock network, we can obtain a substantial power saving. Additionally, the savings will be more when *EN* has a lower probability of being 1, i.e., the clock is mostly disabled.

We can trivially implement a clock gater using an *AND* gate. However, it can lead to glitches on the clock path and associated timing problems. It is illustrated in the following example.

Example 19.9 Consider the clock gating circuit shown in Figure 19.8(a). Let us examine the timing problem associated with this circuit.

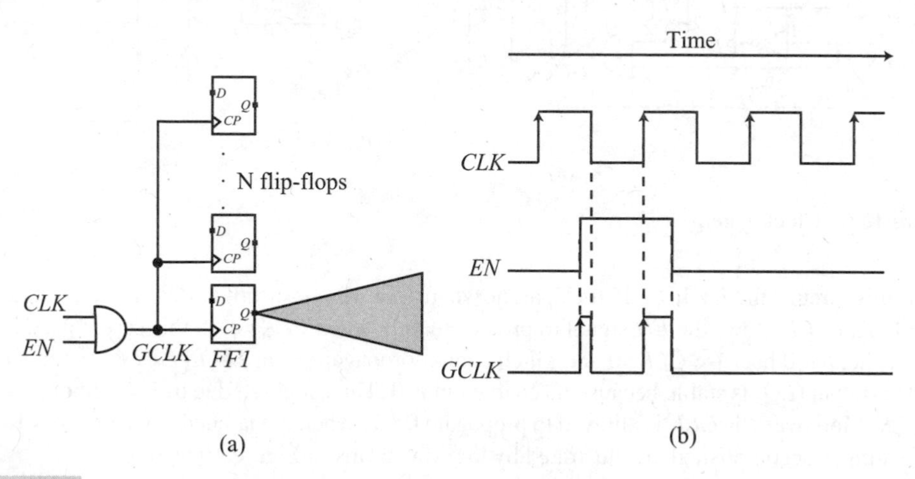

Figure 19.8 Illustration of timing issues in a trivial clock gating (a) implemented using an *AND* gate and (b) waveform of gated clock *GCLK* (contains glitches)

The gated clock *GCLK* will become high whenever both *CLK* and *EN* are high. However, glitches can occur in the gated clock *GCLK* in this implementation due to the change in *EN* during *CLK*=1, as illustrated in Figure 19.8(b). A glitch can spuriously trigger a flip-flop such as *FF1* and lead to the following timing problems:

1. Flip-flops typically require some *minimum pulse-width* to deterministically capture data. However, the width of a glitch can violate this constraint, and the flip-flop *FF1* can go into a metastable state.
2. The data that is launched by a glitch-triggered flip-flop, such as *FF1*, will not get the entire clock period to propagate in its fanout cone. Consequently, timing problems can occur in capturing the data that gets launched by a glitch.

Therefore, we cannot use a simple AND gate for clock gating.

We can implement a glitch-free clock gater using a latch, as illustrated in the following example.

Example 19.10 Consider the circuit shown in Figure 19.9(a). Let us examine how glitches of Ex. 19.9 is avoided in it.

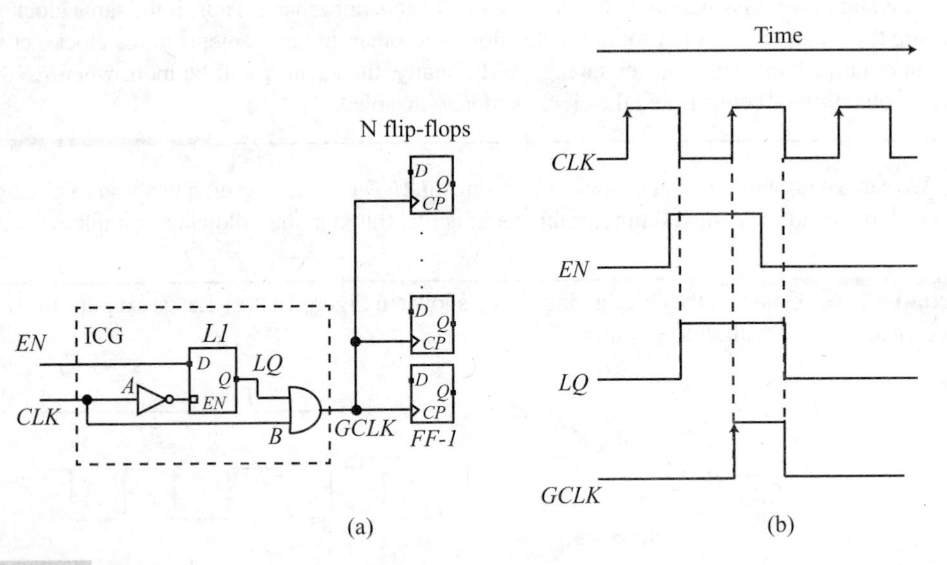

Figure 19.9 Clock gater

In this circuit, the latch *L1* is transparent when *CLK*=0 and it holds the previous state when *CLK*=1. Thus, *L1* allows the *EN* signal to propagate only when *CLK*=0. In this case, the AND gate output is held at 0 because *CLK*=0 and glitch cannot propagate from *EN* to *GCLK*. When *CLK*=1, the latch output (*LQ*) is stable because its enable pin is 0. Thus, a glitch due to *EN* cannot propagate to *GCLK*. Moreover, the *CLK* is allowed to propagate freely when the latched value of *EN* is 1. Thus, clock gating is accomplished, as illustrated by the waveforms of Figure 19.9(b).

19.8.2 Integrated Clock-gating Cells

When we use discrete latch and AND gate instances as a clock gater, we can still observe a glitch in some cases. For example, assume that the clock signal arrive later at the point *A* compared to *B* in the circuit shown in Figure 19.9(a). Assume that *CLK* makes a 0→1 transition. Due to above arrival time difference, for some time duration, the latch will be transparent and *CLK*=1 at the AND gate simultaneously. If *EN* changes during this time duration, we can observe a glitch at *GCLK*. Therefore, when we use discrete latches and AND gates as clock gater, the placement tool should place these discrete elements closely to minimize arrival time differences. Moreover, clock tree synthesis (CTS) should minimize the skew between the latching element (point *A*) and the corresponding AND gate (point *B*).

We can avoid the above problem by designing the entire clock gating circuitry comprising the latch and the AND gate as a separate cell, known as *integrated clock-gating* (ICG) cell. We make ICG cells available in a technology library. Logic synthesis tools instantiate them when an opportunity of saving power by clock gating is detected. We define appropriate *constraints* for the ICG cells such that the internal latch captures data correctly. Additionally, we ensure by the optimal layout design of the ICG that it does not produce glitches when these constraints are honored. Moreover, we ensure that these constraints are honored in the ICG instantiations with the help of a static timing analysis (STA) tool.

It is noteworthy that ICG cells or clock gating elements themselves consume power and area. Therefore, inserting ICG cells for each flip-flop can be counterproductive. When an ICG cell can gate many flip-flops (*N* is large in Figure 19.9), it can save substantial dynamic power. In practice, the state-of-the-art synthesis tools can automatically infer many clock-gating opportunities in an RTL and insert ICG cells appropriately in the netlist. Therefore, clock gating is one of the most popular techniques of saving power in an IC.

19.9 CLOCK DOMAINS

We can reduce the power dissipation in a circuit by employing clock signals of multiple clock frequencies. We can partition a circuit into performance-critical blocks and noncritical blocks. We run performance-critical blocks at higher clock frequencies and the noncritical blocks at lower clock frequencies. Thus, we can reduce the overall power dissipation in a circuit.

A popular energy-efficient design methodology known as globally asynchronous locally synchronous (GALS) employs multiple clock frequencies [30, 31]. A GALS system consists of locally synchronous blocks running at optimal clock frequencies that communicate asynchronously at the top level. We routinely use clock signals with multiple frequencies in low-power and SoC design methodologies. Nevertheless, it poses nontrivial design and verification challenges. We will discuss some of these challenges and their solutions in this section.

19.9.1 Clock Domain Crossing

The flip-flops that are clocked by the same clock source are said to be in the same *clock domain*. In a synchronous circuit, the flip-flops of the same clock domain operate at the same clock frequency and have a fixed phase relationship. However, when data is launched in one clock domain and captured in another clock domain, we say that *clock domain crossing* (CDC) has occurred.

For example, consider a portion of a circuit shown in Figure 19.10(a). Assume that the clock signals *CK1* and *CK2* are asynchronous, i.e., they have different clock frequencies and arbitrary phase relationship. The flip-flop *F1* and *F2* belong to clock domains *CK1* and *CK2*, respectively. The *Data* is launched by *F1* and captured by *F2*. Thus, there exists a CDC in this circuit. When there is a CDC, the circuit behavior can become unpredictable, as explained below.

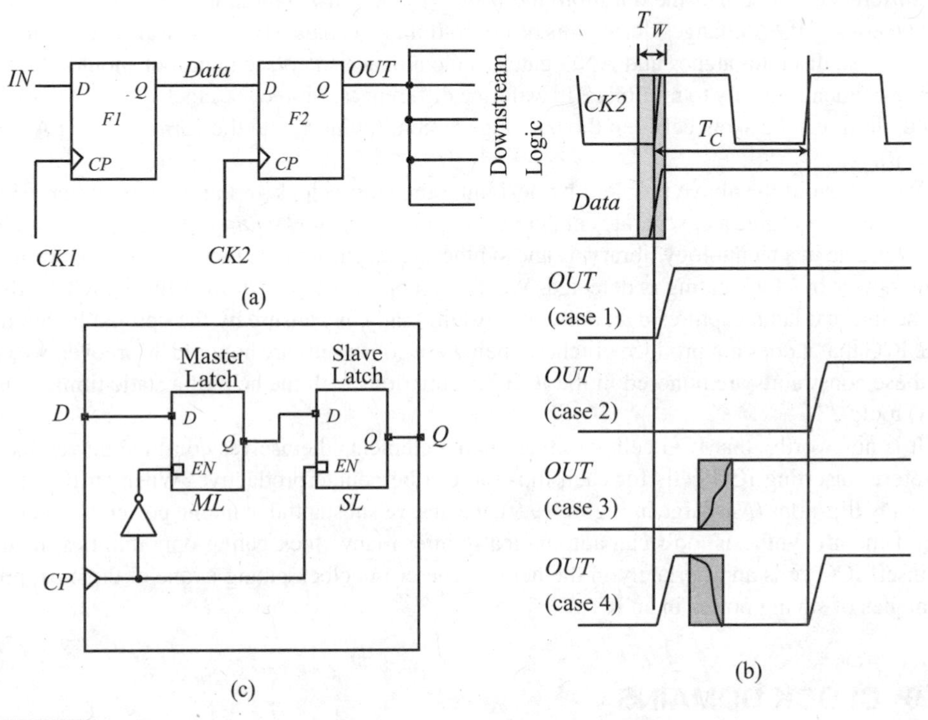

Figure 19.10 CDC (a) circuit condition, (b) some of the possible waveforms, and (c) internal structure of a flip-flop

Assume that *Data* changes very close to the clock edge of *CK2*, as shown in Figure 19.10(b). Thus, the data changes within the *forbidden window* defined by the setup-hold[3] time of *F2*. Let us denote the forbidden window around the clock edge as T_W and refer to the data change within T_W as the *forbidden event*. Normally, STA ensures forbidden event never occurs in a synchronous circuit. However, since *CK1* and *CK2* are asynchronous, forbidden events will occur occasionally. Let us assume that the *Data* change occurs *uniformly distributed* over the clock period T_C of the capture clock *CK2*. Let us denote the frequency of change of *Data* as f_D. The rate of occurrence of the forbidden event f_{FE} can be given as:

$$f_{FE} = \frac{T_W}{T_C} f_D \tag{19.2}$$

Next, let us understand what will be the output of the capture flip-flop *OUT* when the forbidden event occurs. The flip-flop *F2* consists of a master latch *ML* and a slave latch *SL*, as shown in

[3] For setup and hold time constraints, see Section 13.6 ("Modeling constraints").

Figure 19.10(c). When *CK2* is low, the *ML* is transparent, while the *SL* produces previously latched data. When *CK2* goes high, the *ML* latches the data, while the slave simply produces the output of *ML*.

When the forbidden event occurs, *ML* can go into a *metastable state*[4] while closing the latch. However, the time taken to exit the metastable state (settling to 0 or 1) is unpredictable. It depends on the initial voltage latched by *ML* in the metastable state, the device parameters, and the impact of the noise or the environment on the latch [32]. Subsequently, the value latched by *ML* propagates to the output of *SL* through pull-up/pull-down transistors. Note that despite *ML* being metastable, the output voltage of *SL* can still correspond to logic 0/1 due to the pull-up/pull-down transistors between *ML/Q* and *SL/Q*. Some of the possible waveforms that we can expect at *OUT* are as follows (see Figure 19.10(b)):

1. **case 1:** *ML* latches the data properly or it enters into a metastable state and quickly resolves to 1. *SL* latches the data correctly and produces only a stable 1.
2. **case 2:** *ML* misses the data completely or it enters into a metastable state and quickly resolves to 0. *SL* produces only a stable 0. In the next cycle, both *ML* and *SL* produce the data correctly.
3. **case 3:** *ML* goes into a metastable state and resolves to 1 within the first clock cycle. The slave also produces a metastable output initially, but it resolves to 1 within the first cycle. Nevertheless, the $CP \rightarrow Q$ delay is much greater than the nominal $CP \rightarrow Q$ delay.
4. **case 4:** *ML* goes into a metastable state. However, *SL* produces a 1 due to intervening pull-up transistors. Subsequently, the *ML* resolves to 0 and *SL* follows it by toggling its output value. In the next cycle, both *ML* and *SL* produce the data correctly.

Thus, we cannot predict the voltage level of *F2/Q* (or *OUT*). It can lead to the following problems in a circuit:

1. The downstream logic elements can receive and interpret contrasting logic levels for the same metastable output. It can lead to a functional problem in the circuit.
2. Due to the metastable voltage at the gate terminal, the pull-up/pull-down transistors in the downstream logic elements can draw a large current.
3. Due to an unexpected large $CP \rightarrow Q$ delay in *F2*, there can be timing violations in the downstream logic elements.

Therefore, we need to mask the metastable voltage of *F2* from the downstream logic. We can achieve this using *synchronizers*, as explained in the following paragraphs.

19.9.2 Synchronizers

If a latch is metastable at time $t = 0$ and it remains metastable after waiting for $t = S$, we say that a *failure* has occurred. The probability of the occurrence of failure is $e^{-S/\tau}$, where τ depends on the device parameters and the fabrication technology [32]. It is evident from this formula that the probability of failure decreases exponentially with time. It implies that if we wait long enough, the output will settle to 0/1 with a high probability. This observation forms the basis of designing special circuits known as synchronizers which enable proper data transfer between two clock domains.

[4] For details on metastable state, see Section 1.1.6 ("Sequential circuits").

Using Eq. 19.2, we can compute the failure rate FR as [32]:

$$FR = \frac{T_W}{T_C} f_D e^{-S/\tau} \tag{19.3}$$

The reciprocal of the failure rate gives the *mean time between failures* (MTBF):

$$MTBF = \frac{T_C}{T_W f_D} e^{S/\tau} \tag{19.4}$$

We want to increase the MTBF so that the circuit can work for a long time before failing. Note that it is impossible to design a synchronizer that never fails. However, we can increase the MTBF to a very high value (e.g., 10^{10} years, which is the age of the universe) and ensure that it never fails in practice. A synchronizer with an MTBF several orders of magnitude more than the product lifetime is sufficient for many applications.

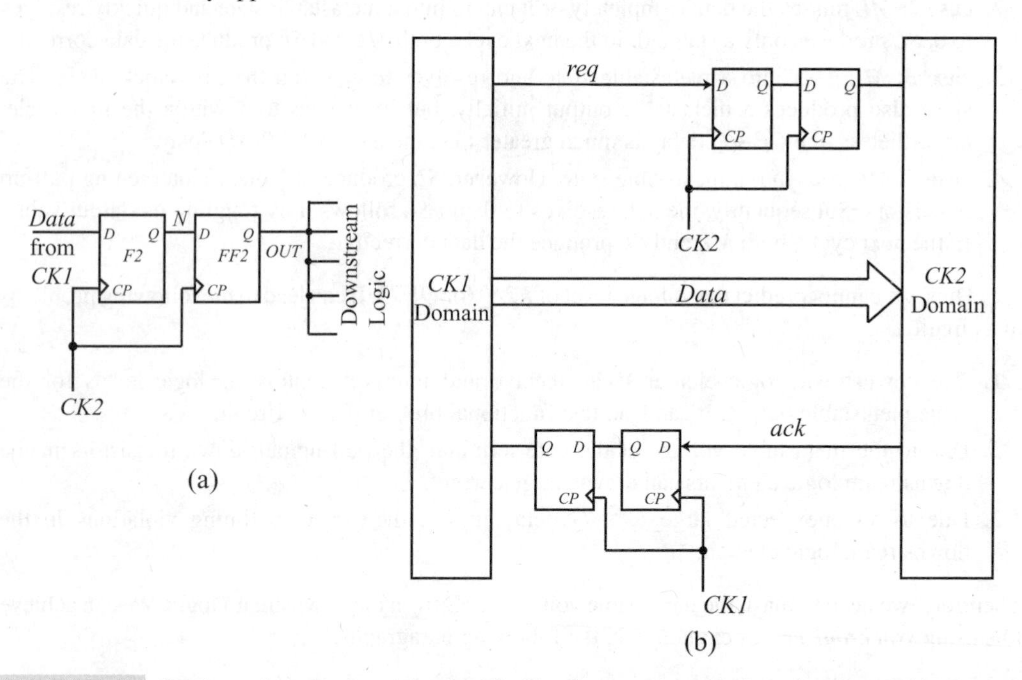

Figure 19.11 CDC synchronization (a) a double flip-flop synchronization circuit and (b) request-acknowledge based synchronization [32]

A popular synchronizer circuit, known as *double flip-flop synchronizer*, is shown in Figure 19.11(a). As explained earlier, the capture flip-flop $F2$ can go into a metastable state due to CDC. To mask the downstream logic from the metastable voltage levels in the clock domain $CK2$, we capture the output of $F2$ in the *next* clock cycle using $FF2$. In the downstream logic, we use the output of $FF2$ instead of the output of $F2$. Thus, we wait and allow the metastable voltage of $F2$ to settle for a longer time S before using it in the downstream logic:

$$S = T_C - t_{cq,F2} - t_{setup,FF2} - t_{wire}. \tag{19.5}$$

where T_C is the clock period of *CK2*, $t_{cq,F2}$ is the $CP \rightarrow Q$ delay of *F2*, $t_{setup,FF2}$ is the setup time of *FF2*, and t_{wire} is the wire delay of the net *N*. By ensuring that *S* is high, we keep the corresponding MTBF acceptable. Thus, *FF2* will have a stable 0/1 (and not a metastable voltage level), except for once every MTBF years [32]. Moreover, the correct *Data* will be available at the output of *FF2* after second clock cycle (cases 1 and 3 in Figure 19.10(b)) or third clock cycle (cases 2 and 4), except once every MTBF years. However, as we explained, if MTBF is high enough, the synchronization provided by the double flip-flop circuit is acceptable.

Note that there is still uncertainty of one clock cycle in the clock domain *CK2*. However, we can handle this functionally by ensuring that the downstream logic of *FF2* works correctly when *FF2* provides *Data* in either the second or the third clock cycle. The primary purpose of the synchronizer is to shield the downstream logic from the metastable voltage levels with a high probability. It is achieved by the double flip-flop synchronizer satisfactorily in most cases. However, it incurs a penalty of one clock cycle in the latency for *Data* propagation and the area increase.

Some practical considerations in employing a double flip-flop synchronization circuit are as follows:

1. We have assumed in Figure 19.10(b) that if *Data* is missed in the first clock cycle or the metastable value settles to the previous state, it gets captured in the second clock cycle. It is guaranteed only when *Data* is held constant for at least two clock cycles at *F2*. To ensure this, we need to have some feedback mechanism from the receiver domain *CK2* to the sender domain *CK1*. Typically, we implement a request-acknowledge protocol between the clock domains, as shown in Figure 19.11(b). The sender does not change *Data* unless it receives an acknowledge signal from the receiver.
2. When we want to send multi-bit data, we cannot synchronize them bitwise. It can lead to some bits being updated one cycle before others. Therefore, the received multi-bit data will be incoherent and can cause functional problems in the downstream logic. Hence, we synchronize the control signal rather than the multi-bit data signal.
3. We should not synchronize a signal from another clock domain using multiple synchronizers in parallel. Due to the uncertainty of one clock cycle, the same signal can be interpreted as both 0 and 1 in the downstream logic and can lead to functional problems.

We can employ other synchronization circuits and methods to tackle specific CDC problems and requirements [33]. Note that CDC poses nontrivial verification challenges also. We cannot verify the synchronization of asynchronous clock domains using STA. We use a combination of topological analyses, pattern matching, and formal techniques to ensure synchronization and adherence to a given CDC protocol.

19.10 RECENT TRENDS

Power optimization techniques, such as clock and power gating, are now routinely applied in a design flow. With the advancement in technology and reduction in threshold voltage, static power dissipation has now become significant. Therefore, shutting off the power supply and using high-V_T cells are prevalent to reduce static power dissipation. The verification of designs that implement low-power methodologies poses nontrivial challenges. For example, verifying that the power intent gets fulfilled, the circuit comes back to a correct state after power restoration, different power domains and clock domains communicate correctly, and DVFS strategies work properly are complex problems. Various verification tools, such as simulators, formal verification tools, and power analyzers, are continuously advanced to tackle these challenges and verify new power-saving strategies. Additionally, many novel approaches are being proposed for design-time and runtime power management at the system level [34]. With power-hungry applications such as big data, artificial intelligence, and exascale computing becoming dominant, developing novel power-saving strategies will become more critical in the future.

REVIEW QUESTIONS

19.1 What are the motivations for reducing power dissipation in an IC?

19.2 How can HW–SW partitioning be done to make the overall system more energy-efficient?

19.3 How can behavior synthesis decisions impact the power dissipation in an IC?

19.4 How can OS make a system more energy-efficient?

19.5 Why is DVFS more effective than simple frequency scaling?

19.6 How can feedback from the core logic to the DVFS controller help in achieving a greater power saving?

19.7 Why can a counter with Gray encoding dissipate less power than with a binary encoding?

19.8 How does path equalization help in reducing power dissipation in a digital circuit?

19.9 What are the complications of power-driven resizing of logic gates?

19.10 How can cells in a multi-V_T library be chosen to reduce power dissipation in a circuit?

19.11 What is power intent?

19.12 How does power intent evolve with the design flow? How can UPF help manage evolving power intent in design flows?

19.13 Why do we need to use level shifters when signals cross power domains?

19.14 Why do we need an isolation cell when a power domain is switched off?

19.15 Why do retention cells have two power supplies?

19.16 How does clock gating help in reducing power dissipation in a digital circuit?

19.17 Why is simple AND gate not used as a clock gater?

19.18 What are the advantages of using an ICG cell over a combination of a discrete latch and AND cell for clock gating?

19.19 What problems can occur in a CDC?

19.20 Why should we avoid reading an output of a flip-flop that is in a metastable state?

19.21 Consider the circuit shown in Figure 19.12. A 32-bit data is transferred from clock domain *CK1* to clock domain *CK2* using parallel synchronizers. What could be a problem in this data transfer? How can this problem be rectified?

19.22 How can we verify that a circuit adheres to a pre defined synchronization protocol?

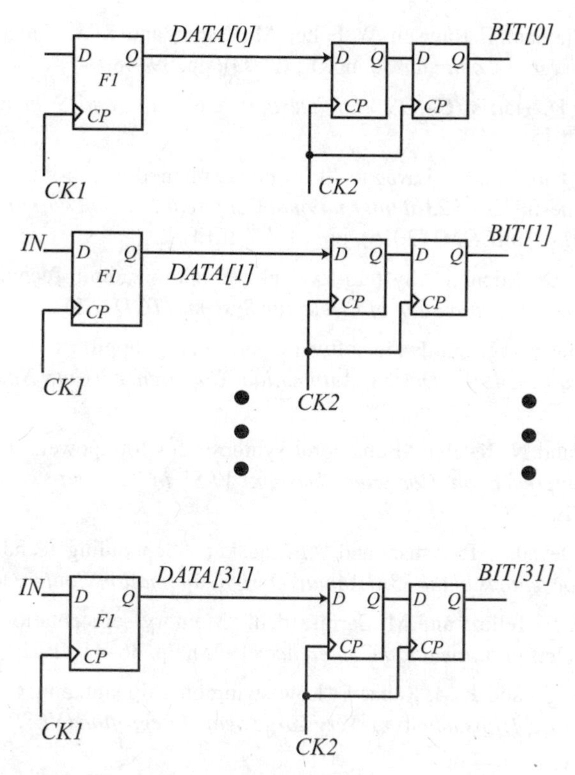

Figure 19.12 Circuit with CDC and incorrect synchronization

TOOL-BASED ACTIVITY

In this activity, use any gate-level power analysis tool, including open-source tools such as OpenSTA [35]. You can use any library, including freely available technology libraries meant for academic purposes [36].

Consider the clock gater shown in Figure 19.9(a). The gain due to clock gating is possible only when the number of flip-flops (N) that can be gated simultaneously is high. Otherwise, the clock path's savings will get offset by the power consumed in the clock gater (CG) itself. Determine the minimum value of N (for your technology library) for which inserting a clock gater will lead to a reduction in total power. You need to write RTL/netlist with CG and without CG. If your library has an ICG cell, then you can directly use it, else use a latch-based clock gater shown in Figure 19.9(a). You need to compare the dynamic power dissipated in the circuit with CG and without CG. By varying N in your circuit, you can determine the minimum value of N, resulting in less power dissipation in the circuit with CG than without CG.

REFERENCES

[1] D. B. Tuckerman and R. F. W. Pease. "High-performance heat sinking for VLSI." *IEEE Electron Device Letters* 2, no. 5 (1981), pp. 126–129.

[2] P. R. Panda, B. Silpa, A. Shrivastava, and K. Gummidipudi. *Power-efficient System Design*. Springer Science & Business Media, 2010.

[3] C. Lefurgy, K. Rajamani, F. Rawson, W. Felter, M. Kistler, and T. W. Keller. "Energy management for commercial servers." *Computer* 36, no. 12 (2003), pp. 39–48.

[4] N. H. Weste and D. Harris. *CMOS VLSI Design: A Circuits and Systems Perspective*. Pearson Education India, 2015.

[5] F. B. Muslim, A. Qamar, and L. Lavagno. "Low power methodology for an ASIC design flow based on high-level synthesis." *2015 23rd International Conference on Software, Telecommunications and Computer Networks (SoftCOM)* (2015), pp. 11–15, IEEE.

[6] L. Benini and G. De Micheli. "System-level power optimization: Techniques and tools." *ACM Transactions on Design Automation of Electronic Systems (TODAES)* 5, no. 2 (2000), pp. 115–192.

[7] J. Henkel. "A low power hardware/software partitioning approach for core-based embedded systems." *Proceedings 1999 Design Automation Conference (Cat. No. 99CH36361)* (1999), pp. 122–127, IEEE.

[8] A. Raghunathan and N. K. Jha. "Behavioral synthesis for low power." *Proceedings 1994 IEEE International Conference on Computer Design: VLSI in Computers and Processors* (1994), pp. 318–322, IEEE.

[9] J. Monteiro, S. Devadas, P. Ashar, and A. Mauskar. "Scheduling techniques to enable power management." *Proceedings of the 33rd Annual Design Automation Conference* (1996), pp. 349–352.

[10] A. H. Farrahi, G. E. Tellez, and M. Sarrafzadeh. "Memory segmentation to exploit sleep mode operation." *32nd Design Automation Conference* (1995), pp. 36–41, IEEE.

[11] H. Zhang, V. George, and J. M. Rabaey. "Low-swing on-chip signaling techniques: Effectiveness and robustness." *IEEE Transactions on Very Large Scale Integration (VLSI) Systems* 8, no. 3 (2000), pp. 264–272.

[12] S. Ramprasad, N. R. Shanbhag, and I. N. Hajj. "A coding framework for low-power address and data busses." *IEEE Transactions on Very Large Scale Integration (VLSI) Systems* 7, no. 2 (1999), pp. 212–221.

[13] M. R. Stan and W. P. Burleson. "Bus-invert coding for low-power I/O." *IEEE Transactions on Very Large Scale Integration (VLSI) Systems* 3, no. 1 (1995), pp. 49–58.

[14] H. Wang, L.-S. Peh, and S. Malik. "Power-driven design of router microarchitectures in on-chip networks." *Proceedings. 36th Annual IEEE/ACM International Symposium on Microarchitecture 2003. MICRO-36* (2003), pp. 105–116, IEEE.

[15] U. Y. Ogras and R. Marculescu. "'It's a small world after all': NoC performance optimization via long-range link insertion." *IEEE Transactions on Very Large Scale Integration (VLSI) Systems* 14, no. 7 (2006), pp. 693–706.

[16] S. Deb, A. Ganguly, P. P. Pande, B. Belzer, and D. Heo. "Wireless NoC as interconnection backbone for multicore chips: Promises and challenges." *IEEE Journal on Emerging and Selected Topics in Circuits and Systems* 2, no. 2 (2012), pp. 228–239.

[17] L. Benini, G. De Micheli, and E. Macii. "Designing low-power circuits: Practical recipes." *IEEE Circuits and Systems Magazine* 1, no. 1 (2001), pp. 6–25.

[18] P. Pillai and K. G. Shin. "Real-time dynamic voltage scaling for low-power embedded operating systems." *Proceedings of the Eighteenth ACM Symposium on Operating Systems Principles* (2001), pp. 89–102.

[19] S. Herbert and D. Marculescu. "Analysis of dynamic voltage/frequency scaling in chip-multiprocessors." *Proceedings of the 2007 International Symposium on Low Power Electronics and Design (ISLPED'07)* (2007), pp. 38–43, IEEE.

[20] T. D. Burd, T. A. Pering, A. J. Stratakos, and R. W. Brodersen. "A dynamic voltage scaled microprocessor system." *IEEE Journal of Solid-State Circuits* 35, no. 11 (2000), pp. 1571–1580.

[21] T. D. Burd and R. W. Brodersen. "Design issues for dynamic voltage scaling." *Proceedings of the 2000 International Symposium on Low Power Electronics and Design* (2000), pp. 9–14.

[22] L. Benini and G. De Micheli. "State assignment for low power dissipation." *IEEE Journal of Solid-State Circuits* 30, no. 3 (1995), pp. 258–268.

[23] K. Roy and S. Prasad. "SYCLOP: Synthesis of CMOS logic for low power applications." *ICCD* 92 (1992), pp. 234–237.

[24] S. Irnan and M. Pedram. "Multi-level network optimization for low power." *IEEE/ACM International Conference on Computer-aided Design* (1994), pp. 372–373, IEEE Computer Society.

[25] C.-Y. Tsui, M. Pedram, and A. M. Despain. "Technology decomposition and mapping targeting low power dissipation." *Proceedings of the 30th International Design Automation Conference* (1993), pp. 68–73.

[26] S. Devadas and S. Malik. "A survey of optimization techniques targeting low power VLSI circuits." *Proceedings of the 32nd Annual ACM/IEEE Design Automation Conference* (1995), pp. 242–247.

[27] M. Alidina, J. Monteiro, S. Devadas, A. Ghosh, and M. Papaefthymiou. "Precomputation-based sequential logic optimization for low power." *IEEE Transactions on Very Large Scale Integration (VLSI) Systems* 2, no. 4 (1994), pp. 426–436.

[28] M. Srivastav, S. Rao, and H. Bhatnagar. "Power reduction technique using multi-V_T libraries." *Fifth International Workshop on System-on-chip for Real-time Applications (IWSOC'05)* (2005), pp. 363–367, IEEE.

[29] "IEEE standard for design and verification of low-power, energy-aware electronic systems." *IEEE Std 1801-2018* (2019), pp. 1–548.

[30] G. Semeraro, G. Magklis, R. Balasubramonian, D. H. Albonesi, S. Dwarkadas, and M. L. Scott. "Energy-efficient processor design using multiple clock domains with dynamic voltage and frequency scaling." *Proceedings Eighth International Symposium on High Performance Computer Architecture* (2002), pp. 29–40, IEEE.

[31] M. Krstic, E. Grass, F. K. Gürkaynak, and P. Vivet. "Globally asynchronous, locally synchronous circuits: Overview and outlook." *IEEE Design & Test of Computers* 24, no. 5 (2007), pp. 430–441.

[32] R. Ginosar. "Metastability and synchronizers: A tutorial." *IEEE Design & Test of Computers* 28, no. 5 (2011), pp. 23–35.

[33] C. E. Cummings. "Clock domain crossing (CDC) design & verification techniques using SystemVerilog." SNUG-2008, *Boston*, 2008.

[34] A. K. Singh, C. Leech, B. K. Reddy, B. M. Al-Hashimi, and G. V. Merrett. "Learning-based run-time power and energy management of multi/many-core systems: Current and future trends." *Journal of Low Power Electronics* 13, no. 3 (2017), pp. 310–325.

[35] T. Ajayi, V. A. Chhabria, M. Fogaça, S. Hashemi, A. Hosny, A. B. Kahng, M. Kim, J. Lee, U. Mallappa, M. Neseem, et al. "Toward an open-source digital flow: First learnings from the OpenROAD project." *Proceedings of the 56th Annual Design Automation Conference 2019* (2019), pp. 1–4.

[36] M. Martins, J. M. Matos, R. P. Ribas, A. Reis, G. Schlinker, L. Rech, and J. Michelsen. "Open cell library in 15nm FreePDK technology." *Proceedings of the 2015 Symposium on International Symposium on Physical Design* (2015), pp. 171–178.

PART THREE

Design For Testability (DFT)

Defects can creep into an IC while fabricating, despite tight process control and sophisticated fabrication technology. The primary purpose of testing is to detect such defects and prevent a defective IC from reaching the end-user.

Though testing is carried out primarily after fabrication, we must consider several aspects of testing during the design phase. We carry out some tasks during designing that simplify the testing process and make it economical. We collectively refer to test-driven design practices as *Design For Testability* (DFT). This part of the book covers essential aspects of DFT.

In Chapter 20, we will describe the basic concepts of DFT and structural testing. In Chapter 21, the scan design technique, the most popular implementation of structured testing, will be explained. In Chapter 22, a methodology to generate test patterns for the most common fault model, i.e., stuck-at fault model, will be presented. Chapter 23 will describe the basic concepts of Built-in Self-Test (BIST), highlighting the advantages and disadvantages of self-testing.

Before going through this part of the book, we suggest that readers become familiar with the basic testing concepts discussed in Chapter 6 ("Testing Techniques").

It is worthy to point out that, in this book, we explain only the essential concepts and principles of testing. To gain further insight into IC testing, we advise readers to go through dedicated books on testing such as [1–4]. Further note that the focus area of this book is the testing of digital circuits. To understand design practices explicitly employed for testing of memories and analog/mixed-signal circuits, readers should refer to books such as [2–4].

REFERENCES

[1] H. Fujiwara, *Logic testing and design for testability*. MIT press Cambridge, MA, 1985.

[2] M. Bushnell and V. Agrawal, *Essentials of electronic testing for digital, memory and mixed-signal VLSI circuits*, vol. 17. Springer Science & Business Media, 2004.

[3] L.-T. Wang, C.-W. Wu, and X. Wen, *VLSI test principles and architectures: design for testability*. Elsevier, 2006.

[4] L.-T. Wang, C. E. Stroud, and N. A. Touba, *System-on-chip test architectures:nanometer design for testability*. Morgan Kaufmann, 2010.

20 Basics of DFT

One should not abandon duties born of one's nature, even if one sees defects in them. Indeed, all endeavors are veiled by some evil, as fire is by smoke.

—*Bhagavad Gita*, Chapter 18, verse 48

In this chapter, we will discuss some of the basic concepts of design for testability (DFT). First, we will introduce the idea of *structural testing* and how it is different from functional testing. Then, we will explain *fault model* and its significance in testing. We will discuss *single stuck-at fault model* in detail. We will highlight the role of structural testing and the single stuck-at fault model in simplifying testing and making it cost-effective. We will also elucidate the problems of controlling and observing signals in a sequential circuit that are encountered during structural testing.

20.1 FUNCTIONAL TESTING VERSUS STRUCTURAL TESTING

Let us assume that we need to test a fabricated circuit that implements a Boolean function with N inputs. We can apply all possible 2^N input combinations and check whether the outputs match the corresponding entries in the truth table. This is known as *functional* testing. However, when N is large (such as 50 or 100), the number of possible input combinations becomes dramatically large, and exhaustive *functional testing* of a circuit becomes infeasible. Therefore, we employ an alternative testing strategy known as *structural testing* for testing a fabricated circuit [1].

In structural testing, we test the *components* or hardware that implements a logic function rather than testing the input–output functionality of the implementation [1]. We can choose components at various abstraction levels such as transistors, switches, logic gates, standard cells, and macros (adders, multipliers, and arithmetic logical units (ALUs)). Nevertheless, we perform structural testing often at the level of logic gates. This testing is known as *structural* testing since it considers the topology and interconnections of logic gates in the implementation. The paradigm of structural testing is widely employed because it reduces the number of test patterns required for good test quality[1] [2].

[1] See Section 6.4 ("Fault coverage and defect level") for metrics to assess the test quality.

Example 20.1 Consider the circuit shown in Figure 20.1 that implements a 16-input AND gate. Let us examine the number of test patterns required for its functional and structural testing.

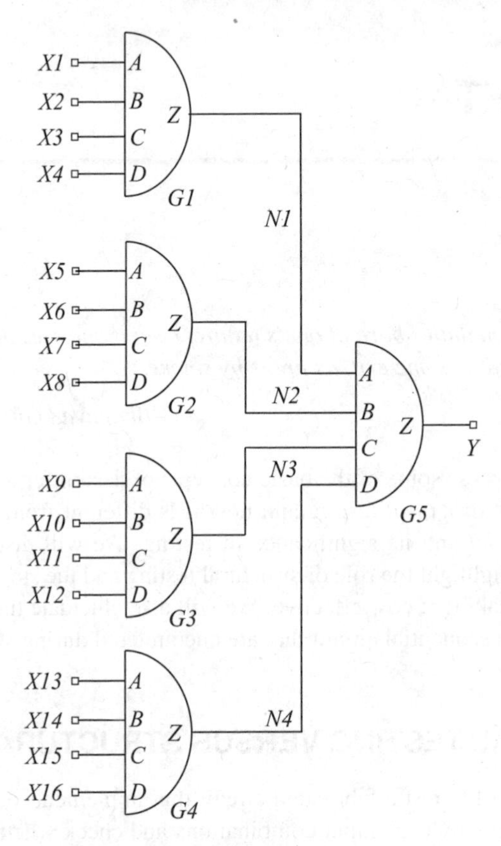

Figure 20.1 A circuit implementing AND of 16 inputs

If we perform functional testing, we need to apply 2^{16}=65536 input combinations and compare the obtained circuit response with the correct response. However, when we perform structural testing, we test individual components of the circuit (i.e., gate *G1*, *G2*, *G3*, *G4*, and *G5*). We need 2^4=16 test patterns to test each four-input AND gate. For example, we apply 16 test patterns at the inputs of *G1* and check the output *G1/Z* for correctness by comparing the obtained response with the expected response for the four-input AND gate. We test all other gates similarly. Hence, a total of 5×16=80 test patterns can test components individually in the circuit. Thus, structural testing requires significantly fewer test patterns compared to functional testing.

Remarks: It is worth highlighting the following assumptions made for structural testing in the above analysis:

1. We can observe the output pins for all the components (such as *G1/Z*). It implies that we can read value at the internal nets (such as *N1*, *N2*, *N3*, and *N4*) of the circuit also.
2. We can write any value at the input pins of all the components (such as *G5/A*). It implies that we can write an arbitrary logic value on any internal net of the circuit. Therefore, to implement component-based testing, we need to make the internal nets *observable* and *controllable*.

3. Despite each component functioning correctly, there could be a problem in their integration. For example, though all the gates *G1*, *G2*, *G3*, *G4*, and *G5* can be functioning correctly, the connection between *G1* and *G5* (through the net *N1*) can be faulty, resulting in a malfunctioning circuit. Therefore, structural testing should test component integration also.

In practice, we enhance the ability to control and observe internal pins by employing some DFT techniques. Additionally, we utilize some *fault models* that work very well in practice. As explained in the following sections, DFT techniques and fault models enable us to make the above assumptions for structural testing and achieve a high test quality.

20.2 FAULT MODELS

Defects are physical and are often not analyzable.[2] Therefore, we represent a defect using a logical or electrical model. These logical or electrical models for defects are known as a *fault models*. A fault model allows us to analyze the impact of a defect using logic or circuit analysis techniques. It transforms the problem of *defect detection* to the problem of *fault detection*. As a result, under the assumptions of a fault model, we can derive *test patterns* algorithmically for detecting a given fault.[3] Furthermore, fault models allow us to make a quantitative assessment of testing effectiveness using metrics such as *fault coverage*. Thus, fault models allow high-quality testing within the structural testing paradigm.

Requirements: We want that the fault model should satisfy the following criteria:

1. It captures the electrical characteristics of a defect realistically.
2. It should allow an efficient derivation of test patterns for a given fault.
3. It should allow fast simulation of a faulty circuit.

Several fault models have been proposed and are employed in practice [3]. Some examples of fault models are the stuck-at fault model (signals stuck to logic 0 or 1), bridging fault model (permanent short circuit between two or more signals), and gate delay fault model (increased delay between an input and an output of a gate) [4]. The most widely used fault model is the stuck-at fault model.

Inconsequential defects: There can be defects of various types in an industrial integrated circuit. It is not feasible to detect all possible defects in a large circuit. A fault model allows us to abstract a huge number of *possible defects* into a smaller number of *possible faults*. However, such an abstract representation can miss detecting some defects. Nevertheless, note that many defects are *inconsequential* and we need not detect them. Moreover, a combination of a few fault models covers most of the defects that cause faulty behavior of a circuit. Hence, fault model-based testing achieves good quality in practice despite its simplicity.

---

[2] See Section 6.1.2 ("Manifestation of defects") for the impact of a defect on a circuit.

[3] See Section 6.3 ("Testing methodology") for the definition of a test pattern.

20.3 STUCK-AT FAULT MODEL

The stuck-at fault model assumes that a defect makes a signal permanently stuck to logic 0 or 1. Therefore, we model a stuck-at fault by tying the corresponding port or pin to a logic 0 or 1. When a port or pin is tied to logic 0, it is said to have a *stuck-at-0* or *SA0* fault. Similarly, when a port or pin is tied to logic 1, it is said to have a *stuck-at-1* or *SA1* fault.

Example 20.2 Consider the NAND gate shown in Figure 20.2(a). When the pin *A* is tied to ground, as shown in Figure 20.2(b), it is said to have stuck-at-0 fault. Similarly, when it is tied to supply voltage, as shown in Figure 20.2(c), it is said to have stuck-at-1 fault.

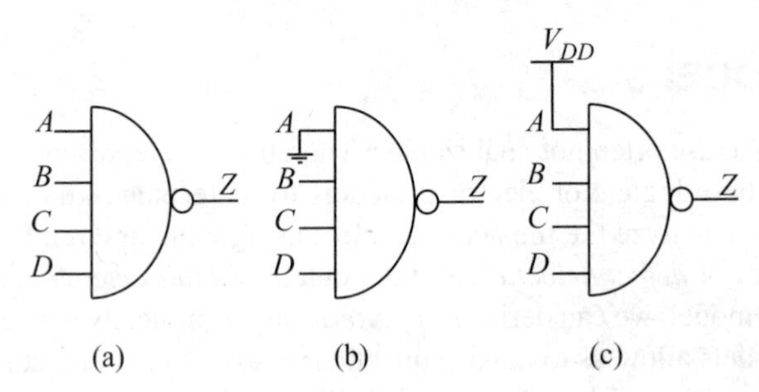

Figure 20.2 Stuck-at faults (a) good circuit, (b) stuck-at 0, and (c) stuck-at 1

For a given circuit, stuck-at fault can occur at various ports and pins. The point where a fault exists or we assume it to exist is known as a *fault site*. At each fault site, two stuck-at faults (SA0 and SA1) are possible. In a circuit simulation, we emulate a stuck-at fault by first disconnecting the corresponding signal source. Subsequently, we tie it to the constant logic (either 0 or 1 depending on the fault type).

Consider a fanout-free net[4] *N1* shown in Figure 20.3(a). A SA0 at its source (at *G1/Z*) is equivalent to a SA0 at its destination (at *G2/A*). Similarly, a SA1 at its source and destination are equivalent. Therefore, we consider only one fault site for fanout-free nets.

For a multiple-fanout net,[5] such as *N2* in Figure 20.3(b), there can be a stuck-at fault at multiple sites. It can occur at its driver (also called the *stem* of the net). A fault at the stem of the net impacts all the pins in its fanout. If a SA0 occurs at *G1/Z*, all pins in its fanout logic (*G2/A*, *G3/B*, and *G4/A*) get stuck to 0. A stuck-at fault can also occur at each *branch* of the net. For example, stuck-at fault can occur at *G2/A*, *G3/B*, or *G4/A*. Note that the stuck-at faults at individual branches of a net are distinguishable, and we need to treat them independently. A SA0 fault at *G2/A* impacts the pins in its fanout, but not the pins *G3/B* and *G4/A*. Therefore, *G2/A* can have a SA0, while *G3/B* and *G4/A* can be fault-free. In circuit analysis, we model SA0/SA1 fault in a branch by disconnecting its driver and connecting 0/1 to the corresponding branch pin. In general, a multiple-fanout net with fanout of *N* (where *N*>1) has (*N*+1) possible fault sites.

[4] A fanout-free net has a fanout of exactly one.

[5] A multiple-fanout net has fanout of more than one.

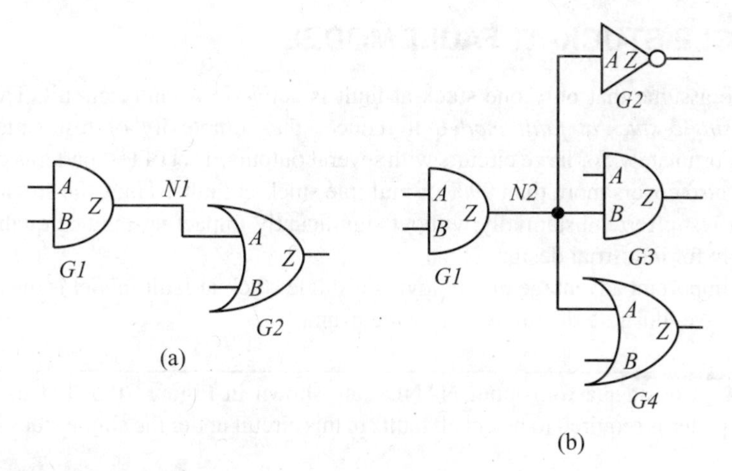

Figure 20.3 Stuck-at fault sites (a) fanout-free net and (b) multiple-fanout net

Example 20.3 Consider the circuit shown in Figure 20.4. Let us examine the number of fault sites in this circuit.

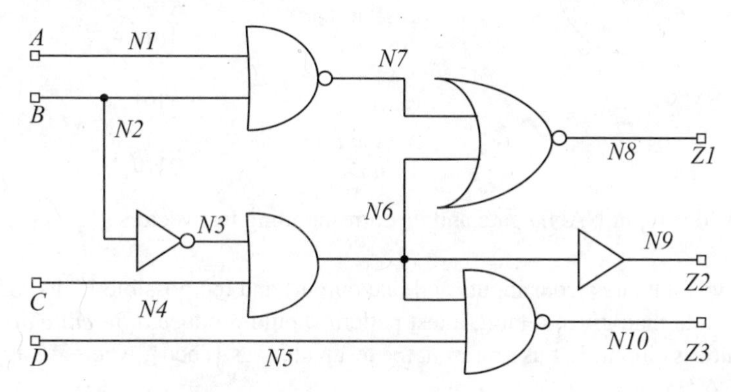

Figure 20.4 A circuit to demonstrate possible stuck-at fault sites

There are eight fault sites for eight fanout-free nets, i.e., *N1*, *N3*, *N4*, *N5*, *N7*, *N8*, *N9*, and *N10*. Additionally, there are three fault sites for the net *N2* and four fault sites for the net *N6*. Thus, there are 15 fault sites and 30 possible stuck-at faults in the above circuit.

In the stuck-at fault model, for a given fault site, there are three possibilities. The site can be fault-free, or it can have SA0 or SA1. Therefore, if we assume that we can have multiple faults simultaneously in a circuit, N fault sites can have faults in $(3^N - 1)$ ways. The term -1 in this expression represents the case when the entire circuit is fault-free. With just 20 fault sites, the number of ways of having one or more faults is more than 3 billion. Therefore, it is impractical to consider multiple stuck-at faults in a circuit.

20.4 SINGLE STUCK-AT FAULT MODEL

In practice, we assume that only one stuck-at fault is active in a faulty circuit. This fault model is known as *single stuck-at fault model.* It reduces the complexity of test pattern generation considerably. Fortunately, for large circuits with several outputs, a set of test patterns covering single stuck-at faults also covers more than 99% of multiple stuck-at faults. Thus, the single stuck-at fault model reduces test effort substantially without significantly impacting the test quality [4]. Hence, we use it widely for industrial designs.

The most important advantage of employing a single stuck-at fault model is the resulting small test pattern set. We illustrate this in the following example.

Example 20.4 Consider a four-input NAND gate shown in Figure 20.5. Let us determine the number of test patterns required to detect all faults in this circuit under the single stuck-at fault model.

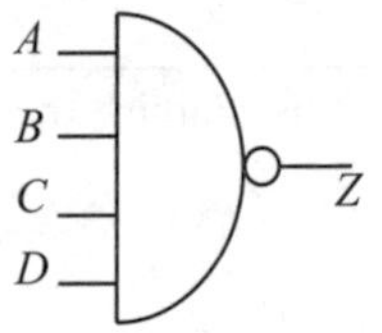

Faults	Test vector
SA0 at *A*, *B*, *C*, *D* or SA1 at *Z*	1111
SA1 at *A* or SA0 at *Z*	0111
SA1 at *B* or SA0 at *Z*	1011
SA1 at *C* or SA0 at *Z*	1101
SA1 at *D* or SA0 *Z*	1110

Figure 20.5 A four-input NAND gate and its corresponding test vectors

There are five fault sites (four inputs and one output) and ten possible faults (SA0 and SA1 at each fault site). Note that, given a fault, a test pattern should produce *different outputs* for a faulty circuit and a faultless circuit. Let us represent the test pattern as {abcd}, where we apply *a* to *A*, *b* to *B*, *c* to *C*, and *d* to *D*.

We can determine the test pattern for SA0 at the input *A* as follows. We consider 16 possible input combinations and pick the one that produces different outputs for a faultless circuit and circuit with an SA0 at *A*. Thus, we arrive at the test pattern {1111}. The output is 0 for the faultless circuit and 1 for SA0 at *A*. We observe that output is 1 for SA0 at other inputs *B*, *C*, and *D* also. Hence, the test pattern {1111} can detect SA0 at other inputs also. Similarly, we determine the test pattern {0111} to detect SA1 at *A* (the output will be 1 for a faultless circuit and 0 for the SA1 at *A*). Similarly, we determine the test patterns for SA1 at *B*, *C*, and *D* as {1011}, {1101}, and {1110}, respectively.

Note that the test patterns {0111}, {1011}, {1101}, and {1110} also test the fault SA0 at the output *Z* (the output will be 1 for faultless circuit and 0 for SA0 at *Z*). Similarly, we can test SA1 at the output *Z* using {1111}. Thus, under the single stuck-at fault model, a test pattern can detect multiple faults. It reduces the number of test patterns required to achieve some fault coverage.

To summarize, for a four-input NAND gate, just five test patterns ({1111}, {0111}, {1011}, {1101}, and {1110}) can detect all possible single stuck-at faults.

Remark: In general, if there are *N*-inputs of a NAND gate, the number of test patterns required to detect all single stuck-at faults is $N + 1$. In contrast, exhaustive functional testing of an *N*-input

NAND gate requires 2^N test patterns. Thus, the complexity of testing all single stuck-at faults in a *gate* is *linear*, while it is *exponential* for exhaustive functional testing. Even in circuits obtained by connecting gates, a single stuck-at fault model helps in reducing the complexity of testing from *exponential* (as in exhaustive functional testing) to *linear* (using the strategy of structural testing).

Let us also consider the test patterns required to test the circuit that implements 16-input AND gate (discussed earlier in Figure 20.1). We can check that the test pattern {0111 1111 1111 1111} can detect SA1 at the internal pin *G1/Z* (the output *Y* is 0/1 for the fault-free/faulty circuit). Similarly, the test pattern {1111 1111 1111 1111} can detect SA0 at the internal pin *G1/Z* (the output *Y* is 1/0 for the fault-free/faulty circuit). Thus, in the single stuck-at fault model, test patterns can detect faults even at the *internal pins* which is essential for structural testing. We will discuss a systematic methodology to derive these test patterns in Chapter 22 ("Automatic test pattern generation").

It is worth pointing out that the circuit shown in Figure 20.1 is simple and used just for illustration. In practice, we have millions of combinational and sequential circuit elements in an integrated circuit (IC). Typically, we need to test all of them. In such cases, detecting faults at internal pins is more challenging, as explained in the following section.

20.5 CONTROLLABILITY AND OBSERVABILITY

We have seen that a fault can occur at any *internal pin* or a net in a circuit. For simplicity, let us first consider detecting faults in a fully combinational circuit. A tester or automatic test equipment (ATE) can apply test patterns and observe their response only using input and output *ports* (i.e., the primary inputs and primary outputs are only available as *external* interfaces). Therefore, detecting faults at a pin that is topologically farther away from the ports of the circuit becomes difficult. Intuitively, the lack of direct contact of a fault site to the input and output ports makes its detection difficult. We describe this aspect of testing using the notion of *controllability* and *observability*, as follows.

Consider a four-input NAND gate located topologically farther away from the input ports, as shown in Figure 20.6(a).

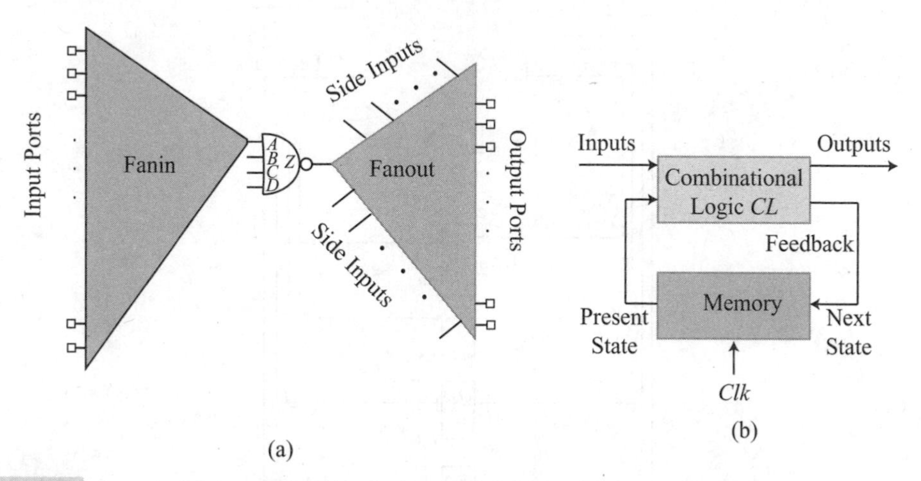

Figure 20.6 A typical (a) combinational circuit and (b) sequential circuit

Controllability: We have seen in Ex. 20.4 that we need five test patterns to detect all possible single stuck-at faults associated with the four-input NAND gate. However, during testing, a testing equipment has only the input ports of a circuit accessible for applying test patterns. The value at the input ports must propagate through the circuit elements, and the requisite bit pattern must appear at the pins A, B, C, and D of the NAND gate. Intuitively, setting a particular bit (0 or 1) at an input pin can be more difficult with more intervening circuit elements. We measure the ease or difficulty of setting a particular signal to a given value (0 or 1) by the *controllability* of that signal. It depends on its topological distance from the input port or the logic depth. The input ports are fully controllable, while a signal lying far away from the input ports is typically less controllable.

Observability: To detect a fault, we need to read the value (0 or 1) obtained at an internal pin at one of the output ports (external interfaces of the circuit). For the NAND gate shown in Figure 20.6(a), its output must propagate through the circuit elements before being read at an output port. Therefore, observing its behavior can be more difficult when there are more intervening circuit elements between the internal pin and the output port. We measure the ease or difficulty of reading a signal value (0 or 1) at an output port by the *observability* of that signal.

Sequential circuits: In a sequential circuit, there are memory elements such as flip-flops that define the *state* of a circuit, as shown in Figure 20.6(b). A sequential circuit transitions to the *next state* on receiving the next clock edge as per the combinational logic CL, the current inputs, and the present state.

To set a given bit (0 or 1) at an internal pin of a sequential circuit, we need to take it to a specific *state* (represented by a combination of zeroes and ones at the Q-pins of the flip-flops). In that state, all memory elements and the input ports would have the right combination of zeros and ones, such that we obtain the desired bit at the internal pin. We illustrate it further in the following examples.

Example 20.5 We first examine the problem of controllability in a sequential circuit. Consider the circuit shown in Figure 20.7.

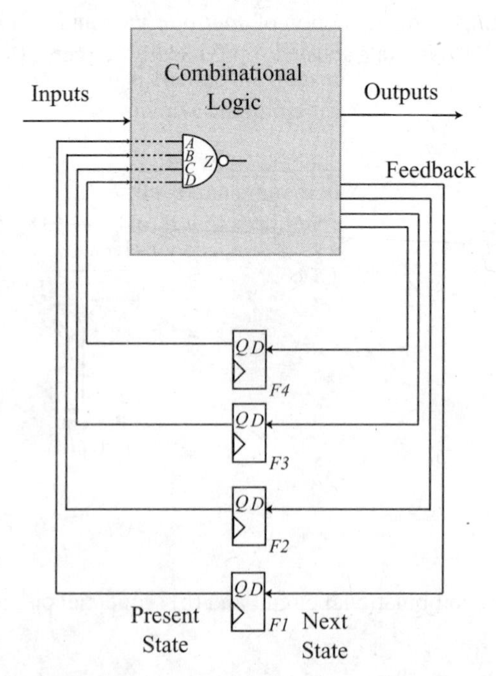

Figure 20.7 Problem of controllability in a sequential circuit

For simplicity, we have assumed that the circuit consists of just four memory elements (flip-flops) *F1*, *F2*, *F3*, and *F4*. We can consider a sequential circuit as a finite state machine (FSM) in which the flip-flops encodes the *states* and the combinational logic implements the *next state function* and the *output function.*[6] In this case, since there are four flip-flops, there can be $2^4 = 16$ states.

In the above circuit, the flip-flops and the input ports drive a four-input NAND gate through some combinational circuit elements. We need to drive the input pins of the NAND gate with the test patterns to detect a single stuck-at fault in that NAND gate. Since the flip-flops lie in their fanin cone, we need to set their Q-pins to the right combination of zeroes and ones such that we obtain the required test pattern at the inputs of the NAND gate. The combination of values at the Q-pins defines a *state* of a sequential circuit. Therefore, we need to take the circuit to a given state to apply a given test pattern at the inputs of the NAND gate (e.g., {1111} to detect SA0 at pin *A*).

In general, given a sequential circuit and an initial state, we can certainly reach a given state only after going through an *exponential* number of intermediate states (exponential in the number of state elements or flip-flops).[7] For example, if the four flip-flops in this example are part of a binary counter that starts with the state {0000}, we will require $2^4-1=15$ transitions to reach the state {1111}. Hence, the problem of controllability is more challenging in a sequential circuit.

Example 20.6 Let us now examine the problem of observability in a sequential circuit. Consider the circuit shown in Figure 20.8.

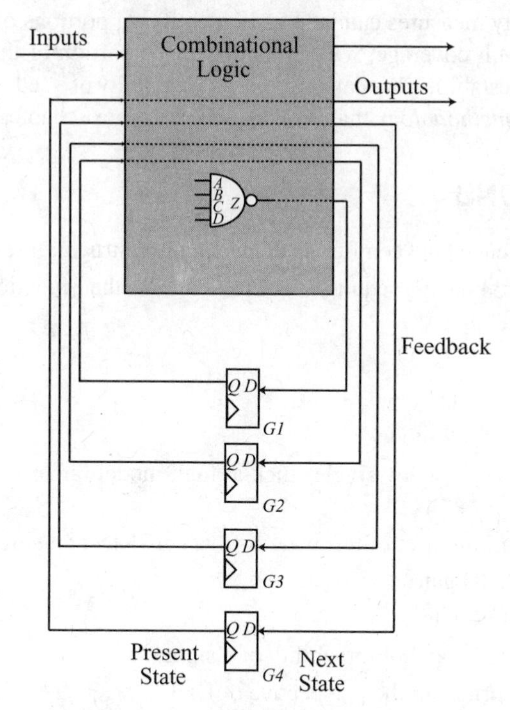

Figure 20.8 Problem of observability in a sequential circuit

[6] See Section 1.1.7 ("Finite state machines") for the basics of FSM and Section 10.6 ("FSM synthesis") for the circuit implementation of an FSM.

[7] We are assuming that the given state is reachable from the initial state.

For simplicity, we again assume that it consists of just four memory elements (flip-flops) *G1*, *G2*, *G3*, and *G4*. Since there are four flip-flops, there can be $2^4 = 16$ states.

The output of the NAND gate propagates to *G1* through some combinational circuit and then through other flip-flops to the output port. In general, the value at an output pin of a sequential circuit depends on the *current state* and the current inputs.[8] Hence, to observe the value of the NAND gate at the output port, we might need to take the sequential circuit to a particular *state*. As explained earlier, in the worst case, we will require state transitions that are exponential in the number of flip-flops to reach a given state from an initial state.[9] Hence, the problem of observability is more difficult in a sequential circuit.

The above examples show that we need to take a sequential circuit through various states to *write* a given value at a pin or *read* its value at a primary output. Therefore, an internal pin is less controllable and observable in a sequential circuit than a combinational circuit.

We can measure the controllability and the observability of a signal in several ways [5, 6]. We can apply random test stimuli at the input ports and compute the probability of that signal being 0 and 1. If the probability of being 1 is low, we infer that it is difficult to make that signal 1 using random test patterns. Therefore, we say that the 1-controllability of the signal is low. Analogously, we can measure 0-controllability.

In general, high controllability and observability of signals in a circuit denote that the circuit is easily testable. If a circuit is easily testable, the effort in devising test methodology and the test cost will be low. The testability measures can help us in identifying portions of a circuit that are hard to test. To achieve a high fault coverage, we can modify those portion of the circuit or insert special hardware to enhance the testability. We can improve the testability of a sequential circuit dramatically by adopting *scan design methodology* that is explained in the next chapter ("Scan design").

REVIEW QUESTIONS

20.1 What is the difference between functional testing and structural testing?

20.2 Why does structural testing require fewer test patterns than functional testing?

20.3 Explain the following:

- **(a)** fault model
- **(b)** stuck-at fault model
- **(c)** single stuck-at fault model.

20.4 Why do we typically employ single stuck-at fault model rather than multiple stuck-at fault model?

20.5 Determine the minimum set of test patterns that can detect all single stuck-at faults in:

- **(a)** a five-input AND gate,
- **(b)** a six-input NOR gate.

20.6 What is controllability and observability of a signal?

20.7 What is a test pattern from the perspective of DFT?

20.8 Why is controlling and observing a signal typically more difficult in a sequential circuit than in a combinational circuit?

---

[8] For Moore type FSM it only depends on the current state.

[9] We are assuming that the given state is reachable from the initial state.

TOOL-BASED ACTIVITY

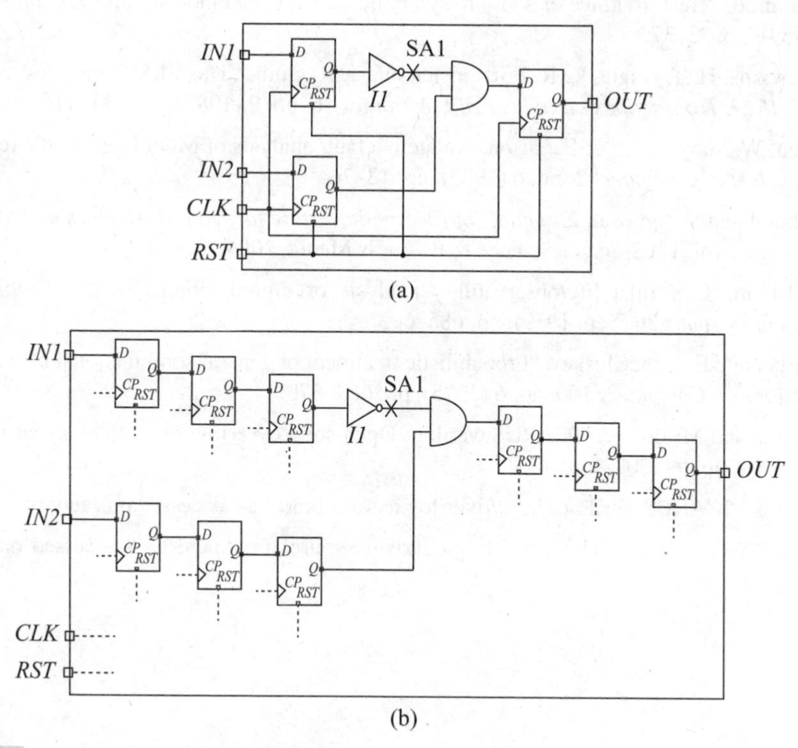

Figure 20.9 Circuits that illustrate difficulty in testing

Consider the circuit shown in Figure 20.9(a). There is an SA1 fault at the inverter *I1*. All flip-flops can be reset by the asynchronous reset port *RST*. Assume that initially, all the flip-flops are in the reset state (output Q-pin is 0). Only the input ports and output ports are accessible to us. We can detect the SA1 (i.e., differentiate between faultless circuit and faulty circuit) at the inverter by making *IN1*=*IN2*=1, de-asserting the *RST* signal and allowing the clock to pass data from the input port to the output port. How many clock cycles will be required to detect the fault (i.e., obtaining different bits at the output port *OUT*). To find this, write a Verilog code and an appropriate testbench for the above circuit, and perform logic simulation. You can use any tool for performing simulation, such as open-source tool ICARUS Verilog [7–9].

Next, consider the circuit shown in Figure 20.9(b). The flip-flops' *CP* and *RST* pins are connected to the input ports *CLK* and *RST*, respectively. They are not shown for the sake of clarity. The circuit has an SA1 fault at *I1*. Similarly, in this case, we can detect the SA1 at the inverter by making *IN1*=*IN2*=1, de-asserting the *RST* signal and allowing the clock to pass data from the input port to the output port. Using logic simulation, find the number of clock cycles required to detect the fault.

Comment on the difficulty in detecting the faults in the given two circuits.

REFERENCES

[1] R. D. Eldred. "Test routines based on symbolic logical statements." *Journal of the ACM* 6 (Jan. 1959), pp. 33–37.

[2] C. F. Hawkins, H. T. Nagle, R. R. Fritzemeier, and J. R. Guth. "The VLSI circuit test problem—A tutorial." *IEEE Transactions on Industrial Electronics* 36, no. 2 (1989), pp. 111–116.

[3] J. P. Shen, W. Maly, and F. J. Ferguson. "Inductive fault analysis of MOS integrated circuits." *IEEE Design & Test of Computers* 2, no. 6 (1985), pp. 13–26.

[4] M. Bushnell and V. Agrawal. *Essentials of Electronic Testing for Digital, Memory and Mixed-signal VLSI Circuits*, vol. 17. Springer Science & Business Media, 2004.

[5] L. Goldstein. "Controllability/observability analysis of digital circuits." *IEEE Transactions on Circuits and Systems* 26 (Sep. 1979), pp. 685–693.

[6] K. P. Parker and E. J. McCluskey. "Probabilistic treatment of general combinational networks." *IEEE Transactions on Computers* 100, no. 6 (1975), pp. 668–670.

[7] S. Williams and M. Baxter. "ICARUS Verilog: Open-source Verilog more than a year later." *Linux Journal* 2002, no. 99 (2002), p. 3.

[8] S. Williams. "ICARUS Verilog." http://iverilog.icarus.com/. Last accessed on January 7, 2023.

[9] T. Bybell. "Welcome to GTKWave." http://gtkwave.sourceforge.net/. Last accessed on January 7, 2023.

21 Scan Design

External nature is only internal nature writ large.

—Swami Vivekananda, *Complete Works of Swami Vivekananda*, Partha Sinha (ed.), 2019

In a sequential circuit, the controllability and observability of signals are low. The state elements (flip-flops) need to transition through several *clock cycles* before the value propagates from the input ports to an internal pin. We encounter a similar problem in observing the value of an internal pin at some output port. In the worst case, the number of clock cycles required to read or write a bit onto an internal pin can be exponential in the number of flip-flops. Even in a circuit with only a hundred flip-flops, the number of required clock cycles can make testing infeasible.

There are several problems associated with the low controllability and observability of signals. First, the tester will require a large number of clock cycles in applying the test pattern sequence. It increases the test execution time and the cost of testing. Second, it becomes difficult for the *automatic test pattern generation* (ATPG) tools to find the sequence of a test pattern for a given fault. The difficulty primarily arises because finding a test pattern in a sequential circuit requires *state exploration* or searching for the appropriate sequence of states. Since, in general, there can be too many states, finding a test pattern sequence in a sequential circuit becomes challenging. Therefore, the test development time becomes dramatically high for a sequential circuit. Consequently, we might need to sacrifice the fault coverage, and the quality of testing can suffer.

We can tackle the above problems by employing the *scan design methodology*. In this methodology, we modify the sequential circuit to improve signal controllability and observability. Consequently, test pattern generation and test pattern application by the tester become extremely efficient. In this chapter, we will explain the scan design methodology in detail.

21.1 BASICS OF SCAN DESIGN TESTING

The basic principle of a scan design methodology is to make design modifications such that controlling and observing the state of memory elements in a design becomes easier than in the original sequential circuit [1].

Salient features: The salient features of scan design methodology are listed below and illustrated in Figure 21.1.

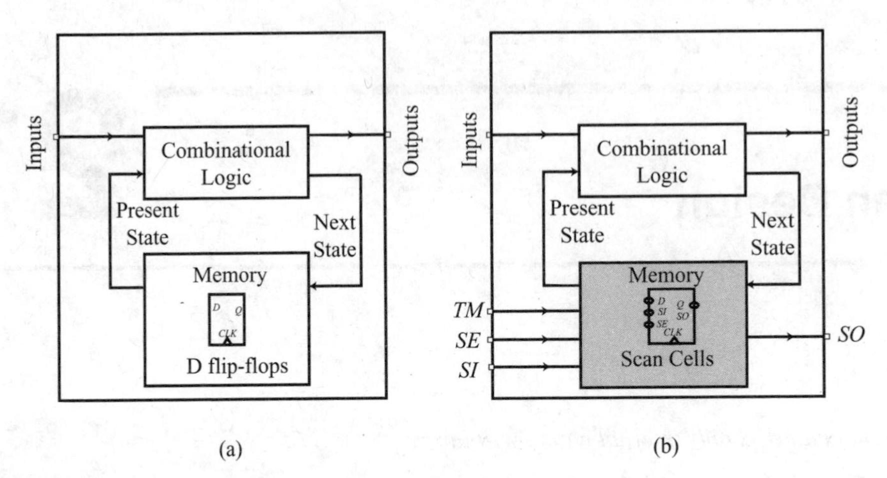

Figure 21.1 Typical modifications made to a design for carrying out scan design test methodology (a) original design and (b) modified design

1. **Test mode:** We operate a circuit in different *modes*. The normal mode of operation is known as the *functional mode*. In this condition, the circuit delivers its main functionality. However, while performing testing (typically using automatic test equipment [ATE]), it is said to operate in the *test mode*. When a circuit goes into the test mode, the test-specific circuit elements become active. These elements increase the controllability and the observability of the signals in this mode.
2. **Extra ports:** We add some extra input and output ports to a design solely for testing. These ports are shown in Figure 21.1(b). The values at the input ports *TM* (test mode) and *SE* (scan enable) define the current operating mode of a circuit.
3. **Scan cells and scan chains:** In scan design methodology, we replace each flip-flop in a design by a circuit element known as *scan cell*. In the functional mode, a scan cell works as a regular flip-flop. We make the same connections for the data inputs (D-pins), clock pins, and outputs (Q-pins) of the scan cells as for the replaced flip-flops. Additionally, we connect the Q-pins of the scan cells to the scan-in (*SI*) pin of another scan cell such that a chain of *shift register* forms in the *test mode* [2]. This chain of shift registers formed during test mode is known as a *scan chain*. Thus, a scan cell works as both a regular flip-flop and a scan chain component in the test mode. We will illustrate scan chains in more detail in the following sections.
4. **Shift-in:** We connect the start of the scan chain to an input port called *SI* (scan in). We can load each scan cell of a scan chain with the required bit (0 or 1) by applying that bit at the *SI* port and allowing the scan chain to *shift-in* data through it for an appropriate number of clock cycles.
5. **Shift-out:** We connect the end of the scan chain to an output port called *SO* (scan out). We can observe the value of a scan cell at the *SO* port by allowing the scan chain to *shift-out* data through it for an appropriate number of clock cycles.

Testability improvements: The combination of values (0 or 1) at the flip-flops' Q-pin represents the *state* of a sequential circuit. If there are N flip-flops in it, there are 2^N possible states. As a result, a normal sequential circuit needs to go through 2^N clock cycles (assuming each clock edge takes it to a new state) to reach a given state in the worst case.[1] For example, a synchronous binary counter with N flip-flops gets to the last count after 2^N clock cycles.

[1] For challenges of testability in a normal sequential circuit, see Section 20.5 ("Controllability and observability").

In a scan design, we organize flip-flops (or scan cells) as a shift register (or scan chain) during testing. We apply a sequence of bits (0 or 1) at the *SI* port and allow it to shift through the scan chain in consecutive clock cycles. Hence, assuming that there are *N* flip-flops in a scan design, we can reach any state in just *N* clock cycles even in the worst case. Thus, to attain a particular state, we require a *linear* number of clock cycles instead of an exponential. It dramatically reduces the test time and the cost of testing for sequential circuits.

The output Q-pin of a scan cell (or the associated flip-flop) is easy to control since it can be taken to any value (0 or 1) using the *SI* port and linear number of clock cycles. Therefore, we can treat the Q-pin of a flip-flop similar to an input port of a design. Hence, we consider the Q-pin of a flip-flop as *pseudo-primary input* (PPI) in testing. Likewise, we can easily observe the value at the D-pin of a scan cell (or the associated flip-flop) at the output port *SO* in a linear number of clock cycles using the scan chain. Therefore, the D-pin can be treated similarly to an output port. Hence, we consider the D-pin of a flip-flop as *pseudo-primary output* (PPO) in testing.

ATPG simplification: With the availability of PPI and PPO in a scan design, the task of ATPG simplifies. It can avoid state exploration in finding test patterns. It just needs to consider the *combinational logic* for test pattern generation and treat the flip-flop's Q-pin and D-pin similar to an input port and an output port, respectively. Thus, in a scan design, the ATPG problem reduces from *sequential ATPG* to *combinational ATPG*. There are efficient methods to solve combinational ATPG problem. We will describe some of them in Chapter 22 ("Automatic test pattern generation").

In the following sections, we will explain the above features of the scan design methodology in more detail.

21.2 SCAN CELLS AND SCAN CHAIN

We can use different kinds of scan cells to implement the scan design methodology. A traditional scan cell implementation, known as *muxed-D scan cell*, is shown in Figure 21.2.[2] It can directly replace an edge-triggered D flip-flop. It consists of a two-to-one multiplexer and an edge-triggered D flip-flop. The *SE* (scan enable) pin drives the select input of the multiplexer. The output of the multiplexer is the value at the D-pin (when *SE*=0) or the SI-pin (when *SE*=1). We refer to the scan cell output pin as *Q* pin or *SO* (scan out) pin. In the functional mode, it works as a regular Q-pin of the replaced flip-flop. In the test mode, it provides input to the next scan cell in the scan chain and works as an SO-pin.

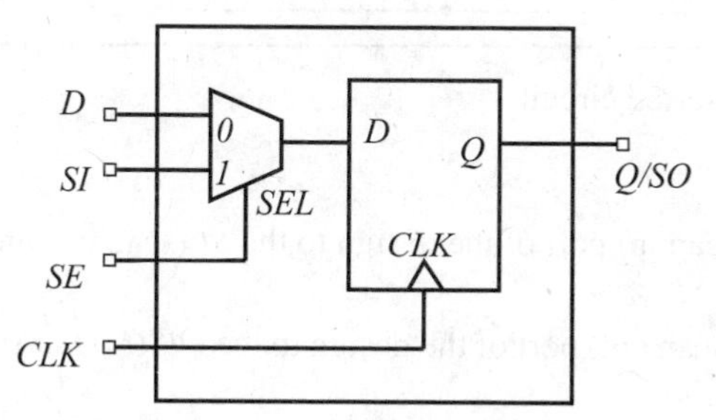

Figure 21.2 Schematic representation of a muxed-D scan cell

[2] The scan cell implementations can have a separate scan clock also, be specifically designed for latch-based designs, and its performance can be improved over the traditional muxed-D scan cell. Interested readers can look into references [2–5] for various other types of scan cells.

In the functional mode, we keep *SE*=0, and the internal flip-flop of the scan cell latches the data present at its *D* pin on receiving the clock signal. When we want the scan cell to work as a shift register, such as in the test mode, we make *SE*=1. Consequently, the scan cell latches the data at the *SI*-pin. Thus, a scan cell latches the data at either the *D* or the *SI*-pin, depending on the *SE* pin value.

We illustrate how scan cells are connected to form a scan chain in the following example.

Example 21.1 Consider the circuit shown in Figure 21.3. Let us carry out scan cell replacement in this circuit.

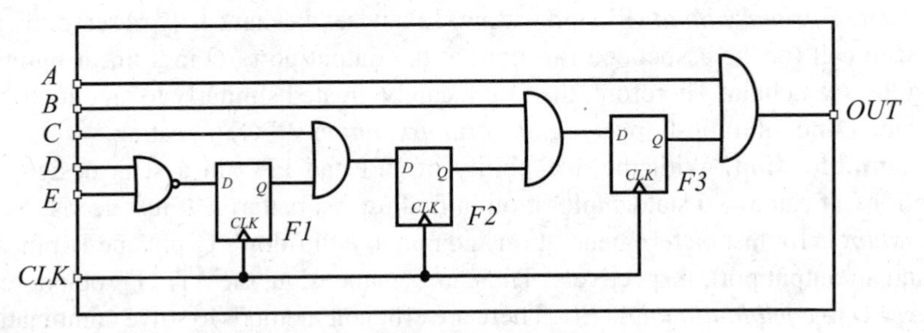

Figure 21.3 Given circuit for scan cell replacement

The circuit post-scan cell replacement circuit is shown in Figure 21.4. The dotted lines represent original connections, while the continuous lines represent new connections required for scan cells. We carry out the following steps to obtain this circuit.

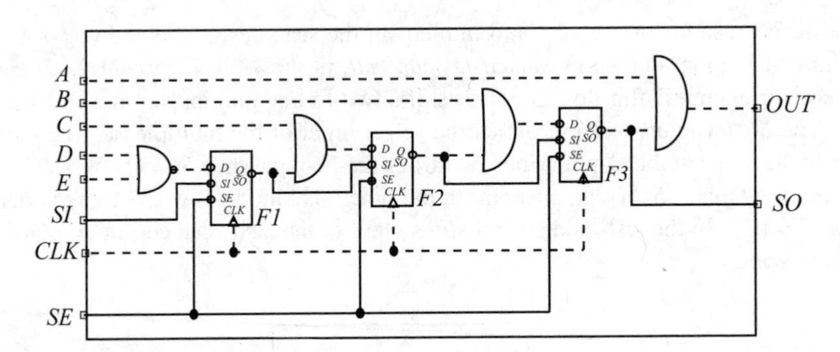

Figure 21.4 Scan cell inserted circuit

1. We connect the *SI* (scan in) port of the design to the *SI* (scan in) pin of the first scan cell in the scan chain.
2. We connect the *SO* (scan out) port of the design to the *Q*/*SO* (Q or scan out) pin of the last scan cell in the scan chain.
3. We connect the *Q*/*SO*-pin of one scan cell to the *SI*-pin of the next scan cell in the scan chain. Thus, data can propagate from one scan cell to the next, operating as a shift register. In Figure 21.4, the scan chain starts at the *SI* port, going through three scan cells *F1/SI* → *F1/Q* or *F1/SO* → *F2/SI* → *F2/Q* or *F2/SO* → *F3/SI* → *F3/Q* or *F3/SO*, and ends at the *SO* port.

4. We connect the *SE* (scan enable) pins of all the scan cells to the *SE* (scan enable) port of the design.
5. Despite replacing D flip-flop with the muxed-D scan cells, we maintain the original connections of the replaced D flip-flop. Thus, we connect the scan cell's D-pin, CLK-pin, and Q/SO-pin as they were connected for the original D flip-flop. In Figure 21.4, these connections are shown by dotted lines. They ensure that, after scan cell replacement, the normal functionality of the circuit does not change for *SE*=0.

21.3 EXECUTION OF SCAN DESIGN TESTING

The modes of operation of a scan design are listed in Table 21.1.

Table 21.1 Modes of operation in a scan design

No.	Mode	TM	SE
1	Functional	0	0
2	Test (shift)	1	1
3	Test (capture)	1	0

A circuit operates normally when *TM*=0. When we need to test, we make *TM*=1, and the circuit moves to the test mode. In the test mode, we can have two separate modes:

1. **Shift mode:** When *TM*=1 and *SE*=1, the circuit goes to the *shift mode*. In this mode, each scan cell receives 1 at the SE-pin. Consequently, the scan cells latch data received at the SI-pin rather than the D-pin. Thus, the scan cells move to the scan chain configuration. The first cell of the scan chain receives data from the *SI* port. Other scan cells of the scan chain receive data input from the preceding scan cell through the SI-pin. The output of the last scan cell becomes available at the *SO* port. Thus, the scan cells form a shift register in the shift mode.
2. **Capture mode:** When *TM*=1 and *SE*=0, the circuit goes to the *capture mode*. In this mode, each scan cell receives 0 at the SE-pin. Consequently, the scan cell latches or captures the data available at the D-pin. Thus, the scan cell works as a regular flip-flop, and the circuit functions normally in the capture mode. It implies that if a scan cell receives a *wrong value* at the D-pin due to some *fault* in the circuit, it will latch or capture that wrong value. We can subsequently read the latched value at the Q/SO-pin at the *SO* port by shifting out the values through the scan chain. By comparing the values obtained at the *SO* port with the golden values for a faultless circuit, we can detect a faulty circuit.

Scan chain testing: In scan design testing, first, we ascertain the integrity of the scan chain. We moved the circuit to the scan mode by making *TM*=1 and *SE*=1. Then we apply a test sequence, such as "0011 0011 0011...", at the *SI* port and allow it to propagate through the scan chain using the system clock. If the scan chain is working properly, then the same sequence is obtained at the *SO* port. The special sequence "0011" is used since it covers all the possible transitions $0 \rightarrow 0$, $0 \rightarrow 1$, $1 \rightarrow 1$, and $1 \rightarrow 0$. If there is an SA0 or SA1 fault in a scan cell, the corresponding transition will be missing in the output sequence. Thus, we can detect a fault in the scan chain.

Shift and capture sequences: After establishing the integrity of the scan chain, we test for faults in the combinational circuit elements. To test them, we apply appropriate test patterns, typically generated by an ATPG tool, as illustrated in Figure 21.5. We can apply test patterns directly at the input ports or primary inputs (PIs) using tester pins. For applying test pattern at the flip-flops' Q-pins or PPIs we take the circuit to the *shift mode*. Then, we stream in test pattern-specific bit sequence through the *SI* port and shift them through the scan chain until it reaches the targeted PPIs. For a scan chain consisting of *N* flip-flops, we will require *N* clock cycles to obtain a given bit sequence at all the PPIs. We denote the bit sequence at PPIs as PP1, PP2, and PP3 in Figure 21.5. While we are loading PPIs with bits corresponding to the test pattern, the values at the PIs other than the test ports (*SI*, *TM*, *SE*, and *CLK*) are irrelevant or don't care. Therefore, we can apply any value at them (as indicated by the shaded region in Figure 21.5).

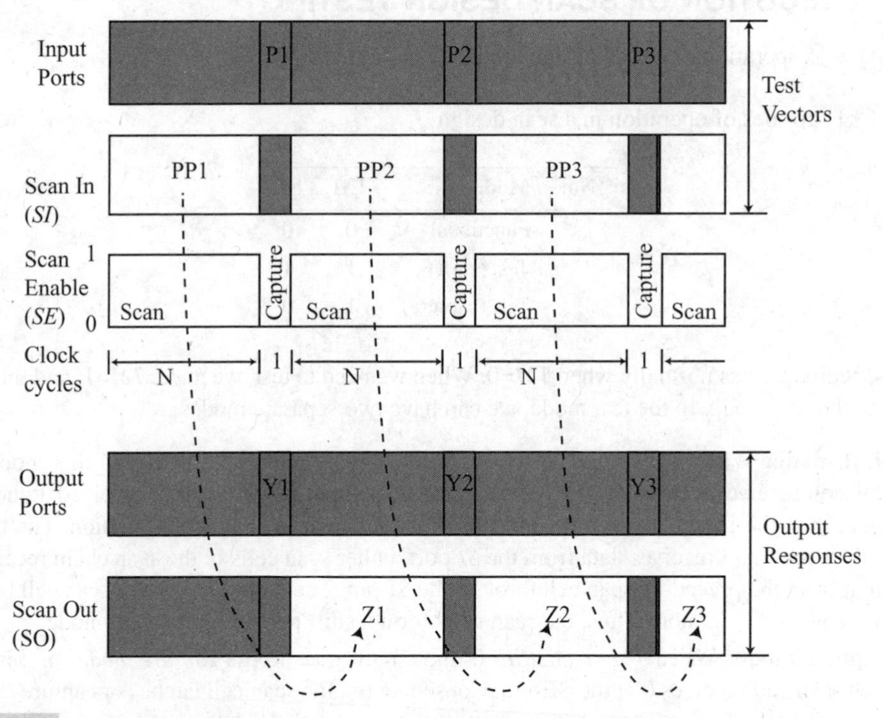

Figure 21.5 Shift and capture operations in testing. The clock cycles are shown along X axis and values at relevant ports are shown along Y axis.

The complete test pattern is loaded when the PIs and the PPIs attain the required values. The values at the PIs are shown as P1, P2, and P3 in Figure 21.5. We can apply patterns P1, P2, and P3 instantaneously at the PIs. Note that {P1, PP1} constitute one test pattern and can simultaneously test many single stuck-at faults in the circuit. Similarly, {P2, PP2} and {P3, PP3} constitute two other test patterns.

After we have loaded the complete test pattern into a circuit, we take it to the *capture mode* for one clock cycle, as shown in Figure 21.5. We capture the circuit response at the output ports or primary outputs (POs). The output response available at the POs, such as Y1, Y2, and Y3, can be directly compared with the expected response. We also capture response at the D-pin of the scan cells or PPOs. However, the output response obtained at the PPOs, such as Z1, Z2, and Z3, requires *N* clock cycles to propagate through the scan chain and become sequentially available at the output ports. Therefore, the circuit is kept in the *shift mode* for N subsequent cycles, as shown in Figure 21.5. We stream out the values obtained at the PPOs in the capture mode to the *SO* port.

Note that in Figure 21.5, the output response corresponding to the test vector {P1, PP1} is shown as {Y1, Z1}, corresponding to {P2, PP2} is shown as {Y2, Z2}, and corresponding to {P3, PP3} is shown as {Y3, Z3}. We compare these responses with the expected response to detect whether a circuit is faulty.

Example 21.2 Let us compute the number of clock cycles needed to test a circuit with the scan test methodology. Assume that the circuit has only one scan chain with N scan cells (N is also called the *length* of the scan chain). Further, assume that we need to apply K test patterns to the circuit.

For one test pattern, we require $(N + 1 + N)$ clock cycles for the following tasks:

1. N cycles for applying bit sequence at *SI* port and shift them in through the scan chain to PPIs,
2. One cycle for capture (and applying pattern at PIs and observing response at POs), and
3. N cycles for shifting the response from PPOs to the *SO* port (for observing and comparing with the expected response).

Note that while we are shifting out the bit sequence Z1 to the *SO* port, we can shift in the PPI component of the next test pattern (i.e., PP2) using the SI port, as shown in Figure 21.5. Hence, for the first $(K - 1)$ test patterns, we can perform the observed response shift out and the next test pattern shift in within the same clock cycle. Consequently, for the first $(K - 1)$ test pattern, we require $\{(K - 1)(N + 1)\}$ clock cycles.

For the last test pattern response shift out, we have nothing to shift in. Hence, it will require $(2N + 1)$ clock cycles. Therefore, the total number of clock cycles in the execution of the shift and capture sequences in a scan design is $(K - 1)(N + 1) + 2N + 1 = K(N + 1) + N$.

The above example shows that the number of cycles required for testing increases with the scan chain length (N). Therefore, we should reduce the scan chain length. We can achieve this is by decomposing long scan chains into multiple small parallel scan chains. However, it will require extra *SI* and *SO* ports (one for each parallel scan chain). Nevertheless, we can avoid creating them by making the following observation from Figure 21.5. When the circuit is in the shift mode, the values at PIs and POs other than *SI*, *SO*, *TM*, *SE*, and *CLK* are *don't care*. In other words, during shift mode, we can set them to an arbitrary value. Therefore, we can use these PIs and POs as *SI* and *SO* for extra scan chains. We can ensure that these ports normally work in the functional mode by using *demultiplexers* and *multiplexers* at the PI and PO, respectively. However, the additional delay of these circuit elements at the PIs and POs can degrade the performance. We should verify that the delay degradations in the input/output (I/O) paths are acceptable [6].

Power dissipation in shift mode: A circuit undergoes a high switching activity as we shift in/out bit sequences through a scan chain in the shift mode. The scan cells toggle values with the changes in the input bits. Additionally, the combinational circuit elements in the fanout of the scan cells also change their values despite these changes being functionally irrelevant in the shift mode. Therefore, a high switching activity and the associated switching power dissipation occurs in the shift mode [7]. We need to constrain this switching power dissipation; otherwise, the circuit can violate the thermal limit and fail. Additionally, there can be an unacceptable voltage drop in the power lines due to the high current drawn by the scan cells and logic gates that are switching too frequently.[3] To alleviate

---

[3] The harmful effects of voltage drop in the power lines in a circuit will be explained in Section 29.3.4 ("Power integrity").

this problem, typically, we reduce the clock frequency in the shift mode to bring the switching power dissipation to a tolerable level, though at the expense of increased test time.

21.4 IMPLEMENTATION OF SCAN DESIGN FLOW

We have described the basics of scan design flow in the previous sections. We need to carry out many tasks in a design flow to implement the scan design flow, and we describe some of them in the following paragraphs [4].

21.4.1 Design Preparation

To implement scan design methodology, the design must follow some guidelines or *rules* [8, 9]. We can check these rules at a higher level of abstraction, such as RTL or netlist. Once we find a rule violation, we can make changes in the design to fix it.

An example of a rule is that the clock pins of all the flip-flops in a design must be controllable from input ports [10]. This rule will ensure that a flip-flop can be part of a shift register after replacing it with a scan cell. The rule will violate if a circuit element blocks the clock signal from reaching the clock pin of a flip-flop. For example, if the clock signal reaches the clock pin of a flip-flop after AND-ing with a *data signal* this rule will violate because the AND gate can prevent that flip-flop from shifting data in the shift mode. To fix this issue, we can move the gate control from the clock path to the data path of the flip-flop [6]. It will ensure that the clock signal can reach the clock pin directly without changing the circuit functionality. Similarly, we need to ensure that the asynchronous set/reset signals of the flip-flops that are not controlled directly by PIs should be disabled during the shift mode [11]. It will ensure that the scan chain can shift the data properly.

We can use RTL *rule checkers* to detect these types of problems early in the design flow.

21.4.2 Scan Synthesis

The scan synthesis processes modify a design to implement the scan design test methodology. These processes are carried out at different stages of a design flow and are typically automated.

Defining configurations: Before carrying out scan synthesis, we need to decide on various attributes or *configurations* of a scan design. For example, we need to choose the number of scan chains, the scan cells from the technology library that we allow to replace flip-flops, the name and the attributes of test ports, and the flip-flop instances that we do not want to change with scan cells. Choosing these attributes typically requires making some trade-offs. For example, to reduce the test time, we can increase the number of parallel scan chains. However, the availability of input and output ports limits the maximum number of scan chains. Moreover, the extra demultiplexers/multiplexers at the PIs/POs can increase the path delay. Therefore, we need to carefully choose the ports that can be the start points or the end points of a scan chain.

We carry out the first stage of scan synthesis *after logic synthesis* and the scan synthesis tool modifies the given netlist according to the given configuration. During scan synthesis, it replaces the flip-flops with specified scan cells. Additionally, it makes appropriate connections in the netlist to build scan chains. It adds some extra test ports to the netlist if required.

Verification: After scan synthesis, the circuit behavior in the *functional mode* should not change. We perform *formal equivalence checking* of a netlist before and after scan synthesis (in the functional mode) to ensure this. However, after scan synthesis, the power, performance, and area (PPA) can degrade. A scan synthesis tool is typically equipped with logic synthesis capability also. Therefore, after scan synthesis, if new timing violations appear, it can perform some *logic re-synthesis* to restore the timing of the circuit. Furthermore, we can provide some hints to the scan synthesis tool by marking timing-critical flip-flops as *dont touch*. In this case, it will avoid replacing *dont touch* flip-flops with scan cells, though at an expense of testability.

Scan cell reordering: The scan synthesis tool connects each scan cell to two other scan cells in a scan chain (one providing input to its SI-pin and the other reading its output from its Q/SO-pin). On the circuit layout, we should place the directly connected scan cells nearby. It makes routing easier. However, scan synthesis tools have no placement information. Therefore, during pre-placement stages, the scan cell sequence in the scan chain does not depend on their location on the layout. Post-placement, we *reorder* scan cells based on their locations such that the closely placed scan cells appear consecutively in the scan chain. After reordering, we connect or *stitch* scan cells together appropriately to form a scan chain. The *scan cell reordering* and *stitching* is carried out during the physical design.[4]

21.4.3 Scan Extraction and Verification

After scan synthesis, for a sanity check, we extract the scan chains from a scan-inserted design. The shift and capture operations are verified on the extracted scan chains using a logic simulator. Furthermore, additional timing and power issues can arise after scan insertion. For example, new setup violations can appear due to scan cells having more delay than the regular flip-flops. Moreover, the timing paths between two scan cells in the shift mode are prone to *hold violations* because no logic gates exist between them. We need to identify and fix these timing issues in a scan design, typically after clock tree synthesis. We should also consider the increased dynamic power dissipation in the shift mode, as discussed earlier.

21.4.4 Test Pattern Generation

We generate test patterns using combinational ATPG tools. These test patterns are applied to a die by an ATE using a *test program*. A typical test program is responsible for overall test operation during ATE testing. It loads test vectors into the chip by setting test ports to a given sequence of zeroes and ones and streaming them through the scan chains. Further, it captures the circuit response, passes the result through the scan chains to the POs, and determines the test result by comparing them with the golden results. Post scan cell reordering, the bit sequence passing through the scan chain for a given test pattern becomes well-defined. Therefore, we can determine the bitstreams that need to be employed by the test program after scan cell reordering.

21.5 OVERHEADS OF SCAN DESIGN TESTING

We have seen that the scan design test methodology simplifies the testing process and makes it economically feasible. However, the benefits of scan design flow come with the following penalty and overheads:

1. **Area overhead**: A scan cell consumes more area than a regular flip-flop due to added multiplexer. Therefore, replacing flip-flops with scan cells increases the total cell area. Furthermore, routing of scan chains and global signals such as *SE* consumes routing resources. Thus, the scan design methodology can increase the overall circuit area.

---

[4] We will discuss scan cell reordering in detail in Chapter 26 ("Placement").

2. **Performance degradation**: In a scan cell, there is an additional multiplexer in the data path, in both the functional mode and the test mode. As a result, the performance of a circuit can degrade in a scan design. Furthermore, the Q/SO-pin of the scan cells are connected to the SI-pin of the following scan cell in the scan chain. Therefore, there is an additional load on the Q/SO-pin of the scan cells. It can further increase the delay of the scan cells.
3. **I/O-pin cost**: We create additional ports in a scan design to support testing (*TM*, *SI*, *SO*, and *SE*), as described in the earlier sections. We can avoid the creation of *SI*, *SO*, and *SE* ports by reusing existing data ports. Nevertheless, we need to create at least one extra port to support the scan design test methodology.
4. **Design effort cost**: To implement scan design test methodology, we need to carry out several additional tasks in a design flow. These tasks consume significant design effort. Nevertheless, considering the benefits of testability and an overall reduction in test development time, we can justify the extra design effort.

21.6 RECENT TRENDS

The scan design methodology has been widely used and has matured over decades. Nevertheless, continuous advancements are required to tackle the increased complexity of designs and the increased number of flip-flops. With increasing design complexity, the scan chains tend to become longer, the power consumed in the test mode increases, and the cost of testing increases. Consequently, we need to adopt new scan architectures to make scan-based testing more efficient and reduce its overheads. For example, we can use scan compression architectures that can reduce the test data volume, modify the test vectors such that toggling reduces in the test mode, and utilize machine learning to arrive at a good scan configuration [12–15].

In recent times, the issue of *security* associated with the scan design has become critical [16]. The scan chains allow us to access internal signals easily. However, sensitive information such as secret keys can be stolen, or a chip can be controlled illegally using these scan chains. Therefore, the scan-based test methodology needs to satisfy both testability and security requirements. We can enhance security by modifying the original scan chains to obfuscate the data stored in them or by restricting access to the scan chain using some authentication method [16].

REVIEW QUESTIONS

21.1 What design changes do we need to make for implementing the scan-based testing methodology in an integrated circuit (IC)?

21.2 How does the timing of a design get affected by scan chain insertion?

21.3 Why can a circuit have a large dynamic power dissipation during scan-based testing?

21.4 How does scan-based testing simplify ATPG?

21.5 What can we do to reduce the length of a scan chain?

21.6 How do we test that a scan chain works fine in a scan-based design?

21.7 Why do we add the port *TM* (test mode) in a scan-based design?

21.8 How is combinational equivalence checking done for a scan-based design?

21.9 In a scan-based design, why is it necessary to ensure that clock pins of all the flip-flops in a design are controllable from input ports?

21.10 Why do we reorder scan cells in a scan chain after placement?

21.11 Why can scan cell insertion lead to numerous hold violations?

21.12 Identify the problem in the circuit shown in Figure 21.6 in implementing scan-based testing. Suggest changes in the circuit that can fix this problem.

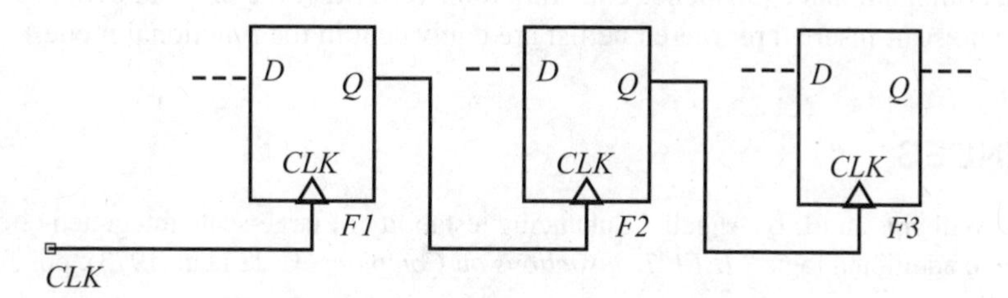

Figure 21.6 Circuit with a problem

21.13 What is the motivation for decreasing the toggle rate in the flip-flops during scan mode operation?

21.14 What are security issues associated with the scan-based testing?

TOOL-BASED ACTIVITY

In this activity, you can use any set of tools that support scan insertion and static timing analysis (STA). You can use any technology library that has scan cells. For example, you can use open-source tools such as OpenSTA [17], and you can also download freely available technology libraries for academic purposes [18].

1. Write a Verilog netlist for the circuit shown in Figure 21.3. Write an Synopsys Design Constraint (SDC) file with a clock period of 800 ps and clock uncertainty of 50 ps. Choose other SDC commands as appropriate. Run STA on the above netlist. Report the worst setup and hold slack for the path between *F1* and *F2*.
2. Perform scan insertion and compare the circuit obtained with Figure 21.4. If you do not have a scan insertion tool, manually write the Verilog netlist as shown in Figure 21.4. Write an SDC file for the functional mode for the scan-inserted netlist (you can add the command `set_case_analysis` to distinguish the modes). Run STA on the scan-inserted netlist using the above SDC file. Report the worst setup and hold slack for the path between *F1* and *F2*. Compare the timing report with the timing report obtained in the previous step. Explain why does the slack change. Specifically, explain the increased impact of load in the scan-inserted netlist.
3. Write an SDC file for the shift mode for the scan-inserted netlist (you can add the command `set_case_analysis` to distinguish the mode). Run STA on the scan-inserted netlist using the above SDC file. Report the worst setup and hold slack for the path between *F1* and *F2*. Compare the timing report with the timing report obtained for the function mode. Explain why does the slack change.

4. Manually reorder the scan chain from SI to SO as (*F3*, *F1*, *F2*) instead of (*F1*, *F2*, *F3*) [i.e., *F3* should be connected to *SI* port and *F2* should be connected to *SO* port]. Note that only scan chain order should change and no other changes in connection is allowed. Write the Verilog file for the scan-reordered netlist.
5. Using a combinational equivalence checking tool, verify that the original netlist, scan-inserted netlist, and scan-inserted reordered netlist are equivalent in the functional mode.

REFERENCES

[1] M. J. Y. Williams and J. B. Angell. "Enhancing testability of large-scale integrated circuits via test points and additional logic." *IEEE Transactions on Computers* C-22 (Jan. 1973), pp. 46–60.

[2] E. B. Eichelberger and T. W. Williams. "A logic design structure for LSI testability." *Papers on Twenty-five Years of Electronic Design Automation* (1988), pp. 358–364.

[3] M. Bushnell and V. Agrawal. *Essentials of Electronic Testing for Digital, Memory and Mixed-signal VLSI Circuits*, vol. 17. Springer Science & Business Media, 2004.

[4] L.-T. Wang, C.-W. Wu, and X. Wen. *VLSI Test Principles and Architectures: Design for Testability*. Elsevier, 2006.

[5] S. Ahlawat, J. Tudu, A. Matrosova, and V. Singh. "A high performance scan flip-flop design for serial and mixed mode scan test." *IEEE Transactions on Device and Materials Reliability* 18, no. 2 (2018), pp. 321–331.

[6] K.-J. Lee, J.-J. Chen, and C.-H. Huang. "Using a single input to support multiple scan chains." *1998 IEEE/ACM International Conference on Computer-aided Design. Digest of Technical Papers (IEEE Cat. No. 98CB36287)* (1998), pp. 74–78, IEEE.

[7] Y. Bonhomme, P. Girard, L. Guiller, C. Landrault, and S. Pravossoudovitch. "A gated clock scheme for low power scan testing of logic ICs or embedded cores." *Proceedings 10th Asian Test Symposium* (2001), pp. 253–258, IEEE.

[8] M. Keating and P. Bricaud. *Reuse Methodology Manual for System-on-a-Chip Designs*. Springer Science & Business Media, 2002.

[9] P. E. Bishop, G. Giles, S. Iyengar, C. T. Glover, and W.-o. Law. "Testability considerations in the design of the MC68340 Integrated Processor Unit." *Proceedings. International Test Conference 1990* (1990), pp. 337–346, IEEE.

[10] M. Bushnell and V. Agrawal. *VLSI Circuits*, vol. 17. Springer Science & Business Media, 2004.

[11] L.-T. Wang, Y.-W. Chang, and K.-T. T. Cheng. *Electronic Design Automation: Synthesis, Verification, and Test*. Morgan Kaufmann, 2009.

[12] P. Wohl, J. A. Waicukauski, J. E. Colburn, and M. Sonawane. "Achieving extreme scan compression for SoC Designs." *2014 International Test Conference* (2014), pp. 1–8, IEEE.

[13] K. Y. Cho, S. Mitra, and E. J. McCluskey. "California scan architecture for high quality and low power testing." *2007 IEEE International Test Conference* (2007), pp. 1–10, IEEE.

[14] B. Kumar, B. Nehru, B. Pandey, V. Singh, and J. Tudu. "A technique for low power, stuck-at fault diagnosable and reconfigurable scan architecture." *2016 IEEE East-West Design & Test Symposium (EWDTS)* (2016), pp. 1–4, IEEE.

[15] A. Zorian, B. Shanyour, and M. Vaseekar. "Machine learning-based DFT recommendation system for ATPG QOR." *2019 IEEE International Test Conference (ITC)* (2019), pp. 1–7, IEEE.

[16] X. Li, W. Li, J. Ye, H. Li, and Y. Hu. "Scan chain based attacks and countermeasures: A survey." *IEEE Access* 7 (2019), pp. 85055–85065.

[17] T. Ajayi, V. A. Chhabria, M. Fogaça, S. Hashemi, A. Hosny, A. B. Kahng, M. Kim, J. Lee, U. Mallappa, M. Neseem, et al. "Toward an open-source digital flow: First learnings from the OpenROAD project." *Proceedings of the 56th Annual Design Automation Conference 2019* (2019), pp. 1–4.

[18] M. Martins, J. M. Matos, R. P. Ribas, A. Reis, G. Schlinker, L. Rech, and J. Michelsen. "FreePDK technology." *Proceedings of the 2015 Symposium on International Symposium on Physical Design* (2015), pp. 171–178.

22 Automatic Test Pattern Generation

... they always purr. 'If they would only purr for "yes" and mew for "no," or any rule of that sort' she had said, 'so that one could keep up a conversation! But how can you talk with a person if they always say the same thing?'

On this occasion the kitten only purred: and it was impossible to guess whether it meant 'yes' or 'no'.

—Lewis Carroll, *Through the Looking-Glass*, Chapter 12, 1871

Test patterns enable testers to distinguish between a faulty circuit and a fault-free circuit. A test pattern is a bit sequence that we can apply at the input ports such that we are able to observe different responses for a faulty circuit and a fault-free circuit. In general, it is difficult to find a test pattern for a given fault manually. Therefore, given a fault model, we generate test patterns for a circuit using some electronic design automation (EDA) tools. The process of automatically finding a set of test patterns is called *automatic test pattern generation* (ATPG). In this chapter, we will discuss ATPG in detail.

22.1 REQUIREMENTS OF ATPG

In Chapter 21 ("Scan design"), we have seen that a given fault of a circuit can be detected using multiple test patterns. Moreover, a given test pattern can detect multiple faults in a circuit. Therefore, given a circuit and a set of faults, different ATPG tools can generate a different set of test patterns. We expect that a good ATPG tool will fulfill the following requirements:

1. **Achieves a high fault coverage:** As the fault coverage increases, the quality of testing improves, and the *defect level* decreases.[1] Therefore, the set of test patterns generated by an ATPG tool should ideally cover *all* (100%) detectable faults for a given fault model. In practice, 98%–100% fault coverage is typically acceptable.
2. **Generates a small test pattern set:** The set of test patterns generated by an ATPG tool is applied to the device under test (DUT) using an automatic test equipment (ATE). Typically, test patterns and their responses are stored in the memory of the ATE and applied to the DUT during testing through some tester channels. For scan-based design, the test application time is directly proportional to the test set size. Therefore, to reduce the recurring cost of testing,

[1] For details, see Section 6.4 ("Fault coverage and defect level").

the number of test patterns (test set size) generated by an ATPG tool for a given fault coverage should be as small as possible. It also helps in reducing the memory requirements of ATE in storing the test patterns and their responses.

3. **Consumes less computational resource:** The computational resource (runtime and memory) required by an ATPG tool to generate a set of test patterns for a given fault coverage should be less.
4. **Meets power constraint during testing:** During testing, the power dissipated in a chip can be significantly higher than the power dissipated during normal circuit operation due to the simultaneous switching of numerous signals. An ATPG tool can help reduce it by generating appropriate test patterns that minimize the switching activities and it should honor peak power constraints for testing [1, 2].

22.2 SEQUENTIAL VERSUS COMBINATIONAL ATPG

The generation of test patterns for a general sequential circuit is known as *sequential ATPG*. Typically, the internal signals of a sequential circuit are less controllable and observable than a combinational circuit. To generate test patterns for those signals we require complex *state exploration* or searching for the appropriate sequence of states. Therefore, sequential ATPG is a difficult problem. We transform the problem of *sequential ATPG* to the problem of *combinational ATPG* using *scan design test methodology*. In a scan design, we replace flip-flops with scan cells and connect them to form scan chains. This design transformation simplifies testing in the following ways:

1. **Improved controllability:** We can obtain any bit (0/1) on the Q-pin of a scan cell (or the replaced flip-flop) in maximum N clock cycles, where N is the number of scan cells in the scan chain. If there was no scan cell insertion, it could take 2^N clock cycles in the worst case (assuming that a new state is reached in each clock cycle). Thus, in a scan cell-inserted design, the Q-pins of scan cells (or the replaced flip-flops) are fully controllable. From the ATPG perspective, we can consider Q-pins of flip-flops as *pseudo-primary inputs* (PPIs).
2. **Improved observability:** We can read the value (0/1) obtained on the D-pin of a scan cell (or the replaced flip-flop) at an output port in maximum N clock cycles, where N is the number of scan cells in the scan chain. Thus, from the ATPG perspective, we can consider D-pins of scan cells (or the replaced flip-flops) as *pseudo-primary outputs* (PPOs). If there was no scan cell insertion, it could take an exponential number of clock cycles to observe the value at an internal pin in the worst case.

Thus, we can generate test patterns in a scan design flow by considering only the combinational circuit elements. We can assume that the Q/D-pins of the scan cells (or corresponding flip-flops) work as input/output ports (PPIs and PPOs). Hence, we can partition a given sequential circuit into multiple combinational circuits. Subsequently, we perform combinational ATPG for them, in which we do not need state-space exploration. Thus, the ATPG problem gets simplified if we implement the scan design test methodology.

Note that the combinational ATPG problem is still an NP-complete problem [3]. However, several efficient algorithms exist for solving the combinational ATPG problem [3]. Commercial ATPG tools employ these algorithms, and we widely use them in testing digital circuits. These tools generate a set of test patterns reasonably fast and often achieve close to 100% fault coverage. Moreover, the test development time (i.e., the time required to implement design for testability (DFT) features in a design flow) becomes predictable when scan design flow and combinational ATPG are employed together. Therefore, scan design flow along with combinational ATPG is the most popular DFT technique for industrial designs. In this chapter, we will look at the combinational ATPG problem in detail.

22.3 TERMINOLOGIES

In this section, we explain terms that are essential for understanding the ATPG algorithm.

1. **Paths in a circuit:** An ATPG tool typically works on a design represented as *netlist*, i.e., interconnections of logic gates. A netlist consisting of combinational circuit elements is shown in Figure 22.1. It has logic gate instances {*G1, G2, G3, ..., G7*}, primary input ports (also called PIs) {*A, B, C, D, E*} and primary output port (also called PO) {*Z*}.

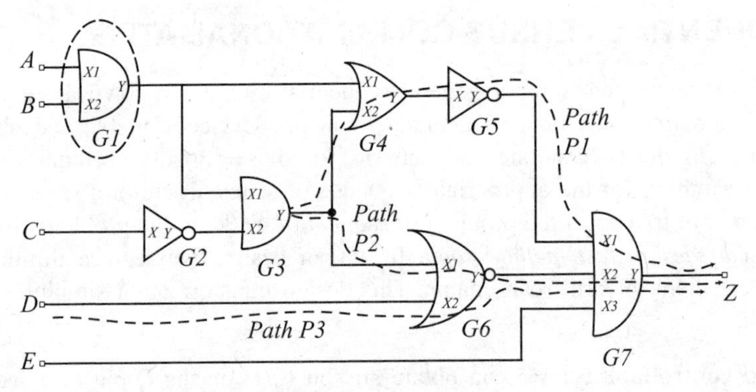

Figure 22.1 Circuit for illustrating various terminologies

During ATPG, we often need to consider different *paths* in a circuit. We define a path as a sequence of pins in topological order. In addition to pins, PIs can be a path start point, and POs can be a path end point. We can denote the paths *P1*, *P2*, and *P3* shown in Figure 22.1 as follows:

(a) *P1*: {*G3/Y, G4/X2, G4/Y, G5/X, G5/Y, G7/X1, G7/Y, Z*}

(b) *P2*: {*G3/Y, G6/X1, G6/Y, G7/X2, G7/Y, Z*}

(c) *P3*: {*D, G6/X2, G6/Y, G7/X2, G7/Y, Z*}

For simplicity, we ignore pins and ports that are bidirectional or inout in our discussions. Furthermore, we assume that there are no loops of combinational circuit elements in the circuit.

2. **Reconvergent fanout:** When a pin or a PI drives more than one fanout, and the paths from these fanouts converge back at some other pin or a PO, then the structure is known as *reconvergent fanout*. In the circuit shown in Figure 22.1, the pin *G3/Y* drives two pins in its fanout, namely *G4/X2* and *G6/X1*. The paths through these pins (*P1* and *P2*) converge at the output port *Z*. Therefore, this is an example of a reconvergent fanout from the pin *G3/Y*.

3. **On-inputs and side-inputs of a path:** For a given path, the input pins lying on that path are known as *on-path inputs* or *on-inputs* of that path. The input pins other than the on-inputs of the instances lying on that path are known as *side path inputs* or *side-inputs* of that path. For example, for the path *P1* shown in Figure 22.1, the on-inputs are {*G4/X2, G5/X, G7/X1*} and side-inputs are {*G4/X1, G7/X2, G7/X3*}. For the path *P2*, the on-inputs are {*G6/X1, G7/X2*} and side-inputs are {*G6/X2, G7/X1, G7/X3*}. For the path *P3*, the on-inputs are {*G6/X2, G7/X2*} and side-inputs are {*G6/X1, G7/X1, G7/X3*}.

4. **Controlling and noncontrolling values:** A *controlling value* of a multi-input combinational gate is the value (0/1) that we can assign to an input pin such that the output is known irrespective of other input pin values. For example, consider an AND gate. If we apply 0 to an input pin, its output becomes 0 regardless of other input values. Therefore, 0 is the controlling value for an AND gate. A *noncontrolling value* of a multi-input combinational gate is the value (0/1) that we can assign to an input pin such that the other input pin values determine its output value. For example, consider an AND gate. If we apply 1 to an input pin, other input pin values decide its output value. Therefore, 1 is the noncontrolling value for an AND gate.

   Table 22.1 shows the controlling and the noncontrolling values of some of the commonly used logic gates. For an XOR/XNOR gate, if we set one of the input pins to 0 or 1, the output can be either 0 or 1, depending on the other input pin value. Therefore, there is no controlling value for an XOR/XNOR gate, and both 0 and 1 are noncontrolling values.

Table 22.1 Controlling and noncontrolling values for different gates

Gate Type	Controlling Value	Noncontrolling Value
AND	0	1
OR	1	0
NAND	0	1
NOR	1	0
XOR	Not defined	Any value
XNOR	Not defined	Any value

22.4 PATH SENSITIZATION METHOD

In this section, we describe a systematic method to generate test patterns in a combinational circuit. Note that an ATPG algorithm must know beforehand the set of faults for test pattern generation. Therefore, we should decide the fault model before invoking the ATPG algorithm. Earlier, we have seen that the *single stuck-at fault model* is simple and the most popular. Therefore, we use the single stuck-at fault model for illustrating the ATPG method. However, note that, in general, for different fault models, different ATPG methods are required.

Though test patterns can be generated in several ways, *path sensitization method* and its variations are widely used [4–7]. For a given fault, the path sensitization method is carried out in three steps, namely, *fault activation*, *fault propagation*, and *line justification*. These steps are described below:

1. **Fault activation:** For a given stuck-at-fault, when we obtain a value *opposite* to the corresponding stuck-at value at the fault site, then the fault is said to be *activated*. For example, consider a stuck-at-0 (SA0) fault at some pin. To activate SA0, we need to obtain 1 at that pin. Similarly, for a stuck-at-1 fault (SA1), we need 0 at that pin. Moreover, during fault activation, we determine values at the PIs such that the fault gets activated. The values at the PIs thus obtained form a part of the generated test pattern.[2] A fault activation ensures that we observe

[2] In the discussion of combinational ATPG algorithms, the role of PIs and PPIs is equivalent. Similarly, the role of POs and PPOs is equivalent.

different behaviors for a faulty circuit and a fault-free circuit at the fault site using the test patterns.

2. **Fault propagation:** Fault propagation allows observing the difference between a faulty circuit and a fault-free circuit at one of the POs. It considers different *paths* from the fault site to the POs. The goal is to find at least one path through which we can observe the faulty behavior. To propagate fault through a path, it is required that we maintain *noncontrolling values* at all the *side-inputs* of that path. Fulfilling this requirement ensures that the on-inputs, rather than the side-inputs determine the output value of each logic gate in that path. Therefore, if we have a distinguishing value at the on-input pin for a faulty circuit, we can observe its effect at its output pin. By maintaining this property for all logic gates in a path, we can propagate the effect of the activated fault up to at least one PO.
3. **Line justification:** The previous step of fault propagation determines the required value at the side-inputs of a path. However, the side-inputs are internal pins, and the PIs determine their values. In the line justification step, we find the values at the PIs that will result in the internal pin values as required by the fault propagation step. The values thus obtained at the PIs constitute another part of the generated test pattern.

We illustrate the path sensitization method of ATPG in the following examples.

Example 22.1 Consider the circuit shown in Figure 22.2. Let us determine a test pattern for the SA1 fault at the pin *G4/X1*.

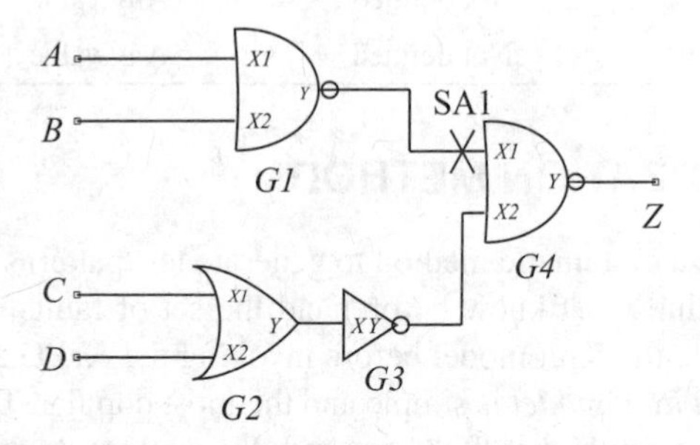

Figure 22.2 Given circuit for test pattern generation

The steps are as follows:

1. **Fault activation:** To activate a SA1 fault, a value of 0 (opposite of the fault value) must be obtained at *G4/X1*. This can be done if *A*=1 and *B*=1.
2. **Fault propagation:** The fault needs to be propagated to the PO *Z* through the path {*G4/X1*, *G4/Y*, *Z*}. The side-input for the path is *G4/X2*, and it must be assigned a noncontrolling value. Since, *G4* is a NAND gate, *G4/X2*=1 is required for fault propagation (see Table 22.1).
3. **Line justification:** Line justification needs to be done such that *G4/X2*=1. To obtain *G4/X2*=1, *G3/X* must be 0, and, therefore, *C* and *D* must be 0. Therefore, the test pattern is: {*A*=1, *B*=1, *C*=0, *D*=0}.

Remark: It is easy to check whether {*A*=1, *B*=1, *C*=0, *D*=0} is a correct test pattern. For the fault-free circuit, this test pattern will produce *Z*=1, while it will produce *Z*=0 for an SA1 at *G4/X1*. Thus, the computed test pattern is correct.

Example 22.2 Consider the circuit shown in Figure 22.3. Let us determine a test pattern for the SA0 at the pin *G4/X1*.

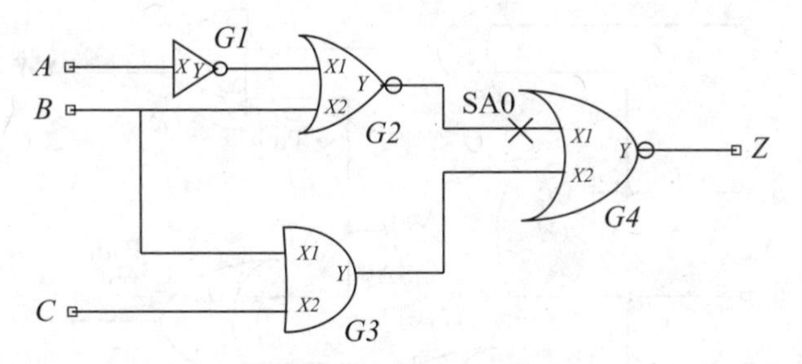

Figure 22.3 Given circuit for test pattern generation

The steps are as follows:

1. **Fault activation:** To activate the fault, a value of 1 must be obtained at *G4/X1*. This can be done if *G2/X1*=0 and *G2/X2*=0. Therefore, *A*=1 and *B*=0 is required.
2. **Fault propagation:** The fault needs to be propagated to the PO Z through the path {*G4/X1*, *G4/Y*, *Z*}. The side-input is *G4/X2*, and it must be assigned a noncontrolling value of 0.
3. **Line justification:** To ensure *G4/X2*=0, at least one of *G3/X1* or *G3/X2* must be 0. Since in fault activation step, we obtained *B*=0, *G3/X1*=0 is already ensured. Therefore, *G3/X2* can be either 0 or 1. Thus, the test patterns are {*A*=1, *B*=0, *C*=0} and {*A*=1, *B*=0, *C*=1}.

Remark: It is easy to check that both the computed test patterns produce *Z*=0 in a fault-free circuit and *Z*=1 in the faulty circuit. Thus, the computed test patterns are valid. Note that the values assigned to the PI (i.e., *B*) during the fault activation and the line justification steps are consistent in the above case. It eases test pattern generation. However, sometimes these values can conflict. In that case, we need to backtrack the decision made during fault activation, or line justification, as explained in the following section.

22.5 BACKTRACKING

In ATPG algorithms, we need to decide the value of signals at different stages. However, the circuit topology and logic functionality define relationships between signals. As a result, fixing the value of a signal implies forcing values on some others. For example, suppose we assign the output of an AND gate to 1. This decision forces all the inputs of that AND gate to take value 1. Moreover, setting a controlling value to a gate's input implies a known value at its output. Thus, deciding the value of a signal triggers many forced assignments. However, some of these assignments can be conflicting, and we need to *backtrack* them.[3] We illustrate this concept in the following example.

[3] Backtracking refers to discarding a part of the current solution if it will not lead to the final solution. It is used in many algorithms. For example, we have discussed backtracking in solving SAT problems (see Section 11.5.2 ("Solving SAT problem")).

Example 22.3 Consider the circuit shown in Figure 22.4. Let us determine a test pattern for detecting the SA0 fault at the pin *G3/Y*.

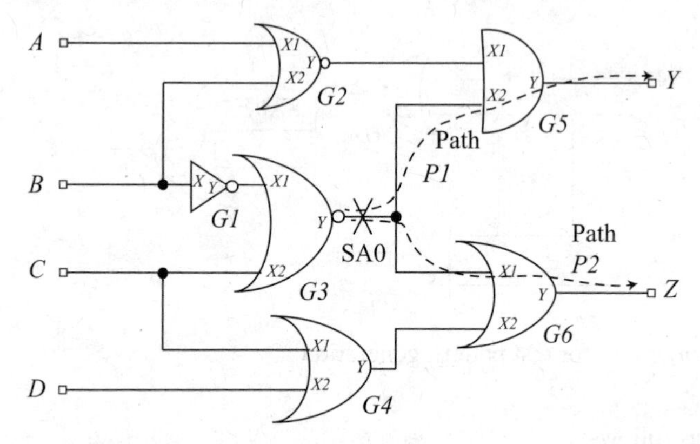

Figure 22.4 Circuit to illustrate backtracking in ATPG

1. **Fault activation:** To activate the fault, *G3/Y* =1. It implies *B*=1 and *C*=0.
2. **Fault propagation and line justification:** There are two paths *P1* and *P2* from the fault site to the output ports *Y* and *Z*, respectively. We can try out both these paths for fault propagation.
   (a) **Through *P1*:** The side-input for this path is *G5/X1*, and it must be assigned a noncontrolling value of 1. However, this is not possible because *B*=1 (obtained in fault activation) forces *G2/Y*=*G5/X1*=0. Therefore, we cannot propagate the fault to the output port *Y*. Hence, we need to *backtrack* the decision to propagate fault to *Y* and try out another path.
   (b) **Through *P2*:** The side-input for this path is *G6/X2*, and it must be assigned a noncontrolling value of 0. Hence, we make *C*=0 and *D*=0. Note that the assignment *C*=0 agrees with the value assigned during fault activation. Therefore, the test pattern is {*A*=X, *B*=1, *C*=0, *D*=0}. The value X at *A* denotes that *A* can be assigned any value (0 or 1). The freedom to choose any value at the port *A* can be exploited in reducing the size of the test pattern set, as we will explain in the following sections.

Remark: In this example, there were only two possible paths for fault propagation. In a typical circuit, there can be many possible paths for fault propagation, and there can be several decisions that require backtracking.

Backtracking is required during ATPG in the following cases:

1. When we need to assign conflicting values (both 0 and 1) to a PI or internal pins.
2. While doing fault propagation, we can find a path blocked (by a controlling value at the side-input) due to previous assignments.

The possibility of backtracking is greater in circuits with reconvergent fanouts. The decision at the fanout point can lead to many implications. Further, the implications can conflict at the reconvergent point. In the circuit shown in Figure 22.4, the signal from *B* drives *G2/X2* and *G1/X*. The paths through these pins reconverge at *Y*. Therefore, the circuit has a reconvergent fanout, and it explains why conflict was observed and backtracking became necessary. In the extreme case, we may need to backtrack all possible decisions, and still, we find no test pattern. It happens for a redundant fault, as described in the following section.

Since backtracking involves discarding some computations that we have already done, it incurs runtime penalties, and we should avoid them, if possible. Most ATPG tools provide a user-defined *backtrack limit*. A *backtrack limit* defines the maximum number of backtracks allowed for a given fault. If the number of backtracks reaches *backtrack limit*, the tool aborts finding a test pattern for that fault. We should define a prudent *backtrack limit* for an ATPG tool. A very high *backtrack limit* can result in too many backtrackings, especially for redundant faults, which can severely increase the runtime. A too low *backtrack limit* can result in many test pattern computations being aborted, and we would obtain a low *fault coverage*.

The preceding description highlights the difficulty of the combinational ATPG problem. The algorithms for combinational ATPG have been studied since the 1960s, and various improvements have been made. The D-algorithm by Roth, the path-oriented decision-making (PODEM) algorithm by Goel, and the fanout-oriented (FAN) algorithm by Fujiwara and Shimono are some of the notable contributions [4, 8, 9]. The key to improving ATPG algorithms is limiting the search space for fault propagation and avoiding backtracking whenever possible.

22.6 REDUNDANT FAULTS

If no test pattern can detect a fault in a combinational circuit, the fault is known as a *redundant fault*. In this case, we cannot distinguish the behavior of a fault-free circuit and a faulty circuit using any combination of bits at the PIs. Therefore, we can use a redundant stuck-at fault to optimize a circuit. We explain this concept in the following example.

Example 22.4 Consider the circuit shown in Figure 22.5. Let us try to determine a test pattern for SA1 at the pin *G3/X1*. We use the path sensitization method for this purpose.

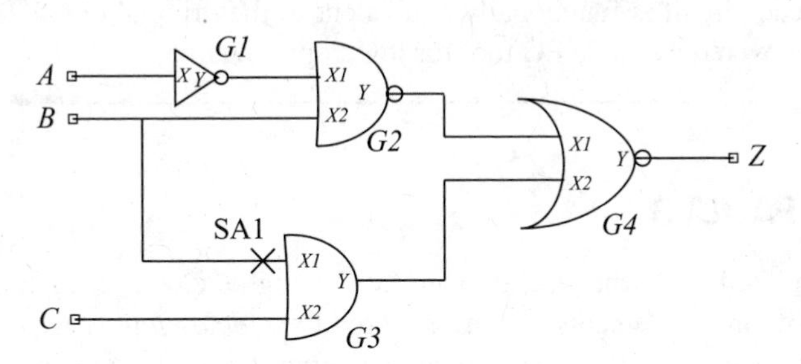

Figure 22.5 Illustration of a redundant fault

1. **Fault activation:** To activate the fault, we require *G3/X1*=0. Hence, *B*=0 is required.
2. **Fault propagation:** The fault needs to be propagated to the PO *Z* through the path {*G3/X1*, *G3/Y*, *G4/X2*, *G4/Y*, *Z*}. The side-inputs for the path are {*G3/X2*, *G4/X1*}, and they must be assigned noncontrolling values. Hence, *G3/X2*=1 and *G4/X1*=0.

3. **Line justification:** To ensure *G3/X2*=1, *C*=1 should be assigned. To ensure *G4/X1*=0, *A*=0, and *B*=1 are required. However, *B*=1 conflicts with the value *B*=0 required for the fault activation. Therefore, we need to backtrack the fault propagation through this path. Since no other path in the circuit exists through which the given fault can propagate, we cannot generate a test pattern for the given fault.

Thus, SA1 at *G3/X1* is a redundant fault. Therefore, the behavior of the fault-free circuit and the faulty circuit is the same. We can utilize this information in optimizing the circuit by assuming that SA1 at *G3/X1* is always present. Hence, *G3/X1*=1. It implies that *G3/Y* =*C*. Therefore, we can remove the gate *G3* and directly connect *C* to *G4/X2*, as shown in Figure 22.6. Thus, we are able to reduce the area of the circuit.

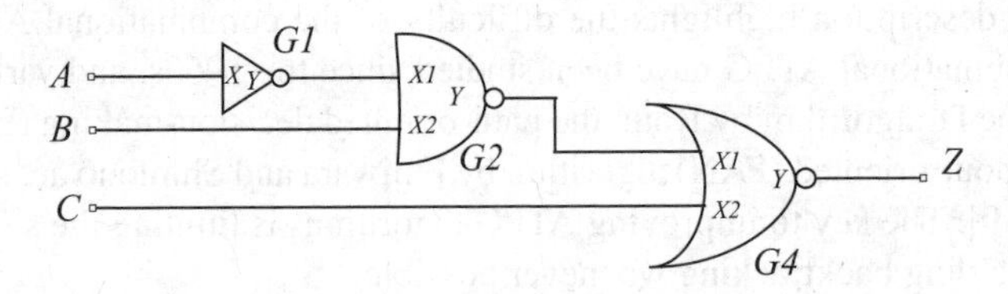

Figure 22.6 Optimized circuit

We can easily check whether the reduced circuit is equivalent to the original circuit by Boolean algebraic manipulations. The functionality of the original circuit is:

$$Z = ((A'.B)' + (B.C))' = (A'B).(BC)' = A'B.(B' + C') = A'BC' \quad (22.1)$$

The functionality of the reduced circuit is:

$$Z = ((A'.B)' + C)' = (A'.B).C' = A'BC' \quad (22.2)$$

Thus, the reduced circuit is functionally equivalent to the original circuit, and we can use the redundant faults discovered by an ATPG tool for logic optimization.

22.7 DELAY FAULTS

Some faults are exhibited only when a chip operates at a higher clock speed. Such faults occur due to increased delay of circuit elements and are referred to as *delay faults*. A popular model for the delay faults is the *transition fault model*. We can have two types of transitions for a signal: rising and falling. Hence, we can have two types of *transition faults* [10]:

1. **Slow-to-rise:** In this case, the transition to 1 is slower than expected. We can consider this fault as a temporary SA0 fault.
2. **Slow-to-fall:** In this case, the transition to 0 is slower than expected. We can consider this fault as a temporary SA1 fault.

For testing a transition fault, we need to create an appropriate transition at the fault site and allow the effect of the transition to propagate to a PO [10]. In general, we need two distinct test patterns *TP1* and *TP2* to test a transition fault. The first pattern *TP1* applies the initial value at the fault site.

The second pattern *TP2* applies the final value at the fault site and propagates the effect of the transition to a PO. We can find the test patterns *TP1* and *TP2* using the ATPG methods described in the previous sections, and use them to detect transition faults.

22.8 TEST SET COMPACTION

For a given fault coverage, we should generate fewer test patterns. It reduces the test application time and testing cost. However, determining the *minimum size* of the test pattern set for detecting all single stuck-at faults in a circuit is an NP-hard problem [11]. Therefore, heuristics have been proposed to obtain a compact test pattern set [9, 12].

A technique to obtain a compact test pattern set is to merge two or more test patterns that are *compatible*. We say that the two test patterns are compatible if they do not require conflicting bits at any PI. Note that a test pattern assigns all PIs one of the three values: 0, 1, or X (X means that we can use either 0 or 1). Two test patterns T_1 and T_2 are compatible if one of the following conditions hold for all PIs:

1. The value is 0 in both T_1 and T_2, or the value is 1 in both T_1 and T_2.
2. The value is X in either T_1 or T_2.

If two test patterns T_1 and T_2 are compatible then they can be merged to a single test pattern T_{1-2} by deriving the values at each PI as follows:

1. If the value is 0 in either T_1 or T_2, then 0 is used
2. If the value is 1 in either T_1 or T_2, then 1 is used
3. If the value is X in both T_1 and T_2, then X is used

Let us illustrate this technique using the following example.

Example 22.5 Consider the circuit shown in Figure 22.7. Let us merge the test pattern obtained for SA1 at *G1/Y* and *G2/Y*. We can derive the test pattern for SA1 at *G1/Y* using the path sensitization method as {*A*=1, *B*=1, *C*=0, *D*=X, *E*=X}. Similarly, the test pattern for SA1 at *G2/Y* can be derived as {*A*=X, *B*=X, *C*=0, *D*=0, *E*=1}. These two test patterns are compatible and can be merged to a single test pattern {*A*=1, *B*=1, *C*=0, *D*=0, *E*=1}. We can check that the merged test pattern can detect both the SA1 faults at *G1/Y* and *G2/Y*.

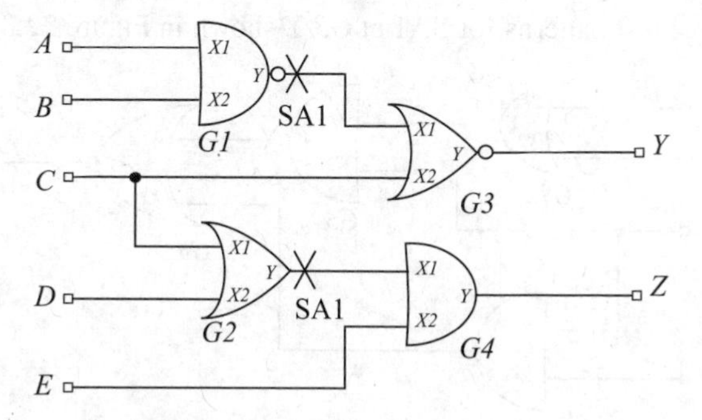

Figure 22.7 Illustration of merging of two test patterns

In the above example, we have carried out merging after test patterns are already generated. This method is known as *static merging*, and it does not impact the fault coverage or the testing capabilities of a given test pattern set. We can carry out merging operations repeatedly over the merged test pattern. Consequently, the size of the given test pattern set can gradually decrease.

Another approach is to consider compaction during the ATPG process. This is known as *dynamic compaction* method. In this method, just after generating a test pattern for a given fault, we intelligently assign the PIs having X to 0 or 1, such that the same pattern can detect stuck-at faults that are not yet covered [13–16].

22.9 RECENT TRENDS

The ATPG techniques have advanced and become more efficient over decades. Besides building novel heuristics over the traditional search-based algorithm, SAT-based approaches are reported to produce test patterns efficiently [17, 18]. Machine learning can also help combine multiple heuristics intelligently, avoid backtracking, and reduce overall runtime [19]. Additionally, with the advancement in technology, tackling other types of faults, such as uncovered multiple stuck-at faults, open net faults, and design-for-manufacturability (DFM) hotspot aware faults, becomes necessary [20–23].

REVIEW QUESTIONS

22.1 What are the challenges of sequential ATPG?

22.2 What is a controlling and a noncontrolling value for a gate?

22.3 Why does an XOR gate have no controlling value?

22.4 Why is backtracking required in the ATPG algorithm?

22.5 Why is it desirable to put a limit on the number of backtracking in an ATPG algorithm?

22.6 What is reconvergent fanout in a circuit? How do reconvergent fanouts make the ATPG difficult?

22.7 In fault propagation, why do we set the side-inputs to noncontrolling values?

22.8 What is a redundant fault?

22.9 How can we detect redundant faults by an ATPG algorithm?

22.10 How can redundant faults be employed to optimize a combinational circuit?

22.11 Determine three test patterns for SA1 at *G3/Y* shown in Figure 22.8.

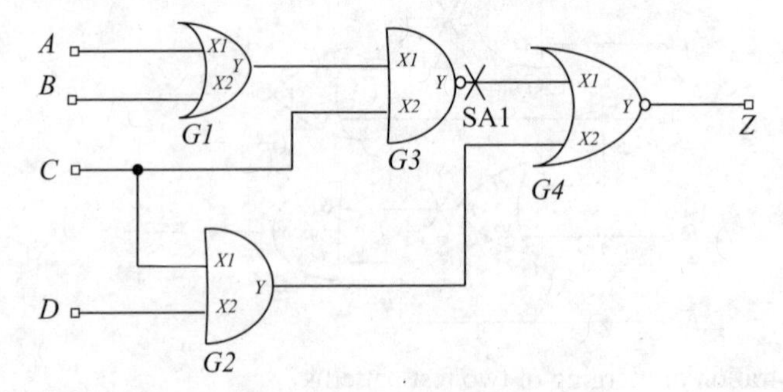

Figure 22.8 Given circuit for ATPG

22.12 Why do we carry out test set compaction?

22.13 What is a compatible pair of test patterns? How can we employ them for test set compaction?

22.14 What is the difference between static test set compaction and dynamic test set compaction?

TOOL-BASED ACTIVITY

Use any ATPG tool, such as Atalanta, for this activity [24].

1. Run the tool and find the test patterns for all possible single stuck-at faults in the circuit shown in Figure 22.8. Make the table as shown in Table 22.2.

Table 22.2 Test pattern for SSA faults

		Test Patterns	
Fault Site	**SA0/SA1**	**Reported by Tool**	**Computed Manually**
..	..	..	..

2. Modify the circuit to remove redundant fault(s) reported by the ATPG tool preserving the Boolean function implemented by the circuit. Report the schematic of the modified circuit. Rerun the ATPG tool. Make the above table again for the modified circuit.

REFERENCES

[1] S. Wang and S. K. Gupta. "ATPG for heat dissipation minimization during scan testing." *Proceedings of the 34th Annual Design Automation Conference* (1997), pp. 614–619, ACM.

[2] S. Ravi. "Power-aware test: Challenges and solutions." *2007 IEEE International Test Conference* (2007), pp. 1–10, IEEE.

[3] M. Prasad, P. Chong, and K. Keutzer. "Why is ATPG easy?" *Proceedings 1999 Design Automation Conference (Cat. No. 99CH36361)* (1999), pp. 22–28, IEEE.

[4] J. P. Roth. "Diagnosis of automata failures: A calculus and a method." *IBM Journal of Research and Development* 10, no. 4 (1966), pp. 278–291.

[5] T. Kirkland and M. R. Mercer. "Algorithms for automatic test-pattern generation." *IEEE Design & Test of Computers* 5, no. 3 (1988), pp. 43–55.

[6] M. Bushnell and V. Agrawal. *Essentials of Electronic Testing for Digital, Memory and Mixed-signal VLSI Circuits*, vol. 17. Springer Science & Business Media, 2004.

[7] L. Scheffer, L. Lavagno, and G. Martin. *EDA for IC System Design, Verification, and Testing*. CRC Press, 2016.

[8] P. Goel. "An implicit enumeration algorithm to generate tests for combinational logic circuits." *IEEE Transactions on Computers,* vol. C-30. no. 3 (1981), pp. 215–222.

[9] H. Fujiwara and T. Shimono. "On the acceleration of test generation algorithms." *IEEE Transactions on Computers* no. 12 (1983), pp. 1137–1144.

[10] J. A. Waicukauski, E. Lindbloom, B. K. Rosen, and V. S. Iyengar. "Transition fault simulation." *IEEE Design & Test of Computers* 4, no. 2 (1987), pp. 32–38.

[11] B. Krishnamurthy and S. B. Akers. "On the complexity of estimating the size of a test set." *IEEE Transactions on Computers* 100, no. 8 (1984), pp. 750–753.

[12] I. Hamzaoglu and J. H. Patel. "Test set compaction algorithms for combinational circuits." *IEEE Transactions on Computer-aided Design of Integrated Circuits and Systems* 19, no. 8 (2000), pp. 957–963.

[13] L.-T. Wang, C.-W. Wu, and X. Wen. *VLSI Test Principles and Architectures: Design for Testability*. Elsevier, 2006.

[14] M. Abramovici, J. Kulikowski, P. Menon, and D. Miller. "SMART and FAST: Test generation for VLSI scan-design circuits." *IEEE Design & Test of Computers* 3, no. 4 (1986), pp. 43–54.

[15] I. Pomeranz, L. N. Reddy, and S. M. Reddy. "COMPACTEST: A method to generate compact test sets for combinational circuits." *IEEE Transactions on Computer-aided Design of Integrated Circuits and Systems* 12, no. 7 (1993), pp. 1040–1049.

[16] B. Ayari and B. Kaminska. "A new dynamic test vector compaction for automatic test pattern generation." *IEEE Transactions on Computer-aided Design of Integrated Circuits and Systems* 13, no. 3 (1994), pp. 353–358.

[17] J. Shi, G. Fey, R. Drechsler, A. Glowatz, F. Hapke, and J. Schloffel. "PASSAT: Efficient SAT-based test pattern generation for industrial circuits." *IEEE Computer Society Annual Symposium on VLSI: New Frontiers in VLSI Design (ISVLSI'05)* (2005), pp. 212–217, IEEE.

[18] S. Eggersglüß, R. Wille, and R. Drechsler. "Improved SAT-based ATPG: More constraints, better compaction." *2013 IEEE/ACM International Conference on Computer-aided Design (ICCAD)* (2013), pp. 85–90, IEEE.

[19] S. Roy, S. K. Millican, and V. D. Agrawal. "Training neural network for machine intelligence in automatic test pattern generator." *2021 34th International Conference on VLSI Design and 2021 20th International Conference on Embedded Systems (VLSID)* (2021), pp. 316–321, IEEE.

[20] P. Wang, C. J. Moore, A. M. Gharehbaghi, and M. Fujita. "An ATPG method for double stuck-at faults by analyzing propagation paths of single faults." *IEEE Transactions on Circuits and Systems I: Regular Papers* 65, no. 3 (2017), pp. 1063–1074.

[21] D. Erb, K. Scheibler, M. Sauer, S. M. Reddy, and B. Becker. "Multi-cycle circuit parameter independent ATPG for interconnect open defects." *2015 IEEE 33rd VLSI Test Symposium (VTS)* (2015), pp. 1–6, IEEE.

[22] A. Sinha, S. Pandey, A. Singhal, A. Sanyal, and A. Schmaltz. "DFM-aware fault model and ATPG for intra-cell and inter-cell defects." *2017 IEEE International Test Conference (ITC)* (2017), pp. 1–10, IEEE.

[23] M. Sauer, A. Czutro, I. Polian, and B. Becker. "Small-delay-fault ATPG with waveform accuracy." *2012 IEEE/ACM International Conference on Computer-aided Design (ICCAD)* (2012), pp. 30–36, IEEE.

[24] H. Lee and D. Ha. "Atalanta: An efficient ATPG for combinational circuits." 1993. https://github.com/hsluoyz/Atalanta. Last accessed on January 8, 2023.

23 Built-in Self-test

I went in search of evil outside. I could not find any. Once I started searching my own heart, I found that none is as big evil as me.

—Kabir Das, 15th century (translated from vernacular Hindi)

In the previous chapters, we have discussed wafer-level testing that uses *automatic test equipment* (ATE). It allows a high degree of automation, achieves good fault coverage, and is cost-effective. Therefore, ATE-based testing is popular and widely adopted for industrial designs. However, ATE-based testing has the following drawbacks or limitations:

1. It requires expensive ATE. Moreover, it requires sophisticated facilities equipped with an ATE. We cannot perform it outside the production testing environment. However, sometimes testing, diagnosis, and repair become necessary for a chip that is already integrated into a system.
2. It employs voluminous test patterns that increase the test time and the cost of testing. Moreover, the number of test patterns increases with the advancement in technology due to increased circuit complexity. Consequently, the cost of testing increases with the advancement in technology.
3. We cannot carry out at-speed testing for high-performance integrated circuits (ICs) due to the impedances associated with the probes of an ATE [1]. However, some types of faults, such as delay faults, can only be detected by at-speed tests [2]. It makes ATE-based techniques inadequate in some situations.

Built-in self-test (BIST) is another testing methodology that addresses the above drawbacks and has become quite popular [3–6]. In this chapter, we will discuss BIST in detail.

We can implement BIST in a design in many ways, depending on the design complexity, target test metrics, and allowed overheads. In this chapter, we describe a typical BIST system to elucidate its basic principles.

23.1 BASICS

BIST is a testing technique in which we incorporate additional hardware and software in a chip that enables it to self-test. During self-testing, an IC can test its *own operation*, both functionally and parametrically. The self-testing does not require an external test pattern. Thus, we eliminate the dependence on an ATE. It allows us to carry out BIST-based testing in the field and even after system integration. Additionally, we can perform at-speed testing because no external signal interfacing is required.

A typical BIST system must include some features for *generating* test patterns internally and *analyzing* the resultant circuit response within the system [3]. However, we should implement these features with minimal hardware overhead. This is achievable in practice due to the following reasons:

1. We can perform adequate testing by just using *random* test patterns.
2. We can detect faults reasonably well using *signature analysis* instead of exact response comparison.

We describe these aspects of testing in the following paragraphs.

23.1.1 Random Test Patterns

BIST relies on the following observation:
It is relatively easy to attain a stuck-at fault coverage of 65%–85% for most combinational circuits just by random test patterns irrespective of the gate count [7].
As an illustration, we show how the stuck-at fault coverage increases with the increasing number of random test patterns in Figure 23.1(a) [6, 8, 9].

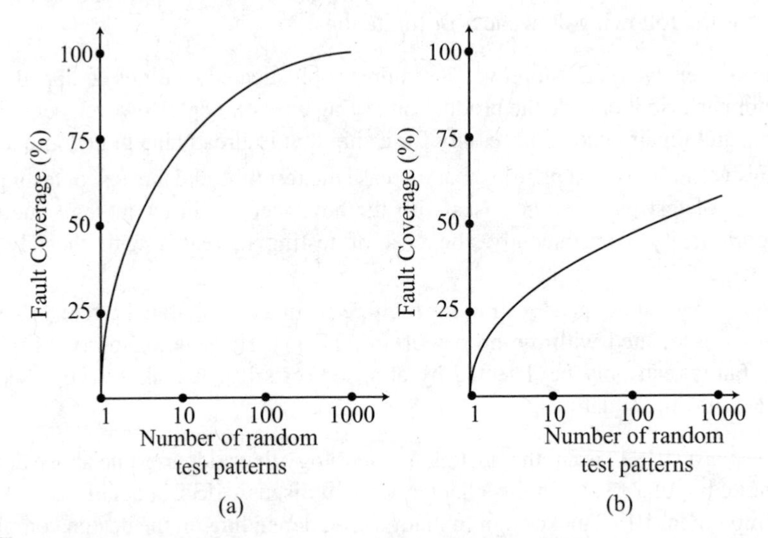

Figure 23.1 Illustration of growth of stuck-at fault coverage with increase in random test pattern (a) a typical combinational circuit and (b) a combinational circuit that is random pattern resistant

We observe that the fault coverage rises rapidly to around 75%, and then at a slower rate [10]. For these types of circuits, we can achieve a reasonable fault coverage just by using *pseudorandom test patterns*. Moreover, we can generate these test patterns using very small circuits such as *linear feedback shift registers* (LFSRs). Generating test patterns using an LFSR eliminates the need to *store* voluminous test patterns inside a chip. We will discuss LFSR in detail, later in this chapter.

For some circuits, we cannot achieve a high fault coverage using random test patterns. These types of circuits are known as *random pattern resistant circuits*. As an illustration, we show the fault coverage for a random pattern resistant circuit in Figure 23.1(b). Even when we use a high number of random test patterns for these kinds of circuits, we do not achieve a good fault coverage. A circuit can become random test pattern resistant in many situations. For example, when the probability of a signal taking a value 0 or 1 is low, it becomes difficult to test such kind of signals using random test pattern. A circuit can also be random pattern resistant when the observability of its signals is low.

Example 23.1 Consider a six-input OR gate. Let us determine the probability of detecting SA1 at the output using random test patterns.

To detect SA1 at the output, we need to apply zero at all the inputs. The probability of obtaining all zeroes in a random test pattern of 6 bits is $= 1/2^6 = 1/64 \approx 1.6\%$. Hence, it is less probable to cover SA1 at the output of a six-input OR gate using random test patterns.

The testability of a combinational circuit using random test patterns primarily depends on the number of primary inputs, the number of gate levels in the longest path, and the average fanin count of gates [11]. To enhance the fault coverage and make a circuit testable by random test patterns, we can insert *control points* and *observation points* in a circuit [12–14]. Moreover, we can augment random test patterns with some deterministic test patterns obtained using automatic test pattern generation (ATPG).

23.1.2 Signature Analysis

We need to apply a large number of test patterns during a BIST operation. Let us assume that we need to apply 1 million test patterns during BIST. Further, assume that a circuit has 100 output ports. Then, we need to store 1 million × 100 = 100 million golden responses to detect failures. Assume that we store each golden response using 1 bit. Thus, we will require a total of $1 \times 10^6 \times 100/8 \approx 12.5$ MB of memory. However, it is not practical to allocate 12.5 MB on a chip to store golden responses. Therefore, we need to devise a method to store these responses compactly.

We can reduce the response size in a BIST system by *compacting* them into *signatures* [12, 15, 16]. We derive a *signature* using some statistical properties of the responses. A signature is typically a number that can be represented using a few bytes. We compute the signature for the obtained responses and compare it with the expected *golden signature*. If they match, we say that the circuit has "passed" the test, else we say it has "failed" the test. Note that we do not compare the actual *responses* with the golden responses. Rather, we compare only the actual *signature* with the golden signature. Hence, we need to internally store only the *golden signature* for BIST which is extremely small. Thus, we eliminate the need to store voluminous golden responses inside a chip.

The above benefit of signature analysis has some downside. It delivers high compression by incurring substantial information loss. It is possible that a *faulty circuit* also generates a signature that matches the golden signature. In such cases, it becomes impossible to distinguish a faulty circuit from a good circuit due to the information loss. Therefore, by comparing signatures instead of exact responses, a faulty circuit can sometimes escape testing. This problem of signature analysis is known as *aliasing* or *masking*. Fortunately, we can keep the probability of aliasing small using techniques described later in this chapter. In practice, we can detect a faulty circuit reasonably well based on signatures and with a modest area overhead.

23.2 BIST ARCHITECTURE

A typical BIST architecture is shown in Figure 23.2. The portion of IC that fulfills its main function is shown as the *circuit under test* (CUT). The primary objective of BIST is to test the CUT. The overall BIST operations are controlled and coordinated by a *BIST controller*.

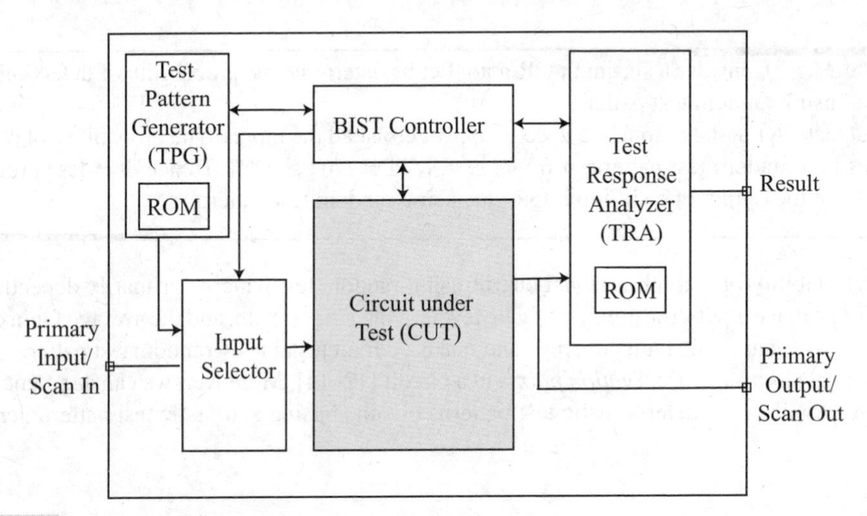

Figure 23.2 A typical BIST architecture

The self-testing can be initiated by a signal given to the BIST controller from the CUT or some primary input port (not shown in Figure 23.2). On receiving such a signal, the BIST controller initiates the operation of the *test pattern generator* (TPG). The TPG generates a sequence of random test patterns and, through the *input selector*, applies them to the CUT. The input selector allows either the external inputs or the test patterns from the TPG to pass through it based on the instruction from the controller. The controller selects the inputs from the TPG during the *BIST operation* (BIST mode) and the external inputs otherwise. Thus, the chip performs its normal operations when BIST is inactive.

A TPG can also provide deterministic test patterns to augment random test patterns and increase the fault coverage. We can use these patterns to test some critical portion of a design that is random pattern resistant. We can store these additional test patterns in a read-only memory (ROM) inside the TPG.

We feed the responses obtained for the applied test patterns to a *test response analyzer* (TRA). The TRA generates a *signature* for the responses and compares it with the golden signature stored in its ROM. Note that the small size of a golden signature helps us to reduce the size of TRA's ROM. After signature analysis, the TRA generates the final BIST result at an output port.

In the following sections, we will describe TPG and TRA in more detail.

23.3 TEST PATTERN GENERATOR

We design a TPG by considering the fault coverage, allowed hardware overhead, and test application time. Ideally, we want a TPG to cover *all* the nonredundant faults for a given fault model at the expense of *minimum* hardware and test application time. To meet these objectives, we typically employ a combination of techniques which we describe in the following paragraphs.

23.3.1 Linear Feedback Shift Register

The most popular technique to generate test patterns for BIST is by obtaining *pseudorandom patterns* using an LFSR. An LFSR is suitable for BIST systems because of its simplicity, ease of design, and low hardware overhead. Additionally, the shift property integrates seamlessly with the scan-based testing. We can feed the output from an LFSR to the internal scan chains in a scan design. Thus, we can reuse test infrastructure developed for scan-based testing for BIST.

To define an LFSR mathematically, we consider an n-bit shift register shown in Figure 23.3 [3, 17]. Let us represent the state of the LFSR in the kth cycle as $(a_0^k a_1^k a_2^k ... a_{n-1}^k)$, where $a_i^k \in \{0, 1\}$. We can compute the state of the LFSR in the $(k+1)$th cycle $(a_0^{k+1} a_1^{k+1} a_2^{k+1} ... a_{n-1}^{k+1})$ as follows:

$$\begin{aligned} a_0^{k+1} &= a_1^k \\ a_1^{k+1} &= a_2^k \\ &\dots \\ a_{n-2}^{k+1} &= a_{n-1}^k \\ a_{n-1}^{k+1} &= f(a_0^k, a_1^k, a_2^k, ..., a_{n-1}^k) \end{aligned} \tag{23.1}$$

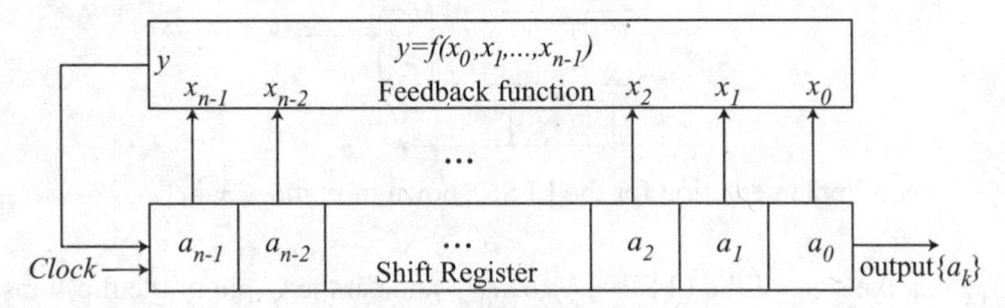

Figure 23.3 Block diagram of a *feedback shift register* (FSR) [17]

The function $f(x_0, x_1, x_2, ..., x_{n-1})$ provides feedback to the shift register. For an LFSR, the feedback function f is a *linear function* that can be represented as:

$$f(x_0, x_1, x_2, ..., x_{n-1}) = \sum_{i=0}^{n-1} c_i x_i \tag{23.2}$$

where the coefficients $c_i \in \{0, 1\}$. If the feedback function f is dependent on a state variable a_i, the corresponding coefficient c_i is taken as 1, else 0. An LFSR is typically built using D flip-flops. The "+" operation in the feedback function f is implemented using an XOR gate. We start the operation of an LFSR from a known initial state (state in the 0th clock cycle) $(a_0^0 a_1^0 a_2^0 ... a_{n-1}^0)$. We refer to the starting state as the *seed* of the LFSR.

We illustrate the implementation and operation of an LFSR in the following example.

Example 23.2 Consider an LFSR of three stages shown in Figure 23.4. Let us understand its operation.

We can infer that the feedback function $f(x_0, x_1) = x_0 + x_1$. We can implement states using flip-flops and "+" using XOR gate as follows:

$$f(x_0, x_1, x_2) = x_0 \oplus x_1$$

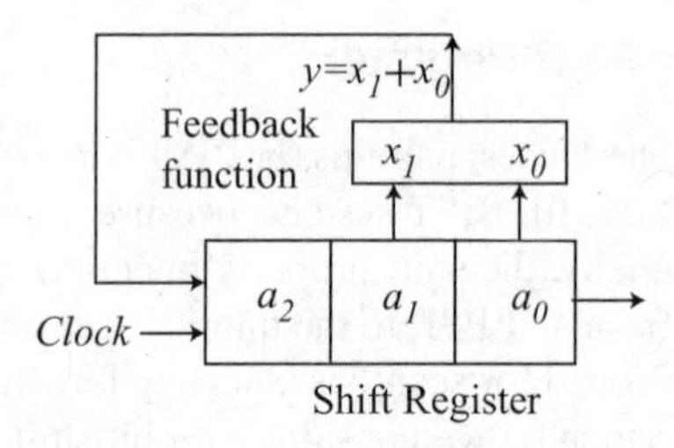

Figure 23.4 Block diagram of an LFSR

Thus, we arrive at the implementation shown in Figure 23.5.

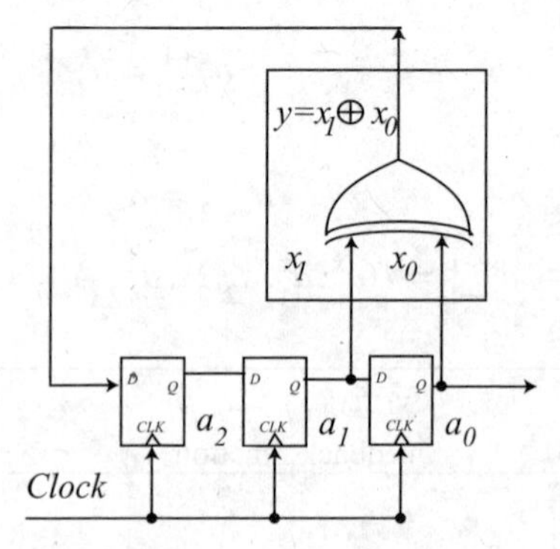

Figure 23.5 Circuit implementation for the LFSR shown in Figure 23.4

Assuming that the state of the LFSR is $(a_2^k a_1^k a_0^k)$ (value at the Q-pin of the flip-flops) in the kth clock cycle, we can compute its next state $(a_2^{k+1} a_1^{k+1} a_0^{k+1})$ (value at the D-pin of the flip-flops) as follows:

$$a_0^{k+1} = a_1^k$$

$$a_1^{k+1} = a_2^k$$

$$a_2^{k+1} = a_0^k \oplus a_1^k$$

Let us assume that the seed is (100). Using the above equations and the seed value, we can compute the successive states as shown in Table 23.1.

We can represent the state transition using a state diagram, as shown in Figure 23.6. The bitstream produced from the output a_0 is 0010111 in successive clock cycles (see a_0^k column for the current state in Table 23.1). Note that if the seed is (000), an LFSR will be stuck in that state forever. Therefore, we avoid the all-zero state in an LFSR. Starting from any other states, the operation of an LFSR is deterministic. Moreover, since the number of states is finite, it must eventually repeat states in the form of a cycle.

Table 23.1 State transition for the LFSR shown in Figure 23.5

Clock	Current State			Next State		
Edge	a_2^k	a_1^k	a_0^k	a_2^{k+1}	a_1^{k+1}	a_0^{k+1}
0	1	0	0	0	1	0
1	0	1	0	1	0	1
2	1	0	1	1	1	0
3	1	1	0	1	1	1
4	1	1	1	0	1	1
5	0	1	1	0	0	1
6	0	0	1	1	0	0

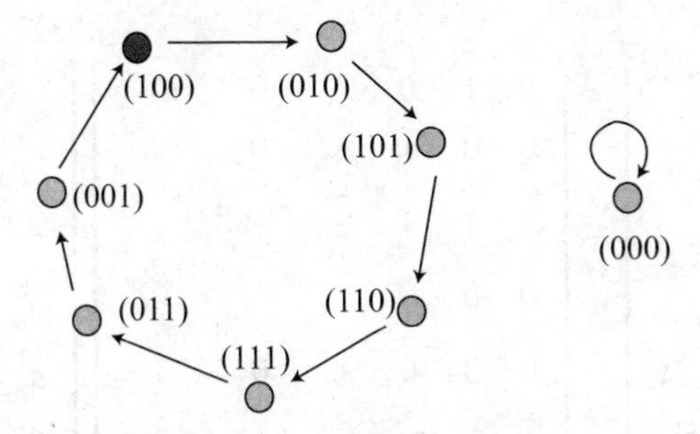

Figure 23.6 State diagram for the LFSR shown in Figure 23.4

By using a suitable feedback function, we can obtain an LFSR that produces states with a lengthy period. For a TPG, a longer test sequence typically delivers a higher fault coverage. The maximum period of an LFSR with n stages is $(2^n - 1)$. In a typical case, we use LFSR to obtain 10^3 to 10^7 patterns, depending on the testability of a circuit and the target fault coverage. For this purpose, an LFSR with a few tens of stages are sufficient [5].

The bits of an LFSR have some advantageous properties. Each bit has almost an equal probability of being 0 or 1. Furthermore, the states of an LFSR appear to be randomly ordered and satisfy most of the properties of random signals [3, 5, 18]. Nevertheless, given an initial state and a feedback function, we can compute these states deterministically. Therefore, the test patterns generated by an LFSR are repeatable and are called *pseudorandom test patterns* [3, 5].

Next, we describe two types of LFSR implementations: *standard LFSR* and *modular LFSR*.

23.3.2 Standard LFSR

The general structure of a standard LFSR is shown in Figure 23.7. It has XOR gates on the external feedback paths. We can tap the output of any flip-flop to provide feedback. If we tap the output a_i of the ith flip-flop in the chain of shift registers, then h_i is taken as 1, else it is taken as 0. The values of

h_i completely specify the architecture of a standard LFSR. If the state of a standard LFSR in the kth cycle is given as $[a_0^k a_1^k \dots a_{n-2}^k a_{n-1}^k]^T$, its state in the next clock cycle $[a_0^{k+1} a_1^{k+1} \dots a_{n-2}^{k+1} a_{n-1}^{k+1}]^T$ can be written in the matrix notation as follows[1]:

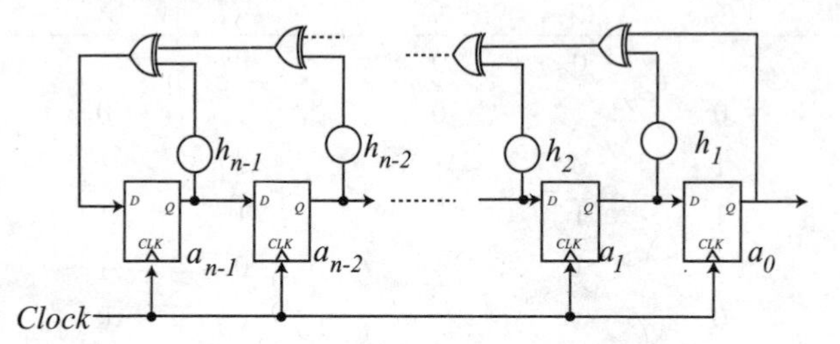

Figure 23.7 General structure of a standard LFSR (external XOR gates)

$$\begin{bmatrix} a_0^{k+1} \\ a_1^{k+1} \\ \vdots \\ a_{n-3}^{k+1} \\ a_{n-2}^{k+1} \\ a_{n-1}^{k+1} \end{bmatrix} = \begin{bmatrix} 0 & 1 & 0 & \dots & 0 & 0 \\ 0 & 0 & 1 & \dots & 0 & 0 \\ \vdots & \vdots & \vdots & \ddots & \vdots & \vdots \\ 0 & 0 & 0 & \dots & 1 & 0 \\ 0 & 0 & 0 & \dots & 0 & 1 \\ 1 & h_1 & h_2 & \dots & h_{n-2} & h_{n-1} \end{bmatrix} \begin{bmatrix} a_0^k \\ a_1^k \\ \vdots \\ a_{n-3}^k \\ a_{n-2}^k \\ a_{n-1}^k \end{bmatrix} \tag{23.3}$$

Note that, $a_i^{k+1} = a_{i+1}^k$ for $0 \le i \le (n-2)$ due to the shift operation. We can derive the feedback function as:

$$a_{n-1}^{k+1} = 1.a_0^k \oplus h_1.a_1^k \oplus \dots \oplus h_{n-2}.a_{n-2}^k \oplus h_{n-1}.a_{n-1}^k \tag{23.4}$$

The "+" obtained in the equation derived from the matrix multiplication corresponds to an XOR gate in the circuit representation. We can concisely represent the above equation as $A(k+1) = T_S.A(k)$, where T_S is the transformation matrix for the standard LFSR, $A(k)$ is the state of the LFSR in the current cycle k, and $A(k+1)$ is the state of the LFSR in the next cycle $(k+1)$.

For the circuit shown in Figure 23.5, the next state can be given as follows:

$$\begin{bmatrix} a_0^{k+1} \\ a_1^{k+1} \\ a_2^{k+1} \end{bmatrix} = \begin{bmatrix} 0 & 1 & 0 \\ 0 & 0 & 1 \\ 1 & 1 & 0 \end{bmatrix} \begin{bmatrix} a_0^k \\ a_1^k \\ a_2^k \end{bmatrix} \tag{23.5}$$

Given an initial state, we can derive the next states of the LFSR using the above matrix equation. We will obtain the state transitions as shown in Figure 23.6. We leave it as an exercise for the readers to verify.

The longest combinational path in a standard LFSR is the feedback path consisting of XOR gates. We need to reduce the delay of this path to increase the maximum operable frequency of the circuit.

[1] We denote the transpose of a matrix A as A^T.

23.3.3 Modular LFSR

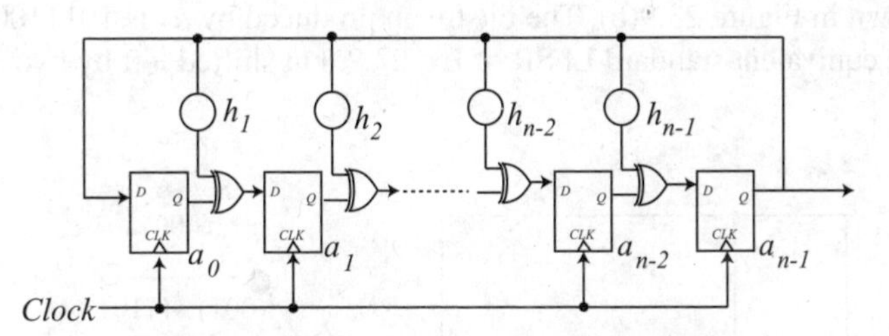

Figure 23.8 General structure of a modular LFSR (internal XOR gates)

We can also implement an LFSR by inserting XOR gates between two adjacent flip-flops in a shift register, as shown in Figure 23.8. This structure is known as a *modular LFSR*. If the *i*th flip-flop receives feedback through an XOR gate, then h_i is taken as 1, else 0. If the state of a modular LFSR in the *k*th cycle is given as $[a_0^k a_1^k \ldots a_{n-2}^k a_{n-1}^k]^T$, its state in the next clock cycle $[a_0^{k+1} a_1^{k+1} \ldots a_{n-2}^{k+1} a_{n-1}^{k+1}]^T$ can be given as follows:

$$\begin{bmatrix} a_0^{k+1} \\ a_1^{k+1} \\ a_2^{k+1} \\ \vdots \\ a_{n-3}^{k+1} \\ a_{n-2}^{k+1} \\ a_{n-1}^{k+1} \end{bmatrix} = \begin{bmatrix} 0 & 0 & 0 & \ldots & 0 & 0 & 1 \\ 1 & 0 & 0 & \ldots & 0 & 0 & h_1 \\ 0 & 1 & 0 & \ldots & 0 & 0 & h_2 \\ \vdots & \vdots & \vdots & \ddots & \vdots & \vdots & \vdots \\ 0 & 0 & 0 & \ldots & 0 & 0 & h_{n-3} \\ 0 & 0 & 0 & \ldots & 1 & 0 & h_{n-2} \\ 0 & 0 & 0 & \ldots & 0 & 1 & h_{n-1} \end{bmatrix} \begin{bmatrix} a_0^k \\ a_1^k \\ a_2^k \\ \vdots \\ a_{n-3}^k \\ a_{n-2}^k \\ a_{n-1}^k \end{bmatrix} \tag{23.6}$$

We can concisely represent the above equation as $A(k+1) = T_M.A(k)$, where T_M is the transformation matrix for the modular LFSR, $A(k)$ is the state of the LFSR in the current cycle k, and $A(k+1)$ is the state of the LFSR in the next cycle $(k+1)$. From Eqs. 23.3 to 23.6, we can note that $T_M = T_S^T$.

We can implement an LFSR as a standard LFSR or its equivalent modular LFSR. We illustrate it using the following example.

Example 23.3 Consider the standard LFSR shown in Figure 23.5. Let us implement it as a modular LFSR.

We take the transpose of the standard LFSR transformation matrix T_S (using Eq. 23.5) to obtain the transformation matrix T_M for the modular LFSR as follows:

$$T_M = T_S^T = \begin{bmatrix} 0 & 0 & 1 \\ 1 & 0 & 1 \\ 0 & 1 & 0 \end{bmatrix}$$

Hence, $h_1 = 1$ and $h_2 = 0$. Thus, we obtain the equivalent modular LFSR in Figure 23.9(a). Starting from the state (001), we can compute the state transitions similar to Ex. 23.2 and obtain the state diagram shown in Figure 23.9(b). The bitstream produced by a_2 is 1011100. This bitstream is the same as in the equivalent standard LFSR of Ex. 23.2, but shifted left by two clock cycles.

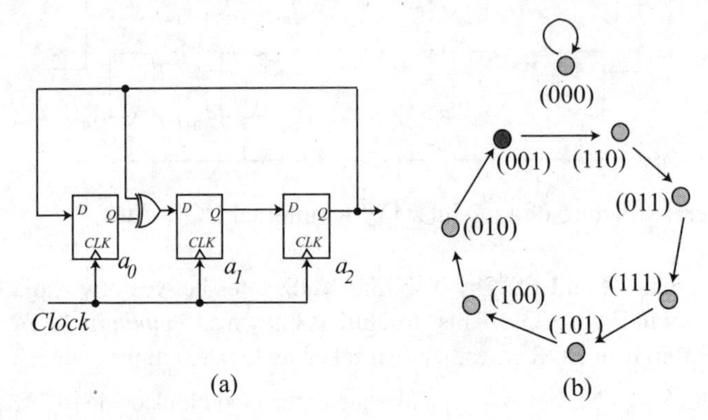

Figure 23.9 Modular LFSR equivalent of LFSR in Figure 23.5 (a) implementation and (b) state diagram

In a modular LFSR, each shift register stage can have, at most, an additional delay of one XOR gate. Hence, a modular LFSR can work at a higher clock frequency than a standard LFSR.

23.3.4 Exhaustive Test Pattern Generation

For a circuit with n inputs, we require 2^n test patterns for an *exhaustive* testing. These patterns ensure that all single and multiple stuck-at faults are covered. For small n, generating an exhaustive set of test patterns is possible using a *binary counter*. However, since the order of application of test patterns is not important, an LFSR can also be used to generate an exhaustive set of test patterns. For the same number of patterns, the area overhead of an LFSR is typically smaller than a binary counter. Therefore, we prefer LFSRs for exhaustive testing using a TPG.

Note that an LFSR does not generate the all-zero pattern. Therefore, some hardware modification is required in a conventional LFSR to obtain a truly exhaustive set of test patterns.

23.3.5 Non-exhaustive Test Pattern Generation

An *exhaustive* generation of test pattern is not feasible when the number of inputs n is large (say $n > 30$). Therefore, we employ a fraction of all possible patterns from an LFSR in testing. However, a non-exhaustive set of patterns may not achieve acceptable fault coverage. Additionally, the number of pseudorandom test patterns required to attain a given fault coverage can be unacceptable. We can employ the following techniques to tackle this problem.

Applying bias: For some circuits, we can attain a better fault coverage by *biasing* test patterns toward 0 or 1 [4, 19, 20]. For example, for a six-input AND gate, if all the inputs have an equal probability of being 1 or 0, then the probability of detecting an SA0 fault is $(0.5)^6 \approx 1.6\%$. However, if all the inputs have a probability of 0.8 to be 1, the probability of detecting an SA0 fault increases to $(0.8)^6 \approx 26\%$. Thus, by biasing the input signal probabilities toward 1, we can achieve better fault coverage for a six-input AND gate. We can bias a set of pseudorandom test patterns toward 0 or 1 by some nonlinear operations on the LFSR bits. For example, when we feed the outputs from two independent bits of an LFSR to a NAND gate, the probability of 1 at the output becomes 75%.

Deterministic test patterns: To enhance the fault coverage, we can augment pseudorandom patterns with some deterministic test patterns in the TPG. We can identify the pseudorandom patterns that do not detect any new fault. Subsequently, we can transform them into deterministic patterns targetting random pattern-resistant faults. This transformation can be done by *bit-flipping* or *bit-fixing* [21, 22].

Alternatively, we can identify faults that the pseudorandom pattern generator does not cover. Subsequently, we can use an ATPG tool to obtain a deterministic set of test patterns for them. We can store these ATPG-based test patterns directly in a ROM and apply them in addition to the test patterns generated by an LFSR.

Multiple seeds: Another strategy to improve fault coverage with a non-exhaustive set of test patterns is starting LFSRs with multiple seeds. We can compute these seeds deterministically and these seeds can be of different lengths. Additionally, for a given LFSR, we can employ a set of feedback functions rather than using a single feedback function [23, 24]. The seeds and the coefficients for the feedback functions can be stored efficiently in the ROM since the size of this data is much smaller than the actual test patterns.

23.4 TEST RESPONSE ANALYZER

A TRA analyzes the responses received from a CUT and decides whether a CUT is faulty or fault-free. It derives signatures by compacting the responses and compares it with the golden signature. We compute the golden signature beforehand by performing simulation of a fault-free circuit and store it in the TRA's ROM.

The most critical consideration in designing a TRA is its ability to distinguish between a faulty circuit and a fault-free circuit. To achieve it, we should keep the probability of *aliasing* small since it reduces the fault coverage. Additionally, it is desirable to implement TRA using minimum hardware. Therefore, compaction schemes must make intelligent trade-offs among aliasing, compaction ratio, and hardware overheads. We explain some of the compaction schemes in the following paragraphs.

23.4.1 Ones Counting

A simple compaction technique is to count the number of ones in the output response. This is also known as *syndrome analysis* [1]. However, any permutation of the correct response has also the same number of ones. Therefore, this compaction technique is highly susceptible to aliasing.

Example 23.4 Consider the function $f(a, b, c) = a \oplus b \oplus c$. Let us consider detecting an SA0 fault at the input a by ones counting and exhaustive set of test vectors.

We have shown the truth table for the fault-free and the faulty circuits in Table 23.2. We observe that there are four ones in the output response, for both the fault-free circuit and the faulty circuit. Hence, the SA0 fault at a gets aliased, and we cannot detect it using ones counting.

Table 23.2 Truth table to evaluate aliasing probability

Input (abc)	Fault-free ($a \oplus b \oplus c$)	Faulty (SA0 at a)
000	0	0
001	1	1
010	1	1
011	0	0
100	1	0
101	0	1
110	0	1
111	1	0

23.4.2 Transition Counting

Another compaction scheme is to count the total number of $0 \rightarrow 1$ or $1 \rightarrow 0$ transitions in the output response [16]. Transition counting checks for the right order of zeroes and ones in the output response to some extent. However, it still suffers from aliasing [5].

Example 23.5 Consider the function $f(a, b, c) = ab + b'c$. Let us consider detecting an SA1 fault at the input a by transition counting. Let the test vectors be generated exhaustively using binary counter starting with (abc)=(000).

Table 23.3 Truth table to evaluate aliasing probability

Input (abc)	Fault-free ($ab + b'c$)	Faulty (SA1 at a)
000	0	0
001	1 ↑	1 ↑
010	0 ↓	1
011	0	1
100	0	0 ↓
101	1 ↑	1 ↑
110	1	1
111	1	1

We have shown the truth table for the fault-free and the faulty circuits in Table 23.3. We observe that the transition count is 3 for both of them, making the fault undetectable by transition counting.

23.4.3 Single-input Signature Register

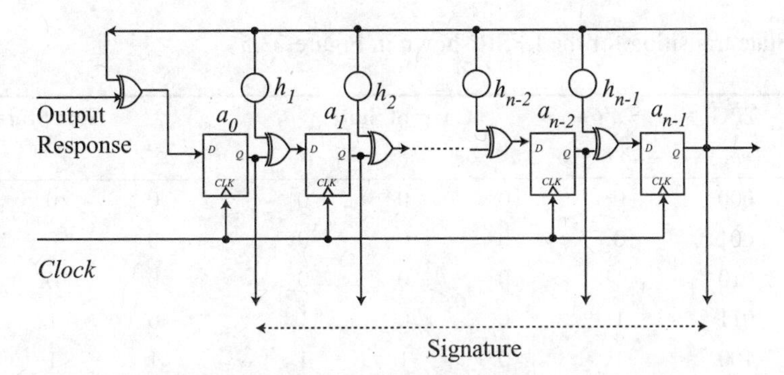

Figure 23.10 A single-input signature register (SISR) implemented in a modular LFSR

We can implement compaction circuits using LFSRs [25]. The output response is fed to a modular LFSR through an additional XOR gate, as shown in Figure 23.10. This structure is known as an SISR. Before compaction, we initialize the LFSR with a seed value. Then, we shift the output bitstream into the LFSR. We consider the final content of the registers as the signature. We compare the resultant signature with the pre-computed golden signature. We expect that the bitstreams produced by a faulty and fault-free circuit would be different. Therefore, the signatures are likely to be different for them. Thus, we can detect faults using signature analysis. We illustrate it in the following example.

Example 23.6 Consider a CUT with the functionality $y = a'b + ab'c'$ shown in Figure 23.11. Assume that the TPG generates test patterns for (abc) using a binary counter in a sequence from (000) to (111). Assume that we start the LFSR with the seed (000). Let us compute the signature for (a) fault-free circuit, (b) circuit with an SA0 fault at a, and (c) circuit with an SA1 fault at y.

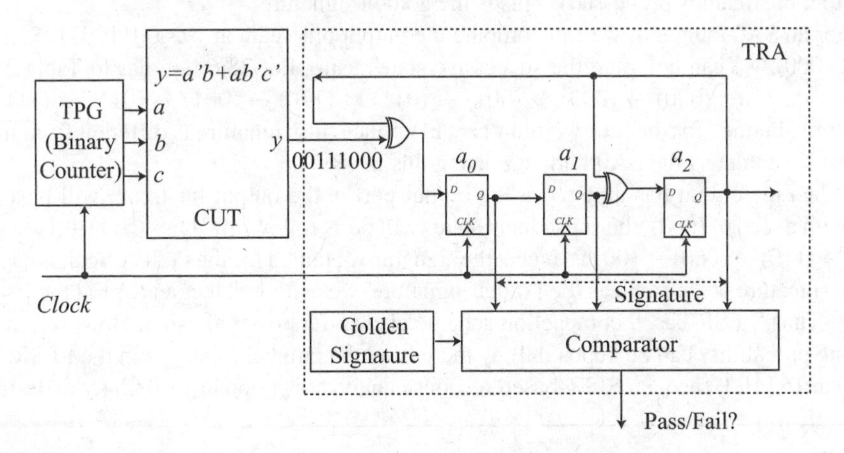

Figure 23.11 Illustration of fault detection using a modular LFSR

(a) For a fault-free circuit, we can compute the output bitstream 00111000 using $y = a'b + ab'c'$, as shown in Table 23.4.

Table 23.4 State transition for the LFSR shown in Figure 23.5

Clock Edge	TPG (*abc*)	$y = a'b+ab'c'$	Current State a_2^k	a_1^k	a_0^k	Next State a_2^{k+1}	a_1^{k+1}	a_0^{k+1}
0	000	0	0	0	0	0	0	0
1	001	0	0	0	0	0	0	0
2	010	1	0	0	0	0	0	1
3	011	1	0	0	1	0	1	1
4	100	1	0	1	1	1	1	1
5	101	0	1	1	1	0	1	1
6	110	0	0	1	1	1	1	0
7	111	0	1	1	0	0	0	1

Let us represent the states of the LFSR in the clock cycle k as $(a_2^k a_1^k a_0^k)$. Then, we can compute the next state $(a_2^{k+1} a_1^{k+1} a_0^{k+1})$ (the value at the D-pin) as follows:

$$a_0^{k+1} = a_2^k \oplus y$$

$$a_1^{k+1} = a_0^k$$

$$a_2^{k+1} = a_1^k \oplus a_2^k$$

We can compute the successive states starting with the seed (000), as shown in Table 23.4. Hence, we obtain the signature (001) for the fault-free circuit (final value in the registers after the entire output bitstream is processed). This is the golden signature.

(b) For an SA0 fault at a, we can compute the output bitstream at y as 00110011. Starting with a seed of (000), we can compute the successive states using Eq. 23.6 (similar to Table 23.4). The successive states are: (000) → (000) → (001) → (011) → (110) → (001) → (011) → (111). Hence, the signature obtained for the faulty circuit is (111). Since this signature is different from the golden signature, we can detect the SA0 fault at a using this SISR.

(c) When the circuit has an SA1 at the output port y, the output bitstream will be 11111111. Starting with a seed of (000), the subsequent states will be as follows (001) → (011) → (111) → (010) → (101) → (110) → (000) → (001). Hence, the signature obtained for the faulty circuit is (001). Since the faulty signature is the same as the golden signature, we cannot detect an SA1 at the port y using this SISR. Thus, LFSR-based compaction scheme also suffers from aliasing. However, in practice, the aliasing probability can be kept small by increasing the number of stages in the LFSR, typically greater than 16 [4]. Hence, an SISR-based response analyzer is popular for industrial designs.

23.4.4 Multiple-input Signature Register

For multiple output circuits, we need to use multiple parallel SISRs. Instead, we can use a compact circuit known as *multiple-input signature register* (MISR). We feed the output responses from different ports to an MISR through multiple XOR gates. An MISR circuit implemented on a modular LFSR is shown in Figure 23.12. We compute the signature in an MISR by shifting in the bitstream from various outputs, similar to an SISR.

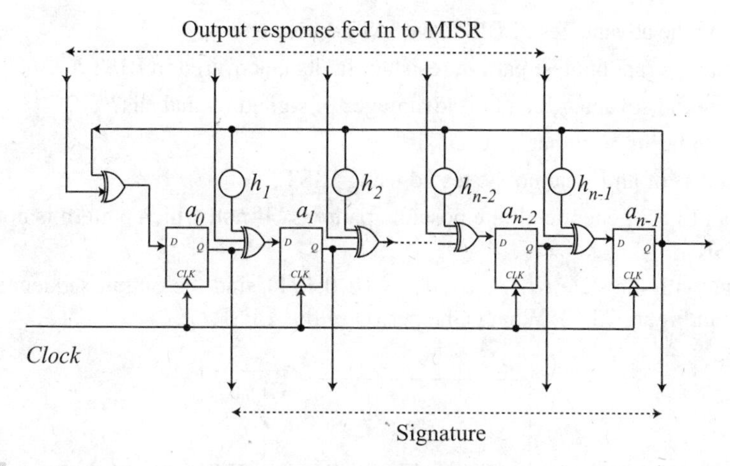

Figure 23.12 An MISR implemented in a modular LFSR

An MISR suffers from aliasing, similar to an SISR. Additionally, an MISR is prone to *error cancellation* [26]. It occurs when two errors coming from different streams at different time instances cancel each other. However, in practice, the probability of aliasing and error cancellation can be made low and tolerable.

23.5 RECENT TRENDS

BIST is a mature technology. It has been used for memories for more than four decades and is currently widely employed in industrial designs [14, 27]. We can carry out several tasks associated with the BIST implementation using commercially available tools. However, to implement BIST, we may need to make some design modifications. For example, control and observe points might be required to be inserted to improve the fault coverage. We need to stop unknown (X) value propagation to the response analyzer by hardware means because it can corrupt the signature. However, anticipating and fixing all of them might be impractical for complex designs. Therefore, techniques of making BIST systems tolerable to unknown (X) values are worth exploring [28].

A demerit of BIST is the area overhead due to additional hardware. It also reduces yield and makes physical design more difficult. Additionally, for some paths, delay can increase due to added logic gates. Moreover, BIST systems increase power dissipation due to excessive switching. Therefore, BIST systems with reduced power dissipation are becoming important [29, 30].

Despite the above drawbacks of BIST, we expect that BIST will continue to be popular. It substantially reduces the test effort, at both the chip level and the system level. It saves maintenance and repair costs that accrue over the complete product life cycle, making it cost-effective in the long run.

REVIEW QUESTIONS

23.1 What are the advantages of BIST over ATE-based testing?

23.2 How can the random test pattern resistant faults be covered in BIST?

23.3 What are the advantages and disadvantages of signature analysis?

23.4 What is aliasing in signature analysis?

23.5 Which nets of an IC cannot be tested using BIST?

23.6 Does an LFSR generate all the possible patterns? If not, which pattern is not generated by an LFSR?

23.7 Starting with the state $(a_3, a_2, a_1, a_0) = (0, 0, 0, 1)$, find the output sequence of the LFSR shown in Figure 23.13. What is the period of this LFSR?

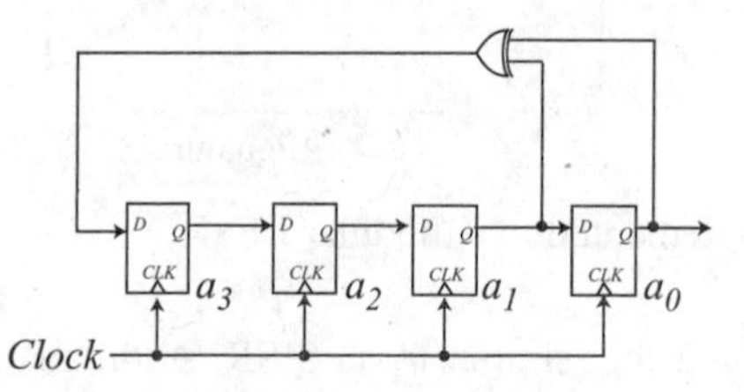

Figure 23.13 A four stage LFSR

23.8 Find the transformation matrix T_S for the standard LFSR shown in Figure 23.13. Derive a modular LFSR whose transformation matrix is the transpose of T_S. What is the output sequence of this modular LFSR and how is it related to the output sequence found in the question 23.7?

23.9 How can you make an LFSR generate all the possible patterns exhaustively?

23.10 Why can a modular LFSR work at higher frequency than a standard LFSR?

TOOL-BASED ACTIVITY

In this activity, you can use any tool for performing simulation, such as the open-source tool ICARUS Verilog [31, 32]. Consider the circuit shown in Figure 23.11. Perform the following tasks:

1. Write a Verilog code for the given circuit.
2. Write a Verilog testbench and simulate to find the golden signature.
3. Introduce an SA0 at *a*. Check whether the SISR can detect the fault.
4. Introduce an SA1 at *y*. Check whether the SISR can detect the fault.
5. Modify the LFSR (by increasing the number of bits or functionality) to detect both the above faults.

REFERENCES

[1] B. Leslie and F. Matta. "Membrane probe card technology (the future for higher performance wafer test)." *IEEE Proceedings International Test Conference* (1988), pp. 601–607, IEEE.

[2] V. Iyengar, T. Yokota, K. Yamada, T. Anemikos, B. Bassett, M. Degregorio, R. Farmer, G. Grise, M. Johnson, D. Milton, et al. "At-speed structural test for high-performance ASICs." *2006 IEEE International Test Conference* (2006), pp. 1–10, IEEE.

[3] E. J. McCluskey. "Built-in self-test techniques." *IEEE Design & Test of Computers* 2, no. 2 (1985), pp. 21–28.

[4] J. Savir and P. H. Bardell. "Built-in self-test: Milestones and challenges." *VLSI Design* 1, no. 1 (1993), pp. 23–44.

[5] V. D. Agrawal, C. R. Kime, and K. K. Saluja. "A tutorial on built-in self-test, Part I: Principles." *IEEE Design & Test of Computers* 10, no. 1 (1993), pp. 73–82.

[6] M. Bushnell and V. Agrawal. *Essentials of Electronic Testing for Digital, Memory and Mixed-signal VLSI Circuits*, vol. 17. Springer Science & Business Media, 2004.

[7] P. Goel. "Test generation costs analysis and projections." *17th Design Automation Conference* (1980), pp. 77–84, IEEE.

[8] V. D. Agrawal and P. Agrawal. "An automatic test generation system for Illiac IV logic boards." *IEEE Transactions on Computers* 100, no. 9 (1972), pp. 1015–1017.

[9] P. Agrawal and V. D. Agrawal. "On Monte Carlo testing of logic tree networks." *IEEE Transactions on Computers* 100, no. 6 (1976), pp. 664–667.

[10] P. Agrawal and V. D. Agrawal. "Probabilistic analysis of random test generation method for irredundant combinational logic networks." *IEEE Transactions on Computers* 100, no. 7 (1975), pp. 691–695.

[11] V. D. Agrawal. "When to use random testing." *IEEE Transactions on Computers*, no. 11 (1978), pp. 1054–1055.

[12] E. B. Eichelberger and E. Lindbloom. "Random-pattern coverage enhancement and diagnosis for LSSD logic self-test." *IBM Journal of Research and Development* 27, no. 3 (1983), pp. 265–272.

[13] J. Savir, G. S. Ditlow, and P. H. Bardell. "Random pattern testability." *IEEE Transactions on Computers* 100, no. 1 (1984), pp. 79–90.

[14] G. Hetherington, T. Fryars, N. Tamarapalli, M. Kassab, A. Hassan, and J. Rajski. "Logic BIST for large industrial designs: Real issues and case studies." *International Test Conference 1999. Proceedings (IEEE Cat. No. 99CH37034)* (1999), pp. 358–367, IEEE.

[15] J. E. Smith. "Measures of the effectiveness of fault signature analysis." *IEEE Transactions on Computers* no. 6 (1980), pp. 510–514.

[16] J. P. Hayes. "Transition count testing of combinational logic circuits." *IEEE Transactions on Computers* no. 6 (1976), pp. 613–620.

[17] G. Gong, T. Helleseth, and P. V. Kumar. "Solomon W. Golomb—Mathematician, Engineer, and Pioneer." *IEEE Transactions on Information Theory* 64, no. 4 (2018), pp. 2844–2857.

[18] S. W. Golomb et al. *Shift Register Sequences*. Aegean Park Press, 1967.

[19] H. D. Schnurmann, E. Lindbloom, and R. G. Carpenter. "The weighted random test-pattern generator." *IEEE Transactions on Computers* 100, no. 7 (1975), pp. 695–700.

[20] H.-J. Wunderlich. “Multiple distributions for biased random test patterns.” *IEEE Transactions on Computer-aided Design of Integrated Circuits and Systems* 9, no. 6 (1990), pp. 584–593.

[21] H.-J. Wunderlich and G. Kiefer. “Bit-flipping BIST.” *Proceedings of International Conference on Computer Aided Design* (1996), pp. 337–343, IEEE.

[22] N. A. Touba and E. J. McCluskey. “Bit-fixing in pseudorandom sequences for scan BIST.” *IEEE Transactions on Computer-aided Design of Integrated Circuits and Systems* 20, no. 4 (2001), pp. 545–555.

[23] B. Koneman. “LFSR-coded test patterns for scan designs.” *Proc. European Test Conf., March 1993* (1993), pp. 237–242.

[24] S. Hellebrand, J. Rajski, S. Tarnick, S. Venkataraman, and B. Courtois. “Built-in test for circuits with scan based on reseeding of multiple-polynomial linear feedback shift registers.” *IEEE Transactions on Computers* 44, no. 2 (1995), pp. 223–233.

[25] N. Benowitz, D. F. Calhoun, G. E. Alderson, J. E. Bauer, and C. T. Joeckel. “An advanced fault isolation system for digital logic.” *IEEE Transactions on Computers* 100, no. 5 (1975), pp. 489–497.

[26] P. Wohl, J. A. Waicukauski, and T. W. Williams. “Design of compactors for signature-analyzers in built-in self-test.” *Proceedings International Test Conference 2001 (Cat. No. 01CH37260)* (2001), pp. 54–63, IEEE.

[27] S. K. Jain and C. E. Stroud. “Built-in self testing of embedded memories.” *IEEE Design & Test of Computers* 3, no. 5 (1986), pp. 27–37.

[28] P. Wohl, J. A. Waicukauski, G. A. Maston, and J. E. Colburn. “XLBIST: X-tolerant logic BIST.” *2018 IEEE International Test Conference (ITC)* (2018), pp. 1–9, IEEE.

[29] J. Rajski, J. Tyszer, G. Mrugalski, and B. Nadeau-Dostie. “Test generator with preselected toggling for low power built-in self-test.” *2012 IEEE 30th VLSI Test Symposium (VTS)* (2012), pp. 1–6, IEEE.

[30] D. Xiang, X. Wen, and L.-T. Wang. “Low-power scan-based built-in self-test based on weighted pseudorandom test pattern generation and reseeding.” *IEEE Transactions on Very Large Scale Integration (VLSI) Systems* 25, no. 3 (2016), pp. 942–953.

[31] S. Williams and M. Baxter. “ICARUS Verilog: Open-source Verilog more than a year later.” *Linux Journal* 2002, no. 99 (2002), p. 3.

[32] S. Williams. “ICARUS Verilog.” http://iverilog.icarus.com/. Last accessed on January 8, 2023.

PART FOUR

Physical Design

Earlier, we discussed dividing the RTL to GDS implementation flow into two stages: logic synthesis and physical design. In part II of this book, we discussed logic synthesis. In this part of the book, we will discuss physical design.

Physical design involves transforming a netlist into a layout. It takes a design from the structural domain to the physical domain. The primary task of physical design is to decide the location of each design element and make their interconnections. During this process, we must ensure that the functionality of a design at the layout level is the same as that delivered by the given netlist. Furthermore, design metrics, such as yield, performance, power, and reliability, must be acceptable. Therefore, physical design implementation and verification are critical in a design flow.

In this part of the book, we will discuss each major physical design task in separate chapters. In Chapter 24 ("Basic Concepts for Physical Design"), we will explain some concepts needed for understanding physical design tasks. In Chapters 25-28, we will discuss essential physical design tasks: chip planning, placement, clock tree synthesis, and routing. In Chapter 29 ("Physical Verification and Signoff"), we will discuss physical verification tasks, including parasitic extraction and signoff. Finally, in Chapter 30 ("Post-silicon Validation"), we will explain the validation tasks performed on the first few fabricated chip samples.

It is worthy to point out that the primary objective of these chapters is to explain essential concepts and principles governing physical design. We have attempted to provide explanations not based on any specific physical design tool or proprietary data format. Therefore, readers can apply these concepts to any tool they choose for physical design, and the learning of these chapters can be relevant despite changes in tools and technology.

In the following chapters, we will also explain some popular algorithms that accomplish physical design tasks. The primary motivation for discussing these algorithms is to illustrate physical design challenges and present a few approaches to tackle them. However, note that we do not analyze these algorithms rigorously in this book. Therefore, readers who are interested in the implementation details of physical design tools should refer to standard textbooks on physical design algorithms such as [1–4].

REFERENCES

[1] M. Sarrafzadeh and C. Wong, *An introduction to VLSI physical design*. McGraw-Hill Higher Education, 1996.

[2] S. M. Sait and H. Youssef, *VLSI physical design automation: theory and practice*, vol. 6. World Scientific Publishing Company, 1999.

[3] C. J. Alpert, D. P. Mehta, and S. S. Sapatnekar, *Handbook of algorithms for physical design automation*. Auerbach Publications, 2008.

[4] N. A. Sherwani, *Algorithms for VLSI physical design automation*. Springer Science & Business Media, 2012.

24 Basic Concepts for Physical Design

A work of art is an ... expression of nature, in miniature ... The standard of beauty is the entire circuit of natural forms,—the totality of nature ... Nothing is quite beautiful alone: nothing but is beautiful in the whole.

—R. W. Emerson, *Nature*, Chapter 1, 1836

A physical design process takes a design from the structural level to the physical level. We start with a *netlist* and finally obtain a *layout* that describes the *location* and *physical connections* of each circuit element in a design. We represent a layout as geometrical patterns required to be fabricated on each layer of the silicon wafer for an integrated circuit (IC).

To appreciate the physical design process, we should be familiar with the properties of various layers in an IC, their organization, and the IC fabrication technology. Additionally, we need to understand the interconnect characteristics and their effects on the circuit parameters, such as delay and signal integrity. In this chapter, we will discuss these concepts that are essential for understanding the physical design tasks described in the following chapters.

24.1 INTEGRATED CIRCUIT FABRICATION

The IC fabrication technology strongly impacts physical design by enforcing constraints on the layout and deciding figures of merit such as performance, power, area, and reliability. Therefore, we need to understand the basics of IC fabrication.

We carry out IC fabrication in two distinct sets of processes: *front end of the line* (FEOL) and *back end of the line* (BEOL). The set of FEOL processes is responsible for fabricating circuit elements such as resistors, capacitors, diodes, and transistors. The set of BEOL processes is responsible for fabricating interconnection layers over the device layers. FEOL processes precede BEOL processes.

24.1.1 FEOL Processing

The FEOL processing starts with a monocrystalline silicon wafer. The starting wafer is typically a lightly doped p-type material. The resistivity, orientation of the crystal, and flatness of the wafer are critical considerations in choosing a wafer.

During FEOL processing, we dope different regions of a wafer with *acceptor* or *donor* materials. Doping with an acceptor creates an excess of holes while doping with a donor creates an excess of electrons. Typically, we use boron/aluminum as acceptors and phosphorous/arsenic as donors.

Ion implantation: We carry out doping using a process called *ion implantation*. During ion implantation, we generate ions by exposing an appropriate gas to an arc discharge. We accelerate and filter these ions using electric and magnetic fields before bombarding onto the substrate. We bombard the substrate through a thin layer of screening silicon dioxide to prevent excessive damage to the crystal and control the depth of the implanted ions. The energy of the ions determines the resultant doping profile on the substrate. The amount of implanted ions is known as the *dose* of the implantation. We can achieve excellent control of the depth, concentration, and doping profile using ion implantation. To selectively dope regions over the substrate, we can use a patterned photoresist as a *soft mask* or a patterned layer of silicon dioxide or other materials as a *hard mask*.

Figure 24.1 illustrates the process of creating an n-well and a p-well for a complementary metal–oxide–semiconductor (CMOS) inverter. A screen oxide is grown over the substrate. Then, the photoresist is applied to the wafer uniformly, exposed to light through an appropriate mask, and developed to obtain the required pattern on the photoresist. Then, the photoresist can act as a soft mask for ion implantation. The patterning and ion implantation is carried out firstly for the n-well and then for the p-well in Figure 24.1.

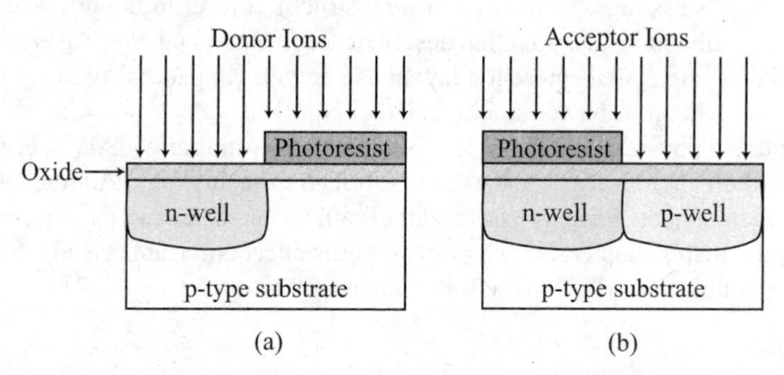

Figure 24.1 Creation of (a) n-well and (b) p-well using ion implantation

After ion implantation, the donor or acceptor atoms get lodged into the substrate, but are still inactive. We need to carry out *annealing* in a drive-in furnace or a high-temperature environment to activate the dopants and create an excess of electrons or holes. Annealing also takes the dopants to the required depth inside the substrate and repairs damages done to the crystalline structure by the bombardment of ions.

Threshold voltage control: In a CMOS circuit, the threshold voltages of the transistors play a crucial role in the circuit operation. Therefore, we need to control the threshold voltage precisely. The threshold voltage of a transistor depends on the doping of the channel, quality of gate–channel interface, dielectric constant of the gate oxide, and its thickness. We can perform additional implantation in the channel region to adjust the threshold voltage, if needed. Depending on the required threshold voltage, we can dope acceptor or donor ions into the channel. We need to control the dose and energy of the threshold adjustment implants tightly, and also use mask to make transistor-specific implants.

Isolation: All transistors are created on the same substrate in an IC during fabrication. Therefore, transistors can interact *through the substrate* if we do not *isolate* them electrically. Isoltaion ensures that the transistors do not interact through the substrate but only through the interconnects. We can achieve isolation by growing a thick layer of silicon dioxide between the transistors using *shallow trench isolation* (STI). We illustrate STI in Figure 24.2.

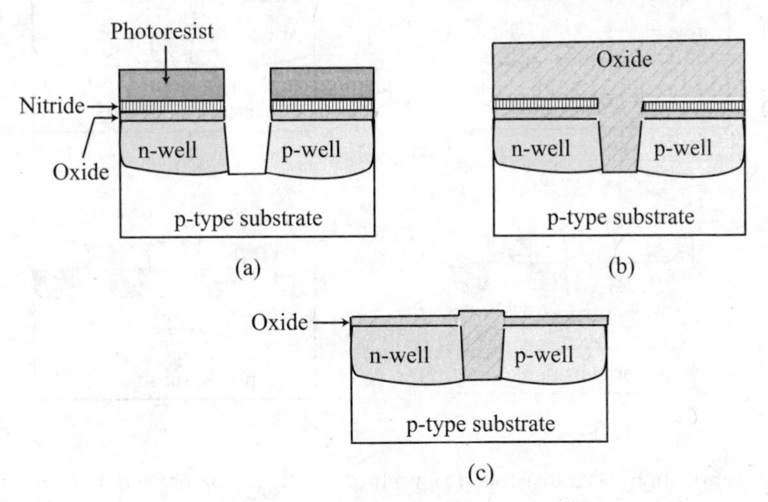

Figure 24.2 STI (a) trench created, (b) trench filled, and (c) oxide planarized and leftover nitride removed

First, we deposit a layer of silicon dioxide and protective silicon nitride. Then, we apply the photoresist and pattern the photoresist layer such that we open up the regions where isolation oxide needs to be deposited. We create trenches using selective plasma etching in the opened-up region, while the photoresist protects the remaining regions (Figure 24.2(a)). Then, we strip off the photoresist layer and grow silicon dioxide over the entire wafer, filling up the trenches (Figure 24.2(b)). Finally, we place the wafer face down in a polishing machine. A high-pH slurry is employed to polish the surface and make it flat. This process is known as *chemical mechanical polishing* (CMP). It removes silicon dioxide from the unwanted regions and also planarizes the wafer surface. We carry out CMP until we reach the silicon nitride layer. We remove the nitride layer post-CMP and obtain the structure shown in Figure 24.2(c).

Fabrication of gate: We need to create gate dielectric layer over the silicon channel in a MOSFET. Since quality of gate dielectric is critical for MOSFET operation, we grow or deposit a high-quality silicon dioxide or high-k oxide such as hafnium oxide over the silicon wafer with precise thickness control.

Post gate oxide deposition, we deposit a layer of polysilicon over the entire wafer, as shown in Figure 24.3(a). We can employ low-pressure *chemical vapor deposition* (CVD) system for this. In a CVD system, we expose the substrate to an appropriate source gas or vapor. For depositing the polysilicon layer, we can use silane (SiH_4) as the source gas. At the surface of the substrate, the vapor reacts or decomposes, leaving behind the required polysilicon layer. We can adjust the work function and the resistivity of the polysilicon layer by unmasked implantation of dopants or adding phosphine or diborane gases in the CVD chamber.

After depositing the polysilicon layer, we pattern the gates of the transistors using a mask and plasma etching, as shown in Figure 24.3(b). We can use polysilicon for short interconnects, which can be patterned using the same mask. After that, we create the source and drain of NMOS and PMOS using ion implantation, as shown in Figure 24.3(c) and (d), respectively.

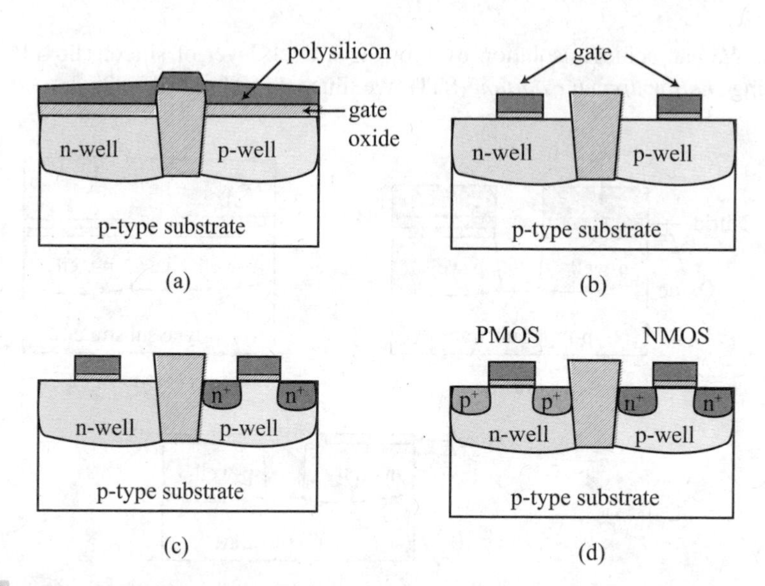

Figure 24.3 Fabrication of transistors (a) high-quality gate oxide and polysilicon deposited, (b) gates of NMOS and PMOS patterned, (c) source and drain of NMOS built, and (d) source and drain of PMOS built

In this section, we have considered some of the basic FEOL process steps. In practice, we employ several other process steps to fabricate advanced devices and incorporate innovative device features. These aspects of device fabrication are outside the scope of this book. We encourage readers to refer to them in the standard textbooks on IC fabrication.

24.1.2 BEOL Processing

BEOL processes make metallic interconnections between devices that were fabricated during the preceding FEOL processing. Typically, copper is used for interconnects because of its low resistivity. Earlier, when we used aluminum interconnects, ICs were more susceptible to a reliability problem known as *electromigration*. Electromigration occurs due to the physical movement of atoms in the interconnect when a continuously high current flows through it. Copper has greater immunity against electromigration compared to aluminum and is preferred for interconnects.

Note that copper atoms easily diffuse through silicon and silicon dioxide layers and can destroy the functionality of a transistor. Therefore, we must ensure that the copper atoms do not penetrate through the dielectric and reach the FEOL layers. We can avoid such contaminations by keeping the BEOL processing environment physically separated from the FEOL processing environment. Additionally, we can insert a *barrier layer* composed of materials such as tantalum, titanium, ruthenium, and metal silicides between the copper layers and the underlying silicon or silicon dioxide layers [1].

Interconnect layers: For making interconnections, we use several layers of interconnects. We separate two different interconnect layers by an *interlayer dielectric* (ILD). Traditionally, we have been using silicon dioxide as the ILD material. However, note that the electric field through an ILD contributes to the load capacitance on the transistors and coupling capacitance between interconnects. Therefore, to enable faster switching, ensure signal integrity, and reduce power dissipation, we need to reduce the capacitance associated with the ILD layer. Hence, materials with dielectric constant lower than silicon dioxide are now used for fabricating ILD layers [2]. We can even use air gaps as an ILD layer at advanced process nodes [3].

Dual damascene process: We fabricate copper interconnect layers using *dual damascene process* [4]. In a dual damascene process, we carve out hollow regions for both vias and interconnecting wires and simultaneously fill them with metal. We illustrate one of the methods to carry out the dual damascene process in Figure 24.4.

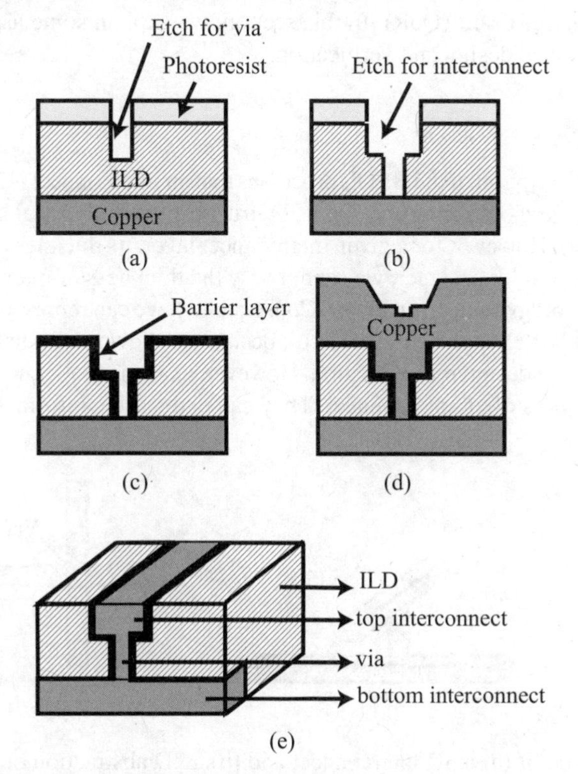

Figure 24.4 Dual damascene process (a) ILD etched for vias, (b) ILD etched interconnects, (c) barrier layer deposited, (d) barrier layer removed from bottom, seed copper deposited, and electroplating done, and (e) final CMP

First, we deposit an ILD layer over a preexisting planarized layer. Then, using masks for the vias and interconnecting wires, we etch hollow regions into the ILD layer (Figure 24.4(a) and (b)). After that, we deposit a diffusion barrier layer over the hollow region (Figure 24.4(c)). Subsequently, we deposit seed copper in the open area using *physical vapor deposition* and fill the trenches with copper using electroplating (Figure 24.4(d)). Finally, we remove the excess copper coating from the top surface using CMP. We stop the CMP process when the ILD layer is reached (Figure 24.4(e)). We repeat this process of fabricating interconnect layers multiple times to create multiple layers of interconnect. At advanced process nodes, we typically employ more than 10 layers of interconnects with varying thicknesses [3].

The above description of the FEOL and BEOL processes is simplistic. In practice, we carry out more than a 100 individual steps sequentially. Despite sophistication, these steps can sometimes produce an unexpected outcome. Therefore, we perform process control tests throughout the fabrication flow, evaluate, and take remedial action as soon as possible.

24.2 INTERCONNECTS

An essential task of physical design is to make interconnections between various components. The nature and properties of interconnects strongly impact the methodologies adopted in physical design and the resultant quality of result (QoR). In this section, we explain some aspects of interconnects that are relevant to physical design and verification.

24.2.1 Thickness

It is evident from the description of the BEOL processes that an interconnect is a 3D structure of metal or polysilicon. Ideally, we should represent it as a 3D structure with polygonal faces or a *polyhedra*, as shown in Figure 24.5(a). However, for a given interconnect layer, its thickness is uniform throughout a wafer. Therefore, as a circuit designer, we cannot vary the thickness of interconnects (it is decided by the fabrication technology for a given layer). Consequently, we can represent interconnects simply as 2D polygons instead of 3D polyhedra in circuit design, as shown in Figure 24.5(b). It simplifies the design process and the design representation. However, analysis tools, such as a delay calculator, need to know the thickness of an interconnect. They can extract this information from the physical libraries.

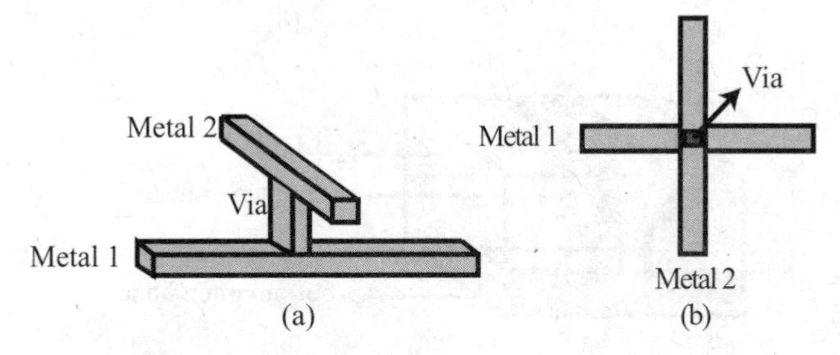

Figure 24.5 Schematic of (a) a 3D interconnect and (b) a 2D abstraction of the 3D interconnect

Example 24.1 Consider the resistor shown in Figure 24.6. It has length L, width W, and thickness T.

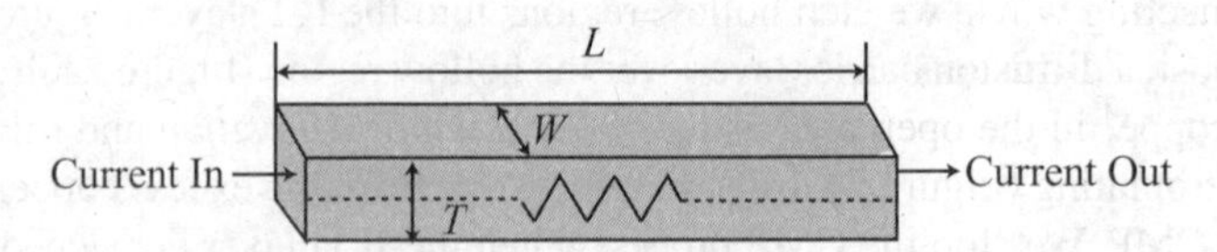

Figure 24.6 Schematic of a section of an interconnect

The resistance R of a 3D interconnect is given as:

$$R = \rho \frac{L}{TW} \tag{24.1}$$

where ρ is the resistivity of the interconnect material. Since the thickness is constant for a given layer, we can write the above equation as follows:

$$R = R_S \frac{L}{W} \tag{24.2}$$

where $R_S = \rho/T$ is known as the *sheet resistance* of the interconnect layer. Typically, the sheet resistance for each layer is defined in the technology-specific physical library. Analysis tools, such as delay calculators, can compute resistance of an interconnect using sheet resistances specified in this library.

The thickness of interconnects in different layers can be different, as illustrated in Figure 24.7. Typically, the thickness of interconnect layers increases from the bottom of an IC toward the top. When we increase the thickness of an interconnect layer, its sheet resistance decreases. We employ thick top metal layers with low sheet resistance for *global interconnect* and in *power delivery network* (PDN) [3, 6]. A low resistance interconnect results in a smaller voltage drop across it. Consequently, thick top metal layers in a PDN facilitate maintaining a uniform supply voltage across a circuit.

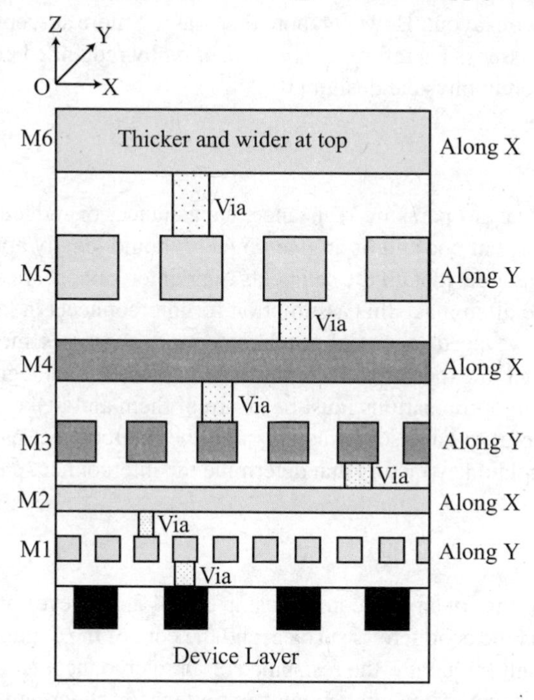

Figure 24.7 Schematic of a six layer metal interconnect (M1–M6) structure [5]

24.2.2 Orientation

We allow interconnects or routes in an IC to run in only two directions: horizontal (along the x-axis) or vertical (along the y-axis). Routing with this constraint is known as *Manhattan routing*. Furthermore, each interconnect layer has an associated *preferred* direction of being laid out: either horizontal or vertical. We allow interconnects to run mostly along the preferred direction. However, sometimes we allow *short* wires, known as *jogs*, to run orthogonal to the preferred direction.

The preferred direction alternate between horizontal and vertical directions for successive layers of interconnects, as illustrated in Figure 24.7 [7, 8]. Note that by relaxing this constraint, we can decrease the overall wire length and improve the QoR of a design. For example, if we allow 45° and 135° interconnect layout, along with the horizontal and vertical directions, we can reduce the wire length [9]. We illustrate it in the following exmaple.

Example 24.2 Assume that two pins are located at the diagonally opposite vertices of a square with sides measuring L micron. Consider that we need to connect these pins. For Manhattan routing, the wire length will be $2L$ micron. If we allow 45° orientation, the wire length will be $\sqrt{2}L$ micron. Hence, we can reduce the wire length by $\frac{2-\sqrt{2}}{2} \times 100 \approx 30\%$ by relaxing the Manhattan routing constraint.
Remark: Despite the reduction in the wire length, we use Manhattan routing and assign a preferred direction for each layer because they greatly simplify the physical design tasks.

A side effect of assigning a direction to each layer is that we need to change the layer and insert a *via* for each bend in the wire layout. However, note that vias are more susceptible to process-induced variations and reliability issues. Therefore, *via minimization* by reducing bends in the wire layout is an important consideration in physical design [10].

24.2.3 Parasitics

Ideally, an interconnect has no parasitic resistance, capacitance, or inductance associated with it. Therefore, a voltage change at one end of an interconnect should ideally appear immediately at the other end. Thus, we can assume that all the segments of an interconnect are at the same potential for an ideal interconnect. We often make this assumption for interconnects in logic synthesis.

In physical design, we need to model the interconnect parasitics more carefully. In reality, parasitics strongly impact the performance, power dissipation, and signal integrity of an IC. Therefore, physical design optimizations must be aware of them and make decisions accordingly.

We often find it helpful to have some insights into the interconnect parasitics during physical design. Therefore, we explain key factors that determine the interconnect parasitics in the following paragraphs.

Resistance

We can compute the resistance of an interconnect using Eq. 24.2. However, at high frequency, current tends to flow primarily on the conductor's surface, and the core of the conductor remains effectively unused. As a result, at high frequency, the resistance of an interconnect tends to increase [11]. This phenomenon is known as *skin effect* and becomes important for wider and thicker wires at the top metal layers. The skin effect is important in clock lines that are generally routed at the top metal layers and work at a high frequency.

Capacitance

Interconnects lie within a dielectric material over a silicon substrate. As a circuit operates, the electric potential of interconnects changes. As a result, the electric field and the stored charges in the surrounding dielectric material change. Consequently, an interconnect exhibits substantial capacitance. The capacitance of interconnects depends on their geometry, environment (the location and geometry of other interconnects), and the property of the surrounding dielectric. The computation of interconnect capacitance is a nontrivial problem. However, to develop an intuitive understanding, we can examine the interconnect capacitance by making some simplifying assumptions.

First, let us consider an interconnect strip over a silicon substrate, as shown in Figure 24.8.

Parallel plate capacitor: Some electric field lines emanate vertically downward from the bottom of the interconnect and terminate at the silicon substrate. In this case, we can model an interconnect as a parallel plate capacitor:

$$C_{pp} = \frac{\epsilon_{di}\epsilon_0}{T_D} WL \tag{24.3}$$

where ϵ_{di} is the relative permittivity of the dielectric material, ϵ_0 is the absolute permittivity of vacuum, T_D is the thickness of the dielectric layer between the bottom of the interconnect and the substrate, W is the width of the interconnect, and L is the length of the interconnect.

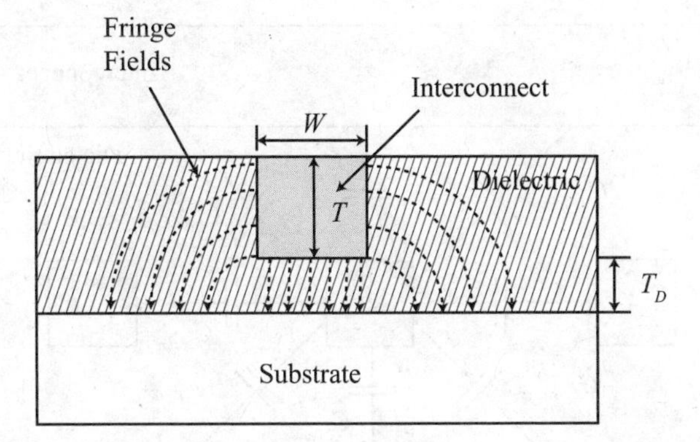

Figure 24.8 Illustration of electric field lines when an interconnect is placed over a substrate

Fringe capacitor: There are field lines that emanate from the sidewalls of the interconnect and terminate on the substrate. These electric fields are known as *fringe fields*. We can account for the fringe fields by adding a fringe capacitor C_{fringe} in parallel with the parallel plate capacitor C_{pp} as follows:

$$C_{strip} = C_{pp} + C_{fringe} \tag{24.4}$$

Depending on the accuracy requirement, we can model C_{pp} and C_{fringe} in various ways [12]. A simplistic model is as follows (assuming $W \geq T/2$) [13]:

$$C_{strip} = C_{pp} + C_{fringe} = \frac{\epsilon_{di}\epsilon_0}{T_D}(W - T/2)L + \frac{2\pi\epsilon_{di}\epsilon_0}{ln\left(1 + \frac{2T_D}{T} + \sqrt{\frac{2T_D}{T}\left(\frac{2T_D}{T} + 2\right)}\right)} L \tag{24.5}$$

where T is the thickness of the interconnect. The first term represents C_{pp}, while the second term represents C_{fringe}. We can observe that the fringe capacitance C_{fringe} increases when T increases. This trend is expected since the increase in the thickness of an interconnect results in a greater sidewall area, and more fringe field lines originate from the sidewalls.

Note that, with scaling, we decrease the width (W) of the interconnects because it allows us to pack more components in a given die area. However, we do not proportionately reduce the interconnects' thickness (T) to avoid sheet resistance becoming too high. Consequently, with scaling, interconnects become taller and narrower. It makes fringe capacitance more significant with scaling.

Capacitance between interconnects: As described earlier, interconnect consists of multiple layers of metals separated by a dielectric material. Therefore, electric field lines emanating from an interconnect are modified by the neighboring interconnects in a complicated manner. Consequently, modeling capacitance between interconnects is a challenging problem. It often requires numerical simulation to obtain a satisfactory accuracy. Neverthless, we can examine its various components qualitatively and analyze their dependencies on the physical parameters.

Figure 24.9 shows three layers of interconnects lying over a reference conductor or ground plane. The conductors are marked 0–5. Let us consider conductor 2, lying between conductors 1 and 3. The conductors 4 and 5 are adjacent to the conductor 2. There are three important components of capacitance for the conductor 2: *overlap capacitance*, *lateral capacitance*, and *fringe capacitance* [14].

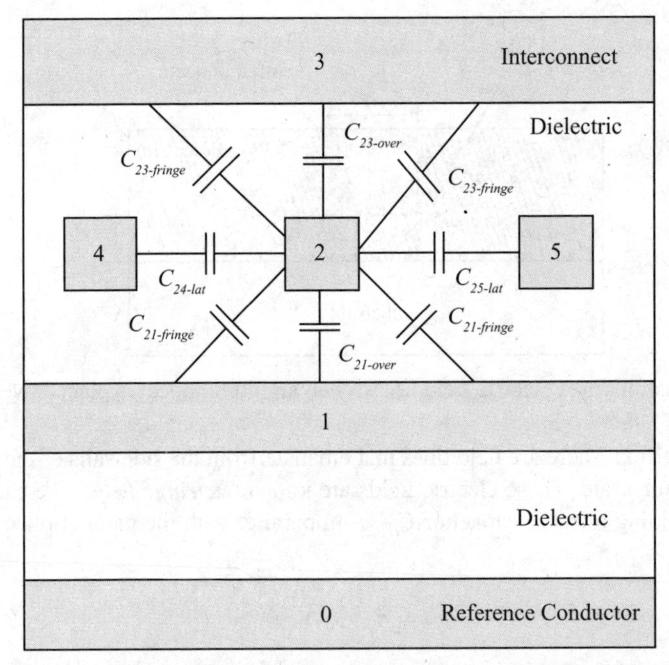

Figure 24.9 Different components of capacitance of an interconnect in an IC. The conductors are marked 0–5. The components of capacitance are shown for the conductor 2 [14]

1. **Overlap capacitance:** The overlap capacitance is due to the overlap between two conductors in different planes. For conductor 2, there are overlaps with conductors 1 and 3. We can model them using overlap capacitors $C_{21-over}$ and $C_{23-over}$.

   We can compute the overlap capacitance using parallel plate approximation, as illustrated in Figure 24.10(a). For example, we can compute $C_{23-over}$ as:

$$C_{23-over} = \frac{\epsilon_{di}\epsilon_0}{T_{23}} A_{23} \tag{24.6}$$

where T_{23} is the thickness of the dielectric layer between the upper edge of conductor 2 and lower edge of conductor 3, and A_{23} is the overlap area between these two conductors.

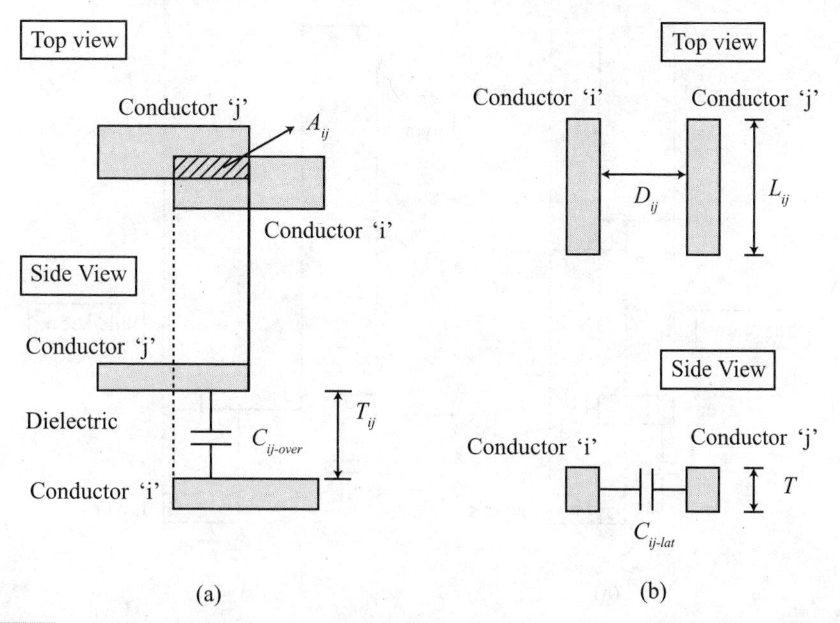

Figure 24.10 Illustration of capacitance between conductors 'i' and 'j' (a) overlap capacitance (the overlapping area A_{ij} is shown as hatched region) and (b) lateral capacitance [14]

2. **Lateral capacitance:** A lateral capacitor is formed by two parallel edges of nonoverlapping conductors in the same plane. For conductor 2, the field lines originate from its sidewalls and terminate at the sidewalls of conductors 4 and 5. We can model them using the capacitors C_{24-lat} and C_{25-lat}.

   The lateral capacitor C_{ij-lat} between two conductors named 'i' and 'j' located in the same layer is illustrated in Figure 24.10(b). The value of C_{ij-lat} depends on the lateral spacing D_{ij}, the length of the parallel edges of the conductors L_{ij}, and the environment between the two parallel edges [14]. As the lateral spacing D_{ij} increases, C_{ij-lat} decreases. For sufficiently large D_{ij}, we can ignore the lateral capacitance. The C_{ij-lat} increases as the length of the parallel edges of the conductors L_{ij} increases. Note that the electric field lines within the parallel edges determine the value of C_{ij-lat}. The existence of other neighboring conductors can modify these electric field lines [14]. Hence, parasitic capacitance extraction becomes complicated, and we need to use numerical techniques to extract them accurately.

3. **Fringe capacitance:** The fringe capacitance represents the coupling between two conductors in different planes due to electric fields originating from the sidewalls. In Figure 24.9, the capacitors $C_{21-fringe}$ and $C_{23-fringe}$, represent the coupling of the conductor 2 through its sidewalls with conductors 1 and 3, respectively.

   The fringe capacitance can exist between conductors with or without overlapping area, as illustrated in Figure 24.11. The magnitude of fringe capacitance depends on the area of the sidewalls, the location of the conducting surface with respect to the edge of the sidewalls, the geometry of the conductors, and the properties of the dielectric layer [14].

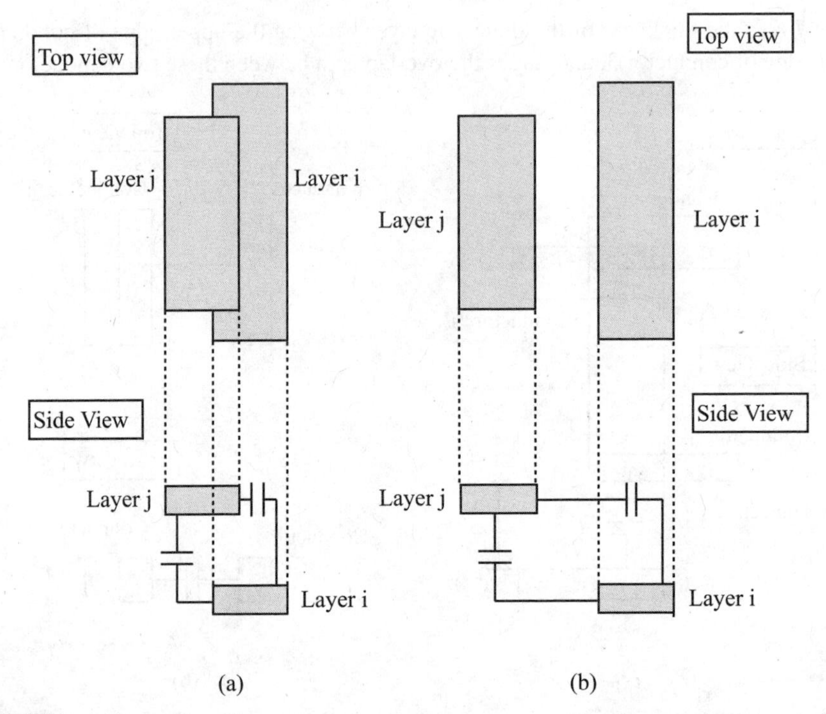

Figure 24.11 Illustration of fringe capacitance in conductors of different layers (a) when there is an overlap and (b) when there is no overlap [14]

Inductance

The inductance of an interconnect is normally insignificant within an IC. However, it becomes important in the following cases:

1. **Top layer metal lines:** At the top level, the interconnects are typically thicker and wider. Therefore, the resistance (R) of these interconnects is smaller. However, the impedance (Z) offered due to the inductance (L) ($Z = j\omega L$, where ω is the angular frequency) becomes significant in comparison to the resistance.
2. **High frequency:** When a signal transitions quickly, the impedance due to the inductance can become comparable to the resistance, and we can no longer ignore the inductance. This condition is especially true for the clock lines.

We can model inductance as follows [15]:

$$cl = \epsilon\mu \tag{24.7}$$

where c and l are capacitance and inductance per unit length, respectively, ϵ and μ are the permittivity and the permeability of the dielectric medium, respectively. For a nonuniform dielectric medium, we can use an *average* value of the dielectric constant to obtain an approximate inductance per unit length [15].

24.2.4 Wire Models

We analyze a physical design using some electrical models for the wires. These models are known as *wire models.*[1] We choose a wire model based on the accuracy requirement and the available computational resource. In the following paragraphs, we describe some commonly used wire models. Note that we assume that the wire is already laid out in the following discussion, and we can extract all the attributes related to the wire layout, if required by the model.

Lumped Capacitance

When the length of an interconnect is short and is located on lower interconnect layers, we can ignore the resistance and the inductance of the interconnect. In this case, we can model the parasitics of the entire interconnect network using a lumped capacitance C_L, as demonstrated in Figure 24.12. Besides capacitance associated with the interconnect, C_L includes the capacitances associated with the pull-up transistor (drain diffusion capacitance C_D) and the capacitances associated with the gates in the fanout of the net. During a rising transition, these capacitances get charged by the pull-up transistor. We can model the pull-up transistor by a linear resistor (R_P) and a voltage source that transitions from 0 V to V_{DD}, where V_{DD} is the supply voltage. In this case, the time taken by a signal to propagate through the interconnect and reach the driven gates (*G1*, *G2*, and *G3*) is insignificant due to negligible resistance of the wire [16]. Under this approximation, we can assume that the voltage levels at all the driven pins are equal, i.e., $v_{out1} = v_{out2} = v_{out3}$. We can compute v_{out1} as follows [17]:

$$v_{out1}(t) = V_{DD}(1 - e^{-t/R_P C_L}) \tag{24.8}$$

The rise time increases as the resistance R_P or the lumped capacitance C_L increases.

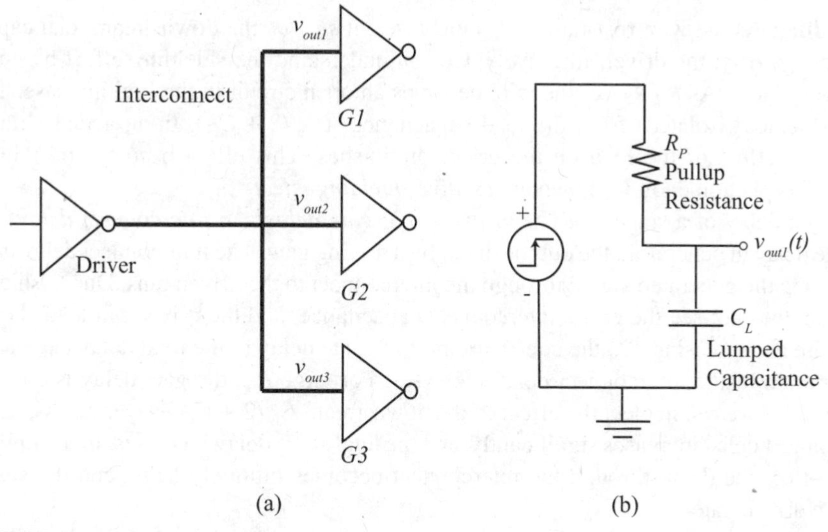

Figure 24.12 Lumped capacitor model of an interconnect

[1] See Section 13.4.6 ("Cells and interconnects") for the application of a wire model in delay calculation.

Resistance–Capacitance Tree

Pi-model: If the interconnect resistance is comparable to the driver's pull-up resistance R_P, we cannot use the lumped capacitance model [17]. For a straight segment of interconnect, the resistance of the wire (R_w) is $R_w = rL$, where r is the resistance per unit length and L is the length of the wire segment. Similarly, the capacitance of the wire (C_w) can be given as $C_w = cL$, where c is the capacitance per unit length. Though the resistance and capacitance are distributed over the length of a wire, we can approximately model them in a combined pi-model, as illustrated in Figure 24.13 [18]. We consider half of the wire capacitance ($C_w/2$) on the driver side and the other half on the load side.

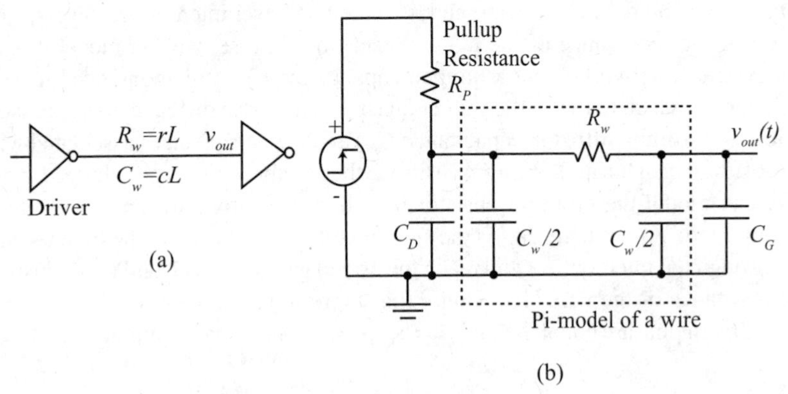

Figure 24.13 Pi-model of an interconnect. The drain diffusion capacitance and the input gate capacitance of the driven gate are shown as C_D and C_G, respectively

Shielding: As the wire resistance (R_w) increases, it *shields* the downstream load capacitances ($C_w/2$ and C_G) from the driver. Intuitively, we can understand the shielding effect by considering R_w to be very large. As $R_w \rightarrow \infty$, the wire becomes an open circuit in the limiting case. The driver becomes shielded (isolated) from the load capacitances ($C_w/2 + C_G$). In an actual circuit, as R_w increases, the effect of the load on the driver diminishes. This effect becomes prominent when $R_w > R_P$. This phenomenon is known as *resistive shielding effect*.

The total delay of a stage consists of two parts: *gate delay* and *interconnect delay*.[2] The gate delay is the delay in generating the output signal by a driving gate. The interconnect delay is the delay in propagating the generated signal through the interconnect to the driven pins. Due to shielding, the driving gate does not see the entire interconnect capacitance, and the driven pin loads [19–21]. As a result, with an increasing R_w, the contribution of the gate delay to the total delay decreases, while the contribution of the interconnect delay increases. For large R_w, the gate delay is due to R_p and $C_D + C_w/2$, and we can neglect the effect of the downstream $C_w/2 + C_G$. Nevertheless, in this case, the interconnect delay increases significantly, and the total stage delay increases. In the limiting case, when $R_w \rightarrow \infty$, the delay through the interconnect becomes infinitely large, and the signal never reaches the driven gate.

Resistance–capacitance tree: In general, when an interconnect has multiple fanouts, the voltage waveforms at different sink pins can be different due to difference in the wire length from the source to the sinks. We cannot model this effect by a single lumped resistance–capacitance (RC)

[2] For details, see Chapter 13 ("Technology library"), section on "Cells and interconnects".

pi-model. Additionally, interconnects can be composed of multiple interconnect segments and have complex layouts.[3] We need to consider them in the wire model.

For interconnects consisting of multiple segments, we can model each segment using an equivalent RC pi-model. Further, we combine them in the form of a *tree*. The circuit thus obtained is referred to as an *RC tree*. A capacitor connects a node to the ground in an RC tree. If more than one capacitors are connected from a node to the ground, we can replace them by an equivalent capacitor. It will reduce the size of an RC tree. Additionally, an RC tree has no floating capacitors, no resistor directly connected to the ground, and no resistor loop.

Example 24.3 Consider a gate *G0* driving three gates *G1*, *G2*, and *G3*, as shown in Figure 24.14.

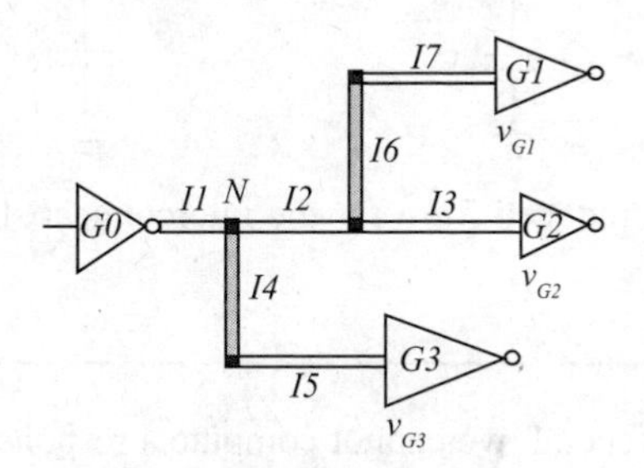

Figure 24.14 Schematic representation of an interconnect. The interconnect consists of seven segments *I1, I2, ..., I7*. The driver gate is *G0*. The driven gates are *G1*, *G2*, and *G3*. The voltage at each driven gate is shown as V_{G1}, V_{G2}, and V_{G3}

When the output of *G0* makes a transition from 0→1, the signal will propagate through the interconnect segments *I1, I2, ..., I7* and reach the input pins of *G1*, *G2*, and *G3* at different time instants. Therefore, a single lumped RC pi-model cannot model this behavior. However, we model different segments of the interconnect using separate pi-model and combine them into a single model. Thus, we obtain an RC tree shown in Figure 24.15.

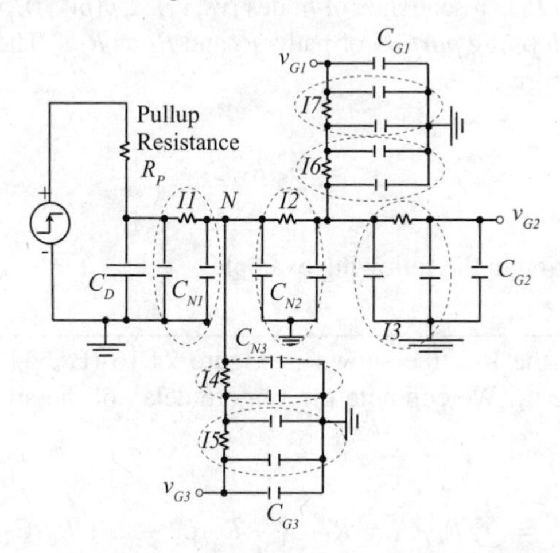

Figure 24.15 RC tree model for the interconnect shown in Figure 24.14. The input gate capacitances at *G1*, *G2*, and *G3* are shown as C_{G1}, C_{G2}, and C_{G3}, respectively

[3] We refer to a straight wire in the layout as *interconnect segment*.

We can reduce the size of the tree by combining capacitors. For example, at the node N, the capacitors C_{N1}, C_{N2}, and C_{N3} are in parallel. We can replace them with an equivalent capacitor with capacitance $(C_{N1} + C_{N2} + C_{N3})$. Similarly, we can combine parallel capacitors at other nodes into one single capacitor. Thus, we obtain a simplified RC tree, as shown in Figure 24.16.

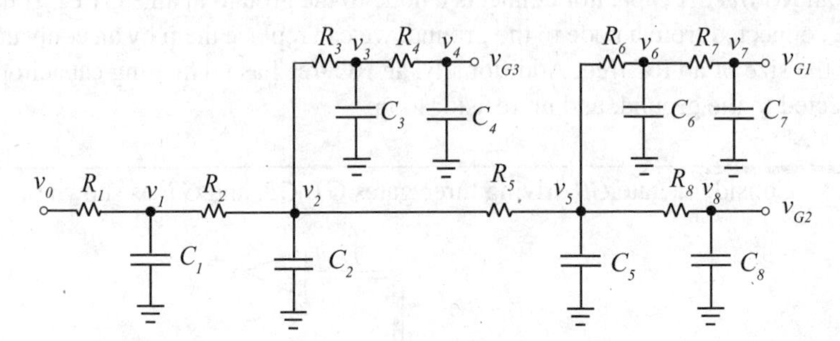

Figure 24.16 The RC tree in reduced form for the interconnect shown in Figure 24.14 (reduced from Figure 24.15)

Elmore delay model: In general, we cannot compute a response in a closed form for a circuit consisting of an RC tree. Therefore, we compute an approximate response with some reasonable accuracy. *Elmore delay model* is a popular model for this purpose [17, 22].

To describe the Elmore delay model, let us assume a generic RC tree with $(N + 1)$ nodes, marked as $v_0, v_1, v_2, ..., v_N$ [21]. We consider the node v_0 as the source voltage node and treat it as the root of the RC tree. Let us assume that a capacitor C_i is associated between each node v_i and the ground for all nodes other than the source node. Let us denote the parent of each node v_i as $p(v_i)$. Note that, since the model is a tree, each node v_i $(1 \leq i \leq N)$ has a *unique parent* $p(v_i)$. Let us denote the resistance between nodes v_i and $p(v_i)$ as R_i.

For a tree, there is a unique path from the root to any node v_i. Let us denote the path from the root to a node v_i as P_i. The path P_i is a sequence of nodes $\{v_0, v_1, ..., p(p(v_i)), p(v_i), v_i\}$. Let us denote the total resistance of the *overlapping portion* of paths P_i and P_k as R_{ik}.[4] The Elmore delay τ_i between the source node v_0 and v_i is given as:

$$\tau_i = \sum_{k=1}^{N} R_{ki} C_k \tag{24.9}$$

We illustrate this computation in the following example.

Example 24.4 Consider the RC tree shown in Figure 24.16 (Ex. 24.3). The input to the gate *G1* is denoted by the node v_7. We compute the Elmore delay of the signal from the source to v_7 using Eq. 24.9:

$$\tau_7 = \sum_{k=1}^{7} R_{k7} C_k = R_{17} C_1 + R_{27} C_2 + ... + R_{67} C_7 \tag{24.10}$$

[4] An overlapping portion of two paths is a set of common nodes in those paths.

The resistance R_{17} is the total resistance of the overlapping portion of the paths P_1 and P_7. Hence, $R_{17} = R_1$. Similarly, $R_{27} = (R_1 + R_2)$. The resistance R_{37} is $(R_1 + R_2)$ because the total resistance in the operlapping portion of the paths P_3 $(\{v_0, v_1, v_2, v_3\})$ and P_7 $(\{v_0, v_1, v_2, v_5, v_6, v_7\})$ is $(R_1 + R_2)$. Similarly, we compute other resistances. Thus, we obtain:

$$\begin{aligned}\tau_7 = R_1C_1 &+ (R_1 + R_2)C_2 + (R_1 + R_2)C_3 + (R_1 + R_2)C_4 \\ &+ (R_1 + R_2 + R_5)C_5 + (R_1 + R_2 + R_5 + R_6)C_6 \\ &+ (R_1 + R_2 + R_5 + R_6 + R_7)C_7 + (R_1 + R_2 + R_5)C_8\end{aligned} \tag{24.11}$$

We can also compute the Elmore delay τ_i of a node v_i *recursively* using the following formula:

$$\tau_i = \tau_{p(v_i)} + R_i(C_i + C_{fo}) \tag{24.12}$$

where $\tau_{p(v_i)}$ is the delay at the parent node $p(v_i)$ and C_{fo} is the sum of capacitances of all nodes in the fanout cone (downstream) of the node v_i. We assume that the delay τ_0 at the source node v_0 is 0.

Example 24.5 Consider the RC tree shown in Figure 24.16. Let us compute the Elmore delay of the signal from the source to v_7 using the above recursive definition, starting from the source ($\tau_0 = 0$). We compute the delay for each node, followed by its child, till we reach the node v_7:

$$\tau_1 = R_1(C_1 + C_2 + C_3 + C_4 + C_5 + C_6 + C_7 + C_8) \tag{24.13}$$

$$\tau_2 = \tau_1 + R_2(C_2 + C_3 + C_4 + C_5 + C_6 + C_7 + C_8) \tag{24.14}$$

$$\tau_5 = \tau_2 + R_5(C_5 + C_6 + C_7 + C_8) \tag{24.15}$$

$$\tau_6 = \tau_5 + R_6(C_6 + C_7) \tag{24.16}$$

$$\tau_7 = \tau_6 + R_7C_7 \tag{24.17}$$

Note that the delay values shown in Eqs. 24.11 and 24.17 are the same. It becomes apparant when we write Eq. 24.11 by keeping all the terms containing the same resistor together, as shown below:

$$\begin{aligned}\tau_7 = R_1(C_1 &+ C_2 + C_3 + C_4 + C_5 + C_6 + C_7 + C_8) \\ &+ R_2(C_2 + C_3 + C_4 + C_5 + C_6 + C_7 + C_8) \\ &+ R_5(C_5 + C_6 + C_7 + C_8) \\ &+ R_6(C_6 + C_7) \\ &+ R_7C_7\end{aligned} \tag{24.18}$$

The recursive definition of Elmore delay sometimes helps in easy algorithmic implementation.

Time constants in an RC tree: An RC tree is a nontrivial linear circuit. We need to solve a set of differential equations to compute its response. For a step input, the output response increases monotonically. However, with a large number of resistors and capacitors, the output response contains several *time constants*. We can compute the response of an RC tree accurately using a circuit simulator such as SPICE. However, given millions of interconnects in a practical design, SPICE-based simulation becomes infeasible. Therefore, we need to make approximations in computing the output response that allows efficient computation and delivers reasonable accuracy.

The Elmore delay approximates the output response with a *single dominant time constant.* The popularity of the Elmore delay model lies in its computational efficiency. We can compute the Elmore delay for an RC tree with n nodes in two $O(n)$ traversals [17, 20]. However, the Elmore delay can be inaccurate in some situations. We illustrate one such case in the following example.

Example 24.6 Consider the RC tree shown in Figure 24.17. It is obtained from the pi-model of Figure 24.13 by combining parallel capacitors.

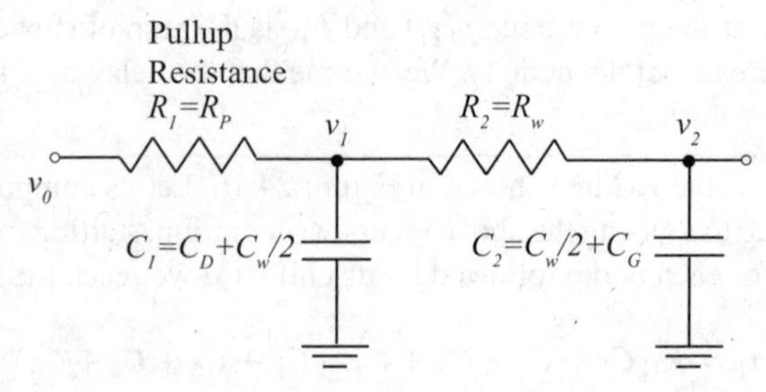

Figure 24.17 The RC tree in reduced form for the interconnect shown in Figure 24.13

The Elmore delay at the node v_1 is:

$$\tau_1 = R_1(C_1 + C_2) \tag{24.19}$$

As explained earlier, when R_2 is very large, $\tau_1 \approx R_1 C_1$ due to *resistive shielding effect*. However, Elmore delay of Eq. 24.19 will consider the effect of C_2 even for large R_2. Thus, the Elmore delay cannot capture the shielding effect, and it can give inaccurate delay for the near-end nodes (i.e., nodes closer to the source) [21]. Therefore, we need to consider alternative delay models to account for more time constants for an RC tree.

Asymptotic waveform evaluation: A generalized technique that approximates the output response of a linear RLC circuit [i.e., circuit consisting of resistor (R), inductor (L), and capacitor (C)] is *asymptotic waveform evaluation* (AWE) [23]. Consider an Nth-order RLC circuit, where N is the number of storage elements (i.e., capacitors or inductors). The AWE-based computation approximates the output response by employing the q most dominant time constants in its impulse response.

In practice, an interconnect model consisting of hundreds of storage elements can have hundreds of time constants. Fortunately, the response is dominated by only a few time constants, and $q \leq 5$ is typically good enough for the AWE model [16]. Furthermore, we can exploit the regularity of an interconnect model to improve the efficiency of the AWE-based delay computation [24]. Nevertheless, due to the computational overhead and the lack of complete interconnect information, we use AWE-based delay models only toward the end of a physical design flow when more accuracy is needed.

Distributed RC and RLC Lines

In reality, the parasitics of an interconnect are distributed along its length. For long interconnects on lower metal layers, we can ignore the inductance and employ a distributed RC model.

Distributed RC model: Let us consider an interconnect of length L whose resistance and capacitance is uniformly distributed, as shown in Figure 24.18(a). Let us represent the resistance per unit length as r and the capacitance per unit length as c. We can represent the voltage v at a distance of x from the source by the following differential equation [15]:

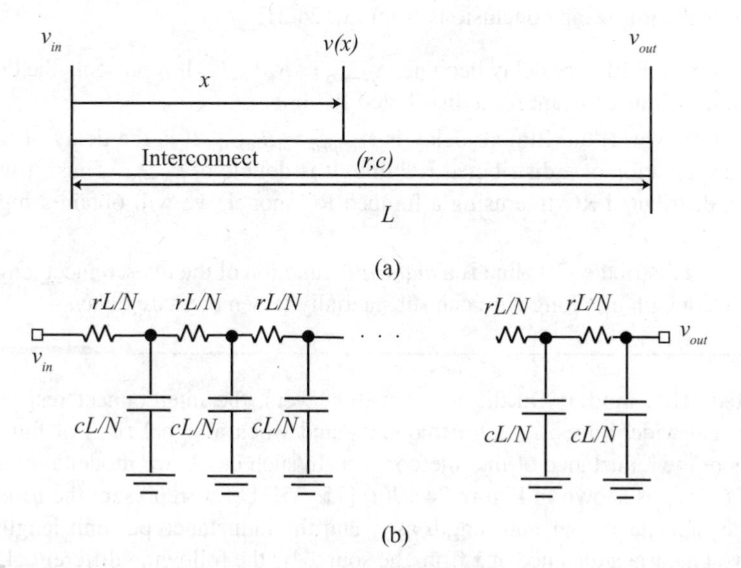

Figure 24.18 RC distributed line (a) an interconnect of length L (resistance per unit length = r and capacitance per unit length = c) and (b) approximation using an N-segment RC-ladder network

$$rc\frac{\partial v}{\partial t} = \frac{\partial^2 v}{\partial x^2} \tag{24.20}$$

There is no closed-form solution for this equation, and we need to solve it numerically. Nevertheless, we can approximate the distributed RC line with an RC-ladder network and estimate the delay using the Elmore delay model. We explain it in the following example.

Example 24.7 Consider a distributed RC line with length L, the resistance per unit length r, and capacitance per unit length c. We divide it into N equal segments. Thus, the total resistance $R_w = rL$ and the total capacitance $C_w = cL$ get equally divided to each segment. The resistance and capacitance of each segment are $R_{segment} = rL/N$ and $C_{segment} = cL/N$, respectively. Thus, we obtain the RC-ladder network as shown in Figure 24.18(b).

We can compute the Elmore delay of the signal as follows:

$$\begin{aligned}
\tau_{v_{out}} &= \frac{cL}{N}.\frac{rL}{N} + \frac{cL}{N}.\frac{2rL}{N} + ... + \frac{cL}{N}.\frac{NrL}{N} \\
&= \frac{rcL^2}{N^2}(1 + 2 + 3...N) \\
&= \frac{rcL^2}{N^2}\frac{N(N+1)}{2} \\
&= \frac{rcL^2(N+1)}{2N} \\
&= \frac{R_w C_w (N+1)}{2N}
\end{aligned} \tag{24.21}$$

We can draw the following conclusions from Eq. 24.21:

1. When $N \rightarrow \infty$, the Elmore delay becomes $\tau_{N\rightarrow\infty} \approx R_w C_w/2$. It represents the delay due to the most dominant time constant for a distributed RC line.
2. When we take $N = 1$, the Elmore delay is $\tau_{lumped} = R_w C_w$. It is the delay of the lumped RC model representation of a distributed RC line. It is double of $\tau_{N\rightarrow\infty}$. Thus, if we estimate the delay of a distributed RC line using a lumped RC model, we will obtain a highly erroneous result.
3. The delay of a distributed RC line is a *quadratic* function of the interconnect length. Therefore, reducing the length of a long wire can substantially lessen the wire delay.

Distributed RLC model: On the upper metal layers, the interconnect resistance decreases due to thicker and wider wires. At high frequency and when a signal rises or falls very quickly, we cannot ignore the inductance of the interconnect. In such cases, we model an interconnect as a *distributed RLC line*, as shown in Figure 24.19(a) [15, 25]. Let us represent the resistance per unit length as r, the capacitance per unit length as c, and the inductance per unit length as l. We can represent the voltage v at a distance of x from the source by the following differential equation [15]:

$$\frac{\partial^2 v}{\partial x^2} = rc\frac{\partial v}{\partial t} + lc\frac{\partial^2 v}{\partial t^2} \tag{24.22}$$

For a short wire, we can consider the line to be lossless ($r = 0$), and the equation reduces to:

$$\frac{\partial^2 v}{\partial x^2} = lc\frac{\partial^2 v}{\partial t^2} \tag{24.23}$$

The above equation is an *ideal wave equation* and suggests that the signal propagates similar to a *wave* in this case.

We can approximate a distributed RLC line also as an RLC-ladder network, as shown in Figure 24.19(b) [25]. However, we cannot employ the Elmore delay model because of the inductance in the network. Nevertheless, we can analyze it using the AWE-based method or SPICE simulation.

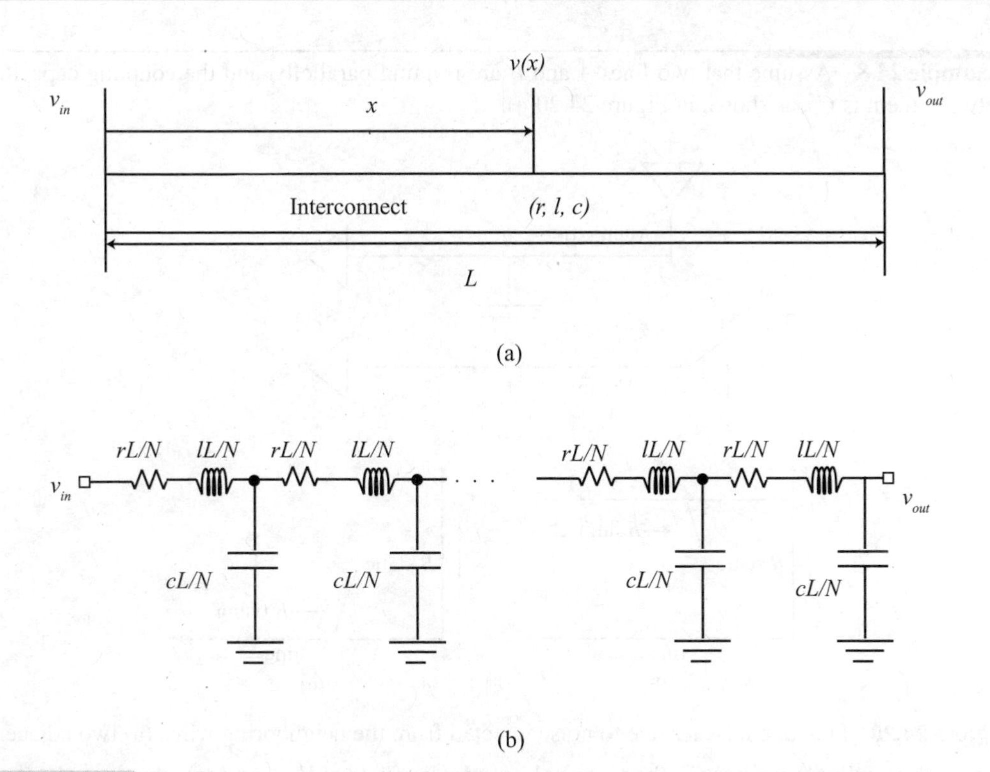

Figure 24.19 RLC distributed line (a) an interconnect of length L (resistance per unit length = r, inductance per unit length = l, and capacitance per unit length = c) and (b) approximation using an N-segment RLC-ladder network

24.3 SIGNAL INTEGRITY

We have discussed that there can be coupling capacitance between two neighboring interconnects due to the overlap and fringe field interactions. As a result, the voltage in one interconnect can impact the voltage in the other. These interactions among neighboring interconnects can create *signal integrity problems* in a circuit. The signal integrity problems get manifested in two forms: *functional issues* and *dynamic delay variations* [26–28]. We explain these problems in the following sections.

24.3.1 Functional Issues

Due to signals interacting through a coupling capacitor, a signal at logic 0 can be interpreted as logic 1 by the gates in its fanout in some situations. Similarly, a 1 can be interpreted as 0. Thus, we can encounter functional failure in a circuit if we do not adequately tackle the problem of signal integrity, as illustrated in the following example.

Example 24.8 Assume that two lines A and B are running parallelly, and the coupling capacitor between them is C_c, as shown in Figure 24.20(a).

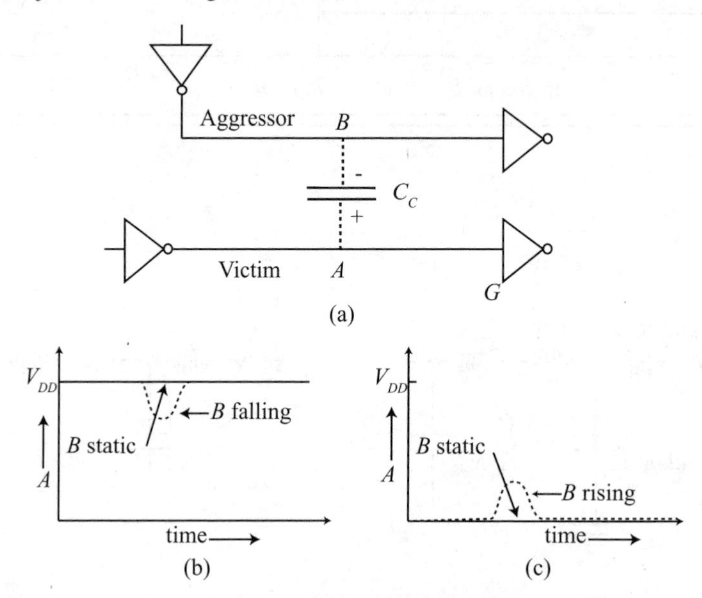

Figure 24.20 Functional issues due to noise injected from the neighboring wires (a) two adjacent lines with coupling capacitor C_c, (b) dip in voltage of a line held at V_{DD}, and (c) bump in voltage of a line held at 0 V [29]

The signal line at which we observe the signal integrity issue is known as the *victim* line (A in this case). Assume that the line A is at a steady state of V_{DD}, as shown in Figure 24.20(b). Assume that the neighboring line B is also at the steady state of V_{DD}. Therefore, the voltage across the coupling capacitor C_c is 0 V.

Next, let the neighboring line B make a quick transition from V_{DD} to 0 V. The capacitor C_c will get charged to V_{DD} due to the difference in the voltage of A and B lines. The driver of the line A will supply this charge, and it will cause a temporary dip in the voltage of the line A, as illustrated in Figure 24.20(b). The line B, which causes the dip in the voltage of the victim line, is known as the *aggressor* line.

Similarly, when the aggressor (line B) makes a 0 to V_{DD} transition, there can be a bump in the voltage of the victim (line A) that was held at logic 0, as illustrated in Figure 24.20(c).

If the glitches or the spikes in the victim line crosses the threshold (i.e., the noise margin) of the driven gate G, the gate G can interpret and propagate a wrong value in its fanout. Thus, coupling of signals can lead to a functional failure in a circuit.

24.3.2 Dynamic Delay Variations

When the victim line and the aggressor line are both switching, the voltage across the coupling capacitor changes dynamically. Consequently, the waveform on the victim line distorts, and the delay becomes a function of the neighboring line switching. The following example illustrates this phenomenon.

Example 24.9 Consider the lines A and B with a coupling capacitor C_c between them, as shown in Figure 24.21(a).

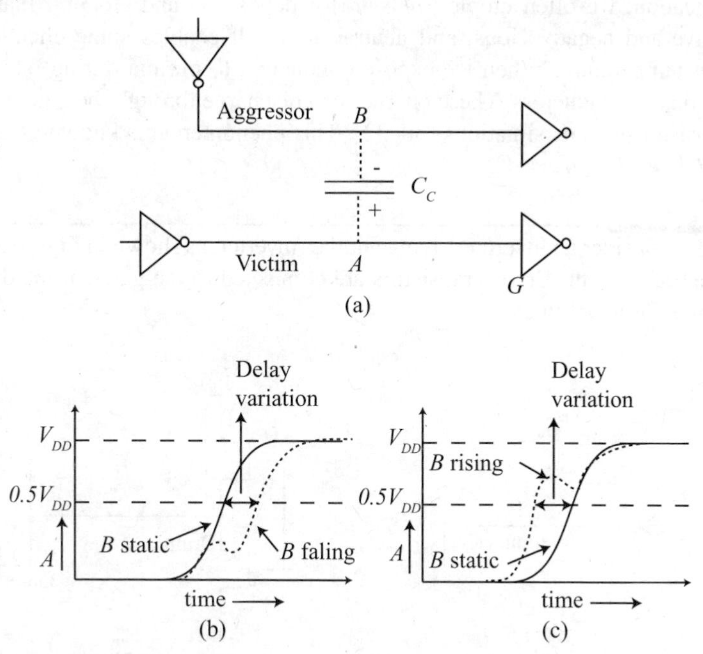

Figure 24.21 Dynamic delay variations (a) two adjacent lines with coupling capacitor C_c, (b) increase in delay due to aggressor switching in the opposite direction, and (c) decrease in delay due to aggressor switching in the same direction [29]

Let the victim line A be at 0 V and the aggressor line B be at V_{DD}. Therefore, the capacitor C_c will be at $-V_{DD}$.

Next, let the victim transition from $0\ V \rightarrow V_{DD}$ and the aggressor transition from $V_{DD} \rightarrow 0\ V$. Therefore, after this transition, the coupling capacitor is charged to $+V_{DD}$. Hence, transitions in lines A and B reverse the voltage across the capacitor C_c. This reversal in voltage requires extra charge being supplied/depleted by the victim during the transition. Therefore, the delay of the victim line A will be impacted by the transition in the aggressor line B. It increases when the aggressor switches in the opposite direction and decreases when it switches in the same direction, as illustrated in Figure 24.21(b) and (c), respectively.

Thus, the delay becomes a function of switching in the neighboring wires. This effect becomes prominent when the victim line has substantial coupling capacitance compared to the capacitance to the ground. Additionally, it becomes prominent when the aggressor line transitions quickly.

The signal integrity issues become important at advanced process nodes because the coupling capacitors between lines increase with the decrease in line spacing. Note that we can accurately analyze signal integrity problems only at the end of a design flow when all interconnects are laid out, and coupling capacitors can be extracted accurately from the layout. Nevertheless, we need to consider and avoid possible signal integrity issues throughout a design flow to achieve design closure quickly.

24.4 ANTENNA EFFECT

During IC fabrication, we often employ *plasma* for deposition and etching. Plasma is a gas of electrons, positive and negative ions, and neutral atoms. It enables some chemical reactions to occur at a lower temperature. When we expose conductors to plasma during fabrication, charges can accumulate on the conductors. These charges can discharge through the gate dielectric and can damage the transistor in some situations [30–32]. This phenomenon is known as *antenna effect* or *plasma-induced gate damage.*

Example 24.10 Consider an inverter driving another inverter, as shown in Figure 24.22(a) and (b). Assume that the drains of the driver transistors are connected to the gates of the driven transistors through a two-layer of metal lines.

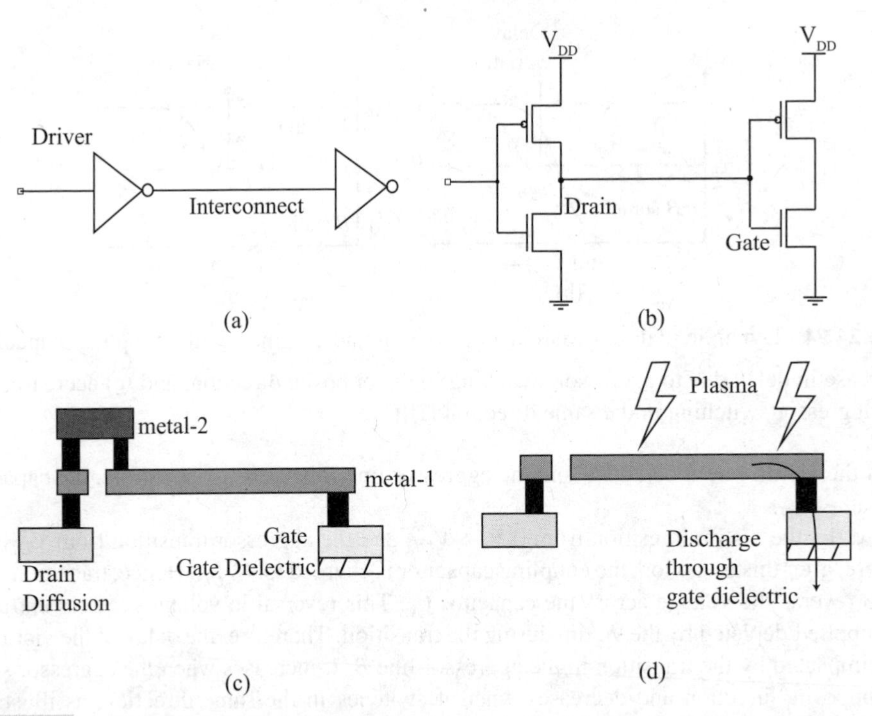

Figure 24.22 Antenna effect (a) an inverter driving another inverter through an interconnect, (b) transistor-level circuit, (c) interconnect to be fabricated in two layers, and (d) plamsa-induced gate damage when only metal-1 is laid

Assume that the drain of the driver and the gate of the driven transistor are directly connected to the metal-1 layer, as shown in Figure 24.22(c) (only one drain and one gate are shown for clarity). Note that the metal-1 line is not continuous but is connected through vias and the metal-2 line.

We fabricate interconnects layer by layer. Hence, the metal-1 layer will be fabricated first. Before fabricating the metal-2 layer, the gate of the driven transistor will be directly connected to the metal-1 layer. Additionally, it will be isolated from the drain diffusion layer due to the gap in the metal-1 layer, as shown in Figure 24.22(d). While fabricating the metal-1 layer, it can get exposed to plasma, and charges can accumulate on the metal lines. These charges will get neutralized by the tunneling current through the gate dielectric (since there is no other low resistance path). If a high tunneling current flows, the gate dielectric can get damaged. Thus, *antenna effect* can damage a gate in some situations.

The net charge collected on the metal lines and the magnitude of the tunneling current depend on the metal area exposed to the plasma. Furthermore, we expect a lower tunneling current if the same charge gets distributed over a large gate area or multiple gates. We can quantify the possibility of plasma-induced gate damage by *antenna ratio* (*AR*), which is defined as:

$$AR = \frac{A_{antenna}}{A_{gate}} \tag{24.24}$$

where $A_{antenna}$ is the area of the metal layer exposed to the plasma and A_{gate} is the area of the gate oxide connected to the exposed metal lines. A high antenna ratio indicates a greater chance of plasma-induced gate damage. For each layer, a constraint on antenna ratio is specified in the physical library as *antenna rules*. During designing, we need to ensure that there are no violations of these antenna rules.[5]

24.5 IMPACT OF PROCESS-INDUCED VARIATIONS

Due to process-induced variations, the attributes of a fabricated interconnect and the ILD layers can vary from their designed value. For example, there can be variations in the thickness and the width of the interconnect, spatial variations in the sheet resistance for a given interconnect layer, and variations in the properties of the ILD layer [33]. These variations impact the parasitics and the delay of a circuit. Hence, we need to account for them in a design flow.

RC corners: We extract parasitic information of interconnects using electronic design automation (EDA) tools. Typically, we extract, store, and share parasitic information among EDA tools using *standard parasitic exchange format* (SPEF) files. We can extract multiple SPEF files for different RC corners to account for process-induced variations. Note that an interconnect with an increased capacitance (due to an increase in the width) can have reduced resistance. Since delay is related to the product of resistance and capacitance, any one RC corner may not exhibit the worst or the best timing behavior. Consequently, we must perform analysis for multiple RC corners.

Multimode multi-corner analysis: In a design flow, we typically consider the impact of process-induced variations and environmental factors by analyzing a design at some discrete set of conditions using a technique called *multimode multi-corner* (MMMC) analysis. We model different process, voltage, and temperature (PVT) conditions using separate technology libraries. We model different modes of a design, such as functional mode, sleep mode, shift mode, and capture mode, using separate constraint (Synopsys Design Constraint, SDC) files. In these SDC files, clock frequencies and values of certain controlling signals typically differ. We create multiple *scenarios* by taking a combination of various PVT corner-based technology libraries, mode-based constraints files, and RC corner-based SPEF files, as illustrated in Figure 24.23 [29]. We can analyze these multiple scenarios simultaneously using MMMC analysis. We make MMMC analysis efficient by avoiding computation of the dominated scenarios and exploiting parallel processing.

[5] We will discuss the techniques to fix antenna rule violations in Chapter 28 ("Routing").

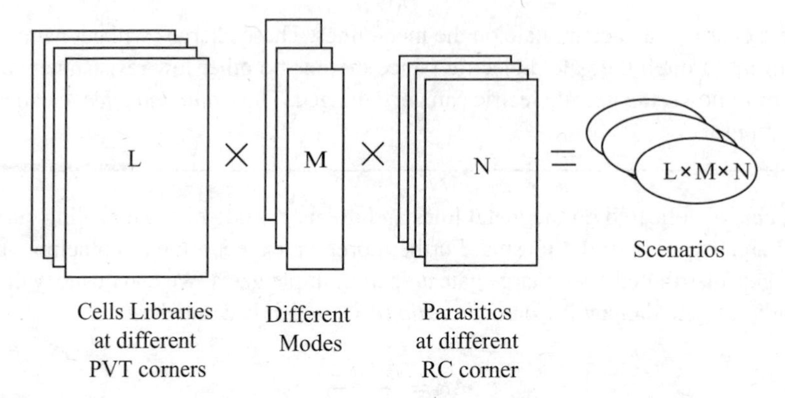

Figure 24.23 Creation of scenarios in the MMMC analysis by combining the PVT corners, modes, and RC corners for a design [29]

Example 24.11 Consider that we need to analyze for 3 PVT corners (Fast, Typical, Slow), 4 modes (sleep, normal, turbo, test), and 4 RC corners (C-worst, C-best, RC-worst, RC-min). We can create 3×4×4=48 scenarios. We perform MMMC analysis that can simultaneously analyze all 48 scenarios. It will be more efficient than analyzing 48 scenarios separately. While performing MMMC analysis, tools can avoid unnecessary computation, share information between scenarios, and exploit parallel processing efficiently.

24.6 LIBRARY EXCHANGE FORMAT

The physical design process needs to know process-dependent parameters such as the number of available layers, their sheet resistance, via dimensions, possible placement locations, and design rules. Additionally, it must know the layout details of the standard cells and macros that are instantiated in the netlist. Though physical design tools can adopt proprietary schemes of acquiring this information, a popular method of storing and retrieving this information is using *library exchange format* (LEF) files. A LEF file is a plain ASCII text file, and it defines elements of process technology and cell models [34].

For easier handling, we generally divide LEF files into two types:

1. **Technology LEF file:** A technology LEF file contains information about the available layers and vias, their properties, and placement sites. It includes sheet resistance and capacitance per unit square[6] for various layers, placement and routing design rules, and antenna rules.
2. **Cell LEF files:** A cell LEF file contains the physical information of the standard cells and macros that we can instantiate in a design. Note that the logic, timing, and power attributes of standard cells and macros are specified in the technology libraries.[7] A cell LEF file contains only abstract information of the cell layout relevant to the physical design, such as cell boundary, list of pins and their locations, and obstructions to placement and routing.

---

[6] Sheet resistance is defined in Ohms per square. Capacitance per unit square is defined in $pF/\mu m^2$ and it models wire-to-ground capacitance.

[7] For details, see Chapter 13 ("Technology library").

REVIEW QUESTIONS

24.1 What are the differences between the FEOL and the BEOL processes?

24.2 How do you carry out region-specific ion implantation?

24.3 Why do we carry out annealing after ion implantation?

24.4 Why is it important to isolate different devices on a silicon substrate? How is isolation typically achieved on a silicon wafer?

24.5 What is CMP? How is it performed?

24.6 What is CVD? How is it performed?

24.7 What is the purpose of the barrier layer in copper interconnects?

24.8 Why is it desirable to keep the dielectric constant of the ILD low?

24.9 What is the dual damascene process?

24.10 Why are copper interconnects not fabricated using conventional deposition and etching?

24.11 Why are interconnect layers typically represented as 2D polygons in the layout of a circuit?

24.12 What is sheet resistance? On what factors does a sheet resistance of an interconnect depend on?

24.13 Why do we typically route power lines and global interconnects on the higher layer of interconnect stack?

24.14 How does the constraint of employing Manhattan routing affects the wire length of a design?

24.15 What is skin effect in an interconnect? In which layers of interconnect do you expect to observe more skin effect?

24.16 What are fringe fields for an interconnect? Why does the capacitance due to fringe fields typically increase with scaling?

24.17 Why is the contribution of resistance to the wire delay increasing with scaling?

24.18 Under what situations can we use a lumped capacitance model?

24.19 How can overlap capacitance and lateral capacitance of an interconnect be reduced?

24.20 The interconnect delay of all the sink pins for a net is the same in the lumped capacitance model. Explain the reason for it.

24.21 Explain the shielding of capacitance by resistors in an RC tree. Explain how does shielding effect is not accounted for in the Elmore delay model.

24.22 In Figure 24.16 determine the Elmore delay from the source v_0 to the sink v_8.

24.23 What is AWE? Explain the trade-off between accuracy and the computational burden in the AWE method of delay computation.

24.24 Explain how we can approximate a distributed RC/RLC line using RC/RLC ladder network.

24.25 Explain how capacitive coupling between two interconnect lines can lead to functional failure in a design? Explain what we can do to alleviate this problem.

24.26 Explain how capacitive coupling between two interconnect lines can lead to delay dependence on the switching in the neighboring lines. How does this problem complicate the static timing analysis?

24.27 What are aggressor and victim lines regarding signal integrity in an integrated circuit? Are the signals on the aggressor lines also impacted by coupling capacitors?

24.28 What is an antenna effect and antenna ratio? What can happen if the antenna ratio for an interconnect exceeds the limit specified by the antenna rule?

24.29 For a given technology, the maximum allowable antenna ratio is 500. A given net has a fanout of 10. Each input pin in its fanout has a gate area of 0.05 μm^2. Calculate the maximum wiring area that we can use for routing this net without causing antenna rule violation.

24.30 What information does a LEF file contain?

24.31 What is MMMC analysis?

24.32 Why is it desirable to account for the behavior of interconnects during logic synthesis?

REFERENCES

[1] T. Standaert, G. Beique, H.-C. Chen, S.-T. Chen, B. Hamieh, J. Lee, P. McLaughlin, J. McMahon, Y. Mignot, F. Mont, et al. "BEOL process integration for the 7 nm technology node." *2016 IEEE International Interconnect Technology Conference/Advanced Metallization Conference (IITC/AMC)* (2016), pp. 2–4, IEEE.

[2] K. Maex, M. Baklanov, D. Shamiryan, F. Lacopi, S. Brongersma, and Z. S. Yanovitskaya. "Low dielectric constant materials for microelectronics." *Journal of Applied Physics* 93, no. 11 (2003), pp. 8793–8841.

[3] S. Natarajan, M. Agostinelli, S. Akbar, M. Bost, A. Bowonder, V. Chikarmane, S. Chouksey, A. Dasgupta, K. Fischer, Q. Fu, et al. "A 14 nm logic technology featuring 2nd-generation FinFET, air-gapped interconnects, self-aligned double patterning and a 0.0588 μm^2 SRAM cell size." *2014 IEEE International Electron Devices Meeting* (2014), pp. 3–7, IEEE.

[4] C. W. Kaanta, S. G. Bombardier, W. J. Cote, W. R. Hill, G. Kerszykowski, H. S. Landis, D. J. Poindexter, C. W. Pollard, G. H. Ross, J. G. Ryan, et al. "Dual damascene: A ULSI wiring technology." *1991 Proceedings Eighth International IEEE VLSI Multilevel Interconnection Conference* (1991), pp. 144–152, IEEE.

[5] J. Ao, S. Dong, S. Chen, and S. Goto. "Delay-driven layer assignment in global routing under multi-tier interconnect structure." *Proceedings of the 2013 ACM International Symposium on Physical Design* (2013), pp. 101–107.

[6] R. H. Havemann and J. A. Hutchby. "High-performance interconnects: An integration overview." *Proceedings of the IEEE* 89, no. 5 (2001), pp. 586–601.

[7] M. Gester, D. Müller, T. Nieberg, C. Panten, C. Schulte, and J. Vygen. "BonnRoute: Algorithms and data structures for fast and good VLSI routing." *ACM Transactions on Design Automation of Electronic Systems (TODAES)* 18, no. 2 (2013), pp. 1–24.

[8] M. Ahrens, M. Gester, N. Klewinghaus, D. Müller, S. Peyer, C. Schulte, and G. Tellez. "Detailed routing algorithms for advanced technology nodes." *IEEE Transactions on Computer-aided Design of Integrated Circuits and Systems* 34, no. 4 (2014), pp. 563–576.

[9] S. L. Teig. "The X architecture: Not your father's diagonal wiring." *Proceedings of the 2002 International Workshop on System-level Interconnect Prediction* (2002), pp. 33–37.

[10] Y. Xu, Y. Zhang, and C. Chu. "FastRoute 4.0: Global router with efficient via minimization." *2009 Asia and South Pacific Design Automation Conference* (2009), pp. 576–581, IEEE.

[11] B. Krauter and S. Mehrotra. "Layout based frequency dependent inductance and resistance extraction for on-chip interconnect timing analysis." *Proceedings of the 35th Annual Design Automation Conference* (1998), pp. 303–308.

[12] E. Barke. "Line-to-ground capacitance calculation for VLSI: A comparison." *IEEE Transactions on Computer-aided Design of Integrated Circuits and Systems* 7, no. 2 (1988), pp. 295–298.

[13] C. Yuan and T. N. Trick. "A simple formula for the estimation of the capacitance of two-dimensional interconnects in VLSI circuits." *IEEE Electron Device Letters* 3, no. 12 (1982), pp. 391–393.

[14] N. D. Arora, K. V. Raol, R. Schumann, and L. M. Richardson. "Modeling and extraction of interconnect capacitances for multilayer VLSI circuits." *IEEE Transactions on Computer-aided Design of Integrated Circuits and Systems* 15, no. 1 (1996), pp. 58–67.

[15] J. M. Rabaey, A. Chandrakasan, and B. Nikolic. "Digital integrated circuit design." *A Design Perspective*. Prentice Hall, 2nd ed., 2002.

[16] J. Cong, L. He, C.-K. Koh, and P. H. Madden. "Performance optimization of VLSI interconnect layout." *Integration* 21, no. 1–2 (1996), pp. 1–94.

[17] J. Rubinstein, P. Penfield, and M. A. Horowitz. "Signal delay in RC tree networks." *IEEE Transactions on Computer-aided Design of Integrated Circuits and Systems* 2, no. 3 (1983), pp. 202–211.

[18] A. Van Genderen and N. Van Der Meijs. "Extracting simple but accurate RC models for VLSI interconnect." *Proc. ISCAS* 88 (1988), pp. 2351–2354.

[19] L. Pileggi. "Timing metrics for physical design of deep submicron technologies." *Proceedings of the 1998 International Symposium on Physical Design* (1998), pp. 28–33.

[20] L. Pileggi. "Coping with RC(L) interconnect induced headaches." *Proceedings of the International Conference on Computer-aided Design* (1995), pp. 246–253.

[21] C. J. Alpert, A. Devgan, and C. V. Kashyap. "RC delay metrics for performance optimization." *IEEE Transactions on Computer-aided Design of Integrated Circuits and Systems* 20, no. 5 (2001), pp. 571–582.

[22] W. C. Elmore. "The transient response of damped linear networks with particular regard to wideband amplifiers." *Journal of Applied Physics* 19, no. 1 (1948), pp. 55–63.

[23] L. T. Pillage and R. A. Rohrer. "Asymptotic waveform evaluation for timing analysis." *IEEE Transactions on Computer-aided Design of Integrated Circuits and Systems* 9, no. 4 (1990), pp. 352–366.

[24] C. L. Ratzlaff and L. T. Pillage. "RICE: Rapid interconnect circuit evaluation using AWE." *IEEE Transactions on Computer-aided Design of Integrated Circuits and Systems* 13, no. 6 (1994), pp. 763–776.

[25] Q. Yu and E. S. Kuh. "Exact moment matching model of transmission lines and application to interconnect delay estimation." *IEEE Transactions on Very Large Scale Integration (VLSI) Systems* 3, no. 2 (1995), pp. 311–322.

[26] D. Sylvester and C. Wu. "Analytical modeling and characterization of deep-submicrometer interconnect." *Proceedings of the IEEE* 89, no. 5 (2001), pp. 634–664.

[27] R. Achar and M. S. Nakhla. "Simulation of high-speed interconnects." *Proceedings of the IEEE* 89, no. 5 (2001), pp. 693–728.

[28] L. Lavagno, I. L. Markov, G. Martin, and L. K. Scheffer. *Electronic Design Automation for IC Implementation, Circuit Design, and Process Technology: Circuit Design, and Process Technology*. CRC Press, 2016.

[29] S. Saurabh, H. Shah, and S. Singh. "Timing closure problem: Review of challenges at advanced process nodes and solutions." *IETE Technical Review* 36, no. 6 (2019), pp. 580–593.

[30] H. Shin, C.-C. King, T. Horiuchi, and C. Hu. "Thin oxide charging current during plasma etching of aluminum." *IEEE Electron Device Letters* 12, no. 8 (1991), pp. 404–406.

[31] S. Fang and J. P. McVittie. "Thin-oxide damage from gate charging during plasma processing." *IEEE Electron Device Letters* 13, no. 5 (1992), pp. 288–290.

[32] Y.-P. Tsai, P.-C. Liou, C. J. Lin, and Y.-C. King. "Plasma charging effect on the reliability of copper BEOL structures in advanced FinFET technologies." *IEEE Journal of the Electron Devices Society* 6 (2018), pp. 875–883.

[33] S. Natarajan, M. A. Breuer, and S. K. Gupta. "Process variations and their impact on circuit operation." *Proceedings 1998 IEEE International Symposium on Defect and Fault Tolerance in VLSI Systems (Cat. No. 98EX223)* (1998), pp. 73–81, IEEE.

[34] Cadence. "LEF/DEF Language Reference, Product Version 5.7, November 2009." http://www.ispd.cc/contests/18/lefdefref.pdf.

25 Chip Planning

While times are quiet, it is easy to take action; ere coming troubles have cast their shadows, it is easy to lay plans ... A journey of a thousand miles began with a single step.

—Lao Tzu, *The Sayings of Lao Tzu* (translated by Lionel Giles, *The Wisdom of the East*, New York: E.P. Dutton and Co, 1904)

The physical design process starts with *chip planning*. As the name suggests, it is the planning phase for creating the layout for a chip. The decisions taken during the chip planning phase strongly impact its final quality of result (QoR). Note that we have a greater degree of freedom in the early stages of physical design. We exploit it during chip planning to make crucial decisions regarding physical design implementation.

The physical design implementation strongly depends on the size of the design and the methodologies adopted. For implementing large designs, we often decompose it into smaller *sub-systems* or *blocks*. This process is known as *partitioning*.

After partitioning, we determine the location of each block and macros. Some blocks can have movable pins. We carry out *pin assignment* to identify their positions on the block. Additionally, we define the location of input and output pads. We also allocate adequate space for creating rows of standard cells. We refer to these tasks collectively as *floorplanning*.

We also decide how to deliver power to all the circuit elements during chip planning. This task is known as *power planning*.

We will discuss partitioning, floorplanning, and power planning in this chapter.

25.1 PHYSICAL DESIGN METHODOLOGIES

25.1.1 Flat Design Implementation

For small or medium size designs, we do not require partitioning. In such cases, the complete design can be *flattened*. During flattening, the hierarchy of design instances gets dissolved, while the library cell instances, including standard cells, are left intact. Subsequently, we can carry out the physical implementation for the flattened design in one go. It is known as *flat design implementation methodology*. However, this methodology becomes challenging for both a designer and the electronic design automation (EDA) tools as the design size increases.

25.1.2 Hierarchical Design Implementation

For large designs, we prefer the *divide-and-conquer* approach. We divide a large design into smaller blocks by partitioning. We can partition blocks recursively such that the leaf-level blocks become small enough to be efficiently managed. An example of partitioning is shown in Figure 25.1. We push various top-level constraints related to timing and physical constraints such as routing and placement blockages to each block. We also push information related to design for testability (DFT), such as the start and the endpoint of the scan chains, to the respective blocks. After that, we implement the leaf-level blocks individually. Then, we combine them hierarchically till we reach the top level. This design methodology is known as *hierarchical design methodology* and is quite popular.

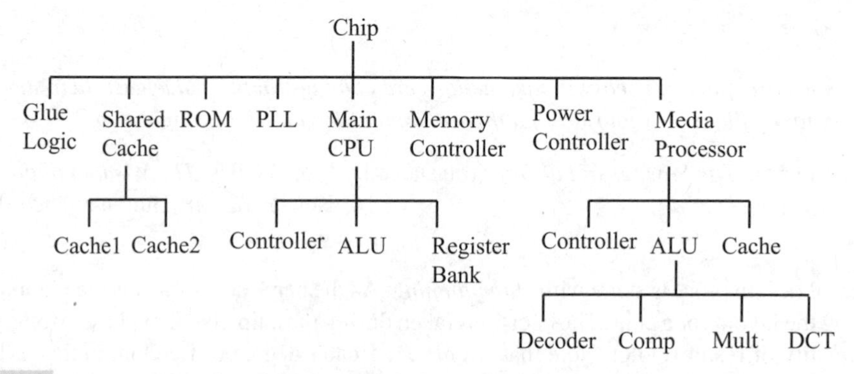

Figure 25.1 An example of partitioning a chip into blocks recursively

Advantages: Hierarchical design methodology has several advantages. In this methodology, we implement each block independently. Since blocks are smaller than the original design, partitioning simplifies the problem for the physical design tools. Thus, we can employ even those tools that cannot handle the complete design in one go. Moreover, different blocks can be designed concurrently by separate individuals or teams. It reduces the overall design time of a chip. However, it is challenging to partition a design *optimally*. Additionally, we can miss some opportunities for inter-block optimizations in the hierarchical design methodology.

Verification challenges: Once we have implemented and verified all the blocks separately, we need to assemble them at the top level and verify. Thus, we need to carry out both block- and top-level verification. We have discussed the challenges in carrying out static timing analysis (STA) in hierarchical implementation methodologies in Section 15.10 ("Constraints for hierarchical design flow"). We need to allocate *timing budget* for each block and ensure that blocks meet their timing budgets. Subsequently, we assemble all the blocks at the top level and carry out STA at the top level. For the top-level STA, we can omit the details of the timing paths *contained* entirely within a block (e.g., a flip-flop to flip-flop timing path fully contained within a block). It will make the top-level STA fast and efficient. However, we still need to analyze the block's timing paths that interact with other blocks or the top. These paths are known as *interface timing paths* of the blocks. Typically, we create an *abstract view* of the blocks for the top-level STA for verifying interface timing paths.

We can create an abstract view of a block in multiple ways. Some of the popular abstract views are *extracted timing model* (ETM) and *interface logic model* (ILM) [1–4]. The ETM and ILM provide a compact block representation and enable efficient STA at the top level. An ETM represents interface timing paths in a technology library format (.lib format). In contrast, an ILM extracts the information of interface timing paths and models them using logic gates.

25.2 PARTITIONING

We can partition a design based on *functionality*. For example, we can partition a design into memory blocks, control units, arithmetic logic units (ALUs), clock generators, power controllers, and network adapters. Alternatively, we can employ the *modules* or hierarchy used in logic synthesis to create partitions. We can group related modules to form a block. If we do not have distinct modules and hierarchies in a design, we can partition the netlist to create various blocks.

During partitioning, we *minimize* the number of interconnections between blocks to simplify routing between blocks and reduce delay of inter-block signals. Additionally, minimizing interconnections between blocks eases physical design task for individual blocks.

25.2.1 Min-cut Partitioning

The task of reducing the number of interconnections between blocks during partitioning can be described by the well-known *min-cut problem* for a graph. Consider a netlist represented as an undirected graph $G = (V, E)$ where V is the vertex set, and E is the edge set.[1]

Let us partition V into two disjoint subsets V_1 and $V_2 = (V - V_1)$ of roughly equal size ($|V_1| \leq \lceil |V|/2 \rceil$ and $|V_2| \leq \lceil |V|/2 \rceil$).[2] When $|V|$ is even then two subsets are of equal size, else they differ by 1. The *edge cut set* for this partition is represented by a subset of edges $E_c \subset E$ such that $E_c = \{(v_{1i}, v_{2j}) : v_{1i} \in V_1, v_{2j} \in V_2\}$, i.e., an edge in E_c has one vertex taken from each partition. A partition represented by the edge cut set E_c is the *min-cut partition* if the number of elements in E_c is minimum.

Example 25.1 Consider the graph shown in Figure 25.2(a). We can partition it in many ways.

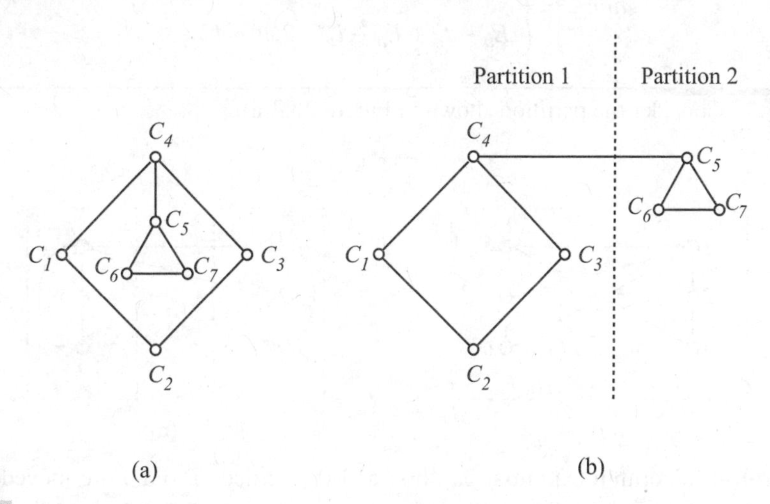

Figure 25.2 Patitioning (a) a given graph and (b) min-cut partition

[1] For the graph representation of a netlist, see Section 16.2 ("Mapping techniques").

[2] We denote the number of elements in a set A as $|A|$.

For example, we can partition as $V_1 = \{C_1, C_2, C_3\}$ and $V_2 = \{C_4, C_5, C_6, C_7\}$. The cut set is $E_c = \{(C_1, C_4), (C_3, C_4)\}$. Thus, the cut size is two (i.e., $|E_c|=2$). It represents the connections or edges crossing these partitions. Another partition is $V_1 = \{C_1, C_2, C_3, C_4\}$ and $V_2 = \{C_5, C_6, C_7\}$. It is shown in Figure 25.2(b). The cut set is $E_c = \{(C_4, C_5)\}$. In this partition, only one edge crosses the partition. We can observe that all other partitions of the graph have more edges crossing the partition. Therefore, the partition shown in Figure 25.2(b) is the min-cut partition for the graph.

There are other formulations of the min-cut partition problem. For example, we can partition a graph into k roughly equal-sized partitions. The min-cut problem is an NP-complete problem [5]. However, there are many efficient heuristics to solve it.

25.2.2 Kernighan–Lin Algorithm

For illustration, let us consider Kernighan–Lin (K–L) algorithm, a popular algorithm for partitioning [6, 7]. Suppose a given graph $G = (V, E)$ is partitioned into two equal partitions $A \subset V$ and $B \subset V$. For each element $a \in A$, we define an internal cost I_a and an external cost E_a. The internal cost I_a is equal to the number of edges for the vertex a that is fully contained within the partition A. The external cost E_a is the number of edges of the vertex a that cross the partition A.

Similarly, we define internal cost I_b and external cost E_b for each vertex $b \in B$.

We can compute the reduction in the number of interconnections obtained after swapping a pair of vertices $a \in A$ and $b \in B$, as follows (by swapping, we mean moving a to B and b to A):

$$gain_{a,b} = \begin{cases} E_a - I_a + E_b - I_b, & \text{if}(a,b) \notin E \\ E_a - I_a + E_b - I_b - 2, & \text{if}(a,b) \in E \end{cases} \tag{25.1}$$

Example 25.2 Consider the partition shown in Figure 25.3(a).

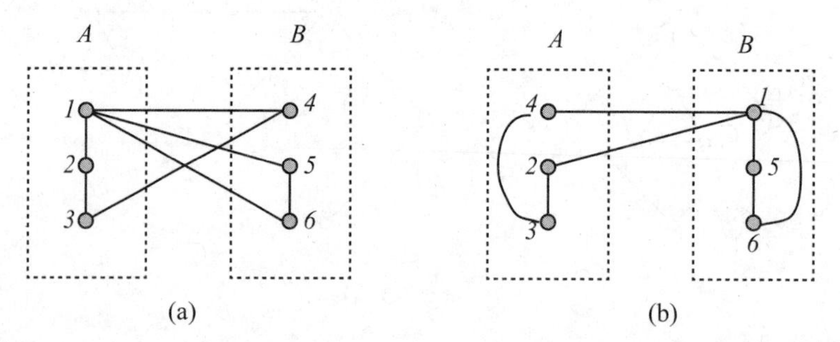

Figure 25.3 K–L algorithm (a) initial partition and (b) vertices *1* and *4* are moved to different partitions

For the vertex *1*, internal cost I_1=1 and external cost E_1=3.
For the vertex *4*, internal cost I_4=0 and external E_4=2.
After swapping the partition of vertices *1* and *4*, we obtain the partitions as shown in Figure 25.3(b). The number of connections between partitions reduced from 4 to 2 by this swapping. The computed

gain is (using the second formula of Eq. 25.1 since $(1, 4) \in E$): $gain_{1,4} = 3 - 1 + 2 - 0 - 2 = 2$. Thus, $gain_{1,4}$=2 represents the decrease in the edge cut size due to moving two vertices across the partition.

Algorithm 25.1 K–L_PARTITIONING

Input: Undirected graph $G = (V, E)$
Output: Two min-cut partitions A and B
Steps:

```
1: Initialize partitions of equal sizes A and B
2: Improvement ← true
3:
4: while (Improvement = true) do
5:     Improvement ← false
6:     Mark all vertices as unlocked
7:     gain-at-k[0] ← 0
8:     k-with-best-gain ← 0
9:
10:    for k ← 1 to ⌊|V|/2⌋ do
11:        Among all unlocked vertices, find pair of vertices a ∈ A and b ∈ B that
       gives the maximum gain_{a,b}
12:        a-move-at-k[k] ← a, b-move-at-k[k] ← b
13:        Mark a and b as locked
14:        gain-at-k[k] ← gain_{a,b} + gain-at-k[k − 1]
15:        if (gain-at-k[k] > gain-at-k[k-with-best-gain]) then
16:            k-with-best-gain ← k
17:        end if
18:    end for
19:
20:    if (k-with-best-gain > 0) then
21:        Improvement ← true
22:        for i ← 1 to k-with-best-gain do
23:            Move a-move-at-k[i] to partition B
24:            Move b-move-at-k[i] to partition A
25:        end for
26:    end if
27: end while
```

The K–L algorithm is shown in Algo 25.1.

Overall algorithm: The algorithm starts with an arbitrary equal-sized partitions A and B (line: 1) and iteratively reduces the number of interconnections by swapping vertices between these partitions until no more improvement is possible (lines: 4–27).

Identifying *k*-best swaps: The algorithm greedily identifies a pair of vertices $a \in A$ and $b \in B$ for which maximum $gain_{a,b}$ can be obtained (line: 11). This step requires exploring the gains between each pair of vertices in both the partitions.

After identifying a and b, it does not immediately swap them. It simply records the pair of vertices that exhibit the best gain in the array *a-move-at-k* and *b-move-at-k* (line: 12). It locks those two vertices and prohibits them from participating in the subsequent swaps in the same pass (line: 13). Additionally, while finding the subsequent pairs with a maximum $gain_{a,b}$, the locked vertices are *considered* to be moved to the other partition without actually moving them.

Avoiding local minima: When k pairs are swapped, the total of all gains in k-swaps are recorded in the array *gain-at-k*[k] (line: 14). Note that the maximum $gain_{a,b}$ can also be negative. However, the algorithm continues identifying successive best possible swaps hoping that the successive swaps can give sufficient positive $gain_{a,b}$ and we can compensate the intermediate negative gains. Thus, the K–L algorithm avoids being stuck in a local minimum, hoping to reach the global minimum finally.

Termination criteria: The algorithm keeps track of the k number of swaps that yielded the maximum sum of gains in the variable *k-with-best-gain* (line: 16). After the algorithm has explored all the $\lfloor |V|/2 \rfloor$ swaps (lines: 10–18), it swaps *k-with-best-gain* vertices if the total gain is positive (lines: 20–26). If the total gain is not positive, we cannot improve further, and the algorithm terminates.

Example 25.3 Consider the partition shown in Figure 25.3(a) as a starting partition. We illustrate the working of K–L algorithm in Table 25.1.

Table 25.1 K–L algorithm operating for the graph shown in Figure 25.3(a)

Pass	k	*a-move-at-k*[k]	*b-move-at-k*[k]	*gain-at-k*[k]	*k-with-best-gain*
I	1	1	4	2	1
I	2	2	5	0	1
I	3	3	6	0	1
II	1	4	1	–2	0
II	2	3	5	–2	0
II	3	2	6	0	0

In the first pass, for $k = 1$, we compute the gains for swapping each pair ($3 \times 3 = 9$ possibilities) and pick the best. The maximum gain is by swapping vertices *1* and *4*. We record this movement (first row in the table) and assume that *1* and *4* are swapped. We lock these two vertices. Similarly, for $k = 2$ and $k = 3$, we compute the gain for swapping each pair of still unlocked vertices and pick the best. After evaluating all moves, we pick k that yields the maximum gain. In this case, we pick $k = 1$. Now, we perform the k number of swaps (swap vertices 1 and 4). We obtain the partition shown in Figure 25.3(b).

We unlock all the vertices in the second pass and repeat the same process as before. However, we do not find any k for which swapping yields positive gain. Consequently, the algorithm terminates. Thus, the K–L algorithm produces the partition shown in Figure 25.3(b).

In practice, the K–L algorithm is quite robust. However, during partitioning, we should consider other metrics also, such as the critical path delay. It is often desirable that the complete critical path is within the same partition. The K–L algorithm can handle such requirements [8]. However, the algorithmic complexity is $O(n^3)$, which may not be acceptable for large designs. Therefore, modifications of the K–L algorithm and other partitioning algorithms have been proposed [9–13].

It is worth pointing out that, though partitioning simplifies the task of physical design, it can sacrifice some QoR. Intuitively, partitioning can prohibit inter-partition optimizations and result in QoR loss. Nevertheless, we prefer the top-down hierarchical design methodology for large designs and tolerate some QoR loss.

25.3 FLOORPLANNING

After partitioning, we undertake the layout planning. We refer to the tasks involved in layout planning collectively as *floorplanning*. We describe the essential elements of floorplanning in the following paragraphs.

25.3.1 Die Size

To create a layout, first, we should define the die size and its aspect ratio. Ideally, the die size should be as small as possible. With a smaller die size, we can fabricate more chips for a given wafer area. Additionally, we can achieve a higher yield for a smaller die.[3] Thus, by reducing the die size, we can reduce the cost of a chip. However, if we choose a too small die size, all the required functionality or circuit components cannot fit into it. A physical design tool, such as a router, will find it tough to produce a feasible solution for a small die. In the worst case, a router can fail to route if the die is too small. Therefore, we should decide the die size by evaluating the trade-offs between the chip cost and the difficulty of finding a feasible solution.

The target *package* for a chip also puts a constraint on the die size and its aspect ratio. Nevertheless, to implement a given functionality, the die should be able to accommodate the following entities [14]:

1. **Blocks:** We need to allocate space on the layout for large instances or blocks or macros. Large instances can be blocks obtained after partitioning or functional blocks such as processors, memories, and analog blocks. We can estimate the total area needed for a block by adding the area of instances within it. Sometimes, we also need to allocate *halos* or empty regions around blocks, as shown in Figure 25.4. Halos can alleviate congestion and routing problems in certain situations and are explained later in this chapter.

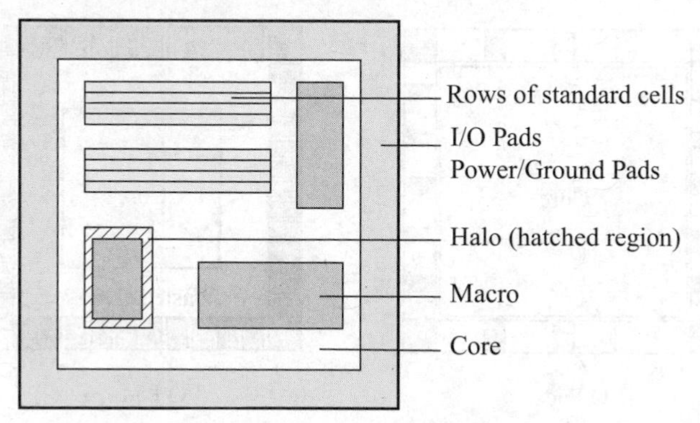

Figure 25.4 A simple floorplan on a die showing various components

[3] For yield–area relationship, see Section 6.2 ("Yield").

2. **Standard cells:** We need to place standard cells on a die. Therefore, we allocate dedicated standard cell rows, as shown in Figure 25.4. The area required for the standard cells can be estimated using the given netlist and the corresponding technology libraries.
3. **Pads:** The signals entering to and leaving from the chip require input/output (I/O) pads. Additionally, power must be delivered to the chip using power and ground pads. The die should have sufficient area to accommodate these pads, as shown in Figure 25.4.
4. **Interconnects:** We should make provisions for the interconnects in the layout. However, at the floorplanning stage, the layout of the interconnects is not yet decided. Therefore, it is challenging to estimate the area we must allocate for the interconnects on the die.

   During physical design, we can account for interconnects using a parameter known as *utilization*. The utilization is defined as:

$$\text{utilization} = \frac{\text{block/macro area} + \text{halo area} + \text{cell area}}{\text{core area}} \tag{25.2}$$

   The *core area* is the die area, excluding the area of the I/O pads and power pads. The above utilization is also known as *core utilization*. Ideally, utilization should be 1, implying that no extra space is allocated for the interconnects, minimizing the die size. In such cases, interconnects can occupy only the additional layers over the cells and devices. Therefore, routing can become infeasible. Hence, we should choose a lower utilization. However, a too low utilization value can waste a large portion of a die and is undesirable. We can choose a good utilization value for a design by using our past experience of working on a similar kind of design (similar in terms of netlist size and functionality). Typically, we target a moderate value for utilization, such as 0.6–0.8, in physical design flows.

An important consideration in deciding the size for a die is whether a design is *core-* or *pad-limited*. A design in which the core area dominates over the area taken by the I/O pads (including pads related to power) is known as a core-limited design. A core-limited design is shown in Figure 25.5(a). To reduce the overall die size in such designs, we should reduce the core area. On the other hand, a design in which the I/O pad area dominates is known as a pad-limited design. In a pad-limited design, we should pack the I/O pads compactly to reduce the overall die size. In some cases, a pad-limited design can waste core area, as shown in Figure 25.5(b).

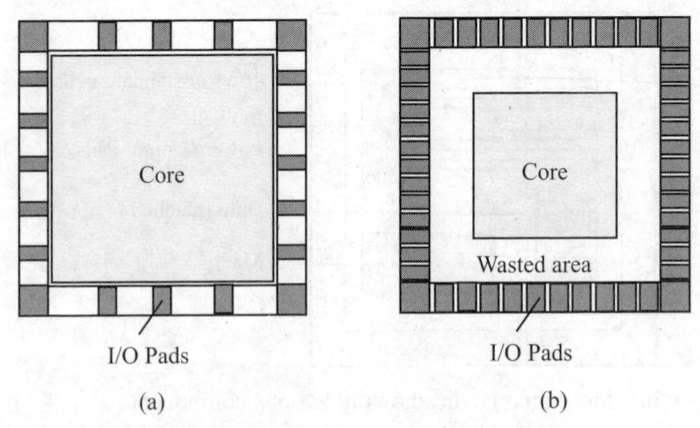

Figure 25.5 Types of design (a) core-limited and (b) pad-limited

25.3.2 Input/Output Cells

A chip communicates with the external world through special circuit elements known as *I/O cells*. An I/O cell can be of type input, output, or bidirectional. Besides providing external connection, an I/O cell can provide other functions, such as capabilities to drive external load, protection against voltage spikes, and voltage-level transformation. Depending on the chip functionality, external environment, and interface signals, we require different kinds of I/O cells. Therefore, we design I/O cells of various functionalities, sizes, voltage levels, drive capabilities, and protection levels. Subsequently, we characterize them and make them available in a technology library (similar to the standard cells for logic gates). We instantiate appropriate I/O cells in the netlist, place them, and create a layout for their connections during physical design.

Example 25.4 Figure 25.6(a) shows a simple bidirectional I/O cell. Let us examine its operation.

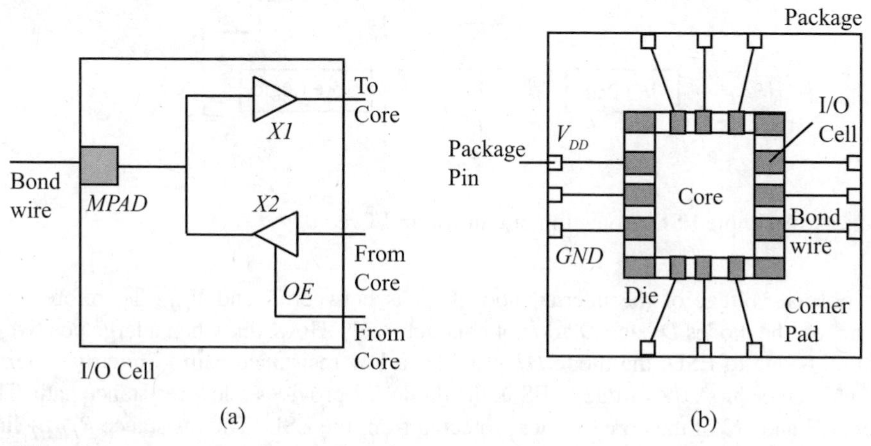

Figure 25.6 I/O cells (a) a bidirectional I/O cell and (b) connection with package pins

There is a metallic pad (*MPAD*) through which a signal enters or leaves. It can be connected to the *package pins* using bonding wires, as shown in Figure 25.6(b). The buffer *X1* is a normal buffer, while the buffer *X2* is a tristate buffer.

When *OE*=0, *X2* is in the high impedance state, and the cell behaves as an input I/O cell. In this case, a signal applied to the package pin passes through *MPAD* and the I/O cell before reaching the core cells.

When *OE*=1, *X2* is enabled, and the cell behaves as an output I/O cell. The signal provided by the core cell passes through the I/O cell and *MPAD* before appearing at the corresponding package pin.

Protection: An important function of an I/O cell is to protect a circuit. Short-duration voltage spikes of several kilovolts, known as *electro-static discharge* (ESD), can damage a circuit if we do not tackle them properly. An ESD occurs when the charge transfers between two oppositely charged bodies (for example, during handling of a chip by human or machine). Therefore, we embed the ESD protection circuit within an I/O cell. When ESD occurs, the ESD protection circuit allows discharge through an alternate path, and the voltage spike does not reach the core cells.

Example 25.5 We show a simple ESD protection circuit for an input pad in Figure 25.7. Let us examine its functionality.

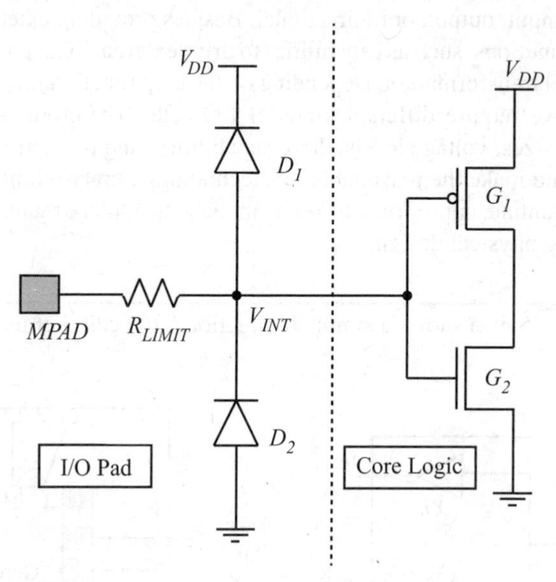

Figure 25.7 A simple ESD protection circuit for an I/O cell

The normal voltage of the internal node V_{INT} is between 0 and V_{DD}. Therefore, in normal circumstances, the diodes *D1* and *D2* do not conduct freely. However, when a large positive voltage occurs at V_{INT} due to ESD, the diode *D1* provides a low resistance path. Similarly, when a large negative voltage occurs at V_{INT} due to ESD, the diode *D2* provides a low resistance path. Thus, the transistors *G1* and *G2* of the core cell get protected from the ESD. The resistance R_{LIMIT} limits the current through the diodes when ESD occurs and protects the diodes from damage.

Remark: The ESD protection circuit shown in Figure 25.7 is simplistic. The design of ESD involves multiple challenges and trade-offs. Interested readers can refer to [15] to delve deeper into this topic.

Voltage transformation: An I/O cell can also transform voltage levels using internal *level shifter*. It becomes essential when the core of a chip and the external circuitry work at different voltage levels. Additionally, an output I/O cell contains an appropriately sized buffer (or a chain of buffers) capable of driving the external load. It is essential because the core cells are typically incapable of driving a heavy external load.

Power pads: There is a specific type of pad or cell, known as a *power pad* or a *power supply cell*, that supplies power to a circuit. In terms of functionality, a power pad provides just a connection to the external power sources using V_{DD} or *GND* lines. The power is not only required for the core cells but also for the I/O cells. However, the power pads that deliver power to the core cells are typically kept separate from those that provide power to the I/O cells. We separate them because I/O cells can draw large transient currents and have larger inductive and capacitive loads. Therefore, we observe more voltage variations in the supply rails connected to the I/O cells. The core cells cannot tolerate such variations in the supply voltage. Additionally, I/O cells often work at higher voltages than the core cells. Therefore, we keep power pads of the core cells and I/O cells separate.

We choose the number of power pads and their locations such that internal voltages are within a *target level* (i.e., it does not drop below a given threshold). Additionally, the current flowing through the pads should be within limits (as defined by the maximum current rating of the power pads) [16, 17]. Power is delivered over the entire die area using a *power delivery network* (PDN). We will describe the design of a PDN later in this chapter.

25.3.3 Bump Pads

The type of interconnection shown in Figure 25.6 is known as *wire bonding*. It is a traditional connection technology in which a die is mounted upright, and fine wires of copper or gold make connections from the package pin to the I/O cell, as explained earlier. Since I/O cells are placed at the die boundary, the number of I/O cells that we can have in a wire-bonded chip is limited.

Another interconnection technology known as *flip-chip* is popular for microprocessors, graphic processors, and high-performance integrated circuits (ICs), despite being expensive. In this technology, connections are made with the help of *bumps*, and we avoid using wires for die–package connections. A bump is a conducting material created on a die, as shown in Figure 25.8(a). The connection is first made from the I/O cells to the bump pads using *redistribution layer* (RDL) during wafer fabrication. RDL is the top metal layer of a die, and we create bump pads on the RDL. Subsequently, we mount the die upside-down (flipped) onto the package substrate, and the die bumps are connected to the bump pads on the substrate by aligning and remelting, as shown in Figure 25.8(b). The bump pads on the substrate provide connections with the package pins or the external devices.

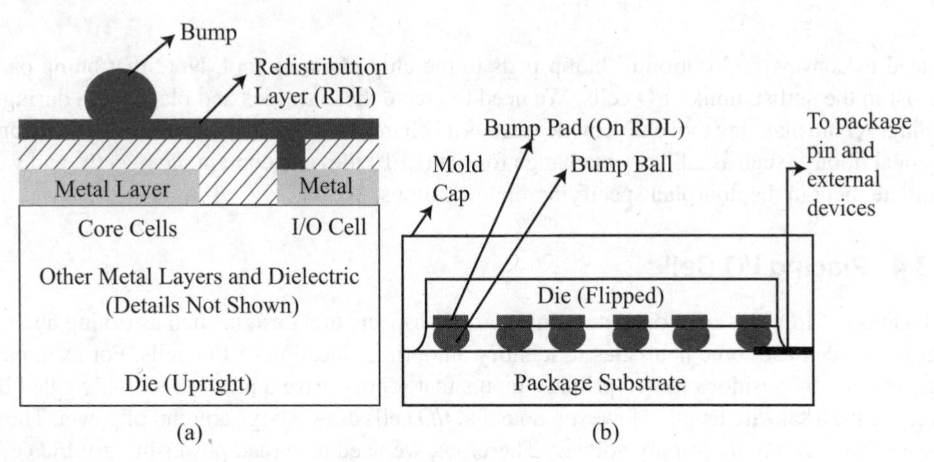

Figure 25.8 Schematic cross-sectional view (a) bump, RDL, and I/O cell; (b) flip-chip connection

The flip-chip technology has the following advantages:

1. It removes the restriction of placing I/O cells only at the periphery of a die. Thus, a chip can have more I/O pins. Additionally, we can place I/O cells closer to the core cells and reduce wire length. It helps in reducing delay and offers flexibility to minimize skews between two signals if required [18].
2. It allows placing power/ground bumps closer to the core cells and helps reduce voltage drop. Additionally, it helps achieve a more balanced power distribution using bumps within a die.

3. It exhibits lower inductance (L) than the wire bonding technology. Hence, the voltage drop (Ldi/dt) due to simultaneous switching of transistors in a circuit gets reduced.

Location of bump pads: We need to decide the location of bump pads on a die during chip planning. It needs to be consistent with the package design and can be of two types [19]:

1. **Peripheral array:** We place bump pads along the boundary of a die, as shown in Figure 25.9(a). The number of bump pads that we can place is limited by the die boundary.
2. **Area array:** We place bump pads on the entire die area, as shown in Figure 25.9(b). It allows placing more bump pads than the peripheral array. Additionally, it allows placing them closer to the core cells and helps derive the full benefits of flip-chip technology.

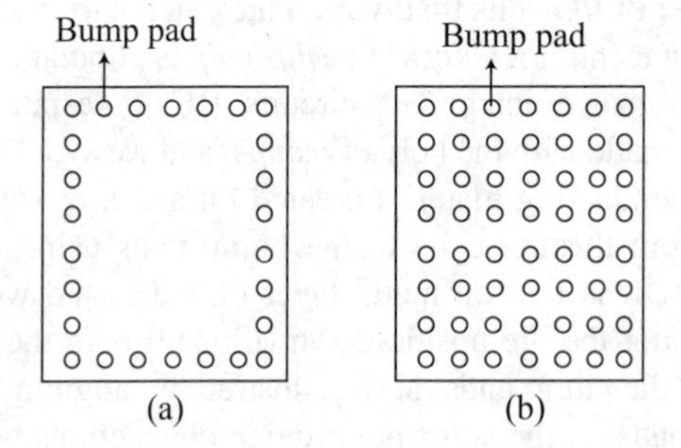

Figure 25.9 Bump pad placement schemes (a) peripheral array and (b) area array

We need to convey the location of bump pads to the chip planning tool. Note that bump pads do not exist in the netlist, unlike I/O cells. We need to create these entities and place them during chip planning. A chip planning tool can provide various mechanisms to accomplish this task. For example, a physical library (such as a library exchange format (LEF) file) can contain *bump cells*, and we can instantiate them in the floorplan specifying their locations.

25.3.4 Placing I/O Cells

The location of I/O cells on a die determines various aspects of a design, such as timing and signal integrity. We can use some heuristics to identify an optimal location of I/O cells. For example, we can assign nearby positions to two primary inputs that jointly drive a multi-input logic gate [20]. It can reduce the total wire length. However, note that I/O cells draw a large amount of power. They can cause a severe drop in the supply voltage. Therefore, we need to spread power-hungry I/O cells all over the die area to avoid creating *voltage drop hotspots* [21]. Additionally, we do not place sensitive signals such as clocks near an I/O cell having a high current transients [22].

In general, we can place I/O cells in the following two ways:

1. **Peripheral I/O:** We can place I/O cells along the boundary of a die. It is the most commonly employed placement scheme for the I/O cells. We can use this placement method for both wire bonding and flip-chip designs. For connecting I/O cells to the bump pads we use RDL routing, as shown in Figure 25.10(a).
2. **Area I/O:** Flip-chip technology allows us to place I/O cells closer to the core cells and obtain shorter wire length [18, 23]. We can place I/O cells as clusters or groups inside the core area.

We use RDL routing for connecting I/O cells to the bump pads, as shown in Figure 25.10(b) [23]. It offers more flexibility in I/O cell placement. Due to reduced wire length, it can help in reducing I/O delays and ameliorate signal integrity issues. However, it impacts core cell placement and leads to complications in achieving timing closure. We also need to carefully consider the higher power delivery requirements of the I/O cells placed in the core area while designing a PDN.

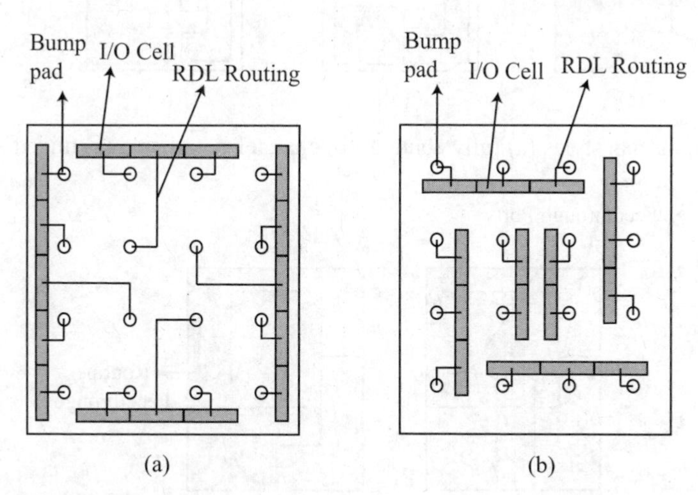

Figure 25.10 I/O cell placement schemes (bumps placed as area array) (a) peripheral I/O and (b) area I/O

We need to establish correspondence between an I/O cell and a bump pad during I/O cell placement. This task is referred to as *bump assignment*. Subsequently, we need to create routing between the I/O cells and the corresponding bump pads using RDL. Sometimes creating routes between I/O cells and bump pads can become infeasible due to bad bump assignment or bad I/O cell placement. Typically, we face fewer issues in RDL routing for area I/O cell placement compared to peripheral I/O cell placement because of its greater flexibility.

25.3.5 Floorplanning Styles

There are three styles of floorplan: *fully abutted*, *channel-based*, and *mixed* floorplan. A fully abutted floorplan has no gap or channel between blocks, a channel-based floorplan has channels between blocks, and a mixed floorplan has some blocks abutted and some blocks having channels between them, as shown in Figure 25.11. We need to choose the floorplan style by considering the pros and cons of each style. The top-level logic and interconnect planning are the crucial factors that determine the floorplan style.[4]

1. **Fully abutted floorplan:** An abutted floorplan has no extra routing resource at the top level. Therefore, entire top-level wiring needs to be done over the macros. We can connect blocks using abutment and pin alignment, or with the help of *feedthroughs*. A feedthrough is an entity which we insert in a macro or partition to allow a signal to pass through it. There are two types of feedthroughs: *feedthrough buffers* and *routing feedthroughs*, as illustrated in Figure 25.12.

[4] By top-level logic, we refer to the logic gates present at the top level and are not contained inside any block.

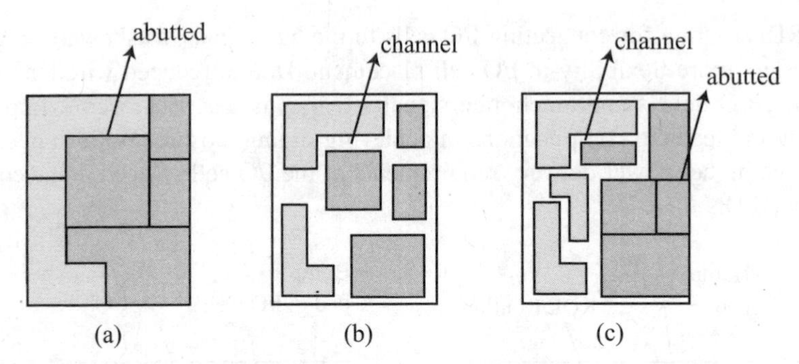

Figure 25.11 Floorplan styles (a) fully abutted, (b) channel-based, and (c) mixed

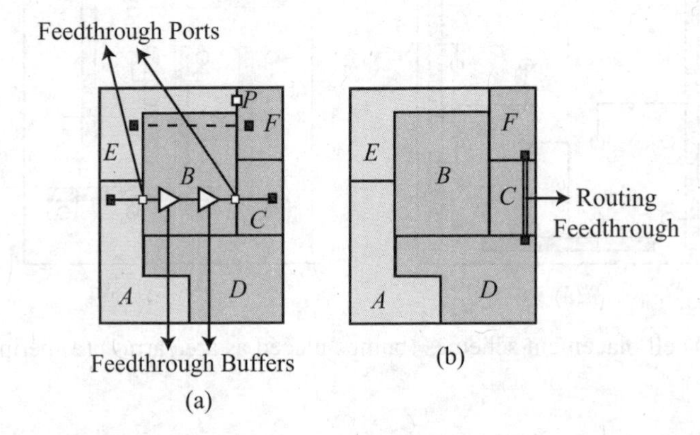

Figure 25.12 Feedthrough types (a) feedthrough buffer and (b) routing feedthrough

(a) **Feedthrough buffer:** A feedthrough buffer connects input ports to output ports of a macro using a buffer or a chain of buffers. Thus, inserting a feedthrough buffer changes the block or partition netlist by creating extra ports and inserting buffers. Note that we can insert feedthrough buffers at various stages of a design flow, such as during register transfer level (RTL) coding, logic synthesis, or physical design. After we have inserted feedthrough buffers and routed the signal through the ports at the block level, any signal can cross that block. In Figure 25.12(a), for connecting blocks A to C, we can use the routes created for the feedthrough inside the block B. Therefore, top-level routing can avoid utilizing routing resources over the block B. However, we should carefully tackle feedthrough buffers in combinational equivalence checking (CEC). The existence of extra ports and feedthrough buffers can lead to mismatches and false failures in CEC.

(b) **Routing feedthrough:** It allows a top-level wire to cross over a block without changing the netlist of that block. In Figure 25.12(b), a routing feedthrough is created in block C. It allows top-level routing to utilize the routing resources over the block C for making connections between D and F. However, it does not change the netlist of the block C. It simply instructs physical design tools to consider the routing feedthrough for the specified layer while routing. Typically, a tool will create routing blockages for routing feedthroughs at the block level. However, it will make those routing resources available for routing at the top level.

Feedthrough management: For fully abutted designs, we need to insert feedthroughs and manage them carefully. We should avoid creating unnecessary feedthroughs since they can consume extra routing resources and degrade the overall QoR. For example, in Figure 25.12(a), consider that we need to connect elements in *E* and *F*. Instead of creating a feedthrough (shown by dashed line), we can assign pins *P* at the common boundary and connect them by alignment. Typically, physical design tools are equipped with capabilities to insert feedthroughs based on the connections and the congestion information gathered by performing coarse routing.

An abutted floorplan style has no space for top-level logic. Therefore, we need to push the logic required between two macros in one of the partitions. Sometimes it can hinder optimizations that need a top-level view. Additionally, a fully abutted design has complexities of feedthrough management and can lead to congestion. Hence, channel-based styles are also considered in floorplanning.

2. **Channel-based floorplan:** In channel-based routing, we keep gaps between blocks to allow top-level routing. These gaps are referred to as *channels*. We can face congestion issues in the channel since all top-level routing goes through it. Additionally, some routes through the channel can be longer than a direct path through another block. Therefore, depending on the criticality of the congestion and timing issues, we can insert some optimized feedthroughs in a channel-based floorplan [24].

Example 25.6 Consider the channel-based floorplan shown in Figure 25.13(a). There are three connections *A1–A2*, *B1–B2*, and *C1–C2* at the top level. Assume that the channels have a capacity of two wires. As a result, we see that there is a congestion between blocks *M1* and *M3*. Additionally, net connecting *C1–C2* needs to take detour that increases its wire length and delay.

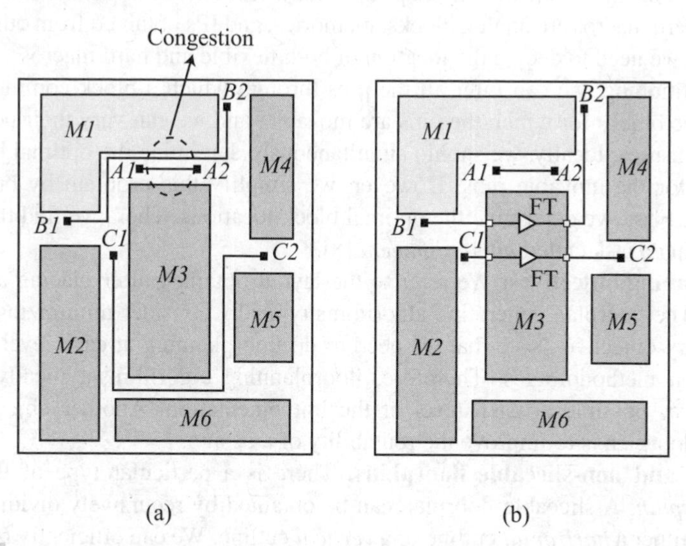

Figure 25.13 Channel-based floorplan (a) no feedthrough and (b) with feedthroughs buffers

We can alleviate congestion and reduce wire length by inserting feedthrough buffers in the block *M3*, as illustrated in Figure 25.13(b).

The channels also allow placing logic cells between blocks. To allow top-level logic optimizations, we should keep the channel wide enough. Nevertheless, creating channels between each pair of blocks may not be needed and can waste floorplan area.

3. **Mixed floorplan:** It derives the benefits of both abutted and channel-based floorplan styles. We abut blocks and create connections by pin alignment wherever possible. We create feedthroughs for some selected nets. We can also create channels where top-level cells are needed between blocks or for special cells such as clock routing buffers.

25.3.6 Placing Large Objects

There are two types of large objects in a design:

1. Blocks obtained after partitioning
2. Directly instantiated design entities such as ROMs, RAMs, ALU, analog blocks, and intellectual properties (IPs).

For the sake of easy description, we refer to all these large objects as macros in this section.

Flexible and hard macros: The area of a block depends on its total cell area. Post partitioning, the cells within a block and the associated block area become fixed. However, we can still vary its shape. These types of macros are known as *flexible macros*. Besides their location, we need to determine the shape of flexible macros during floorplanning.

There are some macros with both shape and size fixed. These macros are known as *hard macros*. Examples of hard macros are analog blocks, memories, and IPs obtained from other vendors. During floorplanning, we need to decide the location of both flexible and hard macros.

After partitioning, we can infer all the pins through which a block communicates externally. There are some blocks for which the pins are movable, and we can vary their positions to improve some QoR measures. Ideally, we should simultaneously determine the optimal block locations and the positions for the movable pins. However, we simplify this problem by breaking it into two separate tasks. First, we determine the optimal block locations. Then, we find the pin locations for each block using a task called *pin assignment* [8].

Floorplanning objectives: We refer to the layout obtained after placing all macros on a die as *floorplan*. The floorplan generating algorithms typically consider minimizing the floorplan area as their primary objective. Note that we need to do floorplanning at each level in the hierarchical implementation methodologies. Therefore, floorplanning algorithms typically pack the macros as compactly as possible at each level of the implementation. Another objective considered in generating a floorplan is to improve the routability of a design.

Sliceable and non-sliceable floorplans: There is a particular type of floorplan known as *sliceable floorplan*. A sliceable floorplan can be obtained by recursively dividing a rectangle into two parts, by either a *horizontal* cutline or a *vertical* cutline. We can efficiently represent a sliceable floorplan using a *binary tree*. The leaf nodes represent the macros. All internal nodes represent a horizontal cutline or a vertical cutline. For sliceable floorplan, efficient techniques are known to derive an optimal floorplan [25–29].

Example 25.7 Consider the schematic of a floorplan shown in Figure 25.14(a).

We can represent the recursive partitioning of the layout by horizontal and vertical lines using the binary tree shown in Figure 25.14(b). A circle with a horizontal line denotes a horizontal cutline, while a circle with a vertical line denotes a vertical cutline. The recursion stops when we reach a layout that is not partitioned.

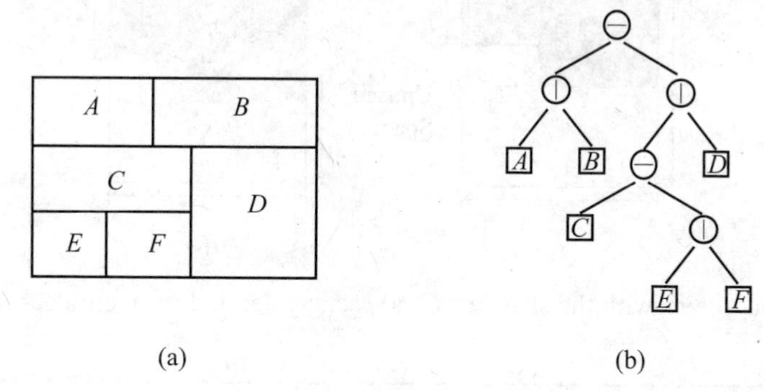

Figure 25.14 A sliceable floorplan (a) layout and (b) binary tree representation

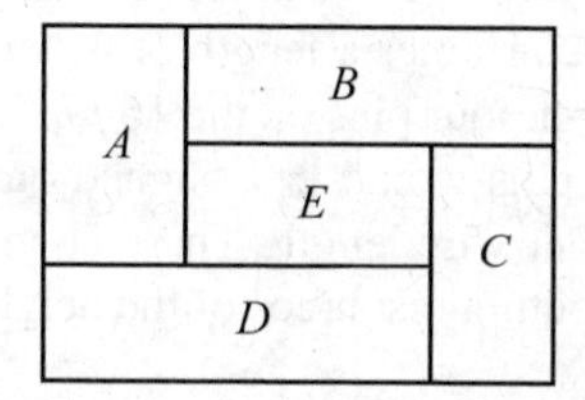

Figure 25.15 A non-sliceable floorplan

There are some floorplans which cannot be obtained by recursive horizontal and vertical cutlines. These floorplans are called non-sliceable floorplans. An example of non-sliceable floorplan is shown in Figure 25.15. Non-sliceable floorplans are more challenging to optimize. The floorplanning problem for the general orientation of macros is an NP-complete problem [28]. Readers can refer to [8, 30] for algorithms on floorplanning.

Rectilinear polygonal shapes: Assuming rectangular shapes for macros simplifies floorplanning algorithms. However, we can often reshape flexible macros to rectilinear polygonal shapes (such as L, T, H, or U). It can help us in reducing area, delay, and congestion in a design [31–33]. We illustrate it in the following examples.

Example 25.8 Consider two *hard macros* A and B of size 3×3 μm^2 and 2×2 μm^2, respectively, and a *flexible macro* C with an area of 12 μm^2. Assume that we need to pack these macros within a rectangle with minimum area.

First, let us assume that C can have rectangular shape only. Hence, we obtain the floorplan shown in Figure 25.16(a). The total floorplan area is 27 μm^2. Note that there is some unused space in this floorplan.

Next, let us assume that C can have a rectilinear polygonal shape. We can now reshape C to utilize the unused space, as shown in Figure 25.16(b). Hence, the total floorplan area reduces to 25 μm^2.

Thus, allowing flexible macros to have a rectilinear polygonal shape can help us reduce floorplan area.

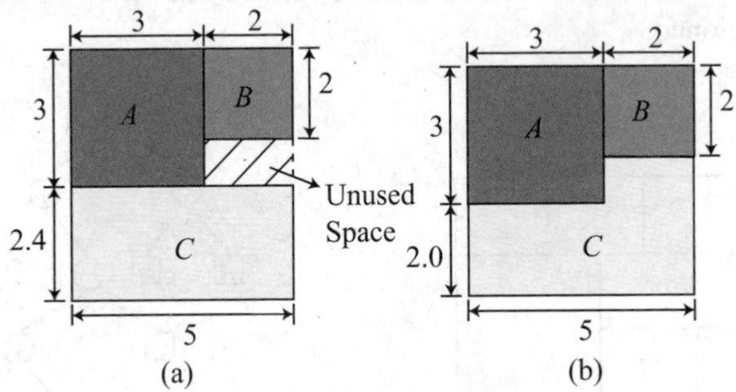

Figure 25.16 Floorplans with the shape of C (a) rectangular and (b) rectilinear polygon (lengths in μm)

Example 25.9 Consider two flexible macros *M1* and *M2* of size 16×16 μm^2 each. Assume that they are adjacent in the floorplan, as shown in Figure 25.17(a). Assume that they have eight connected pins. Further, assume that each pin consumes a length of 4 μm (wire width and spacing included) on the boundary. Hence, we can assign four pins on the 16 μm long common boundary and connect them directly. However, four other pins would be assigned on other edges and their connections (*A–A'*, ..., *D–D'*) would require longer wire length. These connections can increase delay and also lead to congestions by consuming routing resources of the neighboring macros or channels.

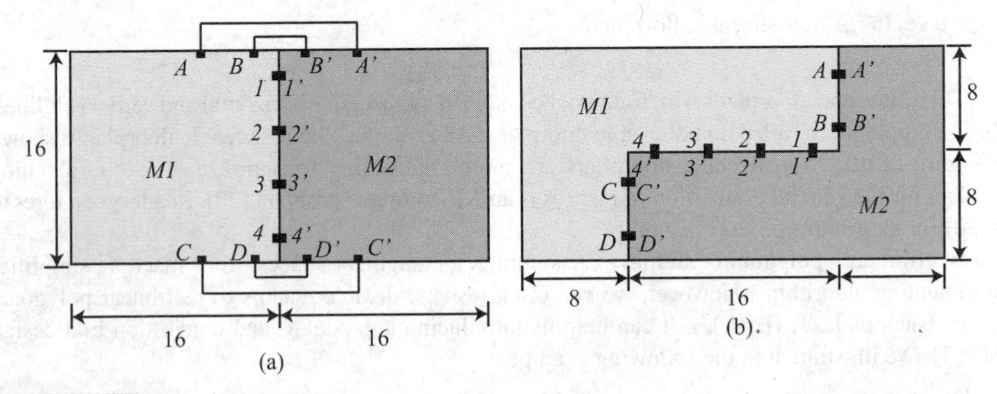

Figure 25.17 Floorplan and pin assignment for (a) rectangular macros and (b) rectilinear polygon macros (lengths in μm)

We can achieve better pin assignment and connectivity by making *M1* and *M2* L-shaped polygon, as shown in Figure 25.17(b) [33]. The length of the common boundary now becomes 32 μm. We can assign all eight pins along this boundary and connect them directly.

Thus, we can reshape macros before pin assignment and improve the QoR of a design.

The above examples are very simple. However, in real designs, we can find many such opportunities of manually reshaping macros and improving some figures of merit. Note that chip planning requires significant manual interventions for meeting tight power, performance, and area (PPA) constraints in an industrial design. As a designer, we can often figure out opportunities to tweak a floorplan to improve some design metrics that are difficult for an automatic tool to determine. Hence, industry-standard chip planning tools provide infrastructure and mechanisms to carry out such manual manipulations easily and efficiently. These mechanisms can be tool-specific commands, options, soft and hard constraints, or graphical user interface (GUI). Readers should refer to them in tool-specific manuals and user guides.

Flylines: We can use the connections between macros to guide us in finding favorable locations for the macros. We place the strongly connected macros nearby. Additionally, we place macros that interact strongly with the external world close to the corresponding I/O cells. We can also take the help of *flylines* in deciding the locations for the macros. A flyline represents the total number of connections between a pair of macros or I/O cells.

Example 25.10 Consider the layout and *flylines*, shown in Figure 25.18. We can observe that there are 52 connections between macros *B1* and *B2*. These two macros have the maximum number of connections for any pair. Hence, we should place these two macros nearby. Similarly, we can decide on good locations for other macros with the help of flylines.

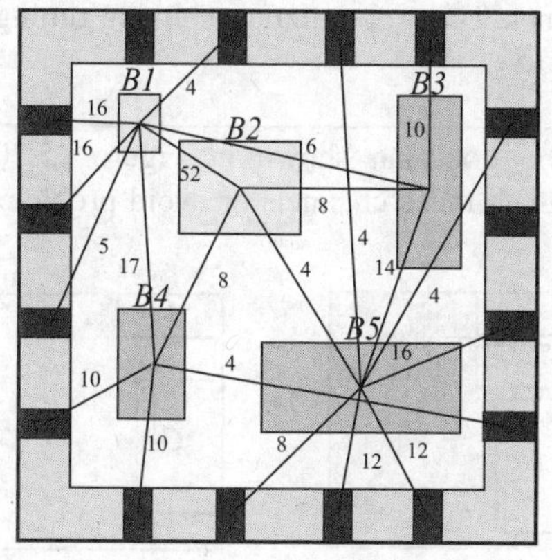

Figure 25.18 Flylines showing connectivity of macros

We can also adopt some guidelines or rules of thumb to determine good locations for macros. A few guidelines are as follows.

1. **Allot contiguous region for standard cells:** We can keep macros at the edges of a floorplan and allot contiguous space for standard cells. Note that we can have more than a million standard cells in an industrial design. We place them with the help of an automatic placement tool. A placement tool works well if the placement region is contiguous rather than fragmented. Therefore, the allocated regions for standard cells should be as contiguous as possible.

Example 25.11 Consider the two floorplans shown in Figure 25.19. The region for standard cell is contiguous in the floorplan shown in Figure 25.19(a), while it is fragmented for the floorplan shown in Figure 25.19(b). Hence, we should prefer the floorplan of Figure 25.19(a).

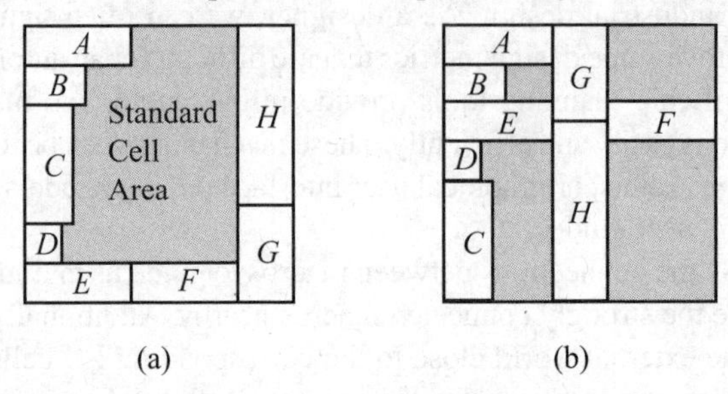

Figure 25.19 Regions alloted for standard cells (shaded) (a) contiguous and (b) fragmented

2. **Avoid narrow channels between macros:** While placing macros, we should avoid narrow channels between them. We should also prevent narrow channels between macros and the edges of the floorplan. A narrow channel has fewer routing tracks available for routing. As a result, we can have congestion in it. Additionally, a placement tool can place some standard cells in these narrow channels. However, there is not much space for upsizing these cells or inserting buffers in a narrow channel. Thus, optimizing or fixing timing problems for these cells can become challenging.

Example 25.12 Consider the floorplan shown in Figure 25.20(a). It has narrow channels between blocks. We can eliminate these channels or avoid problems associated with them in the following ways:

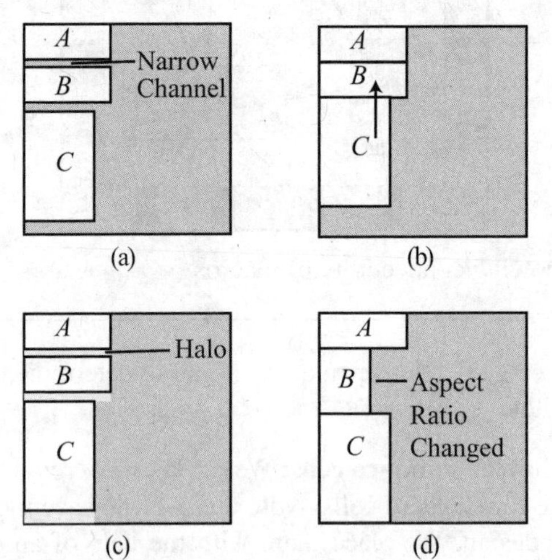

Figure 25.20 Avoiding problem due to narrow channels (a) given narrow channels, (b) channel removal by placement, (c) creating placement blockages (halos) in the channels, and (d) changing aspect ratio of flexible macro

(a) We place macros such that narrow channels do not exist, as shown in Figure 25.20(b).

(b) We add placement blockages or halos in these narrow channels, as shown in Figure 25.20(c). Due to halos, a placement tool will not place any standard cell in these channels, and the problem gets alleviated.

(c) If some macros are flexible, we can modify their aspect ratio to eliminate the narrow channels. If we assume that *B* is flexible, we can modify its aspect ratio as shown in Figure 25.20(d), to get rid of the narrow channels.

---

3. **Add halos around corners:** We often find congestions around the corners of a macro. It is because macros occupy most of the routing tracks above them. Therefore, routing must be done around its periphery to connect cells on different sides of a macro, as shown in Figure 25.21(a). It results in the crowding of wires (congestion) at the corner of the macros. If a placement tool places standard cells at the corners of a macro, the problem aggravates. Therefore, we can add placement blockages or halos around the corners to alleviate this problem, as shown in Figure 25.21(b).

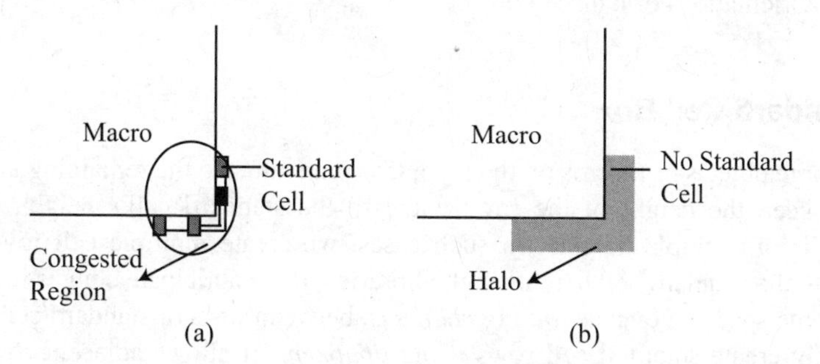

Figure 25.21 Corner of a macro (a) congestion issues and (b) avoiding congestion using halos

25.3.7 Orientation

We can place various objects such as I/O cells, macros, and standard cells in various orientations. If we consider a 2D object and allow horizontal reflection, vertical reflection, and 90° rotations, we can obtain eight different orientations, as illustrated in Figure 25.22. The R0 orientation is the normal orientation of an object. We can reflect it along the X-axis and Y-axis to obtain MX and MY orientations. If we reflect MX along Y-axis or MY along X-axis, we obtain MXY orientation. The MXY orientation is the same as rotating R0 by 180°. We can obtain other orientations by rotating R0 by 90° (R90), R0 by 270° (R270), MX by 90° (MX90), and MY by 90° (MY90). Note that different tools can use other keywords to represent these orientations. Readers should refer to them in the tool-specific user guide or manuals.

We can change the orientation of the placed object to optimize some design metrics. However, we should note that as we change the orientation of an object, the orientation of power and ground lines within that object also changes. Hence, we can allow only those orientations of an object which have internal power and ground line orientations consistent with the given orientation of power and ground lines.

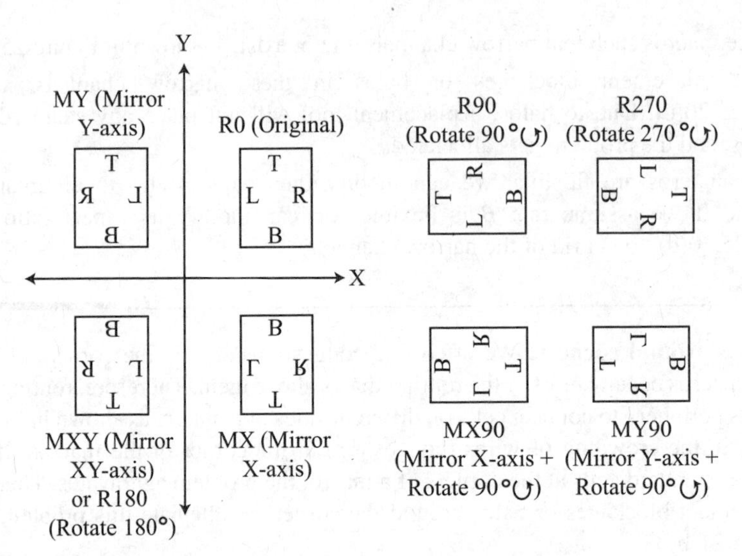

Figure 25.22 Orientations of a placed object

25.3.8 Standard Cell Rows

After placing large objects or macros on the layout, we can allocate the remaining area for standard cell rows. We keep the height of the rows equal to the standard cell's height. Some libraries can contain cells of multiple heights. In such cases, we create multiple cell rows with heights corresponding to the standard cell heights in the libraries. If we anticipate congestion in our design, we can leave some space, known as *routing channels*, between rows of standard cells.

We typically create standard cell rows using *abutment*. It allows adjacent rows to share the supply power rails (V_{DD}) and ground rails (GND) [34]. Thus, we can save routing resources. The following example illustrates it.

Example 25.13 Suppose we want to create four rows of standard cells. We can create them abutted as shown in Figure 25.23. In the R0 orientation, a cell has V_{DD} at the top and ground line at the bottom.

In the first row, we keep V_{DD} at the top and GND at the bottom. Hence, we can use R0 and MY-oriented cells in the first row.

In the second row, we keep V_{DD} at the bottom and GND at the top. Hence, we can use R180 and MX-oriented cells in the second row.

The third and fourth rows are similar to the first and second rows, respectively. Such an abutment of adjacent rows allows us to share the supply power rails (V_{DD}) and ground rails (GND).

Note that R90, R270, MX90, and MY90 orientations will have V_{DD} and GND lines vertically oriented. Hence, they will not align with the horizontally running V_{DD} and GND lines.

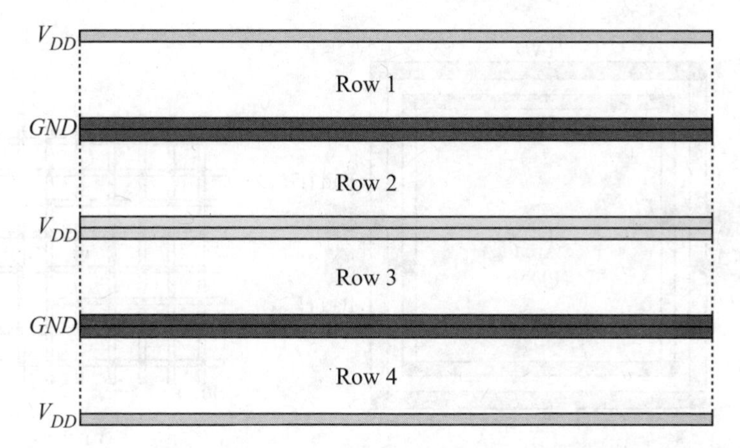

Figure 25.23 Abutment of cell rows

25.3.9 Pin Assignment

We create blocks during partitioning and fix their location on the floorplan. Subsequently, we decide the position of pins of the blocks along the block boundaries. This task is known as *pin assignment*. We provide the following information to the pin assignment tool: the partitioned netlist, the floorplan, the number of pins on each block, and their relative ordering. Pin assignment tool produces pin locations with the primary objective of minimizing the wire length.

Post pin assignment, we can carry out a crude routing. Typical industry-standard physical implementation tool provide mechanism to perform quick placement and quick global routing at the chip planning stage. Using these tools, we can identify problems such as *congestion hotspots* and manually tweak the floorplan to fix various issues. For example, we can change the location of macros, modify the shapes of flexible macros, try out different locations of movable pins, or add halos around macros.

25.4 POWER PLANNING

We discussed earlier that a chip draws external power through *power pads*. From the power pads, power gets delivered to each active element of a circuit using PDN. Typically, we make *power rings* around the periphery of a die, as shown in Figure 25.24(a), and connect appropriate PDN to the power rings.

A popular topology employed for PDN is the *mesh grid topology*. This topology organizes metal lines as uniformly spaced arrays, as shown in Figure 25.24(b). Metal lines run horizontally and vertically in alternate metal layers, and we add vias at their intersections. We refer to the horizontal metal lines as the *power rails* and the vertical metal lines as the *power straps*. The mesh grid topology has a good reliablity. Note that there are multiple paths from the power source to any sink in a mesh grid topology. Consequently, it can deliver power to a sink through an alternate route, even when there is a failure in one of the paths. Additionally, the mesh grid topology offers a low resistance and allows uniform distribution of current [35].

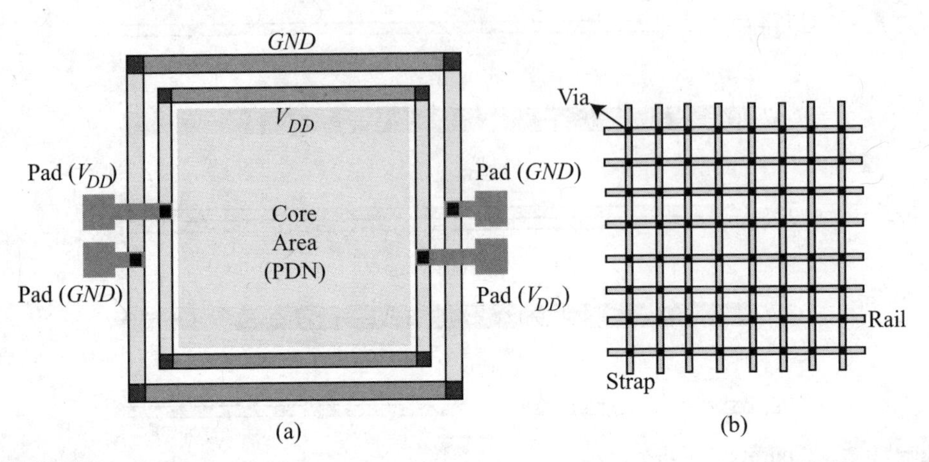

Figure 25.24 Power delivery structures (a) power ring and (b) a mesh grid type PDN

For the mesh grid type PDN, we need to decide the metal layers in which the grid spans. We typically choose top metal layers for the PDN because of their low sheet resistance. Additionally, we need to decide the metal line width and the spacing between them. As we explain in the following paragraphs, these decisions are guided by the fabrication technology and the allowed variation in the voltage levels.

25.4.1 Electromigration

While designing PDN, we should ensure that the current in the metal lines is within some limits. If unidirectional current flows in metal lines for a long duration, metal atoms can move in the direction opposite to the current flow (because electron moves in a direction opposite to the conventional direction of current flow). This effect is known as *electromigration* and can eventually create short- or open-circuit in a wire. Since a unidirectional current flows in a PDN, it is prone to electromigration. Therefore, we need to identify *current density hotspots* (regions where there is high current density) in a PDN using an appropriate EDA tool and make design modifications to fix the issue [36].

25.4.2 Voltage Drop

An ideal PDN should deliver constant voltage level across a die. However, supply voltage varies in different regions of a die because of the following reasons.

1. **Static voltage drop:** Due to the resistance (R) of the intermediate metal lines and current (I) flowing through them, the supply voltage drops before reaching the active circuit elements. This drop in the supply voltage is known as *IR drop*. The IR voltage drop depends on the average current in the circuit and is also referred to as *static voltage drop*. Industry-standard power analysis tools provide mechanism to compute static voltage drop. During early stages of physical design, we can compute static voltage drop and check the robustness of the PDN.

   We can reduce the static voltage drop by increasing the width of the metal lines, which reduces their resistances [37]. The fabrication process imposes a lower limit on the metal line width, while the available routing resources impose an upper limit.

2. **Dynamic voltage drop:** Dynamic voltage drop occurs due to switching activities in a circuit. When a circuit operates, many cells can switch simultaneously and draw current from the power supply. It increases the peak transient current through the supply rails and leads to dynamic voltage drop. Additionally, due to the inductance (L) of the metal lines and the package–die interconnections, there can be a voltage drop due to a sudden current change (high di/dt). We refer to the associated voltage drop as the Ldi/dt voltage drop. The Ldi/dt voltage drop is high when there is a *simultaneous switching* of multiple circuit elements, and there is a sudden high current demand in a circuit. Therefore, supply voltage can change dynamically in a circuit. Note that, we need to account for the switching activities for computing dynamic voltage drop. Hence, computing dynamic voltage drop is computationally more challenging than the static voltage drop.

IR drop hotspots: When the supply voltage drops, the CMOS logic gate delay increases. Consequently, the maximum operable frequency and the noise margin of a circuit decrease. Therefore, we need to perform voltage drop analysis in both static and dynamic conditions [38]. We should identify the localized regions where there is a high voltage drop. We refer to these regions as *IR drop hotspots*. With the increasing wire resistance, decreasing noise margin, and reduced voltage levels at advanced process nodes, the IR drop problems have become extremely critical. If the voltage drop in the IR drop hotspots is unacceptable, we need to fix them.

Decap cells: A popular technique to tackle the dynamic voltage drop problem is inserting an on-chip capacitor between the power (V_{DD}) and the ground lines [39]. These capacitors are known as *decoupling capacitors* or *decap cells* and act as local charge storage reservoirs. During the static condition, decap cells get charged to the supply voltage V_{DD}. When there is a high current transient due to loading, decap cells provide the stored charge and mitigate the associated dynamic voltage drop. We place decap cells on the die at strategic locations, such as close to the IR drop hotspots. Note that a decap cell consumes area on the die and dissipates leakage power also. Therefore, we should carefully choose the number of decap cells in our design and their sizes.

Tackling voltage drop problem: We can tackle the voltage drop problem during various stages of VLSI design flow [38, 40, 41]. A few examples are as follows. We can decide to power-ON different cores in a staggered manner during architectural planning. In the floorplanning stage, we can place power-hungry blocks closer to the power pads. We can add skews in the clock network during clock tree synthesis to avoid simultaneous switching events. Sometimes, we can increase the operating voltage of a circuit to compensate for the voltage drop, though at the cost of increased power dissipation [42]. We can also design circuits that supply boosting current to compensate for the voltage drop.

25.5 RECENT TRENDS

In recent times, design flows have moved from *find and fix* approach to *predict and prevent* approach [35]. Since chip planning is a bridge between logic synthesis and physical design, we often employ it to predict physical design problems during the early design stages. We use the floorplan obtained during the chip planning stage to generate a rough design layout. A logic synthesis tool can compute load capacitance and identify critical paths using the rough layout. Thus, it can quickly predict physical design problems during logic synthesis and prevent them from occurring.

Despite tremendous progress in physical design, chip planning still requires considerable manual interventions. In recent times, machine learning techniques are being explored for floorplanning and have shown good results [43]. Machine learning techniques are also being applied for predicting and preventing problems in power planning [44].

The advent of 3D ICs has opened new frontiers for chip planning. In a 3D IC, we stack multiple layers. Therefore, blocks can directly communicate in the third dimension, we can do pin assignment in the middle of the block, and routing can be more efficient [45]. However, the power density and peak temperature increase in a 3D IC due to the stacking of layers. Thus, 3D ICs modify many traditional 2D physical design problems into 3D problems, and innovative solutions are necessary to tackle them [46, 47].

REVIEW QUESTIONS

25.1 What is the significance of chip planning in VLSI design flow?

25.2 What is the difference between flat and hierarchical design implementation?

25.3 When do we prefer a flat design implementation?

25.4 What are the challenges in timing signoff in hierarchical design implementation?

25.5 How can we derive an abstract view of a block for timing analysis?

25.6 How do we create a timing budget for a block in hierarchical design implementation?

25.7 What is an ETM?

25.8 What are the challenges in ensuring signal integrity in hierarchical design implementation?

25.9 How can we derive partitions for a design for hierarchical implementation?

25.10 What is a min-cut partition for an undirected graph?

25.11 Can the K–L min-cut partition algorithm escape from local minima? If yes, how?

25.12 How do we decide the die size for a chip?

25.13 What is core utilization? What can be the consequences of targetting a very high core utilization (such as 0.9) in a physical design flow?

25.14 What is the difference between a core-limited design and a pad-limited design?

25.15 What is an I/O cell? Describe the functionality of a bidirectional I/O cell.

25.16 What is an ESD? Why should an I/O cell be protected from ESD?

25.17 How is an I/O cell protected from ESD?

25.18 Why is the power supply of core cells and I/O cells kept separate?

25.19 What is a power pad?

25.20 What is the difference between a flexible macro and a hard macro?

25.21 What is a sliceable floorplan?

25.22 Draw the binary tree representation of the floorplan shown in Figure 25.25.

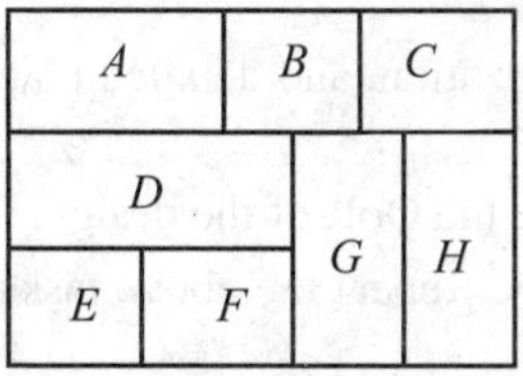

Figure 25.25 Layout of a floorplan

25.23 How can flylines be employed to improve the floorplan of a design?

25.24 Why is it advisable to keep the standard cell area as a contiguous region?

25.25 How can congestion be avoided at the corners of a macro?

25.26 Why is it not a good idea to have a narrow channel allocated between two macros?

25.27 Why is the orientation of standard cells flipped in alternate rows?

25.28 What is the purpose of the task pin assignment?

25.29 What is a PDN?

25.30 What are the advantages of mesh grid topology for a PDN?

25.31 What is electromigration? When can electromigration occur in a PDN? How can it be avoided?

25.32 What is the difference between the static voltage drop and the dynamic voltage drop in a PDN?

25.33 How can we mitigate the voltage drop problem during floorplanning?

25.34 What are decap cells? Where should we place decap cells?

25.35 How can the PPA be affected by inserting decap cells in a design?

25.36 How is chip planning different for wire-bonded and flip-chip packages?

25.37 What are bump pads?

25.38 How is a connection between an I/O cell and a bump pad made?

25.39 What are the different styles of floorplanning?

25.40 What are the advantages of using rectilinear polygonal shapes for blocks instead of rectangular shapes?

25.41 What are the different possible orientations of a placed object in a floorplan?

25.42 How is chip planning of a 3D IC different from a 2D IC?

TOOL-BASED ACTIVITY

Take any chip planning tool that is available to you, including open-source tools proposed in [48]. Load a Verilog netlist with 100–1000 instances and 10–100 flip-flops. You should give the timing constraints, technology library, and physical information (possibly using LEF files) as inputs. Choose clock frequency and I/O constraint such that the worst slack I/O path and register-to-register path have negative slack after placement. Perform the following tasks:

1. Initialize floorplan by assuming some reasonable die size and core size value.
2. Perform pin placement. Observe the location of pins/ports.

3. Perform power planning. Observe the layers used for PDNs. Observe the location of VDD/GND pins.
4. Perform global placement, legalization, and detailed placement. Perform timing analysis after each step.
5. Perform global routing. Analyze the QoR of the design.
6. Change the location of I/O ports. Repeat the above tasks. Observe and analyze the change in the worst slack path.

REFERENCES

[1] C. W. Moon, H. Kriplani, and K. P. Belkhale. "Timing model extraction of hierarchical blocks by graph reduction." *Proceedings of the 39th Annual Design Automation Conference* (2002), pp. 152–157.

[2] S. Saurabh, N. Kumar, and I. Keller. "Method and apparatus for comprehension of common path pessimism during timing model extraction." Jan. 20 2015. US Patent 8,938,703.

[3] S. Saurabh and N. Kumar. "Method and apparatus for efficient generation of compact waveform-based timing models." Aug. 8 2017. US Patent 9,727,676.

[4] S. Saurabh and N. Kumar. "Method and apparatus for concurrently extracting and validating timing models for different views in multi-mode multi-corner designs." Apr. 9 2019. US Patent 10,255,403.

[5] M. R. Garey and D. S. Johnson. *Computers and Intractability*, vol. 174. Freeman San Francisco, 1979.

[6] B. W. Kernighan and S. Lin. "An efficient heuristic procedure for partitioning graphs." *The Bell System Technical Journal* 49, no. 2 (1970), pp. 291–307.

[7] A. E. Dunlop, B. W. Kernighan, et al. "A procedure for placement of standard cell VLSI circuits." *IEEE Transactions on Computer-aided Design* 4, no. 1 (1985), pp. 92–98.

[8] N. A. Sherwani. *Algorithms for VLSI Physical Design Automation*. Springer Science & Business Media, 2012.

[9] C. M. Fiduccia and R. M. Mattheyses. "A linear-time heuristic for improving network partitions." *19th Design Automation Conference* (1982), pp. 175–181, IEEE.

[10] R. Murgai, R. K. Brayton, and A. Sangiovanni-Vincentelli. "On clustering for minimum delay/area." *1991 IEEE International Conference on Computer-aided Design Digest of Technical Papers* (1991), pp. 6–7, IEEE Computer Society.

[11] Y.-C. Wei and C.-K. Cheng. "Towards efficient hierarchical designs by ratio cut partitioning." *1989 IEEE International Conference on Computer-aided Design. Digest of Technical Papers* (1989), pp. 298–301, IEEE.

[12] J.-S. Yih and P. Mazumder. "A neural network design for circuit partitioning." *IEEE Transactions on Computer-aided Design of Integrated Circuits and Systems* 9, no. 12 (1990), pp. 1265–1271.

[13] D. Kolar, J. D. Puksec, and I. Branica. "VLSI circuit partition using simulated annealing algorithm." *Proceedings of the 12th IEEE Mediterranean Electrotechnical Conference (IEEE Cat. No. 04CH37521)*, vol. 1, pp. 205–208, IEEE, 2004.

[14] K. Muller-Glaser, K. Kirsch, and K. Neusinger. "Estimating essential design characteristics to support project planning for ASIC design management." *1991 IEEE International Conference on Computer-aided Design Digest of Technical Papers* (1991), pp. 148–149, IEEE Computer Society.

[15] E. A. Amerasekera and C. Duvvury. "ESD in silicon integrated circuits." John Wiley & Sons, 2nd ed., 2002.

[16] M. Zhao, Y. Fu, V. Zolotov, S. Sundareswaran, and R. Panda. "Optimal placement of power-supply pads and pins." *IEEE Transactions on Computer-aided Design of Integrated Circuits and Systems* 25, no. 1 (2005), pp. 144–154.

[17] R. Zhang, K. Wang, B. H. Meyer, M. R. Stan, and K. Skadron. "Architecture implications of pads as a scarce resource." *2014 ACM/IEEE 41st International Symposium on Computer Architecture (ISCA)* (2014), pp. 373–384, IEEE.

[18] M.-F. Lai and H.-M. Chen. "An implementation of performance-driven block and I/O placement for chip-package codesign." *9th International Symposium on Quality Electronic Design (ISQED 2008)* (2008), pp. 604–607, IEEE.

[19] J.-W. Fang, I.-J. Lin, P.-H. Yuh, Y.-W. Chang, and J.-H. Wang. "A routing algorithm for flip-chip design." *ICCAD-2005. IEEE/ACM International Conference on Computer-aided Design, 2005* (2005), pp. 753–758, IEEE.

[20] M. Pedram, K. Chaudhary, and E. S. Kuh. "I/O pad assignment based on the circuit structure." *1991 IEEE International Conference on Computer Design: VLSI in Computers and Processors* (1991), pp. 314–315.

[21] J. N. Kozhaya, S. R. Nassif, and F. N. Najm. "I/O buffer placement methodology for ASICs." *ICECS 2001. 8th IEEE International Conference on Electronics, Circuits and Systems (Cat. No. 01EX483)*, vol. 1, pp. 245–248, IEEE, 2001.

[22] G. Yasar, C. Chiu, R. A. Proctor, and J. P. Libous. "I/O cell placement and electrical checking methodology for ASICs with peripheral I/Os." *Proceedings of the IEEE 2001. 2nd International Symposium on Quality Electronic Design* (2001), pp. 71–75, IEEE.

[23] J. Xiong, Y.-C. Wong, E. Sarto, and L. He. "Constraint driven I/O planning and placement for chip-package co-design." *Asia and South Pacific Conference on Design Automation, 2006* (2006), pp. 207–212, IEEE.

[24] S. Prasad and A. Kumar. "Simultaneous routing and feedthrough algorithm to decongest top channel." *2009 22nd International Conference on VLSI Design* (2009), pp. 399–403, IEEE.

[25] R. H. Otten. "Automatic floorplan design." *19th Design Automation Conference* (1982), pp. 261–267, IEEE.

[26] P.-N. Guo, C.-K. Cheng, and T. Yoshimura. "An O-tree representation of non-slicing floorplan and its applications." *Proceedings 1999 Design Automation Conference (Cat. No. 99CH36361)* (1999), pp. 268–273, IEEE.

[27] Y.-C. Chang, Y.-W. Chang, G.-M. Wu, and S.-W. Wu. "B*-trees: A new representation for non-slicing floorplans." *Proceedings of the 37th Annual Design Automation Conference* (2000), pp. 458–463.

[28] L. Stockmeyer. "Optimal orientations of cells in slicing floorplan designs." *Information and Control* 57, no. 2–3 (1983), pp. 91–101.

[29] W. Shi. "A fast algorithm for area minimization of slicing floorplans." *IEEE Transactions on Computer-aided Design of Integrated Circuits and Systems* 15, no. 12 (1996), pp. 1525–1532.

[30] M. Sarrafzadeh and C. Wong. *An Introduction to VLSI Physical Design*. McGraw-Hill Higher Education, 1996.

[31] D. P. Mehta and N. Sherwani. "On the use of flexible, rectilinear blocks to obtain minimum-area floorplans in mixed block and cell designs." *ACM Transactions on Design Automation of Electronic Systems (TODAES)* 5, no. 1 (2000), pp. 82–97.

[32] C. C. Chu and E. F. Young. "Nonrectangular shaping and sizing of soft modules for floorplan-design improvement." *IEEE Transactions on Computer-aided Design of Integrated Circuits and Systems* 23, no. 1 (2004), pp. 71–79.

[33] H.-H. Huang, C.-C. Chang, C.-Y. Lin, T.-M. Hsieh, and C.-H. Lee. "Congestion-driven floorplanning by adaptive modular shaping." *48th Midwest Symposium on Circuits and Systems, 2005* (2005), pp. 1067–1070, IEEE.

[34] F. Moraes, L. Torres, M. Robert, and D. Auvergne. "Estimation of layout densities for CMOS digital circuits." *Proceeding International Workshop on Power and Timing Modeling Optimization Simulation (PATMOS'98)* (1998), pp. 61–70.

[35] L. Lavagno, I. L. Markov, G. Martin, and L. K. Scheffer. *Electronic Design Automation for IC Implementation, Circuit Design, and Process Technology: Circuit Design, and Process Technology*. CRC Press, 2016.

[36] X. Huang, T. Yu, V. Sukharev, and S. X.-D. Tan. "Physics-based electromigration assessment for power grid networks." *Proceedings of the 51st Annual Design Automation Conference* (2014), pp. 1–6.

[37] C. J. Alpert, D. P. Mehta, and S. S. Sapatnekar. *Handbook of Algorithms for Physical Design Automation*. Auerbach Publications, 2008.

[38] S. Nithin, G. Shanmugam, and S. Chandrasekar. "Dynamic voltage (IR) drop analysis and design closure: Issues and challenges." *2010 11th International Symposium on Quality Electronic Design (ISQED)* (2010), pp. 611–617, IEEE.

[39] H. Su, S. S. Sapatnekar, and S. R. Nassif. "Optimal decoupling capacitor sizing and placement for standard-cell layout designs." *IEEE Transactions on Computer-aided Design of Integrated Circuits and Systems* 22, no. 4 (2003), pp. 428–436.

[40] M. S. Gupta, J. L. Oatley, R. Joseph, G.-Y. Wei, and D. M. Brooks. "Understanding voltage variations in chip multiprocessors using a distributed power-delivery network." *2007 Design, Automation & Test in Europe Conference & Exhibition* (2007), pp. 1–6, IEEE.

[41] K.-Y. Chao and D. Wong. "Signal integrity optimization on the pad assignment for high-speed VLSI design." *Proceedings of IEEE International Conference on Computer Aided Design (ICCAD)* (1995), pp. 720–725, IEEE.

[42] H. Mair, E. Wang, A. Wang, P. Kao, Y. Tsai, S. Gururajarao, R. Lagerquist, J. Son, G. Gammie, G. Lin, et al. "3.4 A 10 nm FinFET 2.8 GHz tri-gear deca-core CPU complex with optimized power-delivery network for mobile SoC performance." *2017 IEEE International Solid-state Circuits Conference (ISSCC)* (2017), pp. 56–57, IEEE.

[43] A. Mirhoseini, A. Goldie, M. Yazgan, J. W. Jiang, E. Songhori, S. Wang, Y.-J. Lee, E. Johnson, O. Pathak, A. Nazi, et al. "A graph placement methodology for fast chip design." *Nature* 594, no. 7862 (2021), pp. 207–212.

[44] Y. Cao, A. B. Kahng, J. Li, A. Roy, V. Srinivas, and B. Xu. "Learning-based prediction of package power delivery network quality." *Proceedings of the 24th Asia and South Pacific Design Automation Conference* (2019), pp. 160–166.

[45] B. W. Ku and S. K. Lim. "Pin-in-the-middle: An efficient block pin assignment methodology for block-level monolithic 3D ICs." *Proceedings of the ACM/IEEE International Symposium on Low Power Electronics and Design* (2020), pp. 85–90.

[46] T. Ni, H. Chang, S. Zhu, L. Lu, X. Li, Q. Xu, H. Liang, and Z. Huang. "Temperature-aware floorplanning for fixed-outline 3D ICs." *IEEE Access* 7 (2019), pp. 139787–139794.

[47] H. Park, K. Chang, B. W. Ku, J. Kim, E. Lee, D. Kim, A. Chaudhuri, S. Banerjee, S. Mukhopadhyay, K. Chakrabarty, et al. "RTL-to-GDS tool flow and design-for-test solutions for monolithic 3D ICs." *2019 56th ACM/IEEE Design Automation Conference (DAC)* (2019), pp. 1–4, IEEE.

[48] T. Ajayi, V. A. Chhabria, M. Fogaça, S. Hashemi, A. Hosny, A. B. Kahng, M. Kim, J. Lee, U. Mallappa, M. Neseem, et al. "Toward an open-source digital flow: First learnings from the OpenROAD project." *Proceedings of the 56th Annual Design Automation Conference 2019* (2019), pp. 1–4.

26 Placement

System is an arrangement to secure certain ends, so that no time may be lost in accomplishing them ... There must be a place for everything, and everything in its place ...

—Samuel Smiles, *Thrift*, Chapter 2, 1875

A critical task in physical design is to determine the locations of each entity in a design. In Chapter 25 ("Chip planning"), we have discussed how we determine the position of *larger entities* such as macros, memories, analog blocks, and input/output (I/O) pads during floorplanning. However, we also need to decide the locations of numerous *standard cells* in a design. We make these decisions during *placement*. The tool that performs placement is referred to as a *placer*.

The primary goal of placement is to ensure that the design becomes *routable*. This task is challenging because a placer must decide the cell locations without carrying out *actual routing*.[1] Typically, placers use cost measures that are easier to compute, such as the total *wire length* (WL), rather than directly *estimate* the routability of a design. Intuitively, to reduce WL, a placer should place the *connected* cells nearby. Therefore, *minimizing* total WL often improves routability. Moreover, a bad placement solution can make a design unroutable or prevent meeting timing requirements. Hence, other cost measures, such as timing and congestion, are also considered by a placer. The placement methodology that targets improving the slack of the timing paths is known as *timing-driven placement* and is widely employed for industrial designs.

The placement problem is challenging due to the *large* number of standard cells that needs to be placed. Placers often need to tackle more than a million cells in an industrial design. We need to consider this aspect of the problem while developing placement strategies.

Typically, we simplify the placement problem by dividing it into multiple tasks: *global placement* and *legalization*, followed by *detailed placement*.

During the early phases of placement, we try to find approximate cell locations and treat cells as point objects [1]. We *spread* the cells over the layout reducing the cell density and minimizing some cost metrics, such as the total WL. The cell locations are decided just by their connectivity, and we ignore the attributes of a cell, such as its size and pin locations. This task is known as *global placement*. After global placement, cells can overlap and occupy *illegal* positions. We fix these problems during the *legalization* and *detailed placement* stages. Post-placement, we

[1] We cannot determine an optimal placement location for each cell by performing actual routing because routing is a complex problem and will have an unacceptable runtime if we perform routing for each possible cell location.

perform some more optimizations, such as resizing and buffering, to improve the post-placement quality of result (QoR).

In this chapter, we will discuss the above aspects of placement in detail.

26.1 GLOBAL PLACEMENT

A global placer decides the location of a cell by considering it as a point object. Thus, a cell instance C_i is represented by its coordinates (x_i, y_i) on the layout. For a given circuit and its floorplan, the placement solution depends on the following factors:

1. **Location of fixed objects:** A floorplan contains *fixed objects*. For example, I/O ports and macros. A placer is not allowed to change the location of fixed objects. However, it can move cells connected to the fixed objects such that the total WL gets minimized. The optimal placement solution of movable cells will depend on the location of the fixed objects specified in the floorplan and connected to them.

Example 26.1 Consider a movable object *C1* connected to two fixed objects (an input pad *p1* and a macro pin *p2*), as shown in Figure 26.1(a). To minimize WL, we should place *C1* somewhere in the highlighted region (we assume that the wires can be laid out only in horizontal or vertical directions).

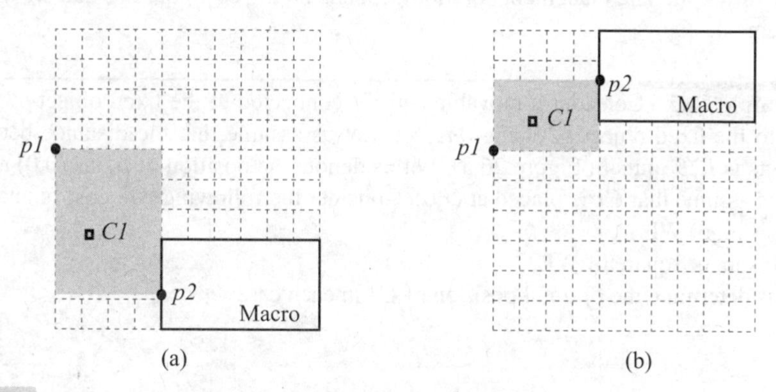

Figure 26.1 Placement solutions for a cell *C1* (highlighted by shaded rectangle) connected with fixed objects *p1* and *p2*

Let us change the location of the macro pin *p2* as shown in Figure 26.1(b). As a result, we need to place *C1* in another region (highlighted by the rectangle) for minimizing the WL.

Thus, the optimal placement solution depends on the location of the fixed objects in a given floorplan.

2. **Cell connectivity:** When a cell is connected to multiple objects, we can minimize the total WL by placing it close to them. Thus, changing the connection in a design will change the placement solution.

Example 26.2 Consider a cell *C1* connected to two fixed objects *p1* and *p2* using two *separate nets*, as shown in Figure 26.2(a). The cell *C1* can be placed anywhere in the highlighted region to minimize the total WL.

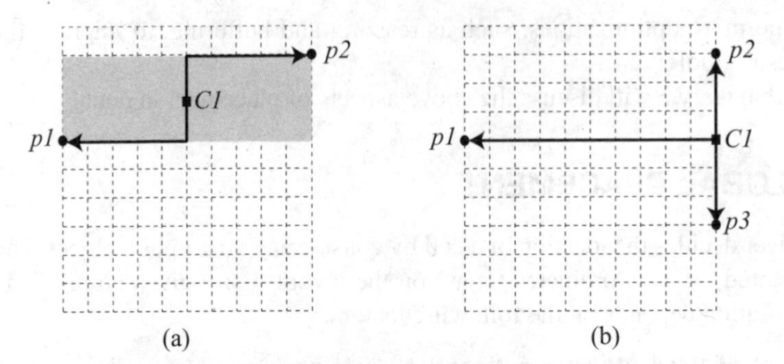

Figure 26.2 Placement solutions when cell *C1* is connected to (a) two fixed objects (solution should be in the shaded region) and (b) three fixed objects

Next, consider that the cell *C1* is connected to three fixed objects *p1*, *p2*, and *p3* using three separate nets, as shown in Figure 26.2(b). To minimize the total WL, the cell should be placed as shown in the figure.

Thus, the placement solution depends on the connectivity of a cell.

3. **Cost function:** The placement solution depends on the *cost function* that we *minimize* during placement.

Example 26.3 Consider a movable cell *C1* connected to the fixed object *p1* using *two* nets and to the fixed object *p2* using *one* net. Let us assume that the distance between the fixed objects is L (9 units in Figure 26.3). Let us denote the position of *p1* as (0,0) and *p2* as (L,0). Let us assume that *C1* is placed at (x,0). Consider the following two cost functions:

(a) sum of WL

(b) sum of square of WL.

Let us determine the optimal position of *C1* in each case.

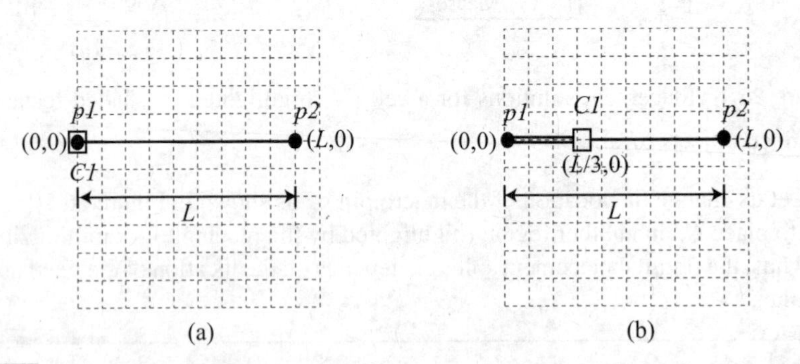

Figure 26.3 Optimal placement solution when the cost function is (a) linear and (b) quadratic

(a) If the cost function is taken as the sum of WL of each net (a linear function of WL), the cost will be $2x + (L - x) = x + L$ (a factor of 2 in the first term exists because it is given that *C1* is connected to *p1* using *two* nets). The cost minimizes for the position (0, 0) (i.e., $x = 0$), as shown in Figure 26.3(a) [2]. The cost in this case is $0 + 0 + 9 = 9$ units. We can check that the cost for the solution in Figure 26.3(b) is more ($3 + 3 + 6 = 12$ units).

(b) If the cost function is taken as sum of the square of WL of each net (a quadratic function of WL), the cost will be $2x^2 + (L - x)^2 = 3x^2 - 2Lx + L^2$. The cost minimizes for the position $(L/3,0)$ (i.e., $x = L/3$), as shown in Figure 26.3(b) [2]. For $L = 9$, the optimal location is (3,0). The cost is $3^2 + 3^2 + 6^2 = 54$ unit^2. We can check that the cost for the solution in Figure 26.3(a) is more ($0^2 + 0^2 + 9^2 = 81$ unit^2).

Thus, the cost function strongly impacts the placement solution.

26.1.1 WL Estimates

The cost function that a global placer seeks to minimize typically depends on the WL. Therefore, a placer needs to compute the WL of each net multiple times while finding a solution. Ideally, the computed WL should match the post-routing WL. However, during placement, the layout of nets is not yet decided. Therefore, the global placer needs to *estimate* the WL of each net.

There are many techniques to estimate the WL during placement. These techniques differ in the computational efficiency and accuracy with respect to the post-routing WL [3–5]. Let us look at a few commonly used WL estimation techniques.

Half-perimeter Wire Length

The half-perimeter wire length (HPWL) estimate is widely used in placement algorithms because it is easy to compute. To compute the HPWL for a given net, we consider a minimum-sized *bounding rectangle* that encloses all the pins of that net. We take half of the perimeter of this bounding rectangle as the HPWL of that net.

Assume that k pins are connected to a net N. Assume that the location[2] of the pins is (x_1, y_1), (x_2, y_2), ... (x_k, y_k). Then, the HPWL can be given as:

$$HPWL(N) = [MAX(x_i) - MIN(x_i)] + [MAX(y_i) - MIN(y_i)] \tag{26.1}$$

where *MAX* and *MIN* functions find the maximum and the minimum value of the coordinates, respectively.

Example 26.4 Consider a set of cells *C1*, *C2*, *C3*, and *C4* whose pins are connected to a net, as shown in Figure 26.4.

The HPWL of the net is the half perimeter of the minimum-sized bounding rectangle, which is 11 units. Alternatively, the HPWL can be computed as $[MAX(x_i) - MIN(x_i)] + [MAX(y_i) - MIN(y_i)] = (6 - 2) + (9 - 2) = 11$ units.

---

[2] Since we treat a cell as a point object, the location of all its pin is the same as its location.

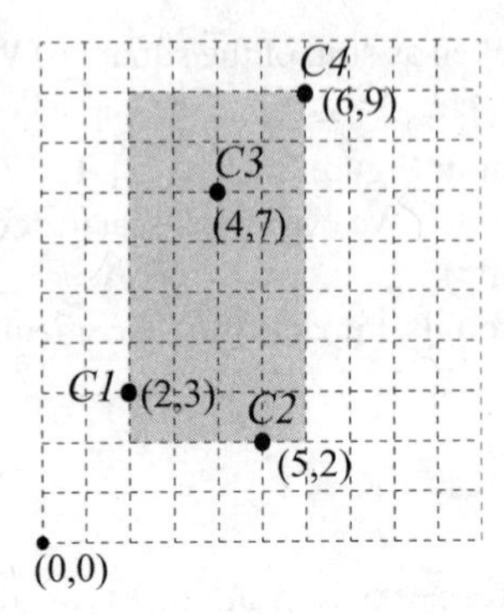

Figure 26.4 Half-perimeter based estimate of WL

Steiner Tree and Spanning Tree

Let us assume that there is a net N connected with k pins whose locations are given as (x_1, y_1), $(x_2, y_2), \ldots, (x_k, y_k)$. *Steiner minimum tree* (SMT) for this net N connects all these pins by a network of *shortest length*. The *length* of this network is defined as the sum of the lengths of all its corresponding net segments.[3] Note that we are allowed to add *extra junctions points* (points where three or more segments meet) in an SMT to minimize the length of the network. These extra junction points are known as *Steiner points*.

Finding SMT for a given set of k points is an NP-complete problem [6, 7]. Therefore, instead of finding SMT, we can use *minimum spanning tree* (MST) to estimate the WL. In contrast to an SMT, *no extra junction points* are allowed in an MST network. The length of an MST network is greater than or equal to the length of an SMT network. However, an MST can be computed in a polynomial time, in contrast to an SMT [6].

Example 26.5 Consider a net that connects pins of the cells *C1*, *C2*, *C3*, and *C4*, as shown in Figure 26.5(a). The network shown is an SMT, and is composed of five net segments *NS1*, *NS2*, ..., *NS5*. Note that extra junction points *S1* and *S2* are added to reduce the total length of the network.

The MST network is shown in Figure 26.5(b). Each net segment in an MST connects only two points. The length of this MST network is greater than the SMT network (see Problem 26.5).

[3] We refer to the portion of a net that connects two points by a straight line in the schematic or a straight wire in the layout as a *net segment*.

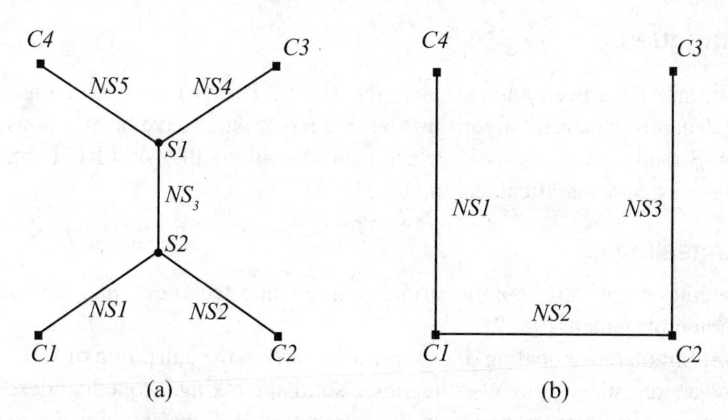

Figure 26.5 Illustration of (a) SMT and (b) MST

Typically, we allow interconnects to run in either horizontal or vertical directions. If the net segments are constrained to be either horizontal or vertical, then the SMT is known as *rectilinear SMT*, and the MST is known as *rectilinear MST*. For WL estimation during placement, we can use the length obtained for the rectilinear SMT and the rectilinear MST since they closely resemble actual wire layouts.

Example 26.6 The rectilinear SMT and rectilinear MST for a set of four points are shown in Figure 26.6(a) and (b), respectively.

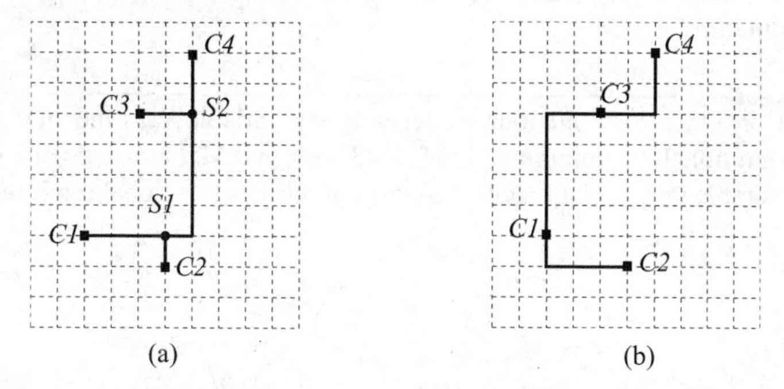

Figure 26.6 Illustration of WL estimation (a) rectilinear SMT. Steiner points *S1* and *S2* are added (b) rectilinear MST

Note that we have added Steiner points *S1* and *S2* in rectilinear SMT in contrast to rectilinear MST. We can see that the rectilinear SMT WL estimate is 13 units, while the rectilinear MST WL estimate is 14 units.

The HPWL estimate is popularly employed by placers because it is computationally simple. Additionally, the HPWL estimate matches the rectilinear SMT estimate for two- and three-pin nets. For nets with four or more pins, the HPWL estimate can be significantly less than the actual WL. The rectilinear SMT estimate can correlate better with the post-routing WL in these cases [4].

26.1.2 Techniques

Placement algorithms have been studied since the 1970s. The general placement problem is NP-complete [6, 8]. Therefore, several algorithms based on heuristics have been proposed over the last several decades. Broadly, we can classify placement algorithms into simulated annealing, min-cut placement techniques, and analytical techniques [1].

Simulated Annealing

Simulated annealing is a general combinatorial optimization technique that we can use for many problems, including placement [9–12].

Inspiration: Simulated annealing draws inspiration from the annealing of solids that forces the material to a low energy state. It involves heating a solid and taking it to a disordered state. When a solid is at a high temperature, more configurations are possible. Subsequently, the material is slowly cooled, allowing it to be at the equilibrium at each temperature. Finally, the solid attains a crystalline or desired ordered state. The final form is determined by the statistical mechanics and the *cooling schedule*.

Approach: In simulated annealing, the algorithm starts with an initial configuration and iteratively moves to new configurations until some stopping criterion is met. At each stage of algorithm, *low-cost* configurations are *accepted,* which is intuitive. However, simulated annealing also accepts *high-cost* configurations with some *probability*. Accepting a high-cost solution differentiates simulated annealing from a simple greedy approach. We keep reducing the probability of acceptance of a *high-cost* solution as the algorithm progresses. In the initial stages of the algorithm, the acceptance of a high-cost solution allows simulated annealing to escape *local minima* and possibly reach the *global minimum.* In the final stages of the algorithm, it accepts only low-cost solutions and thus settles greedily on some local minima. In the best case, this local minimum can be the global minimum.

Example 26.7 Consider a cost function that depends on a variable X, as shown in Figure 26.7. The cost function has multiple local minima M1, M2, M3, M4, and M5. However, the global minimum is M3, and it is desirable that an optimization algorithm reaches the global minimum M3.

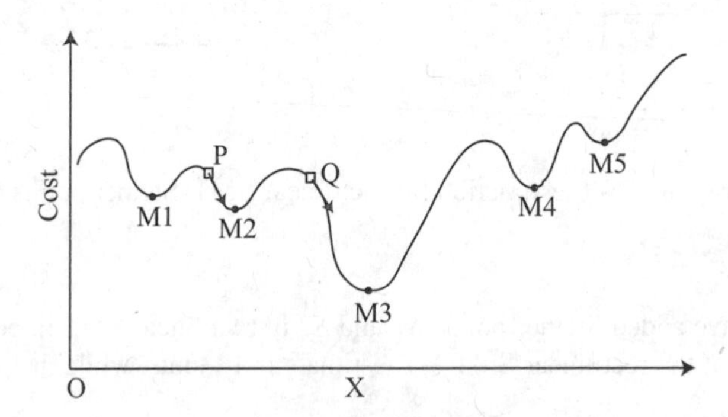

Figure 26.7 A function with multiple local minima

Let us assume that the initial solution starts at the point P. If we accept only low-cost solutions, we can reach the local minima M2. After reaching M2, a greedy algorithm cannot escape it since the cost only increases in its vicinity in both directions.

However, simulated annealing can get out of the local minima M2 and possibly reach the point Q since it can accept high-cost solutions also with some probability. After reaching Q, simulated annealing can greedily settle at the global minimum M3. Thus, simulated annealing is more likely to reach the global minimum than a simple greedy approach.

Algorithm: We explain the simulated annealing algorithm using functions *Update* and *SimAnneal* shown in Algo. 26.1 and Algo. 26.2, respectively. The function *Update* updates the placement configuration at a given temperature T. By placement configuration, we mean a placement solution in which the location of each cell is well-defined. The temperature T is a parameter that mimics the temperature of the solid that is being annealed. Initially, we start with a high value for T and gradually reduce it as the algorithm proceeds.

Algorithm 26.1: *Update*

Input:

- Given configuration CC
- Maximum number of moves M
- Current temperature T

Output:

- Updated configuration CC

Steps:

1: **for** $i \leftarrow 1$ to M **do**
2:     *Generate* a new configuration NC using current configuration CC
3:     $\Delta E \leftarrow Cost(NC) - Cost(CC)$
4:     **if** ($\Delta E < 0$) **then**
5:         $CC \leftarrow NC$
6:     **else**
7:         acceptance probability $P \leftarrow e^{-\Delta E/T}$
8:         *random* $\leftarrow$ *Generate* a random number between 0 and 1
9:         **if** (*random* $< P$) **then**
10:             $CC \leftarrow NC$
11:         **end if**
12:     **end if**
13: **end for**

The function *Update* tries out M different configurations iteratively. It can obtain a new configuration in several ways. For example, it can obtain a new configuration by displacing a cell or swapping the position of two cells. For the new configuration NC, it computes the associated cost *cost*(NC). The cost function can be the estimated total WL or some more sophisticated function. Then, the algorithm computes the cost difference ΔE between the new and the current configurations (line: 3). If the cost decreases, it accepts the new configuration (line: 5); else it accepts the new configuration with a probability $P = e^{-\Delta E/T}$ (line: 10). If it accepts the new configuration *NC*, it updates the current configuration *CC* with *NC*.

Algorithm 26.2: *SimAnneal*

Input:

- Given configuration C_0
- Initial temperature T_0
- Maximum number of moves at a given temperature M

Output:

- Updated configuration

Steps:

1: Current configuration $CC \leftarrow C_0$
2: Current temperature $T \leftarrow T_0$
3: **while** (!*Terminate*(*T*)) **do**
4:     *Update*(*CC*, *M*, *T*)
5:     $T \leftarrow Cool(T)$
6: **end while**
7: **return** *CC*

The top-level algorithm for simulated annealing *SimAnneal* is shown in Algo. 26.2. It takes an initial placement configuration C_0 and iteratively improves the configuration. We can choose C_0 to be some random initial configuration or construct it heuristically. We choose the starting effective temperature T_0 to be high enough such that a random search of the configuration space is possible. We choose the number of moves M tried by *Update* at a given temperature based on the design size and the allowed runtime.

The function *SimAnneal* optimizes the configuration at a particular temperature by calling the function *Update* (line: 4), and then it reduces the temperature using the function *Cool* (line: 5). The cooling can be done simply by updating the temperature as $T = \alpha T$, where α is a positive constant less than 1. The quality of the solution depends heavily on the cooling schedule. When the cooling is gradual, the runtime is high, and the solution quality can improve. As $T \rightarrow 0$, the probability of accepting the high-cost solution $P = e^{-\Delta E/T}$ moves toward zero. We can see that when $T \approx 0$, P becomes ≈ 0 for $\Delta E > 0$. Hence, when $T \rightarrow 0$, simulated annealing starts behaving as a greedy search algorithm.

The simulated annealing algorithm continues until it meets the termination condition, i.e., the function *Terminate* returns *true*. We can choose to terminate the algorithm when the temperature reduces below a predefined threshold. Alternatively, we can terminate if the runtime exceeds a predefined limit or we obtain a negligible improvement in a few successive iterations.

The simulated annealing algorithm has been widely used for global placement since the 1980s, though most industry-standard placement tools now employ other techniques as described in the following paragraphs [11].

Min-cut Placers

Min-cut placers attempt to place the components of densely connected sub-circuits nearby [13–15]. The algorithm repeatedly partitions a given circuit (netlist) into densely connected sub-circuits to achieve this goal. It determines densely connected sub-circuits using a *min-cut partition* algorithm. There are efficient algorithms, such as the *Kernighan–Lin algorithm* and the *Fiduccia–Mattheyses algorithm*, which can be employed to obtain min-cut partition [16–18].[4]

In addition to finding the min-cut partition for a circuit/sub-circuit, a min-cut placer divides the layout area into *bins* by a horizontal or vertical line. Subsequently, it places each sub-circuit or partition in different bins. A min-cut placer repeats the above process *recursively*. The recursion terminates when each sub-circuit consists of only one cell, and its position in the layout is uniquely defined. We illustrate this technique in the following example.

Example 26.8 Consider a graph shown in Figure 26.8(a).

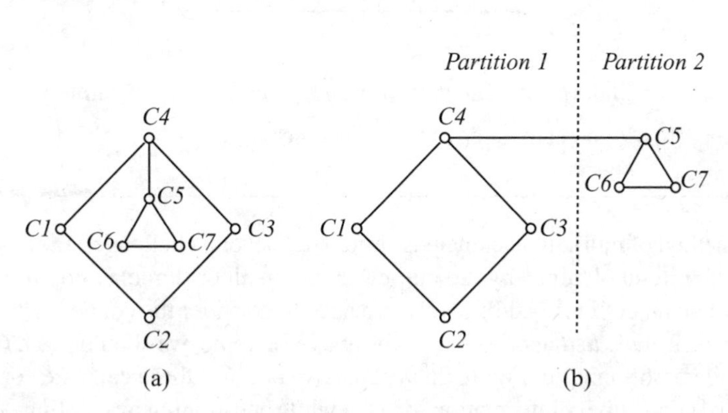

Figure 26.8 Illustration of min-cut partition (a) given graph with vertex set {*C1*, *C2*, ..., *C7*} and (b) min-cut partition with two sets {*C1*, *C2*, *C3*, *C4*} and {*C5*, *C6*, *C7*}

The min-cut partition for this graph is shown in Figure 26.8(b). In this partition, only one edge crosses the partition. When we try to find other possible partitions, we observe that all other partitions have more edges crossing the partition. Therefore, the partition shown in Figure 26.8(b) is the min-cut partition.

After finding the min-cut partition, a min-cut placer divides the layout area into two bins by a vertical line, as shown in Figure 26.9(a). Subsequently, it places the two min-cut partitions in them. In the next step, it divides the layout by a horizontal line and places the corresponding min-cut partitions in them, as shown in Figure 26.9(b). Next, it divides the layout by vertical lines and places the corresponding min-cut partitions, as shown in 26.9(c). Then, the recursion terminates because each sub-circuit consists of only one cell.

[4] For Kernighan–Lin algorithm, see Section 25.2 ("Partitioning").

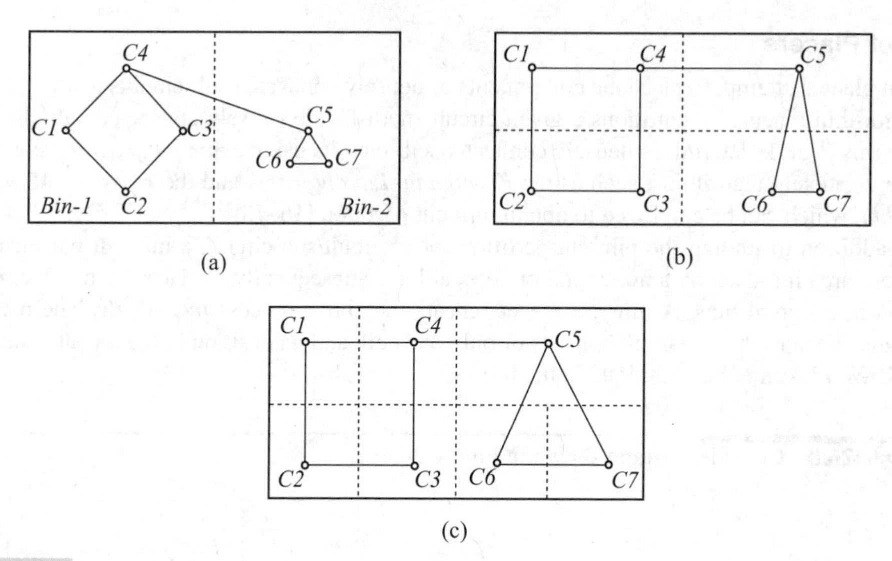

Figure 26.9 Steps for finding min-cut partition and placing them separately in the partitioned layout (a) first step, (b) second step, and (c) final placement

The above method of min-cut placement is simplistic. Several challenges arise in practical cases. For example, a sub-circuit obtained by the min-cut partition algorithm may not fit the bins obtained by vertical or horizontal cutlines. Additionally, we need to consider the connectivity of the cells in a partition with the cells *outside* that partition. In the above example, we should place *C4* and *C5* in the adjacent bins while partitioning in Figure 26.9(b) and (c) because these cells are connected. We can preserve the global connectivity information for a net while partitioning by creating *dummy terminals* on the partition boundary. The dummy terminals bias the placement algorithm such that the cells connected to other partitions get placed closer to the boundary and reduce the total WL. Interested readers can refer to [18] to know more about this aspect of the min-cut placement algorithm.

Practically, min-cut placers have been quite successful and have typically outperformed simulated annealing-based global placers for industrial designs since the mid-1990s [1]. However, most industry-standard placement tools now employ analytic placement algorithms.

Analytical Placers

Analytical placers capture the placement problem in a concise set of equations and use efficient mathematical solvers to solve those equations [1, 8, 19–22]. Analytical placement techniques have been studied since the 1970s [19, 23]. Analytical placers matured with time and could outperform min-cut placers on large industrial designs by 2005 [1].

Analytical placers consider cells as point objects with coordinates (x_i, y_i) on the layout. Thus, given N cells, an analytical placer needs to determine the matrices $X = [x_1, x_2 ... x_N]$ and $Y = [y_1, y_2 ... y_N]$. While determining these matrices, an analytical placer minimizes some cost functions and honors the given constraints, such as the location of fixed objects. We can mathematically model the cost function for analytical placement in many ways.

Quadratic cost function: Let us assume that the cost of connecting two cells C_i and C_j is represented by an element c_{ij} in a two-dimensional $N \times N$ *connection matrix* C. The element c_{ij} can be computed by counting the number of nets between C_i and C_j.

A popular quadratic cost function of WL for a given placement solution is [19]:

$$\Phi(X, Y) = \frac{1}{2}\sum_{i,j=1}^{N} c_{i,j} \times [(x_i - x_j)^2 + (y_i - y_j)^2]. \quad (26.2)$$

We can express the above quadratic cost function in another form such that determining X and Y for the minimum cost solution becomes easier. Let us represent the sum of all the elements in the ith row of the connection matrix as s_i. Let us define a diagonal matrix D as follows:

$$d_{i,j} = \begin{cases} 0, & i \neq j \\ s_i, & i = j \end{cases} \quad (26.3)$$

If we define a matrix $B = D - C$, then it can be shown that the cost function $\Phi(X, Y)$ can also be written as [19][5]:

$$\Phi(X, Y) = XBX^T + YBY^T \quad (26.4)$$

We illustrate the computation of the above cost function in the following example.

Example 26.9 Consider the graph shown in Figure 26.10(a) representing connections between cells.

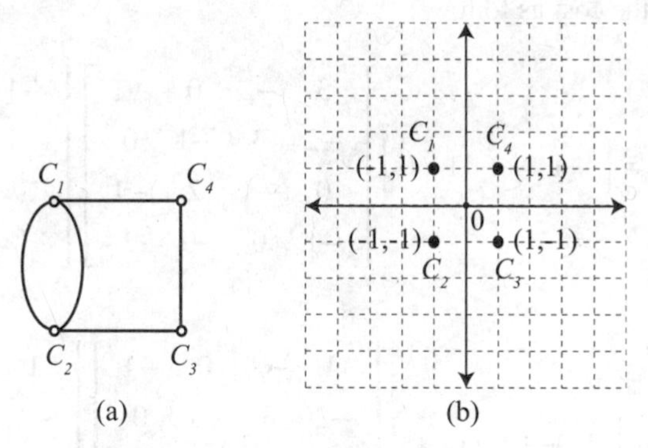

Figure 26.10 Given cell (a) connections and (b) placement

We can determine its connection matrix as:

$$C = \begin{bmatrix} 0 & 2 & 0 & 1 \\ 2 & 0 & 1 & 0 \\ 0 & 1 & 0 & 1 \\ 1 & 0 & 1 & 0 \end{bmatrix}$$

Note that, since there are two nets between cells C_1 and C_2, $c_{12} = c_{21} = 2$.

[5] We denote the transpose of a matrix A as A^T.

Next, consider the placement of these cells in Figure 26.10(b). Let us first compute its cost using Eq. 26.2.

$$\Phi(X, Y) = \frac{1}{2}[2 \times \{(-1-(-1))^2 + (1-(-1))^2\} + ...] = \frac{1}{2} \times 40 = 20$$

Next, let us compute the cost using Eq. 26.4. We first derive matrix D using the connection matrix C and Eq. 26.3:

$$D = \begin{bmatrix} 2+1 & 0 & 0 & 0 \\ 0 & 2+1 & 0 & 0 \\ 0 & 0 & 1+1 & 0 \\ 0 & 0 & 0 & 1+1 \end{bmatrix} = \begin{bmatrix} 3 & 0 & 0 & 0 \\ 0 & 3 & 0 & 0 \\ 0 & 0 & 2 & 0 \\ 0 & 0 & 0 & 2 \end{bmatrix}$$

Then, we compute B as:

$$B = D - C = \begin{bmatrix} 3 & -2 & 0 & -1 \\ -2 & 3 & -1 & 0 \\ 0 & -1 & 2 & -1 \\ -1 & 0 & -1 & 2 \end{bmatrix}$$

Then, we compute the cost as follows:

$$XBX^T = \begin{bmatrix} -1 & -1 & 1 & 1 \end{bmatrix} \begin{bmatrix} 3 & -2 & 0 & -1 \\ -2 & 3 & -1 & 0 \\ 0 & -1 & 2 & -1 \\ -1 & 0 & -1 & 2 \end{bmatrix} \begin{bmatrix} -1 \\ -1 \\ 1 \\ 1 \end{bmatrix} = 8$$

$$YBY^T = \begin{bmatrix} 1 & -1 & -1 & 1 \end{bmatrix} \begin{bmatrix} 3 & -2 & 0 & -1 \\ -2 & 3 & -1 & 0 \\ 0 & -1 & 2 & -1 \\ -1 & 0 & -1 & 2 \end{bmatrix} \begin{bmatrix} 1 \\ -1 \\ -1 \\ 1 \end{bmatrix} = 12$$

$$\Phi(X, Y) = XBX^T + YBY^T = 8 + 12 = 20.$$

It is same as computed using Eq. 26.2.

Minimizing cost function: The quadratic cost functions, such as given by Eq. 26.4, are an indirect measure of the total WL. Nevertheless, analytical placers commonly use them because mathematical solvers can efficiently minimize a *quadratic* cost function for large problems [22]. A trivial solution that minimizes the cost function is $X = Y = 0$. We add constraints $XX^T = 1$ and

$YY^T = 1$ to yield a nonzero solution. We can minimize the cost function represented by Eq. 26.4 if we choose the smallest eigenvalues of the matrix B [19]. The corresponding eigenvectors give the placement solution X and Y. Interested readers can refer to [19] for details.

The above formulation of analytical placement is one of the several techniques used by analytical placers. The cost functions are typically modified to include fixed objects and avoid the high density of cells over the layout. Nevertheless, the solution obtained using an analytical placer typically has cell overlaps, and legalization becomes necessary.

26.2 LEGALIZATION AND DETAILED PLACEMENT

The location of a cell produced by a global placer can be *illegal* due to the following reasons:

1. Two or more cells overlap.
2. A cell occupies illegal sites, such as between cell rows.
3. Location is not aligned to the power rails and routing tracks.

Therefore, after global placement, we need to perform legalization.

Legalization: It removes overlap between cells and snaps a cell to a legal site with minimum impact on costs such as WL, timing, or congestion. Note that the task of legalization is not to spread cells over the layout. Typically, the spreading of cells is undertaken during global placement by including it in its cost function.

Legalization makes only incremental changes in the placement solution such that the new location is legal, as illustrated in Figure 26.11. It finds a legal position for an illegally placed cell or a group of cells by making the slightest movement [1, 24].

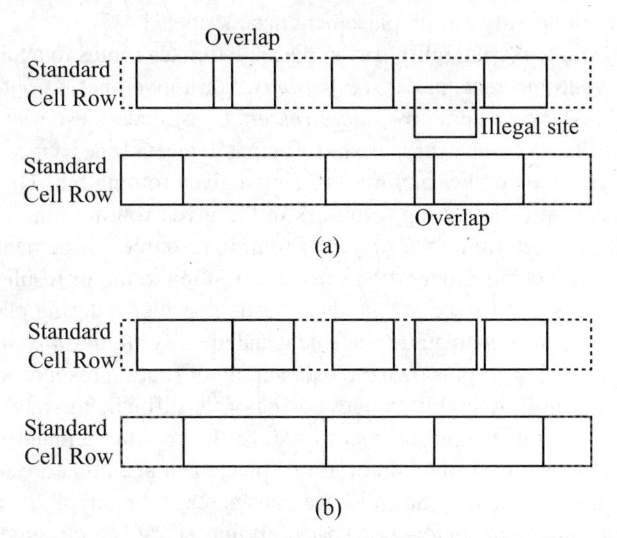

Figure 26.11 Placement (a) before legalization and (b) after legalization

Detailed placement: Post-legalization, we can improve some QoR measures by incrementally changing the location of a group of cells. This task is performed by a *detailed placer*. Note that a detailed placer is expected to preserve the legality of the given placement solution. Moreover, it should not create extra violations of congestion and placement density. Typically, a detailed placer improves WL or routability by swapping neighboring cells, redistributing free sites, and moving cells in an unused portion of the layout [1, 25].

26.3 TIMING-DRIVEN PLACEMENT

During global placement, in addition to the total WL, we can consider other cost metrics. In performance-critical designs, we want a placer to minimize the delay of the critical path.

A timing-driven placer internally performs static timing analysis (STA) and can identify *critical nets* in a design. It attempts to minimize delay metrics, such as total negative slack (TNS) and worst negative slack (WNS), by placing cells that are on the critical nets close-by [26–29]. It can achieve this by assigning more *weights* or cost to nets that are timing-critical. The extra weights on the timing-critical nets bias the placement solution such that timing-critical objects get placed nearby.

Challenges: A crucial consideration in net weight-based timing-driven placement is the stability of the placement algorithm. Instability can occur if fixing a critical path makes some other paths more critical. Consequently, a net can become noncritical and then critical in successive steps. Thus, the convergence of the placement algorithm can become difficult [30]. This problem can be avoided by carrying out *incremental STA* and maintaining a history of modifications. The industry-standard placers are typically timing-driven.

26.4 ROUTABILITY-DRIVEN PLACEMENT

Routability depends on several factors. Despite minimizing the total WL during placement, a design can become unroutable due to a shortage of available routing resources in some regions. Therefore, explicitly considering routability during placement has merits [31–35].

Detecting congestion: A routability-driven placer estimates routes to avoid routing failures. It divides the circuit layout into rectangular *routing bins*. For a given net, a *route* is defined by a set of adjacent routing bins. We can employ *global routers* to estimate these routes during placement because global routers have become very fast and efficient in recent times. Once we have done global routing, we can compute the number of wires crossing a given routing bin. This information allows us to determine the demand for routing resources in the given routing bin. The available routing resources in a routing bin determine the *supply* of routing resources. If demand is high (compared to supply), congestion can occur. Congestion can cause routing to fail or result in timing issues due to *detour*. Therefore, we should avoid potential congestion problems during placement.

We can use a *congestion map* to guide placement and identify *congestion hotspots*. A congestion map highlights regions with a surplus demand over supply of routing resources.[6]

Alleviating congestion: A highly congested region is difficult to route. Therefore, we can move out cells from the congestion hotspots to reduce the routing demand. We can incorporate the congestion metrics in the cost function of global placement such that the solutions with higher congestion are penalized more. Alternatively, we can increase the supply of routing resources in a routing bin by allocating *white space* (i.e., space left unused during placement) in the congested region. We can introduce white space in a region by temporarily inflating cells during placement. Note that allocating white space to relieve congestion improves routability at the expense of longer WLs [34].

[6] We will discuss the precise definition of congestion in Chapter 28 ("Routing").

If we know possible congestion regions in our design, we can guide a placer to alleviate congestion. For example, if corner regions of a macro are prone to congestion, we can create *halos* or *placement blockages* around that macro. Alternatively, we can specify low *utilization* locally in the region where we expect congestion. A placer typically honors user-specified utilization constraints. Therefore, it will place fewer cells in the low-utilization regions, and will be able to avoid potential congestion problems. Similarly, if we notice that there are *voltage drop hotspots* in some regions due to high cell density, we can specify low utilization locally in those regions.

26.5 POST-PLACEMENT TRANSFORMATIONS

Placement assigns locations to design entities. As a result, the information content for a design increases after placement. We can carry out some design transformations that utilize the newly added placement information to identify appropriate targets and improve post-placement QoR.

26.5.1 Standard Cell Optimizations

Post-placement, we can estimate WL and net capacitances with a greater accuracy. Therefore, we can carry out standard cell optimizations such as *resizing* to fix timing violations and satisfy *maximum load* and *maximum slew* constraints for a net. Additionally, we can insert *buffers* according to the estimated capacitance for large fanout nets. We can also add *repeaters* for nets connecting far-located pins. Note that we need to perform incremental placement and legalization after making changes to the cells (or adding or deleting cells).

Post-placement, we can choose cells with an appropriate threshold voltage (V_T). For example, we can replace a cell that is timing noncritical (i.e., cells having a high positive slack) with a high-V_T cell. Though a high-V_T cell is slower, it dissipates less leakage power. Therefore, employing a high-V_T cell can reduce leakage power dissipation. We can employ low-V_T cells to improve performance in timing-critical paths at the expense of leakage power.

26.5.2 Scan Cell Reordering

As described in Chapter 21 ("Scan design"), we carry out the scan insertion before physical design. It involves replacing flip-flops with scan cells, connecting them to form scan chains, and making connections for scan-enable pins. Hence, scan design requires significant routing resources. We can reduce the routing resource requirement of a scan chain by *reordering scan cells* based on their locations. We cannot carry out such reordering before placement because we do not know the locations of the scan cells (or the corresponding flip-flops) on the layout.

Example 26.10 Consider a simple scan chain consisting of six flip-flops, as shown in Figure 26.12.

Assume that the preplacement scan chain order is: *FF1*, *FF2*, *FF3*, *FF4*, *FF5*, and *FF6*, as shown by the flylines in Figure 26.12(a).

Assume that scan cells are reordered as: *FF1*, *FF3*, *FF5*, *FF2*, *FF4*, and *FF6*, as shown by the flylines in Figure 26.12(b).

We can observe that by reordering scan cells, we can reduce the total WL and improve the routability of a design. This routability improvement is the primary motivation of carrying out scan cell reordering after placement.

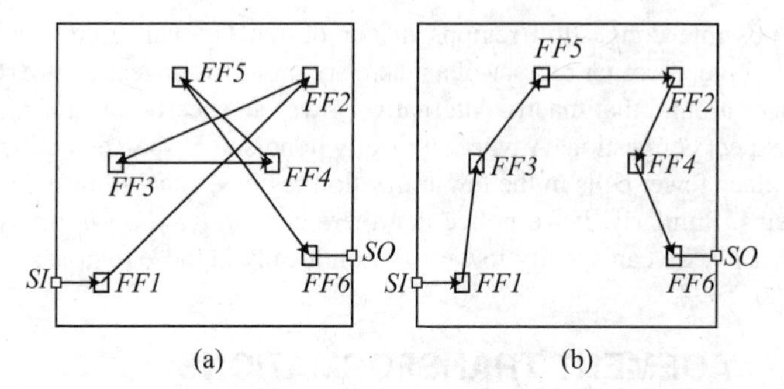

Figure 26.12 Scan cell order (a) before reordering and (b) after reordering

In a typical design flow, we ignore the scan chain connections *while* performing placement. It relaxes scan chain routing constraints and can improve the placement solution. Post-placement, we perform scan chain reordering based on the scan cell locations [36–38]. Typically, we reorder such that scan cells that are nearby form consecutive flip-flops in the scan chain. We can consider other objectives also to reorder scan cells, such as reducing the number of transitions in shift operations and dynamic power dissipation.

26.5.3 Spare Cell Placement

As design complexity increases, some bugs can escape design verification and are detected only in the post-silicon stages. Fixing post-silicon bugs often requires costly and time-taking *respins*. A mask set, consisting of 20–40 individual masks, contributes greatly to the cost of respins [39]. Typically, the cost of device layer masks (lower level masks) is more than the metal layer masks (upper layer masks) due to smaller feature sizes in the device layers [39]. We can reduce the cost of respin by reusing device layer masks and making necessary fixes *only* on the metal layer masks. This is the primary motivation of inserting *spare cells* in a design [39–41].

A spare cell is a standard cell and exists in a manufactured die but is left unconnected to any logic signal. However, if some extra standard cells are needed during respin, we connect spare cells appropriately. It will require modifications in the masks of a few metal layers only, and the device layer masks need not change. Thus, we can reduce the cost of respin using spare cells.

Placement of spare cells: Ideally, we should place a spare cell in the neighborhood of signals that we need to rewire. Additionally, the functionality of the spare cell should match the functional requirement of the respin. However, when designing, we cannot anticipate the fixes that will be required in the future. Therefore, we choose spare cells that are capable of generating a rich set of Boolean functions. We place them in regions of possible fixes. If there are no specific regions where we anticipate fixes, we place them randomly in the empty spaces over the layout.

Note that spare cells are left unconnected to any logic signal in a design. Therefore, design implementation tools can inadvertanly remove them to minimize the total cell area for a design. We may need to inform implementation tools to retain the spare cells that were inserted in a design.

26.6 RECENT TRENDS

As technology advances, the number of cells in a design increases. Consequently, placement becomes more challenging and time-consuming. We typically employ multi-threading and parallelism to reduce runtime of a placer [42, 43]. Additionally, we can consider exploiting machine learning (ML) techniques for efficient placement [44]. There are some similarities between minimizing WL in placement and minimizing error function in an artificial neural network (ANN). In the former, we obtain optimal locations, while in the latter, we obtain optimal weights of the ANN. We can exploit this similarity to design ML-based placement techniques.

As technology advances, standard cell libraries change. For example, in advanced technology libraries, cells can span multiple rows, and a single library can contain cells of different heights [45, 46]. We should consider these changes in cells during legalization and detailed placement. Additionally, as new technologies, such as 3D-IC, are becoming popular, we need to adapt conventional 2D placers to handle emerging requirements [47].

REVIEW QUESTIONS

26.1 What is the role of placement in the design flow? Why is it challenging to perform placement on a large industrial design?

26.2 What is the difference between global placement and detailed placement?

26.3 Why does a placement tool attempt to minimize the total WL in a design?

26.4 On what factors does the placement solution depend?

26.5 Four cells are placed at the corners of a square of length $2L$, as shown in Figure 26.13. These cells need to be connected using a network of wires. Find θ such that total WL is minimum in Figure 26.13(a). Find the value of the minimum WL. Show that the total WL in (b) is longer than in (a).

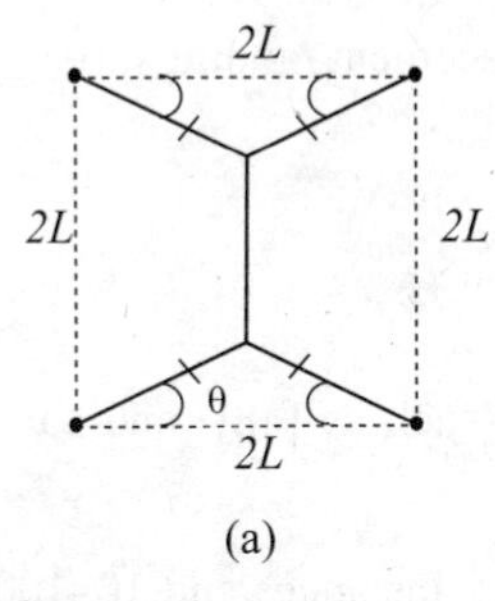

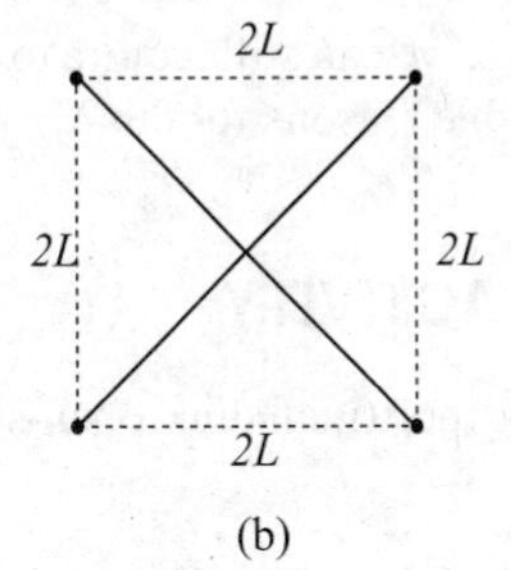

Figure 26.13 Problem 26.5

26.6 Assume that there are three nets in a design: *N1*, *N2*, and *N3*. The net *N1* is connected to three cells *C1*, *C2*, and *C3*. The net *N2* is connected to two cells *C4* and *C5*. The net *N3* is connected to five cells *C6*, *C7*, *C8*, *C9*, and *C10*. A placement tool has identified the following locations for the cells:

C1(0, 0), *C2*(10, 4), *C3*(2, 6),
C4(0, 10), *C5*(0, 20),
C6(0, 30), *C7*(0, 35), *C8*(20, 35), *C9*(35, 35), and *C10*(70, 75).

(a) Compute the total WL of the design, based on HPWL approximation. The unit of length is micron.

(b) A placement tool is using a simulated annealing algorithm to find the optimum solution. The acceptance probability of the new solution is given as $P = e^{-\Delta E/T}$ when $\Delta E > 0$. Here, ΔE is the increase in the total WL due to the new solution and $T = 40$ micron. The placement engine perturbs the above solution by changing the location of *C10* (all other cells are at the same location as mentioned above). Find ΔE and the acceptance probability if the new locations of *C10* are as follows:
(i) *C10*(70, 45)
(ii) *C10*(90, 95)

26.7 How does the aspect ratio of a standard cell affect the legalization problem?

26.8 Why does a detailed placer attempt to redistribute free sites?

26.9 A timing-driven placer assigns more weights to critical nets. Why can this strategy hamper the convergence of the placement algorithm?

26.10 Why does a timing-driven placer attempt to reduce TNS in a design?

26.11 Why is minimizing total WL during placement not sufficient to ensure the routability of a design?

26.12 How can congestion-driven placement be in conflict with WL-driven placement?

26.13 A designer knows congestion hot spots in a design. What steps can s/he take during placement to avoid congestion?

26.14 Post-placement, a designer notices a few nets having maximum capacitance violations. What steps can s/he take to fix this problem?

26.15 A portion of a design is found to have a very high positive slack. Using this information, can some steps be taken to reduce leakage power dissipation in the design?

26.16 How can scan cell reordering be done to improve the routability of a design?

26.17 What is the motivation of adding spare cells in a design?

26.18 Post-routing, we can still require to do some placement modifications. What could be some of the possible reasons for this?

TOOL-BASED ACTIVITY

Take any placement tool (including open-source tools such as [48]) and carry out the following activities:

1. Choose a relatively small netlist (such as 100–1000 instances and 10–100 flip-flops) that can be analyzed easily. You should give the constraints and physical information (possibly using library exchange format (LEF) files) as inputs. Initialize floorplan by assuming some reasonable value for the various parameters (such as utilization factor and aspect ratio). Perform pin placement and power planning.
2. Perform timing-driven global placement, legalization, and detailed placement. Perform timing analysis after this step. Examine the path with the WNS and understand how the tool computes various values in the timing report.
3. Perform routability-driven global placement, legalization, and detailed placement. Perform timing analysis after this step. Examine the path with the WNS and understand how the tool computes various values in the timing report.

4. Change the utilization factor of the core and repeat steps 2 and 3. Observe and analyze the changes in the results.

REFERENCES

[1] I. L. Markov, J. Hu, and M. Kim. "Progress and challenges in VLSI placement research." *Proceedings of the IEEE* 103 (Nov. 2015), pp. 1985–2003.

[2] G. Sigl, K. Doll, and F. M. Johannes. "Analytical placement: A linear or a quadratic objective function?" *Proceedings of the 28th ACM/IEEE Design Automation Conference* (1991), pp. 427–432.

[3] A. E. Caldwell, A. B. Kahng, S. Mantik, I. L. Markov, and A. Zelikovsky. "On wirelength estimations for row-based placement." *IEEE Transactions on Computer-aided Design of Integrated Circuits and Systems* 18, no. 9 (1999), pp. 1265–1278.

[4] J. A. Roy and I. L. Markov. "Seeing the forest and the trees: Steiner wirelength optimization in placement." *IEEE Transactions on Computer-aided Design of Integrated Circuits and Systems* 26, no. 4 (2007), pp. 632–644.

[5] M. Pedram and B. Preas. "Interconnection length estimation for optimized standard cell layouts." *ICCAD* (1989), pp. 390–393.

[6] N. A. Sherwani. *Algorithms for VLSI Physical Design Automation*. Springer Science & Business Media, 2012.

[7] M. R. Garey and D. S. Johnson. "The rectilinear Steiner tree problem is NP-complete." *SIAM Journal on Applied Mathematics* 32, no. 4 (1977), pp. 826–834.

[8] K. Shahookar and P. Mazumder. "VLSI cell placement techniques." *ACM Computing Surveys (CSUR)* 23, no. 2 (1991), pp. 143–220.

[9] N. Metropolis, A. W. Rosenbluth, M. N. Rosenbluth, A. H. Teller, and E. Teller. "Equation of state calculations by fast computing machines." *The Journal of Chemical Physics* 21, no. 6 (1953), pp. 1087–1092.

[10] S. Kirkpatrick, C. D. Gelatt, and M. P. Vecchi. "Optimization by simulated annealing." *Science* 220, no. 4598 (1983), pp. 671–680.

[11] C. Sechen and A. Sangiovanni-Vincentelli. "TimberWolf3. 2: A new standard cell placement and global routing package." *23rd ACM/IEEE Design Automation Conference* (1986), pp. 432–439, IEEE.

[12] R. A. Rutenbar. "Simulated annealing algorithms: An overview." *IEEE Circuits and Devices Magazine* 5, no. 1 (1989), pp. 19–26.

[13] M. A. Breuer. "A class of min-cut placement algorithms." *Proceedings of the 14th Design Automation Conference* (1977), pp. 284–290.

[14] X. Yang, M. Sarrafzadeh, et al. "Dragon2000: Standard-cell placement tool for large industry circuits." *IEEE/ACM International Conference on Computer Aided Design. ICCAD-2000. IEEE/ACM Digest of Technical Papers (Cat. No. 00CH37140)* (2000), pp. 260–263, IEEE.

[15] D. J.-H. Huang and A. B. Kahng. "Partitioning-based standard-cell global placement with an exact objective." *Proceedings of the 1997 International Symposium on Physical Design* (1997), pp. 18–25.

[16] B. W. Kernighan and S. Lin. "An efficient heuristic procedure for partitioning graphs." *The Bell System Technical Journal* 49, no. 2 (1970), pp. 291–307.

[17] C. M. Fiduccia and R. M. Mattheyses. "A linear-time heuristic for improving network partitions." *19th Design Automation Conference* (1982), pp. 175–181, IEEE.

[18] A. E. Dunlop, B. W. Kernighan, et al. "A procedure for placement of standard cell VLSI circuits." *IEEE Transactions on Computer-aided Design* 4, no. 1 (1985), pp. 92–98.

[19] K. M. Hall. "An r-dimensional quadratic placement algorithm." *Management Science* 17, no. 3 (1970), pp. 219–229.

[20] A. B. Kahng, S. Reda, and Q. Wang. "APlace: A general analytic placement framework." *Proceedings of the 2005 International Symposium on Physical Design* (2005), pp. 233–235.

[21] J. Lu, P. Chen, C.-C. Chang, L. Sha, J. Dennis, H. Huang, C.-C. Teng, and C.-K. Cheng. "ePlace: Electrostatics based placement using Nesterov's method." *2014 51st ACM/EDAC/IEEE Design Automation Conference (DAC)* (2014), pp. 1–6, IEEE.

[22] N. Viswanathan and C.-N. Chu. "FastPlace: Efficient analytical placement using cell shifting, iterative local refinement, and a hybrid net model." *IEEE Transactions on Computer-aided Design of Integrated Circuits and Systems* 24, no. 5 (2005), pp. 722–733.

[23] L. Sha and R. W. Dutton. "An analytical algorithm for placement of arbitrarily sized rectangular blocks." *22nd ACM/IEEE Design Automation Conference* (1985), pp. 602–608, IEEE.

[24] P. Spindler, U. Schlichtmann, and F. M. Johannes. "Abacus: Fast legalization of standard cell circuits with minimal movement." *Proceedings of the 2008 International Symposium on Physical Design* (2008), pp. 47–53.

[25] M. Pan, N. Viswanathan, and C. Chu. "An efficient and effective detailed placement algorithm." *ICCAD-2005. IEEE/ACM International Conference on Computer-aided Design, 2005* (2005), pp. 48–55, IEEE.

[26] W. Swartz and C. Sechen. "Timing driven placement for large standard cell circuits." *Proceedings of the 32nd Annual ACM/IEEE Design Automation Conference* (1995), pp. 211–215.

[27] T. Kong. "A novel net weighting algorithm for timing-driven placement." *Proceedings of the 2002 IEEE/ACM International Conference on Computer-aided Design* (2002), pp. 172–176.

[28] B. Halpin, C. R. Chen, and N. Sehgal. "Timing driven placement using physical net constraints." *Proceedings of the 38th Annual Design Automation Conference* (2001), pp. 780–783.

[29] W. E. Donath, R. J. Norman, B. K. Agrawal, S. E. Bello, S. Y. Han, J. M. Kurtzberg, P. Lowy, and R. I. McMillan. "Timing driven placement using complete path delays." *Proceedings of the 27th ACM/IEEE Design Automation Conference* (1991), pp. 84–89.

[30] J. Cong, J. R. Shinnerl, M. Xie, T. Kong, and X. Yuan. "Large-scale circuit placement." *ACM Transactions on Design Automation of Electronic Systems (TODAES)* 10, no. 2 (2005), pp. 389–430.

[31] U. Brenner and A. Rohe. "An effective congestion-driven placement framework." *IEEE Transactions on Computer-aided Design of Integrated Circuits and Systems*, 22, no. 4 (2003), pp. 387–394.

[32] P. N. Parakh, R. B. Brown, and K. A. Sakallah. "Congestion driven quadratic placement." *Proceedings of the 35th Annual Design Automation Conference* (1998), pp. 275–278.

[33] W. Hou, H. Yu, X. Hong, Y. Cai, W. Wu, J. Gu, and W. H. Kao. "A new congestion-driven placement algorithm based on cell inflation." *Proceedings of the 2001 Asia and South Pacific Design Automation Conference* (2001), pp. 605–608.

[34] C. Li, M. Xie, C.-K. Koh, J. Cong, and P. H. Madden. "Routability-driven placement and white space allocation." *IEEE Transactions on Computer-aided Design of Integrated Circuits and Systems* 26, no. 5 (2007), pp. 858–871.

[35] M.-C. Kim, J. Hu, D.-J. Lee, and I. L. Markov. "A SimPLR method for routability-driven placement." *2011 IEEE/ACM International Conference on Computer-aided Design (ICCAD)* (2011), pp. 67–73, IEEE.

[36] S. Ghosh, S. Basu, and N. A. Touba. "Joint minimization of power and area in scan testing by scan cell reordering." *IEEE Computer Society Annual Symposium on VLSI, 2003. Proceedings* (2003), pp. 246–249, IEEE.

[37] M. Hirech, J. Beausang, and X. Gu. "A new approach to scan chain reordering using physical design information." *Proceedings International Test Conference 1998 (IEEE Cat. No. 98CH36270)* (1998), pp. 348–355, IEEE.

[38] Y. Bonhomme, P. Girard, L. Guiller, C. Landrault, S. Pravossoudovitch, and A. Virazel. "Design of routing-constrained low power scan chains." *Proceedings Design, Automation and Test in Europe Conference and Exhibition* 1 (2004), pp. 62–67, IEEE.

[39] K.-H. Chang, I. L. Markov, and V. Bertacco. "Reap what you sow: Spare cells for post-silicon metal fix." *Proceedings of the 2008 International Symposium on Physical Design* (2008), pp. 103–110.

[40] Y.-P. Chen, J.-W. Fang, and Y.-W. Chang. "ECO timing optimization using spare cells." *2007 IEEE/ACM International Conference on Computer-aided Design* (2007), pp. 530–535, IEEE.

[41] Z.-W. Jiang, M.-K. Hsu, Y.-W. Chang, and K.-Y. Chao. "Spare-cell-aware multilevel analytical placement." *2009 46th ACM/IEEE Design Automation Conference* (2009), pp. 430–435, IEEE.

[42] T. Lin, C. Chu, and G. Wu. "POLAR 3.0: An ultrafast global placement engine." *2015 IEEE/ACM International Conference on Computer-aided Design (ICCAD)* (2015), pp. 520–527, IEEE.

[43] C.-X. Lin and M. D. Wong. "Accelerate analytical placement with gpu: A generic approach." *2018 Design, Automation & Test in Europe Conference & Exhibition (DATE)* (2018), pp. 1345–1350, IEEE.

[44] Y. Lin, S. Dhar, W. Li, H. Ren, B. Khailany, and D. Z. Pan. "DREAMPlace: Deep learning toolkit-enabled GPU acceleration for modern VLSI placement." *Proceedings of the 56th Annual Design Automation Conference 2019* (2019), pp. 1–6.

[45] Y. Lin, B. Yu, X. Xu, J.-R. Gao, N. Viswanathan, W.-H. Liu, Z. Li, C. J. Alpert, and D. Z. Pan. "MrDP: Multiple-row detailed placement of heterogeneous-sized cells for advanced nodes." *IEEE Transactions on Computer-aided Design of Integrated Circuits and Systems* 37, no. 6 (2017), pp. 1237–1250.

[46] C.-H. Wang, Y.-Y. Wu, J. Chen, Y.-W. Chang, S.-Y. Kuo, W. Zhu, and G. Fan. "An effective legalization algorithm for mixed-cell-height standard cells." *2017 22nd Asia and South Pacific Design Automation Conference (ASP-DAC)* (2017), pp. 450–455, IEEE.

[47] J. Lu, H. Zhuang, I. Kang, P. Chen, and C.-K. Cheng. "ePlace-3D: Electrostatics based placement for 3D-ICs." *Proceedings of the 2016 on International Symposium on Physical Design* (2016), pp. 11–18.

[48] T. Ajayi, V. A. Chhabria, M. Fogaça, S. Hashemi, A. Hosny, A. B. Kahng, M. Kim, J. Lee, U. Mallappa, M. Neseem, et al. "Toward an open-source digital flow: First learnings from the openroad project." *Proceedings of the 56th Annual Design Automation Conference 2019* (2019), pp. 1–4.

27 Clock Tree Synthesis

...imperfection is in some sort essential to all that we know of life ... No human face is exactly the same in its lines on each side, no leaf perfect in its lobes, no branch in its symmetry. All admit irregularity as they imply change ... All things are literally better, lovelier, and more beloved for the imperfections ...

—John Ruskin, *The Stones of Venice*, Vol. II, Chapter 6, 1853

In a synchronous design, the *clock signal* orchestrates all its operations. It is responsible for the launch and the capture of *data* at the flip-flops. During logic synthesis, we assume that an *ideal clock* signal triggers all the flip-flops. An ideal clock has the same *waveform* (voltage vs. time relationship) at all the points in the circuit. It implies that, for a given clock signal, all the flip-flops launch and capture data at the same time instant. In other words, there is no *clock skew* or difference in the arrival time of the clock signal at the launch and the capture flip-flops.

In physical design, the assumption of an ideal clock does not hold. Clock signals are generated by some clock source (either external or internal) and distributed throughout a circuit. When a clock signal propagates through a circuit, it encounters path-dependent delay due to parasitic resistances and capacitances. Therefore, delivering an ideal clock signal to the clocked circuit elements is impossible for practical designs.

Given that we cannot deliver an ideal clock signal physically, *clock tree synthesis* (CTS) aims to implement a clock distribution network that delivers a clock signal that is *similar* to an ideal clock. Specifically, CTS aims to *minimize* the clock skew. To achieve this, CTS employs special routing topologies and circuit elements such as *clock buffers* to obtain a balanced delay in the clock distribution network. Since clock distribution networks can be responsible for 25–70% of the total dynamic power dissipated in a circuit, CTS also attempts to reduce the clock network's power dissipation [1–3]. Moreover, post-CTS new timing violations can appear due to the nonideal clock behavior. CTS tries to fix these timing violations also.

Traditionally, achieving zero clock skew is the goal of CTS. However, sometimes system performance can be improved by maintaining *useful skews* on specific clock paths. Therefore, CTS inserts useful skews by building an unbalanced clock network wherever it finds such opportunities.

In this chapter, we will explain the above aspects of CTS in detail.

27.1 TERMINOLOGIES

Consider a clock distribution network shown in Figure 27.1. Using this network, we describe a few commonly used terms related to CTS.

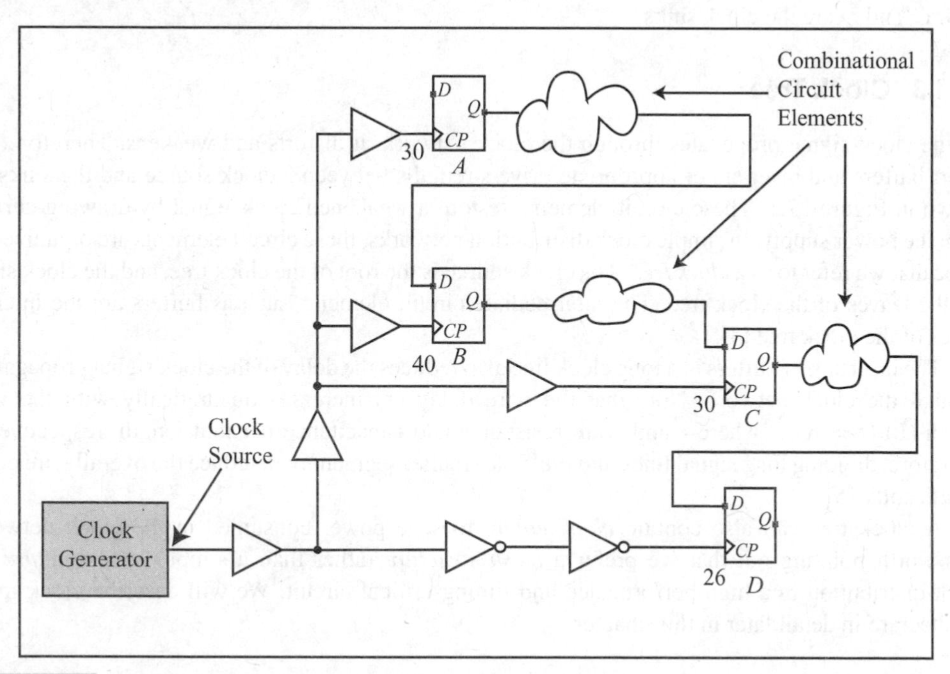

Figure 27.1 An illustration of a clock network. The numbers alongside the clock-pin (*CP*) indicate the arrival time of the clock signal in picoseconds. Assume that the clock signal is generated at the clock source at $t = 0$

27.1.1 Sequentially Adjacent Flip-flops

A flip-flop *F1* is sequentially adjacent to flip-flop *F2* if the output data of *F1* can propagate through at least one *combinational path* and reach the input data pin of *F2*. Note that a combinational path consists of only combinational circuit elements. In the simplest case, a combinational path can just be a wire.

We refer to the flip-flop *F1* that launches data as the *launch flip-flop* and the flip-flop *F2* that captures data as the *capture flip-flop*. In Figure 27.1, *A* is the launch flip-flop and *B* is the capture flip-flop, *B* is the launch flip-flop and *C* is the capture flip-flop, and *C* is the launch flip-flop and *D* is the capture flip-flop. Note that, a flip-flop can be a launch flip-flop for one path and a capture flip-flop for another path.

27.1.2 Clock Source and Sinks

The starting point of a clock signal is referred to as the *clock source*. A clock source can be an internal clock generator, as shown in Figure 27.1. Alternatively, the clock source can be external to a chip, and the clock signal can enter through some input port. In this case, the corresponding input port is considered as the clock source. Note that for multi-clock designs, there can be multiple clock sources. CTS tool typically infers clock sources from the *clock constraints* defined in the SDC file.[1]

The final receiving endpoints of the clock signal are referred to as the *clock sinks*. Typically, the clock pin of a sequential circuit element is a clock sink. In Figure 27.1, the clock pins of the flip-flops *A*, *B*, *C*, and *D* are the clock sinks.

27.1.3 Clock Tree

As the clock signal propagates through the clock network, it distorts and weakens. Therefore, we insert buffers and inverters of appropriate drive strengths between a clock source and the sinks, as shown in Figure 27.1. These circuit elements restore a weakened clock signal by drawing current from the power supply. In simple clock distribution networks, these circuit elements are organized as a tree that we refer to as a *clock tree*. The clock source is the root of the clock tree, and the clock sinks are the leaves of the clock tree. The intermediate circuit elements such as buffers are the internal nodes of the clock tree [4].

The insertion of buffers in a long clock line also reduces the delay of the clock signal propagating through the clock network. Note that the wire delay (τ) increases quadratically with the wire length (L) ($\tau \approx rcL^2$, where r and c are resistance and capacitance per unit length, respectively). Therefore, dividing long signal lines into multiple smaller segments can reduce the overall wire delay significantly [5].

A clock tree can also contain *clock gaters* to save power consumed in the clock network. It is worth pointing out that we prefer a *mesh structure* rather than a simple *tree structure* for clock distribution in a high-performance and timing-critical circuit. We will describe clock mesh architecture in detail later in this chapter.

27.1.4 Insertion Delay

The time taken by the clock signal to propagate through a clock tree and reach a clock sink A is referred to as its insertion delay d_A. It depends on the cell delay and the wire delay along the clock path.

Assume that the clock signal is generated (or arrives) at the clock source at time $t = 0$. Then, the *arrival time* of the clock signal t_A at a sink A will be the same as its insertion delay d_A. In Figure 27.1, the insertion delays and the arrival times of the clock signal are: t_A=d_A=30 ps, t_B=d_B=40 ps, t_C=d_C=30 ps, and t_D=d_D=26 ps.

For a given clock network, the sink where the clock signal arrives last represents the worst-case behavior of the clock network. The worst-case insertion delay can be denoted as $d_{worst} = MAX(d_X)$, where X is any sink in the clock network. A clock routing algorithm often considers minimizing d_{worst} as one of its objectives.

The insertion delay is sometimes referred to as the *clock latency*. However, the term clock latency is preferably used in the pre-CTS stage to represent the expected post-CTS insertion delay.[2]

[1] For details on clock constraints, see Chapter 15 ("Constraints").

[2] For details, see Section 15.3.1 ("Latency").

27.1.5 Clock Skew

Given two clock sinks S_1 and S_2, the difference in the clock arrival times $(t_{S_1} - t_{S_2})$ is known as the clock skew $\delta_{S_1S_2}$. In Figure 27.1, $\delta_{AB} = -10$ ps, $\delta_{AC} = 0$, $\delta_{AD} = 4$ ps, $\delta_{BC} = 10$ ps, $\delta_{BD} = 14$ ps, and $\delta_{CD} = 4$ ps.

The clock skew between two sinks is relevant only for sequentially adjacent flip-flops [6]. For two sinks that are not sequentially adjacent (such as *A* and *C*), the existence of clock skews between them does not directly affect the performance or reliability of a circuit.

Example 27.1 Consider a circuit that contains *N* flip-flops. Theoretically, data can be launched by any flip-flop among *N* flip-flops and it can be captured by the remaining $(N-1)$ flip-flops. Thus, theoretically, we can define $N(N-1)$ clock skews. Nevertheless, practically, only a subset of flip-flops in a design are sequentially adjacent. Therefore, out of $N(N-1)$ clock skews, only a few of them are practically relevant.

In Figure 27.1, only three clock skews ($\delta_{AB} = -10$ ps, $\delta_{BC} = 10$ ps, and $\delta_{CD} = 4$ ps) out of 12 possible skews are practically relevant.

Impact of skews on the circuit timing: For the deterministic operation of a synchronous circuit, the clock skew should satisfy setup and hold constraints. These requirements are discussed in detail in Chapter 14 ("Static timing analysis"). However, for the sake of convenience, we are presenting them again. The setup constraint is:

$$T_{period} > \delta_{lc} + T_{clk-ql} + T_{data_{max}} + T_{setup-c} \tag{27.1}$$

where T_{period} is the clock period, δ_{lc} is the clock skew between the launch flip-flop and the capture flip-flop, T_{clk-ql} is the delay from the clock pin to the Q-pin for the launch flip-flop, $T_{data_{max}}$ is the maximum delay in the data path between the launch flip-flop and the capture flip-flop, and $T_{setup-c}$ is the setup time of the capture flip-flop. Note that, in this book, we have followed the convention that the clock skew is *positive* when the clock arrival time at the launch flip-flop is *greater* than the clock arrival time at the capture flip-flop (i.e., the clock arrives later at the launch flip-flop than at the capture flip-flop). From Eq. 27.1 we can infer that the existence of a positive clock skew results in an increase in the minimum operable clock period and decrease in the maximum operable frequency. The hold constraint is:

$$\delta_{lc} + T_{clk-ql} + T_{data_{min}} > T_{hold-c} \tag{27.2}$$

where $T_{data_{min}}$ is the minimum delay of a signal between the launch flip-flop and the capture flip-flop and T_{hold-c} is the hold time of the capture flip-flop. From this equation we can infer that the existence of positive clock skew δ_{lc} relaxes the hold constraint.

Global clock skew: In a clock distribution network, the maximum value of the clock skew between any two pairs of sinks is referred to as the *global clock skew* Δ_{global}. Hence, $\Delta_{global} = MAX|\delta_{XY}|$, where *X* and *Y* are any two sinks. In Figure 27.1, $\Delta_{global} = \delta_{CB}$=14 ps.

While evaluating global clock skew, we consider system-wide maximum possible clock skew. Thus, the global clock skew targets constrain CTS even for sinks that are not sequentially adjacent. Therefore, global skew can correspond to two clock sinks that do not directly impact the circuit's performance. However, a clock routing algorithm often considers minimizing the global clock skew as one of its objectives since it is easier to model and estimate global clock skew.

27.2 CLOCK DISTRIBUTION NETWORKS

The clock distribution network is typically organized as *global clock network* and *local clock network*.

Global clock network: A global clock network distributes the clock signal to the various parts of a chip and covers a *large area*. It delivers the clock signal from a clock source to the starting terminals of the local clock network. The global clock distribution network attempts to reduce the clock skew and keep it close to zero. Therefore, it consists of a symmetric structure. Moreover, since it delivers a clock signal over a large area, it consists of multiple buffers of high drive capabilities. We carefully choose the architecture of a global clock network and often make manual interventions to achieve desired skew targets.

Local clock network: Local clock networks distribute clocks over a *smaller area* and are scattered all over the layout. They distribute the clock signal from a point where the global clock network terminates to the clocked elements (flip-flops) in their neighborhood. Therefore, the layout of a local clock network depends strongly on the location of the flip-flops. Since the location of flip-flops is not necessarily symmetric, local clock networks can have higher clock skews than global clock networks. A local clock network can have a few buffers, typically of low drive capabilities. Since, a local clock network contains numerous clock sinks, the total capacitance driven by the local clock network and the associated power dissipation can be higher than the corresponding global clock network [1].

27.3 CLOCK NETWORK ARCHITECTURES

There are two basic structures employed in a global clock distribution network: trees and meshes [7]. We can also employ hybrid structures that derive the benefits of both trees and meshes. We discuss these structures in the following paragraphs.

27.3.1 Trees

To minimize the clock skew, we choose a tree architecture such that the distances from the clock source to *all* the terminating points are equal.

H-tree and X-tree: In a symmetric clock tree, we first route the clock signal to the center of the chip. From the center, we feed the clock signal to the centroid of the H-structure or X-structure, as shown in Figure 27.2. Subsequently, the endpoints of the H-structure or the X-structure feed the clock signal to the centroid of the next level of smaller H-structure or smaller X-structure.

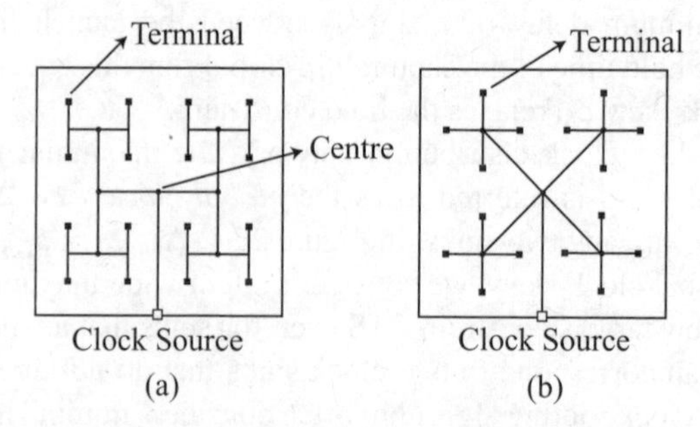

Figure 27.2 Symmetric tree structures: (a) H-tree and (b) X-tree [6]

The hierarchy of progressively smaller symmetric structures continues for a few levels and finally terminates. From the terminating points, flip-flops can be clocked, or a local clock network can be driven. We have shown two-level H-tree and X-tree in Figure 27.2. Nevertheless, we can continue to have more levels in the tree. Note that we can employ the X-tree architecture only in those designs that do not restrict routing to be rectilinear.

The length of paths from the clock source to all the terminals is equal in a symmetric tree. We reduce the width of the conductor progressively in an H-tree in lower levels of the hierarchy [8–10]. It helps in reducing inductive reflections of the clock signal at the branching points and reduces overall insertion delay. To boost the drive strength, we can add identical buffers or repeaters at symmetric locations. However, the addition of buffers in the clock distribution networks makes them more susceptible to process-induced variations.

Trade off between skew and insertion delay: The symmetric structures such as H-tree require more wire length and exhibit greater capacitance for the clock distribution network compared to a simple clock tree, such as shown in Figure 27.1 [11]. Therefore, the insertion delay of the clock signal typically increases in a symmetric clock tree. Thus, we are able to minimize the skew in a symmetric clock tree at the expense of the insertion delay.

Prone to process, voltage, and temperature (PVT) variations: Ideally, the skew between any two terminals in a symmetric tree is zero. However, due to PVT variations, the delay of circuit elements and interconnects on the clock path can deviate from their nominal values. Consequently, skews are observed even in symmetric tree-based clock distribution networks.

Possible solutions: One of the techniques to tackle variations in clock networks is to employ a *non-tree topology*. A simple method to obtain a non-tree topology is by adding crosslinks in an existing clock tree, as shown in Figure 27.3 [12, 13].

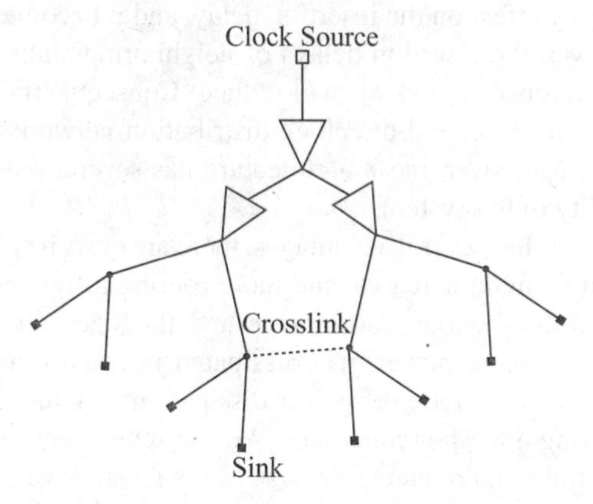

Figure 27.3 Crosslinks in a clock tree to tackle impact of PVT variations [12, 13]

The clock source gets connected to the sinks through *multiple paths* in a crosslinked tree. It results in a stronger correlation among insertion delays due to shared paths. Consequently, insertion delay variations reduce, and the clock distribution network becomes more tolerant to the PVT variations. However, crosslinks increase the total wire length of the clock network [12].

27.3.2 Meshes

Another popular non-tree architecture, known as mesh, is often employed in clock networks to tackle PVT variations, especially in timing-critical circuits. We can implement a clock mesh in multiple ways. A simple mesh architecture is illustrated in Figure 27.4 [14]. A metal wire mesh is laid over the entire layout and driven by *mesh drivers* at different locations. The clock sinks are connected to the nearest point on the clock mesh.

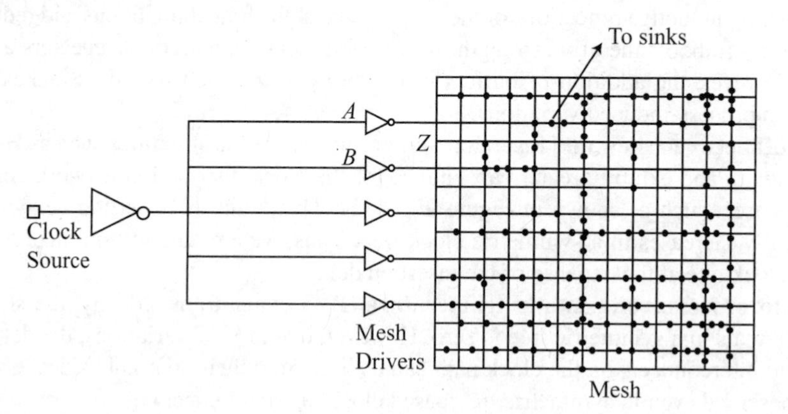

Figure 27.4 A clock mesh (small circles on the mesh indicate the connection points for clock sinks)

The mesh architecture ensures that multiple paths exist between the mesh drivers and the clock sinks. It leads to an averaging effect on the insertion delay, and it becomes less sensitive to process-induced variations. Moreover, the insertion delays of neighboring sinks become correlated due to shared paths, and their differences (clock skews) reduce. Consequently, we can obtain extremely small skews in a mesh architecture, and the clock distribution network becomes more tolerant to PVT variations [1, 15–17]. Moreover, mesh architecture has several redundant wires and devices, which increase the reliability of the system.

Though mesh architecture has several advantages, there are costs involved in adding redundancy to the circuit. The total wire length increases, and more routing resources are consumed compared to tree architectures. Additionally, total capacitance and the size of clock drivers increase in a mesh architecture. Therefore, more power gets dissipated in a mesh architecture compared to a tree architecture. Thus, we need to trade off power dissipation to achieve higher immunity against PVT variations for employing a mesh architecture. We can reduce power dissipation overheads by avoiding some noncritical links and reducing the size of the mesh drivers [16, 18].

Note that a significant portion of the power dissipated in a clock mesh is due to the short-circuit power dissipated in the driving buffers [19]. We explain it in the following example.

Example 27.2 Consider the mesh drivers *A* and *B* of Figure 27.4. We can represent their connections using a simple circuit shown in Figure 27.5(a).

The outputs of the inverters are tied together, while their inputs *A* and *B* receive the clock signal. Due to the PVT variations, the clock signal can arrive at *A* and *B* at different time instants. As a result, a skew δ_{AB} exists between the inputs of these inverters, as shown in Figure 27.5(b).

During the time interval δ_{AB}, the clock signal at A is 1, while the clock signal at B is 0. Consequently, *NMOS-1* and *PMOS-2* are switched on simultaneously inside the inverters, as shown in Figure 27.5(c). Therefore, a low-resistance path between V_{DD} and *GND* opens up as shown in the figure. Hence, a high short-circuit power can be dissipated in a clock mesh drivers even due to small skews in the clock signal.

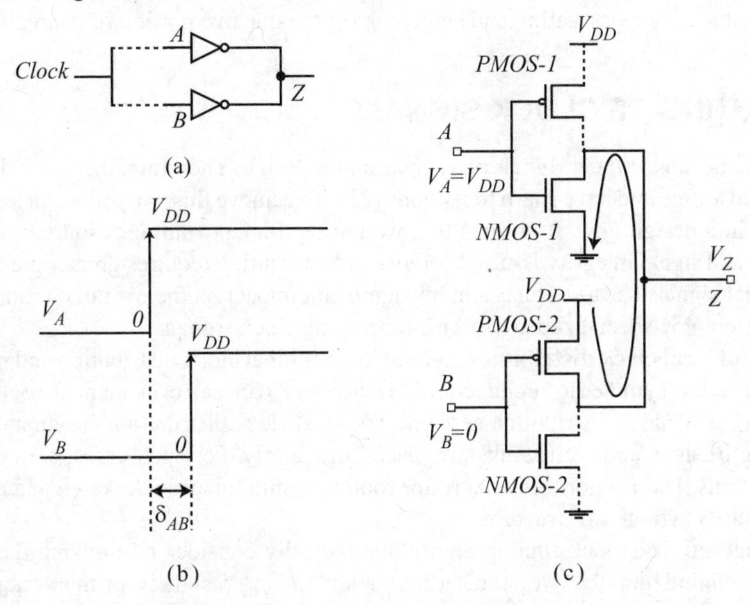

Figure 27.5 A pair of drivers for a mesh: (a) circuit, (b) skew at the inputs, and (c) short-circuit current path

---

We can drive a clock mesh in multiple ways. A popular technique is to drive it at multiple points using a top-level symmetric H-tree, as shown in Figure 27.6 [1, 20, 21].

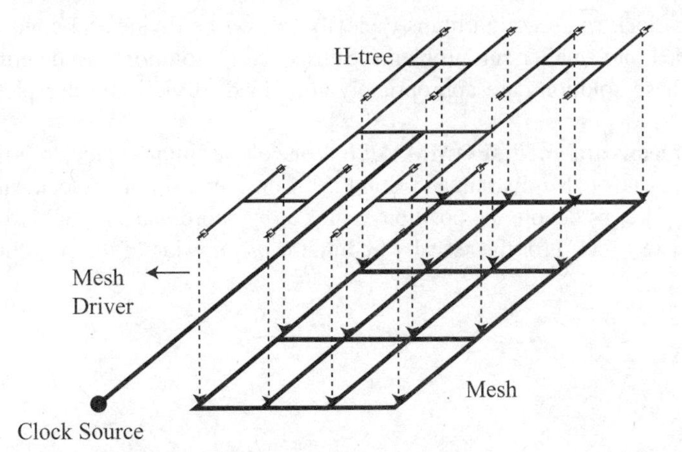

Figure 27.6 Hybrid clock architecture: top-level H-tree driving a mesh. The dashed lines represent drivers from the H-tree to the mesh

The terminals of the H-tree drive the mesh using buffers at the mesh intersection points. This architecture derives the merits of both trees and meshes. Trees achieve low skew over a long distance, offer low capacitances, and exhibit low insertion delays. Moreover, trees utilize routing resources efficiently and dissipate less power. Meshes provide a uniform clock network design irrespective of the load capacitances and achieves smaller local clock skews. Often, we employ multiple meshes in a design to enable easy clock gating and switching off the inactive portion of a circuit [22].

27.4 ROUTING OF CLOCK SIGNALS

The clocks are the most critical signals of a synchronous design. Therefore, the clock signals must be particularly noise-free and have sharp transitions [23]. To achieve this, we route clock signals before other signals in a design flow. It ensures that we have sufficient routing resources during CTS to avoid detours and signal integrity issues. Moreover, clock routing requires greater precision than the routing of other signals because it has a more significant impact on the overall circuit performance. Therefore, we employ specialized algorithms for routing clock signals.

Global and local clock distribution network: For global clock distribution and timing-critical circuits, we employ symmetric architectures. Hence, we can perform manual routing for some portion of the global clock distribution network. For local clock distribution, a symmetric clock tree is not feasible because numerous sinks are placed irregularly. For such cases, we employ special routing algorithms. These algorithms determine routing to minimize clock skews, and their solutions are not necessarily symmetric structures.

Cost function: The clock routing algorithms typically consider minimizing the global skew (Δ_{global}) and minimizing the worst insertion delay (d_{worst}) as their primary objectives [24]. Furthermore, they can consider other secondary objectives, such as minimizing the total wire length and the impact of coupling noises.

Many algorithms for routing clock signals have been proposed in the literature [24–27]. For simplicity, clock routing algorithms sometimes consider wire length as a simple approximation for the insertion delay. Hence, they simply equalize the wire length of the clock paths to all the sinks [25, 27]. Other wire delay models, such as the Elmore delay model, can also be employed during clock routing [6].

Approach: Clock routing algorithms typically follow the divide-and-conquer approach. The problem is divided into smaller sub-problems. Subsequently, solutions are determined for the sub-problems, and these solutions are appropriately combined to yield the complete solution to the given problem.

Method of means and medians (MMM): It is one of the simplest clock routing algorithms and was proposed by Jackson, Srinivasan, and Kuh [25]. Consider a set of N clock sinks represented as $S = \{s_1, s_2, ..., s_N\}$. Let us denote the position of sink s_i by coordinates (x_i, y_i). We can compute the center of mass $(x_c(S), y_c(S))$ for the set S by finding the *mean* value of the coordinates as follows:

$$x_c(S) = \frac{\sum_{i=1}^{N} x_i}{N} \tag{27.3}$$

$$y_c(S) = \frac{\sum_{i=1}^{N} y_i}{N} \tag{27.4}$$

We first route the clock from the clock source to the above center of mass.

Then, we partition S into two approximately equal partitions S_L and S_R based on x-coordinate. We arrange the given set of points in a nondecreasing order of x-coordinate. Subsequently, we take the first half ($\lceil N \rceil$) points in one partition S_L (left partition) and the rest in another partition S_R (right partition). Thus, partitioning is accomplished around the *median* of given points. After partitioning, we route the center of the mass of the set S to the centers of mass of the two partitions S_L and S_R.

Next, we partition sets S_L and S_R based on the y-coordinates. We arrange the points in S_L in the nondecreasing order of y-coordinate. Subsequently, we take approximately half of them in the bottom half S_{BL} and the rest in the top half S_{TL}. Then, we route the center of mass of S_L to centers of mass of S_{BL} and S_{TL}.

We recursively carry out the above partitioning and routing steps until there is only one point in each set. We illustrate this algorithm in the following example.

Example 27.3 Consider the set of sinks $S = \{s_1, s_2, s_3, s_4, s_5, s_6\}$ shown in Figure 27.7. We carry out MMM algorithm in the following steps.

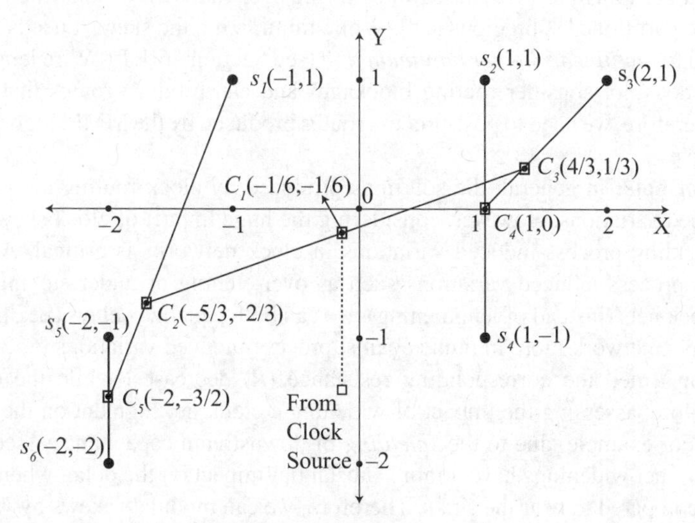

Figure 27.7 Clock routing using MMM algorithm

1. We compute the center of mass of the set S as follows:

$$x_c(S) = \frac{\sum_{i=1}^{N} x_i}{N} = \frac{-1+1+2+1-2-2}{6} = -\frac{1}{6}$$

$$y_c(S) = \frac{\sum_{i=1}^{N} y_i}{N} = \frac{1+1+1-1-1-2}{6} = -\frac{1}{6}$$

Hence, the center of mass of S is $C_1(-1/6, -1/6)$. We apply the clock source at C_1 or route it to C_1.

2. We partition the set into the left half $S_L = \{s_5, s_6, s_1\}$ and the right half $S_R = \{s_4, s_2, s_3\}$ based on x-coordinates. The centers of mass of S_L and S_R are $C_2(-5/3, -2/3)$ and $C_3(4/3, 1/3)$, respectively. We route C_1 to C_2 and C_3 as shown in Figure 27.7. Note that, we have shown the routing using straight lines, though we can use Manhattan routing also.
3. We partition S_L based on y-coordinates. Thus, we obtain $S_{BL} = \{s_5, s_6\}$ and $S_{TL} = \{s_1\}$. Then, we route the center of mass of S_L (C_2) to the centers of mass of S_{BL} ($C_5(-2, -3/2)$) and S_{TL} ($s_1(-1, 1)$). Note that when a set contains only one point then its center of mass is the same point. Similarly, we perform partitioning and routing for the right partition S_R. The partitioned sets are $S_{BR} = \{s_4, s_2\}$ and $S_{TR} = \{s_3\}$, and their centers of mass are $C_4(1, 0)$ and $s_3(2, 1)$, respectively.
4. Finally, we partition and route S_{BL} and S_{BR} because they contain two points each. After that, the algorithm terminates because each set contains only one point. Thus, we obtain the routing shown in Figure 27.7.

The MMM algorithm minimizes the skew by trying to equalize the distance of the center of mass of a region to the partitioned subregions [25]. While minimizing the skew, it needs to trade off wire length compared to *rectilinear Steiner minimum tree* (see Section 26.1.1 ("Wire length estimates")). Additionally, it does not consider routing blockages and can produce routes that intersect (as in Figure 27.7). Therefore, we need to post-process routes produced by the MMM algorithms to legalize and avoid obstacles.

Widening of nets: In general, the solutions produced by clock routing algorithms are prone to process-induced variations. However, considering the huge impact of clock skews on the system performance, tackling process-induced variations in clock networks is critical. A widened net is less sensitive to process-induced variations such as over-etching or under-etching. Therefore, by widening the clock nets (instead of lengthening), we can both equalize delays (i.e., minimize skews) and make the clock network more immune against process-induced variations [6, 28].

By widening a net, the corresponding resistance (R) decreases, while the capacitance (C) increases. Therefore, assessing the impact of widening a clock net segment on the delay ($\tau \sim RC$) is complicated. For example, due to the *shielding* of downstream capacitances (see Section 24.2.4 ("Wire models")), net widenings have a more substantial impact on the delay when we introduce it near the source compared to near the sinks. Therefore, we can modulate skews by widening various net segments of a given clock net by different amounts. Another advantage of reducing skews by widening net segments is that it decouples the skew optimization process from the routing of the clock network. Thus, the overall task of routing and skew reduction becomes easier.

Signal integrity issues: We design clock networks such that the transition of the clock signal is very sharp. It makes a clock line a *strong aggressor*, and it can introduce coupling noise to the adjacent wires. We can shield clock lines by running adjacent parallel wires to avoid this problem. Alternatively, we can increase the spacing between the clock lines and the adjacent signal lines. However, we must also ensure that these solutions do not introduce congestion by consuming excessive routing resources.

The clock network typically employs buffers of large size and drive strength. These buffers draw large currents from the power supply during charging and discharging. Consequently, these buffers can introduce power supply noise and a significant *instantaneous voltage drop*. We can surround the clock buffers with *decap cells* of appropriate sizes to avoid this problem.

27.5 POWER DISSIPATION IN CLOCK NETWORK

The capacitances in the clock network are charged and discharged every clock cycle, irrespective of circuit operation. Therefore, the dynamic power dissipated in the clock network is a significant component of the total power dissipated in a circuit.

Factors that impact power dissipation: The dynamic power dissipated in a clock network can be given as $P_{network} = C_{eff}V_{DD}^2 f$, where C_{eff} is the effective capacitance associated with the clock network (it takes into account all cell and interconnect capacitances), V_{DD} is the supply voltage, and f is the frequency of the clock. Therefore, to reduce the power dissipated in a clock network, we typically consider reducing the clock network's capacitances and supply voltage. Note that the performance goals determine the clock frequency, and we cannot reduce it.

Reducing power dissipation: In general, the effective capacitance of a clock distribution network depends on the clock architecture, the drivers employed in the clock networks, and the attributes of the interconnects such as their length, width, and resistivity [2]. The skew considerations primarily drive the architectural decisions of a clock network. Nevertheless, by *minimizing* the wire length of a clock network, the effective capacitance C_{eff} and the associated power dissipation can be reduced. Similarly, we can reduce power dissipation by keeping the clock drivers of *minimum size* that meet the delay and the slew targets.

Note that if drivers of the clock network are too small or too few, the delay and the slew targets of a clock signal can be violated [29]. Moreover, when the slew of a clock signal increases, the short-circuit power dissipation in the clock network also increases.[3] It is worth pointing out that though zero clock skew is desirable, nonzero clock skew can be tolerated if there are sufficient margins on the data path. In such situations, we can employ smaller drivers and shorter wires in the clock network to reduce the C_{eff} of the clock network. Thus, we can reduce the power dissipated in the clock network at the expense of clock skews and without compromising the clock frequency [30].

Since dynamic power is a quadratic function of V_{DD}, by making a clock network operate at a reduced supply voltage (e.g., 50% of V_{DD}), we can obtain a substantial saving in the power dissipation. However, on reducing the supply voltage, the cell delay increases. Nevertheless, we can compensate for this increase in delay by using faster clock buffers and pipelined registers (we also need to tolerate some capacitance increase). Overall, this technique allows us to save more than 50% of the clock network's dynamic power without compromising the overall circuit performance, even at reduced supply voltages [31, 32].

Clock gating: Another effective and popular technique to tackle the power dissipated in a clock network is to employ *clock gating*.[4] However, we should carefully consider *glitches* in the clock network that can get introduced due to clock gating. When *integrated clock-gating* (ICG) cells are employed for clock gating, then glitches are easily avoided. However, if individual latches and AND gates are used as clock gating elements, the placement should ensure that these cells are placed nearby. Moreover, CTS should minimize the skew between the latching element and the corresponding AND gate to avoid glitches.

---

[3] If slew increases, the PMOS and the NMOS in a CMOS inverter remain simultaneously ON for a longer duration, leading to a higher short-circuit power dissipation.

[4] For details, see Chapter 19 ("Power-driven optimizations").

27.6 USEFUL SKEWS

Though attaining zero skews is considered as the primary objective of CTS, introducing well-controlled skews can improve the system performance [33]. If a significant difference in the slack exists on the two sides of a flip-flop, we can allocate the excess margin on one side of the flip-flop to the more critical path on the other side of the flip-flop.

Motivation: From Eq. 27.1, we can observe that a *negative* clock skew ($\delta_{lc} < 0$) can relax the *setup requirement* and increase the maximum operable frequency. Therefore, we design a clock network such that we obtain a negative clock skew on the *critical path* (the worst setup slack path) at the expense of positive clock skew on the adjacent path. It allows data to propagate for a longer time on the critical path and increases the *maximum operable frequency* of the circuit.

The skews introduced between the launch flip-flop and the capture flip-flop that improve the system performance are called *useful skews*. Note that when we introduce *negative clock skew* on a path, the *hold requirement* becomes stricter. Therefore, while introducing useful skew, we should also consider the minimum delay of the data path and the corresponding hold constraints.

Example 27.4 Consider a portion of a circuit shown in Figure 27.8.

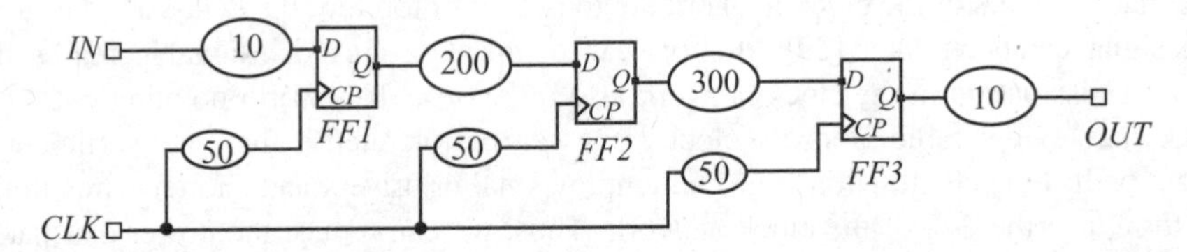

Figure 27.8 Circuit with zero skew. The delays of circuit elements are in picoseconds. Ovals represent combinational block

For simplicity, let us assume that the flip-flops are ideal (setup time, hold time, and $CP \rightarrow Q$ delay are zero). Assume that we ignore the path from the primary input to *FF1* and from *FF3* to the primary output due to sufficiently high slacks in these paths. In Figure 27.8, the clock skew is zero $\delta_{lc} = 0$ for each launch–capture flip-flop combination. It can be observed that the critical path is between *FF2* and *FF3*. The constraint on the clock period from Eq. 27.1 is: $T_{period} > 300$ ps. Thus, the maximum operable frequency is f_{max}=1000/300=3.33 GHz.

We can notice that adjacent to the critical path *FF2*→*FF3*, there is a noncritical path *FF1*→*FF2* with some margins. Therefore we can introduce some useful skews.

In Figure 27.9, we introduce useful skews by increasing the insertion delay to *FF1* and *FF3*, while keeping the insertion delay to *FF2* unchanged.

Now, the setup constraint for the data path between *FF2* and *FF3* is as follows:

$$T_{period} > 50 - 100 + 300 \Rightarrow T_{period} > 250$$

Similarly, the setup constraint for the data path between *FF1* and *FF2* is as follows:

$$T_{period} > 100 + 200 - 50 \Rightarrow T_{period} > 250$$

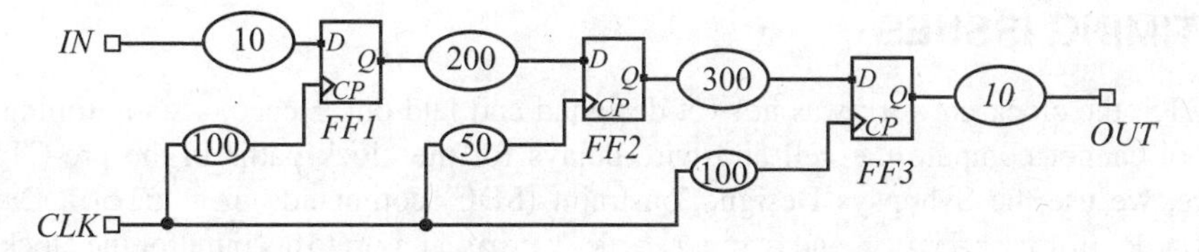

Figure 27.9 Circuit with useful skew in the clock path. The delays of circuit elements are in picoseconds. Ovals represent combinational block

The constraint imposed on T_{period} by the above two inequations are the same. Thus, the operable frequency f_{max}=1000/250=4 GHz. Thus, we can achieve higher maximum operable frequency by introducing useful skews on the clock paths.

The negative clock skew in the *FF2→FF3* path helped in transferring the surplus slack from the *FF1→FF2* side to the *FF2→FF3* side. Note that if we make the clock skew between *FF2→FF3* more negative, the path between *FF1→FF2* will become more critical, and the overall performance of the system will reduce. Thus, in general, we can maximize the circuit performance by choosing an optimum set of clock skews.

Technique: For a given circuit, we can determine a set of nonzero localized clock skews for each flip-flop such that we achieve the *maximum operable clock frequency* for the circuit. The set of localized clock skews thus obtained forms the *clock schedule* for maximizing the performance of the circuit. We need to consider both the *minimum* and the *maximum* delays of the combinational paths between each launch–capture flip-flop pair to meet both hold and setup constraints. We can formulate the problem of finding the optimal clock schedule as a *linear programming problem* and solve it using efficient linear programming solvers [6].

Corresponding to the optimal clock schedule, we can compute the minimum insertion delay for each flip-flop. Subsequently, a CTS tool can create an unbalanced clock distribution network to meet the targeted insertion delays [34]. Industry-standard CTS tools are typically equipped to insert useful skews in the clock network.

Other benefits: The useful skews can also help in reducing the *IR drop problems*. By introducing clock skews, we avoid the simultaneous triggering of flip-flops. Consequently, the peak current drawn by the power supply gets reduced, and the dynamic IR drop problem is ameliorated.

Comparison with retiming: It is instructive to compare useful skews with *retiming*.[5] Both techniques exploit the difference in the slacks on the input and the output sides of a flip-flop to achieve higher circuit performance.

The retiming transformation requires moving logic in the *data path*, while introducing useful skew requires changes in the *clock path*. Since we need to move a complete logic function or logic gate during retiming, a retiming transformation achieves coarser control of slack compared to inserting useful skews. Additionally, retiming poses problems in *combinational equivalence checking* (CEC) due to the combinational logic changes between flip-flops, while inserting clock skews do not create such problems.

---

[5] For details of retiming, see Section 17.3.5 ("Retiming").

27.7 TIMING ISSUES

Before CTS, the clock network was not yet designed and laid out. Hence, a static timing analysis (STA) tool cannot compute the cell and wire delays for the clock paths in the pre-CTS stages. Therefore, we use the Synopsys Design Constraint (SDC) commands `set_clock_latency`, `set_clock_uncertainty`, and `set_clock_transition` to account for the clock network in the pre-CTS timing analysis. Post-CTS, STA tools can compute cell delay and wire delay on the clock path using appropriate delay models. To instruct an STA tool to compute actual delays rather than estimates of the pre-CTS stage, we use the SDC command `set_propagated_clock` in the SDC file for the post-CTS timing analysis.

Post-CTS timing violations: Post-CTS, new timing violations can emerge when STA is done using a *propagated clock* signal: setup violations can occur due to additional positive clock skews, and hold violations can occur due to negative clock skews. A high transition time or slew in the clock path can also cause timing violations. Moreover, CTS can displace some data path elements from their original position to accommodate buffers for the clock network. Consequently, the delay of the data paths can also change and result in a few timing violations.

Another reason for timing violations in the post-CTS stage is the *pessimistic* approach taken by the STA tools in analyzing the *common circuit elements* in a *launch* clock path and the corresponding *capture* clock path. The setup requirement is determined by considering the *late launch* clock path (maximum arrival time) and the *early capture* clock path (minimum arrival time). However, this approach can become overly pessimistic when the launch clock path and the capture clock path have common circuit elements. We illustrate this concept in the following example.

Example 27.5 Consider a portion of a circuit shown in Figure 27.10.

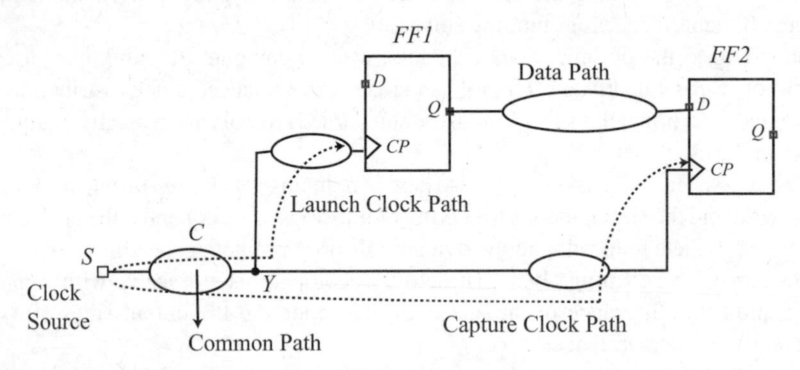

Figure 27.10 Illustration of common path pessimism in STA (ovals represent combinational block)

The launch clock path is from the clock source *S* to *FF1/CP* and the capture clock path is from *S* to *FF2/CP*.

For setup requirements, an STA tool considers the maximum arrival time of the path *S*→*FF1/CP*, and the minimum arrival time of the path *S*→*FF2/CP*. Note that the path from *S*, through *C*, to *Y* is common to both the launch clock path and the capture clock path. Thus, assuming different delay values for the same circuit elements in the common clock path is unrealistic and introduces pessimism in the timing analysis.

An STA tool can compute different delay for the same circuit element *C* while accounting for PVT variations. For example, assume that we have specified the following on-chip variation (OCV) derate factors[6]:

Late path derating factor=1.1
Early path derating factor=0.9

In setup analysis, it will scale the delay of *C* in the clock launch paths by a factor of 1.1, while it will scale it by a factor of 0.9 in the clock capture path.

A similar pessimism occurs in the computation of the hold requirement also.

Common path pessimism removal (CPPR): Industry-standard STA tools are equipped with CPPR algorithm to determine common path segments between the launch clock path and the capture clock path and remove pessimism arising due to the difference in the delay of the common circuit elements.

In the post-CTS timing analysis, we should enable CPPR in the STA tool to remove unnecessary pessimism. Moreover, a clock tree implementation tool should try to share circuit elements between the launch clock path and the capture clock path to ameliorate the problem of PVT variations.

Fixing post-CTS timing violations: We can fix the new timing violations obtained after CTS using timing-driven post-placement transformations explained in Chapter 26 ("Placement"). We can tackle post-CTS hold violations arising from large negative clock skews by introducing delay elements in the data path.

Post CTS, once we have verified that the circuit is timing safe using an STA tool and ensured the *logical equivalence* of the pre-CTS and the post-CTS netlist using a CEC tool, we freeze the clock network and allow only minimal engineering change order (ECO) fixes at the end of the design flow.

27.8 RECENT TRENDS

CTS has evolved over the decades as it needs to keep pace with the advances in technology and handle newer effects that become prominent. For example, at advanced process nodes, it needs to consider the greater impact of process-induced variations, clock waveform degradation, and high power dissipation [35–40].

Though historically, CTS has been skew and wire length-driven, in recent times, power-driven, timing-driven, and slew-driven CTS are becoming essential. Moreover, new circuit architectures such as 3D integrated circuits (ICs), require CTS algorithms to meet stringent power dissipation constraints and efficiently route long clock lines using through-silicon vias (TSVs).

REVIEW QUESTIONS

27.1 Why is the routing of the clock signal done before routing other signals?

27.2 What are sequentially adjacent flip-flops? Why is clock skew for sequentially adjacent flip-flops significant for timing verification?

27.3 How does dividing long clock lines into smaller segments and adding buffers between them help reduce the clock lines' overall delay?

27.4 Why are the clock skews typically higher for a local clock network than the global clock network?

---

[6] See Section 14.9.3 ("On-chip variation") for OCV.

27.5 How do symmetric tree structures such as H-tree and X-tree help minimize clock skews?

27.6 How do crosslinks help in ameliorating the problem of PVT variations in a symmetric tree clock distribution network?

27.7 What are the advantages and disadvantages of a mesh architecture compared to a tree architecture for clock distribution?

27.8 Why can the short-circuit power dissipation be high in a mesh-based clock distribution network? How can it be reduced?

27.9 Why should the slew of a clock signal be tightly controlled in a clock distribution network?

27.10 The critical portion of a circuit is shown in Figure 27.11.

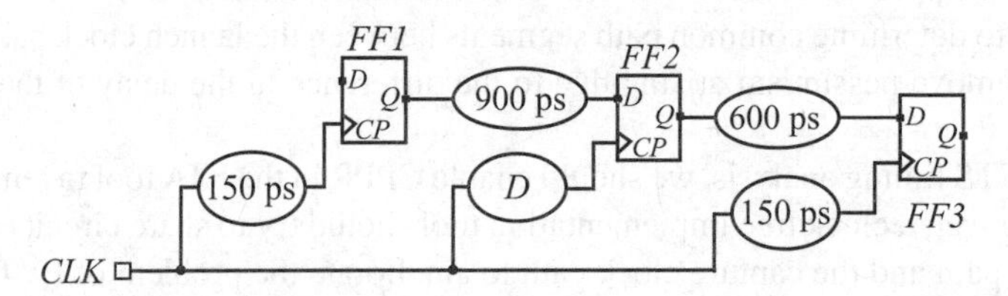

Figure 27.11 A given portion of a circuit

Assume that all the flip-flops are ideal (setup time, hold time, and $CP \rightarrow Q$ delay are zero). Find the delay D of the circuit element in the clock path such that the circuit operates at the maximum clock frequency satisfying the setup constraint. Ignore the delay of the wires and the hold constraints. Find the corresponding clock period.

27.11 A portion of a sequential synchronous circuit is shown in Figure 27.12. The following attributes are valid for both the flip-flops *FF1* and *FF2*: setup time=25 ps and $CP \rightarrow Q$ delay=25 ps. The delay of the combinational block in the data path is 900 ps. The frequency of the *Clock* is 1 GHz. Ignore the delay of all the wires.

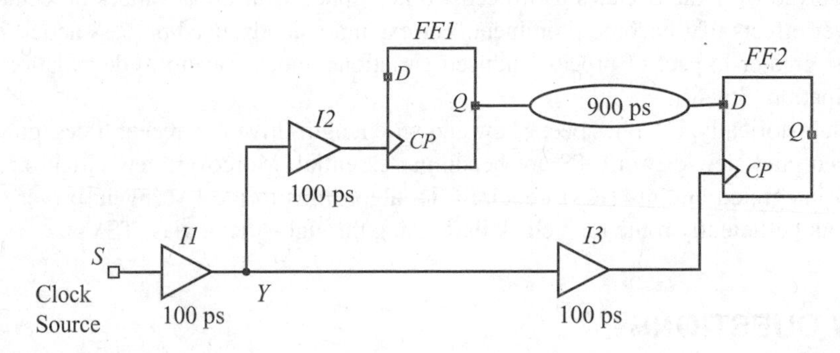

Figure 27.12 A given portion of a circuit

Find the setup slack under the following conditions:

(a) Nominal delays, as shown in the figure, are considered.

(b) Assume that an OCV derate factor of 1.1 is added to the delay for the late paths, and 0.9 is added to the delay for the early paths.

(c) Assume that an OCV derate factor of 1.1 is added to the delay for the late paths, and 0.9 is added to the delay for the early paths. Furthermore, consider CPPR in the computation.

TOOL-BASED ACTIVITY

Take any CTS tool (including open-source tools proposed in [41]) and carry out the following activities:

1. Load a design in which placement is already done. Choose a design with 100–1000 instances and 10–100 flip-flops for easy analysis. The design obtained after placement in the activity of the previous chapter can also be used.
2. Perform pre-CTS STA.
3. Carry out CTS.
4. Perform post-CTS STA. Analyze the worst setup slack path and the worst hold slack path.
5. Compare the timing reports of post-CTS STA with pre-CTS STA. Specifically, examine the clock paths.
6. Analyze the impact of CPPR on the worst setup slack.

REFERENCES

[1] P. J. Restle, T. G. McNamara, D. A. Webber, P. J. Camporese, K. F. Eng, K. A. Jenkins, D. H. Allen, M. J. Rohn, M. P. Quaranta, D. W. Boerstler, et al. "A clock distribution network for microprocessors." *IEEE Journal of Solid-State Circuits* 36, no. 5 (2001), pp. 792–799.

[2] R. Chen, N. Vijaykrishnan, and M. Irwin. "Clock power issues in system-on-a-chip designs." *Proceedings. IEEE Computer Society Workshop on VLSI'99. System Design: Towards System-on-a-chip Paradigm* (1999), pp. 48–53, IEEE.

[3] M. A. El-Moursy and E. G. Friedman. "Exponentially tapered H-tree clock distribution networks." *IEEE Transactions on Very Large Scale Integration (VLSI) Systems* 13, no. 8 (2005), pp. 971–975.

[4] E. G. Friedman and S. Powell. "Design and analysis of a hierarchical clock distribution system for synchronous standard cell/macrocell VLSI." *IEEE Journal of Solid-State Circuits* 21, no. 2 (1986), pp. 240–246.

[5] H. Bakoglu and J. D. Meindl. "Optimal interconnection circuits for VLSI." *IEEE Transactions on Electron Devices* 32, no. 5 (1985), pp. 903–909.

[6] E. G. Friedman. "Clock distribution networks in synchronous digital integrated circuits." *Proceedings of the IEEE* 89, no. 5 (2001), pp. 665–692.

[7] A. L. Sobczyk, A. W. Luczyk, and W. A. Pleskacz. "Power dissipation in basic global clock distribution networks." *2007 IEEE Design and Diagnostics of Electronic Circuits and Systems* (2007), pp. 1–4, IEEE.

[8] F. H. A. Asgari and M. Sachdev. "A low-power reduced swing global clocking methodology." *IEEE Transactions on Very Large Scale Integration (VLSI) Systems* 12, no. 5 (2004), pp. 538–545.

[9] J. Fishburn and C. A. Schevon. "Shaping a distributed-RC line to minimize Elmore delay." *IEEE Transactions on Circuits and Systems I: Fundamental Theory and Applications* 42, no. 12 (1995), pp. 1020–1022.

[10] S. Tam, S. Rusu, U. N. Desai, R. Kim, J. Zhang, and I. Young. "Clock generation and distribution for the first IA-64 microprocessor." *IEEE Journal of Solid-State Circuits* 35, no. 11 (2000), pp. 1545–1552.

[11] D. Somasekhar and V. Visvanathan. "A 230-MHz half-bit level pipelined multiplier using true single-phase clocking." *IEEE Transactions on Very Large Scale Integration (VLSI) Systems* 1, no. 4 (1993), pp. 415–422.

[12] A. Rajaram, J. Hu, and R. Mahapatra. "Reducing clock skew variability via crosslinks." *IEEE Transactions on Computer-aided Design of Integrated Circuits and Systems* 25, no. 6 (2006), pp. 1176–1182.

[13] A. Rajaram and D. Z. Pan. "Variation tolerant buffered clock network synthesis with cross links." *Proceedings of the 2006 International Symposium on Physical Design* (2006), pp. 157–164.

[14] M. P. Desai, R. Cvijetic, and J. Jensen. "Sizing of clock distribution networks for high performance CPU chips." *33rd Design Automation Conference Proceedings, 1996* (1996), pp. 389–394, IEEE.

[15] S. M. Reddy, G. R. Wilke, and R. Murgai. "Analyzing timing uncertainty in mesh-based clock architectures." *Proceedings of the Design Automation & Test in Europe Conference*, vol. 1, pp. 1–6. IEEE, 2006.

[16] X.-W. Shih, H.-C. Lee, K.-H. Ho, and Y.-W. Chang. "High variation-tolerant obstacle-avoiding clock mesh synthesis with symmetrical driving trees." *2010 IEEE/ACM International Conference on Computer-aided Design (ICCAD)* (2010), pp. 452–457, IEEE.

[17] M. Mori, H. Chen, B. Yao, and C.-K. Cheng. "A multiple level network approach for clock skew minimization with process variations." *ASP-DAC 2004: Asia and South Pacific Design Automation Conference 2004 (IEEE Cat. No. 04EX753)* (2004), pp. 263–268, IEEE.

[18] G. Venkataraman, Z. Feng, J. Hu, and P. Li. "Combinatorial algorithms for fast clock mesh optimization." *IEEE Transactions on Very Large Scale Integration (VLSI) Systems* 18, no. 1 (2009), pp. 131–141.

[19] G. Wilke, R. Fonseca, C. Mezzomo, and R. Reis. "A novel scheme to reduce short-circuit power in mesh-based clock architectures." *Proceedings of the 21st Annual Symposium on Integrated Circuits and System Design* (2008), pp. 117–122.

[20] D. W. Bailey and B. J. Benschneider. "Clocking design and analysis for a 600-MHz alpha microprocessor." *IEEE Journal of Solid-State Circuits* 33, no. 11 (1998), pp. 1627–1633.

[21] H. Chen, C. Yeh, G. Wilke, S. Reddy, H. Nguyen, W. Walker, and R. Murgai. "A sliding window scheme for accurate clock mesh analysis." *ICCAD-2005. IEEE/ACM International Conference on Computer-aided Design, 2005* (2005), pp. 939–946, IEEE.

[22] G. R. Wilke and R. Murgai. "Design and analysis of "Tree + Local Meshes" clock architecture." *8th International Symposium on Quality Electronic Design (ISQED'07)* (2007), pp. 165–170, IEEE.

[23] C. Yeh, G. Wilke, H. Chen, S. Reddy, H. Nguyen, T. Miyoshi, W. Walker, and R. Murgai. "Clock distribution architectures: A comparative study." *7th International Symposium on Quality Electronic Design (ISQED'06)* (2006), pp. 7–91, IEEE.

[24] N. A. Sherwani, *Algorithms for VLSI Physical Design Automation*. Springer Science & Business Media, 2012.

[25] M. A. Jackson, A. Srinivasan, and E. S. Kuh. "Clock routing for high-performance ICs." *27th ACM/IEEE Design Automation Conference* (1990), pp. 573–579, IEEE.

[26] R.-S. Tsay. "An exact zero-skew clock routing algorithm." *IEEE Transactions on Computer-aided Design of Integrated Circuits and Systems* 12, no. 2 (1993), pp. 242–249.

[27] A. Kahng, J. Cong, and G. Robins. "High-performance clock routing based on recursive geometric matching." *Proceedings of the 28th ACM/IEEE Design Automation Conference* (1991), pp. 322–327.

[28] S. Pullela, N. Menezes, and L. T. Pillage. "Reliable non-zero skew clock trees using wire width optimization." *Proceedings of the 30th International Design Automation Conference* (1993), pp. 165–170.

[29] G. E. Tellez and M. Sarrafzadeh. "Minimal buffer insertion in clock trees with skew and slew rate constraints." *IEEE Transactions on Computer-aided Design of Integrated Circuits and Systems* 16, no. 4 (1997), pp. 333–342.

[30] J. L. Neves and E. G. Friedman. "Minimizing power dissipation in non-zero skew-based clock distribution networks." *Proceedings of ISCAS'95-International Symposium on Circuits and Systems*, vol. 3, pp. 1576–1579, IEEE, 1995.

[31] E. De Man and M. Schobinger. "Power dissipation in the clock system of highly pipelined ULSI CMOS circuits." *Proceedings of International Workshop on Low Power Design* (1994), pp. 133–138.

[32] H. Kojima, S. Tanaka, and K. Sasaki. "Half-swing clocking scheme for 75% power saving in clocking circuitry." *IEICE Transactions on Electronics* 78, no. 6 (1995), pp. 680–683.

[33] J. P. Fishburn. "Clock skew optimization." *IEEE Transactions on Computers* 39, no. 7 (1990), pp. 945–951.

[34] J. L. Neves and E. G. Friedman. "Design methodology for synthesizing clock distribution networks exploiting nonzero localized clock skew." *IEEE Transactions on Very Large Scale Integration (VLSI) Systems* 4, no. 2 (1996), pp. 286–291.

[35] W. Liu, C. Sitik, E. Salman, B. Taskin, S. Sundareswaran, and B. Huang. "SLECTS: Slew-driven clock tree synthesis." *IEEE Transactions on Very Large Scale Integration (VLSI) Systems* 27, no. 4 (2019), pp. 864–874.

[36] T. Lu and A. Srivastava. "Low-power clock tree synthesis for 3D-ICs." *ACM Transactions on Design Automation of Electronic Systems (TODAES)* 22, no. 3 (2017), pp. 1–24.

[37] M. Lin, H. Sun, and S. Kimura. "Power-efficient and slew-aware three dimensional gated clock tree synthesis." *2016 IFIP/IEEE International Conference on Very Large Scale Integration (VLSI-SoC)* (2016), pp. 1–6, IEEE.

[38] T.-J. Wang, S.-H. Huang, W.-K. Cheng, and Y.-C. Chou. "Top-level activity-driven clock tree synthesis with clock skew variation considered." *2016 IEEE International Symposium on Circuits and Systems (ISCAS)* (2016), pp. 2591–2594, IEEE.

[39] D. K. Oh, M. J. Choi, and J. H. Kim. "Thermal-aware 3D symmetrical buffered clock tree synthesis." *ACM Transactions on Design Automation of Electronic Systems (TODAES)* 24, no. 3 (2019), pp. 1–22.

[40] D. Oh, M. Choi, and J. Kim. "Symmetrical buffered clock tree synthesis considering NBTI." *2020 IEEE International Symposium on Circuits and Systems (ISCAS)* (2020), pp. 1–5, IEEE.

[41] T. Ajayi, V. A. Chhabria, M. Fogaça, S. Hashemi, A. Hosny, A. B. Kahng, M. Kim, J. Lee, U. Mallappa, M. Neseem, et al. "Toward an open-source digital flow: First learnings from the OpenROAD project." *Proceedings of the 56th Annual Design Automation Conference 2019* (2019), pp. 1–4.

Routing

28

... I rather pride myself in finding my way where there is no path, than in finding it where there is ... half the time they don't know the difference between a trail and a path, though one is a matter for the eye, while the other is little more than scent.

—J. F. Cooper, *The Pathfinder*, Chapter 1, 1840

Routing involves making physical interconnections between different components of a design. We perform routing after floorplanning and placement. Therefore, a router knows the location of all the pins, input/output (I/O) pads, and macros in a design. Moreover, it gathers information on connectivity of instances from the netlist that we provide. A router determines the complete layout of interconnects based on the connectivity and tries to honor timing constraints and design rules.

We perform routing after creating a power delivery network and clock tree synthesis. Therefore, power lines and clock signals are already routed, and the primary responsibility of a router is to route the data signals. Since power lines and clock signals consume a significant amount of routing resources, a router can utilize only the leftover routing resources.

Routing is a very complicated and time-consuming task. One of the reasons for the complication is the large number of components handled by a router. A design can contain more than a million pins that must be routed. Additionally, there are tight routing resource constraints, design rules, timing requirements, and signal integrity constraints.

We divide routing into two phases, *global routing* and *detailed routing*, to make it more manageable. The global routing is the planning stage of routing. It determines the regions through which a given net passes. The detailed routing determines the actual layout of each net in the pre-assigned routing regions. After detailed routing, a router makes local changes to a design to improve its quality of result (QoR). Additionally, it fixes timing violations, signal integrity issues, and design rule violations. In this chapter, we will discuss all these tasks in detail.

28.1 GLOBAL ROUTING

A global router determines *global routes* of each net in a design at a coarser level. Subsequently, a detailed router produces the final layout of the interconnects using the global routes. Therefore, routing models, data structures, and algorithms of the global router and detailed router are interdependent. The quality of global routing greatly influences the QoR of detailed routing and the figures of merit of the final chip.

28.1.1 Problem Formulation

We typically model the global routing problem using a *grid graph*. A grid graph is obtained by dividing a layout or a routing region into rectangular grids, as shown in Figure 28.1(a). Each rectangular region in the layout is called a *global bin* (GB). It is also referred to as a *global tile* or a *global bucket*.

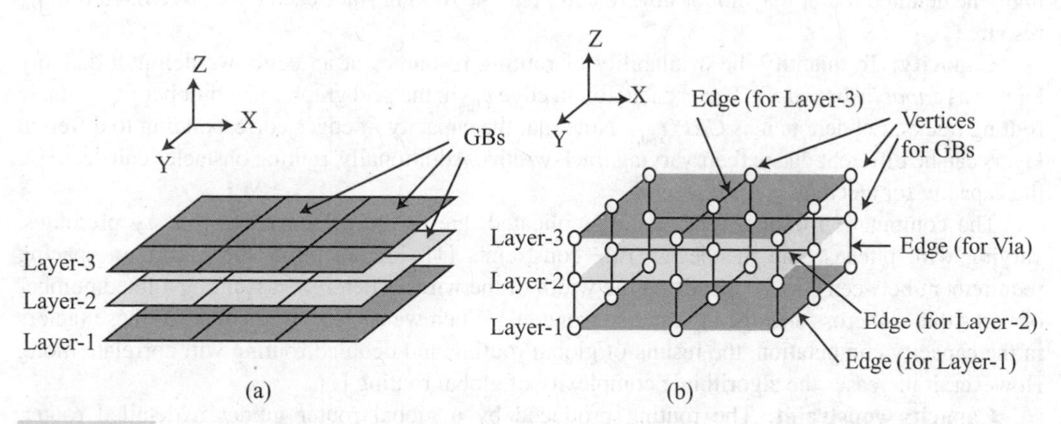

Figure 28.1 Deriving grid graph for three layers of interconnect (a) partitioning of a layout into GBs and (b) corresponding grid graph

Grid graph: We obtain a grid graph using the GBs in the rectangular grid. A grid graph $G = (V, E)$ is an undirected graph that consists of a vertex set V and an edge set E. Each vertex $v \in V$ corresponds to a GB. For integrated circuits (ICs) in which interconnects can span multiple layers, we create vertices corresponding to each layer. Figure 28.1(b) shows a grid graph for a three layers of interconnect.

An edge $e_{i,j} \in E$ corresponds to the boundary between two adjacent GBs associated with the vertices $v_i, v_j \in V$. In a typical IC, each layer has a preferred direction of the wire layout. In Figure 28.1(b), we have assumed that layer-1 and layer-3 have preferred direction along the y-axis, and layer-2 has preferred direction along the x-axis. We create edges between two adjacent vertices only if they lie along the preferred direction of routing. Additionally, we create edges between two vertices that lie vertically adjacent. These edges represent the vias of a layout.

Global routing problem: A global routing algorithm works on a netlist that consists of k nets $\{n_1, n_2, n_3, ..., n_k\}$. A net n_i consists of m pins $\{p_{(i,1)}, p_{(i,2)}, p_{(i,3)}, ..., p_{(i,m)}\}$. We know the location of each pin, since we perform routing after placement. Therefore, we can associate a pin $p_{(i,j)}$ with a vertex $v \in V$ in the grid graph using the GB in which the pin is placed. For simplicity, we assume that all pins in a GB lie at the center of that GB.

A routing solution for a net n_i requires finding a tree T_i in the grid graph G such that all its constituent pins $p_{(i,j)}$ belong to T_i. The branches of the tree T_i represent the layout of the net n_i. The complete global routing solution requires finding a set of optimal trees T_i for a design. The optimization criteria can be to achieve the minimum wire length. Additionally, a global router needs to consider the following measures.

Overflow and Routability

Usage: When a net crosses the boundary of two GBs corresponding to $v_i, v_j \in V$, we say that the net has *utilized* the corresponding edge $e_{i,j} \in E$. The same edge $e_{i,j}$ can be utilized by multiple nets or interconnects. The number of interconnects that utilize an edge $e_{i,j}$ is called the *usage* or *demand* of that edge. Let us denote the usage of an edge $e_{i,j}$ as $USE(e_{i,j})$. It is evident that if $USE(e_{i,j})$ is very high, the detailed router may not be able to complete the routing since each edge has a fixed routing resource.

Capacity: To quantify the availability of routing resources at an edge, we define a quantity known as *capacity* or *supply*. The capacity of an edge $e_{i,j}$ in the grid graph is the number of available routing tracks. Let denote it as $CAP(e_{i,j})$. Note that the capacity of edges corresponding to different layers can be different due to their varying track widths. Additionally, routing obstacles can decrease the capacity for an edge.

The computation of $CAP(e_{i,j})$ gets complicated due to the presence of arbitrary preroutes, varying wire pitches, and the design rule constraints [1]. For example, the minimum spacing requirement between wires depends on the width of the wires. Therefore, assuming a fixed number of wires that can cross an edge $e_{i,j}$ can be erroneous. When we accurately account for these factors in the capacity computation, the results of global routing and detailed routing will correlate more. However, it increases the algorithmic complexity of global routing [1].

Capacity constraint: The routing produced by a global router guides a detailed router. Therefore, the usage of an edge should not exceed its capacity, i.e.,

$$USE(e_{i,j}) \leq CAP(e_{i,j}) \tag{28.1}$$

Overflow: If $USE(e_{i,j}) > CAP(e_{i,j})$, then the edge is said to *overflow*. In this case, we define overflow $OF(e_{i,j})$ as:

$$OF(e_{i,j}) = USE(e_{i,j}) - CAP(e_{i,j}) \tag{28.2}$$

If $USE(e_{i,j}) \leq CAP(e_{i,j})$, there is no overflow, and we say $OF(e_{i,j}) = 0$. Since a global router should ensure the routability for a detailed router, it attempts to find a solution such that $OF(e_{i,j}) = 0$ for all $e_{i,j} \in E$. In this endeavor, it often sacrifices other optimization criteria, such as achieving minimum wire length [2].

Overflow is a popular metric to quantify routability. Nevertheless, it is sensitive to the approximations of the routing models, similar to many other metrics that quantify routability [1].

Congestion

Even when there is no overflow, the relative difficulty in producing a detailed routing solution can be measured using a parameter known as *congestion* or *density*. The congestion $CG(e_{i,j})$ of an edge $e_{i,j} \in E$ is defined as:

$$CG(e_{i,j}) = \frac{USE(e_{i,j})}{CAP(e_{i,j})} \tag{28.3}$$

If $CG(e_{i,j})$ is high in a region for a layout, it implies that many wires are passing through that region, and that region is said to be congested. It is better to avoid congestions in a layout since it can fail a detailed router (i.e., a detailed router will be unable to find a feasible solution). We will explain the other problems associated with congestion later in this chapter.

Timing

At advanced process nodes, the wire delay contributes significantly to the total path delay. Therefore, we need to consider wire delay and the timing of the critical path during global routing. Note that merely minimizing the wire length is not sufficient to ensure good timing. We explain it in the following example.

Example 28.1 Consider the net n_1 shown in Figure 28.2. It consist of pins $p_{(1,1)}, p_{(1,2)}, p_{(1,3)}$, and $p_{(1,4)}$. Let us assume that the timing arc between the pins $p_{(1,1)}$ and $p_{(1,4)}$ is in the critical path.

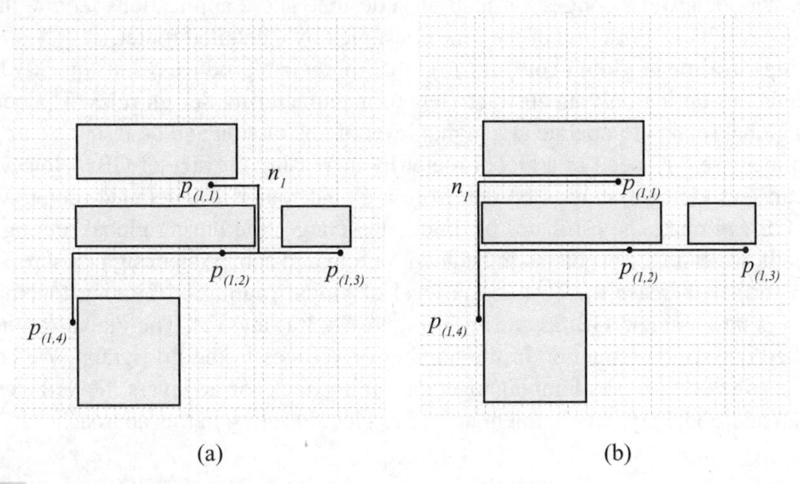

Figure 28.2 Global routing for a net n_1 consisting of pins $p_{(1,1)}$, $p_{(1,2)}$, $p_{(1,3)}$, and $p_{(1,4)}$ (rectangular routing grids are of unit length) (a) routing for the minimum wire length and (b) routing for the best timing between $p_{(1,1)}$ and $p_{(1,4)}$

A route that will have the minimum wire length to connect these pins is shown in Figure 28.2(a). The total wire length is 48 units, and the wire length between critical pins $p_{(1,1)}$ and $p_{(1,4)}$ is 39 units.

Next, we route n_1 such that the wire length between critical pins $p_{(1,1)}$ and $p_{(1,4)}$ is minimum, as shown in Figure 28.2(b). The total wire length in this case is 58 units, while the length of the wire segment between critical pins $p_{(1,1)}$ and $p_{(1,4)}$ is 29 units. Therefore, this route is better for the timing of the critical path despite having a higher total wire length.

This example illustrates that the solution exhibiting the minimum wire length does not necessarily achieve the best timing. Therefore, timing-driven global routing becomes necessary in high-performance designs.

We can use various types of delay models for the interconnects in global routing. As a first approximation, we can use a linear delay model in which delay is proportional to the wire length. For greater accuracy, we can also use the Elmore delay model.

Runtime

Global routing is the planning phase of creating wire layouts. We expect global routing to be significantly faster than detailed routing. The routes produced by a global router simplify the problem for a detailed router. Nevertheless, a detailed router still needs to determine wire layouts and consumes a very high runtime. Moreover, a detailed router can sometimes fail to produce a feasible solution after consuming much runtime due to high congestion in the design, less routing resource availability, incorrect tool usage, or limitation of the detailed router. Therefore, a global router needs to anticipate problems of detailed routing and report it quickly. Then, a designer can analyze and fix the reported problem before moving to detailed routing.

We also employ global routers during prototyping, floorplanning, and placement for estimating wire length, wire delay, and congestion map for a design. These applications require that a global router produce results in a fraction of the time consumed by a detailed router.

Reducing runtime of global routing: A global router can produce results quickly by working with GBs and considering only an abstract view of manufacturing design rules. The coarseness of GBs allows global routers to operate at a higher level of abstraction and be fast.

We can increase the speed of a global router by increasing the size of GBs. However, a large-sized GB transfers the computation burden from a global router to a detailed router. Moreover, a large-sized GB can make the estimates of timing and congestion during global routing inaccurate [3]. As a result, we can face problems in detailed routing and achieving design closure.

We can further increase the abstraction level of global routing by binning together wires of different layers into a single GB, as shown in Figure 28.3(a) and (b). The global routing problem gets simplified due to the reduction in the number of vertices in the grid graph. We compute the capacity of an edge $CAP(e_{i,j})$ by lumping together routing tracks on all layers. Nevertheless, we need to perform an additional layer assignment task after global routing in this approach.

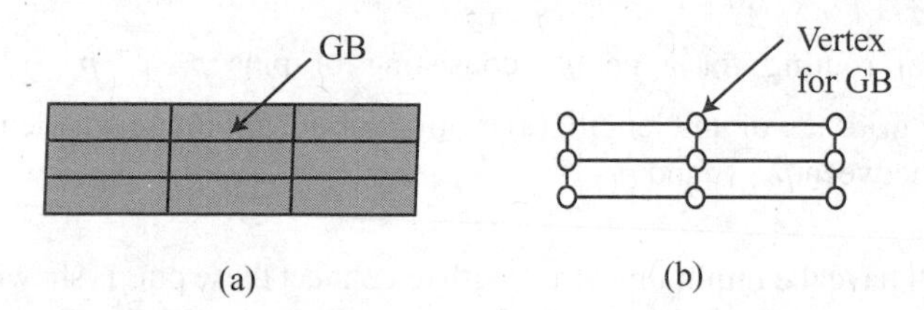

Figure 28.3 Reduced grid graph for the routing layers shown in Figure 28.1 (a) partitioning of a layout into GBs and (b) corresponding grid graph

28.1.2 Methodologies

In general, a global router needs to solve three types of problems: (a) routing a two-pin net, (b) routing a net with more than two pins, and (c) routing multiple nets. We will discuss the techniques to solve each of them in the following paragraphs.

Routing Two-pin Nets

A basic task in many routing algorithms is finding the shortest path between a pair of pins on a grid graph having obstacles. A well-known algorithm that accomplishes this task is *Lee–Moore algorithm* or *maze routing algorithm* [4, 5].

In the maze routing algorithm, one of the pins is called the source pin (p_{src}), and the other pin is called the sink pin (p_{sink}). We refer to the grid that has the source pin as the source grid. We mark it with a *tag* of 0. The tag indicates the smallest *distance* of a grid from the source grid. We refer to a grid at a distance i from the source grid as *i-distant* grid. The maze routing algorithm progresses in two phases: (a) tagging or labeling the grids with the smallest *distance* and (b) backtracking the shortest path from the sink to the source grid.

We tag grids with the smallest *distance* similar to a propagating wave starting from the source grid. If a grid has been tagged i, we tag all its *adjacent* untagged grid with $(i + 1)$. We say two grids are adjacent only if they are horizontally or vertically touching each other. We do not consider the grids touching only at the corners as adjacent because we allow only rectilinear paths (Manhattan routing) on the maze.

We ensure that all the i-distant grids are always tagged before the $(i + 1)$-distant grids during tagging. In other words, we perform breadth-first traversal of grids. We continue tagging until we reach the sink node. We illustrate tagging in the following example.

Example 28.2 Consider the source pin (p_{src}) and sink pin (p_{sink}) shown in Figure 28.4(a). Let us find the shortest path between them using maze routing. Note that there are obstacles in the grid through which no path can cross.

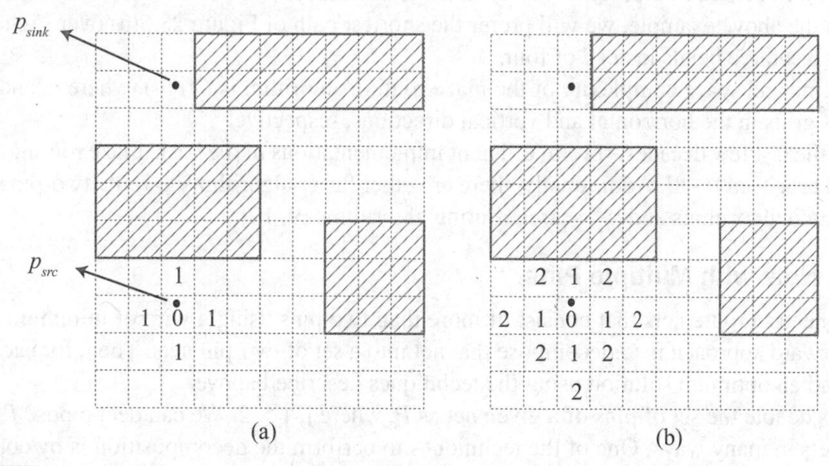

Figure 28.4 Progress of maze routing algorithm (the obstacles are shown as hashed grids) (a) tagging of 1-distant grids and (b) tagging of 2-distant grids

We mark the source grid with tag 0. Then, we mark the four adjacent grids around the source grid with tag 1, as shown in Figure 28.4(a). These are 1-distant grids. Then, we tag all 2-distant grids (excluding the grids with obstacles), as shown in Figure 28.4(b). We continue tagging until we reach the sink node, as shown in Figure 28.5.

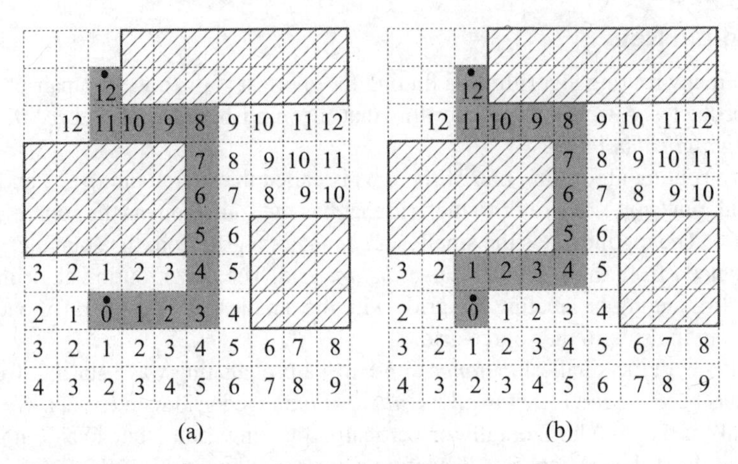

Figure 28.5 Tags in maze routing algorithm (continued from Figure 28.4) and backtracking (path shown as shaded grids) (a) one of the shortest paths and (b) another shortest path

In the second phase of the maze routing algorithm, we backtrack from the sink grid to the source grid by following tags in a strictly decreasing order. It traces the shortest path between these pins. Note that by following the tags in decreasing order, we can reach the source grid through multiple paths, as illustrated in Figure 28.5(a) and (b). Nevertheless, all such paths will be shortest and of the same length. In general, we prefer a path with the least number of bends since it reduces the number of vias. In the above example, we will prefer the shortest path of Figure 28.5(a) over Figure 28.5(b) because it has three bends instead of four.

The time and space complexity of the maze routing algorithm is $O(mn)$, where m and n are the number of grids in the horizontal and vertical directions, respectively.

Over the last few decades, several efficient implementations of the basic maze routing algorithm have been proposed [6–8]. Additionally, there are other faster algorithms to route two-pin nets, such as line-search algorithms and A*-search routing algorithms [9, 10].

Routing Nets with Multiple Pins

We often need to route nets that consist of more than two pins using a wire of minimum length. A straightforward approach is to decompose that net into a set of two-pin nets. Then, for each two-pin net, we find an optimum solution using the techniques described above.

Let us denote the set of pins of a given net as P, where $|P| > 2$. We can decompose P into two-element sets in many ways. One of the techniques to perform the decomposition is by constructing a *minimum spanning tree* (MST) over the set P.[1] Nevertheless, the routing produced by this decomposition may not be optimum. We can reduce the wire length by introducing a set of virtual pins S for a given net at appropriate locations and finding MST over the set $P \cup S$. These virtual pins are known as *Steiner points*.

[1] See Section 26.1.1 ("Wire length estimates") for MST.

Example 28.3 Consider the global routing for a net consisting of pins $p_{(1,1)}$, $p_{(1,2)}$, and $p_{(1,3)}$, as shown in Figure 28.6.

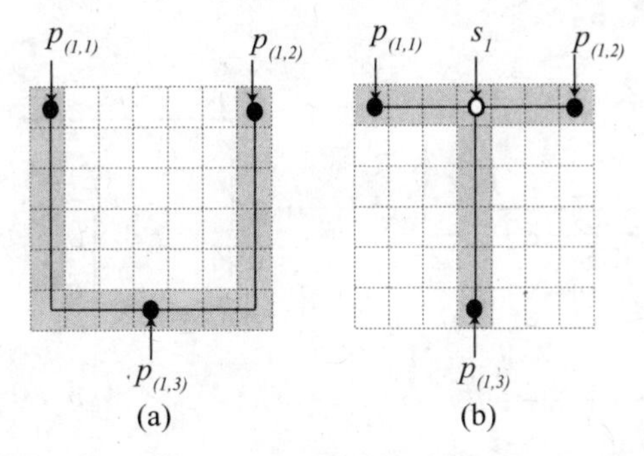

Figure 28.6 Two different routing for a net (a) using rectilinear MST and (b) using extra Steiner point (each grid is of unit length)

Figure 28.6(a) shows routing using MST. The total wire length is 17 units.

Figure 28.6(b) shows a route in which we add a Steiner point s_1. The resultant wire length is 12 units. It is less than the wire length obtained using MST.

We are typically interested in routes with vertical and horizontal net segments. Therefore, we are interested in finding a *rectilinear* Steiner minimum tree (SMT) for a given set of pins[2].

Finding rectilinear SMT for a set of pins is an NP-complete problem [11]. Therefore, we use various heuristics to solve this problem. One of the approaches is to employ the *Hanan grid* for finding an appropriate set of Steiner points S [12]. We construct a Hanan grid by extending vertical and horizontal lines through all the pins P [13]. Hanan has shown that the set of Steiner points S would be contained in the Hanan grid. Therefore, we can consider the points in the Hanan grid as potential Steiner points while solving the rectilinear SMT problem [14, 15].

Routing Multiple Nets

We typically perform global routing using a method known as *sequential routing*. In this method, we sequentially route the given set of nets by connecting their pins using the shortest paths in the grid graph. However, as routing progresses, the edges of the grid graph start exhibiting overflow and congestion. Therefore, we must also handle the problem of congestion during sequential routing.

One of the methods of handling congestion is to start by routing each net independently, ignoring the existence of other routes in the routing region. Once routing is done for all the nets, we create a congestion map for the complete layout and identify regions of high congestion. The nets that cross a congested region are ripped up and re-routed through alternate or less congested regions. It is evident that by changing the sequence of *rip-up and re-route* of nets, the solution of routing will change. Therefore, the final QoR of routing depends strongly on the sequence of re-routing.

Another routing technique is to consider the existing routes in a region while computing the optimal routes for the first time. Nevertheless, the final QoR still depends on the sequence in which the nets get routed.

[2] See Section 26.1.1 ("Wire length estimates") for SMT.

Example 28.4 Consider a layout shown in Figure 28.7. It contains two nets n_1 and n_2. The net n_1 contains pins $p_{(1,1)}$ and $p_{(1,2)}$. The net n_2 contains pins $p_{(2,1)}$ and $p_{(2,2)}$. For the sake of simplicity, let us assume that each edge of the grid graph has a capacity of 1 wire.

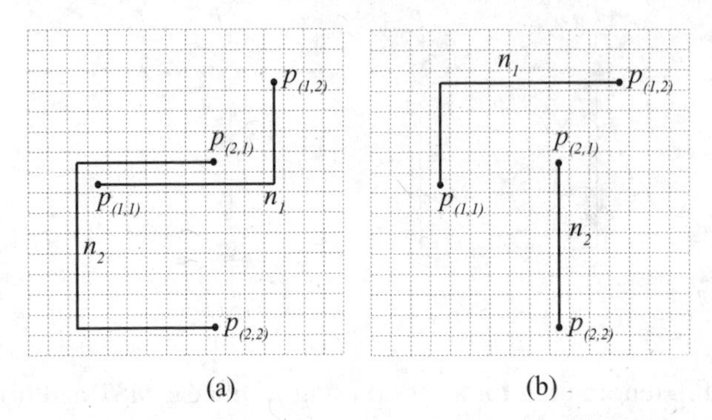

Figure 28.7 Sequential routing (grids are of unit length) (a) routing of n_1 followed by n_2 and (b) routing of n_2 followed by n_1

If we route n_1 followed by n_2, we obtain the route shown in Figure 28.7(a). In this case, the net n_2 takes a longer route because it finds edges fully *used* by the routes of n_1.

Next, we route n_2 followed by n_1. We obtain the route shown in Figure 28.7(b). In this case also, we encounter edges that are fully *used* by the routes of n_2 while routing n_1. However, there is an alternative route for n_1 with the same cost as in Figure 28.7(a). Therefore, the total wire length is smaller for the layout shown in Figure 28.7(b) than in Figure 28.7(a).

Thus, the total wire length is strongly dependent on the sequence in which we route the nets.

We can also compute routes *concurrently* for all the nets. A methodology in which the routing problem is formulated as a *zero–one integer linear program* is an example of concurrent routing [12, 16]. In this technique, we represent a route by a pattern of zeroes and ones in the grid graph. We determine the global routing solution by minimizing the maximum congestion in the grid graph. However, a zero–one integer linear program is an NP-complete problem and this method is typically not feasible for large problem sizes.

28.2 DETAILED ROUTING

A detailed router determines the exact layout of each net, including all the attributes of wire segments such as width and location. It utilizes the solution provided by a global router as a guide. Typically, a detailed router expects that the solution provided by a global router satisfies the capacity constraints. However, industry-standard detailed routers can tolerate a few capacity constraint violations [17].

An essential requirement for a detailed router is to reproduce the connectivity specified in the given netlist. Therefore, a detailed router should not produce any extra open- or short-circuit. Additionally, it must honor design rules for interconnects. A detailed router also handles signal integrity issues using various techniques. For example, it can avoid laying out two sensitive signals (the aggressor and the victim) in parallel tracks longer than a fixed length. It can also insert grounded lines between sensitive signals to reduce the crosstalk.

28.2.1 Problem Formulation

The most basic operation for a detailed router is to find an optimal path between any two given points in the available *routing space* with the constraint that the design rule must be honored. In general, this problem can be tackled by modeling the routing space in two ways: *gridless* model and *grid-based* model.

Gridless model: A gridless detailed routing model does not consider a separate grid for routing. It considers the *manufacturing grid* as feasible points on which the central line of a wire can pass through.[3] The manufacturing grid is determined by the resolution of a given technology and is typically very dense. Therefore, directly working on the manufacturing grid is often infeasible.

In a gridless model, the location of pins, nets, and vias can be defined at a much finer granularity. Additionally, it allows defining the width of wires at much smaller dimensions. Therefore, a gridless router can exploit greater freedom, though computation in a gridless model can be more complex than a grid-based model.

Grid-based model: During detailed routing, we need to pack wires compactly. Therefore, we typically restrict wires to run parallelly in each layer in its preferred direction. These restricted lines of parallel wires are commonly known as *routing tracks* or simply *tracks*. The spacing of the routing tracks in each metal layer is known as the *minimum pitch* or the *routing pitch*. We can make most of the net connections for a given design by keeping the distance between wires and the width to their minimum values without violating design rule constraints. We can compactly represent and efficiently manipulate these connections and routing paths using a grid-based model that employs routing tracks.

Detailed routing grids: The detailed routing grids are similar to the global routing grids explained in the previous section. However, detailed routing grids are fine-grained, and the routing resources are modeled in greater detail. For simplicity, we keep the distance between adjacent grid points uniform, equal to the routing pitch. Moreover, we employ suitable data structures to store and efficiently manipulate densely packed wires.

For a given layer, all the vertices in the grid graph that lie in a straight line in the preferred direction constitute a routing track. A router typically creates layout for a wire on these routing tracks. However, pins can have multiple blockages around them and can also be misaligned. Therefore, wires can be laid out in the direction orthogonal to the preferred direction for accessing pins. These orthogonally oriented wires are known as *jogs*. The jogs and vias are modeled appropriately as edges in the grid graph.

While defining routing pitch for a detailed routing grid, we need to consider the following two conflicting requirements:

(a) The routing pitch should be kept small to ensure that the wires can be packed compactly

(b) The routing pitch should be kept sufficiently large to avoid creating design rule check (DRC) violations.[4]

We explain these requirements in the following example.

[3] The manufacturing grid is used for geometry alignment in the layout, and its dimension is defined in the LEF file.

[4] We will explain DRC in Section 29.2.3 ("Design rule check").

Example 28.5 Consider that we define the routing pitch based on the minimum design rule spacing requirement between two parallel wires (i.e., minimum line-on-line spacing), as shown in Figure 28.8.

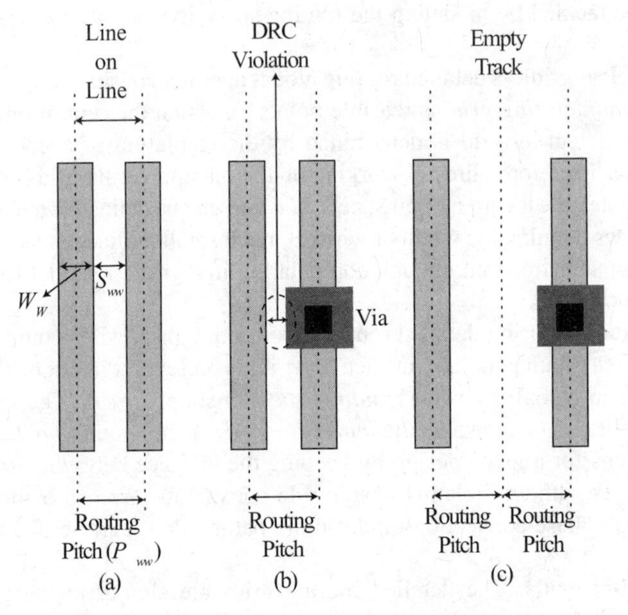

Figure 28.8 Routing pitch defined by line-on-line spacing (a) two minimum width wires placed on adjacent tracks, (b) DRC violation due to a wire and a via, and (c) fix DRC violation using an empty track

In this case, the routing pitch P_{ww} can be given as:

$$P_{ww} = W_w + S_{ww} \tag{28.4}$$

where W_w is the minimum width of the wire and S_{ww} is the required minimum spacing between the wires. Two minimum width wires placed on different routing grids will exhibit no DRC violation, as shown in Figure 28.8(a). However, if we increase the width of these wires or place a via on a line, we will obtain a DRC violation, as shown in Figure 28.8(b).

To fix the above DRC violations, we must increase the spacing of wires by keeping an empty track between them as shown in Figure 28.8(c) or consider an alternate route for these lines. These solutions can result in congestion because each via effectively occupies two tracks on the routing grid.

To avoid problems illustrated in the above example, we can define routing pitch based on the via-on-via minimum spacing requirement, as illustrated in Figure 28.9(a). In this case, the routing pitch P_{vv} can be defined as:

$$P_{vv} = W_v + S_{vv} \tag{28.5}$$

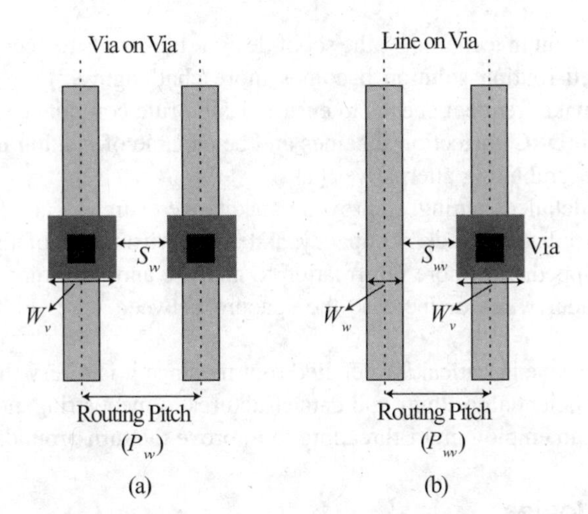

Figure 28.9 Routing pitch defined by (a) via-on-via spacing and (b) line-on-via spacing

where W_v is the minimum width of a via and S_{vv} is the required minimum spacing between two adjacent vias. It will avoid DRC violations for vias, though considerable routing area can get wasted for typical routes in which two vias need not be adjacent.

We can obtain a better trade-off by defining routing pitch based on the line-on-via minimum spacing requirement, as illustrated in Figure 28.9(b). In this case, the routing pitch P_{wv} can be defined as:

$$P_{wv} = W_w/2 + W_v/2 + S_{wv} \tag{28.6}$$

where W_w is the minimum width of a wire, W_v is the minimum width of a via, and S_{wv} is the required minimum spacing between a line and a via.

28.2.2 Detailed Routing Constraints

A detailed router needs to consider various constraints while producing a solution. Some of these constraints are described in the following paragraphs.

Design rules: A design must satisfy several technology-dependent *design rules* to ensure the manufacturability of a circuit. An essential requirement for detailed routing is to produce an interconnect layout that satisfies these design rules [18].

For example, the routing pitch is decided by the rule that defines the minimum distance between shapes (layout pattern) belonging to different nets, as discussed above. The minimum distance depends on the width of the wires and the parallel run-length of the shapes [19]. In multi-patterning technologies, the required minimum distances for the shapes of the same color are larger than the shapes of different colors because the same color shapes use the same mask and are patterned together during photolithography [19].[5]

For the shapes belonging to the same net also, some configurations are prohibited. For example, rules define the minimum length of the edge of the shapes and minimum shape area. These *same-net rules* should also be accounted for by a detailed router.

[5] See Section 7.2.3 ("Double patterning or multi-patterning") for multi-patterning.

With the advancement in technology, the set of design rules tends to become more complicated, and producing detailed routing solution becomes more challenging. Therefore, while routing, a detailed router must make frequent checks to ensure design rule compliance. Additionally, a router must be equipped with DRC correction schemes and be capable of making intelligent trade-offs in selecting one of the several fixing alternatives [1].

Timing: During detailed routing, improved capacitance estimates are possible due to a more accurate modeling of adjacent track occupancy and the environment of the wires. Therefore, a detailed router is equipped with more information to analyze and fix timing violations. A detailed router can employ wider wires or increase the spacing between wires to meet aggressive delay targets [18].

Runtime: The runtime is critical for detailed routing since it is a very time-consuming task. A detailed router uses efficient algorithms and data structures to meet stringent runtime requirements [18]. Additionally, it can employ multi-threading to improve the turn-around time [20].

28.2.3 Methodologies

Traditional algorithms for detailed routing operate in isolated regions of the layout. We can classify detailed routing into: *channel routing*, *switchbox routing*, and *over-the-cell (OTC) routing*. With the advancement in technology and availability of many layers for routing, we can apply the graph-based techniques that were discussed for global routing to detailed routing algorithm also, though with a finer granularity of layout, resources, and delay models [17]. We can also apply techniques such as rip-up and re-route to detailed routing algorithm [1]. In the following paragraphs, we discuss different types of detailed routing.

Channel Routing

When only a few layers of metals are available for routing, most of the routing resources get used up by the standard cells. In such cases, we reserve space between two rows of standard cells for routing. These spaces between standard cell rows are known as *channels*.

The input to a channel router is channel boundaries and a set of pins on the channel boundaries, as shown in Figure 28.10. We give tags to pins to identify the net with which they are associated. Two pins with the same tag belong to the same net. We can make a connection of nets using a horizontal metal segment known as *trunk* or a vertical metal segment known as a *branch*. A net that consists of more than one trunk is said to be composed of *doglegs*, as shown in Figure 28.10. The number of routing tracks used in the channel is known as the *channel height*. We assume that the channel height is flexible. A channel router aims to produce a DRC clean routing such that the channel height gets minimized. A small channel height results in a smaller die area.

In the classical channel routing problem, only two layers of metal are allowed. Furthermore, each pin has precisely one branch, and each net has exactly one trunk. Therefore, nets that have overlapping intervals must be laid out on different tracks. This is known as *horizontal constraint*. Additionally, the branch emanating from pins lying at the same longitudinal position but belonging to different nets must change layer before they overlap. This is known as *vertical constraint*. We illustrate these constraints in the following example.

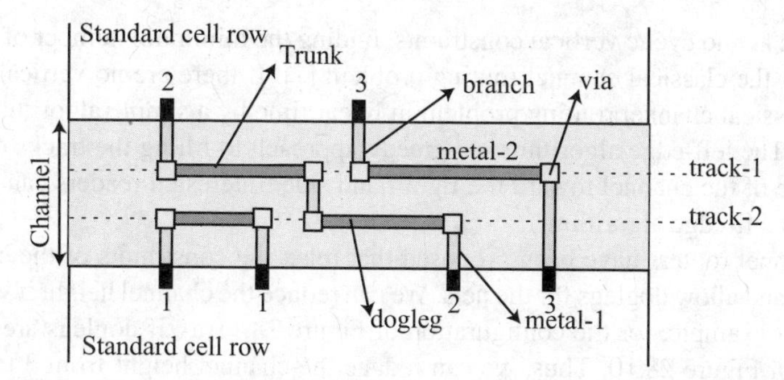

Figure 28.10 Channel routing features (metal-1 is vertical and metal-2 is horizontal)

Example 28.6 Consider the layout shown in Figure 28.11(a). Nets 2 and 3 have an overlapping interval. Hence they must be laid out on different tracks. This is a horizontal constraint. The pins 1 and 2 near the left edge emanate from the same longitudinal position and belong to different nets. Hence they must change layers before overlapping. This is a vertical constraint.

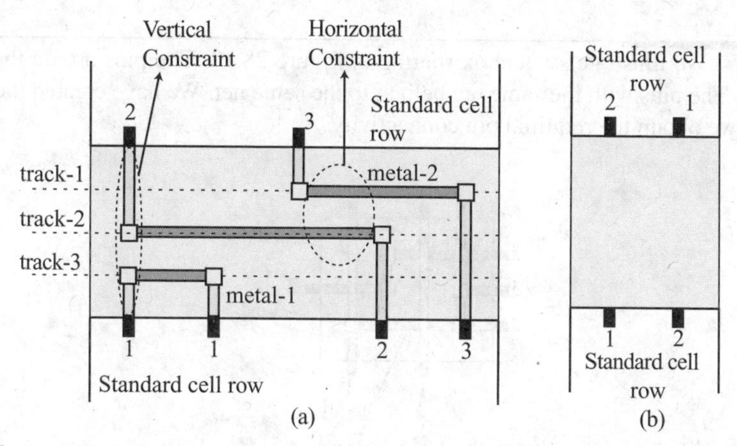

Figure 28.11 Classical channel routing (a) simple example and (b) unroutable configuration

Next, consider the layout shown in Figure 28.11(b). In this case, pin 1 lies below pin 2, and pin 2 lies below pin 1. The branch emanating from pin 2 near the left edge needs to change the layer to avoid a short circuit with pin 1. However, for connecting with pin 2 near the right edge, it needs to change the layer again, creating a short-circuit with the pin 1 connections. Hence, we cannot complete the routing in the classical channel routing model for this layout. The layout, such as shown in Figure 28.11(b), is said to have a *cyclic vertical constraint.*

Even if there are no cyclic vertical constraints, finding the minimum number of tracks is an NP-hard problem for the classical channel routing problem [1]. If there are no vertical constraints, we can solve the classical channel routing problem in linear time by a simple algorithm known as *left-edge algorithm*. The left-edge algorithm is a greedy approach to filling the tracks one by one from the left-hand side of the channel toward the right-hand side. Interested readers can refer to [21, 22] for details on the left-edge algorithm.

Various channel routers have been proposed that relax the constraints of the classical channel routing problem and allow doglegs for the nets. We can reduce the channel height if we allow doglegs in the layout. For example, for the configuration of Figure 28.11(a), if doglegs are allowed we can use the routing of Figure 28.10. Thus, we can reduce the channel height from 3 tracks to 2 tracks with the help of doglegs.

In the present technologies, we have more than 10 layers of interconnects. Hence, the channel router has lost its earlier significance [1].

Switchbox Routing

A *switchbox* has pins on all its four sides. We can consider it as a generalization of a channel. Since the pin locations are fixed, the routing area is fixed. Therefore, the main objective of switchbox routing is to route all nets adhering to the design rules.

Example 28.7 We illustrate switchbox routing in Figure 28.12. The pins are on the boundary of the switchbox. The pins with the same tag belong to the same net. We have created the layout of the nets such that we obtain the required pin connectivity.

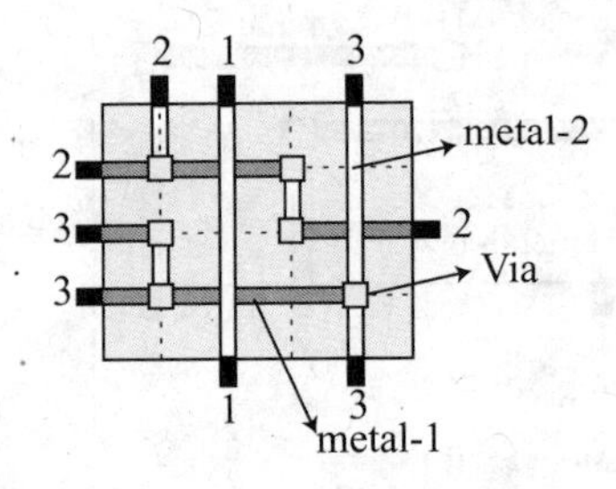

Figure 28.12 A switchbox routing using two metal layers

The switchbox routing problem is considered more complex than the channel routing problem [22]. Consequently, various heuristics have been developed to tackle it [23–25]. For example, we can use a greedy approach, such as employed in channel routing, for switchbox routing also [23, 24]. Moreover, we can prioritize the nets based on their pin locations while assigning tracks. We can also allow incremental modifications and backtracking if no routing solution can exist in the current search space. [25].

OTC Routing

The increased number of available routing layers in current fabrication technologies has shifted detailed routing from channel-based approaches to OTC-based approaches. The metal layers that are not obstructed by standard cells can be used for OTC routing.

Example 28.8 Consider the layout of wires shown in Figure 28.13. Assume that metal-1 and metal-2 are fully utilized over the standard cell rows.

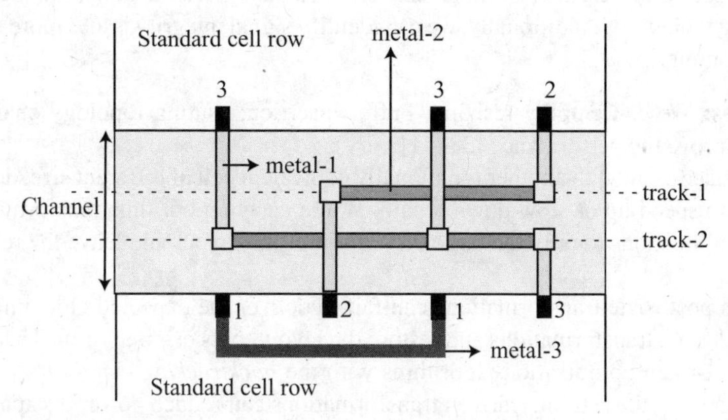

Figure 28.13 An OTC routing

The pins marked 1 can be OTC-routed using metal-3, as shown in the figure, since metal-3 is not obstructed by the standard cells.

The OTC routing allows a reduction in the height of the channel. With sufficient metal layers available for routing, we can eliminate channels from the circuit layout.

Similar to the classical channel routing problem, OTC routing is an NP-hard problem [22]. Therefore, several heuristics have been proposed to solve it [22, 26–29]. We can also apply general routing techniques such as maze routing to solve the OTC routing problem. We can consider the complete die as a routing region and also adopt a divide and conquer strategy to solve the OTC routing problem [12, 30, 31].

28.3 POST-ROUTING OPTIMIZATION

The routing process considers several aspects of a circuit besides finding a feasible design rule-compliant routing. For example it can consider congestion, delay, signal integrity, and power dissipation during routing [32–37]. Nevertheless, we can improve some of these QoR measures even after routing, as explained below.

Opportunity for timing optimization: Some of the reasons why detailed routing can leave some scope for further timing optimization are as follows:

1. The primary objective of a global router is to ensure the routability of a design. It uses wire length and overflow to measure the quality of a solution. As illustrated in Ex. 28.1, for a multi-fanout net, minimizing wire length does not necessarily minimize the critical path delay [32].
2. A router tries to minimize congestion in a region. However, reducing congestion and reducing delay are often competing objectives. For example, to avoid congestion, some wires can get detoured, increasing the delay of the signal associated with that wire [35].

3. Once detailed routing is done, the layout of a design is well-defined. The parasitics of interconnects, including coupling capacitances, can be extracted accurately only after detailed routing. Therefore, we can perform precise timing analysis after detailed routing and determine real timing violations. Additionally, we can identify signal integrity issues more accurately after detailed routing.

Techniques: We can employ resizing, buffer insertion, routing topology changes, and wire widening to improve the performance [38–41].

We can replace a cell with another functionally equivalent cell of different size, delay, and power characteristics to speed up or slow down a path. While carrying out this transformation, we make *incremental* modifications to only the targeted portion of the layout and leave the rest of the layout unchanged.

Sometimes post-route transformations can fail to deliver the expected QoR improvement. We need to roll back such transformations and restore the layout to its original form. Therefore, we need to equip the post-route optimization algorithms with the *backtracking capabilities*. It is worthy to point out that some of these timing-driven transformations can reduce power dissipation also.

Wire width and spacing adjustments: One of the post-route optimization techniques is to adjust the width and spacing of the wires. By increasing the width of a wire, the resistance will decrease. It can help in reducing delay. However, increasing the width of a wire increases the coupling capacitance and the crosstalk effects. Thus, we must consider two conflicting objectives while determining the width of a wire [38]. We can reduce the coupling capacitance and the crosstalk noise by adjusting the spacing between wires.

Both wire widening and increasing the spacing between wires require additional routing resources. We can combine both these techniques to reduce delay and crosstalk effects [39]. Typically, after routing, some tracks can remain unused. We can utilize them for wire widening and increasing the spacing between wires. In practice, we can adjust the wire width and spacings only in discrete steps defined by the given manufacturing technology.

28.4 RELIABILITY ISSUES

During routing, we need to consider some reliability issues also. In this section, we explain some of these issues and possible solutions.

28.4.1 Antenna Effect

In Section 24.4 ("Antenna effect"), we have explained the antenna effect. We have seen that if a lower-level metal is directly connected to the gate, the gate dielectric can be damaged due to plasma-induced charges. We can avoid this issue by adhering to the *antenna rules*.

Example 28.9 Suppose we have antenna rule violation for the routing shown in Figure 28.14(a). The area of lower-level metal *M1* directly connected to the gate is greater than the antenna rule limit. We can avoid antenna rule violation using the following routing techniques.

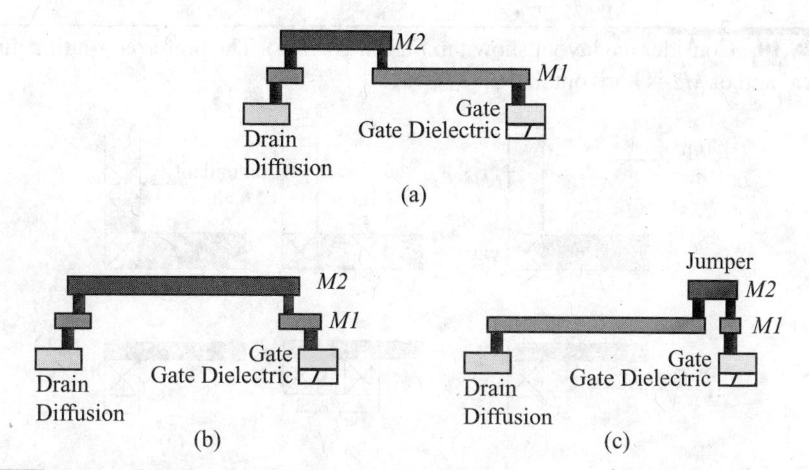

Figure 28.14 Techniques to avoid antenna rule violations (a) a routing with antenna rule violation, (b) using higher layer metal for routing instead of lower layer metal, and (c) using jumpers

1. We allow the major portion of the net to be in the higher-level metal (*M2*), as illustrated in Figure 28.14(b). Consequently, we can reduce the area of the lower-level metal (*M1*) directly connected to the gate and meet the given antenna rule requirements.
2. We can add a jumper that avoids a continuous lower-level metal (*M1*) being directly connected to the gate, as illustrated in Figure 28.14(c). This technique is useful when we do not have sufficient routing resources available in the higher-level metal (*M2*).

28.4.2 Redundant Vias

During routing, we need to insert a via whenever there is a change in layer for a net connection. A router typically minimizes the number of vias. There are two main reasons for it. First, the resistance of a via is typically higher than the corresponding metal layers. Consequently, vias contribute significantly to the interconnect delays. Second, vias are prone to failure due to various reasons such as mask misalignment and thermal stresses. These failures can lead to open circuit defects and the consequent loss in yield and reliability.

Techniques: A popular technique to increase the yield is the insertion of *redundant via* adjacent to a single via [42]. It allows a design to tolerate single via failure and improves the reliability of a design.

Redundant via insertion is typically undertaken after detailed routing or while performing detailed routing [43]. We can insert redundant vias either on-track or off-track, as explained in the following example.

Example 28.10 Consider the layout shown in Figure 28.15(a). The preferred routing direction for *M1* is vertical and of *M2* is horizontal.

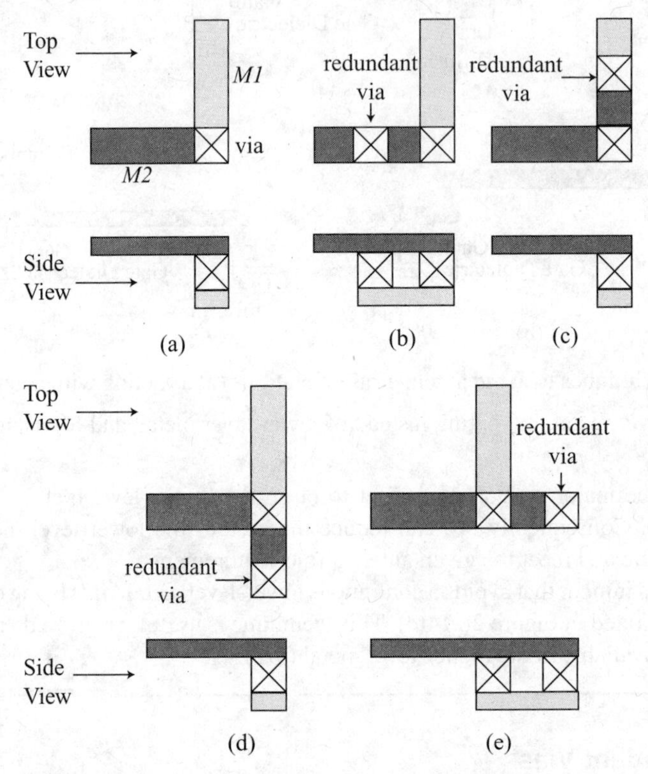

Figure 28.15 Insertion of redundant vias (a) original layout, (b, c) on-track redundant via, and (d, e) off-track redundant via

We can insert on-track redundant vias by extending wire in one of the layers in the non-preferred direction along the occupied track. In Figure 28.15(b), we extend *M1* in the non-preferred direction, while in Figure 28.15(c) we extend *M2* in the non-preferred direction.

We can insert off-track redundant vias by extending wires in both layers. In Figure 28.15(d), we extend *M1* in the preferred direction and *M2* in the non-preferred direction. In Figure 28.15(e), we extend *M1* in the non-preferred direction and *M2* in the preferred direction.

Typically, on-track redundant vias consume less routing resources than off-track redundant vias and also exhibit better electrical characteristics [42].

We must decide the location of inserting redundant via carefully because inserting a via blocks some routing resources and can inhibit other redundant via insertion in its neighborhood [42].

28.5 MANUFACTURABILITY

The interconnects and routing play a vital role in determining the manufacturability of a design. To a large extent, compliance with design rules ensures manufacturability and profitable yield. However, there are more techniques to improve the manufacturability of a design and we regularly apply them after routing.

Uneven surface during manufacturing: Due to the difference in the hardness of metal such as copper and an interlayer dielectric (ILD) such as SiO_2, chemical mechanical polishing (CMP) can produce irregular surface topography if there is a considerable variation in the metal density over the layout. We illustrate it in the following example.

Example 28.11 Consider the cross-section of an IC shown in Figure 28.16(a). There is a high metal density on the left side, while the right side has only ILD. As a result, CMP etches more on the right side where metal density is low. Hence, we obtain an uneven surface, as shown in Figure 28.16(b).

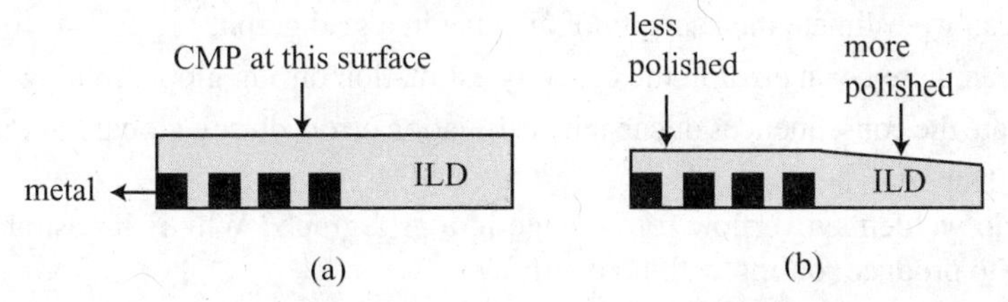

Figure 28.16 Irregular surface created due to difference in metal density (a) before CMP and (b) after CMP

Dummy metal fills: An uneven surface after CMP reduces yield and creates significant performance variations in a circuit. Therefore, we add dummy metal fills in regions where there are fewer wires to obtain a uniform metal density over the layout [44].[6]

The existence of dummy metal fills in the vicinity of current-carrying wires can affect the coupling capacitance. Therefore, dummy metal fills can impact the timing of a circuit, despite being floating. Hence, we need to also consider the timing impact of dummy metal fills in a design flow [45].

28.6 RECENT TRENDS

At advanced technology nodes, routing becomes highly complicated. Producing a DRC-clean detailed routing can take days of runtime for sub-10 nm IC designs [46]. Therefore, significant effort is directed toward addressing the runtime and memory scalability of detailed routers [47–49]. The runtime problem is widely addressed by exploiting the multi-threading framework.

One of the primary reasons responsible for making routing challenging is the complexity of the design rules at the advanced process nodes [50]. Some of these rules must be checked and complied with *during* routing because it can become very difficult to fix their violation after routing.

[6] Foundries typically provide rules for dummy metal fills. We can provide those rules to a DRC tool, and a DRC tool typically generates data for dummy metal fills, which we can merge with the design layout.

Additionally, ignoring these rules by a global router creates a large miscorrelation between the estimates of a global router and a detailed router. Accounting for design rules complicates the capacity estimation significantly for a global router [46, 51]. One of the techniques to alleviate these problems is to predict DRC violations early in the design flow and fix them before carrying out detailed routing. Machine learning and artificial intelligence framework can be tried out for this prediction [52, 53].

REVIEW QUESTIONS

28.1 What are the merits and demerits of carrying out routing in two steps (global and detailed routing) instead of one step?

28.2 What is a grid graph and GBs?

28.3 What trade-offs should we consider while choosing the size of a GB?

28.4 What is the capacity of an edge in a grid graph?

28.5 How can we estimate the capacity of an edge in a grid graph?

28.6 Why can there be an error in the capacity estimation during global routing?

28.7 What are the consequences of capacity estimation errors during global routing on the design convergence?

28.8 How do we define overflow for an edge in a grid graph? Why is it desirable for a global router to produce routing with no overflow?

28.9 During global routing, overflow was found. Suggest some techniques to reduce the overflow.

28.10 How do we define congestion for an edge in a grid graph?

28.11 What are the consequences of high congestion produced by a global router?

28.12 Suggest some techniques to alleviate the problem of congestion encountered during global routing.

28.13 What is the need to perform timing optimization after routing?

28.14 How does the maze routing algorithm deal with the obstacles in the grid?

28.15 In Figure 28.17, find the minimum wire length path between p_{src} and p_{sink} using maze routing algorithm. The grid size is one unit length.

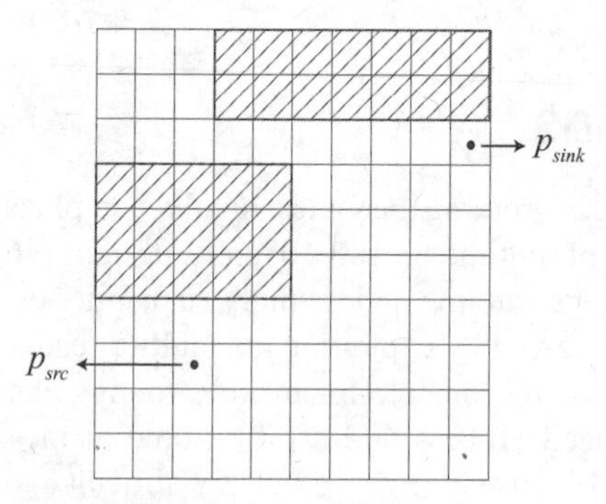

Figure 28.17 Given maze with obstacles

28.16 What are the trade-offs involved in deciding the routing pitch for detailed routing?

28.17 What are the differences in the interconnect delay model in global routing and detailed routing?

28.18 What are the differences between grid-based routing and gridless routing? What are the advantages and disadvantages of each of them?

28.19 Name a few post-route optimization methods.

28.20 How can the chances of plasma-induced gate damage be reduced during routing?

28.21 Why do we insert redundant vias in a design?

28.22 Why can insertion of dummy CMP-fills create new timing violations?

TOOL-BASED ACTIVITY

Take any routing tool that is available to you and carry out the following activities:

1. Load a placed design in which power routing and clock tree synthesis are already done. For easy analysis, choose a small design (say with 100–1000 instances and 10–100 flip-flops). You can use the design you obtained after clock tree synthesis in the activity of Chapter 27 ("Clock tree synthesis").
2. Perform global routing and detailed routing. Perform timing analysis. Analyze the worst slack path.
3. Observe the difference in the delay estimates of the global routing and the detailed routing for the same net (take a few nets with a varying number of connected pins).

REFERENCES

[1] C. J. Alpert, D. P. Mehta, and S. S. Sapatnekar. *Handbook of Algorithms for Physical Design Automation*. Auerbach Publications, 2008.

[2] Y.-J. Chang, Y.-T. Lee, and T.-C. Wang. "NTHU-Route 2.0: A fast and stable global router." *2008 IEEE/ACM International Conference on Computer-aided Design* (2008), pp. 338–343, IEEE.

[3] W.-T. J. Chan, P.-H. Ho, A. B. Kahng, and P. Saxena. "Routability optimization for industrial designs at sub-14 nm process nodes using machine learning." *Proceedings of the 2017 ACM on International Symposium on Physical Design* (2017), pp. 15–21, ACM.

[4] E. F. Moore. "The shortest path through a maze." *Proc. Int. Symp. Switching Theory, 1959* (1959), pp. 285–292.

[5] C. Y. Lee. "An algorithm for path connections and its applications." *IRE Transactions on Electronic Computers*, no. 3 (1961), pp. 346–365.

[6] S. B. Akers. "A modification of Lee's path connection algorithm." *IEEE Transactions on Electronic Computers*, no. 1 (1967), pp. 97–98.

[7] F. Hadlock. "A shortest path algorithm for grid graphs." *Networks* 7, no. 4 (1977), pp. 323–334.

[8] J. Soukup. "Fast maze router." *Proceedings of the 15th Design Automation Conference* (1978), pp. 100–102, IEEE Press.

[9] D. W. Hightower. "A solution to line-routing problems on the continuous plane." *Proceedings of the 6th Annual Design Automation Conference* (1969), pp. 1–24.

[10] P. E. Hart, N. J. Nilsson, and B. Raphael. "A formal basis for the heuristic determination of minimum cost paths." *IEEE Transactions on Systems Science and Cybernetics* 4, no. 2 (1968), pp. 100–107.

[11] M. R. Garey and D. S. Johnson. "The rectilinear Steiner tree problem is NP-complete." *SIAM Journal on Applied Mathematics* 32, no. 4 (1977), pp. 826–834.

[12] H.-Y. Chen and Y.-W. Chang. "Global and detailed routing." *Electronic Design Automation* (2009), pp. 687–749, Elsevier.

[13] M. Hanan. "On Steiner's problem with rectilinear distance." *SIAM Journal on Applied Mathematics* 14, no. 2 (1966), pp. 255–265.

[14] A. B. Kahng and G. Robins. "A new class of iterative Steiner tree heuristics with good performance." *IEEE Transactions on Computer-aided Design of Integrated Circuits and Systems* 11, no. 7 (1992), pp. 893–902.

[15] M. D. Moffitt, J. A. Roy, and I. L. Markov. "The coming of age of (academic) global routing." *Proceedings of the 2008 International Symposium on Physical Design* (2008), pp. 148–155.

[16] M. Cho and D. Z. Pan. "BoxRouter: A new global router based on box expansion and progressive ILP." *IEEE Transactions on Computer-aided Design of Integrated Circuits and Systems* 26, no. 12 (2007), pp. 2130–2143.

[17] C. J. Alpert, Z. Li, M. D. Moffitt, G.-J. Nam, J. A. Roy, and G. Tellez. "What makes a design difficult to route." *Proceedings of the 19th International Symposium on Physical Design* (2010), pp. 7–12.

[18] M. Gester, D. Müller, T. Nieberg, C. Panten, C. Schulte, and J. Vygen. "BonnRoute: Algorithms and data structures for fast and good VLSI routing." *ACM Transactions on Design Automation of Electronic Systems (TODAES)* 18, no. 2 (2013), pp. 1–24.

[19] M. Ahrens, M. Gester, N. Klewinghaus, D. Müller, S. Peyer, C. Schulte, and G. Tellez. "Detailed routing algorithms for advanced technology nodes." *IEEE Transactions on Computer-aided Design of Integrated Circuits and Systems* 34, no. 4 (2014), pp. 563–576.

[20] F.-K. Sun, H. Chen, C.-Y. Chen, C.-H. Hsu, and Y.-W. Chang. "A multithreaded initial detailed routing algorithm considering global routing guides." *2018 IEEE/ACM International Conference on Computer-aided Design (ICCAD)* (2018), pp. 1–7, IEEE.

[21] A. Hashimoto and J. Stevens. "Wire routing by optimizing channel assignment within large apertures." *Proceedings of the 8th Design Automation Workshop* (1971), pp. 155–169.

[22] N. A. Sherwani. *Algorithms for VLSI Physical Design Automation*. Springer Science & Business Media, 2012.

[23] W. K. Luk. "A greedy switch-box router." *Integration* 3, no. 2 (1985), pp. 129–149.

[24] J. P. Cohoon and P. L. Heck. "BEAVER: A computational-geometry-based tool for switchbox routing." *IEEE Transactions on Computer-aided Design of Integrated Circuits and Systems* 7, no. 6 (1988), pp. 684–697.

[25] H. Shin and A. Sangiovanni-Vincentelli. "A detailed router based on incremental routing modifications: Mighty." *IEEE Transactions on Computer-aided Design of Integrated Circuits and Systems* 6, no. 6 (1987), pp. 942–955.

[26] J. Cong and C. Liu. "Over-the-cell channel routing." *IEEE Transactions on Computer-aided Design of Integrated Circuits and Systems* 9, no. 4 (1990), pp. 408–418.

[27] N. D. Holmes, N. A. Sherwani, and M. Sarrafzadeh. "New algorithm for over-the-cell channel routing using vacant terminals." *Proceedings of the 28th ACM/IEEE Design Automation Conference* (1991), pp. 126–131.

[28] T. Her and D. Wong. "On over-the-cell channel routing with cell orientations consideration." *IEEE Transactions on Computer-aided Design of Integrated Circuits and Systems* 14, no. 6 (1995), pp. 766–772.

[29] H.-P. Tseng and C. Sechen. "Multi-layer over-the-cell routing with obstacles." *Proceedings of CICC 97-Custom Integrated Circuits Conference* (1997), pp.565–568, IEEE.

[30] J. Cong, J. Fang, and Y. Zhang. "Multilevel approach to full-chip gridless routing." *IEEE/ACM International Conference on Computer Aided Design. ICCAD 2001. IEEE/ACM Digest of Technical Papers (Cat. No. 01CH37281)* (2001), pp. 396–403, IEEE.

[31] J. Cong, M. Xie, and Y. Zhang. "An enhanced multilevel routing system." *Proceedings of the 2002 IEEE/ACM International Conference on Computer-aided Design* (2002), pp. 51–58.

[32] X. Hong, T. Xue, J. Huang, C.-K. Cheng, and E. S. Kuh. "TIGER: An efficient timing-driven global router for gate array and standard cell layout design." *IEEE Transactions on Computer-aided Design of Integrated Circuits and Systems* 16, no. 11 (1997), pp. 1323–1331.

[33] J. Huang, X.-L. Hong, C.-K. Cheng, and E. S. Kuh. "An efficient timing-driven global routing algorithm." *Proceedings of the 30th International Design Automation Conference* (1993), pp. 596–600.

[34] S.-W. Hur, A. Jagannathan, and J. Lillis. "Timing-driven maze routing." *IEEE Transactions on Computer-aided Design of Integrated Circuits and Systems* 19, no. 2 (2000), pp. 234–241.

[35] J. Hu and S. S. Sapatnekar. "A timing-constrained algorithm for simultaneous global routing of multiple nets." *IEEE/ACM International Conference on Computer Aided Design. ICCAD-2000. IEEE/ACM Digest of Technical Papers (Cat. No. 00CH37140)* (2000), pp. 99–103, IEEE.

[36] H.-P. Tseng, L. Scheffer, and C. Sechen. "Timing- and crosstalk-driven area routing." *IEEE Transactions on Computer-aided Design of Integrated Circuits and Systems* 20, no. 4 (2001), pp. 528–544.

[37] T.-Y. Ho, Y.-W. Chang, S.-J. Chen, and D.-T. Lee. "Crosstalk- and performance-driven multilevel full-chip routing." *IEEE Transactions on Computer-aided Design of Integrated Circuits and Systems* 24, no. 6 (2005), pp. 869–878.

[38] N. Hanchate and N. Ranganathan. "A linear time algorithm for wire sizing with simultaneous optimization of interconnect delay and crosstalk noise." *19th International Conference on VLSI Design held jointly with 5th International Conference on Embedded Systems Design (VLSID'06)* (2006), pp. 8. IEEE.

[39] J.-A. He and H. Kobayashi. "Simultaneous wire sizing and wire spacing in post-layout performance optimization." *Proceedings of 1998 Asia and South Pacific Design Automation Conference* (1998), pp. 373–378, IEEE.

[40] Y.-M. Jiang, A. Krstic, K.-T. Cheng, and M. Marek-Sadowska. "Post-layout logic restructuring for performance optimization." *Proceedings of the 34th Annual Design Automation Conference* (1997), pp. 662–665.

[41] T. Xiao and M. Marek-Sadowska. "Gate sizing to eliminate crosstalk induced timing violation." *Proceedings 2001 IEEE International Conference on Computer Design: VLSI in Computers and Processors. ICCD 2001* (2001), pp. 186–191, IEEE.

[42] K.-Y. Lee and T.-C. Wang. "Post-routing redundant via insertion for yield/reliability improvement." *Proceedings of the 2006 Asia and South Pacific Design Automation Conference* (2006), pp. 303–308.

[43] H.-Y. Chen, M.-F. Chiang, Y.-W. Chang, L. Chen, and B. Han. "Novel full-chip gridless routing considering double-via insertion." *Proceedings of the 43rd Annual Design Automation Conference* (2006), pp. 755–760.

[44] G. Y. Liu, R. F. Zhang, K. Hsu, and L. Camilletti. "Chip-level CMP modeling and smart dummy for HDP and conformal CVD films." *arXiv preprint cs/0011014*, 2000 https://arxiv.org/abs/cs/0011014, Last accessed on January 11, 2023.

[45] Y. Chen, P. Gupta, and A. B. Kahng. "Performance-impact limited area fill synthesis." *Proceedings of the 40th Annual Design Automation Conference* (2003), pp. 22–27.

[46] D. Park, D. Lee, I. Kang, C. Holtz, S. Gao, B. Lin, and C.-K. Cheng. "Grid-based framework for routability analysis and diagnosis with conditional design rules." *IEEE Transactions on Computer-aided Design of Integrated Circuits and Systems* 39, no. 12 (2020), pp. 5097–5110.

[47] S. Mantik, G. Posser, W.-K. Chow, Y. Ding, and W.-H. Liu. "ISPD 2018 initial detailed routing contest and benchmarks." *Proceedings of the 2018 International Symposium on Physical Design* (2018), pp. 140–143.

[48] W.-H. Liu, S. Mantik, W.-K. Chow, Y. Ding, A. Farshidi, and G. Posser. "ISPD 2019 initial detailed routing contest and benchmark with advanced routing rules." *Proceedings of the 2019 International Symposium on Physical Design* (2019), pp. 147–151.

[49] S. Dolgov, A. Volkov, L. Wang, and B. Xu. "2019 CAD contest: LEF/DEF based global routing." *2019 IEEE/ACM International Conference on Computer-aided Design (ICCAD)* (2019), pp. 1–4, IEEE.

[50] S. M. Gonçalves, L. S. Rosa, and F. S. Marques. "DRAPS: A design rule aware path search algorithm for detailed routing." *IEEE Transactions on Circuits and Systems II: Express Briefs* 67, no. 7 (2019), pp. 1239–1243.

[51] D. Shi, E. Tashjian, and A. Davoodi. "Dynamic planning of local congestion from varying-size vias for global routing layer assignment." *IEEE Transactions on Computer-aided Design of Integrated Circuits and Systems* 36, no. 8 (2017), pp. 1301–1312.

[52] W.-T. Hung, J.-Y. Huang, Y.-C. Chou, C.-H. Tsai, and M. Chao. "Transforming global routing report into DRC violation map with convolutional neural network." *Proceedings of the 2020 International Symposium on Physical Design* (2020), pp. 57–64.

[53] W. Zeng, A. Davoodi, and Y. H. Hu. "Design rule violation hotspot prediction based on neural network ensembles." *arXiv preprint arXiv:1811.04151*, 2018. https://arxiv.org/abs/1811.04151, Last Accessed January 11, 2023.

29 Physical Verification and Signoff

'Rule Forty-two. All persons more than a mile high to leave the court.' ... said the King.

—Lewis Carroll, *Alice's Adventures in Wonderland*, Chapter 12, 1865

In the previous chapters, we have discussed various physical design tasks, viz. chip planning, placement, clock tree synthesis, and routing. After performing these tasks, we obtain a layout that contains sufficient information for fabricating a chip. However, we need to perform some post-layout verification tasks on the final layout before sending it to the foundry for fabrication. These verification tasks ensure that the final layout delivers the desired functionality, meets the power, performance, and area (PPA) requirements, adheres to rules provided by the foundry, and is free from signal integrity (SI) issues and other electrical problems. Note that we carry out some of these tasks in the earlier stages of the design flow also. However, now we perform these tasks more rigorously because this is the last opportunity to fix design problems before fabrication. Additionally, the final layout allows us to perform these tasks more accurately because it contains complete design information.

The post-layout verification involves three major tasks:

1. **Layout extraction:** The layout describes the polygons or shapes required on each fabricated layer. We need to extract various information from the layout so that other verification tools can work with the extracted information. The problem for a verification tool is greatly simplified if it works with the extracted information rather than the original layout.
2. **Physical verification and signoff checks:** We need to check whether a layout adheres to manufacturability *rules*, matches the functionality of the initial netlist, and is free from electrical and timing problems. We refer to the checks that we carry out on the final layout before sending it to a foundry as *signoff checks*.
3. **Engineering change order (ECO):** If we need to make some minor changes to a design almost at the end of a design flow, we make these changes and carefully verify them using an *ECO process*.

In this chapter, we will discuss the above post-layout verification tasks in detail.

29.1 LAYOUT EXTRACTION

There are two major steps in layout extraction: *circuit extraction* and *parasitic extraction*. The circuit extraction involves extracting devices and their interconnections to reconstruct the circuit that the layout is supposed to represent. Additionally, we extract parasitic information from the layout that can be used in static timing analysis (STA), SI analysis, and power integrity analysis. We discuss these tasks in more detail in the following paragraphs.

29.1.1 Circuit Extraction

The framework for circuit extraction is shown in Figure 29.1. Before carrying out circuit extraction, we need to obtain the complete layout database for a design. Note that, during implementation, the layout contains macros and standard cells with an *abstract layout view*. However, this view of the layout is insufficient for layout extraction because it does not contain the *full layout details* of the instantiated macros and standard cells. Therefore, we need to *merge* the graphical database system (GDS) of the top-level design layout with the GDS files of the macros and standard cells contained within it. Typically, implementation tools provide some mechanism to create a *merged GDS file*. We use the merged GDS file for circuit extraction.

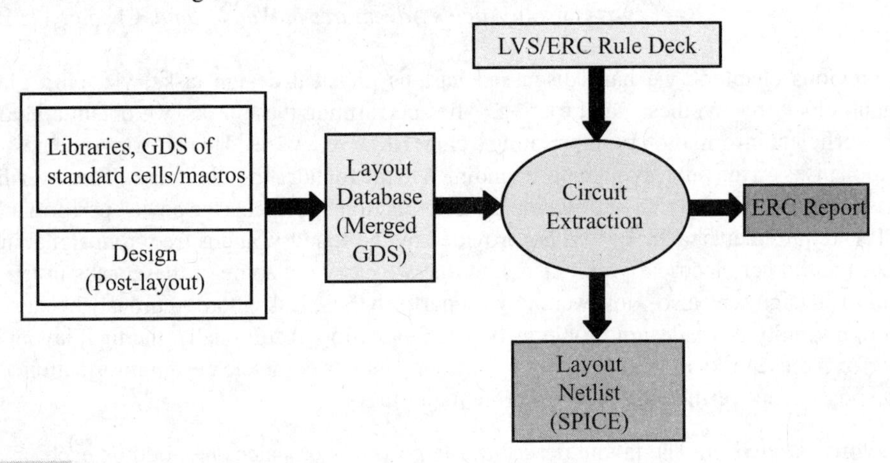

Figure 29.1 Framework for circuit extraction

The circuit extraction task involves extracting devices and connections from the layout [1]. The circuit extraction tool identifies devices and connections by comparing features in the layout with the given set of patterns or *extraction rules*. These rules are technology-specific and are defined by the foundries.

For a given technology, a foundry identifies a set of devices, such as transistor, capacitor, resistor, and diode, that can be fabricated. Then, it determines the set of extraction rules or patterns to identify and extract these devices from the layout database. These rules are encoded in a format that a physical verification tool can comprehend.

Layout versus schematic (LVS) rule deck: Typically, we carry out circuit extraction using LVS checking tools. Hence, the rules for circuit extraction are represented as instructional code in files that are commonly referred to as *LVS rule deck*. Typically, a foundry bundles the LVS rule deck with the process design kits (PDKs). While extracting the circuit from a layout database, we need to provide an LVS rule deck to the tool. The LVS rule deck has the following information for extraction:

1. **Layer definition rules:** These rules help identify the various physical layers in the given design layout.
2. **Layer derivation rules:** These rules help derive additional layers by using existing layers. These rules can employ *logical operations* on polygons or shapes.

3. **Connectivity rules:** These rules help identify electrical connectivity between various layout layers. For example, a connectivity rule can define the electrical connection between two metal layers using a via layer.
4. **Device and parameter extraction rules:** These rules specify how to extract devices and their terminals from the layout layers, typically using a sequence of polygonal logical operations [2].

We illustrate the device extraction process in the following example.

Example 29.1 Consider the simplistic layout of an NMOS shown in Figure 29.2(a). Let us understand how we can extract various layers for this device with the help of an LVS rule deck.

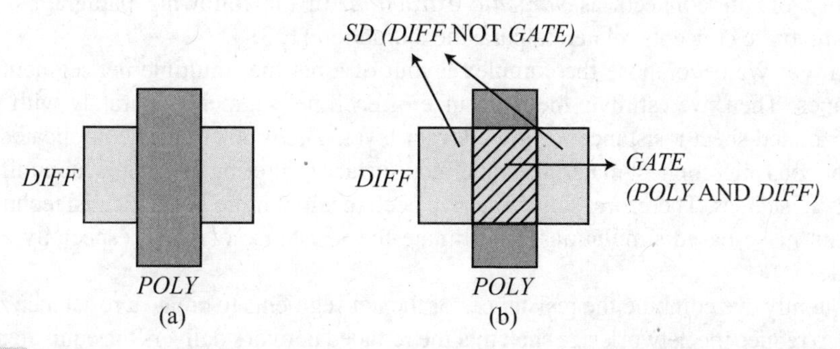

Figure 29.2 Device extraction: (a) layout of a MOSFET and (b) derivation of various layers

The physical layers of the layout would be defined in the LVS rule deck. Assume that the layer *DIFF* for the n-type diffusion layer and the layer *POLY* for the polysilicon layer are defined. Then, with the help of the LVS rule deck, we can derive the *GATE* layer by taking the common region between the *DIFF* and the *POLY* layer. We can consider this operation as logical AND applied to sets of points contained in these two polygons, as shown in Figure 29.2(b). We can define the *SD* (source and drain) layer by considering the *DIFF* layer that is not shared with the *GATE* layer. We can view this operation as NOT operation of *GATE* layer over *DIFF* layer.

The parameters of the extracted MOSFET can be derived from the associated polygons. For example, the gate length and width can be extracted from the *GATE* layer, while the source and drain area can be extracted using the *SD* layer.

A circuit extraction tool can identify devices in the given layout using the LVS rule deck, as demonstrated in the above example. Subsequently, from the remaining portion of the layout, it will determine distinct *nets* that connect these devices.

A net consists of electrically connected objects in the layout. A net can be in the same layer created by the overlap of polygons or abutment. A net can also connect objects in different layers using vias and contacts.

During net extraction, the circuit extraction tool also determines the device terminals connected by the nets. It also assigns a unique identifier or name to each extracted net. Once the devices (transistors, diodes, capacitors, resistors, etc.) and their terminals are extracted, and connections are made between them using extracted nets, the circuit extraction is complete. The extracted circuit is typically represented as a SPICE netlist and is referred to as *layout netlist*.

We also perform electrical rule checks (ERCs) during circuit extraction, as shown in Figure 29.1. Physical verification tools typically provide options to enable ERC during circuit extraction and select the electrical rules of our interest from the rule deck.

29.1.2 Parasitic Extraction

We need to determine the electrical characteristics of the interconnects from the final layout to perform STA, SI analysis, IR drop analysis, and electromigration analysis. Ideally, interconnects should simply connect devices in the layout. However, they exhibit undesirable resistance (R), capacitance (C), and inductance (L). Hence, we refer to the task of determining the electrical characteristics of interconnects as *parasitic extraction*. In the following paragraphs, we briefly describe some basic concepts related to parasitic extraction [1, 3].

Resistance: We decompose the complex layout of a net into multiple net segments that have simpler shapes. Then, we estimate the resistance of each net segment separately with the help of its shape, size, and sheet resistance for the relevant layer. The problem gets complicated due to the nonhomogeneous interconnect material, skin effect, surface scattering of carriers, nonuniform width, and contact resistances. Therefore, sometimes we need to adopt more sophisticated techniques, such as discretized mesh-based simulations, to estimate the resistance of a net, especially at advanced process nodes.

Subsequently, we combine the resistances of the net segments to create a resistance network for a net. We also reduce the network size such that the reduced network delivers the equivalent electrical characteristics. Such network reductions are application-specific and try to lessen the computational burden of the applications, such as delay calculators and SI analyzers.

Capacitance: We have discussed various components of interconnect capacitance (*overlap*, *lateral*, and *fringe*) in Chapter 24 ("Basics concepts for physical design").[1] Accounting for all these components in parasitic extraction is computationally challenging due to complex electrostatic interactions among interconnects.

Field solvers: We can extract capacitance of a net using a *field solver* [4]. A field solver calculates parasitics for the given structure by numerically solving Maxwell's equation. The numerical simulation-based techniques allow field solvers to achieve greater accuracy than analytical techniques, especially for complicated geometries.

Field solvers can use *finite element method* in which we create a discretized mesh for the entire space in consideration. Subsequently, it formulates physics-based matrix equations and invokes an efficient matrix equation solver to produce the result. Alternatively, we can use *floating random walk* methods that employ statistical techniques to derive capacitance. Floating random walk-based field solvers have matured over the last two decades and are widely used for capacitance extraction.

A field solver can compute parasitic capacitance accurately in a 3D space but can take long runtime. Therefore, we do not use a field solver for the full-chip parasitic extraction.

Chip-level parasitic extraction: We perform chip-level parasitic extraction using *pattern matching* technique. In this technique, we divide the parasitic extraction process into the following two tasks [3]:

1. **Technology pre-characterization:** First, we compute and store the capacitance models for a set of structural patterns for a given technology. We need to carry out this task once for a given technology, and this task is referred to as *technology pre-characterization*. The major steps involved in technology pre-characterization are shown in Figure 29.3.

[1] See Section 24.2.3 ("Parasitics") for various components of capacitance.

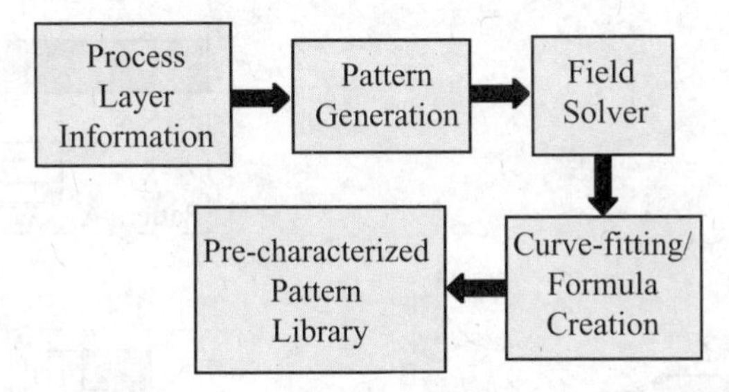

Figure 29.3 Major steps in technology pre-characterization

For a given process, we enumerate millions of sample geometries or structures by choosing various combinations and arrangements of interconnect layers. We use the technology layer stack information and the characteristics of the layer stack to derive these structures. For example, we use the permittivities of the dielectric layers and the spacing rules for deriving these structures.

Then, we compute the capacitance of the above structures using an accurate field solver [4, 5]. The capacitances computed by the field solver are stored in lookup tables, or empirical formulae are derived using curve fitting techniques [6, 7]. Thus, we create a *pre-characterized pattern library* for parasitic extraction. We also group together structures with similar formulas and try to reduce the number of patterns in a pattern library. Note that the task of technology pre-characterization is very timing consuming. However, we need to perform this task only once for a given technology. Therefore, the cost and effort of technology pre-characterization get distributed over multiple designs that use that technology.

2. **Pattern matching:** For extracting capacitance from a given layout, we first partition the layout into smaller structures enclosed by a window. Then, we match each window with the patterns in the pre-characterized library and compute the capacitance with the help of lookup tables or empirical formulae. We use the actual geometries measured from the layout in these formulas for the capacitance calculation.

We illustrate the pattern matching-based capacitance extraction technique in the following example.

Example 29.2 Consider the arrangement of wires of two metal layers *M1* and *M2* shown in Figure 29.4(a). The *M1* layer is running along x-axis and *M2* layer along y-axis. These layers are separated by a dielectric layer of thickness *D*. These wires will exhibit overlap and fringe capacitance among themselves. Let us examine how capacitance can be extracted using pattern matching.

Consider the 2D cross-section of the structure taken using the Y–Z plane along the cutline AA'. It will match pattern *A* shown in Figure 29.4(b). Similarly, the 2D cross-section of the structure taken using the X–Z plane along the cutline BB' will match pattern *B*. We can compute the total capacitance by combining the capacitances of patterns *A* and *B*. However, note that there are two overlap capacitors associated with regions where the wires of *M2* layer lie directly over the wire of *M1* layer. The overlap capacitance will be considered in both pattern *A* and pattern *B*. Hence, we need to subtract the overlap capacitances (shown by two patterns *C*) from the total capacitance.[2]

[2] The patterns shown in this example are for illustrative purposes only. A parasitic extraction tool can use its own set of patterns and employ various compensation techniques for the overlapped region.

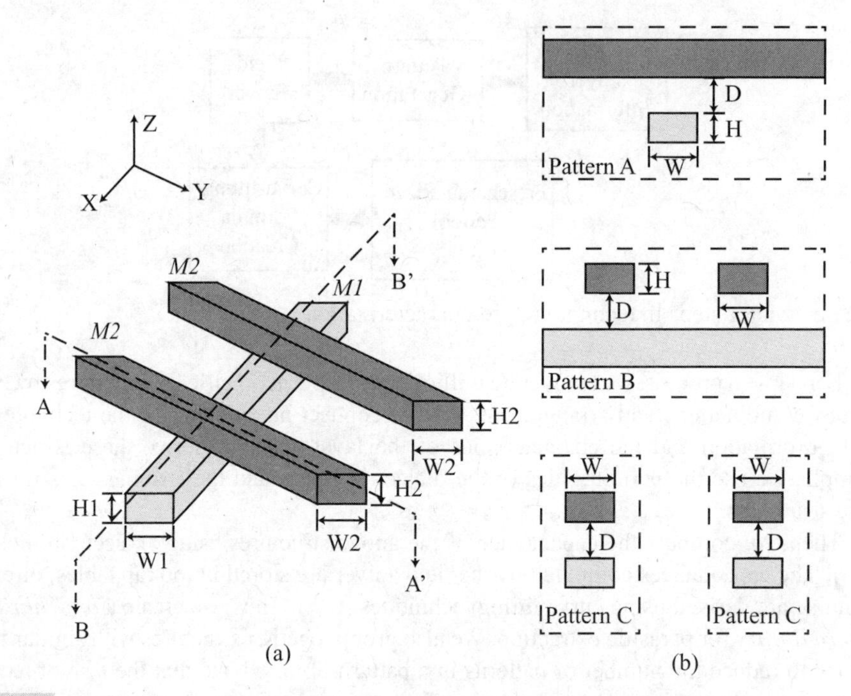

Figure 29.4 Capacitance extraction: (a) given arrangement of wires in the design and (b) patterns matched [8]

Remark: In this example, we have extracted capacitances by considering 2D shapes along two orthogonal planes. Therefore, this technique of capacitance extraction is known as 2.5D capacitance extraction. In this technique, we scan the given layout in the X and Y directions and match the corresponding 2D cross-sectional geometry with the pre-characterized pattern library, as described above.

The pattern matching-based capacitance extraction technique is very efficient for large circuits and typically delivers results with acceptable accuracy. However, for the critical nets, we can re-extract capacitance using field solvers to achieve a higher accuracy.

Inductance: The extraction of inductance is more complicated than extracting resistance and capacitance. We define inductance using the current in a *closed loop*. Therefore, we need to know the *return current path* for the inductance computation. However, determining the return current path is challenging because the current takes the path of least impedance. Hence, the return current path can change with frequency and the circuit state. The return current can also be in the form of a *displacement current*, which makes the computation even more complicated.

The magnetic field due to an interconnect can terminate over a large area (depending on the return current path), and we cannot use a simple rule-based approach for inductance calculation. Therefore, we use specialized numerical techniques for inductance extraction. However, these techniques can take excessive computational resources because of the need to consider a larger area.

We can simplify the inductance extraction problem by extracting it only when necessary. Fortunately, in most situations, we can ignore inductance extraction. The inductance effect is insignificant in narrow wires in which the resistance dominates. The inductance extraction becomes important for wider wires on the upper metal layers. We can also simplify the inductance extraction problem by imposing some restrictions on the interconnect structure that allows us to easily identify the return current path.

Network reduction: The straightforward extraction of parasitics produces an extensive network of resistors, capacitors, and inductors. Consequently, parasitic-included netlists become too large, and applications such as STA, SI analysis, and IR drop analysis become inefficient on such netlists. Therefore, parasitic extraction tools need to reduce the extracted network. The reduction can be application-specific. For example, a parasitic extraction tool can reduce network size by ensuring that the signal delays remain fairly correct. It can also reduce the network size by attempting to preserve the total capacitance and resistance of the network. Alternatively, it can preserve a few dominant time constants of the network. Typically, extraction tools also provide some mechanism to control network reduction using threshold values for the extracted quantities.

Framework for parasitic extraction: The chip-level parasitic extraction framework is shown in Figure 29.5. We provide the layout and physical libraries (library exchange format (LEF) files) as inputs to the extraction tool. Before carrying out parasitic extraction, we perform some basic connectivity checks (such as short circuit and open circuit) on the layout. We also provide the process technology information to the extraction tool. An extraction tool needs this information to know the sheet resistance of each layer for resistance extraction and capacitance models for capacitance extraction. This information is provided by foundries for a given extraction tool and is bundled with the PDKs.

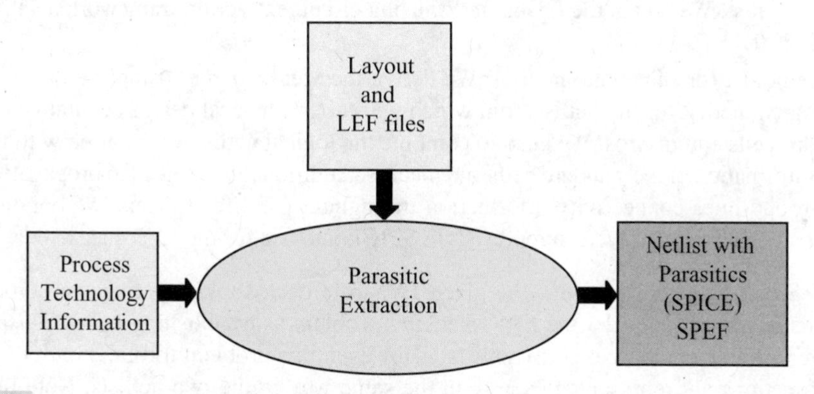

Figure 29.5 Framework for parasitic extraction

We can instruct the parasitic extraction tool to produce the type of parasitic model based on our requirements. For example, we can extract parasitics as *lumped capacitance*, *distributed resistance*, or *distributed resistance and capacitance* with/without *coupling capacitance*. We can use different types of parasitic models based on the requirement of the application, such as STA, SI analysis, or IR drop analysis.

A parasitic extraction tool can produce outputs in various forms. Depending on the application, we can instruct a parasitic extraction tool to produce a SPICE netlist or parasitics in *detailed standard parasitic format* (DSPF) or *standard parasitic exchange format* (SPEF). A SPEF file is an ASCII file (readable using an editor) and standardized by IEEE [9].

The electrical properties of the fabricated interconnects can vary due to the variations in parameters such as line width, thickness, and spacing. To account for these variations, we need to extract parasitics at different *corners*. Industry-standard parasitic extraction tools provide mechanisms to extract parasitics from the layout for various corners. The extraction rule deck also needs to support multiple corner extraction. Some tools can have capabilities to extract parasitics simultaneously for multiple corners. A simultaneous extraction can be more efficient than extracting separately for multiple corners because it can reuse computation among different corners.

29.2 PHYSICAL VERIFICATION TASKS

There are three major physical verification tasks that we need to carry out on the layout:

1. LVS check
2. ERC
3. Design rule check (DRC)

We will discuss these tasks in this section.

29.2.1 LVS Check

The framework of LVS check is shown in Figure 29.6. It establishes the equivalence of the following two entities:

1. Layout netlist: We obtain the *layout netlist* using circuit extraction framework, as illustrated in Figure 29.1.
2. Source netlist (or schematic netlist): We derive the *source netlist* using the design database. The logical netlist (or the netlist from which we started physical design) contains instances of standard cells and macros. We need to combine the logical netlist information with the device-level information of the standard cells and macros. Additionally, we need to propagate the power and ground lines connectivity information throughout the circuit. Thus, we obtain a device-level source netlist that we can represent in SPICE and use for the LVS check.

An LVS-checking tool employs the given LVS rule deck for comparing the layout and the source netlists. We can consider the LVS-checking problem as establishing *isomorphism* between the graph representations of these two netlists.[3] However, this problem formulation will work only when the functions are represented exactly in the same way in the two netlists. Note that we can represent a Boolean function as a graph in multiple non-isomorphic ways. Hence, checking for graph isomorphism can be too restrictive for the LVS check. Therefore, LVS tools augment the graph isomorphism detection techniques with formal methods for establishing the equivalence of two netlists.[4]

An LVS-checking tool starts by finding the correspondence between the ports of two netlists, typically using their port names. Subsequently, it tries to find a correspondence between the circuit elements of the layout netlist and the source netlist. If it can find such a correspondence for the entire circuit, the two netlists are said to pass the LVS check. If the tool finds any discrepancies, it reports them, along with some hints that can help us locate the root problem.

---

[3] For isomorphism, see Section 11.4.3 ("Reduced ordered binary decision diagram").

[4] For formal methods, see Section 11.8 ("Equivalence checking").

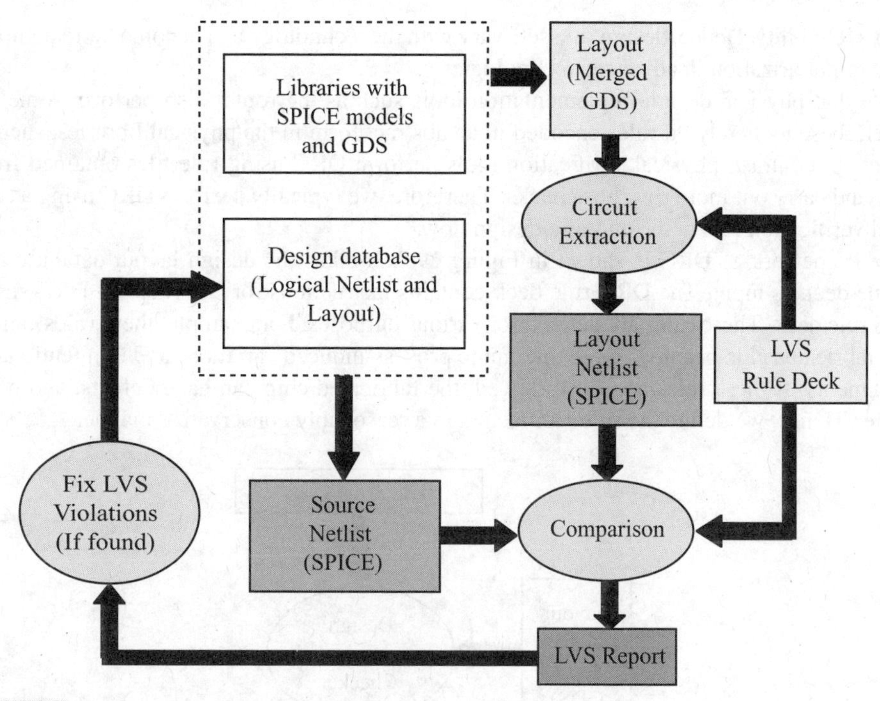

Figure 29.6 Framework for LVS checking

Some of the common LVS errors are the short circuit and open circuit of nets. If nets get short circuited (or open circuited) in the layout, the total number of nets reported by the LVS tool for the layout netlist will be less (or more) than the number reported for the source netlist. We can use the number of reported nets as a hint in debugging LVS errors.

An LVS tool typically improves performance by processing a repeated block only once. This technique is known as *hierarchical processing*. Further, an LVS tool can achieve additional runtime improvements by multi-core and distributed processing.

29.2.2 Electrical Rule Check

An ERC ensures that the layout does not contain any problematic electrical connection. As explained earlier, ERC is typically performed while extracting the netlist from the layout.

Some of the problems that we can detect using ERC are unintentional short circuit or open circuit, wrong power (V_{DD}) or ground (GND) connections, floating gates, undriven nets, and n-well/p-well not connected to power/ground lines. For complex chips that integrate many IPs and have multiple power domains, ensuring the sanity of power/ground lines becomes critical. We can also detect the existence of *electrostatic discharge* (ESD) protection components in a design during ERC. Commercial physical verification tools provide mechanisms to customize ERC based on design needs. These rule checks are referred to as *programmable ERCs*.

29.2.3 Design Rule Check

We carry out DRCs to ensure that the layout meets the constraints required for manufacturing a chip. These constraints or rules are defined by the respective foundry to achieve a good yield and improve the chip's reliability. Hence, design rule sets vary with the technology and become more complicated with the miniaturization in advanced technologies.

Note that physical design implementation tools such as the router also perform some DRCs. However, these tools rely on rules encoded in an abstract form in the physical libraries, such as the LEF files. In contrast, physical verification tools perform DRC using rule files obtained from the foundry and carry out more rigorous checks. Therefore, we typically perform DRC using a separate physical verification tool at the end of a design flow.

The framework of DRC is shown in Figure 29.7. It takes the design layout database and the DRC rule deck as input. The DRC rule deck contains instructions for carrying out DRC similar to the LVS rule deck. These rules are defined for various purposes. For example, these rules can ensure that the fabricated chip becomes more tolerant to process-induced variations and unintentional mask misalignments. If these rules are not followed, the fabricated chip can have defects, and the yield can suffer. Hence, we define the rule parameters in a reasonably conservative manner.

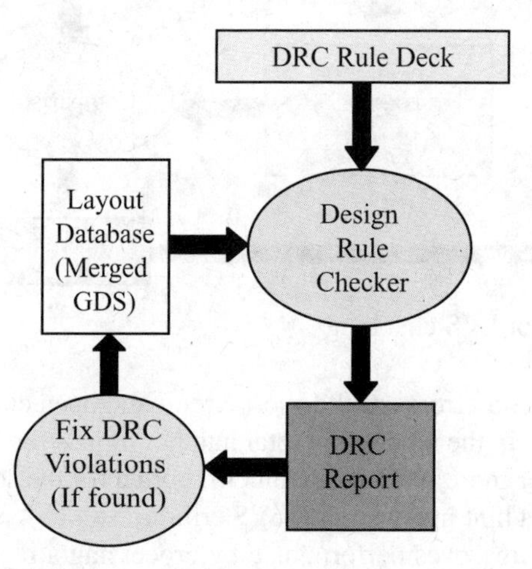

Figure 29.7 Framework for DRC

DRC rules can be of several types, such as *size rules*, *spacing rules*, *enclosure rules*, and *extension rules*, as illustrated in Figure 29.8.

1. **Size rules:** A rule can define the minimum dimension for the shapes in the layout. For example, a foundry can define the minimum width for the shapes of various layers, such as diffusion, polysilicon, metal, and contact.
2. **Spacing rules:** A rule can define the minimum spacing between two shapes on the same or different layers. For example, a foundry can define the minimum spacing between two polysilicon lines in the same layer.
3. **Enclosure rules:** A rule can define the minimum spacing between the edges of a shape enclosed within another shape. For example, a foundry can define the minimum spacing between the edge of the polysilicon line and the enclosed contact.
4. **Extension rules:** A rule can define the minimum extension length between two shapes. For example, a foundry can define the minimum length by which the polysilicon line should extend beyond diffusion.

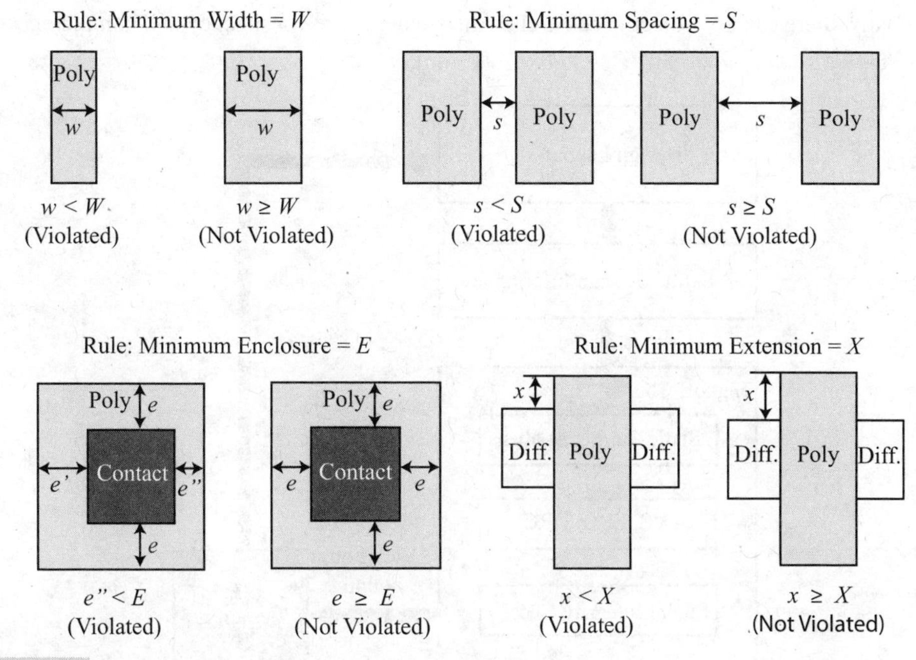

Figure 29.8 Examples of various types of design rules

A DRC tool searches for the patterns specified by the rule in the given layout database. On finding such patterns, it reports rule violations on the layout. Then, we need to analyze and fix those violations in the design.

The industry-standard DRC tools have capabilities to make design changes to fix DRC violations. For example, with the help of foundry-provided rules, a DRC tool can identify regions on the layout that can have issues in *chemical mechanical polishing* (CMP). Further, we can use that tool for inserting dummy metal fills. Similarly, we can use DRC tools for finding and fixing violations of *antenna rules* and *design for manufacturing* (DFM) related rules such as multi-patterning. After these layout changes, we need to generate a merged GDS file for further processing and signoff.

29.3 SIGNOFF

A design becomes ready for fabrication once we have obtained its layout. However, before sending the layout to a foundry, we need to ensure that the layout delivers the intended functionality and meets various figures of merit, such as performance, power, manufacturability, yield, and reliability. This task is known as design *signoff*. It is a critical task in a design flow because any bug that escapes signoff will find its way into the fabricated chip.

Figure 29.9 shows the framework of design signoff.[5] We start with the finished layout. We extract various information about the design using this layout, such as net parasitics and connectivity. Then, we carry out multiple signoff checks related to timing, foundry-defined rules, SI, and power integrity.

[5] The sequence of signoff tasks shown in the figure need not be followed in a design project. We often carry out these tasks parallelly by distributing them among multiple teams to meet the stringent design schedule.

Additionally, there can be project-specific checklists that ensure consistency and completeness of a design signoff.

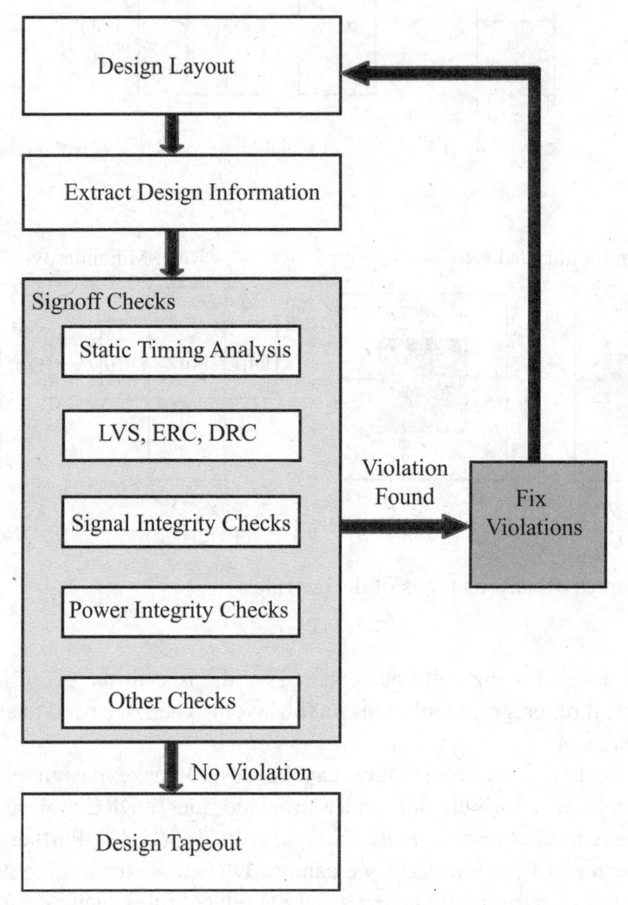

Figure 29.9 Framework of signoff

At the signoff stage, we have all the information about the layout. Therefore, we can carry out the signoff checks with greater accuracy than during design implementation. Since signoff is the last chance to detect and fix design problems, we need to analyze a design more accurately, even if we incur extra computational resources. Moreover, we typically perform signoff checks using separate signoff tools without depending on the tools we used in the design implementation. It helps us discover design problems that escaped detection during implementation.

If we find violations during a signoff stage, two possibilities can occur after analysis. First, we find that the violation is tolerable. In that case, we can simply waive the violation. Second, we find the violation is a genuine design problem. In that case, we need to fix that violation by making some design changes. After making these design changes, we need to perform the signoff checks again to verify the correctness of the fix and also ensure that the fix did not produce any new violations. Hence, a signoff check is an iterative process. To minimize the number of iterations, we try to make only local or incremental changes to a design during signoff.

After a design passes all signoff checks, we send the corresponding layout database to a foundry for fabrication. This task is referred to as *tape-out*. It is an occasion to celebrate for a design team since tape-out marks the end of an arduous task of designing an integrated circuit (IC).

In the following paragraphs, we describe some of the essential signoff checks.

29.3.1 Static Timing Analysis

We perform signoff STA at a higher level of accuracy. Therefore, the following considerations are critical for signoff STA:

1. **Parasitic extraction:** We need to extract parasitics more accurately for signoff. We extract parasitics using the parasitic extraction tool in a suitable format, such as DSPF and SPEF and provide them as inputs to the STA tool. When we use both the parasitic extraction tool and STA tool from the same electronic design automation (EDA) vendor, we can extract parasitics in a *binary format* (a machine-readable format that can be loaded quickly) and provide them as inputs to the STA tool. It can save runtime in loading the voluminous parasitics information for large designs.

   To improve the accuracy of signoff parasitic extraction, we need to consider the following:

   (a) For critical nets, we can instruct parasitic extraction tools to extract capacitance more accurately by invoking a field solver.

   (b) We can instruct parasitic extraction tool to include inductance for nets carrying high-frequency signals and having large width on the upper metal layers. There can be overshoots, undershoots, or ringing due to the inductance effect in these nets. Sometimes, these distortions can trigger hold time violations and chip failure.

   (c) We should include metal fills at the signoff level parasitic extraction because their existence can significantly change neighboring wires' capacitances.

   (d) At the signoff level, we extract parasitics for a design for multiple corners to account for process, voltage, and temperature (PVT) variations.

2. **Library models:** Besides accurate models for parasitics, we need to use accurate models for library cells at the signoff level. We can use advanced delay models, such as composite current source (CCS) or effective current source model (ECSM), for signoff STA. We can also use waveform propagation methods, typically available in industry-standard STA tools, to account for waveform distortion, especially at advanced process nodes.

3. **Process-induced variations:** We need to account for *intra-die variations* in signoff STA. We can adopt various techniques depending on the technology node.[6] For older technologies (65 nm and above), we can use on-chip variation (OCV) derate-based techniques. For advanced technology nodes, we need to adopt more sophisticated techniques such as advanced on-chip variation (AOCV) or parametric on-chip variation (POCV).

   We also need to account for *inter-die variations* in signoff STA. We use multimode multi-corner (MMMC) analysis for this purpose. However, an exhaustive set of corners and modes for advanced process nodes creates too many scenarios. It increases the complexity of signoff STA. The industry-standard STA tools can reduce the turnaround time of MMMC analysis by exploiting the commonalities among scenarios and distributed computing.

[6] See Section 14.9 ("Accounting for variations") for OCV, AOCV, and POCV.

4. **Reducing pessimism:** We need to be slightly conservative during signoff. However, we can face issues in achieving timing closure during signoff if we are overly pessimistic. Extra pessimism during signoff can lead to unnecessary timing fixes and hurt the design schedule and quality of result (QoR) (e.g., area and power dissipation increase). Therefore, we try to eliminate extra pessimism in STA during signoff. We can use *path-based analysis* (PBA) for the failing endpoints to reduce some pessimism.[7]

We can fix the timing violations obtained during the signoff stage using various techniques, such as gate resizing, adding or removing buffers, re-routing, wire-widening, a minimal adjustment to placement, and minimal logic changes. At this stage, we can also utilize positive timing slack on some paths for final *leakage power recovery* without impacting the timing of a circuit.

29.3.2 Physical Verification

As explained in the previous sections, we need to carry out LVS, DRC, and ERC on the final layout. We need to fix the violations detected by these physical verification tasks before tape-out.

29.3.3 Signal Integrity

During signoff, we need to consider the SI aspects of a design very carefully.[8] We need to ensure that the SI issues do not lead to noise-induced functional failures. Additionally, we need to ensure that we do not have timing failures due to dynamic delay variations in a design.

We need to extract coupling capacitance between interconnects to carry out SI analysis. We might need to instruct a parasitic extraction tool to extract coupling capacitances since it may not be doing that by default. Additionally, it is prudent to re-check the existence of coupling capacitances in the extracted parasitic files before carrying out an SI analysis. There can be too many coupling capacitances for real designs, and considering all of them can be computationally costly. Hence, we can filter out the coupling capacitances that will have an insignificant impact on the circuit behavior. For example, we can filter out the coupling capacitances with values below some threshold or those having values a few orders less than the total net capacitance.

An SI analysis tool examines the impact of coupling capacitances for: (a) *glitch* or *noise* and (b) *dynamic delay variations*. Typically, we carry out SI analysis using tools that perform an integrated STA–SI analysis. Let us understand how these SI effects are accounted for by an integrated STA–SI analysis tool.

1. **Glitch:** A glitch can occur in a victim net held at constant logic value due to a transition in the aggressor net. Depending on the value of the victim net and the type of transition at the aggressor nets, four types of glitches are possible:
   - **(a)** Logic 0 has a rise glitch,
   - **(b)** Logic 1 has a fall glitch,
   - **(c)** Logic 0 has an undershoot,
   - **(d)** Logic 1 has an overshoot.

[7] See Section 14.6.4 ("Slew propagation") for PBA.

[8] See Section 24.3 ("Signal integrity") for the basic concepts related to SI.

These types of glitches are illustrated in Figure 29.10. These glitches are also referred to as *crosstalk noises* by the EDA tools. If the magnitude of the glitch is above a threshold, it can propagate to a sequential circuit element. Consequently, a wrong value can get latched (0 instead of 1 or 1 instead of 0), leading to a circuit failure.

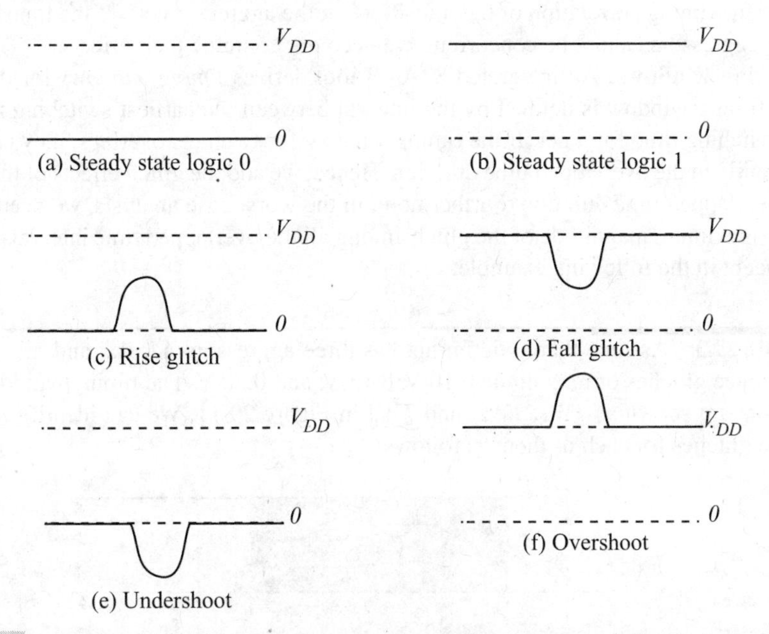

Figure 29.10 Types of glitches at the victim: (a, b) steady state 0 and 1, (c, d) rise and fall glitches due to aggressor rising and falling, respectively, and (e, f) undershoot and overshoot due to aggressor falling and rising, respectively

An SI analysis tool computes the magnitude of a glitch with the help of *noise models* in the given technology library. The magnitude of the glitch depends on the following factors:

(a) Slew of the aggressor net: glitch increases when the aggressor transitions quickly.

(b) Coupling capacitance between the aggressor and the victim net: glitch increases with the increase in the coupling capacitance.

(c) Ground capacitance of the victim net: glitch decreases as the ground capacitance of the victim increases.

(d) Strength of the driver of the victim net: glitch decreases as the strength of the driver of the victim net increases.

The glitch received at the input of a gate will propagate to its output based on its attribute and the circuit conditions. A glitch with a height less than some threshold will not cause a circuit problem. For example, consider a CMOS inverter having input as logic 0. Ideally, the input should have a voltage of 0 V. If the input rises to $V_{IL_{MAX}}$ (the maximum threshold for the input in the low state), the CMOS inverter will still produce a logic 1 at its output. Hence, a rise glitch with a height less than $V_{IL_{MAX}}$ cannot lead to a functional failure in the CMOS inverter.

A narrower glitch (i.e., of shorter duration) and a higher load on the driver of the victim net can stop the propagation of glitches. Therefore, an SI analysis tool propagates glitches in a circuit by considering the width and height of the input glitches and the load at the output pin.

Multiple aggressors: When a victim receives glitches due to multiple aggressors, we can compute the effect of each aggressor separately and then add their effects. However, simply adding the magnitude of individual glitches is *safe* but too *pessimistic* because the victim can receive glitches from different aggressors at different time instants. Therefore, we need to consider the timing correlation of the transitions in the aggressor nets. If the transitions in two or more aggressors cannot be concurrent, we need not combine their effects.

Timing windows: An integrated STA–SI tool derives *timing windows* for the aggressor nets. A timing window is defined by the interval between the earliest switching time and the latest switching time for a net. If the timing window for two nets overlaps, they can transition concurrently in the overlapped time duration. Hence, we add the glitch effects of the aggressors in the overlapped time duration. Furthermore, in the worst-case analysis, we need to consider only the maximum magnitude of the glitch among all the overlapped time intervals. We explain this concept in the following example.

Example 29.3 Assume that a victim net has three aggressors *A1*, *A2*, and *A3* for a net and they produce glitches of magnitude 0.10 V, 0.15 V, and 0.20 V. The timing windows for these aggressors are shown as *TW1*, *TW2*, and *TW3* in Figure 29.11. We can identify five bins and compute glitches for each of them as follows:

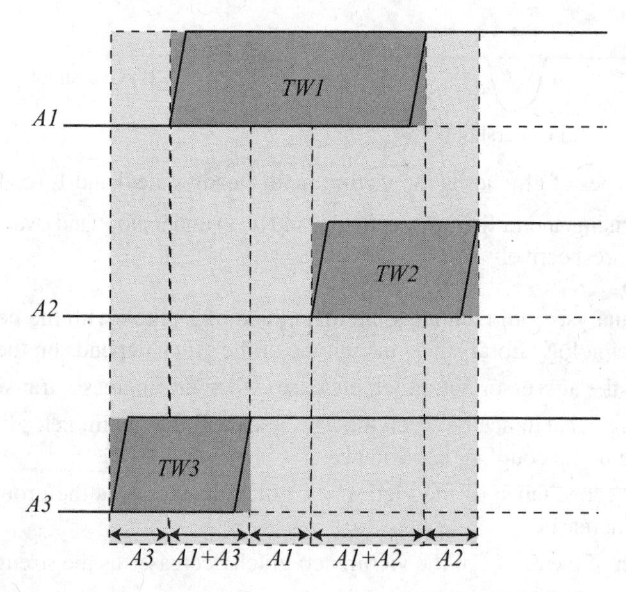

Figure 29.11 Timing window for three aggressors *A1*, *A2*, and *A3*

(a) Only *A3* transitions: Magnitude of glitch = 0.20 V
(b) *A1* and *A3* transition concurrently: Magnitude of glitch = 0.10+0.20 = 0.30 V
(c) Only *A1* transitions: Magnitude of glitch = 0.10 V
(d) *A1* and *A2* transition concurrently: Magnitude of glitch = 0.10+0.15 = 0.25 V
(e) Only *A2* transitions: Magnitude of glitch = 0.15 V

Hence, the worst-case glitch has a magnitude of 0.30 V (when *A1* and *A3* transition concurrently). Note that if we did not consider the timing window, we would have computed the worst-case glitch as 0.10+0.15+0.20=0.45 V, which would have been significantly pessimistic.

2. **Dynamic delay variations:** Besides producing glitches, the transitions in the neighboring nets can change the delay of a circuit element. We describe this effect in the following example.

Example 29.4 Consider the portion of a circuit shown in Figure 29.12. Assume that the gate *G1* (driver of the victim net *A*) makes a 0→1 transition at the output. The driver *G1* provides charge for the ground capacitor C_G to change from 0 to V_{DD}.

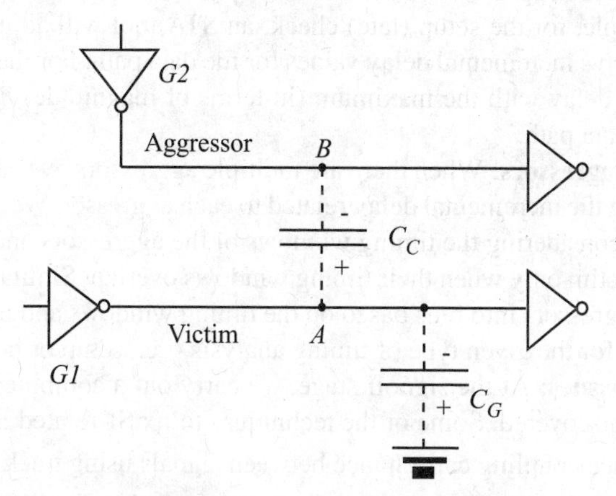

Figure 29.12 A portion of circuit that illustrates the impact of coupling capacitance

First, let us assume that the net *B* is held constant to 0. Hence, the driver *G1* needs to provide a charge for the coupling capacitance C_c to change from 0 to V_{DD}. Hence, *G1* needs to charge $C_G + C_c$.

Next, let us assume that the net *B* is held constant to 1. Hence, initially the voltage across C_c is $-V_{DD}$ and finally when the 0→1 transition completes, the voltage across C_c is 0 V. Hence, *G1* needs to charge $C_G + C_c$. Thus, when the neighboring line *B* is held constant, *G1* needs to provide charge for $C_G + C_c$ for both cases (*B*=0 and *B*=1). When we carry out STA (without SI analysis), we assume that the neighboring nets are held steady, and the capacitance $C_G + C_c$ is considered for the delay calculation. The delay obtained using this assumption is known as the *base delay*.

Next, let us assume that the net *B* also makes a 0→1 transition while *A* was transitioning. In this case, the voltage across C_c is 0 V, both initially and finally. Hence, less charge needs to be provided by *G1* for charging C_c, and the delay of *G1* will decrease.

Next, let us assume that the net *B* makes the opposite transition, i.e., 1→0 transition while *A* was transitioning. In this case, the voltage across C_c is $-V_{DD}$ initially and $+V_{DD}$ finally. Hence, in effect, $2 \times C_c$ needs to be charged, and the delay of *G1* will increase.

The above variations in delay due to coupling capacitors are also referred to as *crosstalk delay* by the EDA tools.

The above example shows that, depending on the transition direction of the aggressor, the delay of the victim net can decrease or increase. An SI analysis tool makes a negative or positive incremental adjustment to the base delay to account for these dynamic delay variations. Moreover, these adjustments can be different for the rise and the fall transitions of the victim. Therefore, an integrated STA–SI analysis tool computes the following information for the victim:

a) maximum rise positive incremental delay

b) maximum fall positive incremental delay

c) maximum (in terms of magnitude) rise negative incremental delay

d) maximum (in terms of magnitude) fall negative incremental delay.

Further, it uses appropriate incremental delay values depending on the path and analysis type. For example, for the setup (late) check, an STA tool will adjust the base delay with the maximum *positive* incremental delay values for the data path. For the hold (early) check, it will adjust the base delay with the maximum (in terms of magnitude) *negative* incremental delay values for the data path.

Multiple aggressors: When there are multiple aggressors, we can estimate their combined effect by adding the incremental delay related to each aggressor. We can make this analysis less pessimistic by considering the timing windows of the aggressors and the victim. An aggressor can impact a victim only when their timing windows overlap. Similar to the glitch analysis, we can separate aggressors into bins based on the timing windows and use the bin that exhibits the worst behavior for the given type of timing analysis (i.e., setup or hold).

Fixing SI issues: At the signoff stage, we carry out a combined STA–SI analysis and fix issues that are discovered. Some of the techniques to fix SI-related issues are as follows [10]:

1. We can reduce coupling capacitance between signals using track re-assignment.
2. We can upsize the driver of the victim net and downsize the driver of the aggressor nets.
3. We can shield the victim net with a ground net if routing resource is available.
4. We can add a buffer to the long victim nets. However, we should consider the possible increase in delay and power dissipation.

The above discussion shows that the crosstalk delay depends on the timing windows, and the timing windows depend on the crosstalk delay. Hence, STA and SI analysis are interdependent, and we often need to carry out SI analysis multiple times to achieve timing closure.

29.3.4 Power Integrity

We need to ensure the power integrity of a chip during signoff. As a chip operates, it draws electrical energy through the supply rails and dissipates it as heat energy. Power integrity issues can arise in a chip due to excessive power dissipation and manifest in several forms. For example, high power dissipation can lead to unacceptable temperature rise, performance loss due to voltage drop in the power lines, and electromigration. We need to ensure that a chip is free from these power integrity issues by performing a set of checks. In the following paragraphs, we describe some of the analyses that need to be carried out during signoff.

1. **Dynamic power analysis**: We perform *dynamic power analysis* by gathering the time-dependent current and voltage profile of various nets in a design. We provide the value change dump (VCD) file as input to the power analysis tool.[9] The power analysis tool uses the VCD file to extract the switching information for each instance in the design. Since a VCD file typically contains data for several clock cycles, we might need to identify a smaller time interval in which the circuit has higher switching activity. We can perform dynamic power analysis for only that time interval to save computational resources. The power analysis tool employs the *power arcs* in the given technology library, the output load, and the input slew to compute the power and current profile for each instance in the design. This approach of power analysis is known as *vector-based approach*.

   We use the current and voltage profiles obtained in dynamic power analysis in the subsequent power integrity checks, as described below.
2. **IR drop analysis**: During the early stages of physical design, we perform IR drop analysis to assess the robustness of the power delivery network (PDN). Therefore, we can perform *static IR drop analysis* by considering the *average current* drawn through the supply rails. A static IR drop analysis does not require switching activities and can detect some problems in a PDN during the early stages of a design flow. However, IR drop is a time-dependent phenomenon. Therefore, we need to perform *dynamic IR drop analysis* during signoff because it can discover more problems associated with the IR drop [11].

   The dynamic IR drop analysis requires estimating the maximum instantaneous current flowing through the supply rails. The instantaneous current through the supply rails depends on the input stimuli and the delay of the circuit elements. It increases if more signals switch simultaneously. Typically, we can carry out dynamic IR drop analysis with the help of the dynamic power analysis framework described above. However, we should also consider the transient IR drop due to parasitics (resistance, capacitance, and inductance) in the *package* pins, bond wires, and bumps (in the flip-chip packages).

   The dynamic IR drop analysis provides information about the local *IR drop hotspots* in a design. After performing IR drop analysis, we can make design changes to fix these IR drop hotspots. For example, we can insert decoupling capacitors, move cells out of the high-density region, downsize cells, and add more power/ground stripes if routing resources are available. We often need to conduct dynamic IR drop analysis and hotspot fixes iteratively until all IR drop-related problems are resolved.

   Impact of IR drop on timing: An IR drop problem can impact the timing of a circuit [12]. For example, consider an inverter *G1* shown in Figure 29.13(a). Assume that the supply voltage reduces to $(V_{DD} - \Delta V)$, where ΔV is the IR drop. Further, assume that it receives an input voltage V_{IN} that changes from $1 \rightarrow 0$, as shown in Figure 29.13(b). Due to the IR drop, the output of *G1* will change from 0 to $(V_{DD} - \Delta V)$ instead of 0 to V_{DD}. Additionally, the load at the output of *G1* will charge slowly. Hence, the delay and the output slew of *G1* can increase from their nominal values, as shown in Figure 29.13(b). The increase in the output slew will further increase the delay of *G2* and other gates in the fanout of *G1*. Thus, IR drop can impact the timing behavior of an entire circuit. Therefore, during signoff, we need to ensure that the IR drop does not create any timing problem.[10]

---

[9] See Section 18.3.2 ("Estimation of activity") for details on VCD.

[10] The gate delay becomes more sensitive to the supply voltage variations with the advancement of technology and decreasing supply voltage [13]. Hence, considering the timing impact of IR drop becomes necessary at advanced process nodes.

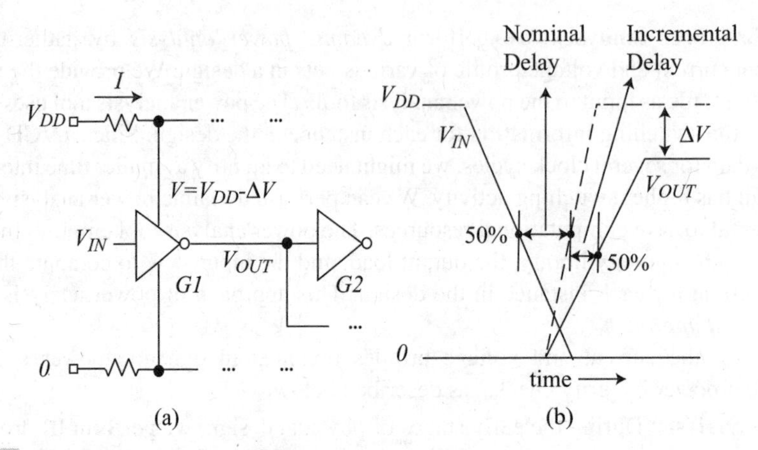

Figure 29.13 Illustration of impact of IR drop on timing: (a) a portion of a circuit with inverter *G1* having some IR drop and (b) change in output waveform due to IR drop (the dashed lines show output waveform for *G1* with no IR drop)

We can use an integrated STA-rail analysis tool framework, as shown in Figure 29.14, to perform IR-drop-aware STA. First, we carry out the IR drop analysis for the entire circuit. Thus, we can obtain the range of supply voltages (minimum and maximum operating voltage) for each instance in a design. Subsequently, a delay calculator can compute the incremental changes in the delay due to the IR drop for each instance. Then, STA can use the IR-drop-adjusted delay in the timing analysis instead of the nominal delay. Note that, depending on the path type, the STA tool will choose an appropriate pessimistic value of the delay. For example, for the setup (late) check, the STA tool will use the delay corresponding to the minimum operating voltage (corresponding to the maximum delay) for the data path. For the hold (early) check, it will use the delay corresponding to the maximum operating voltage (corresponding to the minimum delay) for the data path. Moreover, IR drop-induced delay changes on the *clock path* can also lead to timing violations due to changes in the clock skews. Hence, we need to consider the impact of IR drop for both setup and hold analyses.

We should note that IR drop analysis and timing analysis are interdependent. The IR drop computation depends on the delay and the switching time of the signals computed during STA. Likewise, the computation of STA (delay, slew, and arrival time) depends on the IR drop values, as discussed above. Therefore, we require iterations to converge on the rail voltage and timing attributes of a circuit.

3. **Electromigration checks:** We need to ensure during signoff that the designed circuit is tolerant to electromigration.[11] From the perspective of electromigration, we need to consider three classes of nets [14]. First, the power/ground nets that carry high current with a dominant DC component. For such nets, the average current magnitude can be very high. Second, the clock nets that carry current in both directions. In such nets, the bidirectional current flow allows some *self-healing* and the electromigration effect gets ameliorated. However, the root mean square (RMS) current for such nets can be very high. Third, the other signal nets that carry pulsed and asymmetric current with a low RMS current value. We need to use different strategies for each class of nets.

[11] See Section 25.4.1 ("Electromigration") for the basics of electromigration.

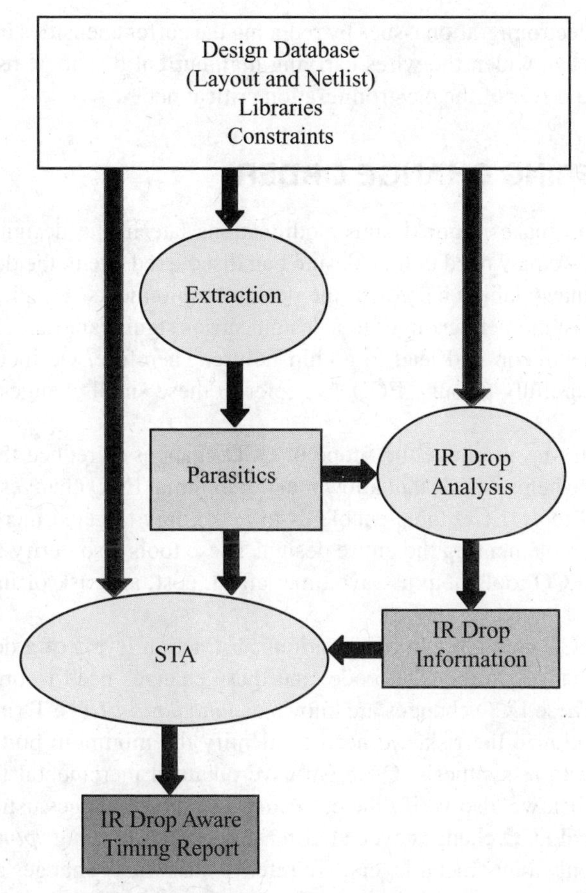

Figure 29.14 A framework for integrated STA-rail analysis

The self-healing effect is negligible in power/ground nets because of the unidirectional current flow. Moreover, power/ground nets carry a high current. Hence, power/ground nets are more susceptible to electromigration.

During signoff, we need to ensure that the *electromigration rules* are followed for the metal wires, vias, and contacts in a design. Typically, these rules are provided by a foundry and they set limits on the current densities for each layer. The current density limits for a metal line can depend on the width, length, number of vias, and temperature. Note that electromigration worsens at an elevated temperature. Therefore, we need to consider the operating conditions for electromigration rules checking.

The industry-standard electromigration checking tools use three kinds of current density limits to identify violations: average current, RMS current, and peak current. The average current is a good measure for identifying electromigration issues in the power/ground nets. We can use static IR drop analysis for computing the average current flowing through these nets and then compare it with the average current limit. The RMS current limit is conservative because it ignores the self-healing effect. However, it can account for thermal effects due to Joule heating. The peak current limit can be used to protect against short-pulsed events such as ESD that can damage the metal due to electrical overstress [14].

We can fix electromigration issues by reducing the current densities in the electromigration critical nets. We can widen the wires carrying high current densities, use short-length wires, and downsize the driver of the electromigration critical net.

29.4 ENGINEERING CHANGE ORDER

Sometimes we need to make minor design modifications late in the design flow due to various reasons. For example, we may need to fix a design bug discovered late in the design flow, or we may need to incorporate unavoidable changes in the design requirements. Despite these modifications being small, making last moment changes to a design carries significant risk. It can introduce errors that will escape to the silicon and lead to a chip failure. Therefore, we incorporate last moment changes to a design carefully through ECO. We refer to these small changes made to a design as *ECO changes*.

ECO tools: We try to minimize the scope of ECO changes to reduce the risk of introducing new errors. We take the help of tools that allow making minimal ECO changes to a design. We refer to these tools as *ECO tools*. ECO tools enable us to make only targeted incremental changes to a design rather than re-implementing the entire design. These tools also verify the correctness of the ECO changes. Thus, ECO tools help us save time, effort, cost, and risk of introducing a new bug into a design.

ECO changes: ECO changes can be at various abstraction levels of a design. Sometimes we need to make ECO changes in the RTL code, and these changes need to propagate to the netlist and the final layout. These ECO changes are known as *functional ECO*s. To make quick functional ECO changes and minimize the risk, we need to identify the minimum portion of the RTL code that requires change and re-synthesis. Once we have obtained incremental functional changes in the corresponding netlist, we also verify the correctness of these changes using formal techniques. We can make functional ECO changes even after mask preparation using *spare cells.*[12] These ECO changes will modify only a few metal layers. We refer to these ECO changes as *metal-mask ECO*.

We can make ECO changes in the layout to fix setup/hold time violations, SI-related issues, and design rule violations. To enable making localized changes to a layout, physical design tools such as placer and router work in the *incremental mode*. In this mode, physical design tools limit the scope of transformation to a small layout region and cause minimal disruption to a design. These features of an ECO tool allow us to make ECO changes with minimal risk and meet the design schedule. Therefore, an ECO tool is a critical component of a robust design flow.

29.5 RECENT TRENDS

With the advancement in process nodes, the physical verification tasks become more challenging. The physical verification tools need to continuously support newer features, such as multiple patterning, 3D ICs, and advanced devices. Additionally, these tools need to handle larger designs and produce results more efficiently. Industry-standard physical verification and signoff tools take advantage of distributed and multi-threaded processing to reduce the runtime of physical verification. At the signoff level, improving the accuracy of the analysis and accounting for the effects that become dominant at smaller geometries are continuously required. Moreover, the interdependence of timing, power, SI, IR drop, electromigration, yield, reliability, and turnaround time poses challenges in developing signoff strategies. In recent times, researchers are trying to exploit machine learning to address some of these challenges [15–19].

[12] See Section 26.5.3 ("Spare cell placement"), for details on spare cells.

REVIEW QUESTIONS

29.1 What information does an LVS rule deck contain?

29.2 What is a field solver?

29.3 Why do we not use a field solver for the complete chip parasitic extraction?

29.4 What is the role of technology pre-characterization in capacitance extraction?

29.5 What is the difference between 2D, 2.5D, and 3D capacitance extraction?

29.6 For which kinds of interconnects extracting parasitic inductance becomes necessary?

29.7 What are the challenges in extracting inductance from a given layout?

29.8 Why does a parasitic extraction tool try to reduce the extracted network?

29.9 What is the significance of LVS checks in a design flow?

29.10 What are some of the common electrical rules we need to check in a design flow?

29.11 What are the various types of DRCs?

29.12 Why do we need to extract parasitics more accurately during signoff than during detailed routing?

29.13 How can the timing window help in reducing the pessimism in the analysis of crosstalk noise and crosstalk delay?

29.14 Consider a victim net *V* and four aggressor nets *A1*, *A2*, *A3*, and *A4*. The incremental crosstalk delay impact of *A1*, *A2*, *A3*, and *A4* are 5 ps, 10 ps, 15 ps, and 20 ps. The timing windows for these nets are shown in Figure 29.15. Compute the worst-case incremental delay impact of the aggressors on the victim.

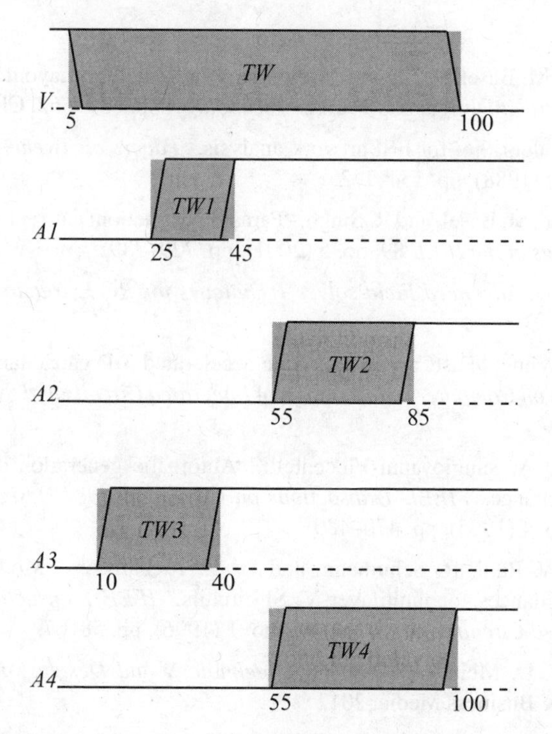

Figure 29.15 Timing windows for the victim and aggressor nets (unit of time on the X-axis is picosecond)

29.15 How can we fix crosstalk delay-induced timing violations in signoff?
29.16 Why do we need to account for the switching activity in the dynamic IR drop analysis?
29.17 How can IR drop lead to a hold violation in a circuit?
29.18 How can we fix IR drop-induced timing violations for a cell instance?
29.19 Why do we need to check for electromigration during signoff?
29.20 Why are power/ground nets more susceptible to electromigration?
29.21 How can we fix violations of electromigration during signoff?
29.22 What are functional ECO changes and metal-mask ECO changes?
29.23 How does a good ECO tool help achieve design closure quickly?

TOOL-BASED ACTIVITY

Take any **signal integrity analysis** tool and carry out the following activities:

1. Extract the parasitics in the post-routed design using a suitable tool and save it appropriately (possibly in SPEF).
2. Load the design and associated information, including parasitics. Carry out SI analysis.
3. Analyze how SI issues can be manifested as (a) change in delay and (b) glitches in the circuit.
4. Understand how increased activities and simultaneous switching of signals cause SI issues.

REFERENCES

[1] W. Kao, C.-Y. Lo, M. Basel, R. Singh, P. Spink, and L. Scheffer. "Layout extraction." *EDA for IC Implementation, Circuit Design, and Process Technology*, pp. 637–662, CRC Press, 2018.

[2] H. S. Baird. "Fast algorithm for LSI artwork analysis." *Papers on Twenty-five Years of Electronic Design Automation* (1988), pp. 154–162.

[3] W. H. Kao, C.-Y. Lo, M. Basel, and R. Singh. "Parasitic extraction: Current state of the art and future trends." *Proceedings of the IEEE* 89, no. 5 (2001), pp. 729–739.

[4] W. Yu and X. Wang. *Advanced Field-solver Techniques for RC Extraction of Integrated Circuits*. Springer, 2014.

[5] K. Nabors and J. White. "FastCap: A multipole accelerated 3-D capacitance extraction program." *IEEE Transactions on Computer-aided Design of Integrated Circuits and Systems* 10, no. 11 (1991), pp. 1447–1459.

[6] U. Choudhury and A. Sangiovanni-Vincentelli. "Automatic generation of analytical models for interconnect capacitances." *IEEE Transactions on Computer-aided Design of Integrated Circuits and Systems* 14, no. 4 (1995), pp. 470–480.

[7] N. D. Arora, K. V. Raol, R. Schumann, and L. M. Richardson. "Modeling and extraction of interconnect capacitances for multilayer VLSI circuits." *IEEE Transactions on Computer-aided Design of Integrated Circuits and Systems* 15, no. 1 (1996), pp. 58–67.

[8] J. A. Davis and J. D. Meindl. *Interconnect Technology and Design for Gigascale Integration*. Springer Science & Business Media, 2012.

[9] "IEEE standard for Integrated Circuit (IC) Open Library Architecture (OLA)." *IEEE STD 1481-2009* (2009), pp. c1–658.

[10] M. Becer, R. Vaidyanathan, C. Oh, and R. Panda. "Crosstalk noise control in an SoC physical design flow." *IEEE Transactions on Computer-aided Design of Integrated Circuits and Systems* 23, no. 4 (2004), pp. 488–497.

[11] S. Nithin, G. Shanmugam, and S. Chandrasekar. "Dynamic voltage (IR) drop analysis and design closure: Issues and challenges." *2010 11th International Symposium on Quality Electronic Design (ISQED)* (2010), pp. 611–617, IEEE.

[12] A. Vakil, H. Homayoun, and A. Sasan. "IR-ATA: IR annotated timing analysis, a flow for closing the loop between PDN design, IR analysis & timing closure." *Proceedings of the 24th Asia and South Pacific Design Automation Conference* (2019), pp. 152–159.

[13] T. Okumura, F. Minami, K. Shimazaki, K. Kuwada, and M. Hashimoto. "Gate delay estimation in STA under dynamic power supply noise." *IEICE Transactions on Fundamentals of Electronics, Communications and Computer Sciences* 93, no. 12 (2010), pp. 2447–2455.

[14] J. Lienig and M. Thiele. *Fundamentals of Electromigration-aware Integrated Circuit Design*. Springer, 2018.

[15] S.-S. Han, A. B. Kahng, S. Nath, and A. S. Vydyanathan. "A deep learning methodology to proliferate golden signoff timing." *2014 Design, Automation & Test in Europe Conference & Exhibition (DATE)* (2014), pp. 1–6, IEEE.

[16] N. Chang, A. Baranwal, H. Zhuang, M.-C. Shih, R. Rajan, Y. Jia, H.-L. Liao, Y.-S. Li, T. Ku, and R. Lin. "Machine learning based generic violation waiver system with application on electromigration sign-off." *2018 23rd Asia and South Pacific Design Automation Conference (ASP-DAC)* (2018), pp. 416–421, IEEE.

[17] S.-Y. Lin, Y.-C. Fang, Y.-C. Li, Y.-C. Liu, T.-S. Yang, S.-C. Lin, C.-M. Li, and E. J.-W. Fang. "IR drop prediction of ECO-revised circuits using machine learning." *2018 IEEE 36th VLSI Test Symposium (VTS)* (2018), pp. 1–6, IEEE.

[18] Y.-C. Lu, S. Nath, S. S. K. Pentapati, and S. K. Lim. "A fast learning-driven signoff power optimization framework." *2020 IEEE/ACM International Conference on Computer Aided Design (ICCAD)* (2020), pp. 1–9, IEEE.

[19] M. Agarwal and S. Saurabh. "An efficient timing model of flip-flops based on artificial neural network." *2021 ACM/IEEE 3rd Workshop on Machine Learning for CAD (MLCAD)* (2021), pp. 1–6, IEEE.

30 Post-silicon Validation

... To expect the unexpected shows a thoroughly modern intellect ...

—Oscar Wilde, *An Ideal Husband*, Third Act, 1895

In the earlier chapters, we have discussed various tasks involved in a design flow. The design process culminates in sending the final layout to a foundry for fabrication or design *tape-out*. However, we do not start with the *mass production* of a chip immediately. Rather we check the first few samples of the fabricated chip by running actual applications under realistic operating conditions. This task is known as *post-silicon validation*. If we find problems in post-silicon validation, we need to debug and fix them before moving ahead with the mass production of a chip. In this chapter, we will discuss the basic concepts related to post-silicon validation.

30.1 NEED FOR POST-SILICON VALIDATION

During the pre-silicon phase, we perform verification on an *abstract design model* in a *virtual environment* using simulation, formal verification, and other techniques described in the earlier chapters. For complex designs and system-on-chips (SoCs), some verification gaps inevitably remain in the pre-silicon phase. Consequently, a fabricated chip can produce an unexpected or wrong response on an actual application. We need to detect such unexpected behavior of the fabricated chip and fix them before it reaches the end-users on a large scale.

Despite tremendous progress in the pre-silicon design verification techniques, we obtain errors in a majority of newly manufactured SoCs [1]. The primary reasons why verification gaps remain in the pre-silicon stage are as follows:

1. **Inadequate functional coverage:** The speed of simulation and other functional verification methods is orders of magnitude slower than obtaining results using a fabricated chip. The increasing complexity of integrated circuits (ICs), the exponential growth in the design space, and the tighter design schedule leave room for undetected functional issues in the pre-silicon stage. In contrast, we can verify the functionality of a *fabricated* chip by running software, operating systems (OS), and other real-world applications at full speed and for a longer duration, which is not possible using simulation-based verification. For example, we can execute the OS boot sequence on a fabricated chip in a few seconds, while simulating it using a register transfer level (RTL) model will require several years [2]. Hence, post-silicon validation allows us to cover those scenarios and operating conditions in the post-silicon validation stage that were missed during the pre-silicon functional verification.

2. **Inadequate screening of electrical issues:** A fabricated die can have electrical issues due to crosstalk, IR drop in the power supply lines, thermal effects, and process-induced variations. It is infeasible to check for all the possible electrical issues in the pre-silicon stage. In the post-silicon validation, we can subject a chip to electrical, thermal, and physical stress, operate under a realistic environment, and check for robustness to electrical noises and security risks. As a result, we can detect electrical and other issues that were missed in the pre-silicon verification.

Due to the above reasons, post-silicon validation is an essential activity before the mass production of a chip. It is one of the most time-consuming activities of the new chip development cycle and requires a lot of manual interventions.

Note that post-silicon validation is different from *manufacturing test*. The purpose of manufacturing test is to detect *manufacturing defects* and avoid faulty chips from reaching the end-user. We perform manufacturing tests for each manufactured chip when we are into mass production. Depending on the yield of a process, manufacturing test will detect faults in some fabricated dies and prevent those dies from reaching the end-user. Part III ("Design for testability") of this book discussed design techniques that made manufacturing tests more efficient.

On the other hand, post-silicon validation needs to be done before moving into mass production. The functional and electrical problems that we detect in post-silicon validation typically originate due to some design bug. If we manufacture a chip without fixing that bug, all the chips will have the corresponding functional or electrical problem. Once we have fixed that bug, the subsequently manufactured chips will be free from it. Therefore, post-silicon validation is critical for ensuring the quality of complex chips and SoCs.

30.2 MAJOR TASKS

There are four major tasks involved in post-silicon validation.

1. **Preparation:** The quality of post-silicon validation and the effort required in identifying and fixing problems heavily depend on the design infrastructures. Therefore, we need to plan for the post-silicon validation during the early stages of design implementation. We need to create a *post-silicon test plan* during the early phases of design implementation. Subsequently, we need to create design infrastructures that support the post-silicon test plan and enable efficient debugging. This task can involve making design changes. We refer to the design tasks focused on increasing the design's debuggability as *design-for-debug* (DFD) [1]. We need to consider DFD during the early stages of the design implementation because it can become impossible to incorporate new DFD features after we find them missing during the post-silicon validation stage. The DFD features incorporated in the early stages, such as during system-level design, permeate throughout a design flow.

   The preparation for post-silicon validation includes developing *debug software* that will provide infrastructure support for the post-silicon validation and debugging. The debug software can provide features such as reading/writing from/to internal registers, transporting data from the chip to an external database, and analyzing the data produced by the chip.

2. **Applying test:** We subject a chip to various use cases, inputs, and environmental conditions during validation. Subsequently, we observe its behavior and compare it with the expected functionality. If the chip's functionality deviates from the expected functionality, we need to diagnose and fix the problem.

3. **Root cause identification:** After detecting a post-silicon validation issue, we need to diagnose its root cause. This task can be tedious and time taking.
4. **Bug fixing:** Once we have diagnosed a problem, we need to fix it. Alternatively, we can find some workaround that allows us to meet stringent schedule requirements.

In the following sections, we will discuss the above tasks in more detail.

30.3 DESIGN-FOR-DEBUG

DFD is a critical component of post-silicon validation because diagnosing the root cause of a bug is challenging [3]. It requires controlling and observing millions of internal signals with the help of a limited number of input/output (I/O) ports available on the fabricated chip. To address this challenge, DFD focuses on developing infrastructures that increase the controllability and observability of the internal signals. The primary design features that facilitate debugging are: scan chains, trace buffers, and customized debug features. We explain these features in the following paragraphs.

1. **Scan chains:** We create *scan chains* in a design for facilitating manufacturing test.[1] In the *shift mode*, we can write values to the registers (or scan cells) by streaming in the required bit pattern through the primary input (*SI* port) and allowing them to shift through the scan chain. Similarly, we can read the values of the registers by shifting them through the scan chain and streaming out to the primary output (*SO* port). Thus, scan chains improve the controllability and observability of a design and greatly simplify the manufacturing test. We can use the scan chain infrastructure for post-silicon debugging also.

   While debugging, if we want to read values at the Q-pin of a scan cell (or the corresponding flip-flop) at any given clock cycle, we first allow the scan cell to *capture* (or latch) the required value. Subsequently, we take the circuit to the *shift mode* and allow that value to propagate to the respective primary output (*SO* port) through the scan chain. This method of reusing scan chains for debugging is popular because it does not require extra hardware (scan chains are already created for the manufacturing test) [4]. However, when we move a circuit to the scan mode, we disrupt the normal functionality of the circuit. Therefore, we prefer using a scan chain for debugging when we want to know the entire state of a system *at a given clock cycle*.
2. **Trace buffers:** When we want to observe the value of a small set of signals for *several* clock cycles, we store them in a dedicated internal memory known as *trace buffers*. Subsequently, we read the values from the trace buffers with the help of the I/O ports for debugging. Note that we cannot directly read the values of the internal signals at the I/O ports because the internal circuit typically operates at a significantly higher speed than the speed at which values can be read from the I/O ports. Moreover, we have a limited number of I/O ports in a circuit. Therefore, storing the values of the internal signals during at-speed operation and subsequent transfer of that data for offline debugging when needed is more efficient for post-silicon validation.

   Figure 30.1 shows a simplistic implementation of a debug infrastructure that employs trace buffers. The trace buffers store the value of M signals (M is also referred to as the *width* of the trace buffers) for N clock cycles (N is also referred to as the *depth* of the trace buffers) [3]. Hence, the total size of the trace buffer is $M \times N$ bits. The L internal signals reach the trace

[1] See Chapter 21 ("Scan design") for details on scan chains.

buffers through a multiplexer logic. The multiplexer logic selects M out of L signals based on the value in the control registers. The value of the control registers can be changed dynamically based on the debug requirements. Other components that control the read/write operations of the trace buffers are not shown in this figure.

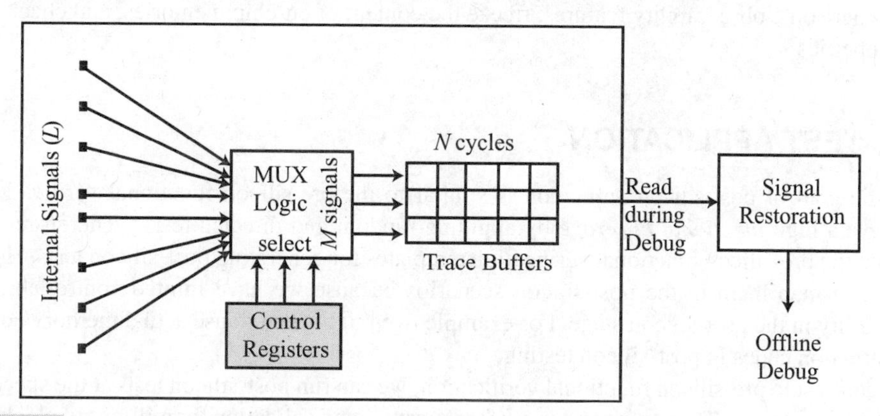

Figure 30.1 A simple scheme for signal tracing

Trace signal selection: The area and power overheads of the trace buffers increase with the increase in M and N. Therefore, we cannot afford to keep M (number of trace signals) more than a few hundred. It imposes a severe restriction on signal tracing.

During *design implementation*, we need to select signals we want to trace. This task is challenging because a typical industry-scale design contains millions of signals. We do not know which signals will be needed for debugging until we have moved into post-silicon validation. Therefore, we need to employ various strategies to select trace signals during design implementation. Traditional methods rely on the designer's experience and often select signals that will increase the observability of a design for the error-prone scenarios. Various sophisticated techniques have been proposed recently that allow maximum restoration of states from the set of traced signals [2, 5–8].

The selected trace signals need to be routed to the multiplexer logic and the trace buffers. Since too many signals can converge to the inputs of the trace buffers, we can face congestion and other physical design issues in developing signal tracing infrastructures. Therefore, we need to consider these aspects of design implementation early in the design flow.

Debugging: During debugging, the values in the trace buffers are streamed out through some I/O interface, such as the standard *joint test action group (JTAG)* interface and *universal serial bus (USB)* interface. Additionally, to read voluminous debug data from a chip more efficiently, we can use strategies such as *on-chip compression* of the debug data [9].

After reading debug data from a chip, we can restore the values of other signals in a circuit. For example, if we know that the input of an OR gate is 1, we can infer that its output is 1, irrespective of the other inputs. Such kind of restorations from the input to the output is referred to as *forward restoration*. Similarly, if we know that the output of an OR gate is 0, we can infer that all its inputs are 0. Such kind of restorations from the output to the input is referred to as *backward restoration*. Starting with the values of the traced signal, we can perform forward and backward restoration iteratively until we can restore no more new signal.

After signal restoration, we use the values of the traced and the restored signals to carry out debugging and diagnosis.

3. **Customized features:** Sometimes we add customized features that help in debugging. For example, we can add features to collect runtime statistics (such as the number of cache misses), trigger some observability features, freeze the content of on-chip memories, and check system properties.

30.4 TEST APPLICATION

The basic goal of post-silicon validation is similar to the pre-silicon functional verification, i.e., to achieve a high *functional coverage* by applying random and directed tests.[2] Therefore, we can start with the pre-silicon functional verification templates for generating post-silicon tests. However, we need to map them to the post-silicon scenarios because we have limited controllability and observability in the post-silicon stage. For example, we might need to use actual memory addresses and instruction codes in post-silicon testing.

In contrast to pre-silicon functional verification, we can run post-silicon tests at the speed of the hardware being tested. Therefore, we can perform post-silicon validation for millions of clock cycles. It allows us to run end-user applications such as OS and complex calculations that are impossible to run during pre-silicon functional verification due to extremely high runtime.

On-chip test generation: We can develop *on-chip test generation infrastructures* for post-silicon validation to tackle the problem of low controllability. However, these infrastructures are distinct from BIST infrastructures because the goal of post-silicon validation is to perform functional verification rather than the structural testing done for detecting manufacturing faults.[3] Therefore, purely random stimuli are of not much help in post-silicon validation. We need to constrain the stimuli to be compliant with the system functionality [10].

We also need to consider the problem of observability in post-silicon validation. The erroneous behavior may remain unobservable for a long duration before its effect becomes observable. This problem makes debugging more difficult because the trace buffers store values only for the last few clock cycles. Therefore, the trace buffers can get overwritten by the time the problem gets manifested, and debugging becomes more difficult. Hence, techniques for *quick error detection* (QED) become essential in post-silicon validation [11].

Hardware assertions: One of the techniques to reduce the latency of bug manifestation is by adding *hardware assertions*. We can insert *hardware monitors in a design* corresponding to the assertions for post-silicon validation. These hardware monitors will perform on-chip assertion checking during post-silicon validation. If the assertion fails during validation, the hardware monitor will raise a flag or signal that the assertion has failed. We can appropriately connect the output of these monitors to observe the assertion failures immediately and reduce the latency of bug manifestation.

We normally encode assertions in the RTL code for functional verification.[4] We can reuse these assertions in post-silicon validation by synthesizing them. By default, RTL synthesis tools do not synthesize assertions. Therefore, we need to instruct the synthesis tool to synthesize the assertions or use special tools meant for synthesizing assertions [12, 13].

---

[2] See Section 9.5.2 ("Coverage-driven verification") for coverage.

[3] See Chapter 23 ("Built-in self-test") for BIST.

[4] See Section 9.5.3 ("Assertion-based verification") for details on the use of assertions in pre-silicon functional verification.

Example 30.1 Consider the following SystemVerilog assertion (SVA) (from Ex. 9.12).

```
ready_check: assert( ready -> !reset);
```

It checks that whenever the signal *ready* is 1, the signal *reset* must be 0. We can synthesize this assertion as shown in Figure 30.2.

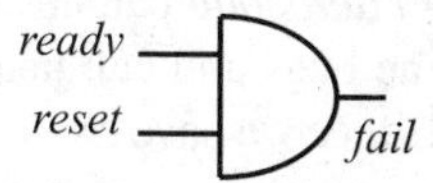

Figure 30.2 Synthesized assertion

The output *fail* becomes 1 when *ready*=1 and *reset* also becomes 1. Hence, if we insert the above hardware in the circuit, the signal *fail* will indicate whether the assertion has failed. Moreover, if we can directly observe the signal *fail* during post-silicon validation, we can know the status of the above assertion immediately.

The above example shows a simple assertion. In general, assertions can contain logical operators, temporal operators, and regular expressions. Therefore, an assertion can be synthesized into a complicated sub-circuit consisting of combinational and sequential circuit elements. In such cases, we also need to consider the area and power penalty of the synthesized assertions. We need to intelligently trade off these figures of merit to improve the design debuggability and select only the most useful set of assertions for hardware synthesis.

30.5 ROOT CAUSE IDENTIFICATION

After we have observed an issue in the post-silicon testing, we need to *reproduce* it. Reproducing an issue involves finding the sequence of steps by which we can deterministically observe the problem. This task is essential because we need to subsequently observe the problem multiple times for bug fixing, validation of the fix, and quality assurance purposes. Reproducing an issue can be tedious and can involve many trials and errors. We can utilize the signal traces and take the help of the simulation setup, field programmable gate array (FPGA) prototype, or virtual machines to reproduce an issue.

After reproducing a problem, we need to identify whether it is a *bug* that can impact the system or a behavior that can be tolerated. We also need to ensure that the bug is not a manifestation of a manufacturing defect and is a genuine design problem. Subsequently, we need to localize the root cause of that bug in the design with the help of debug infrastructures. We can also take the help of formal techniques to localize the root cause of a problem [14–16].

30.6 BUG FIXING

The post-silicon validation is carried out under a strict schedule. Therefore, we cannot detect and fix bugs *sequentially*. Instead, when we discover a bug, we try to find a workaround and move ahead with the rest of the validation tasks while another designer or group works on the bug fixing [3]. This strategy helps us achieve a shorter turnaround time for completing the post-silicon validation tasks. However, in practice, it is challenging because of the uncertainty in finding an efficient workaround, bug reproducibility, and finding the root cause of a bug.

The bug fixing method will depend on the design type and the kind of bug. For many microprocessors and SoCs, we can use *microcode* patches to fix the problem and avoid costly re-spins. We can use spare cells to fix some bugs and can take the help of engineering change order (ECO) tools for making design changes. Once we have fixed a bug, we need to validate its correctness and also ensure that the bug fix did not introduce any new problem.

30.7 RECENT TRENDS

With the increasing design complexities, the task of post-silicon validation is becoming more complex, costly, and schedule-critical. Therefore, it has attracted a lot of attention from researchers recently. Several sophisticated techniques are being developed to enhance the debuggability of a design. However, while adding debugging features to a design, we can inadvertently expose its details and make it vulnerable to security attacks [17]. Therefore, interactions between debuggability and security of a design have now become a critical consideration.

In this chapter, we have presented a rudimentary description of post-silicon validation. In practice, many project-specific features are added to a design to carry out these activities more efficiently [3]. Often these features are manually developed or borrowed from other designs on an ad-hoc basis. With increasing design complexities, these approaches will become unscalable. Therefore, we need to develop a more structured framework for post-silicon validation and provide electronic design automation (EDA) tool support for carrying out these activities.

REVIEW QUESTIONS

30.1 Why do we need to carry out post-silicon validation?

30.2 How is post-silicon validation different from manufacturing test?

30.3 How is post-silicon validation different from pre-silicon functional verification?

30.4 Why do we need to prepare a design for post-silicon validation right from system-level design?

30.5 How can we use scan chains to debug a design during post-silicon validation?

30.6 How do we use trace buffers to debug a design during post-silicon validation?

30.7 How do we decide the width and depth of the trace buffers in a design?

30.8 Consider the circuit shown in Figure 30.3 for debugging.

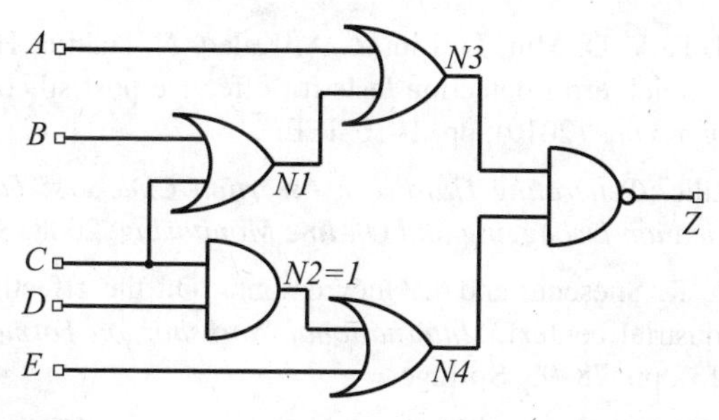

Figure 30.3 A given circuit

Assume that we know that the value at the net $N2$=1. Compute the values at the nets $N1$, $N3$, $N4$, and Z by iterative forward and backward restoration.

30.9 Why do we need QED in post-silicon validation?

30.10 How can we use SVAs for post-silicon validation?

30.11 How can debuggability and security conflict in a design?

REFERENCES

[1] M. Abramovici, P. Bradley, K. Dwarakanath, P. Levin, G. Memmi, and D. Miller. "A reconfigurable design-for-debug infrastructure for SoCs." *Proceedings of the 43rd Annual Design Automation Conference* (2006), pp. 7–12.

[2] P. Mishra, R. Morad, A. Ziv, and S. Ray. "Post-silicon validation in the SoC era: A tutorial introduction." *IEEE Design & Test* 34, no. 3 (2017), pp. 68–92.

[3] P. Mishra and F. Farahmandi. *Post-silicon Validation and Debug*. Springer, 2019.

[4] B. Vermeulen and S. K. Goel. "Design for debug: Catching design errors in digital chips." *IEEE Design & Test of Computers* 19, no. 03 (2002), pp. 37–45.

[5] H. F. Ko and N. Nicolici. "Automated trace signals identification and state restoration for improving observability in post-silicon validation." *2008 Design, Automation and Test in Europe* (2008), pp. 1298–1303, IEEE.

[6] X. Liu and Q. Xu. "Trace signal selection for visibility enhancement in post-silicon validation." *2009 Design, Automation & Test in Europe Conference & Exhibition* (2009), pp. 1338–1343, IEEE.

[7] K. Rahmani, P. Mishra, and S. Ray. "Efficient trace signal selection using augmentation and ILP techniques." *Fifteenth International Symposium on Quality Electronic Design* (2014), pp. 148–155, IEEE.

[8] K. Rahmani, P. Mishra, and S. Ray. "Scalable trace signal selection using machine learning." *2013 IEEE 31st International Conference on Computer Design (ICCD)* (2013), pp. 384–389, IEEE.

[9] P. R. Panda, M. Balakrishnan, and A. Vishnoi. "Compressing cache state for postsilicon processor debug." *IEEE Transactions on Computers* 60, no. 4 (2010), pp. 484–497.

[10] X. Shi and N. Nicolici. "On-chip cube-based constrained-random stimuli generation for post-silicon validation." *IEEE Transactions on Computer-aided Design of Integrated Circuits and Systems* 35, no. 6 (2015), pp. 1012–1025.

[11] T. Hong, Y. Li, S.-B. Park, D. Mui, D. Lin, Z. A. Kaleq, N. Hakim, H. Naeimi, D. S. Gardner, and S. Mitra. "QED: Quick error detection tests for effective post-silicon validation." *2010 IEEE International Test Conference* (2010), pp. 1–10, IEEE.

[12] M. Boulâe and Z. Zilic. *Generating Hardware Assertion Checkers: For Hardware Verification, Emulation, Post-fabrication Debugging and On-line Monitoring*. 2008. Springer.

[13] L. Pierre, F. Pancher, R. Suescun, and J. Quévremont. "On the effectiveness of assertion-based verification in an industrial context." *International Workshop on Formal Methods for Industrial Critical Systems* (2013), pp. 78–93, Springer.

[14] A. Smith, A. Veneris, M. F. Ali, and A. Viglas. "Fault diagnosis and logic debugging using Boolean satisfiability." *IEEE Transactions on Computer-aided Design of Integrated Circuits and Systems* 24, no. 10 (2005), pp. 1606–1621.

[15] F. M. De Paula, M. Gort, A. J. Hu, S. J. Wilton, and J. Yang. *Backspace: Formal Analysis for Post-silicon Debug*. 2008, IEEE.

[16] C. S. Zhu, G. Weissenbacher, and S. Malik. "Post-silicon fault localisation using maximum satisfiability and backbones." *2011 Formal Methods in Computer-aided Design (FMCAD)* (2011), pp. 63–66, IEEE.

[17] S. Ray, J. Yang, A. Basak, and S. Bhunia. "Correctness and security at odds: Post-silicon validation of modern SoC designs." *Proceedings of the 52nd Annual Design Automation Conference* (2015), pp. 1–6.

Answers

CHAPTER 2

2.6 176, 397, $5.68, $3.02

2.13 (a) 150000 (b) FPGA technology (c) cell-based technology

CHAPTER 3

3.5 (a) Run F6 and F8 on dedicated hardware, $180, 505s (b) Run F6, F8, F3, and F5 on dedicated hardware, 280s, $260

CHAPTER 6

6.4 (a) 89.4%, 3.96 cents (b) 97.6%, 3.62 cents (c) 91.3%, 15.53 cents

CHAPTER 8

8.9 At time 0 a=0 b=0
At time 10 a=1 b=0
At time 20 a=1 b=1
At time 25 a=0 b=0

8.15 At time 0 clk=0 en=0 a=0 b=0
At time 2 clk=1 en=0 a=0 b=0
At time 3 clk=1 en=1 a=1 b=0
At time 4 clk=0 en=1 a=1 b=1
At time 5 clk=0 en=1 a=1 b=0
At time 6 clk=1 en=1 a=1 b=1
At time 7 clk=1 en=1 a=1 b=0
At time 8 clk=0 en=0 a=0 b=1

CHAPTER 9

9.12 *top2* has a race condition. Depending on which always block executes first, *c* will get old or updated value of *b*.

9.18 The property *casrt* specifies that whenever *request*=1, *grant* should be made 1 within five clock ticks.

CHAPTER 10

10.14 The state diagram is shown in Figure A-10.14:

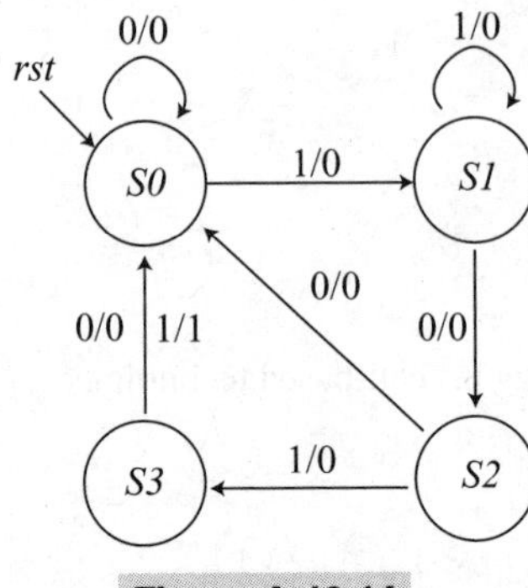

Figure A-10.14

CHAPTER 11

11.3 The ROBDDs are shown in Figure A-11.3:

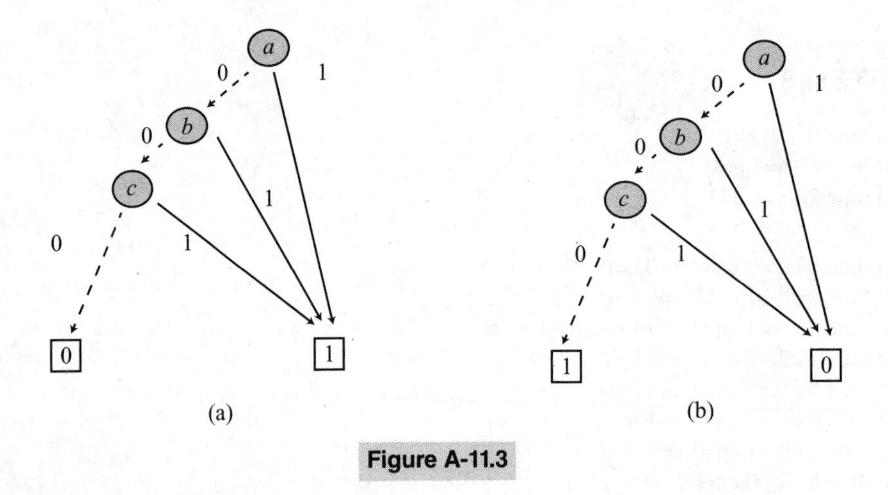

Figure A-11.3

11.4 (a) $a + bc$ (b) $a'(b' + c')$

11.5 The terminal node with logic value "1"

11.14 The miter is shown in Figure A-11.14. The patterns at inputs that satisfy the miter (or make the models inequivalent) are: $A = 0, B = 1$ and $A = 1, B = 0$.

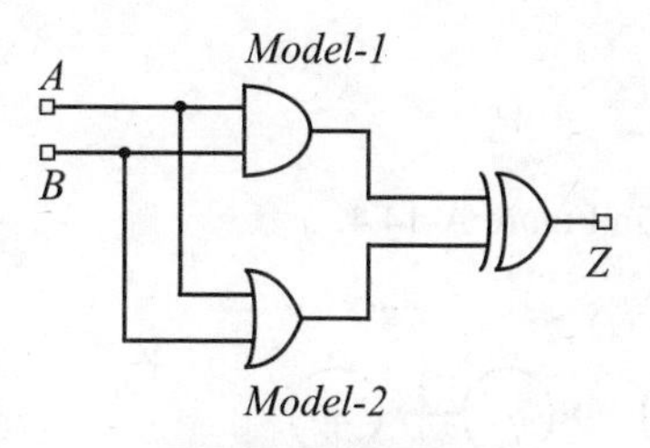

Figure A-11.14

11.15 The failing miter is shown in Figure A-11.15. The patterns at inputs that satisfy the miter (or make the models inequivalent) are: $F1/Q = 0, F2/Q = 0$ and $F1/Q = 1, F2/Q = 1$. No, these are invalid patterns, because F1/D is the complement of F2/D. Therefore, the possible values of {F1/Q,F2/Q} are {01} and {10}. Thus, it is a false failure.

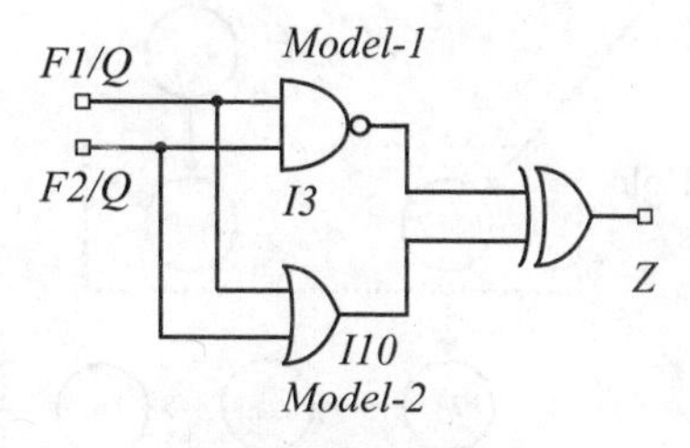

Figure A-11.15

CHAPTER 12

12.14 {Kernel, co-kernel} pairs are: $\{e + cd, a\}$, $\{e + cd, b\}$, $\{a + b, cd\}$, $\{a + b, e\}$, and $\{ae + be + acd + bcd, 1\}$.

12.18 ODC condition is a. The optimized logic network is shown in Figure A-12.18.

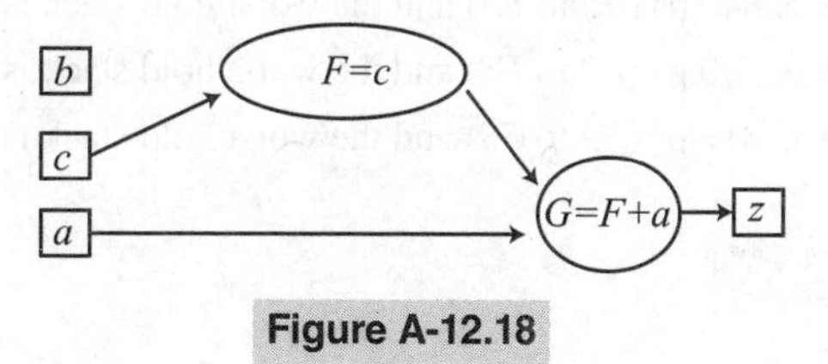

Figure A-12.18

CHAPTER 13

13.4 (a) 8 ps and (b) 4 ps

CHAPTER 14

14.3 978 ps

14.4 The timing graph is shown in Figure A-14.4.

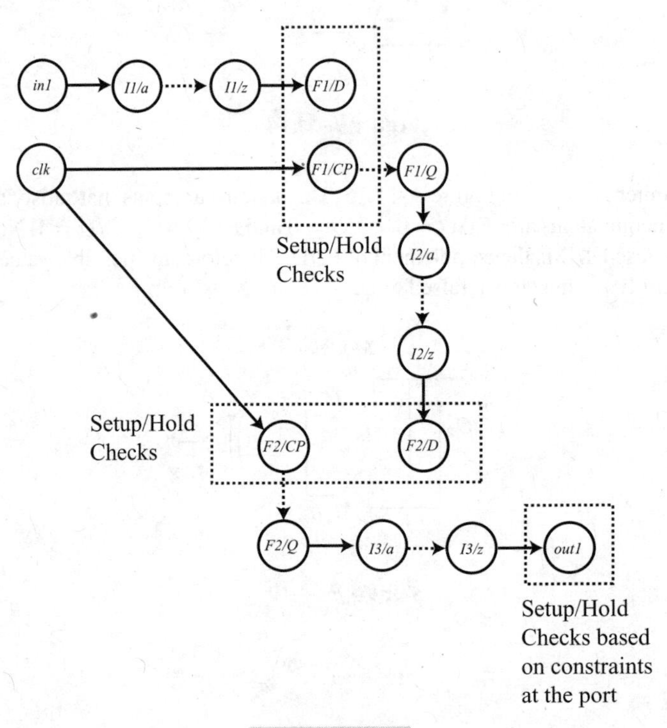

Figure A-14.4

14.7 (a) $A_{Z,max}$=42 ps, $S_{Z,max}$=20 ps; (b) $A_{Z,max}$=26 ps, $S_{Z,max}$=5 ps; (c) $A_{Z,min}$=15 ps, $S_{Z,min}$=5 ps; and (d) $A_{Z,max}$=26 ps, $S_{Z,max}$=5 ps.

14.9 The worst setup slack is 840 ps (*F2* to *F3*) and the worst hold slack is 95 ps (*F1* to *F3*)

14.10 The worst setup slack is 790 ps (*F2* to *F3*) and the worst hold slack is 45 ps (*F1* to *F3*)

14.11 The worst setup slack is 816 ps (*F2* to *F3*) and the worst hold slack is 73.5 ps (*F1* to *F3*)

CHAPTER 15

15.1
```
current_design MYDES
create_clock -name Clock -period 12 -waveform{0 3} [get_ports clk]
```

15.2 100 MHz and 33.33 MHz.

15.5
```
current_design MYDES
create_clock -name Clk -period 10 [get_ports CLK]
set_output_delay 0.350 -max -clock [get_clocks Clk] [get_ports OUT]
```

CHAPTER 16

16.3 (a–c) The solutions are shown in Figure A-16.3.

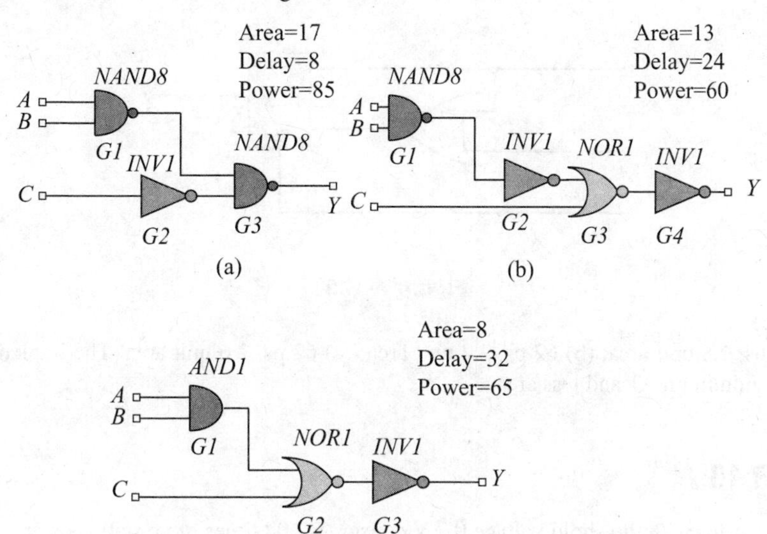

Figure A-16.3

16.4 (a, b) The solutions are shown in Figure A-16.4.
(c) Possible matches: N1: INV1
N2: INV1
N3: NOR1, AND1
N4: NOR1 Y: INV1
For smallest area: N1, N2, N3 covered by AND1, N4 covered by NOR1, and Y covered by INV1.

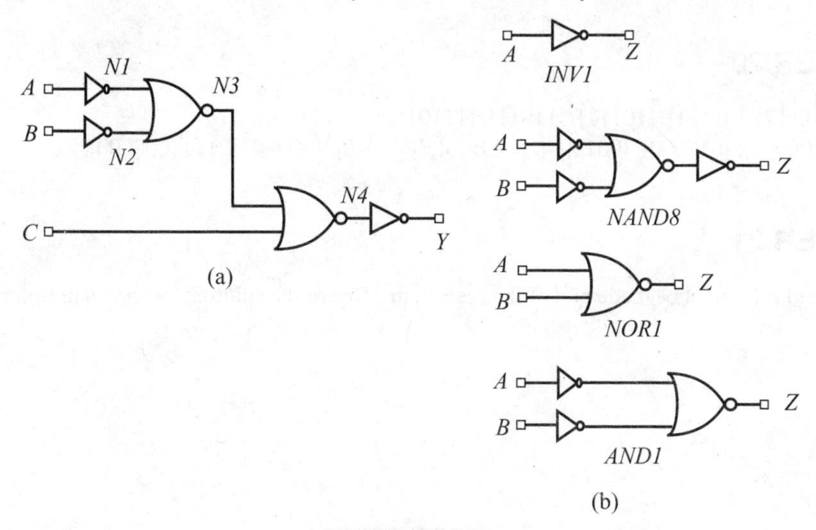

Figure A-16.4

CHAPTER 17

17.9 (a) 145 ps and (b) 130 ps. The restructured circuit is shown in Figure A-17.9.

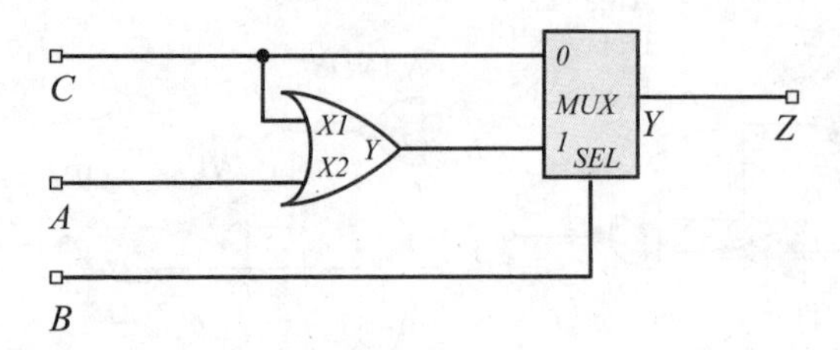

Figure A-17.9

17.11 (a) 75 ps, 1.8 unit area; (b) 62 ps, 2.2 unit area; (c) 62 ps, 2.6 unit area. The implementation (b) has the minimum AT and less area.

CHAPTER 18

18.3 The transistor with threshold voltage 0.2 V consumes 608 times more static power.

18.10 0.1 μW.

CHAPTER 19

19.21 The individual bits of *BIT[0:31]* can be incoherent (some bits can be sampled in one clock cycle, while others are sampled in the next clock cycle). To avoid this, individual bits should not be synchronized. One of the techniques to achieve this is by synchronizing the control signal as in Figure 19.11.

CHAPTER 20

20.5 (a) {11111}, {01111}, {10111}, {11011}, {11101}, and {11110}
(b) {100000}, {010000}, {001000}, {000100}, {000010}, {000001}, and {000000}.

CHAPTER 21

21.12 *F2* and *F3* cannot be made part of the scan chain. One of the solutions is shown in Figure A-21.12.

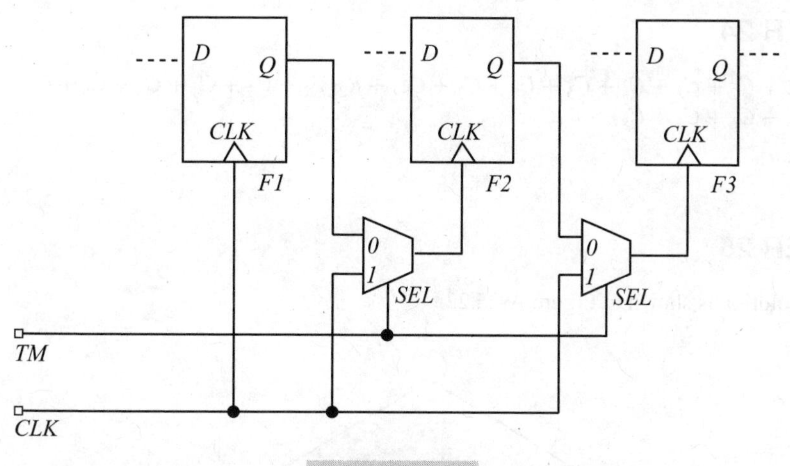

Figure A-21.12

CHAPTER 22

22.11 {A=0, B=1, C=1, D=0}, {A=1, B=0, C=1, D=0}, and {A=1, B=1, C=1, D=0}.

CHAPTER 23

23.7 The sequence is as follows:
{**0001**}
{1000}
{0100}
{0010}
{1001}
{1100}
{0110}
{1011}
{0101}
{1010}
{1101}
{1110}
{1111}
{0111}
{0011}
{**0001**} .... The period is 15.

CHAPTER 24

24.22 $R_1(C_1 + C_2 + C_3 + C_4 + C_5 + C_6 + C_7 + C_8) + R_2(C_2 + C_3 + C_4 + C_5 + C_6 + C_7 + C_8) + R_5(C_5 + C_6 + C_7 + C_8) + R_8C_8$

24.29 250 μm^2

CHAPTER 25

25.22 The solution is shown in Figure A-25.22.

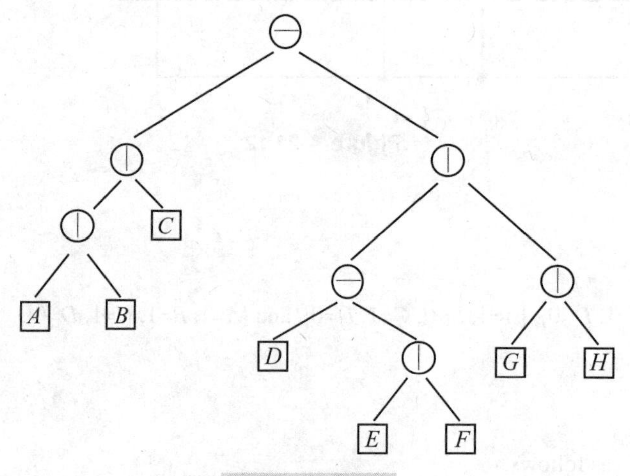

Figure A-25.22

CHAPTER 26

26.5 30°, $2L(1 + \sqrt{3})$

26.6 (a) 141 (b) (i) 1 (ii) 1/e

CHAPTER 27

27.10 300 ps, 750 ps

27.11 (a) 50 ps (b) −82.5 ps (c) −62.5 ps

CHAPTER 28

28.15 12 units. The solution is shown in Figure A-28.15.

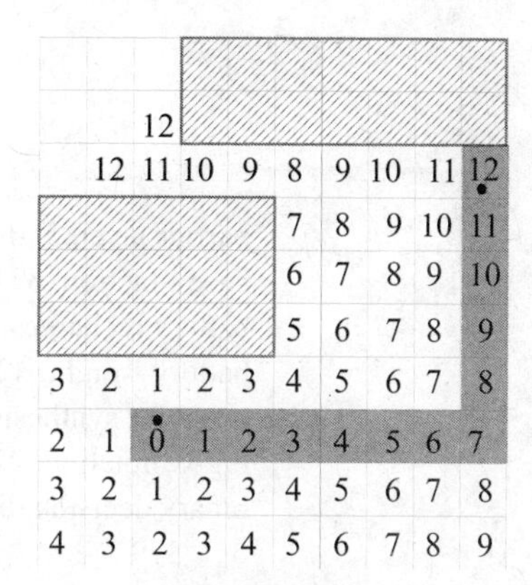

Figure A-28.15

CHAPTER 29

29.14 30 ps

CHAPTER 30

30.8 *N1*=1, *N3*=1, *N4*=1, and *Z*=0.

Index